y Press has long been a pioneer in the reissuing of
n its own backlist, producing digital reprints of books
er by scholars and students but could not be reprinted
aditional technology. The Cambridge Library Collection
a wider range of books which are still of importance to
sionals, either for the source material they contain, or as
ry of their academic discipline.

orld-renowned collections in the Cambridge University
ner libraries, and guided by the advice of experts
ambridge University Press is using state-of-the-art
ts own Printing House to capture the content of each
sion. The files are processed to give a consistently clear,
oks finished to the high quality standard for which the
und the world. The latest print-on-demand technology
will remain available indefinitely, and that orders for
s can quickly be supplied.

ry Collection brings back to life books of enduring
ng out-of-copyright works originally issued by other
e range of disciplines in the humanities and social
and technology.

CAMBRID

This series includes
expeditions to the A
in such endeavours
which, if successful
Cartographers and
their work made in
botany and zoolog
about the indigenc
of the American co
accounts of the ha
isolation in bygon

Man
and Physics

This 1875 manua
Arctic environme
it was prepared f
to inform and in
information, ran
northern lights,
(1831–1915), th
reached. Thoug
for the furthest
expedition's ga
manual clarifie
polar research
of the journey

Cambridge Universit
out-of-print titles fro
that are still sought a
economically using t
extends this activity t
researchers and profe
landmarks in the hist

Drawing from the
Library and other par
in each subject area, C
scanning machines in
book selected for inclu
crisp image, and the b
Press is recognised aro
ensures that the books
single or multiple copi

The Cambridge Libr
scholarly value (includi
publishers) across a wi
sciences and in science

Manual of the Natural History, Geology, and Physics of Greenland and the Neighbouring Regions

Prepared for the Use of the Arctic Expedition of 1875, under the Direction of the Arctic Committee of the Royal Society

EDITED BY T. RUPERT JONES

CAMBRIDGE
UNIVERSITY PRESS

University Printing House, Cambridge, CB2 8BS, United Kingdom

Cambridge University Press is part of the University of Cambridge.
It furthers the University's mission by disseminating knowledge in the pursuit of education, learning and research at the highest international levels of excellence.

www.cambridge.org
Information on this title: www.cambridge.org/9781108071918

This edition first published 1875
This digitally printed version 2014

ISBN 978-1-108-07191-8 Paperback

MANUAL

OF THE

NATURAL HISTORY, GEOLOGY, AND PHYSICS

OF

GREENLAND

AND THE NEIGHBOURING REGIONS;

PREPARED FOR THE USE OF THE ARCTIC EXPEDITION OF 1875, UNDER THE DIRECTION OF THE ARCTIC COMMITTEE OF THE ROYAL SOCIETY,

AND EDITED BY

PROFESSOR T. RUPERT JONES, F.R.S., F.G.S., &c., &c., &c.,

TOGETHER WITH

INSTRUCTIONS

SUGGESTED BY THE ARCTIC COMMITTEE OF THE ROYAL SOCIETY

FOR THE USE OF THE EXPEDITION.

PUBLISHED BY AUTHORITY OF THE LORDS COMMISSIONERS OF THE ADMIRALTY.

LONDON:
PRINTED BY GEORGE E. EYRE AND WILLIAM SPOTTISWOODE,
PRINTERS TO THE QUEEN'S MOST EXCELLENT MAJESTY.
FOR HER MAJESTY'S STATIONERY OFFICE.

1875.

36122.

INSTRUCTIONS

FOR THE USE OF

THE SCIENTIFIC EXPEDITION

TO

THE ARCTIC REGIONS,

1875.

SUGGESTED BY THE ARCTIC COMMITTEE OF THE ROYAL SOCIETY.

PREFACE.

THE President and Council of the Royal Society were informed by a letter from the Secretary of the Admiralty dated 4th December 1874, that it was their Lordships' intention to despatch an expedition, in the spring of 1875, to endeavour to reach the North Pole, and to explore the coast of Greenland and adjacent lands; and were invited to offer any suggestions which "might appear to them de- " sirable in regard to carrying out the scientific conduct " of the voyage."

This letter was referred to a Committee consisting of the President and Officers, Prof. J. C. Adams, Dr. Carpenter, Capt. Evans, Mr. J. Evans, Mr. F. Galton, Dr. Günther, Prof. Ramsay, Sir H. Rawlinson, and Mr. Scott, with power to add to their number. The Committee decided that it was desirable to prepare (1) a Manual of Scientific Results already obtained in Arctic Expeditions, (2) Instructions for future observations.

To prepare the instructions, the Committee divided the branches of science which were to be represented among several sub-committees, consisting partly of its own members, partly of other scientific persons who were specially conversant with the respective branches. It has been thought best that the several portions into which the instructions are divided should appear with the names of those by whom they were individually drawn up in the first instance, and who, after consultation with their colleagues, are held responsible for their final form.

In a few instances it will be found that the same subject has been mentioned by more than one of those concerned in the preparation of the instructions. In such cases it has been thought best not to incorporate the instructions relating to the same subject, but to leave them under the names of those who drew them up; partly to maintain the principle of individual responsibility, partly because the observers would more readily enter into the spirit of the instructions if the different points of view from which the same subject was regarded were separately presented to their minds.

The preparation of the Manual of Scientific Results was entrusted to the care of Prof. Rupert Jones, F.R.S., as Editor in chief, who also undertook himself to compile the part relating to Zoology, Botany, Geology, and Mineralogy. The Committee appointed Prof. W. G. Adams, F.R.S., Sub-editor, to assist Prof. Rupert Jones by compiling the part relating to Physics. The Manual will be found at the end of the Instructions.

INSTRUCTIONS FOR THE USE OF THE ARCTIC EXPEDITION, 1875.

CONTENTS.

PART I.—PHYSICAL OBSERVATIONS.

I.—Astronomy.

II.—Terrestrial Magnetism.

III.—Meteorology.

IV.—Atmospheric Electricity.

V.—Optics.

VI.—Miscellaneous Observations.

PART II.—BIOLOGY.

I.—Zoology.

II.—Botany.

III.—Geology and Mineralogy.

INSTRUCTIONS.

PART I.—PHYSICAL OBSERVATIONS.

I.—ASTRONOMY.

1. Astronomical Data. (Eclipses of the Sun and Occultations.) By J. R. Hind, F.R.S.

Eclipse of the Sun, 1876, March 25.

In longitude 60° West and latitude 82°, this eclipse commences March 25 at $4^h\ 12{\cdot}1^m$ local mean time, 139° from the Sun's north point towards the west, and ends at $6^h\ 8^m$. Magnitude of eclipse, 0·67.

For any position not far from the above, the longitude of which is λ (taken *negatively*) and geocentric latitude *l*, the Greenwich time *t*, of commencement of the eclipse, may be found from the following formulæ :—

$$\text{Cos}\ \omega = 1{\cdot}11253 - [0{\cdot}19595].\sin l + [9{\cdot}95510].\cos l.\cos(\lambda + 36°\ 19{\cdot}4')$$
$$t = 8^h\ 3^m\ 50^s - [3{\cdot}57030].\sin\omega + [3{\cdot}52333].\sin l - [3{\cdot}76690].\cos l.\ \cos(\lambda - 150°\ 2{\cdot}0'),$$

and applying the longitude expressed in time to *t*, thus found, the local mean time of first contact is obtained. The quantities within square brackets are logarithms.

The distance of the point of contact on the Sun's limb from his north point reckoning towards the west $= \omega + 29°\ 45'$.

Eclipse of the Sun, 1877, August 8.

In longitude 60° West, and latitude 82°, this eclipse commences August 8 at $12^h\ 17{\cdot}3^m$ local mean time, 16° from the Sun's north point towards the west, and ends at $13^h\ 21^m$. Magnitude of eclipse, 0·21.

For any position not far from the above, the Greenwich time *t* of commencement will be found from the formulæ :—

$$\text{Cos}\ \omega = 2{\cdot}52151 - [0{\cdot}22120]\sin l + [9{\cdot}93713].\cos l.\cos(\lambda + 289°\ 47{\cdot}1')$$
$$t = 17^h\ 22^m\ 57^s - [3{\cdot}49064]\sin\omega - [3{\cdot}30948]\sin l - [3{\cdot}73517].\cos l.\cos(\lambda - 26°\ 24{\cdot}9').$$

The distance of the point of contact on the Sun's limb from his north point reckoning towards the west $= \omega - 21°\ 36'$.

Note.—The *north point* is here to be distinguished from the Sun's upper point or vertex.

As an example of the application of the above formulæ of reduction of time of commencement, the calculation may be made for the position assumed in the original calculation upon which the equations are founded, viz., longitude 60° or 4^h West, and latitude (geographical) 82°. To reduce the geographical to the geocentric latitude (*l*), a correction is to be applied which may be taken from the Table at p. 57 of "Appendices to various "Nautical Almanacs between the years 1834 and 1854;" with argument 82°, this correction is 3·1′ to be subtracted from the geographical latitude, and hence $l = 81°\ 56{\cdot}9'$.

In the case of the eclipse of 1876, March 25, the computation is then as follows:—

Constant	−0·19595	Constant	+9·95510	λ	− 60° 0′·0
sin l	+9·99570	cos l	+9·14633	Constant	+ 36 19·4
	−0·19165	cos A	+9·96181	A	− 23 40·6
No.	−1·55472		+9·06324		
		No.	+0·11567		
		Constant	+1·11253		
			+1·22820	λ	− 60 0·0
			−1·55472	Constant	−150 2·0
		cos ω	−0·32652	B	−210 2·0
		or log. cos ω	−9·51391		
		ω	+109° 3·5′		

Constant	−3·57030	Constant	+3·52333	Constant	−3·76690
sin ω	+9·97552	sin l	+9·99570	cos l	+9·14633
	−3·54582		+3·51903	cos B	−9·93738
No.	−3514·2s	No.	+3303·9s		+2·85061
				No.	+ 708·9s
					+ 3303·9
					+ 4012·8
					− 3514·2
					+ 498·6

	°	′
ω =	109	4
Constant =	29	45
Angle from N. Point at first contact =	138	49

	H.	M.	S.
or +		8	19
Constant	8	3	50
	8	12	9
Subtract Long.	4	0	0
Local time of commencement	4	12	9

which agrees with the direct calculation.

Occultations.

The list appended includes those stars of the *Nautical Almanac* catalogue, which may possibly be occulted in 82° of north latitude; but in order to ascertain with certainty whether any star is occulted, and the circumstances of the occultation, supposing the position of the point where the observation is to be made approximately known, the formulæ given at p. 134 of *Appendices, &c.*, cited above, may be employed. An example of the application of these formulæ for Greenwich is given at p. 145; but to further illustrate the method of computation, the occultation of the planet Mars 1876, January 31, is here calculated for longitude 4^h West, latitude 82°.

The following are the circumstances for this position of the principal occultations visible in 1876 and 1877.

		Immersion. H.	M.	°	Emersion. H.	M.	°
Mars	1876, Jan. 30–31	23	36	86	0	28	333
Venus	„ Oct. 13	4	4	12	4	40	294
Regulus	„ Dec. 6	1	57	50	2	47	262
„	1877, Jan. 2	9	52	70	10	46	240
„	„ Jan. 29	21	22	141	21	35	169
„	„ Feb. 26	8	10	95	8	57	214

The above are local mean times, and the angles from N. point are reckoned as usual in the *Nautical Almanac*.

Nautical Almanac Office, J. R. Hind.
1875, February 10.

OCCULTATIONS of Stars to the 5th magnitude inclusive that may be visible in or near 82° N. Lat. and 60° W. Long.

Date.	Star's Name.	Magnitude.	Local Mean Time of ☌ in R. A. of ☾ and ✳.	Date.	Star's Name.	Magnitude.	Local Mean Time of ☌ in R. A. of ☾ and ✳.
1875.			h. m. s.	1876.			h. m. s.
Sept. 21	136 Tauri	5	11 32 37	Nov. 9	c Leonis	5	21 9 42
Oct. 16	ζ Arietis	4½	6 34 23	29	17 Tauri	4	23 46 26
16	τ′ Arietis	5	9 4 30	30	23 Tauri	5	0 20 22
18	136 Tauri	5	17 48 23	30	η Tauri	3	0 47 4
Nov. 12	ζ Arietis	4½	17 7 15	30	27 Tauri	4	1 26 1
12	τ′ Arietis	5	19 34 13	Dec. 1	136 Tauri	5	22 39 49
15	136 Tauri	5	2 36 23	3	κ Geminor.	3½	14 44 4
20	χ Leonis	5	12 21 59	6	α Leonis	1½	2 8 21
Dec. 10	ζ Arietis	4½	4 33 39	6	ρ Leonis	4	13 27 33
10	τ¹ Arietis	5	7 1 47	27	17 Tauri	4	10 49 5
12	136 Tauri	5	13 42 23	27	23 Tauri	5	11 23 23
15	γ Cancri	4½	3 29 12	27	η Tauri	3	11 50 23
17	χ Leonis	5	19 33 5	27	27 Tauri	4	12 29 45
1876.				29	136 Tauri	5	9 45 32
Jan. 6	ζ Arietis	4½	14 31 24	31	κ Geminor.	3½	1 5 42
6	τ¹ Arietis	5	17 4 27	1877.			
9	136 Tauri	5	1 1 58	Jan. 2	α Leonis	1½	10 34 11
11	γ Cancri	4½	14 36 32	23	ε Arietis	4½	1 26 14
Feb. 2	ζ Arietis	4½	21 42 55	23	17 Tauri	4	20 35 39
5	136 Tauri	5	10 16 10	23	19 Tauri	5	20 43 17
8	γ Cancri	4½	1 20 48	23	20 Tauri	5	20 58 28
Mar. 1	ζ Arietis	4½	3 11 8	23	23 Tauri	5	21 11 6
1	23 Tauri	5	16 6 30	23	η Tauri	3	21 39 1
1	27 Tauri	4	17 15 12	23	27 Tauri	4	22 19 41
3	136 Tauri	5	16 45 45	25	136 Tauri	5	20 52 41
6	γ Cancri	4½	9 47 14	27	κ Geminor.	3½	12 41 18
				29	α Leonis	1½	21 27 30
Oct. 6	17 Tauri	4	6 7 22	Feb. 19	ε Arietis	4½	8 7 24
6	23 Tauri	5	6 42 48	20	17 Tauri	4	3 49 1
6	η Tauri	3	7 10 40	20	19 Tauri	5	3 56 53
6	27 Tauri	4	7 51 21	20	20 Tauri	5	4 12 32
8	136 Tauri	5	7 25 48	20	η Tauri	3	4 54 21
13	ρ Leonis	4	1 49 1	22	136 Tauri	5	5 49 38
13	c Leonis	5	15 7 58	23	κ Geminor.	3½	23 2 33
Nov. 2	17 Tauri	4	13 44 18	26	α Leonis	1½	8 50 24
2	23 Tauri	5	14 18 47	Mar. 18	ε Arietis	4½	13 33 48
2	η Tauri	3	14 45 54	19	17 Tauri	4	9 19 26
2	27 Tauri	4	15 25 29	19	19 Tauri	5	9 27 23
4	136 Tauri	5	13 45 13	19	20 Tauri	5	9 43 8
6	κ Geminor.	3½	7 7 43	19	η Tauri	3	10 25 15
9	ρ Leonis	4	7 37 32	21	136 Tauri	5	12 10 35

OCCULTATION ☌ 1876, JANUARY 31.

Long. 4^h W.

	°	′	″
Lat. N.	82	0	0
Correction		−3	7
l	81	56	53

ρ	9·99860		
$\cos l$	9·14633		
$\phi^{(1)}$	9·14493		9·14493
$\cot l$	9·15063	const.	9·41916
$\phi^{(2)}$	9·99430	$\phi^{(3)}$	8·56409

	H.	M.	S.
☾'s R.A.	0	38	41·1
✳'s R.A.	0	39	25·8
time	−		44·7
arc	−11′ 10·5′		

	H. M. S.		H. M. S.		M. S.
Local sidereal time at Greenwich noon	16 40 14·8			α_1	time 1 51·2
Mars R.A.	− 0 39 25·8				arc 27′ 48·0″
	−7 59 11·0		− 7 59 11·0		° ′ ″
		T	+ 4 2	D ☾'s Dec.	+5 13 50·8
			+ 39·8	δ ✱'s Dec.	+4 10 13·3
			− 3 56 31·2	D−δ	+1 3 37·5
		h	− 59° 7·8′	D_1	+14 56·8

P 57′ 23·0″	+3·53694		+3·53694		+3·53694
$\phi^{(2)}$	+9·99430	cos h	+9·71019	sin h	−9·93366
	+3·53124		+3·24713		−3·47060
cos δ	+9·99885	sin δ	+8·86166	sin δ	+8·86166
{	+3·53009		+2·10879		−2·33226
	+56 29·2	$\phi^{(1)}$	+9·14493	$\phi^{(3)}$	+8·56409
	+ 17·9		+1·25372	{	−0·89635
	+56 11·3				− 0 7·9
D−δ	+63 37·5			D_1	+14 56·8
x	+ 7 26·2			x_1	+15 4·7
{ α	−11 10·5			{ α_1	+27 48·0
	−2·82640		15 39·8		+3·22220
cos δ	+9·99885		57 17·7	cos δ	+9·99885
{	−2·82525			{	+3·22105
	−11 8·7				+27 43·6
P sin h	−3·47060	P	3·53694	P cos h	+3·24713
$\phi^{(1)}$	+9·14493	const.	9·43677		+8·56409
{	−2·61553	Δ′	2·97371	{	+1·81122
	−6 52·6				+ 1 4·7
{ y	−4 16·1			{ y_1	+26 38·9
	−2·40841				+3·20382
x	+2·64953			x_1	+2·95650
tan S	−9·75888	S	−29 51·2	cot ι	+0·24732
cos S	+9·93817	ι	+29 30·1	cos ι	+9·93969
W	+2·71136			W	+2·71136
cos[−(S+ι)]	+9·99999	−(S+ι)	+0 21·1		3·55630
n	+2·71135				+6·20735
Δ′	2·97371			H	+3·00353
cos ω	+9·73764	ω	+56 52·1	cos ω	+9·73764
c	+3·26589	a	−56 31·0	c	+3·26589
sin a	−9·92119	b	+57 13·2	sin b	+9·92467
	−3·18708				+3·19056
t_1	−0 25·6m			t_2	+0 25·9m
	4 2				4 2
	3 36·4				4 27·9
Long.	4 0·0 W				4 0·0 W
Imm. Jan. 30d	23 36·4	(Loc. Mean T.)		Em. Jan. 31d	0 27·9
(−ι)	−29·5			(−ι)	−29·5
ω	+56·9			ω	+56·9
	−86·4				+27·4

Note.—In this particular example, T ($4^h\ 2^m$) was taken from a previous calculation, but it may be obtained with a sufficient degree of approximation, by the method described at p. 129 of Appendices to the Nautical Almanac.

If the angles from the Sun's vertex are required, the parallactic angles must be computed and applied to the above.

2. Suggestions for Observations of the Tides to be made by the North Pole Expedition, by the Rev. Samuel Haughton, M.D., F.R.S.

I.—Summary of Arctic Tidal Observations already made.

The tidal wave enters the Arctic Polar Basin by three distinct channels :—

1. By Behring's Strait.
2. By Davis' Strait.
3. By the Greenland Sea and Barentz Sea.

As to the first two of these tidal waves, I can offer some useful observations, but I know little of the third wave, beyond the fact, recorded by Captain Markham, that the tide wave No. 2, entering Smith Sound and Kennedy Channel, meets at Cape Frazer (Grinnell Land), Lat. 80° N. with a tidal wave coming from the north, which I believe to be the wave No. 3, which has travelled round the north coast of Greenland, thus proving it to be an island.

1. *Behring Strait Tidal Wave.*

Observations on this tidal wave have been made at—

1. Port Clarence - - Captain Moore.
2. Point Barrow - - Captain Rochfort Maguire.
3. Walker Bay - - Captain Collinson.
4. Cambridge Bay - Captain Collinson.

All these observations lead to the result that this tidal wave is a simple lunar semi-diurnal tide, without any complication of solar or of diurnal tide, which seem, from some unknown cause, unable to enter the Arctic Basin through Behring's Strait, although the diurnal tide is well developed in many parts of the North Pacific Ocean. This tide has been traced eastwards as far as Victoria Strait, where it meets the Davis' Strait tide No. 2, entering Victoria Strait, from the north, through Bellot Strait and Franklin Strait.

[The Franklin expedition perished at the meeting of these two tides, which forms a line of still water and immoveable pack ice. In fact the "Erebus" and "Terror," having become beset in September 1846, were abandoned in April 1848, having moved only 15 miles during the 18 months.]

It is extremely probable that the Behring Strait tidal wave enters Banks' or Maclure Strait and passes as far eastward as the Bay of Mercy, where Maclure's Expedition was abandoned, in 1853, after two years ineffectual attempts to enter Melville Sound from the West. I am persuaded that this failure was due to the meeting of the Behring Strait and Davis' Strait tidal waves at the western outlet from Melville Sound. Unfortunately this important fact cannot be determined with certainty in consequence of the apparent loss of the tidal observations made by Maclure in the Bay of Mercy in 1851–52–53; and by Kellett in Bridport Inlet in 1852–53. If these tidal observations could be discovered they would throw much light on the theory of the tidal motion of this part of the American Arctic Archipelago.

2. *Davis' Strait Tidal Wave.*

This tidal wave is much better known than that of Behring Strait. Observations upon it have been made at—

1. Fredericksdal	-	Missionary Asboe.
2. Godthaab -	-	Dr. Rink.
3. Holsteinborg	-	Director Elberg.
4. Próven -	-	Assistant Bolbroe.
5. Frederickshaab	-	——
6. Port Leopold	-	Sir James Ross.
7. Bellot Strait	-	Sir Leopold McClintock.
8. Beechey Island	-	"Resolute" and "Assistance", "North Star."
9. Griffith Island	-	"Resolute" and "Assistance."
10. Refuge Cove	-	Sir E. Belcher.
11. Northumberland Sound.		Sir E. Belcher.

This tidal wave, in passing Cape Farewell has a luni-tidal interval of $6^h\ 22^m$, which is increased (Inglefield) to $11^h\ 0^m$ at Upernavik, and to $11^h\ 50^m$ at Van Rennselaer Harbour (Kane). The diurnal element is well developed along the Greenland coast. On reaching the head of Baffin's Bay, the tidal wave moves northward through Smith's Sound, and (according to Captain Markham) meets another tide at Cape Frazer.* The tidal wave flows also through Lancaster Sound to the westward to Port Leopold, where it divides into three branches, through—

a. Barrow Strait (westward).

b. Wellington Channel, Queen's Channel, and Penny Strait (northward).

c. Prince Regent Inlet (southward).

The progress of the tidal waves may be thus estimated by the luni-tidal intervals :—

		H.	M.
(*a*).	Port Leopold - - - -	11	44
	Griffith Island (Admiralty Tide Tables) -	0	15
	Dealy Island (" ") -	1	48
	Bay of Mercy, not given (Admiralty Time Tables).		
	(The range is given at 2 ft. in the Bay of Mercy, and at 4 ft. at Dealy Island; this circumstance, and the presumed difficulty of fixing the time of high water is in favour of the tide at Mercy Bay being the Behring Strait tide.)		
(*b*).	Port Leopold - - - -	11	44
	Penny Strait - - - -	0	15
(*c*.)	Port Leopold - - - -	11	44
	Bellot Strait - - - -	11	48

* Captain Markham's remarks show that the diurnal element is well developed in the tidal wave south of Cape Frazer.

All these tidal waves are complex, and consist of four well marked waves.

1. Lunar semidiurnal.
2. Solar semidiurnal.
3. Lunar diurnal.
4. Solar diurnal.

This tidal wave cannot, therefore, for a moment be confounded with the Behring Strait wave, which is a simple lunar semidiurnal wave.

The western branch moves (as I believe) across Melville Sound, and meets the Behring Strait tidal wave in Maclure Strait. The northern branch proceeds regularly through Penny Strait to lat. 76°, showing no sign of meeting an opposing tide although it would probably meet the Behring Strait tide somewhere about 80°. The southern branch, as I have proved meets the Pacific tide at the north entrance of Victoria Channel, where the Franklin expedition was abandoned. If the statement of the meeting of two tidal waves in Kennedy Channel be confirmed, it will diminish the chance of reaching the North Pole by that route, even though the northern tidal wave be not the Behring Strait wave which is highly improbable.

It is not at all unlikely that the Behring Strait tidal wave may meet the united Atlantic waves to the north of Greenland, and at this side of the Pole ; in which case it is probable that sledges will do more work than ships.

As it may be of use to determine quickly the character of the tidal wave, I now give a method of doing so.

II.—Method of determining quickly the existence of a Diurnal Tide.

Hourly observations of the height of the tide made for 48 successive hours, will determine accurately the diurnal tide for every hour of the middle 24 hours. Let h_1, h_2, h_3, be three heights of tide separated from each other by intervals of 12 hours, then the diurnal tide, at the period corresponding to the middle observation h_2 is given by the formula :—

$$D=\frac{h_1-2h_2+h_3}{4} \quad (1.)$$

The time selected for making the 48 hours observations should be when the Moon's declination is great (either north or south) because the diurnal tide vanishes with the declination of the Moon or Sun respectively. The expression for the diurnal tide is of the form,—

$$D=M \sin 2\mu \cos (m)+S \sin 2\sigma \cos (S) \quad (2.)$$

Where μ=Moon's declination.

σ=Sun's declination.

m=An angle that goes through all its changes in a lunar day.

s=An angle that goes through all its changes in a solar day.

At the time of equinox $\sigma=0$, and hence the 48-hour observation, if made at this time, and also when $\mu=0$, would show the non-existence of a diurnal tide, although there might be really a large one. The form of equation (2.) shows the reason for directing the observations to be made when the Moon's declination is great.

As a rule the diurnal tide is of considerable amount both lunar and solar, in all the branches of the Davis' Strait tidal wave; and in some cases the solar diurnal tide is actually greater than the lunar diurnal tide.

III.—General Rules for Tidal Observations.

Much valuable time has been often misspent on tidal observations of little value, and great disappointment felt at the small results produced by most laborious and carefully conducted observations; whereas at other stations, a simple month's observations properly made have given results of great value, although the observations themselves did not cost one-tenth part of the labour of other observations which gave but little result.

I offer the following suggestions for tidal observations made for a lengthened period.

1. Hourly observations of height should be made for one month at the times of solstice and equinox.

2. At the intervening periods, in order to save the labour of the observers as much as possible, it is recommended (instead of noting the time and height of high and low water each day) that the height of the tide should be registered every *four* hours of *mean solar time*. This would correspond with the times of striking bells, which would ensure punctuality and accuracy as to the time of observation, and the observation itself could be made in one minute. I should prefer observations made every *four* hours, for this reason among others, that the diurnal and semi-diurnal tides could be at once separated, and discussed independently of each other.

3. The times of observation must be carefully kept to, but whether the exact hours, or a fixed number of minutes after the exact hours, may be decided according to the convenience of the observers.

4. Remark carefully that the times of observation must be according to *mean solar time*, not according to apparent solar time.

---

3. Pendulum Observations. By Prof. Stokes, M.A., Sec. R.S.

It must be remembered that pendulum observations are of little value unless very accurately made.

The pendulum station will of course be adjacent to the ship's winter quarters. It must *if possible* be on land, chiefly because the clock's rate at the time of observation must be determined by transits, and we have no guarantee that ice covering the sea, however apparently firm, may not be subject to small motions in azimuth, which would vitiate the transits.

Clear weather should be chosen for the observations, that transits may be observed.

The observers are assumed to be already acquainted with the mode of making pendulum observations, and therefore it will only be necessary to mention some precautions.

It is recommended that great care be taken as to the mode of illuminating the bright patch on the clock-pendulum. Sir George Airy found a gold-leaf surface of an oblique section of a cylinder projecting from the bob towards the observer to be best. The light is then to be lateral, and may be distant.

As even an astronomical clock cannot be trusted to go for short intervals of time with a rate equal to its mean rate for 24 hours, it is desirable to take a series of consecutive swings extending over 24 hours, which would have the further advantage that the mean temperature of the pendulum would more accurately correspond to the mean indication of the thermometers. The time chosen for commencement should be about the middle of the time most favourable for transits. As a swing may be expected to last about four hours, and it is sufficient to observe two or three coincidences at the beginning and end of each swing, the observer would have time enough to take transits and to rest in the intervals between observing coincidences. The observer must remember, however, that he is responsible for the number of coincidences that have taken place, and therefore he would do well to take at first, or in preliminary trials, one or two intermediate coincidences, merely as counters not intended for reduction, and not leave off this practice till he has convinced himself that it may be safely dispensed with.

In observing coincidences the observer must, of course, register both the disappearance and the reappearance of the mark. But as it is somewhat perplexing to observe and register four events which succeed one another at intervals of a few seconds, namely, the two disappearances (those of the right and left edges of the mark) and the two reappearances, the observer (unless he can thoroughly depend upon himself to record the four events without confusion) is advised to be careful in the adjustment of the mark and diaphragm, so as to secure the two disappearances or the two reappearances taking place on consecutive seconds even when the pendulum is swinging in the smallest arc that will be observed with, in which case it will, of course, suffice to observe one disappearance and one reappearance for each coincidence.

The barometer and the thermometers hung near the pendulum should be read at the beginning and end of each swing. Should there be much variation of temperature, the thermometers should also be read at noted times once or oftener during the swing. It is to be remembered that what we want to know is, not the exact temperature at the moment of coincidence, but the mean temperature during the swing.

In one of the swings, or, if more convenient, in a preliminary or subsequent special swing taken for this sole object, and in which coincidences need not be attended to, the arc should be carefully observed five or six times distributed over the swing, or, which would probably be found more accurate, the clock times should be noted when the arc attains definite values, beginning with an arc slightly greater than the greatest used for coincidences, and going on till it is reduced to about one-third of its initial value. The barometer and thermometer should be read at the same time. The object of this is to determine the law of decrease of the arc, and thereby render it possible, in the subsequent reduction of the observations for time, to correct for the arc without assuming that it decreases *strictly* in geometric progression.

The geographical position, latitude especially, of the pendulum station must be found, and the height above the sea level. The geological character of the formation on which the pendulum observatory is built should be stated. Should it be found impracticable to erect the observatory on land, it may be built on the ice, provided there be no sensible change of level of the ice, and no motion of any kind, the alteration of which is not extremely gradual, and provided also, that means can be employed for checking the clock's rate by astronomical observations. Should the pendulum be swung on ice, the depth of the sea at the place must be measured.

Twenty-four hours' observation with each pendulum would give an excellent result, provided the weather permit of a trustworthy determination of the clock's rate. The days on which the two pendulums are swung need not be consecutive.

---

4. On the DETECTION of METEORIC (COSMICAL) DUST in the SNOW of ARCTIC REGIONS. By PROF. H. E. ROSCOE, F.R.S.

It has been shown by Nordenskiöld* that pure snow collected in the northern regions far distant from any source of dust, contains small black particles left behind when clean snow is melted. These black particles consist mainly of iron, but contain distinct quantities of cobalt, thus proving their non-terrestrial character. It would be very interesting to confirm these observations of the wide-spread depositions of fine cosmical dust by a repetition of the process adopted by Nordenskiöld, which consisted in collecting a large quantity of apparently pure snow, and allowing the same to melt, placing it, for this purpose on a clean sheet, spread out, and arranged so that the water should drain away, leaving the black particles on the sheet. These should then be carefully collected, when the greater part of the snow was melted, by placing the remaining snow in a bottle or glass, and allowing it to melt completely, when the black particles will sink to the bottom and the clear water can be poured off. Or the black particles can

* Poggendorff's Annalen, 151, p. 154.

be brushed off the sheet by means of a feather as soon as all the snow has melted. They must then be carefully preserved in a tube or stoppered bottle and brought home for analysis.

The same observer noticed that the black magnetic particles were frequently seen in the "firn" or granular old snow above which several layers of recent snow had accumulated; it would therefore be well to look out for the black grains below the snow.

II.—TERRESTRIAL MAGNETISM.

1. MEMORANDUM on DETERMINATION of ELEMENTS and use of MAGNETICAL INSTRUMENTS, by PROFESSOR J. C. ADAMS, M.A., F.R.S., and CAPTAIN EVANS, C.B., F.R.S.

The determination anywhere in the Arctic regions of the elements, by means of which the earth's magnetic force is usually expressed (*Declination, Inclination and Intensity*) will be valuable:—if made within the limits of former voyages, by affording the means of determining the approximate amount of the secular changes by comparison with earlier observations;—if made beyond the limits of former explorations, by materially adding to our knowledge of the distribution of the magnetic force over the earth's surface, and thereby contributing towards the perfection of the theory of Terrestrial magnetism.

The multiplication of the observations to be made in the Arctic Expedition being so much dependent on circumstances and climate, no definite suggestions can be offered on this head; it may, however, be borne in mind that the several elements above mentioned must be considered as possessing an equal importance, and that the value of each new station is proportional to its distance from those where observations have already been made.

The Article on Terrestrial Magnetism in the Admiralty Manual of Scientific Enquiry, by Sir Edward Sabine, K.C.B., with its appendices on the use of the principal instruments furnished to the Expedition, will be found an excellent guide for observers, and should be carefully consulted. In lieu of the maps therein referred to, provisional maps of the Magnetic Elements suitable to the requirements of the Expedition are appended to this memorandum.*

The instruments furnished comprise—

1. Portable unifilar magnetometers (in duplicate) for determining the absolute horizontal intensity at a fixed station.

2. Barrow's Circles (in duplicate) for determining the inclination. These circles are further provided with additional needles for the

* Magnetical instructions (in duplicate) for the use of portable instruments adapted for magnetical surveys and portable observatories, &c. by Lieut. C. J. B. Riddell, R.A., F.R.S., 1844, are also furnished.

purpose of determining the total force by Dr. Lloyd's method, which is independent of any changes in the magnetic moments of the needles employed. It should be observed that the poles of the *additional* needles so furnished *are never to be reversed or disturbed.**

3. Azimuth compasses (in duplicate), fitted with Admiralty Standard circle, special needles, and levelling foot screws, for determining the absolute declination. The declination as determined by this instrument at a fixed station is to be considered as the zero to which the observations made with the differential declinometer (mentioned in the following paragraph) are to be referred.

4. Portable declination magnetometer, for differential observations only at a fixed station.

5. Mr. Fox's apparatus for observing the inclination and force (in triplicate) for use in sledge or travelling parties. [This instrument is generally known as Fox's Circle].

6. Three-inch prismatic compasses (nine in number) for observing the declination (variation of the compass), to be used by sledge or travelling parties.

At winter quarters, and in an observatory established at a distance from the ship, so as to be free from the disturbance of her iron, it is assumed that the declination magnetometer will be firmly secured on its pedestal; and pedestals or stands arranged for the reception of the unifilar magnetometer and Barrow's circle, at suitable distances apart, to avoid inter-disturbance among the magnets of the several instruments. The necessary observations should, if possible, then be made as follows :—

It may be expedient frequently to determine the absolute declination with the azimuth compass specially furnished for the purpose before winter darkness sets in, so that a reliable zero may be obtained for the differential observations of declination. From 6 to 9 a.m., and from 4 to 6 p.m., will probably be found the best times for observing, on the assumption that the declination is then at or near the mean daily value.

The line of detorsion of the declination magnetometor should be carefully adjusted at the outset.

The inclination by Barrow's circle, and also the total (relative) force by means of Lloyd's needles to be observed once a week. Occasionally, it is desirable that the observations with Lloyd's needles should be repeated several times in the day, in order to find the approximate amount of the diurnal changes.

The absolute horizontal intensity to be determined once a month, avoiding days of unusual disturbance as denoted by the declination magnetometer. At the same time, or nearly so, observations should be made with Lloyd's needles employed as deflectors at different distances, in order to obtain absolute results.†

* See Admiralty Manual, Appendix 2 B., p. 105.
† See Lloyd's Treatise on Magnetism, Art. 97, p. 99.

Following the course pursued at Point Barrow, in H.M.S. Plover in 1852–4, and at Port Kennedy in the Fox, Sir Leopold McClintock 1858–9, it is anticipated that hourly observations of the declination can be continuously made, and if so, the observations should be made at the commencement of each hour local mean time.

On all occasions of marked disturbance of the magnet and especially during the occurrence of an aurora, extraordinary observations should be made. If the changes are rapid and irregular, instead of recording the observations at stated intervals of time, observe the extreme readings of the scale and the times at which the magnet commences its return movement towards an opposite extreme, so as to determine the extent and duration of the movements in opposite directions.

At stations where the stay of the ship is only of a few days duration, the same observations for the absolute declination, inclination, and intensity, should be made, omitting the differential observations of declination.

Observations to be made by Sledge or Travelling Parties.

The apparatus devised by Mr. R. W. Fox, F.R.S., for observing inclination and force [Fox's circle] proved of such great value under exceptional conditions in the Antarctic Magnetic Survey, performed in H.M. ship Erebus and Terror (1840–3) under Sir James Ross, that it has been deemed expedient to furnish it to the present Arctic Expedition. The instrument from its construction, will, it is considered, be found invaluable to travelling parties as it can be carried safely (and manipulated) under circumstances which would be fatal to more delicate instruments.

Preparatory to travelling parties setting out from winter quarters (or from a fixed station when absolute determinations have been made) comparative observations for inclination and force, should be made with the Fox circles which accompany them, and the same repeated on return.

The comparative observations for inclination are requisite to determine the index errors of the Fox needles as they are *not reversible.* The observations made with the deflectors and the weights should be sufficiently extended to embrace the circumstances of all possible observations whilst travelling, so as to ensure that no travelling observation be lost for want of comparative observations. Experience will soon determine when travelling the extent to which deflectors and weights may be conveniently used.

The small prismatic compasses furnished for the use of travelling parties are exclusively intended for observing the declination (or variation of the compass). On account of the large changes in declination, consequent on a small change of geographical position in the regions of high latitudes which will be explored, these observations will, as a matter of necessity, be frequently required.*

* At the magnetic pole, the horizontal force vanishes and consequently the direction of the magnetic meridian is indeterminate. At the geographical

Reference to the maps of the magnetic elements accompanying this memorandum will show that these changes of declination do not arise from magnetical causes, as the direction of the needle with reference to the north magnetic pole remains the same and the inclination and force values remain comparatively unaltered. These compasses should be preserved with care, to ensure accuracy in the observations, and in aid thereof, small extra travelling compasses (fitted so as to be corrected for the declination if necessary), are furnished to perform the rougher work of the steering compass for sledge parties or travellers.

The several constants, index and temperature corrections of the various instruments have been determined at Kew and the Admiralty compass observatories, and will be furnished to the Expedition.

III.—METEOROLOGY.

1. Meteorological Instructions. By Robert H. Scott, F.R.S., Director of the Meteorological Office.

The meteorological observations to be made during the Arctic Expedition will, in the main, be similar to those made on any voyage, and accordingly the instructions for the management of the instruments and for taking the readings will in great measure be identical with those furnished to observers at sea under ordinary circumstances. The latest copy of these instructions is contained in the Report of the Conference on Maritime Meteorology in 1874, of which copies are furnished to the Expedition.

There are, however, certain points which require special notice, and particularly so as it is hoped that for a considerable period of time observations will be taken regularly at fixed stations or even on shore.

Hours of Observation.—In Sir J. C. Ross's expedition to the Antarctic regions the observations were taken hourly, but the expedition was always at sea, so that the number of available observers was never reduced by sledging expeditions, &c.

It is certain that observations will be taken as frequently as practicable on the present occasion, but it must be remembered that in all cases *quality of observations is of much greater importance than quantity*, so that if it be impossible to give *correct*

---

pole, the direction of the magnetic meridian is determinate, but that of the geographical meridian is indeterminate, every different meridian, as defined by its longitude from Greenwich, having its corresponding variation of the compass. At all points near the geographical pole the statement of the variation of the compass should be accompanied by a statement of the longitude of the meridian to which the variation is referred.

LINES OF EQUAL HORIZONTAL FORCE IN ABSOLUTE MEASURE. (Approximate 1875.)

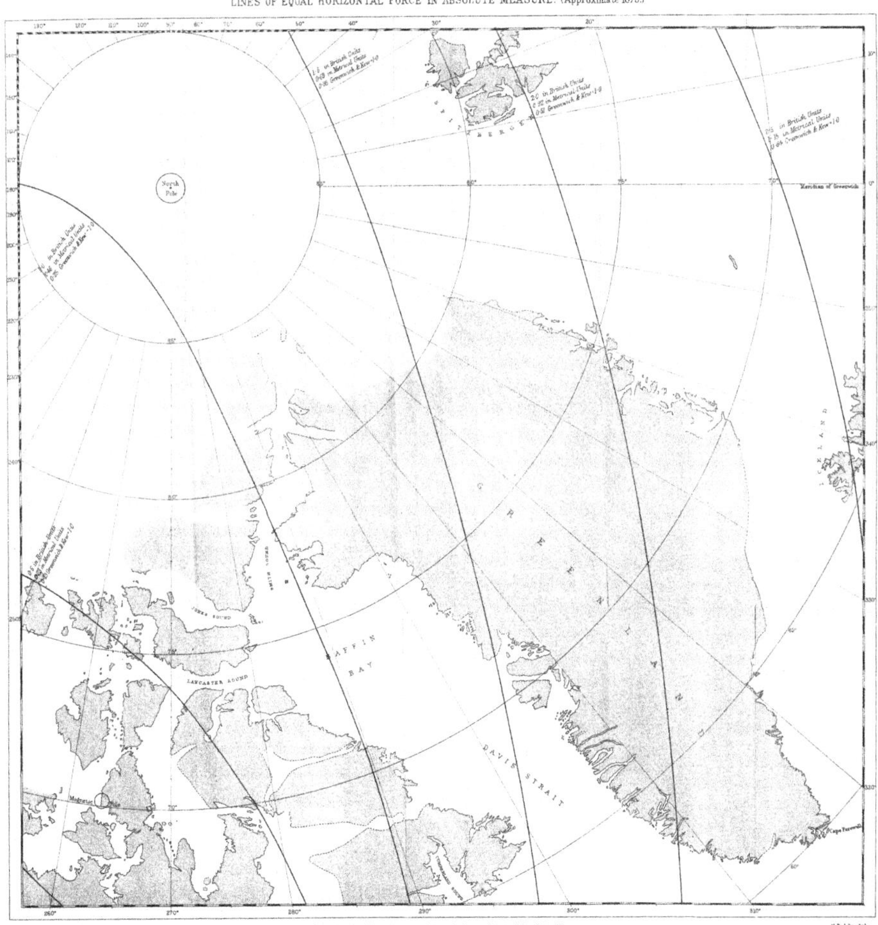

The material originally positioned here is too large for reproduction in this reissue. A PDF can be downloaded from the web address given on page iv of this book, by clicking on 'Resources Available'.

LINES OF EQUAL INCLINATION. (Approximate 1875.)

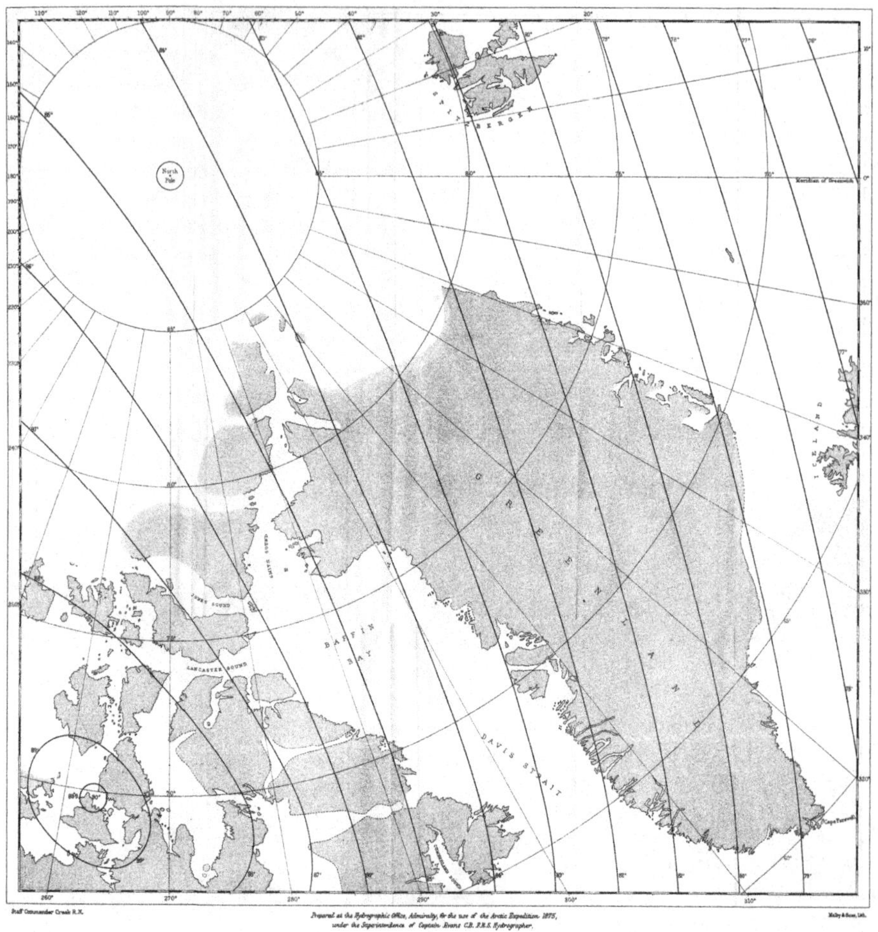

The material originally positioned here is too large for reproduction in this reissue. A PDF can be downloaded from the web address given on page iv of this book, by clicking on 'Resources Available'.

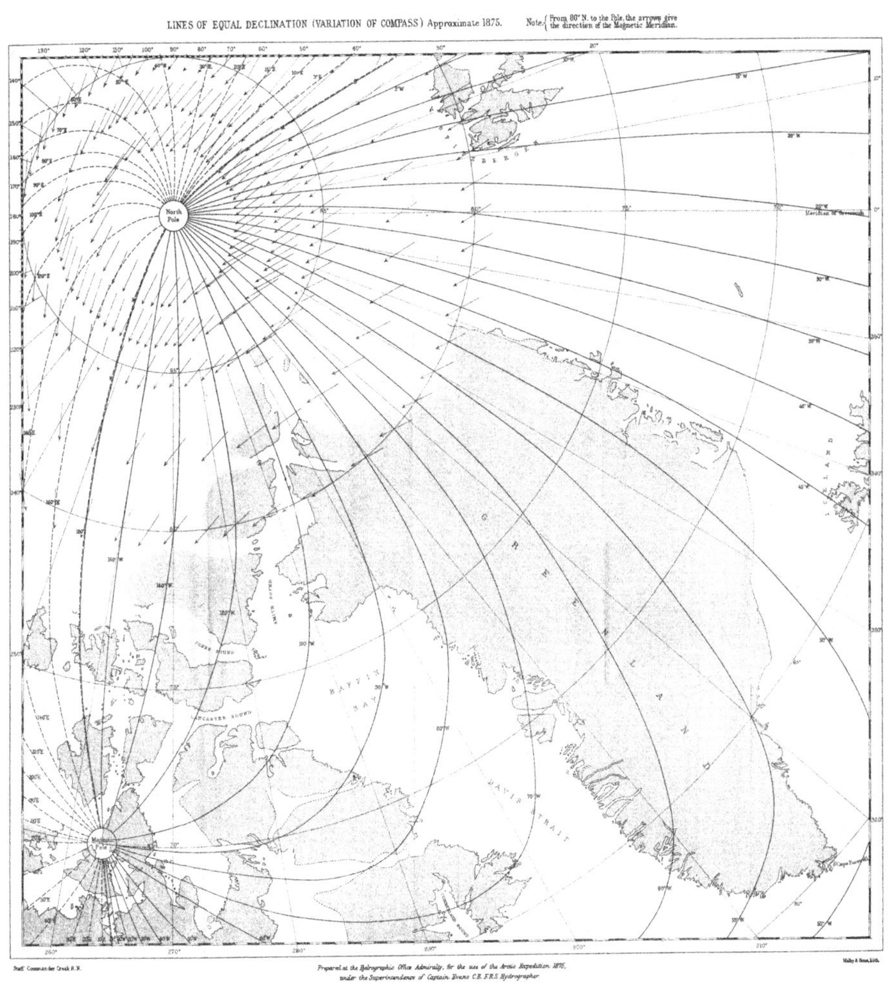

The material originally positioned here is too large for reproduction in this reissue. A PDF can be downloaded from the web address given on page iv of this book, by clicking on 'Resources Available'.

hourly observations, two-hourly, or in default of these four-hourly readings may be registered. They should always be at equidistant intervals of time.

The logs supplied are ruled for two-hourly observations.

Barometers.—The barometers supplied are of two kinds, marine barometers and aneroids. The latter are supplied for use on sledge expeditions and for determination of heights.

The mercurial barometers are the only barometers to be used for the regular observations. They should be suspended in some place where they will be shielded from sudden changes of temperature.

The aneroids must be carefully protected from blows or falls, which might seriously affect their action. Whenever opportunity offers, especially before and after their employment on any service, *e. g.*, an exploring expedition, the aneroids used should be most carefully compared with the mercurial barometers, giving the temperatures at each comparison, and the results of such comparisons most carefully noted in the log.

It may, perhaps, not be out of place to remark that the omission to make such entries in the logs of comparison of barometers, &c., and of the distinguishing marks and numbers on the instruments actually used, &c., will most seriously depreciate the scientific value of the registers though otherwise most carefully kept.

Temperature.—The thermometers should be suspended in some position where they will show the true temperature of the open air, and will be affected as little as possible by the warmth of the ship. For this reason, while the Expedition is in winter quarters, they should if possible be placed in an observatory at some distance from the ship.

Two thermometer screens are supplied for these observations, besides the ordinary small screens for use on board ship. They should be erected on posts, so that the bottom of the screen should be at a level of 4 feet above the surface of the snow ; but, of course, experience will alone show whether a height of 4 feet above the snow is sufficient.

The screens are furnished with doors at the back as well as in front, so as to allow of the reception of several thermometers, which may thus be placed back to back.

If any rime is deposited on the instruments it should be carefully removed some time (if feasible about half an hour), before the observation.

The maximum and minimum thermometers should be placed in the screen and read regularly at the latest observing hour in the evening, if not at midnight.

In addition other thermometers, both ordinary and registering, are supplied for various experiments.

As regards the pocket thermometers, it will be found a very good plan, in order to ascertain speedily the true temperature of the air, to pass a strong string through the hole at the top of the mounting and swing the thermometer round the head for about half a minute.

By this method, even in full sunshine, a very close approximation to the true temperature of the air may be made.

Radiation.—The radiation thermometers should be placed in the clips sent with them, and attached to an upright post at the height of 4 feet above the ground. Alongside of the black bulb thermometer *in vacuo* should be placed a bright bulb thermometer also *in vacuo*, which should be read at the same time, so as to obtain a measure of the radiation by the difference between the simultaneous readings of the two instruments.

It will further be interesting to erect a black board, say 2 feet square, at a level of 4 feet from the ground, and to place a black bulb *in vacuo* a few inches above this board, which will thus intercept all heat radiated from the ground.

For *terrestrial radiation*, a board about 2 feet square should be placed upon the ground, its upper surface being fully exposed to the sky. A slight groove in the board should mark the place where the bulb of the thermometer, a minimum thermometer, should be placed.

Hygrometry.—Two hair hygrometers are supplied, which should be erected in the thermometer screen. They should be read at the same hour as the ordinary instruments.

The principle of these instruments is as follows, the hygrometrical condition of the air is given by the elongation or contraction of a hair, according as the air is moist or dry.

They bear two scales on the same arc.

The *lower* scale (Saussure's) divided into *equal* parts 0—100 gives the so-called scale of absolute humidity ranging from perfect dryness to saturation. The *upper* scale, divided into *unequal* parts, 0—100 gives the relative humidity in per-centages. *The upper scale is that which is alone to be used.*

The following is the mode of setting the instrument; inasmuch as it easily gets out of adjustment in carriage, the following rules are to be observed: On a day of heavy rain or thick fog, when the air is perfectly saturated with moisture, the screw at the top is to be turned gently until the index comes to stand at 100 on the scale. If perfect saturation does not occur, the index should be brought by turning the screw to the point on the scale which gives the per-centage of relative humidity shown by careful observation of the wet and dry bulb hygrometer.

The instrument is to be suspended at some little distance, say 4 inches, in front of the upright to which it is attached. When the instrument has been originally set to be correct at 100 or at any definite degree of relative humidity, its indications should be carefully checked for a few days, and the position of the index corrected by means of the screw, when it will soon attain the requisite degree of consistency in its behaviour.

The regulations for observing the instrument, as in force at the Russian stations, where hair hygrometers are generally in use in winter, are as follows: Whenever the temperature is above 32°, the hygrometrical observations entered in the register, are to be recorded by means of the wet and dry bulb hygrometer *exclusively*, and the degree of humidity given in the Tables (Glaisher's 5th

Edition) for each observation is to be compared with the reading of the hair hygrometer taken at the same time which should be entered in the "Remarks" space. The mean of these observations for a few weeks when the temperature is above 32° is to be taken, and thereby the correction for the readings of the hair hygrometer, from time to time, is to be ascertained.

The following is an instance :—

Mean degree of Humidity by Wet and Dry Bulb Hygrometer.	Mean Reading of Hair Hygrometer.
73.	77.

correction to all readings of the hair hygrometer is accordingly —4.

Every possible opportunity must be taken at first to determine these corrections for the hair hygrometer, inasmuch as it is not to be expected that during the severe weather many such opportunities will occur. Whenever the Hair Hygrometer is readjusted, a note to that effect should be inserted in the register. It will be seen from the above that the object of reading the two different classes of instruments above 32°, is to determine a correction for the hair hygrometer while the wet and dry bulb themometer is acting properly. For readings below 32°, the hair hygrometer may be supposed to be a more trustworthy instrument than the wet and dry bulb hygrometer, but both instruments should be read together, and their indications entered in the register, as thereby very valuable experience of the comparative utility of the two instruments in the Arctic regions may be gained.

In addition a pair of mercurial thermometers are to be mounted as a dry and wet bulb hygrometer. The greatest care will be requisite to ensure that these observations are correct when the temperature is below 32°. Inasmuch as it is to be expected that for a long period the moisture on the wet bulb will be frozen, it will be necessary to see that the muslin covering of the bulb is always provided with a *very thin* coating of ice. This is effected by moistening it with pure water at least once a day, and not less than half an hour before an observation. The moisture on the bulb will freeze at once, and the film of ice so formed will in most cases last for 24 hours, inasmuch as at the very low temperatures which will probably prevail the evaporation from ice is very slight.

Care should be taken that the water employed is always fresh water; no addition of spirit or of salt, in order to lower its freezing point, is on any account to be admitted.

In case of severe cold setting in, *i.e.*, of temperatures below —20° F., it will be best to take in the mercurial thermometers and make all the observations with spirit thermometers, but it should never be forgotten that spirit thermometers are very deceptive, unless great care is taken to keep them in order. They should not be used if the spirit column be broken in any part by bubbles or shortened by condensation of spirit at the top of the tube.

Wind.—In addition to the ordinary observations of the estimated direction and force of the wind, an electrical anemometer is supplied, which should be erected on a high pole, or on the masthead when the ship is in winter quarters. The description of this instrument and its management and registration will be supplied with it.

Two small anemometers are also supplied; these must be read at least once daily.

Clouds.—The directions for observing clouds are contained in the ordinary Instructions for Observers at Sea, but there are certain observations of much interest which may be made on clouds as indicating the motion of upper currents in the atmosphere.

It is therefore of importance that, whenever possible, the direction and apparent rate of motion of the different strata of clouds should be recorded, whether these agree with the wind at the surface of the earth or not.

There is especial interest in the question of the motion of the upper clouds near the region of greatest cold, as it is believed by most meteorologists that these upper currents flow towards the districts of great cold, where the air descends to the surface of the earth.

Hydrometeors or Weather Observations.—These should be recorded as directed in the "Instructions in the Use of Meteorological Instruments," of which copies are supplied to the Expedition.

As regards the Aurora, Prof. Stokes has furnished certain notes which are appended to these instructions. The greatest care should be taken to note and register the most striking particulars as to its appearance and modifications, especially as to its manifestations in any quarter as regards the existence of open water, &c., in that direction. The subject of magnetic observations or of spectroscopic examination of the Aurora does not fall within the scope of meteorological observations.

As regards Halos, Parhelia, and Paraselenæ, &c., the variety in these appearances is so great that no definite rules can be laid down for observing them. The phenomena observed should be carefully sketched, and wherever possible measurements should be taken by sextant.

In conclusion, the attention of the observers should be specially directed to the fact that many of the storms which pass over the extreme north of Europe are connected with areas of barometrical depression which follow tracks lying within the Arctic Circle; they have been traced almost wherever the whaling and sealing ships have advanced between Spitzbergen and Greenland, and there is reason to believe that they will be observed at the head of Smith Sound also.

It will therefore be a matter of the greatest interest to notice whether or not these areas of barometrical depression appear to be formed near the spot where the Expedition is in winter quarters.

At all times of sudden barometrical oscillations the observations should be made as frequently as possible, as by this means it may be rendered possible to ascertain whether the centre of the disturbance has passed north or south of the ship.

A copy of Capt. Hoffmeyer's weather charts is supplied, and it will be evident from an inspection of these charts of what importance accurate observations in high latitudes will be for the subsequent study of the weather of Europe during the period embraced by the stay of the Expedition in the Arctic seas.

2. Note on Auroral Observations. By Prof. Stokes, Sec. R.S.

The frequency of the Aurora in Arctic regions affords peculiar facilities for the study of the general features of the phenomenon, as in case the observer thinks he has perceived any law, he will probably soon and repeatedly have opportunities of confronting it with observation. The following points are worthy of attention.

Streamers.—It is well known that, at least as a rule, the streamers are parallel to the dipping-needle, as is inferred from the observation that they form arcs of great circles passing through the magnetic zenith. It has been stated, however, that they have sometimes been seen curved. Should anything of this kind be noticed, the observer ought to note the circumstances most carefully. He should notice particularly whether it is one and the same streamer that is curved, or whether the curvature is apparent only, and arises from the circumstance that a number of short, straight streamers start from bases so arranged that the luminosity as a whole presents the form of a curved band.

Have the streamers any lateral motion, and if so, is it from right to left or left to right, or sometimes one and sometimes the other, according to the quarter of the heavens in which the streamer is seen, or other circumstances? Again, if there be lateral motion, is it that the individual streamers move sideways, or that fresh streamers arise to one side of the former, or partly the one and partly the other? Do streamers, or does some portion of a system of streamers, appear to have any uniform relation to clouds, as if they sprang from them? Can stars be seen immediately under the base of streamers? Do streamers appear to have any definite relation to mountains? Are they ever seen between the observer and a mountain, so as to appear to be projected on it? This or any other indication of a low origin ought to be most carefully described.

When streamers form a corona, the character of it should be described.

Auroral arches.—Are arches always perpendicular to the magnetic meridian? If incomplete, do they grow laterally, and if so, in what manner, and towards which side? Do they always move from north (magnetic) to south, and if so, is it by a southerly motion of the individual streamers, or by new streamers springing up to the south of the old ones? What (by estimation, or by reference to known stars) may be the breadth of the arch in different positions in its progress? Do arches appear to be nothing but congeries of streamers, or to have an independent existence? What relation, if any, have they to clouds; and if related, to what kind of clouds are they related?

Pulsations.—Do pulsations travel in any invariable direction? What time do they take to get from one part of the heavens to another? Are they running sheets of continuous light, or fixed patches which become luminous, or more luminous, in rapid succession; and if patches, do these appear to be foreshortened streamers? Are the same patches luminous in successive pulsations?

Sounds (?)—As some have suspected the Aurora to be accompanied by sound, the observer's attention should be directed to this question when an Aurora is seen during a calm. If sound be suspected, the observer should endeavour by changing his position, brushing off spicules of ice from the neighbourhood of his ears, his whiskers, &c., to ascertain whether it can be referred to the action of such wind as there is on some part of his dress or person. If it should clearly appear that it is not referable to the wind, then the circumstances of its occurrence, its character, its relation (if any) to bursts of light, should be most carefully noted.

These questions are proposed merely to lead the observer to direct his attention to various features of the phenomenon. Answers are not demanded, except in such cases as definite answers can be given; and the observer should keep his attention alive to observe and regard any other features which may appear to be of interest. It is desirable that drawings should be made of remarkable displays.

Observations with Sir William Thomson's electrometer would be very interesting in connexion with the Aurora, especially a comparison of the readings before, during, and after a passage of the Aurora across the zenith.

IV.—ATMOSPHERIC ELECTRICITY.

1. INSTRUCTIONS for the OBSERVATION of ATMOSPHERIC ELECTRICITY. By PROFESSOR SIR WILLIAM THOMSON, LL.D., F.R.S.

The instrument to be used is the portable electrometer described in Sir Wm. Thomson's reprint of "Papers on Electrostatics and Magnetism," §§ 368–378.* Full directions for keeping the instrument in order, preparing it for use, and using it to make observations of atmospheric electricity, are to be found in sections 372–376; these are summarized in the following short practical rules:—

I. The instrument having been received from the maker with the inner surface of the glass, and all the metallic surfaces within,

* A copy of this book has been sent by the author for the use of the officer or officers to whom the observations of atmospheric electricity are committed.

clean and free from dust or fibres, and the pumice dry. To prepare it for use:

(1.) Remove from the top the cover carrying the pumice. Drop upon the pumice a small quantity of the prepared sulphuric acid supplied with the instrument, distributing it as well as may be over the whole surface of the stone. There ought not to be so much acid as to show almost any visible appearance of moisture when once it has soaked into the pumice. Replace the cover without delay, and screw it firmly in its proper position, and then leave the instrument for half an hour or an hour, or any longer time that may be convenient to allow the inner surface of the glass to be well dried through the drying effect of the acidulated pumice on the air within.

(2.) Turn the micrometer screw till the reading is 2,000. (There are 100 divisions on the circle which turns with the screw on the top outside, and the numbers on the vertical scale inside show full turns of the screw. Thus each division on the vertical scale inside corresponds to 100 divisions on the circle; and 20 on the vertical scale is read "2,000.") Introduce the charging rod and give a charge of negative electricity by means of the small electrophorus which accompanies the instrument. When enough has been given to bring the hair a little below the middle of the space between the black dots, give no more charge; but remove the charging rod and close the aperture immediately. If now the hair is still seen a little below the middle of the space between the black dots, turn the screw head in such a direction as to raise the attracting disc, and so diminish the attraction till the hair is exactly midway between the dots. Watch the instrument for a few minutes, and if the hair is seen to rise, as it generally will (because of the electricity which has been given, spreading over the inner surface of the glass), turn the micrometer screw in the direction to lower the attracting plate, so as to keep the hair midway between the dots.

(3.) The insulation will generally improve for several hours, and sometimes for several days, after the instrument is first charged. The instrument may be considered to be in a satisfactory state if the earth reading does not diminish by more than 30 divisions per 24 hours. If the maker has been fortunate with respect to the quality of the substance of the glass jar, the earth reading may not sink by more than 30 divisions per week, when the pumice is sufficiently moistened with strong and pure sulphuric acid. Recharge with negative electricity occasionally so as to keep the earth reading between 1,000 and 2,000.

II. To keep the instrument in order. Watch the pumice carefully, looking at it every day. If it begins to look moist, remove the cover, take out the screws holding the lead cup, remove the pumice and dry it on a shovel over the galley fire. When cool put prepared sulphuric acid on it, replace it in the instrument, and reelectrify according to No. I.

Never leave the pumice unwatched, in the instrument, for as long as a week. WHEN THE INSTRUMENT IS TO BE OUT OF USE FOR A WEEK OR LONGER TAKE THE PUMICE OUT OF IT.

III. To use the portable electrometer for observing atmospheric electricity:

(1.) The place of observation, if on board ship, must be as far removed from spars and rigging as possible. In a sailing ship or rigged steamer the best position for the electrometer generally is over the weather quarter when under way, or anywhere a few feet above the tafferel when at anchor. On shore or on the ice a position not less than 20 yards from any prominent object (such as a hut or a rock or mass of ice or ship), standing up to any considerable height above the general level, should be chosen. Whether on board ship or in an open boat or on shore or on the ice, the electrometer may be held by the observer in his left hand while he is making an observation; but a fixed stand, when conveniently to be had, is to be preferred, unless in the case of making observations from an open boat.

(2.) To make an observation in ordinary circumstances the observer stands upright and holds or places the electrometer in a position about five feet above the ground (or place on which he stands), so as to bring the hair and two black dots about level with his eye. The umbrella of the principal electrode being *down* to begin with (and so keeping metallic connection between the principal electrode and the metallic case of the instrument), the observer commences by taking an "earth reading." * The steel wire, with a match stuck on its point, being in position on the principal electrode, the match is then lighted, the umbrella lifted, and the micrometer screw turned so as to keep the hair in the middle between the black dots. After the umbrella has been up and the match lighted for 20 seconds or half a minute, a reading may be taken and recorded, called an "air reading." A single such reading constitutes a valuable observation. But a series of readings taken at intervals of a quarter of a minute, or half a minute, or at moments of maximum or minimum electrification during the course of two or three minutes, the match burning all the time, is preferable. In conclusion, remove the match if it is not all burned away, lower the umbrella home, and take an earth reading.

(3.) The electric potential of the air at the point of the burning match is found by subtracting the earth reading from the air reading at any instant. When the air reading is less than the earth reading the air potential is negative, and is to be recorded as the difference between the earth reading and the air reading with the sign – prefixed. The earth reading may be generally taken as the mean between the initial and final earth readings. But the actual earth readings and air readings ought all to be recorded carefully, and the full record kept.

(4.) Note and record the wind at the time of each observation, also the character of the weather.

IV. Observations to be made:

(1.) At the commencement of the Expedition, in the course of the northward voyage, observations of atmospheric electricity

* Electrostatics and Magnetism, § 375.

ought to be taken regularly three or four times a day; also occasionally during the night to give the observer some practice in the use of a lantern for reading the divisions on the circle and of the vertical scale.

(2.) When stationary in winter quarters observations should be made three times a day at intervals of six hours; for example, at 8 a.m., 2 p.m., and 8 p.m., or at 7.30 a.m., 1.30 p.m., and 7.30 p.m. Whatever times are most convenient may be chosen provided they be separated by intervals of six hours.

(3.) It is very desirable that hourly observations should be made, if only for a few days, in winter and in summer. If possible arrangements to do so at least for six consecutive days in winter, and for six consecutive days in summer should be made. The results will be very interesting as showing whether there is a diurnal or semi-diurnal period in either the Arctic winter or summer, as we know there is at every time of year in places outside the Arctic circle.

(4.) Make occasionally special observations when there is anything peculiar in the weather, especially with reference to wind.

V. Special precautions:

(1.) In the Arctic climate more care may be necessary than in ordinary climates as to earth connections. Therefore put a piece of metal on the stand on which the electrometer is placed during an observation on board ship, and keep this in metallic communication with the ship's coppers or lightning conductors. If the electrometer is held in the hand with or without a glove, a fine wire ought to be tied round the brass projection which carries the lens, or otherwise attached to the outer case of the electrometer, and by this wire sufficient connection maintained with the earth during an observation. The connection will probably be sufficient if a short length of the wire is laid on the ice and the observer stands on it. Enough, however, is not yet known as to electric conductivity of ice: and to make sure it *may* be necessary to have a wire or chain let down to the water through a hole in the ice, and metallic connection kept up by a fine wire between this and the electrometer case during an observation.

(2.) The observer's cap (particularly if of fur) and his woollen clothing, and even his hair if not completely covered by his cap, will be apt in the Arctic climate to become electrified by the slightest friction, and so to give false results when the object to be observed is atmospheric electricity. A tin foil cover for cap and arms, kept in metallic communication by a fine wire with the hand or hands applied to the case of the electrometer or to the micrometer screw head, should therefore be used by the observer (and assistant, if he has an assistant to carry lanthorn, or for any other purpose), unless he has made sure that there is no sensible disturbance from those causes, without the precaution.

VI. Instruments, stores, and appliances for observation of atmospheric electricity sent with the Expedition:

1. Two portable electrometers, Nos. 35 and 36, each with one steel wire for carrying match, one charging rod, and one electrophorus for charging the jar.

2. Six spare steel wires (three to go with each instrument).

3. Supply of matches ready made. (The slower the match burns, the better. If those supplied burn too fast, steep them in water and dry them again.)

4. White blotting paper and nitrate of lead to make more matches when wanted. (Moisten the paper with weak solution of nitrate of lead, and roll into matches with thin paste made with a very little nitrate of lead in the water.)

5. Six spare pumices (three for each electrometer); india rubber bands to secure pumice in lead case.

6. Eight small stoppered bottles of prepared sulphuric acid (four for each electrometer).

7. Tin foil and fine wire.

V.—OPTICS.

1. Spectroscopic Observations. By Prof. G. G. Stokes, Sec. R.S.

(1.) Observations of the Spectrum of the Sun with a view to Terrestrial Absorption.

It has long been known that when the Sun is near the horizon additional lines and bands are seen in its spectrum, which are either not observed when the Sun is high, or are found to be much narrower. These are referrible to terrestrial absorption, and maps have been made of them by Brewster and Gladstone, by Ångström, and by Hennessey, which are sent with the Expedition. Recent researches appear to show that the greater part at any rate of these additional lines are due to watery vapour, but it is still a question whether some of them may not be due to some other constituent of the earth's atmosphere, to some substance present in the atmosphere in such minute quantity as to elude chemical tests.

In the extreme cold of the Arctic regions the quantity of water present in the elastic state in the atmosphere must be comparatively small, and consequently the absorption due to aqueous vapour at a given small altitude of the Sun, might be expected to be considerably reduced as compared with what is observed in warmer, and especially in tropical countries; and, as we have no reason to suppose that the other absorbing constituent or constituents of the atmosphere, if such there be, would be similarly affected by cold, the comparison of the absorption-spectrum obtained in Arctic countries with that observed in more temperate climates might afford means of detecting bands of absorption, if such there be, of other than aqueous origin, and thereby perhaps, by subsequent researches at home, of leading to the discovery of some other constituent of the atmosphere present in quantity too minute to admit of direct detection. Besides, the length of time that the Sun remains at a low altitude every day shortly after he makes his first appearance in the spring affords time for more deliberate observations than can be made during the few minutes he remains at a low altitude in places in comparatively low latitudes.

The Expedition will be furnished with a spectroscope of high dispersion, and with maps of the spectrum showing the additional bands seen when the Sun is low.

The best time for observation will be in the spring, shortly after the first appearance of the Sun, or, if circumstances should allow, in the autumn, shortly before he disappears for the winter, inasmuch as in either case the Sun will remain for a comparatively long time at a very low altitude.

As the appearance of the spectrum changes a good deal with the degree of detail in which it is seen, the spectrum should, in a preliminary trial, be compared with the maps, and that map more especially worked with which best matches the object. The power of the spectroscope might even be reduced by the removal of one or two of the prisms, if a better match as to degree of detail can thus be obtained.

For the actual observations, days should be chosen when the Sun is clear down to the horizon. The object is to be compared with the map or maps selected, going regularly through a portion of the spectrum each day as time permits, and making memoranda of the accordance or otherwise of the object with the map. Measurements need not in general be taken, except when the identification of a line is doubtful. Should the relative strength of any terrestrial line as compared with its fellows appear distinctly different in the object from what it is in the map, such line should be marked for re-examination.

The lines so marked should subsequently be re-examined at various low altitudes of the Sun. It is to be remembered that the various lines or bands of absorption as seen under otherwise given conditions and at a given place, do not all increase in the same proportion as the Sun approaches the horizon; so that the apparent abnormal strength of any particular line as compared with its fellows which had been noted in the first instance *might* be due to its having been seen at a different altitude of the Sun from that to which the map relates.

The probability of the discrepancy between the object and the map being thus explicable is to be judged of by the result of the re-examination. Should it appear at all probable that the result is *not* thus to be explained, the line should be noted by reference to the map, or if its identification should be at all doubtful, its distances from two neighbouring easily identifiable lines right and left of it should be measured, and its breadth measured or estimated with reference to some other line, the appearance of which agrees with its representation in the map.

Should any of the lines, even after the re-examination remain apparently discrepant from the map, it would be well, if leisure permit, to examine them again further on in the season under different conditions of temperature, direction of wind, &c., and note whether any change is observed.

(2.) Spectrum of the Aurora.

The spectrum of the aurora contains a well-known conspicuous bright line in the yellowish-green, which has been accurately observed. There are also other bright lines of greater refrangibility, the determination of the positions of which is more difficult account of their faintness, and there are also one or more lines in the red in red auroras.

Advantage should be taken of an unusually bright display to determine the position of the fainter lines. That of the brightest line, though well known, should be measured at the same time to control the observations. The character of the lines (*i.e.*, whether they are strictly lines, showing images of the apparent breadth of the slit, or narrow bands, sharply defined or shaded off) should also be stated.

Sometimes a faint gleam of light is seen at night in the sky, the origin of which (supposed from the presence of clouds) is doubtful. A spectroscope of the roughest description may in such cases be usefully employed to determine whether the light is auroral or not, as in the former case, the auroral origin is detected by the chief bright line. The observer may thus be led to be on the look out for a display which otherwise might have been missed.

It has been said, however, that the auroral light does not in all cases exhibit bright lines, but sometimes, at least in the eastern and western arch of the aurora, shows a continuous spectrum. This statement should be confronted with observation, special care being taken that the auroral light be not confounded with light which, though seen in the same direction, is of a different origin, such, for example, as light from a bank of haze illuminated by the moon.

Sir Edward Sabine once observed an auroral arch to one side (say north) of the ship, which was in darkness. Presently the arch could no longer be seen, but there was a general diffuse light so that a man at the mast head could be seen. Later still the ship was again in darkness, and an auroral arch was seen to the south.

Should anything of the kind be observed, the whole of the circumstances ought to be carefully noted, and the spectroscope applied to the diffuse light.

---

2. Polarisation of Light. By W. Spottiswoode, M.A., LL.D., Treas. R.S.

The fact of atmospheric polarisation, and the laws which regulate it, are already well known. And it is therefore not probable that observations upon it, although made under somewhat unusual circumstances, will add materially to our knowledge. At the same time, as the instruments are extremely portable, and the observations readily made, it appears quite worth while to repeat some of them.

The main features of polarisation in light from the sky are described in the book which accompanies the instruments; and they may be observed with a Nicol's prism and biquartz, with a Savart's polariscope, or even better with a Nicol, or a double-image prism, alone. The statement that the polarisation is due to the scattering of light from small globules of water suspended in the atmosphere in the shape of mist must be regarded perhaps rather as a suggestion than as an established fact; and any observations made under different atmospheric conditions capable of being brought to bear on this question will be valuable.

It is known that the light coming from a rainbow is polarised. It will be worth while to examine whether the same is the case with that from *halos*, &c. If this be so, observe the positions of the Nicol, or double-image prism, in which the light is extinguished (or most enfeebled) at different parts of the phenomenon.

It has been suggested that the *Aurora*, inasmuch as it presents a structural character, may afford traces of polarisation. Having reference to the fact that the striæ of the electric discharge in vacuum tubes presents no such feature, the probability of the suggestion may be doubted. But it will still be worth while to put the question to an experimental test.

If traces of polarisation be detected, it must not at once be concluded that the light of the Aurora is polarised; for the Aurora may be seen on the background of a sky illuminated by the moon, or by the sun, if not too far below the horizon, and the light from either of these sources is, in general, more or less polarised; therefore, if the light of the Aurora be suspected to be polarised, the polariscope should be directed to an adjacent portion of clear sky, free from Aurora, but illuminated by the moon or sun as nearly as possible similar, and similarly situated, to the former portion; and the observer must then judge whether the polarisation first observed be merely due to the illumination of the sky.

The light from the *Ice blink* should be also examined for traces of polarisation.

The presence of polarisation is to be determined:

(1.) With a Nicol's prism, by observing the light through it, by turning the prism round on its axis, and by examining whether the light appears brightest in some positions and least bright in others. If such be the case, the positions will be found to be at right angles to one another. The direction of "the plane "of polarisation" will be determined by that of the Nicol at either of these critical positions. The plane of polarisation of the light transmitted by a Nicol is parallel to the longer diagonal of the face; and, accordingly, the plane of polarisation, or partial polarisation, of the observed light is parallel to the longer diameter of the Nicol when the transmitted light is at its greatest intensity, or to the shorter when it is at its least.

(2.) The observation with a double-image prism is similar to that with a Nicol. This instrument, as its name implies, gives the images which would be seen through the Nicol in two rectangular positions, both at once, so that they can be directly compared; and when in observing polarised light the instrument is turned so that one image is at a maximum, the other is simultaneously at a minimum. Both these methods of observation, (1) and (2), are especially suitable for faint light; because in such a case the eye is better able to appreciate differences of intensity than differences of colour.

(3.) The observation with a biquartz differs from (1) only by holding a biquartz (a right-handed and a left-handed quartz cemented side by side) at a convenient distance beyond the Nicol, and by observing whether colour is or is not produced. If the Nicol be so turned that the two parts of the biquartz give the same colour (choose the neutral tint, teint de passage, rather than the yellow), we can detect a change in the position of the plane of polarisation by a change in colour, one half verging towards red, the other towards blue. This observation is obviously applicable to a change in the plane, either at different parts of the phenomenon at the same time, or at the same parts at different times.

(4.) We may use a Savart's polariscope, which shows a series of coloured bands in the field of view. For two positions at right angles to one another, corresponding to the two critical positions of a Nicol, these bands are most strongly developed; for two positions midway between the former the bands vanish. In the instruments here furnished the plane of polarisation of the observed light will be parallel to the bands when the central one is light, perpendicular to them when the central band is dark.

The instruments supplied will enable the officer charged with them to repeat all the principal experiments in the polarisation of light. After a little practice the observer will be able to make his own selection of apparatus, and to apply the method of observation best suited to the particular circumstances in which he finds himself placed.

---

3. INSTRUCTIONS in the use of the SPECTROSCOPES supplied to the ARCTIC EXPEDITION. BY J. NORMAN LOCKYER, F.R.S.

The instruments supplied are as follows :—

A.—An automatic 6-prism spectroscope of large dispersion for observations of the Sun.

B.—A direct-vision spectroscope by Merz for observations of the sun and of the aurora.

C.—A direct-vision spectroscope by Browning for observations of the aurora.

D.—A miniature spectroscope for observations of the aurora.

Instructions for use of Instrument A.

The stand, with its train of prisms, should be taken out of the box and placed on a firm support. The prisms if dusty should then be wiped with a soft brush or leather, care being taken, if the leather be used, to move it in one direction as little as possible, to prevent scratches. The tube furnished with a focussing screw should, after the adjustment referred to in the next paragraph, be screwed into the ring which moves along the arc. The other tube should be fixed into the other fixed ring.

To focus the observing telescope insert the eye-piece to be used, and obtain an image of a distant object, using a piece of glass, of the colour of that part of the spectrum to be observed, between the eye-piece and the eye. When this has been done and both telescope and collimator fixed in position, the image of the slit should be brought into focus by using the sliding tube of the collimator.

The slit plate and slipping piece should be so inserted into the collimator tube that the adjusting screw of the slit will be on the side nearest the observer.

The whole train of prisms should be covered with black velvet. This may be supported by a piece of thick cardboard (blackened) resting on the two telescopes, and train of prisms in order to prevent it from intercepting any of the light.

Coloured glasses are supplied in order to prevent the observations of each part of the spectrum being interfered with by the presence of stray light of any other colour than that under observation; these should be placed in front of the slit.

It will be well to commence all series of observations of the region B—A at the more refrangible side of B, noting micrometer reading, both number of revolutions and $\frac{1}{100}$ths, in each case. B will then serve as a zero. It will be found that the slit will bear opening without much loss of definition as the less refrangible region between *a* and A is approached.

The less refrangible line of D and *b*, and F and C themselves, may be used in the same manner for observations in the regions of the spectrum near these lines.

To render the light more intense lenses are supplied to throw an image of the sun or an intense parallel beam through the slit: the sun can be followed by the tangent screw movement fixed to the support. The height of the sun should be recorded against each observation, and this can be facilitated by keeping the collimator carefully levelled, marking degrees on the slit plate and observing where the image of the sun falls.

It will be observed that in Ångström's maps the gradations of the effect of absorption from maximum (above) to minimum (below) are shown. In Mr. Hennessey's maps these gradations are absent. It will be well to prepare a wide spectrum, on a scale about equal to that employed by Mr. Hennessey, showing the gradations and the variations, if any, from those observed by Ångström.

Instructions for use of Instrument B.

Three sets of direct-vision prisms are supplied with this spectroscope. When all three sets are used the dispersion will be sufficient to enable it to be used for the Sun in the same manner as instrument A (except for the least refrangible portion), when some of the foregoing instructions will apply.

It has been, however, provided chiefly for observations of the Aurora, and when used for this purpose only one set of prisms (as packed in the box two sets are mounted) and the lower power eyepiece should be employed.

One method of measurement which can be employed by this instrument is as follows:—The adjusting screw with scale nearest the eye, for reference called the "reading screw," is set to the central division marked 10 on the paper scale. The other similar adjusting screw furthest from the eye, for reference called the "bending screw," should be adjusted so that the needle point visible in the field of view is coincident with some definite line in the green, say the green line in the aurora.

Measures of auroral lines right and left of this line can then be made and recorded, and these measures can be compared with those of the lines of other spectra, either simultaneously by means of the reference prism, or afterwards.

A variation of this method will be to set the reading screw to 10, and the bending screw so that the solar line *b* is adjusted to the needle point. Solar positions can then be read off before the auroral observations are made, and a map prepared on which the positions of auroral lines can be at once recorded.

SPECTROSCOPIC WORK.

Scales prepared on Mr. Capron's plan, together with forms for recording positions, also accompany the instrument. In using these carefully insert the principal solar lines in their places on the forms, as taken with a fine slit, and keep copies of this scale for use. If the slit opens *only on one* side, note on scale in which direction the lines widen out, whether towards red or violet. Also fill up some of these forms with gas and other spectra, as taken at leisure *with the same instrument* and scale.

When observing close the slit (after first wide opening it), as much as light will permit, and then with pen or pencil record the lines as seen upon the micrometer scale on the corresponding part of the form, and note *at once relative intensities* with Greek letters, α, β, &c. (or numbers).

Reduce at leisure line-places on scale to wave lengths, and note as to each line the *probable limits of instrumental error*.

In case the auroral spectrum is so faint that the needle point or micrometer scale is invisible, half of the field of view may be covered with tinfoil, with a perfectly straight smooth edge running along the diameter of the field, in perfect focus, and parallel to the lines of the spectra. The reading screw being set to 10, the bending screw should then be adjusted so that the green line of the aurora is just eclipsed behind the blackened edge of the tinfoil. A similar eclipse of other lines will give their positions.

In this instrument the reference prism is brought into action by turning the slipping piece to which is fixed the two terminals. Care should be taken that the prism itself is adjusted before commencing observations, as it may be shaken out of position on the voyage. The tubes provided for the reference spectra may be either fastened to the terminals or arranged in some other manner. The air spectrum may also be used as a reference spectrum. To get this two wires should be screwed into the insulators, their ends being at such a distance apart and in such a position that the spectrum is well seen.

Instruments C *and* D.

These being of the well-known forms require no special instructions.

*General Observations regarding the Spectrum of the Aurora.**

Note appearance, colour, &c. of *arc*, *streamers*, *corona*, and *patches* of light.

Get compass positions of principal features, and *note any change of magnetic intensity*. If corona forms, take its position and apparent height.

Look out for *phosphorescence* of aurora and adjacent clouds. Listen for reported sounds. Note any peculiarity of cloud scenery prior to or pending the aurora.

Sketch principal features of the display, and indicate on this sketch the parts spectroscopically examined.

Examine line in *red* specially in reference to its assumed connexion with *telluric* lines (little *a* group), and note *as to its brightening in sympathy with any of the other lines*.

Examine line in yellow-green (Ångström's) as to *brightness*, *width*, and *sharpness* (*or nebulosity*) at the edges. Notice as to a peculiar *flickering* in this line sometimes seen; note also whether this line is *brighter* (or the reverse) *with a fall of temperature*. Note *ozone* papers at the time of aurora.

Note whether the auroræ can by their spectra be classed into distinct types or forms, and examine for *different spectra* as under—

- *α*. The auroral *glow*, pure and simple.
- *β*. The white arc.
- *γ*. The streamers and corona.
- *δ*. Any phosphorescent or other patches of light or light-cloud in or near the auroræ.

---

* In these observations some suggestions made by Mr. Capron have been incorporated.

The information collected together in the "Manual" should be carefully consulted, and the line of observations suggested by Ångström's later work followed out. To do this, not only record the positions of any features you may observe in the spectrum, but endeavour to determine if any, and if so which, of the features vary together. Compare, for instance, the two spectra of nitrogen in the Geissler tube supplied, by observing first the narrow and then the wider parts of the tube. It will be seen that the difference in colour and spectrum results simply from an addition to the spectrum in the shape of a series of channelled spaces in the more refrangible end in the case of the spectrum of the narrow portion.

Try to determine whether the difference between red and green auroras may arise from such a cause as this, and which class has the simpler spectrum.

See whether indications of great auroral activity are associated with the widening or increased brilliancy of any of the auroral lines.

Remember that if auroral displays are due to gaseous particles thrown into vibration by electric disturbance, increased electric tension may either (1) dissociate those particles and thus give rise to a new spectrum, the one previously observed becoming dimmer; or (2) throw the particles into more intense vibration without dissociation, and thus give rise to new lines, those previously observed becoming brighter.

Careful records of auroral phenomena from both ships may enable the height of some observed from both to be determined. It will be very important that those the heights of which are determined by such means should be carefully observed by the spectroscope, in order to observe whether certain characteristics of the spectrum can be associated with the height of the aurora.

VI.—MISCELLANEOUS OBSERVATIONS.

1. On SALINE MATTER in ICE. By DR. RAE, F.R.G.S.

I find in the note-book of my visit to Repulse Bay in 1853–4:— Bay frozen over for some miles on 22nd September 1853.

					ft.	in.
20th December 1853,	{ 89 days after bay froze over }		ice thick	-	4	7
24th January 1854,	35	days' interval	,,	-	5	9
25th February ,,	32	,, ,,	,,	-	7	0¾
25th April ,,	59	,, ,,	,,	-	7	11½
25th May ,,	30	,, ,,	,,	-	8	1½

In the first three (December, January, and February) of the above examinations, carefully made by cutting holes in the ice, the ice was tough (what Weyprecht calls *leathery*), opaque, that is translucent, certainly not transparent like fresh-water ice, and always so salt that the water from it was not fit to drink.

In March I find no examination was made, my seven men and myself being either away or very busy that month, but the men who measured the ice in April and May assured me that in cutting the holes they found the ice in the same state (tough and opaque) as it had been on the three previous occasions. Being absent in April and May on a long sledge journey, I could not test the ice personally.

I am told that the "*rough, old*, wasted ice" mentioned by me as *almost* always giving good drinking water when thawed, has been confounded with what is called "rotten ice." Now the two are quite distinct and different. Rotten ice, means ice that is worn away whilst lying "in situ," generally early in spring, by certain currents of the sea (usually where there is shallower water between two deeper places) acting on its under surface, whilst the temperature of the air is still much too low to have any effect on the upper part.

This "rotten ice" becomes spongy and dangerous to travel over. It is very common in Smith's Sound, according to Kane and Hayes.

The wasted old ice I spoke of, generally breaks up into detached floes when quite solid and several feet thick, and is gradually worn away, sometimes into all sorts of curious shapes, by the combined action of the sea and atmosphere.

In the very excellent scientific report of the Austro-Hungarian Arctic Expedition given in "Nature" of the 11th March, Weyprecht tells us, "that in 24 hours ice a foot thick was formed on " the sea by a temperature of minus 37°·5 to 50° C., and that the " salt of the sea water had not time to be displaced entirely, as " the formation of the ice went on too quickly," &c.

I found that ice formed on the sea by the gradually lowering temperature of early winter did not eliminate the salt any more than in the case of quick freezing given above.

To quote again from Weyprecht, "The melted water (from sea " ice) at the end of summer is therefore almost free from salt, " and has a specific gravity of 1·005."

This distinguished officer may be quite right, but I do not go so far as he does.

My idea is that at the end of summer the portion of the floe that floats above water will be found fresh or nearly so in most cases but that the submerged part will be decidedly briny. I am not, however, certain of this.

The following are the mean temperatures of the months in which the ice was measured, corrected as nearly as practicable for error of thermometer ascertained by freezing mercury :—

		Below zero.
Mean temperature of December 1853	- -	25·5 F.
„ „ January 1854	- -	32·4 „
„ „ February „	- -	38·4 „
„ „ March „	- -	20·5 „
„ „ April „	- -	1·7 „

2. Hints towards Observations in the Arctic Regions. By Prof. J. Tyndall, LL.D., F.R.S.

I beg to recommend the following subjects for observation :—

1. The formation of snow crystals ; their shapes, sizes, and the atmospheric conditions under which they occur. In connexion with this point it will be useful to consult Scoresby's Arctic Regions, Vol. I.

2. In water contained in vessels surrounded by cold brine, I once observed the formation of small hexagonal, and stellar crystals. They were formed at some depth below the surface, and rose to the surface. Water in the Arctic regions could be rapidly exposed so as to render a verification and extension of this observation possible.

3. By permitting a thermometer to be frozen in water, and exposed to a varying temperature, a rough notion of the rapidity of the conduction of heat through ice might be obtained. The experiment would be more valuable if a prism of ice, at a very low temperature, were warmed at one end, and the flux of heat determined by the observation of thermometers, sunk in the ice at different distances from that end. Only, however, when there is plenty of time at the observer's disposal should this observation be made.

4. The rate at which the ends of some of the Arctic glaciers advance into the sea ought to be determined by a theodolite.

5. If possible, it would be desirable to compare this terminal motion with the motion at some distance from the sea.

6. The question whether the glaciers break off to form icebergs through being lifted by the water underneath their snouts, or through the gravity of their overhanging ends would be worthy of decision. In the former case the surface of the glacier would be in a state of longitudinal compression, and no crevasses would be formed ; in the latter case the surface would be in a state of longitudinal strain, and crevasses might be expected.

7. The height of some of the tallest icebergs ought to be accurately determined.

8. The moraine matter carried down by the glaciers and transported by icebergs, would be worthy of observation.

9. The condition of the rocks and hills adjacent to existing glaciers ought to be examined, with a view to decide whether the glaciers, in former times, extended beyond their present limits.

10. The veining of the ice, at the ends of the glaciers, ought to be sketched and described.

11. Observations might be made on the colour of the ice. It would also be interesting to determine the colour of the sky on different days by a cyanometer ; and how the colour varies with the zenith distance. Various cyanometers are described by Dr. Hermann Schlagintweit in the Philosophical Magazine for 1852, vol. iii. p. 92.

12. The polarization of the sky, and the determination of the neutral points in the Arctic firmament, might also be made an interesting subject of observation.

13. The presence or absence of germs in the Arctic air might be ascertained by experiments similar to those of Pasteur upon the Mer de glace.

14. The range of a sound of a definite character on different days, and at different hours of the same day, ought to be determined. I have myself derived much instruction from experiments made—

1. With a dog-whistle.
2. With an open organ-pipe producing from 300 to 400 waves a second ; a pistol fired with a definite charge would also be useful. It would also be easy to fit up a bell with a hammer to deliver upon it a stroke of constant strength. Mr. Tisley would prepare such a bell immediately. In all cases the state of the weather, when experiments on sound are made, ought to be noted.

15. The aerial echoes ought to be observed ; here the sound of a cannon will be necessary. The echoes of a cannon, fired to windward, ought to be compared with those of the same, or of a similar cannon, fired to leeward.

16. The range of two sounds differing in pitch, say an octave apart, ought to be determined; the experiments ought to be repeated on different days, with the view of determining whether the same sound has, at all times, the greatest range.

I beg to enclose with these suggestions :—

1. A copy of a paper on the Physical Properties of Ice ;*

2. A copy of a paper on the Atmosphere as a Vehicle of Sound ;

3. A copy of a book entitled "Forms of Water ;"

from which various hints and suggestions may be derived.

---

* The internal structure of the Arctic ice ought to be explored by concentrated sunbeams, in the manner indicated in this paper.

INSTRUCTIONS
FOR
THE NATURALISTS ENGAGED IN THE ARCTIC EXPEDITION.

PART II.—BIOLOGY.

I.—ZOOLOGY.

1. INSTRUCTIONS for MAKING OBSERVATIONS on, and COLLECTING SPECIMENS of, the MAMMALIA* of GREENLAND. By DR. ALBERT GÜNTHER, F.R.S.

To obtain information on the present state of our knowledge of the Mammalian Fauna of Greenland, the Naturalists ought to acquaint themselves with, and if possible to take with them copies of, the following publications :—

1. Richardson, "Fauna Boreali-Americana." (Part containing the Mammals.) 1829. 4to.

2. Brown, R. "On the Mammalian Fauna of Greenland," in the "Proceed. Zool. Soc." 1868, pp. 330–362.

3. Brown, R. "Notes on the history and Geographical relations "of the Pinnipedia frequenting the Spitzbergen and Greenland "Seas," in the "Proceed. Zool. Soc." 1868, pp. 405–440.

4. Boyd Dawkins, W. "The British Pleistocene Mammalia." Part V. *Ovibos moschatus*. Lond. 1872.

5. "Die zweite Deutsche Nordpolarfahrt in den Jahren 1869 "und 1870," under Karl Koldewey. Leipzig, 1874. Containing numerous observations on Mammals scattered in the body of the work, chapter 13 being entirely devoted to zoology.

The number of Greenland Mammalia is so small that most of the desiderata can be specified under the heads of the several species ; the following general remarks, however, may be given for the guidance of the naturalists :—

1. One of their most important tasks is to ascertain all facts bearing upon the distribution or possibly gradual disappearance of Mammalian life in the direction towards the Pole.

2. For this purpose attention is to be paid not merely to such animals as may be met with in a living state, but also to any osseous remains or fragments which may be found on the shore ; and such remains are to be brought home, if practicable.

* Not including Cetacea.

3. Specimens of every species met with north of 80° should be preserved, if found in any way to vary from more southern races.

4. No opportunity should be neglected of examining the uterus of female animals, partly to ascertain the period of procreation, partly to obtain fœtal specimens for future examination.

5. To enable future inquirers to institute comparisons as regards the numerical increase or decrease of certain Mammals, within the Arctic circle, registers ought to be kept in which the numbers of individuals of every species seen during the voyage, are entered from day to day: this refers more especially to the bear, musk-ox, reindeer, and walrus.

With regard to the single species the following points deserve particular attention:—

1. *Polar Bear.*—Ascertain the proportion of the number of males to females, noting the age (whether full or not full grown) and the time of the year at which the individuals have been observed. The accounts of a partial hybernation of this animal appear to refer to the females only, which probably at some time in the winter retire into secluded spots to give birth to the young. Gather information respecting the condition of the cubs before they are able to follow the mother, and preserve the skins and skulls of such young examples.

2. *Wolverine.*—Occurs, according to Fabricius, in South Greenland; but if Fabricius was not mistaken in his determination, it must be limited to parts of the interior where reindeer can subsist. Obtain if possible a skin or other positive evidence of the existence of this animal by offering a reward to the natives.

3. *Weasel* and *Stoat* have hitherto not been found in Greenland, but if the *Lemming*, which has been met with in East Greenland only, should reappear further north or on the Crest coast, it is possibly accompanied by some species of *Mustela* which feeds on the Lemming. Specimens (skins and skeletons) or even mere fragments of them should be earefully preserved.

4. *The Arctic Fox.*—The blue and white varieties are said to occur promiscuously at certain localities (one or the other being predominant, and to be found even in the same litter.) Accurate observations should be made upon this subject. Is the colour permanent in the same individual all the year round? Are any cases known of an individual having changed the colour of its coat? Is the diversity of colour at all in connexion with their food and the mode of obtaining it? It might be surmised that the white-coloured variety is better able to approach hares or ptarmigan than the blue which would obtain its food chiefly from the nests of birds, any animal offal, shells, crustaceans, and from the caches they have been observed to prepare for the dearth of winter. Is the sense of smelling as well developed as in its European congener?

Skins and skeletons of both varieties to be preserved.

5. The *Eskimo Dog.*—If unfortunately opportunity should occur, the phenomena connected with the Arctic Dog-madness

should be a matter for serious observation. The subject has been treated by Dr. W. L. Lindsay in the "British and Foreign "Medico-chirurgical Review" for January 1870 and July 1871. The points requiring special attention are: 1. The probable causes (generally believed to be cold and darkness). Does the disease show itself first in dogs of the pure Eskimo-breed, or in crosses between native and European dogs? 2. The symptoms and course of the disease (are they identical in both races?). 3. The means by which the contagion is conveyed.

The time of gestation of the *pure native breed* (said to be a direct descendant of the American Wolf, *Canis occidentalis*) ought to be ascertained; further information collected as regards the statements, according to which it readily reverts into the wild state; and a number of skulls and skeletons and some skins obtained.

6. *Domestic Cat.*—In localities where no recent importation of the cat has taken place, it will be of interest to ascertain whether, in the course of some generations, any change in the closeness and colour of the fur and in the fertility of the species has been observed.

7. *Hare.*—A series of good skins obtained at different seasons of the year (with the skulls) as well as some skeletons are required.

8. *Musk Ox.*—Every fact adding to our knowledge of its actual geographical range, as well as of the changes that have taken place in its distribution in time, is of great interest. It seems that evidences of its former existence on the West coast are not scarce; and the skulls and other parts of the skeleton which may be found, ought to be preserved, with careful observations on the conditions of the locality. As many skins, skulls, and skeletons of animals of both sexes and of all ages, as can be conveniently procured, prepared, and packed, should be preserved.

A careful dissection of the soft parts ought to be made, and some of them, such as the brain, a gravid uterus (if not too far advanced in pregnancy), and the intestinal tract preserved, the first two in spirits, the last in strong brine. Finally, it would be most desirable to make an attempt to bring young animals to Europe; and as it cannot be expected that the transport could be effected on board the exploring vessels, the co-operation of whalers or European residents might be secured by holding out the prospect of a fair pecuniary reward.

9. *Reindeer.*—Obtain a series of skulls of adult animals, with and without horns, and if possible one or two skeletons, the Greenland Reindeer being considered to be a distinct variety. Any differences in size and colour, and in the shape of the horns observed in animals from different localities should be noted.

10. *Walrus.*—Although perfect skins and skeletons of adult individuals are great desiderata in our museums, they can be acquired from whaling vessels; and the naturalists of an exploring expedition will be satisfied to preserve skulls of extraordinary size or the exceptionally tuskless skulls of females, and particularly the heads of newly-born or fœtal animals, which are to be preserved in

strong spirit. It will be also useful to obtain faithful sketches of the heads of adult animals (in different views) and of the attitudes assumed by them during life.

11. *Seals.*—So much remains to be done towards elucidating the life-history of the Seals (of which six species have hitherto been found off the coasts of Greenland), that the naturalists should never neglect an opportunity of collecting further materials on any point referring to the occurrence, habits, propagation, migration, variation, &c., and note their observations, be they confirmatory of, or at variance with, the statements of previous observers. All perfect skins, skeletons, or skulls which can be spared for scientific purposes, should be preserved; and in obtaining these specimens the collector ought to be particularly anxious (1.) to obtain skin and skeleton (or at least skull) of the same individual; (2.) to obtain specimens out of the same flock or family and to mark distinctly the examples thus related to each other; (3.) to secure and prepare the mother with the young.

---

2. Instructions for making Observations on and Collecting Specimens of the Cetacea of the Arctic Seas. By Prof. W. H. Flower, F.R.S.

The study of the habits and structure of the Cetacea is beset by so many difficulties that every accurately observed and carefully recorded fact relating to them will be of value to science. Fortunately the work of the numerous naturalists who have devoted themselves to this group, during the last few years, has done much to clear away the main sources of confusion and error in all the earlier accounts of their anatomical characters, habits, and geographical distribution, and at length, at least as regards the Northern species, we have been able to arrive at a tolerably satisfactory knowledge of the principal distinctive characteristics of all the common species, and of their relations to each other. The ground having been so far cleared, and a definite framework based on solid fact having been raised, future observers will be in a far better position, than was possible till very recently, to fill in all the required details for completing our knowledge of this interesting order of Mammals.

A list is appended of the species which may be met with in the seas to be traversed by the Expedition, with their principal distinctive characters, an outline of what is now known of their geographical distribution, and notes on the chief points in their history still requiring elucidation. It is probable that many of these are not truly Arctic, but in the absence of satisfactory information as to the limits of their range in that direction, it seems best to include all species known to inhabit the North Atlantic. It may, however, be mentioned generally that the appearance in the sea of every Cetacean should be noted, the correct specific designation being, if it can possibly be made out, assigned to it, the vague designations of seamen, such as "whales," "bottlenoses," "porpoises," &c., being avoided. If in "schools," the number of individuals, as nearly as they can be estimated, the direction in which they are swimming, the character of the "blowing," the average duration of the intervals between each expiration, &c., will also be subjects for observation.

If any animal is actually captured, or found dead, abundant opportunity for observation will be afforded, as our knowledge of most of the species is derived chiefly from skeletons, badly preserved portions of the soft parts, and imperfect descriptions of their external form. Good drawings, made to scale and accurately coloured, of the external appearance of nearly all the species are still desiderata. Careful measurements, devoid of the (often unconscious) exaggeration which vitiates so many of those already given by voyagers, are also required, especially of the larger species. The extreme length should always be given, if possible, in a straight line from the tip of the nose to the notch between the flukes of the tail, as measurements following the curves of the body give a very erroneous idea of the actual size. Any parasites which may be found attached to the external surface or contained within the animal should be carefully described, and, if practicable, preserved.* The contents of the stomach should always be noted, with a view to ascertain the natural food of the animal.

Collecting will probably be limited to smaller or rarer specimens, as it will not be possible to occupy valuable space by such bulky objects as are the skeletons of most of the Northern Cetaceans. If, however, whole skeletons cannot be preserved, certain portions of them might be removed and brought home without difficulty, especially the pelvic bones and rudiments of the hinder extremity, which are nearly always wanting in the skeletons in museums; next to these, the skull, the cervical vertebræ, the hyoid bones, the sternum, and the fore limb or paddles are the most characteristic parts. Brains of any of the larger species are much wanted, if they can be obtained in a tolerably fresh condition and carefully preserved in spirit. If any fœtuses are met with in dissecting, they should, if possible, be preserved entire, with the uterus and membranes, in spirit or strong brine.

List of Cetacea of the North Atlantic.

I.—WHALEBONE WHALES. (*Mystacoceti.*)—Easily recognised by the baleen or "whalebone" with which the palate is furnished, and by the double openings of the blow-holes on the top of the head.

Genus *Balæna.* The one species inhabiting the Arctic regions is *B. mysticetus*, Linn., the Greenland Right-Whale, distinguished from the other Whalebone Whales of the same seas by the very large size of the head (one-third, or even more, of the entire length), by the great length of the baleen, the absence of longitudinal furrows in the skin of the throat, and the absence of

* On this subject *see* Van Beneden, "Les Cétacés, leur Commensaux et leur Parasites," Bull. de l'Acad. royale de Belgique, 2me série, tome XXIX., No. 4, p. 347, 1870.

a dorsal fin. This is by far the most important of the Northern Cetacea to man, being the animal which yields train oil and whalebone in greatest quantity and finest quality. It appears to have a regular seasonal migration, wintering in the southern portions of Davis Straits, Hudson Strait, and the coast of Labrador, though never coming farther south, but the extent of its northern range in the summer yet remains to be ascertained. For full accounts of its habits and geographical distribution, *see* the works of Martens, Zorgdrajer, and Scoresby; R. Brown, "Notes on the History and "Geographical Relations of the Cetacea frequenting Davis Strait "and Baffin's Bay," Proc. Zool. Soc., 1868, p. 533; and especially the elaborate monograph by Eschricht and Reinhardt "On the "Greenland Right-Whale (*Balæna mysticetus*)," translated from the Danish in "Recent Memoirs on the Cetacea," Ray Society, 1866.

A really good drawing of an adult Greenland Whale is still a desideratum, also accurate statements as to the size it attains. There is no evidence from specimens in any of the European museums that it exceeds 55 feet in length (in a straight line), but voyagers generally give much larger dimensions.

A Right Whale, with a smaller head and shorter whalebone (*Balæna biscayensis*), was formerly abundant in the temperate portions of the North Atlantic, but is now nearly extinct. Any information about the species would be extremely valuable. Is it identical with the American Black Whale (*B. cisarctica*, Cope)?

See Eschricht and Reinhardt, *op. cit.* Van Beneden and Gervais, "Ostéographie des Cetacés." Fischer, "Documents pour servir "a l'histoire de la Baleine des Basques (*Balæna biscayensis*)," Annales des Sciences Nat., 1871.

All the remaining Whalebone Whales of the Northern seas have a small dorsal fin, and the skin of the throat and breast marked with deep longitudinal furrows.

Megaptera boops (O. Fabricius)=*M. longimana* (Rudolphi). Hump-backed Whale, *Keporkak* or *Krepokak* of the Greenlanders, German *Buckelwall.* Known externally from the true Rorquals by the low and obtuse dorsal fin, and especially by the great size of the pectoral fins, which are more than one-fourth of the entire length of the animal (from 45 to 50 feet). Colour black above, and black and white below in streaks and patches; the pectoral fins wholly white, baleen black. As far as is known at present there is but one species of Megaptera in the Northern seas, which has a far more extensive range than the Greenland Right-Whale, passing in the winter as far south as Bermuda, and in the summer up to the Greenland coast (66° North). For the fullest account of this species, *see* D. F. Eschricht, "Untersuchungen "über die Nordischen Wallthiere" (1849), which extremely valuable work contains an exhaustive bibliography of the Northern Cetacea up to date of publication.

Genus *Balænoptera.* The Rorquals, or Fin Whales, are known from the last by the pectoral fin not exceeding one-sixth of the length of the animal, and by the falcate form of the dorsal fin. Four Northern species are now generally recognised.

B. musculus, Companyo = *Physalus antiquorum*, Gray. The Common Rorqual or Razor-back, *Keporkarnak* of the Greenlanders. Black above, white below; flippers black; baleen slate-colour on the outer edges, streaked longitudinally with yellow, and yellowish-white on the inner fibrous edges. Length of adult 60 to 70 feet. This is the commonest of the Rorquals of the temperate Atlantic and Mediterranean. Its Northern range has not been well ascertained, as it has till lately been confounded with the next species. For external characters and figure, *see* W. H. Flower, Proc. Zool. Soc., 1869, p. 604, and pl. xlvii., with references to previous figures.

B. Sibbaldii, Gray. Sibbald's Rorqual, *Tonnolik* of the Greenlanders, *Steypireythr* of the Icelanders. Black above, shading into slate-grey below, more or less varied with dashes or spots of white; flippers black above and whitish below; baleen uniform deep black. Is of larger size (70 to 80 feet) and has a more Northern range than the last. *See* W. Turner, "An Account of "the Great Finner Whale (*Balænoptera Sibbaldii*) stranded at "Longniddry," Trans. Roy. Soc. Edinburgh, vol. xxvi. A. W. Malm, "Monographe illustrée du Baleinoptère," Stockholm, 1867.

B. laticeps (Gray). Rudolphi's Rorqual.

Only four of five specimens of this species, which is chiefly distinguished by its osteological characters, have hitherto been met with, and but little is known of its external appearance or geographical distribution. It does not appear to exceed 40 feet in length. All the specimens referred with certainty to this species have occurred in the North Sea, between the North Cape and the Dutch coast.

Balænoptera rostrata (O. Fabricius). The Lesser Rorqual, *Tikagulik* of the Greenlanders, *Vaagevhal* of the Norwegians. Black above, white below; flippers black, with a broad white band across the middle; baleen yellowish-white. Length not exceeding 30 feet. This is the smallest and perhaps the best known of all the Rorquals, having an extensive range in the North Atlantic.

The genus *Agaphelus* (Cope, Proc. Acad. Nat. Sc. Philadelphia, 1868, p. 221) is founded on the imperfect remains of a Whale cast ashore on the coast of New Jersey, indicating the existence of a species in the North Atlantic resembling *Balæna* in the absence of pectoral cutaneous furrows and of dorsal fin, but having the elongated form of body, tetradactylous hand, and general osteological characters of *Balænoptera*. Further indications of this animal are much needed.

II. Toothed Whales (*Odontoceti*). No baleen. Nostrils (in nearly all) united to form a single median crescentic opening or "blow-hole."

Physeter macrocephalus, Linn. The Cachalot or Sperm Whale. An inhabitant of the tropical and warmer temperate seas, and only met with as an accidental straggler in the North Atlantic.

Hyperoodon rostratus (Chemnitz). The Common Beaked Whale or Bottlenose. No teeth in the upper jaw; one or two pairs in the front end of the lower jaw, small and concealed in the

1. Collect specimens of every species, with the exception of the common Wolf-fish (*Anarrhichas lupus**), the species of the genus *Gadus*, viz., Cod-fish, Whiting, Pollack, Coal-fish, Ling, and Torsk, the Halibut and the Capelin.† Beyond lat. 71° N. collect all fishes without exception; generally three or four specimens of each kind will be sufficient.

2. The smaller kinds, that is, specimens which can be packed in tin boxes 2½ feet long, preserve in spirits 20° over proof; the larger specimens can be skinned and preserved dry; skins of sharks are best preserved in very strong brine.

3. To judge from the collections brought home by previous travellers there must be, at suitable localities, an abundance and a great variety of small shore-fish, such as Father-lashers (Cottoids), Sticklebacks, Blennies, etc., which may be obtained by the usual means or by employing natives or residents.

4. The absence of Fish life at or near the surface is no proof that fishes are not abundant at a greater depth; and whenever circumstances permit, long hand-lines should be tried. These hand-lines should differ from the ordinary cod-line in being much longer, upwards of 80 fathoms, and (in the fashion of a pater-noster) provided with hooks for about 20 fathoms from the sinker. In order to allow the lines to be in the water for several hours, the hooks ought to be fastened to the snood by a number of open strands of soft twine about three inches long, so that the fish cannot bite through the line. I have no doubt that in this manner those extraordinary Arctic forms which we know from isolated examples only, can be obtained; nearly all of them are evidently very voracious fishes.

5. Although the Sharks are well represented in the Arctic Seas, our knowledge of them is extremely incomplete. Scarcely the outlines of their specific characters are known, and absolutely nothing of their life-history. No instance is on record of a young Basking Shark (a species by no means uncommon) having ever been seen. Therefore, all observations regarding them ought to be collected; and specimens of a manageable size preserved. Should the Naturalists have an opportunity of examining very large examples, an exact outline drawing of the entire animal showing the exact position of the eyes, nostrils and blow-holes (spiracula) should be made, and the jaws cut out and preserved. The species of Sharks which are regularly killed for the sake of the oil extracted from their liver, ought to be determined; and some of the European residents on the coast be induced to prepare the skins of full grown examples (from 25 to 35 feet in length) for sale to the Museums in Europe.

6. We know of but one family of fishes inhabiting fresh waters of the Arctic region, viz., Salmonoids. Trout were caught during

---

* There is a second species in Greenland, *Anarrhichas denticulatus*, with much smaller teeth and an unspotted blackish-brown body; this species is a desideratum.

† Greenlandic: Angmaksak; Sennersulik (male with villous excrescences), Sennersuitsut (male without). Eskimo: Angmaggeuck.

the voyage of the "Fox" in 72° (Bellot Strait), and in Spitzbergen they occur several degrees further north; so that it is probable that fishes of this family live in still higher latitudes. It is of the highest interest to ascertain the extreme limits at which fresh-water species can exist; their existence being dependent on the presence of food and on the conditions necessary for the development of the spawn. Charr can live in Alpine pools which are free from ice for a few weeks only in favourable seasons.

a. In collecting specimens of this family take large individuals in preference to small ones, as young examples of less than 8 inches long, are but rarely suitable for specific determination. If the specimens cannot be brought away, ascertain whether they have or have not teeth along the body of the vomer, that is, whether they are Charr or Trout.

b. Has the locality in which they are found, a communication with the sea? and is there any reason to believe the fish to be migratory?

c. If possible ascertain the depth of lakes inhabited by fish, and whether the water is likely to freeze to the bottom. As some fish of temperate regions (*Cyprinoids*) can endure for a considerable period complete enclosure in ice, it would be of interest to prove experimentally that certain Arctic fish (marine or freshwater) are endowed with a similar tenacity of life, and to see for what period they can survive.

d. Examine the stomachs of all Salmonoids, and note their contents, some of which may be worth preserving.

---

5. Instructions for making Observations on, and Collecting the Mollusca of, the Arctic Regions. By J. Gwyn Jeffreys, F.R.S.

Exactly two centuries ago Frederic Martens, of Hamburg, first noticed the Mollusca which he met with in his voyage to Spitzbergen and Greenland. These were the *Clione papilionacea* of Pallas and *Limacina arctica* of Fabricius; the former a naked or shell-less mollusk, and the latter a smaller shell-bearing species, both being of the Pteropod order, which inhabit the surface of the Arctic ocean in countless numbers, and are usually (but questionably) supposed to constitute the food of whales. Since that time Linné, Müller, Fabricius, Chemnitz, Leach, Gray, Broderip and Sowerby, Möller, Torell, Mörch, Lovén, Forbes, Reeve, Albany Hancock, Davidson, and several others, have described or noticed species from high northern latitudes; besides Sars and his Scandinavian fellow-workers, who have so carefully and laboriously investigated the Mollusca of the Norwegian Coasts within the Arctic circle. No nation has done, especially of late, so much as Sweden to advance our knowledge of the Arctic Mollusca. In 1857 and 1858, Professor Torell, aided and encouraged by the Academy of Sciences at Stockholm, explored by dredging the seas of Spitzbergen and Iceland; the greatest depth reached by him was 280 fathoms. In 1861 a second Swedish expedition was made to Spitzbergen, when a species of *Cylychna* was recorded by Professor Lovén as having been dredged at a depth of over 1,000 fathoms. A third Swedish expedition in the same direction was made in 1868; and by the kindness of Professor Lovén I was favoured with an opportunity of examining at Stockholm some of the results. The dredgings and soundings appear to have extended from 5 to 2,600 fathoms; and, at the last-named depth, in N. lat. 78° a living Crustacean (*Cuma*), and a valve of a Mollusk (*Astarte compressa*) were obtained. Again, in 1871, the Swedish frigate "Eugenia," was dispatched on a scientific voyage to Greenland; when Mr. Josua Lindahl, who had assisted us in the "Porcupine" expedition of the previous year, was the naturalist in charge. The results of this last expedition, as regards the Mollusca, have not yet been published; but I was informed by Mr. Lindahl, that in Davis's Straits he dredged a species of *Pecchiolia* (or *Verticordia*), apparently *acutecostata*;* and Professor Lovén told me that a *Marginella* (which is a southern form) and a *Limopsis* were dredged also in Davis's Straits at a depth of 900 fathoms.

The importance of such investigations cannot be too highly estimated, especially in a geological point of view. The palæontological basis of the glacial epoch consists mainly in the identification of certain species of Mollusca, which inhabit the Polar seas and are fossil in Great Britian and even as far south as Sicily. But such species may owe their present habitat and position to other than climatal causes, viz., to the action of marine currents. Certain small Spitzbergen species (*e.g. Lega frigida*, and *L. abyssicola*), have been lately found everywhere in the depths of the North Atlantic as well as in the Mediterranean; and the question naturally arises what is the home of these species, or where did they originate? That question cannot be answered for want of sufficient information. It is likewise quite premature to assume that Arctic Mollusca comprise very few species, although they may abound in individuals. We know very little about them, because of the difficulty of investigation. The researches of Professors Torell and Sars induce us to believe that these Mollusca are not less varied than numerous.

It is hoped that each of the vessels to be fitted out for the Polar expedition will have a donkey engine, by which the dredges can be lifted; and that a sufficient supply of necessary apparatus will be provided, regard being of course had to the limited space allowed for such a secondary object. The great experience of Capt. Nares renders any suggestions as to dredging quite superfluous.

---

* This species is fossil in the Coralline Crag of Suffolk, and the Zanclean formation of Sicily; and I dredged it at considerable depths in the Bay of Biscay. It has been also dredged by Mr. Arthur Adams in the seas of Japan, and by Count Pourtales in the Gulf of Mexico.

One difficulty will be the preservation of any soft or shell-less Mollusca in high latitudes, where spirit of wine would freeze; but this may be obviated by having accurate drawings of the animals made on the spot. Perhaps one of the medical staff in each vessel could do this. In lower latitudes such Mollusca might be kept in methylated spirit in thick glass jars, space being allowed for expansion by subsequent freezing.

The larger shell-bearing Mollusca may be wrapped up in wool or paper and placed in wooden covered trays, such as were used in the "Porcupine" and "Challenger" expeditions; and the smaller shells could be kept in wooden boxes or stout pill boxes. The rest of the dredged material should be carefully kept in canvass bags.

In every case, it is of the greatest consequence that the latitude and longitude of the place of capture, as well as the depth, should be recorded by means of linen labels, prepared so as to make the print and writing ineffaceable by sea water.

Where the ground or sea-bottom is muddy, a fine sieve may be used in a large tub of water to get rid of the impalpable mud; or the "globe sieve" may be worked overboard for the same purpose. A descriptive account of these contrivances will be found in the preliminary report of the "Porcupine" exploration in 1869, in No. 121 of the proceedings of the Royal Society, p. 415. I would strongly recommend also the "scoop-sieve" (*loc. cit.*) for catching Pteropods, Cuttle-fishes, and other animals on the surface of the sea. Towing-nets of muslin or fine gauze may be used for the capture of small Pteropods and Oceanic Crustacea; and the stomachs of fishes and star-fishes should be examined for shells. See "Hints for Collecting" in "British Conchology," vol. v. The crops of sea-birds occasionally contain shells; but these shells would be of littoral species and therefore not of much scientific value.

The rocks and seaweeds on the coast should be diligently searched for species of *Littorina*, *Lacuna*, *Purpura*, *Buccinum*, and other littoral shells. And it should be borne in mind that too many specimens of different ages cannot be collected. The mischievous practice of species-making would not have been carried to such an extent if naturalists had before them a suite of specimens to show the range of variation, instead of being restricted to two or three specimens, and sometimes only to what is called a "unique" specimen.

Land and fresh-water shells, when procurable, would be extremely interesting, and would serve to elucidate one of the difficult problems of geographical distribution. The former may be looked for under loose stones and among mosses; and the latter in pools of water, during the summer and autumn. Möller described several species from Greenland.

I may add that all fossil shells should be collected, and their position accurately noted, especially the height above the present level of the sea. The former conditions and climate of the Polar region may be thus ascertained, and a new chapter opened in the history of our globe.

---

6. Instructions on the Collection and Preservation of Hydroids and Polyzoa. By G. J. Allman, M.D., P.L.S., F.R.S.

Hydroida.

In the notes on the towing net reference is made to the Hydroid Medusæ as among the objects most frequently captured in the net; and it is stated that these almost always originate as buds from rooted plant-like zoophytes. These hydroid zoophytes grow like sea-weeds beneath the sea; the buds, which when mature detach themselves and spend the remainder of their lives as free-swimmers in the open sea, are the sexual portion of the hydroid colony, and are destined to give origin to generative elements, male and female, by which the species is perpetuated, while the rooted plant-like portion consists for the most part of a multitude of little hydranths or polypites destined not for generation but for nutrition. These are minute flower-like bodies, each with a mouth occupying the central point in the flower, and surrounded by contractile tentacles which are mostly so disposed as to resemble the petals and other verticils of a regular flower; the whole of the hydranths are organically united into a composite rooted colony. In many cases, however, the generative buds do not detach themselves at any period of their existence, and then they form, like the hydranths, a permanent plant of the common colony. The rooted or nutritive portion of the colony is the trophosome; the assemblage of generative buds whether permanently fixed or destined to become free is the gonosome. When these buds become free, as in the case of the medusæ above referred to, they may be named planoblasts. A few words on the modes of collecting the trophosomes may here be added to what has already been said regarding the planoblasts.

While the planoblasts must be sought for by the towing net, the rooted trophosomes are obtained by means of the dredge or by examining the rocks left exposed by the retiring tide. There is scarcely any depth at which these beautiful organisms may not be encountered by the dredge, but they are most abundant in moderately deep water. They here fix themselves to masses of rock or to old shells, or to sea-weeds or other bodies affording a sufficient surface for attachment, and in some rare cases they root themselves in the sandy sea bottom. The region between tide marks will at low water often afford a rich harvest to the collector. Here the hydroids will be found rooted to the sides of rock pools or fixed in the clefts or under the projections of the rocks, or spreading over the surface of exposed sea-weeds. Large loose stones lying between tide marks ought to be turned over, and the under surface examined for hydroid trophosomes many of which are chiefly found in such situations, especially on stones which lie near the extreme limit of low water. If any portion of the stones or rocks on which they grow can be broken off and carried away with them so much the better; if not, they are to be detached by means of a broad knife blade passed under their point of attachment. When growing on shells they ought not to be separated from these. If the shells be too large these may be broken and the fragments preserved with the hydroids attached to them. When growing over the surface of sea-weeds the portion of the plant to which the hydroids have attached themselves ought also to be retained. When thus secured no time should be lost in placing the specimens in spirit, for when exposed to the air they very soon dry up and lose some of their most important characters.

Polyzoa.

The Polyzoa often resemble the hydroid trophosomes so closely that the inexperienced collector can scarcely be expected to distinguish them, and indeed on this point he need not trouble himself, for the determination of their distinguishing characters may well be left to future careful examination. Like the hydroids they form for the most part plant-like colonies, sometimes giving rise to branching colonies like tufts of sea-weed, and sometimes spreading like lichens over the surface of stones, shells, and algæ. The circumstances under which they occur are almost entirely the same as in the case of the hydroid trophosomes, and they are to be collected and preserved in the same way.

---

7. Instructions on the Construction and Method of Using the Towing Net, and Notes on the Animals which may be obtained by its employment. By G. J. Allman, M.D., P.L.S., F.R.S.

Construction of the Towing Net.

The towing net is a small bag made of some material open enough in its texture to allow of the water easily passing through it, and yet sufficiently close to retain within it such minute bodies as it may encounter in its passage through the sea. Its mouth is kept open by a strong ring, and it is towed behind the vessel by means of a line fastened to the ring.

The bag may be made of fine straining-linen or of new bunting; and in the size which will be found most generally useful it may have a depth of about 18 inches, and a width across its mouth of about a foot. The ring which surrounds the mouth may be a wooden hoop; or it may be made of brass rod strong enough to resist the tendency to become bent when the net is being drawn through the water.

Three pieces of strong line about two feet in length are to be fastened at equal intervals to the ring, and tied together securely at their opposite ends. To the point of their union the towing line is to be attached. With a net of the size here suggested, a reel of strong "hake-line" will make the best towing line for all the ordinary velocities at which towing may be most advantageously practised.

Mode of Using the Net.

When thus rigged the net may be used from a row boat, or from a sailing vessel or steamer under moderate way. It may be thrown out from the stern; and in surface-towing sufficient way must be given to keep the mouth of the bag close to the surface of the water. Many of the small objects which may be floating on or near the surface of the sea in the way of the net as it is thus towed behind the vessel, will necessarily pass into it; and after it has been allowed to remain out for a period varying with the abundance of surface life in the sea at the time, it is to be hauled in and examined.

Though the richest results are usually obtained by using the towing net close to the surface of the sea, it will frequently be found important to employ it at various depths, in order to obtain information regarding the organisms which either habitually or temporarily inhabit zones other than the most superficial one.

For this purpose the net is to be weighted; the weight attached to it depending on the depth to which it is desired to sink it, and on the velocity of the ship. Care should be taken that while the net is out the motion of the vessel be not interrupted, and that sufficient way be given to keep the net constantly distended in its passage through the water.

It will generally be advisable to employ two nets at the same time, one working close to the surface, and the other sunk to some determined depth below it.*

In the directions now given, the towing net is supposed to be towed behind the vessel in open water; but the Arctic explorer should be reminded that some of his richest fields will be found in places where the ice is for short distances discontinuous, and where small portions of unfrozen water will be thus exposed. Here oceanic forms will congregate in rich profusion attracted by the light and air. In the smaller spaces so exposed we may use with most advantage a towing net similar to that here described, but, instead of being provided with a towing line, it should be fixed to the end of a pole, and worked with the hand.

Another mode of using the towing net, which is often attended with the best results, consists in leaving it suspended from the ship while at rest in the tideway or in the course of currents. It may be so left for several hours, and then hauled in for examination. A net used in this way, however, will be found most effective if it be constructed somewhat differently from the ordinary one. A piece of the same material as that of which the rest of the net is composed should be sewed within its mouth so as to form a sort of diaphragm in the shape of an inverted cone with an open apex, as shown in the annexed figure. This serves to retain whatever has once made its way into the net. The fundus of the bag is closed by simply tying a cord round it, and its contents are to be examined from the bottom by untying the cord and washing out the bag in the way to be presently described.

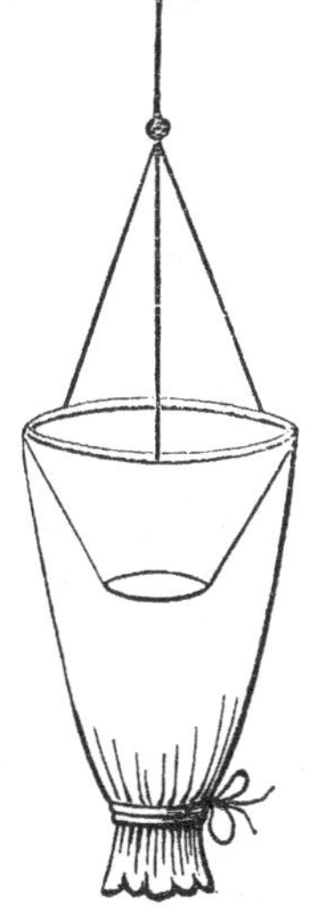

* I am informed by Captain Nares that this plan was commonly adopted during the voyage of H.M.S. "Challenger."

The chief difficulty which the collector will here have to contend against will be found in the presence of floating refuse matter which is being constantly discharged from the ship, and which when the vessel is under way is generally carried clear of the net by the force of the water thrown off from the ship's sides. In order to avoid as much as possible this source of annoyance, the towing line may be attached to the extremity of a long pole fixed at right angles to the side of the ship.

Freeing the Net of its Contents.

The ordinary towing net immediately on being hauled on board is to be carefully turned inside out into a vessel containing some sea-water in which it is to be moved about in order to wash off such minute organisms as may be adhering to its surface. The kind of vessel best suited for this purpose will be found to be a white glazed earthenware pan provided with a lip, such as are used in dairies for holding milk.

From the washings of the net the larger objects are now to be removed, and quickly transferred to clear glass jars of sea-water for further examination; while the water with the remaining organisms should be poured from the pan into one or more such jars, each capable of holding about half a pint.

These smaller organisms are frequently so colourless and transparent that it is at first difficult to see them in the jars; a little practice, however, will enable the observer to recognise them, and he must now transfer to other jars, containing sea-water, such as he wishes to keep and observe further in a living state, for if left crowded together, even for a few hours, the water will become vitiated, and the delicate, frequently gelatinous organisms become decomposed and worthless for observation.

This separation and transference is best effected by glass dip tubes.

Results obtained.

The objects captured in the towing net are very numerous and various, and are among the most beautiful and interesting of the more simply organised inhabitants of the sea. The towing net has been hitherto used almost exclusively in the temperate and equatorial latitudes, and we, as yet, know very little of what may be obtained by it in the Arctic Seas. The following account of its results applies, therefore, directly only to those seas where the naturalist has used it, but it will nevertheless serve as a guide to the Arctic explorer, and suggest to him what he ought to keep in view.

Plants.

The members of the vegetable kingdom which will find their way into the towing net will chiefly consist of the very low groups constituting the orders *Diatomaceæ* and *Oscillatoriæ,* the former provided with siliceous cases and retaining indefinitely their external form; the latter destitute of any firm support, and speedily decomposing and losing all their important characters. In some seas these low forms of vegetable life abound to such an extent as to discolour the water over very wide areas, and they not unfrequently seriously interfere with the work of the towing net by rendering the washings of the net so turbid as to hide the small animals taken at the same time, while the rapid decomposition of their soft parts speedily vitiates the water and destroys such animals as have been allowed to remain with them.

To preserve them the washings of the net should be thrown on a filter in order to get rid of the superfluous water, and the matter which remains should while still moist be transferred to glass tubes containing spirit.

Protozoa.

Among this lowest group of the animal kingdom the collector should be on the look-out for ***Radiolariæ*** and ***Foraminiferæ***. Though the hard siliceous cases and framework of the Radiolariæ and the calcareous shells of the Foraminiferæ will usually retain their forms after the destruction of the soft parts, it is far better to transfer to the spirit the whole organism at once by means of the dip tube. Other microscopic Protozoa, such as ***Noctiluca*** and ***Peridinium***, as well as the true ***ciliate Infusoria***, ought also to be watched for. Among these last are the ***Dictyocystidæ***, a group of pelagic Infusoria having a close resemblance to certain Radiolariæ by their elegant siliceous bell-shaped shells perforated in the manner of lattice-work. In most cases, however, these microscopic organisms are so minute as to render impracticable the separation and transference to spirit of any great quantities of them. We must then be contented with putting up such as can be obtained, on microscope slides or in cells, a process, however, which takes time and labour, and needs some practice in the art of microscopic mounting.

Cœlenterata.

It is in this group that some of the most abundant and important results follow the use of the towing net.

Among the most striking and interesting inhabitants of the surface zone of the sea in all latitudes are the ***Hydroid Medusæ***, clear gelatinous, more or less bell-shaped or umbrella-shaped organisms, which mostly originating as buds from plant-like hydroids (or zoophytes) fixed to the sea bottom, free themselves after a time from their supporting stems, and spend the rest of their lives in a state of activity at the surface, where they swim by the expansion and contraction of their gelatinous bells. When possible, drawings ought to be made of these beautiful animals, for no means have yet been discovered of preserving their forms after death with anything like satisfactory results. Many, however, may be fairly preserved by placing them in methylated spirit; and as this seems to be the best method we possess of preserving their zoological characters, it should never be neglected.

Nearly allied to the Hydroid Medusæ, and with very similar habits, are the ***Siphonophora***. These usually form long garland-like series of transparent, gelatinous, variously shaped bodies, frequently ornamented with spots of bright colour, scarlet or orange. They may be easily detected swimming with a rhythmical repetition of impulses near the surface of the sea. They require great care in their capture, being formed of numerous zooids or more or less independent buds, which by rough handling are easily separated from one another. Though less easily broken to pieces than some other groups of associated zooids, such as the chains of Salpæ, to be presently referred to, those taken in the towing net will often be found injured, and a perforated or brass wire gauze ladle slipt under them while swimming will generally be found the best way of removing them from the sea. Here again drawings ought to be made when possible, as no known method of preservation will satisfactorily retain their characters. The best is that above recommended for the Hydroid Medusæ. Other forms of Siphonophora, such as *Physalia* (Portuguese Man of War), *Velella* and *Porpita*, obtained abundantly in the more equatorial latitudes, have firm supports of the soft parts, and are therefore much more easily preserved. None of these have yet been taken in Arctic seas. They are best preserved by being placed in methylated spirit.

The *Discophorous Medusæ* present a general resemblance to the Hydroid Medusæ in their gelatinous umbrella-shaped swimming disc, by the rhythmical contraction of which they impel themselves through the sea. They attain, however, for the most part a much larger size. The smaller forms may be treated as has been recommended for the preservation of the Hydroid Medusæ and Siphonophora; but we should scarcely recommend any attempt to preserve the larger ones, which frequently attain a diameter of one or more feet, and which would need (to obtain at best very unsatisfactory results) more preserving liquid and space than could be afforded them. Here again the aid of the draughtsman is indispensable.

Both Hydroid and Discophorous Medusæ are commonly known by the name of "jelly fish." The observations of Scoresby and of the naturalists attached to Kotzebue's voyages have proved that both forms abound in high latitudes.

The *Ctenophora* (*Beroe*, *Cydippe*, &c.) form a large part of the surface life of the sea. They are often of considerable size, are constant products of the towing net, and ought to be preserved in the way indicated for the Medusæ and Siphonophora.

The other division of the Cœlenterata, exemplified by Sea Anemones and Corals, have few free-swimming representatives. One of these, however, *Arachnactis*, may be expected to occur in the Arctic seas.

Echinodermata.

The larval forms of most of the *Echinodermata* (Sea Urchins Star-fishes, &c.) consist of minute free-swimming organisms which are among the most frequent captures of the towing net. They should be removed by the dip tube from the washings of the net and transferred to spirit.

Vermes.

Among the Vermes (Sea Worms, &c.), we have some remarkable pelagic free-swimming forms which frequent the uppermost zone of the sea.

The curious *Sagitta*, a little clear crystalline stylette-like body from half an inch to an inch in length with a delicate quadrilateral tail-fin, and moving by a succession of rapid jerks, is sometimes very abundant in the British seas, and would probably be found further north. So also the beautiful *Tomopteris*, a little animal of the purest transparency, attaining a length of about an inch and provided with a series of transparent double paddle-like fins which run down each side of the body, is another pelagic form of the Worms which ought to be met with in high latitudes. Both these animals will be best preserved in spirit.

A great many of the Sea Worms, however sedentary they may be in their adult stages, are in their larval condition free-swimming pelagic forms. They are full of interest in the light they throw upon the phenomena of development, and on the affinities of distant groups of the animal kingdom. They should, therefore, be always carefully noted, removed by the aid of the dip tube from the washings of the net, and preserved as far as possible by immersion in spirit.

Arthropoda.

Among the invertebrate life which abounds in Arctic seas, and which from the concurrent testimony of Arctic voyagers constitute a characteristic feature of their fauna will be found the *Amphipodous Crustacea*. These are small active animals, most familiarly known to us by the " sand hoppers " of our own shores. In Arctic regions they are often attracted in countless multitudes by fragments of offal thrown into the sea. To such an extent do they abound there that the carcass of a seal has been in a few hours reduced by them to the condition of a clean skeleton. They frequent various depths from the surface downwards, and may be all well preserved in spirit.

Among the *Isopod Crustacea* (*Idotea*, &c.), we also find active free-swimming species which frequent the surface zone of the sea and are constantly captured in the towing net. They may be preserved like the amphipoda in spirit.

Minute *Entomostracous Crustacea*, especially those belonging to the group of the *Copepoda*, are often captured in amazing quantities in the towing net. As with other microscopic forms the attempt to separate them from the washings of the net with the view of preserving them is troublesome and difficult. If however the collector has an abundant stock of patience he may here use his dip tube and phials of spirit with advantage. At all events other animals which it may be desirable to preserve for any length of time in a living state should never be left along with these little Crustacea when they are contained in the jars in any considerable quantity, for the Entomostraca rapidly decompose and render the water unfit for other forms of life.

The Crustacea are also rich in larval forms which abound in the most superficial zone of the sea, where their development is favoured by the more intensified conditions of light and aeration to which they are there exposed. Among those larval forms are the free-swimming young of the higher Crustacea, especially those known to the older observers under the name of Zoea, at a time when these immature organisms were regarded as completely developed and independent forms. Also the larvæ of the *Barnacles*, a low section of the Crustacea, which, though absolutely fixed in their adult state, spend the early period of their lives as free-swimmers in the open sea. These are all active creatures of singular, and often grotesque aspect, and are among the most frequent captives of the towing net. Many of them are of great interest in their bearing on the laws of development and on the affinities of groups. They admit of being well preserved in spirit.

Mollusca.

Among the most abundant and striking pelagic forms are the Salpæ belonging to the low molluscoid group of the *Tunicata*. They are of a somewhat oval or prismatic shape, attaining a length of from half an inch to two or even three inches; they are of crystalline transparency, with usually a large brown, reddish, or purple globular body visible within them near one end, and caused by the location at this spot of some of their more important viscera. They swim in jerks near the surface of the sea, either singly or united into long chain-like groups. Being of considerable consistence notwithstanding their clear gelatinous appearance, they can generally be very well preserved in spirit; but great care must be taken not to separate the components of the chain-like series which are very easily detached from one another. Indeed when taken in the towing net these are very often found to be broken up, and the safest way of capturing them is by gliding under them as they swim past the vessel a perforated ladle, as has been recommended in the case of the Siphonophora, and then carefully transferring the whole chain to spirit.

Appendicularia is another tunicate also frequently taken in the towing net. It is a minute clear oviform creature, of about the size of a millet seed, and easily recognized by a rapidly vibrating transparent ribbon-shaped swimming organ, somewhat resembling the tail of a tadpole, and springing from a point near one end of the body. It may be transferred to spirit by means of the dip tube.

Holding a much higher position among the Mollusca are the *Pteropoda*. These are free-swimming animals provided with a pair of wing-like appendages by the aid of which they flit through the superficial zone of the sea. They are usually clear-bodied, and either colourless or tinged with some shade of purple, and generally attain a length of from half to three quarters of an inch. Some of them are said to be diurnal in their habits, sinking into the deeper regions during the night, while others are believed to be nocturnal and to withdraw themselves from observation during the day. These statements, however, require confirmation. Some of them are provided with a delicate transparent shell, others are quite naked. Some of the naked forms (*Clio*) have long been associated in the accounts of Arctic voyage with the fauna of high northern seas, where they occur in immense numbers, and are believed by the whale fishers to form the principal food of the whale. They are easily taken in the towing net, are of considerable consistence, and can be well preserved in spirit.

The ***Heteropoda*** (*Carinaria, Atlanta, Firola,* &c.), another free-swimming form of the higher Mollusca, are also either naked or provided with an external shell, which may be large enough to enclose the entire animal, or be only sufficient for the protection of the respiratory and reproductive organs. They swim by means of a vertical fin, which projects from the ventral surface. They are abundant in the warmer temperate, and equatorial seas. Their habits resemble those of the Pteropoda, and they may be captured and preserved in the same way.

Some of the *Nudibranchiate Gasteropodous Mollusca* also possess pelagic habits. It is rare, however, to find among them free-swimming species, and they are mostly indebted for their pelagic life to floating sea weed (Gulf weed, &c.) on the fronds of which they habitually live, and by which they are carried about from place to place in the open sea. Floating sea-weed, indeed, ought always to be carefully examined. It frequently affords a rich storehouse of rare animals, which are for the most part easily preserved in spirit.

Among the Mollusca are also many free-swimming larval forms. These are all minute animals, generally furnished with a pair of conspicuous wing-like swimming organs, and with a little nautilus-like shell. Though proceeding from more or less sedentary parents, their life in this stage is entirely that of the free-swimming Pteropods, and they become easy captives of the towing net. They must be removed by the dip tube, and preserved in spirit.

Vertebrata.

We can hardly expect to meet with vertebrate animals among the contents of the towing net. Occasionally, however, small fishes (Syngnathidæ, &c.) frequent the most superficial zone, and will be captured in the net. Small fishes of pelagic habit are not unfrequently taken among floating sea weed. All these should be preserved in spirit. The occurrence of floating fish-eggs should be noted, and specimens reserved.

PHOSPHORESCENCE.

It is now well ascertained that the Phosphorescence of the sea is mainly due to living animals which frequent by night the more superficial zones; and no opportunity of carefully observing this phenomenon ought to be neglected. It is of importance to know the various species to which the light-giving function must be assigned, and to determine the conditions which may aid the luminosity or interfere with it. The collector should always make a note of the possession of this property by any animals in which he may have observed it. On occasions when the luminosity of the sea may be exceptionally intense, or when on the other hand this phenomenon may be exceptionally feeble, the temperature of the sea and the meteorological conditions present should be carefully noted.

Times of using the Towing Net.

The hours during which the towing net may be employed with the best results are various. In the temperate and equatorial seas some of the surface-dwellers remain habitually in the deeper regions during the day, and come to the surface only in the evening and during the night, while others will be found near the surface only in the daytime. In such latitudes the surface-life of the sea is usually found most abundant about sun rise, and again shortly after sunset. In Arctic regions, however, with the very different distribution of light and darkness, the habits of marine animals may be something quite different. These can be learned only by careful observation, and we as yet know little or nothing of them.

Preserving Liquids.

In the above directions the only preserving liquid mentioned is alcohol. This is certainly the most generally useful one, and will probably be found the only one practically available in high polar latitudes. It may be used in the form of methylated spirit of the ordinary commercial strength.

Schulze recommends for the preservation of very small Medusæ and other small delicate organisms, that they be placed, while alive, in a watch-glass with sea water, and then rapidly killed by dropping into the water a 1 per cent. solution of osmic acid. After lying some minutes in the osmic acid they are to be immersed in pure water, and from this transferred to spirit.

---

8. Supplementary Instructions. By Professor Huxley, Sec. R.S.

The authors of the preceding pages have so fully covered the ground of zoological instruction, that I have but few observations to offer.

It is desirable that no opportunity of seeking for Insecta, Arachnida, Myriapoda, and Annelida, on land or in fresh water, should be neglected. As in the Swiss glaciers, insects may occur in pools on land ice. Considering how few such specimens are likely to be obtained, it will be well to preserve any that may be found in spirits. The contents of the crops of birds will be worth examination on the chance of finding remains of such animals. Minute fresh-water entomostracous Crustacea and Infusoria are particularly worthy of notice and preservation. The latter may be preserved in spirit if previously treated with osmic acid.

The external and internal parasites of mammals, birds, and fishes should be sought for and preserved in spirit, the organs of the animal from which they are obtained being carefully noted. It would be interesting to know if the Arctic Canidæ are liable to be infested with *Pentastomum*, a large-sized vermiform parasite which occurs in the frontal sinuses of the dog. The abdominal cavity of fishes of the cod tribe and other deep-water fishes may yield specimens of the worm-like Myxioid fishes, of which only very few forms are at present known.

The experience of previous voyagers shows that amphipod and isopod Crustacea may be captured in great numbers by letting down a piece of meat into an ice-hole; and the exploration of the contents of the stomachs of fishes, and especially of any of the whalebone whales, will probably yield a harvest of pelagic Crustacea and Molluscs.

It is needless to remark on the importance of dredging whenever opportunity offers, and on the value of all specimens of stalked Crinoids that may be obtained. The rare and singular Ascidian *Chelyosoma*, the test of which is covered with polygonal plates, may possibly be met with, and, if so, should be carefully preserved in spirit.

One of the most interesting points to which the naturalists can direct their attention, however, is the obtaining of materials for the determination of the nature of the microscopic surface Fauna and Flora, and the comparison of it with the sea bottom in the same localities. The latter will, of course, be obtained by sounding. The former may be secured and preserved in the manner adopted by Dr. Hooker in the Antarctic expedition. This method consists simply in filtering a certain quantity of sea-water, taken at the surface and free from obvious impurities, through fine filter paper. After a sufficient filtrate has been obtained, the square of filtering paper may either be folded up with the filtrate inside, the latitude and longitude being written with a dark black pencil on the outside, and simply dried; or better still, it may be put, while still damp, into strong spirit. Perhaps even water, strongly impregnated with creosote, might suffice to preserve such collections; but it will be desirable not to trust to this without first trying the effect of maceration in such a fluid on the paper.

Salpæ are excellent collectors of surface organisms, and whenever they are met with in numbers it will be worth while to preserve a good many for the sake of the microscopic organisms contained in the alimentary canal. In the case of the larger *Salpæ*, the end, usually coloured, which contains the stomach may be cut off from a number of specimens, and preserved for the same purpose.

The stomachs of Lamellibranchs obtained by the dredge will give equally valuable information respecting the minute organisms at the bottom.

II.—BOTANY.

1. Instructions in Botany. By Dr. J. Dalton Hooker, C.B., President of the Royal Society.

There are many observations to be made on the habits and distribution of Arctic plants, and important collections to be formed illustrative of the local conditions of the climate and geological character of the regions they inhabit. A reference to the account of the Greenland Flora, republished in the Manual prepared for the use of the Expedition, shows how complicated is the problem of the migration of Arctic plants, and how much there is still to be learned from mere collections of specimens, provided these be complete for each locality, well preserved, and carefully ticketed. Quite as much also is to be learned of the life-history of Arctic plants; a field of research in which nothing has been observed, and one so wide that but a few indications as to what may be done can here be given. In this particular branch of inquiry the observations must for the most part be suggested by the observer himself; and an original and inquiring mind may find many paths to discovery even in the study of the poorest flora under its most unpromising aspects.

Flowering Plants.

There is reason to suppose that certain of the species of Arctic genera freely hybridise, especially those of Draba, Saxifraga, and Salix. I can account in no other way for the number of intermediate forms that are found in all extensive collections, and this between plants so distinct in other countries as the white and yellow-flowered Drabas. Hybridisation may also account in some degree for various supposed species rarely fruiting, though this is more probably due to the sudden accession of snow or other meteorological causes at the period of fertilization.

In connexion with the above subject, the pollen of the various species should be carefully examined, and observations made as to whether it is carried by the wind or by insects from flower to flower; and whether the surface of the stigma is viscid, or papillose, or clothed with hairs; and whether the flowers secrete honey on the petals, disk, or elsewhere. All association of insects with plants should be carefully noted, and their effects watched. It is doubtful if any annual flowering plant attains a very high latitude; the haunts of land animals, as the musk ox, &c. should be searched for such. Specimens of flowering plants should be abundantly collected both in flower and fruit, and this in all localities, keeping a very careful look out to secure all the species of such Families as grasses and sedges which resemble one another so much, and to secure both sexes of the willows.

Selected specimens of extreme forms and varieties of species should be sedulously collected, in order to show the limits of variation in a given area, and all circumstances that seem to influence variation should be noted.

Any modification of the *facies* of the vegetation in the various localities should be noted, as also the relative abundance and variety of the ubiquitous as well as of the scarcer species, their luxuriance of growth, &c.

Soil collected on icebergs, or on transported masses of ice, should be searched for seeds, roots, and remains of plants, and if practicable spread out and kept moist, till any seeds it may contain should have germinated. The number of kinds that germinate under such circumstances should be noted.

Mosses and Hepaticæ.

These have never been collected with the care they deserve in the Arctic regions. They are much more numerous than a casual observer, or one who attends to flowering plants only, would suppose, and can only be satisfactorily collected by close attention. Not unfrequently several species grow together in one tuft, and the hepaticæ especially are often found threading the tufts of mosses as solitary individuals. When collected, the tufts, if they have to be carried far, should be wrapped singly in paper, as their leaves and organs of fructification are liable to be injured. To preserve them the tufts should be broken up by the hand into fan-shaped specimens and pressed; such specimens indicate the habit of growth; one tuft will thus supply many instructive specimens. In the case of mosses in fruit, the calyptra and operculum should be carefully sought for, and if fugacious put in a little fold of white paper by the specimen. The male organs are often minute and obscure, and should be diligently sought for, using the pocket lens in the field if necessary. Many species have the sexes in different tufts; and in the Arctic regions the male plants are probably more frequent than the female. Of some species indeed the male inflorescence is only known.

Lichens.

These have not been collected with any method in the Polar American Islands or in the high latitudes of Greenland. Many of the larger species that grow on rocks or on the earth have indeed been brought home, especially by Lyall and Walker, but of the minute kinds that inhabit the bark of shrubs, and possibly the leaves of various plants in those regions little is known; nor of the crustaceous kinds that adhere to stones, and which cannot be removed without pieces of the rock or stone on which they grow. To remove them a hammer and a chisel are necessary, and the specimens should be trimmed so as to take as little bulk as possible, consistently with preserving the whole specimen. Besides drying between papers the branched and leafy kinds, bags of them should be brought home in a rough state for chemical analysis.

Fungi.

This family of plants is rare in the Polar regions, and a few Agarics and Pezizæ are the forms most frequently to be met with. Search should be made for the minute species which are parasitic on the branches and leaves of woody plants. In the case of Agarics the spores should be collected on white paper and their colour noted, and the plant itself preserved in spirits. In all cases the colours should be noted, or, better still, the plants should be drawn. It is well also to note of Agarics, &c. whether the stalk is solid or hollow, and the top dry or viscid.

Algæ.

Marine Algæ may be found between tide-marks attached to rocks and stones, or rooting in sand, &c.; those in deeper water are got by dredging, and many are cast up after storms; small kinds grow on the larger, and some forming fleshy crusts on stones, shells, &c. must be pared off by means of a knife.

The more delicate kinds, after gentle washing, may be floated in a vessel of fresh water, upon thick and smooth writing or drawing paper; then gently lift out paper and plant together; allow some time to drip; then place on the sea-weed clean linen or cotton cloth, and on it a sheet of absorbent paper, and submit to moderate pressure. Many adhere to paper but not to cloth; then change the cloth and absorbent paper till the specimens are dry. Large coarser kinds may be dried in the same way as land plants; or are to be spread out in the shade, taking care to prevent contact of rain or fresh water of any kind; when sufficiently dry, tie them loosely in any kind of wrapping paper. Those preserved in this rough way may be expanded and floated out in water at any time afterwards. A few specimens of each of the more delicate Algæ ought to be dried on mica or glass. A note of date and locality ought to be attached to every species.

Delicate slimy Algæ are best prepared by floating out on smooth-surfaced paper (known as "sketching paper"); then allow to drip and dry by simple exposure to currents of air without pressure.

Very little information exists regarding the range of depth of marine plants. It will be very desirable that observations should be made upon this subject, as opportunity from time to time presents itself.

Professor Dickie remarks, and the caution should be borne in mind:—"When the dredge ceases to scrape the bottom, it becomes "in its progress to the surface much the same as a towing net, "capturing bodies which are being carried along by currents, "and therefore great caution is necessary in reference to any

" marine plants found in it. Sea-weeds are among the most
" common of all bodies carried by currents near the surface or
" at various depths below, and from their nature are very likely
" to be entangled and brought up."

Carefully note and preserve Algæ brought up in the dredge at moderate depths, under 100 fathoms, or deeper. Preserve specimens attached to shells, corals, &c., which would indicate their being actually in situ.

The following observations in the methods of collecting Diatomaceæ are extracted from the Flora Antarctica, vol. ii., p. 504, and apply to Rhizopods and many other minute oceanic organisms:

"The various means employed for selecting the species varied according to circumstances, as the following enumeration of the processes pursued will show. I. Sea-water was filtered through closely woven bibulous paper (filter paper), which latter was folded, dried, and carefully put away. If a certain measure of water be always thus treated, an approximate knowledge of the abundance and scarcity of the various species and genera occurring at different positions may be gained. II. The scum of the ocean almost invariably contains many species entangled in its mass; it was preserved in small phials well secured. III. A tow-net of fine muslin used when the vessel's rate does not exceed two or three knots secures many kinds, which may be washed off the muslin and collected on filter paper. IV. The stomachs of Salpæ and other (especially of the naked) Mollusca, invariably contain Diatomaceæ, sometimes several species. These Salpæ were washed up in masses on the pack ice, and in decay they left the snow covered with animal mattter impregnated, as it were, with Diatomaceæ; the *reliquiæ* were preserved in spirits. V. The dirt and soil of the Penguin rookeries, and especially their guano abound in Diatomaceæ, perhaps originally swallowed by the Salpæ and Cuttle-fish, which themselves become the prey of the Penguins. VI. Ice encloses Diatomaceæ; they are deposited on the already formed ice by the waves, or frozen into its substance during calm weather when the upper stratum of water rapidly congeals. Ice so formed generally breaks up by the swell of the sea into thin angular masses which become orbicular by attrition, whence the name pancake-ice. The pancake-ice was often seen a few hours after a calm, covering leagues of ocean, and uniformly stained brown from the abundance of these plants. It was taken in buckets, and when removed from the water appeared perfectly pure and colourless. On melting, however, it deposited a pale red cloudy precipitate, excessively light, consisting wholly of Diatomaceæ. This precipitate was bottled on the spot, and proved more rich in species than any of the other collections. The specimens were also the best preserved, for Professor Ehrenberg observes that some thus obtained appeared as if still alive, though collected three years previous to his examination, and subjected to many vicissitudes of climate. The snow sometimes falls on the surface of the still ocean-water and does not freeze, but floats a honey-like substance, often called brash-ice;

treated in the same way as the pancake-ice it yielded an abundant harvest. VII. The mud and other soundings from the bottom of the ocean, when brought up on the arming of the deep-sea lead, or the chlam or dredge, generally contain the siliceous skeletons or coatings of many species, with the markings of their surface retained. VIII. The fresh and salt waters and muddy estuaries of the Falkland Islands, and similar localities present us with species, occurring under circumstances altogether similar to what accompany their allies in Europe, and are caught by the dredge as it comes up."

Note Algæ on ships, &c., with the submerged parts in a foul condition; also preserve scrapings of coloured crusts or slimy matter, green, brown, &c.

Observe Algæ floating, collect specimens, noting latitude and longitude, currents, &c.

Examine loose floating objects, drift-wood, &c., for Algæ, if no prominent species presents itself, preserve scrapings of any coloured crusts, note as above.

It might be useful to have a few moderate sized pieces of wood, oak, &c., quite clean at first, attached to some part of the vessel under water to be examined, say, monthly. The larger or shorter prominent Algæ should be kept and noted, and crusts on such examined and preserved, with notes of the vessel's course.

Various instances have been mentioned by travellers of the coloration of the sea by minute Algæ, as in the Straits of Malacca by Harvey; and cases of this kind would be worth special attention.

The calcareous Algæ (*Melobesia*, &c.) are comparatively little known and are apt to be overlooked, they probably do not inhabit the Polar area, but their northern distribution is undetermined.

Fresh-water Algæ should be collected as occasion presents. Prof. Dickie states that they may be either dried like the marine kinds, or preserved in a fluid composed of three parts alcohol, two parts water, one part glycerine, well mixed.

Cases are recorded of the presence of Algæ in hot springs. If such are met with the temperature should be noted and specimens preserved.

Mr. William Archer, of Dublin, has supplied the following notes on collecting Diatoms, &c. in Ireland, and which probably apply equally to various Polar localities where water stagnates.

"In these lands the nicest Algæ are those found in peaty districts, not in peat-bogs, but in uncultivated spots with a peaty bottom, on the edges of springs, and this in spring and summer when the species are found in conjugation or in fruit. Desmidieæ are met with far more abundantly in such spots either mixed with confervæ or crowded into cloudy masses on dead leaves, or (the larger kinds) forming a mucous stratum. The collecting a little of such material from many sites offers the best means of obtaining many and rare species. To obtain fruiting specimens choose in most cases the least green or attractive looking specimens, and even brown and dead looking patches afford the best chance of obtaining fruiting specimens of the conjugatæ. This follows from the fact that the cells empty in conjugation, and the patches hence consist for the most part of empty siliceous coats. Scytonematous and Palmellaceous Algæ grow upon wet rocks; the former keep well enough folded in paper and left to dry, after which they can be re-moistened, but the latter should be bottled."

The Arctic Expedition affords excellent opportunities to the naturalist for making observations on the power of seeds to resist cold whilst retaining their vitality. To this end samples of various seeds will be supplied for conducting the experiment, which is of the simplest description. Certain fixed numbers of any one kind of seed should be exposed to a low temperature, and sown, side by side with as many that have not been so exposed, in pans of earth kept moist, and the time required for the germination of every seed noted. Such kinds as survive the degree of cold to which they have been exposed should then be referred to successively longer periods of cold and to greater cold till the power of germination is lost. Variations of this experiment will suggest themselves to the naturalists; equal numbers of small and light seeds, and larger and heavy may be compared, which may be selected by weight or by measurement. The germinated plants also may be exposed to successively greater degrees of cold and the results noted. Seeds may be immersed in fresh and in salt water of different temperature artificially raised, with a view of testing their power of resisting their influence, as may also roots of Polar plants, the hybernacula or buds of such plants as *Saxifraga cernua*. The seeds supplied are mustard, cress, radish, turnip, pea, bean, sweet pea, wheat, barley, oats, maize.

Observations are wanting of the temperature to which Arctic plants are exposed during the winter when covered with snow. This can be approximately ascertained by sinking a tube of wood or other non-conducting material containing a thermometer attached to the base of a rod of wood through the snow into the soil in which the plants grow. The base of the tube should be made of a conducting material which would take the temperature of the soil, and the bulb of the thermometer should be covered with wool, so that it may retain the temperature during withdrawal for reading off.

III.—GEOLOGY AND MINERALOGY.

1. General Instructions for Observations in Geology. By Prof. A. C. Ramsay, LL.D., F.R.S., Director-General of the Geological Survey, &c., and John Evans, F.R.S., President of the Geological Society.

The instruments and other appliances necessary for daily or occasional work are few.

1. A hammer blunt at one end for breaking rocks, and flattened and somewhat sharp on the edge at the opposite end for splitting purposes when in search of fossils, &c. It is well to be provided with a spare hammer or two in case of loss, and one or two of smaller size for trimming specimens. The most convenient way of carrying a hammer is to sling it in a flat piece of leather through which the handle is passed to a belt buckled round the waist.
2. Two or three small steel chisels.
3. A measuring tape is sometimes useful.
4. A pocket compass, in which there may be a clinometer.
5. A larger pendulum clinometer, which is also a foot-rule.
6. A common ivory protractor.
7. A leather satchel slung across the shoulder in which to carry specimens, fossils, &c.
8. Gummed labels with printed numbers to stick on rock specimens and fossils collected.
9. Some cotton to pack delicate specimens in.
10. A supply of packing paper, and small canvas bags for special specimens.
11. Note-books, which may also serve for sketch-books.
12. A box of colours and some drawing paper.
13. An aneroid barometer for the measurement of heights.
14. For the purpose of refreshing the memory in case it should be at fault, a late manual of geology should be provided, such as Jukes' Student's Manual, and Lowry's Figures of Fossils stratigraphically arranged. This gives an excellent idea in a compendious form of the different forms of fossils which may be expected in the various formations. Also Ramsay's Physical Geology and Geography of Great Britain, which explains the connexion of Geology with Physical Geography in a somewhat condensed manner, and is more or less applicable to many countries besides Great Britain.

Stratified and Igneous Rocks.

The following are the chief preliminary points to which the attention of the observer should be drawn, assuming him not to have had much experience in geological work in the field, but to be already acquainted with the elementary principles of the stratification of sediments by water; if not, it will be necessary for him to consult a manual, like that of Jukes and Geikie.

On a voyage such as that contemplated it is probable that sections, whether of solid rocks or of looser superficial detritus, will chiefly be seen on lines of coast cliffs, or indicated by the outcropping of strata on occasional flat spaces between high and low water mark.

a. Do the rocks lie in horizontal layers or beds, and are they all

Fig. 1.

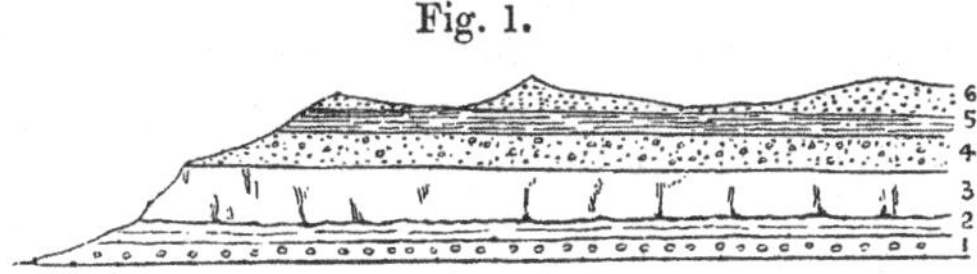

of the same lithological character, such as of Conglomerate, Sandstone, Limestone, or Shale; or do they vary, such as 1 Conglomerate, 2 Shale, 3 Limestone, 4 Sandstone mixed with pebbles, 5 Shale again, and 6 Sandstone; or are they of mixed character partaking of two or more kinds of material. If Conglomerate, are the included stones rounded and water-worn, or angular; of what kind or kinds of material do the fragments consist; of what size are the largest stones; and the site of the parent rock from which they were derived should be noted when it can be ascertained.

Do all, or any, of the beds contain fossil shells, or other kinds of organic remains. If so of what genera, and if possible try by genera and species to determine to what part of the scale of the geological formations they belong, such as Silurian, Carboniferous, Oolitic, and so on. The fossils from each separate bed should also be numbered when collected.

Fig. 2.

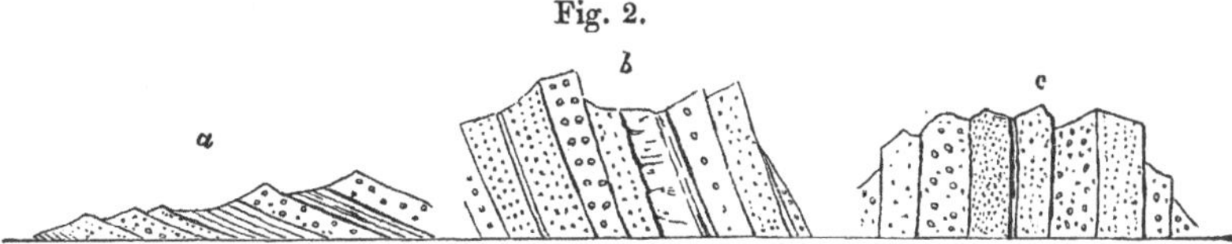

b. Are the strata inclined (*a*, *b*, fig. 2) or vertical (*c*., fig. 2.)

If inclined, at what angles do they *dip*. This is ascertained by the use of the clinometer. The point of the compass towards which the strata dip should also be accurately registered.

Fig. 3.

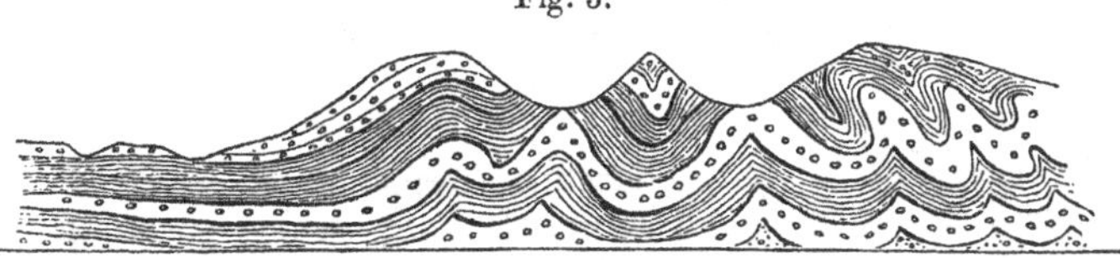

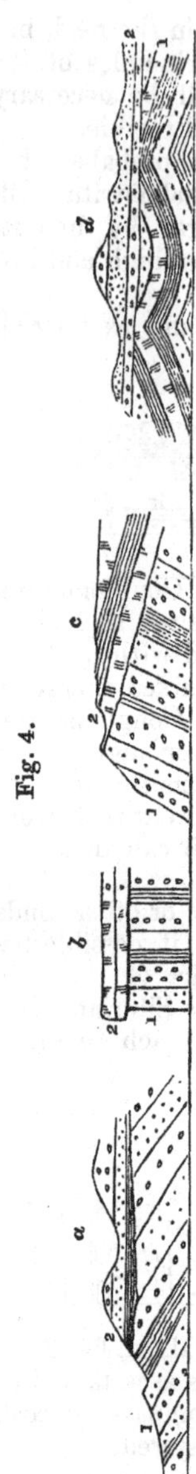

Fig. 4.

c. It may happen that the strata do not dip regularly in any given direction but are bent and contorted, and this should be noted. In all cases observations of their fossil and other contents should be registered similar to those mentioned with regard to the supposed horizontal strata, that is to say, in serial order.

d. It is of great importance to notice if all the strata in a given section lie *conformably* on each other as they do in the preceding diagrams, or if *unconformable stratification* is obvious, or may be inferred; cases of which are shown in the diagrams, Figs. 4, *a*, *b*, *c*, *d*.

If so then it is probable that the strata numbered 1 will be found to belong to a much earlier period of geological time than those numbered 2, and if the strata are fossiliferous that some of the genera and most or all of the species will be distinct in the formations that lie unconformably to each other.

e. Interstratified with common sedimentary strata there are often beds of coal, lignite, gypsum, rock salt or other minerals to be found, and they sometimes contain pseudomorphous crystals of rock salt in marly, shaley, or sandy bases, for these often remain where no solid beds of salt are found. These circumstances are of importance as indicating terrestrial surfaces where the plants, the fossilized remains of which form coal and lignite, grew, and in the case of gypsum and salt, to the probable existence of inland lagoons and salt lakes in which gypsum and salt were deposited.

Note and collect any other minerals in the rocks that seem to be of scientific importance.

It is also very important to note the colour of the rocks, grey, green, brown, blue, or red, &c., as the case may be, and also the effect of weathering on the surface of the rocks.

Igneous Rocks.

f. Are any of the ordinary stratified formations associated with igneous rocks such as bosses of granites, syenites, quartz-porphyries, diorites, &c. Are they pierced by trap dykes, passing more or less across the planes of stratification as in Fig. 5, and if so of what kinds.

Fig. 5.

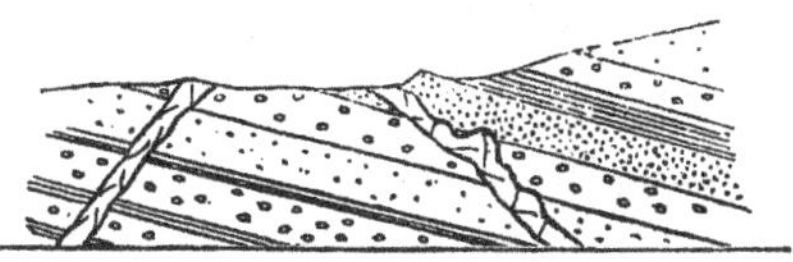

Are common sedimentary strata ever associated with *interbedded lavas* and volcanic ashes and tuffas, in such a way as to show that they were poured and spread out under water at intervals during the accumulation of the strata, and what is the mineral character of the igneous rocks. In the case of lavas that have been poured out over ancient sea bottoms, it may often be noted that the sedimentary stratum *underneath* has been altered or *baked* by the overlying melted mass of lava, while the sedimentary bed that *overlies* the lava remains unaltered by heat, the underlying lava having cooled before the deposition of the sediment that succeeds it in the section exposed. This is one way to distinguish between such lava beds, and *sheets of melted matter that have been forcibly injected between ordinary sedimentary strata.* Beds of coal, or of lignite, underlying sheets of igneous rocks, should be examined to see whether any portion has become altered, possibly into graphite.

Fig. 6.

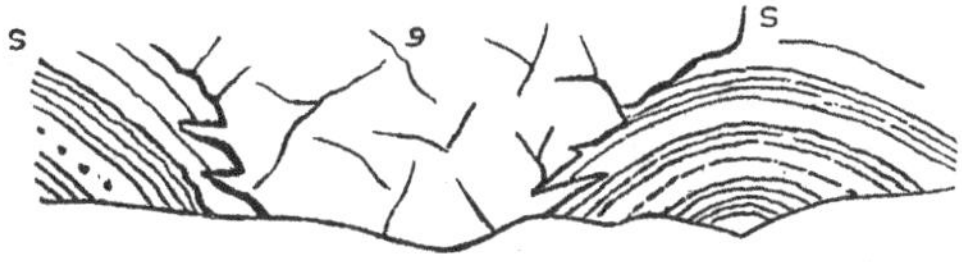

Do igneous rocks, such as lavas and volcanic ashes, show signs of having accumulated on land in successive layers. If so, are there any signs of soils and plant-bearing beds between them, or of other strata that may have been formed in fresh water, bearing bivalve Crustacea, such as Cypris, &c., or any other kinds of organic remains, such as fish, terrestrial mammalia, &c. &c.

g. Besides strata that have merely been hardened into rocks and show all the signs of ordinary stratification, it is probable that tracts of metamorphic rocks may be met with, such as slaty beds merely altered by slaty cleavage, also gneiss of various kinds, mica schist, chlorite schist, hornblende schist, serpentines, &c. For the theory of those alterations of common formations that resulted in the production of metamorphic rocks still showing traces of stratification, the observer must refer to any good manual of geology. It is sufficient now that he should be able to distinguish their leading varieties.

It must be understood that gneiss and other metamorphic rocks are not necessarily of the greatest geological antiquity.

In Europe, America, and Asia there are metamorphic rocks of all geological ages ranging between the Laurentian and the Eocene formations. It is, therefore, important to discover or surmise to what formation or set of formations any metamorphic series of rocks may belong, should any data be available for that purpose. In the absence of this, the observer must be content to register the character of the rocks and their modes of occurrence.

More Special Observations.

h. With regard to ordinary stratified rocks, it is important to discover whether the organic remains they contain are marine, estuarine, or freshwater mollusca, fishes, reptilia, &c.

Also whether terrestrial mammalia, insects, and land plants occur, and if so to what genera and species they belong; and, if possible, to collect a sufficient quantity of all kinds of fossil remains to be examined and described by the best authorities on the return of the voyagers.

i. In connexion with this they should endeavour to determine in any given section, or in sections of rocks more or less apart, whether more than one geological formation or set of formations is present, as, for example, strata that lithologically or palæontologically can be compared to the European or American Silurian and Carboniferous rocks, or to the Liassic and Oolitic series, or to the Cretaceous series, or to the Wealden, Eocene, or to the Miocene strata, &c. &c.

Should this be practicable, it is important to endeavour to show in drawn sections their order of succession and superposition in the manner given in ordinary geological sections, and also the way in which they are affected by faults or dislocations, either visible, or that may reasonably be inferred, as for example, by later deposits, No. 1 in Fig. 7, seeming to dip under older strata 2, in what may be called an unnatural manner without the visible

Fig. 7.

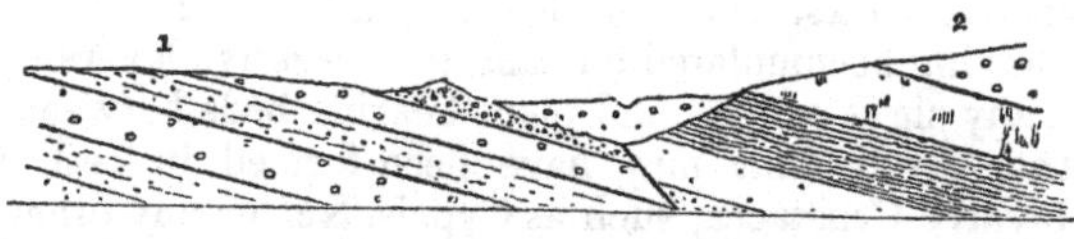

intervention of a fault or faults, or of partial inversion of the strata by contortion of the masses as in Fig. 8.

1, Secondary strata; 2, Palæozoic strata; *f*, Fault.

Fig. 8.

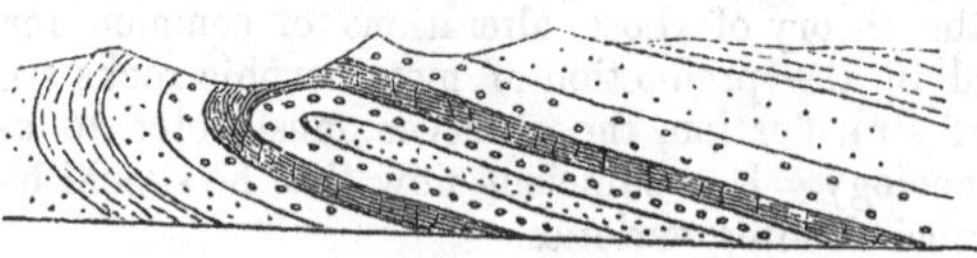

Inverted strata.

In connexion with fractures, faults, and wide joints in the rocks, mineral lodes may be looked for, such as lead, copper, tin, gold, Cryolite, phosphate of lime, and other minerals; and, if the observer is in doubt as to their nature, if possible, let him bring away specimens.

There are also some special points that ought to be attended to which may occur in these northern regions, such as:—

To gather additional information respecting the Oolitic fauna in any newly discovered area similar to those already known at Cook's Inlet in latitude 60° N. (M'Clintock), and of the Liassic fauna found by Sir Edward Belcher, and in another by the Swedish Expedition in latitude 78° 30′. Also to ascertain if the Carboniferous flora occurs in any continental lands or islands resembling that found in Bear Island, lat. 70° 30′, or possibly in Discoe Island, where loose blocks were found containing Sigillaria and Stigmaria, or again similar to the Carboniferous strata of Melville Island.

Also special attention should be paid to new areas containing a Miocene flora, such as has been collected by Nordenskiöld, Sir R. M'Clintock, Sir R. Maclure, Colomb, Inglefield, Dr. Brown, and Whymper at Atanakerdluk in the Waigat, and at other places near and in the island of Disco, in Greenland. A similar flora is also known in the Miocene rocks of Spitzbergen.

In connexion with this latter subject, the explorers in the late Austrian Expedition mention that many great sheets of basaltic lavas were seen, in the new archipelago which they discovered, to overlie, unconformably, masses of gneiss, in a manner that conveys the idea that the overlying igneous rocks consist of vast masses of horizontal sheets of lava. The description reminds the writer of the manner of occurrence of the Miocene igneous rocks in and near Disco in Iceland, the Faroe Islands, and of some of the Inner Hebrides. Should these or other islands be visited which are more or less composed of such like sheets of lava, it is important to notice if terrestrial surfaces occasionally occur between them, showing signs of terrestrial soils and the remains of land plants, or if freshwater beds occur between the igneous rocks bearing relics of land plants and of freshwater or terrestrial animals, such as Crustacea of the genus Cypris (found in Mull along with leaves of land plants by the Duke of Argyll), Insects, and Mammalia, and, if so, specimens should if possible be preserved. Any notes of this kind will be of great value, as throwing much light, not only on changes of climate, but also on the subject of a great continental extension of land during the Miocene epoch into far northern regions, as suggested by Dr. Robert Brown, much of which still remains—as is also indicated by Iceland, the Faroe Islands, the Inner Hebrides, the North of Ireland, the Madeira Islands, and other Atlantic isles—as surmised by Mr. Judd.

As, in some places, it may be impossible to obtain access to rocks *in situ*, it will be well to examine any pebbles on the beach, and any moraine boulders for organic remains, noting in each case the direction from which the pebbles or boulders appear to have travelled. In searching for fossils the rocks should be broken along the planes of stratification rather than across them.

Glacial Observations.

The observer must be supposed to be already acquainted with the phenomena of ordinary European glaciers such as those of the Alps, and, by reading, with those of Greenland, and with the subject, generally, of the ordinary glacial boulder clay and occasional marine deposits holding shells, &c., so widely spread over the North of Europe and America. In Greenland and any other land he may visit it is important to notice :—

a. Are linear surface moraines corresponding in direction with what may be called the trend of the flow of the existing glacier streams common or occasional, and are they similar to those on the Alpine glaciers. If so, are any cliffs or bare slopes observable from which the débris could have fallen from which such moraines were derived. If cliffs bordering the glaciers are not visible, the existence of such moraines would indicate their existence further inland.

b. Are there any glaciers in high northern latitudes that do not descend to the level of the sea, and, if so, what are the forms, extent, and height of the terminal moraines that accumulate at their ends.

c. Observations, if possible, to be made on moraine matter under the glacier ice, that is to say, between the glacier and the rocky floor over which it flows (*moraines profondes*) ; the possible extent and thickness of such moraine matter, and its coarseness, fineness, other general characters, and mode of occurrence. Are the ordinary phenomena of scratched stones common under such circumstances, and especially are large boulders found there. Are they ice-scratched.

d. Can the thickness of the ice of certain glaciers be ascertained which pass seaward beyond the shore or lines of sea-cliff, and which at their ends may be supposed to grate along the sea-bottom. This may be done by soundings at the ends of such glaciers, in conjunction with estimates of the height of the surface of the glaciers above the level of the sea.

e. Where glaciers protrude out to sea and there expand after the manner of the Rhone glacier, where at its lower end it protrudes and expands in a wide valley, is it possible to form an idea of the shape of the ground on either side of the valley through which the thicker mass of the glacier ice descends to the sea. Is it likely to be merely undulating ground somewhat higher than the valley, or hilly or even mountainous, the whole region being more or less smothered in ice. In connexion with this it may be asked, is the so-called continental ice of western Greenland to a great extent an ice sheet formed independently of mountains that bound deep valleys, the slopes of the bottoms of these valleys being westward or in the far north, is it an exaggeration of a system of confluent glaciers generated by high mountains on the east side of the continent or elsewhere.

f. What is the appearance of the surface of such glaciers. Are they *crevassed* in the interior of the country, that is, traversed by fissures, large and small, like the Alpine glaciers. Are there any crevasses traversing the surface of the ice in cases where it passes out far seaward.

g. To make, if possible, observations on the temperature of the ice at the surface, and at various depths below the surface, for the purpose of discovering to what depth the ice is affected by the external temperature of the air. It is usually stated that all glacier ice below a shallow variable depth is just about the temperature of 32° Fahr., and therefore in part always passing into the state of fluid water. This by some has been doubted with regard to the Swiss glaciers in winter. It is stated that streams of water flow all the year round from underneath the ends of Greenland glaciers, which are charged with glacier mud, and so to speak, boil up with the freshwater from the ends of glaciers that pass out to sea all the year. If so, does the quantity of freshwater and mud seem to decrease in winter.

h. It has been stated by Dr. Sutherland that the surface ice of the Greenland glaciers of Melville Bay, &c., for a depth of 8 or 10 feet is more solidly frozen than the underlying strata of ice, because of the influence at the surface of the cold air; and that the underlying ice, having the temperature of ordinary deep glacier ice (about 32°), flows faster than the overlying thoroughly frozen stratum, and that this upper stratum, adhering to and being dragged unwillingly onward by the underlying more rapidly moving ice, decrepitates and is shattered because of its solidity and power of resistance to the onward motion of the underlying more rapidly moving body of melting ice.

Further observations on this point are desirable.

i. Measurements actual or approximate of the size of boulders on glaciers are desirable, and notes of the various kinds of rocks that form surface moraines. Sketches of such boulders would be sometimes of value.

k. It has been stated that the solid Greenland rocks which form the surface of the country are not grooved and striated like the rocks affected by old and modern Alpine glaciers, or like the rocks of Scotland, the North of England, and Wales, which, like much of modern Greenland, are believed to have once been buried under universal thick sheets of glacier ice of what is called the Glacial Epoch. The reason given for this is that the whole or most of Greenland, being or having been entirely covered by glacier ice, no moraine matter from bare cliffs fell on the surface of the glaciers, and that therefore no stones and other glaciers débris found its way from the sides of glaciers and through crevasses to the bottom of the ice, by means of which their rocky floors could be grooved and scratched because of the great superincumbent pressure of the moving ice-flow. The more northern rock surfaces of Greenland have therefore been said to be ice-polished and *moutounée*, but not grooved and scratched. Is this the case.

l. What is the state of the bare rocky cliffs described by Kane in the far north of Greenland as regards ice-markings.

m. Should new islands be discovered further to the north, what are their physical characters as regards height and form. If mountainous ordinary glaciers may be expected. If so, have they any special characteristics, and of what kinds.

If low or flat, is true glacier ice formed on them, and if so of what thickness, and what is its general *behaviour*. If in such cases the rocks are sometimes bare of ice, are they smoothed, polished, grooved, and scratched as if by the action of glacier ice that once was there, and if so in what direction do the striations run, and if in more than one direction, which appears to have been the prevailing one.

n. Specially to observe the *ice-foot* or flat fringe of ice that adheres to the shore for a time after the main masses of the ice-floes have become detached from it. Note the quantity of detritus that falls on its surface from the adjoining cliffs, and its subsequent flotation seaward into deeper water, and the scattering of boulders thereby over the sea bottom as the ice melts.

o. Observe all icebergs of importance. Note if possible their length, breadth, and probable circumference. Observe their shapes, whether tabular, or serrated and peaky. If tabular endeavour to determine their heights above the level of the sea, for this may serve to indicate the thickness of the glaciers from which they broke, since in tabular masses of ice, the mass above the water bears a definite proportion to the mass submerged.

Note if possible whether or not they are aground, and if so, in what depth of water.

Observe if any icebergs are laden with masses of moraine rubbish, and if so, try to estimate its amount after the manner of Scoresby. Are boulders ever seen encased in the ice far below the surface. Is it probable that grounding icebergs are capable of attaching submarine boulders, gravel, sand, and mud, and carrying them on as they float and melt in other areas. Is it likely that grounding icebergs polish, round, groove, and striate the rocks over which they grate, both on the side on which they first impinge, and on the opposite side as they are forced over the opposing mass of rock.

p. In connexion with floating ice generally, endeavour to indicate the direction of the flow of marine currents. Does floating coast ice uproot and transport boulders, &c., and does it smooth and striate rocks, and help to produce *roches moutounées.* Are grounding icebergs, and other kinds of floating ice likely to contort the soft strata of the sea bottoms on which they impinge.

q. It is stated by Dr. Rae that "in the Arctic regions ice is "sometimes by great pressure forced up on shore many feet above "high-water mark, and carries with it or pushes before it stones; "and these are left in such varied forms as to cause the belief that "the work has been done by human hands. If the shores of the "Arctic Sea are gradually rising, stones thus pushed up by ice may "be found at a very considerable height above the sea." Such observations are valuable, and note ought to be taken as to mounds of shore gravel having been pushed landward by the pressure of sea ice packed and forced up above high-water mark. It is possible that such observations may throw some light on the ridges known as Eskers in Ireland and Kames in Scotland. These are long mounds of gravel believed to be intimately connected with the *Glacial Epoch*, and by some supposed to have been formed on the shores of the icy sea of that period.

r. In connexion with the subject of sea coasts it is of importance to observe if there are traces or lines of raised sea-beaches running in a terrace or in lines of terraces at different heights more or less parallel to the present sea-shores. Also whether or not they contain sea-shells and other marine remains. Also at what height above the present sea-level each individual beach or terrace lies. Note also, if possible, the direction from which shore pebbles may have come and larger boulders, and if they have any relation to prevalent winds and marine currents.

s. A good deal has been written about the occurrence of meteorites (meteoric iron) in Greenland. Should such be observed, their position and size should be noted, and if possible, specimens collected.

---

2. Instructions for making Observations on, and Collecting Mineralogical Specimens. By Prof. N. Story Maskelyne, F.R S.

A.—Minerals and Rocks.

In offering advice as to the mode of setting about collecting minerals and mineralogical facts in a land that is peculiarly rich in rare and curious mineral species, while it is clad in an iron-mail of ice, one cannot lose sight of the fact that the special experience which the Arctic voyager must soon acquire will fit him better than any instructions for the tasks of exploration and collecting. Nevertheless some notice of the sort of localities that may repay research, and of the observations that it would be worth recording, may not be out of place, while a short summary of the mineral objects that are to be looked for, and of the modes of determining something about their characters are of the first importance. The instructions on the subject of geology will certainly impress on the scientific observer the great importance of carefully noting and laying down in profile, and where possible, in plan, all important lines of mountain chain or protruding rock, and of collecting specimens of every distinct kind of rock, and further, of fastening to or carefully enrolling with all specimens, labels that can hardly err in the fulness with which they state the circumstances and the position of the spot at which they are obtained. To the mineralogist rock specimens have a special interest as being aggregates of minerals and often containing crystals in cavities or otherwise distributed through them, from the presence of which the history and associations of the rock itself may be gathered. Hence a judiciously made collection of rocks has the character of an index to the petrology of a whole country.

It is among igneous rocks that the Arctic mineralogist will probably find his chief occupation. The important minerals that occur under other conditions, such as where they are found lining the fissures which carry mineral lodes, may indeed be accidentally met with—perhaps among the weathered masses at the foot of a cliff on the section of which the mineral vein may be recognised from which they came; and any minerals so found that from their metallic lustre, their weight, or some other striking character may appear peculiar should be preserved in specimens, so that their characters may be determined at leisure. But the rocks fruitful of minerals for the Expedition, north of Upernavik, will in all probability be of a different kind. It is among the minerals that belong to or are associated with the occurrence of igneous rocks developed on a large scale that the Expedition will be able probably most effectively to deal. For the conditions of Arctic travel are hardly consistent with the close and careful search needed for the discovery of the rarer kinds of minerals; nor are the characters of the country and climate such as to expose such minerals to view under favourable conditions for finding them, as for instance in the beds of torrents. However uncertain may be the early accounts that recorded the existence of an active volcano and fumerole action in the south of Greenland, it is quite possible that volcanic forces may still be in action in the regions of a remoter north. Should this be found to be the case there will be an ample field provided for all that enterprise and observation can do in collecting materials for the description of such a district. Some notes drawn up by an experienced observer, Mr. J. W. Judd, are appended to these instructions, and they deal with the more important petrological questions that arise in such a neighbourhood.

In the case of rocks of the plutonic class being met with, it is less easy to offer as precise injunctions as in the former case in regard to the methods of observation. A collection of well selected specimens of the rocks themselves is in all cases the first requisite; and next, it is important to gather illustrations of any special peculiarity in mineral associations that these rocks furnish. In cases where plutonic rocks have intruded into other formations it is desirable that specimens of the adjacent rock should, where practicable, be collected from points at different distances from the intruding mass. And in the case of rocks of the granitic class it may be that irregular cavities may be met with in which the crystals of the minerals forming the rock are distinctly developed; and these are sometimes associated with other minerals of interest such as beryl, topaz, tourmaline, &c.

Among the rocks of the volcanic class the trachytes will often be found rich in interest for the mineralogist, as well from the varied forms which they assume, including a porphyritic structure on the one hand and a glassy structure on the other, as from the various minerals that they include. And the doleritic class on the other hand, including basalt, presents a special interest in the amygdaloidal cavities with which such rocks teem, and which are so often found to be the home of minerals of great variety and interest, well repaying a careful search.

While passing near a coast, as, for instance, along the strip of ice-foot, the rocky cliffs and bluffs should be carefully scrutinised where these form the coast-line; while the talus that conceals their feet may yield specimens weathered out from the rest of the mass which it would be well to search for where any peculiarity is presented by the face of the rock itself. In collecting specimens carrying crystals that are at all delicate, it would be well to place them first in a fold of tissue paper and then to cover them with some soft material like cotton wool before finally packing them in an outer paper; and it would be better to insert the label next to the tissue paper.

Where delicate crystals present salient points it is best to secure them by packing them in chip boxes, into which they should be wedged by plugs of cotton, and, when opportunity offers, by subsequently fastening them to the box by a little glue on their under side.

It would seem to be preferable to an attempt to condense into a few pages descriptions of the more important minerals (which to a person familiar with mineralogy will have little value) that, for the use of collectors not well versed in the science, a very small series of such minerals carefully selected as representing their more important characteristics should form part of the equipment of the Expedition. By reference to such a small cabinet comprising perhaps some 50 specimens, the collector will not only familiarise his eye with their aspect, but may compare with them on his return to the ship the specimens which he has collected during a temporary expedition. Thus, the large number of minerals referable generally to the group of augites or those of which hornblende is a type, or again those forming the group of garnets, though widely differing in the case of each group in respect of colour and even of habit, yet present such general mineralogical resemblances that with the aid of a treatise on discriminative mineralogy and a few implements, the collector might go far towards identifying many of the minerals he has collected, should he not be content with merely storing them for investigation at home. And with this view, it will be well to mention two or three handy books by the use of which, and by the aid of a few experiments he may find for himself and practically apply all the information which he immediately requires. Such books are either Dana's smaller Manual or his larger treatise on mineralogy, preferably the latter, Frazer's translation of Weisbach's Tables for the Determination of Minerals (Philadelphia, 1875); to which may be advantageously added, for the details of results with the blowpipe, Brush's Manual of Determinative Mineralogy (New York, 1875), and a Treatise on Rocks, by Cotta, translated by

Lawrence (Longman & Co., 1866). And included with the small series of minerals above recommended, a few samples of the more important igneous rocks should be taken for the purpose of comparison.

With regard to tools and instruments requisite for obtaining mineral specimens and for recognising them when obtained, besides the personal companionship of a small portable hammer of the best steel and of not too hard a temper, at least one more massive hammer, and two or three large chisels and wedges should form a part of the equipment that accompanies an exploring party; and doubtless means of blasting masses of rock in special cases, by methods involving comparatively little labour will not be wanting to the Expedition. Tools of large size are requisite in order to obtain good pieces even of small magnitude of tough igneous rocks. The specimens of rock themselves need not be larger than four inches by three, and one inch thick; but, where many have to be carried, in the case of ordinary-looking rocks a size of about 3 in. × 2 in. must be deemed sufficient.

But it is before all important that, where possible, the specimens secured should not be merely the weathered outside of a protruding rock, but a piece of the rock with fresh fracture from the interior of such a mass.

The instruments of observation requisite for determining the direction and inclination of ridges and of the faces of rock-masses belong rather to instructions in geology than to those for collecting minerals, and will doubtless be provided for the Expedition. For the actual scrutiny of the minerals themselves the following apparatus should be provided for each collector. An ordinary pocket lens (and one or two in reserve in case of loss), with a moderately high and a low power; a small strong stoppered bottle for containing dilute hydrochloric acid; a not too elaborate set of blow-pipe apparatus, including a lamp for colza oil, or a supply of large sized stearine candles; two or three small fine three-cornered steel files; a small collection of ten mineral specimens representing the degrees of hardness; a magnetized needle, and a small hammer and two or three little steel chisels for trimming specimens.

B.—Meteorites.

There is no spot in the world around which so much interest has gathered in connexion with the subject of meteorites as that to the N.W. of Disco Fjord in the island of Disco, from which Prof. Nordenskiöld first brought to Europe large masses of iron, which he announced as having been embedded in Miocene times in the basaltic rocks that there overlie to a vast thickness the gneissoid formations of the island. This spot is Ofivak, and from it an expedition in 1871, a year after Prof. Nordenskiöld's return, brought to Stockholm a mass of iron weighing nearly 20 tons, and others only inferior to it in size.

The great interest of the discovery lay, however, not even in the acquisition of these masses of apparently meteoric iron, but in the fact that they were found in close proximity to a ridge of basalt, entangled in which other specimens of native iron similar in character and associated with a kind of pyrites (Troilite, Fe. S.), only met with in meteorites, were found.

Such an ingredient could only have found its way into the basaltic dyke in one of two ways: it might have fallen into the basalt in the very remote epoch when that rock was yet in a plastic condition, or there is the possibility that it might have been terrestrial iron borne upwards with the melted rock mass from the interior of the globe.

It is, therefore, a matter of much interest that this place should be again thoroughly explored, and the point in question settled. The best means for this end will be to ascertain by careful inspection of the site how far the basaltic ridge from which Nauckhoff in the Swedish expedition separated the specimens of iron and troilite extends, and whether iron can be found in it in other places than that immediately investigated.

Experiments with a dip-magnet in the neighbourhood may lead to the discovery of such masses.

Under any circumstances it is important that portions of the basaltic rock itself and of the so-called basalt wacké (or decomposed basalt) on either side of it, should be blasted from the mass and brought home.

And it will be of much interest in connexion with the subject of meteorites, that any specimens of iron in use by the Esquimaux, indicating rude hammering or workmanship, should be secured and all possible information obtained as to the sources whence the metal is obtained. There is good reason for believing that meteoric iron has been habitually used by these people.

Thus, Sir John Ross records at p. 104 of his narrative that the natives in the neighbourhood of Cape Melville and Prince Regent's Bay obtained their iron for their implements from masses of iron that occurred in the Sowallick or Iron Mountains that rise at the back of that bay. And he mentions that they reported "that one of the iron masses, harder than the rest, was a part of "the mountain; that the others were in large pieces above ground "and not of so hard a nature; that they cut it off with a hard "stone and then beat it flat into pieces of the size of a sixpence, "but of an oval shape."

The locality was stated to be some 25 miles distant from the place in Prince Regent's Bay where the interview with the natives was held.

The masses of iron from Ofivak have a great tendency to undergo a sort of spontaneous corrosion, due to the presence of soluble chlorides enclosed within them.

The only available way of arresting or retarding this action seems to be to keep the meteoric masses either in completely dry air, or in a liquid that is closed as much as possible from the air. Probably putting them in a closed cask filled with fresh water would be the best means of effecting it.

There is another point of no small interest to which the attention of the observer in snowy latitudes should be drawn, in connexion with the meteoric matter that reaches the earth from space. It has been asserted that snow when collected under conditions that seemed, by the remoteness of the locality from human habitations, to be secure from contaminations due to the agency of man, has, when melted, yielded among other products metallic, and therefore, probably, meteoric iron. Nordenskiöld collected the dust thus distributed through snow in Spitzbergen, and proved its meteoric character by finding iron and cobalt in it. The circulation of the winds no doubt carries other forms of dust, including volcanic and desert sands, to enormous distances, but meteoric iron can be distinguished amidst these without difficulty. And meteoric iron is not the only—it, in fact, must be but a single ingredient among several constituting the meteoric dust falling through the atmosphere from the regions of space. Any discolouration of the snow that has not an evident cause in neighbouring sources of contamination should be at once suspected of having an origin thus foreign; and any steps feasible at the time should be taken for preserving as much as possible of the discolouring or otherwise foreign material. Some cubic yards of such snow will yield, probably, barely enough material for a satisfactory examination; and the melting of this quantity and the collecting, drying (at not too high a temperature), and preserving the small amount of residue mixed with it, without contamination from utensils employed in the process, involve care and precautions that will suggest themselves to the observer who may find himself in a position to avail himself of such an opportunity of aiding science.

Favourable places in which the residue from the melting of snows during summer months might have collected without contamination from impurities of local origin may, however, in all probability be found by the observant traveller, and this residuary material may so, perhaps, be secured in appreciable amount.

The icelike snow underlying the more recent or the melted snow may be found in some cases more richly charged by accumulation with the foreign dust in question. The matter is one of such great interest that it is well worth some trouble to endeavour to collect appreciable amounts of this cosmical dust.

---

3. INSTRUCTIONS on the OBSERVATIONS which should be made in case VOLCANOES or EVIDENCES of VOLCANIC ACTION should be met with. By J. W. JUDD, F.G.S.

Should any volcanic rocks be met with, the following suggestions may aid the observer in directing his attention to the most important points in connexion with them.

I. *If the rocks have a fresh appearance, and are of comparatively recent origin,* the following circumstances concerning them should be particularly noted :—

A. LAVA STREAMS. Concerning these should be recorded,—

a. Dimensions. Distance from point of origin ; breadth at various parts of course ; thickness, so far as it can be determined, and especially as affected by the accidents which the current meets with in its flow.

b. Slope over which they flow. This should be measured at various points with a clinometer if possible, and, at points where the inclination suddenly changes, any variations in the dimensions or other characters of the current should be carefully noted.

c. Surfaces of lavas. Attention should be paid to the features presented by these, whether smooth and "ropy," or bristling and scoriaceous.

d. Texture. Note especially if the rock of the current be porphyritic, compact, globular, concretionary, pumiceous, glassy, sphærulitic, or coarsely crystalline. If the rock presents ribboned or banded structures, observe, if possible, the relations of these to the direction of flow of the stream. When the rock exhibits transitions from one texture to another, collect series of specimens, illustrating the gradation. Note especially changes between the surface and interior of current, or those taking place at different points of its course.

e. Structure. All peculiarities of jointed, and especially of columnar, structures are worthy of being recorded. Note the features presented by the upper and lower part of the current, and any changes in its course; also if columns be divided by transverse joints, and the features presented by these, &c.

f. Chemical and mineralogical constitution. If the appearance of the rock does not suggest at once the class to which it belongs, and the component minerals cannot be detected with a lens, recourse may sometimes be had to a determination (even roughly) of its specific gravity.

g. Sometimes lavas contain large masses of *included minerals.* These are very interesting, and should be carefully collected.

h. Cavities, or air bubbles, in comparatively recent lavas, are frequently found coated with beautifully crystallised minerals. And when the rocks are of older date, the similar cavities may be lined or filled with crystals of zeolites and other minerals.

B. NATURE OF BEDS LYING BETWEEN LAVA CURRENTS.

These are of the utmost interest and value to the geologist, but unfortunately the ordinary mode of weathering of volcanic rocks is such as greatly to obscure the interbedded deposits by a talus of fallen fragments. The best opportunities for their study are afforded by sea-cliffs, and deep ravines or river-gorges, which should therefore be carefully examined. In such situations we may expect to find—

a. Burnt soils (Laterites of Lyell), usually of a brick-red colour, and affording various evidences of their modes of origin.

b. Coal or Lignite seams. These are very frequently observed. Note if they rest upon an "underclay" (an old soil with roots), and if they contain wood, leaves, or other plant remains, with recognisable structure.

c. Ash-beds. These are sometimes composed of such impalpable dust as to constitute a matrix in which delicate leaves, shells, and even insect remains are exquisitely preserved.

d. Stratified tuffs. Note especially the degrees and nature of their stratification; also whether they are loose or indurated. They may contain shells and plants of terrestrial or marine origin. Record the elevations at which the latter are found.

e. Gravels or other deposits. Note their characters and materials, and, if possible, define their mode of origin.

C. CONES, CRATERS, &c. Wherever the lavas present a fresh appearance, an attempt should be made to trace them up to their points of origin.

a. If any great *volcanic mountain* be met with, all details concerning the lava streams, fragmentary matters, and dykes of which it is built up will be of great interest. Failing these, however, sketches of the mountain, and of specially interesting portions of it, accompanied by such rock specimens as can be obtained, will be of service to geologists.

b. Cinder-cones on the flanks of a volcano, or scattered around it, should be examined and sketched. Note if they originate streams of lava.

c. The craters, both of volcanic mountains and of cinder-cones, should be examined. Note if they are breached by lava streams, or contain bosses of lava in their interiors, or buttress-like masses adhering to their sides.

d. Note especially the *arrangement* of the smaller and larger cones in respect to one another. Furnish, if possible, plans to illustrate this point, or failing these, as many general outline sketches as possible.

e. In and around the craters look for *fumaroles*, and, if possible, record the nature of the gases evolved from them. Collect the interesting minerals found in the crusts which are deposited round the vents, and in the rocks traversed by the vapours and gases.

f. Hot springs, geysers, &c., often occur in the vicinity of active or recently extinct volcanoes. These the observer should be on the look-out for (their vapours often render them conspicuous at great distances), and their phenomena should be carefully recorded. Specimens of hot and mineral water should be sealed up in bottles, and brought home for examination and analysis.

g. Deposits of siliceous sinter, travertine, &c. These, besides yielding interesting varieties of minerals and illustrations of their mode of formation, often contain incrusted or mineralised remains of plants or animals which may be of great interest.

(In the event of the observer being so fortunate as actually to witness an eruption of a volcano, every detail that he can supply may be of scientific value. Especially should he note the appearances presented by the ascending column of vapour and fragmentary materials issuing from the crater, the height to which this rises, the nature, sequence, and rate of the explosions to which it is due, and the sounds which accompany them. All earthquake shocks and tremblings of the ground should of course be recorded. If lava streams are seen flowing, their rate of motion and attendant phenomena should be carefully noted.)

II. *If the volcanic rocks have evidently been subjected to great denudation*, the following points should be more particularly attended to:—

A. The composition, textures, and various structures of the different lavas should be carefully observed, and all zeolites or other minerals in their included cavities collected.

B. If the igneous rocks be found alternating with sedimentary ones, all fossil remains which can be obtained from the latter will have a *double* value, as throwing light on the age both of the aqueous and the volcanic rocks. But it will be especially necessary to notice whether the igneous masses be *interbedded* and *contemporaneous* with the aqueous deposits, or *intrusive* and *subsequent* to them. In seeking to determine this point, it must be borne in mind that,—

Lava streams have slaggy or scoriaceous upper and under surfaces, and that they only alter the rocks upon which they rest.

Intrusive sheets, on the other hand, are seldom scoriaceous, and alter the rocks both below and *also above* them. They moreover occasionally cross the lines of bedding of the strata, and send off dykes or veins into them.

C. If possible, the lavas of the district should be traced up to central masses of *intrusive rocks.* The forms assumed by these in weathering should be sketched, and specimens illustrating the different characters which they assume and the minerals they contain be collected.

D. All the phenomena of metamorphism exhibited by the stratified rocks in the vicinity of intrusive masses, whether dykes, sheets, or bosses, should be looked for, and their nature and extent recorded. In connexion with this subject, it should be remembered that very many interesting minerals are developed near the junction of igneous rocks with those which they traverse. Series of rock specimens illustrative of a gradual change in characters will be very valuable.

E. If masses of tuffs and volcanic agglomerates be met with, they will frequently be found to contain crystals (more or less perfect) of various volcanic minerals. Not unfrequently they also yield fragments of rock which have been ejected from a volcanic vent. These, if of aqueous origin, may be searched for fossils; in all cases, however, they frequently exhibit signs of having undergone changes by the action of heat, acid vapours, &c. upon them. Such masses should be broken up and carefully examined, for they frequently enclose in their cavities some of the most beautifully crystallised varieties known to the mineralogist.

The general instructions as to the instruments best adapted for the purpose of the geological observer, and of the tools used for obtaining rock specimens and minerals, are of course equally applicable to the student of vulcanology. But as igneous rocks are in many cases especially liable to change by weathering, the greatest efforts should be made to obtain specimens as fresh and little altered as possible. In those cases, however, where the rock assumes any peculiar features in consequence of meteoric actions upon it, specimens both of the unaltered and of the altered rock are desirable.

The work of reference which will be found most serviceable to the traveller who may come across volcanic districts is Mr. G. Poulett Scrope's "Volcanos" (second edition, revised and enlarged, 1872), published by Longman and Co. In this work detailed descriptions of the interesting phenomena of volcanic action are given, and ample illustrations of most of the points adverted to in these notes will be found.

MANUAL

OF

THE NATURAL HISTORY, GEOLOGY, AND PHYSICS

OF

GREENLAND

AND THE NEIGHBOURING REGIONS.

The Zoology, Botany, Geology, and Mineralogy compiled by
T. Rupert Jones, F.R.S., F.G.S.,
Professor of Geology, Royal Military and Staff Colleges, Sandhurst:

The Physics compiled by W. G. Adams, M.A., F.R.S., F.G.S., F.C.P.S.,
Professor of Natural Philosophy and Astronomy
in King's College, London.

EDITED BY

Professor T. RUPERT JONES, F.R.S., F.G.S., &c., &c., &c.,

UNDER THE DIRECTION OF THE ARCTIC COMMITTEE OF THE ROYAL SOCIETY.

PREFACE.

THE subject-matter of this "Manual" is arranged under the two headings of—Part I. Biology and Zoology, and Part II. Physics. The former is subdivided geographically into—§ I. Davis Strait, Baffin's Bay, and the coasts continuing northwards, under the general term of "West Greenland." § II. The great Arctic-American Archipelago, including the Parry Islands. § III. East Greenland, with Spitzbergen and Franz-Joseph Land.

The short time allowed for preparation limited more especially the treatment of the last section, and has caused also an irregularity in the arrangement of some earlier articles, such as Nos. X., LIX., LXXI., LXXVI., and CVIII.

Part II. (Physics), having had a still shorter time of preparation, is without an Index, the Table of Contents serving that purpose. Indeed § III. of Part I. is not fully represented in the Index.

This "Manual" consists mainly of Reprints and Excerpts from Transactions, Proceedings, Journals, Magazines, and Foreign Books, and from English "Voyages," and their "Appendices" when the treatment of a subject required correlative matter to be brought forward.

Perfect uniformity in the printing of geographical and natural-history names has not been attempted. The plans and systems of different authors, the fashions of their day, the inaccuracies of some, the peculiarities of others, and the pressure of circumstances, have affected the style of reproduction in the several articles.

It has not been practicable in many cases to refer direct to the authors of the reprinted papers, even of late date, or to submit proofs for their inspection; several, however, have had the advantage of revision by their authors.

Thanks are due especially to Dr. Lütken and Dr. Mörch, of Copenhagen, for revisions of their previously published Catalogues of Arctic Animals, and to Dr. Lütken for new catalogues of several groups of animals living in Greenland and the neighbouring seas. The Editor has received much information about Greenland from Dr. R. Brown, of Campster, some of whose Greenland Memoirs are reprinted in the "Manual." Among others also, who have aided him with information and books, are R. H. Scott, Esq., E. Whymper, Esq., C. E. De Rance, Esq., and Count Marschall. Lieut. C. Cooper King, R. M. Art., F.G.S., has materially aided him in his work; and W. S. Dallas, Esq., F.L.S., has kindly revised a few of the lists. The officers of the Royal Geographical and other Societies in London have helped him in their Libraries; and the Secretaries of the Royal Dublin, the Edinburgh Botanical, and the Glasgow Geological Societies have courteously conveyed permission to reprint papers, and in some cases have presented the memoirs wanted.

Capt. F. J. Evans, C. B., has also most obligingly aided the Editor.

In the Index, geographical names have not been made a prominent feature, but natural-history terms, and the names of writers and explorers, have been chosen as the most convenient and useful keys to the many subjects of the "Manual." As time and space were limited, the name or term alone is usually given, and its various relations and associations are indicated by the often numerous references to pages that follow it.

T. R. J.
May 7, 1875.

TABLE OF CONTENTS.

MANUAL OF THE NATURAL HISTORY, &c. OF GREENLAND AND NEIGHBOURING LANDS.

PART I.—BIOLOGY AND GEOLOGY.

§ I.—West Greenland, including Davis' Strait, Baffin's Bay, Smith Sound, and Kennedy Channel.

§ II.—Parry Islands and East-Arctic America, &c.

---

§ III.—East Greenland, Spitzbergen, Franz-Joseph Land, &c.

PART II.—PHYSICS.

I.—Meteorology.

II.—Temperature of the Sea, &c.

III.—Physical Properties of Ice.

IV.—Tides and Currents.

V.—Geodesy and Pendulum Experiments.

Measuring the Motion of Glaciers.

VI.—Observations on Refraction and on Air.

Observations and Experiments on Sound.

VII.—Terrestral Magnetism.

VIII.—Aurora Borealis.

Additional Errata.

Page iv., line 3, *for* Zoology *read* Geology.
,, vi., line 8 from bottom, *for* Views *read* Veins.
,, 732, line 4, *for* Zonolrichia *read* Zonotrichia.

MANUAL, &c.

PART I.—BIOLOGY AND GEOLOGY.

§ I.

West Greenland; including Davis Strait, Baffin's Bay, Smith Sound, and Kennedy Channel.

I.—On the Mammalian Fauna of Greenland. By Dr. Robert Brown, F.L.S., F.R.G.S., &c.

[Reprinted by Permission from the Proceedings of the Zoological Society of London, 28 May 1868. With Corrections and Annotations by the Author, March 1875.]

Contents.

1. *History of the Subject.*

In entering upon a review of the Greenlandic species of Mammalia, it may be a matter of surprise to some that anything remains to be said concerning the larger animals of a country so comparatively near home, and regarding which so much has been written, where Egede, Fabricius, Vahl, and Rink lived, and regarding which we possess the remarks of such excellent naturalists as the acute authors of the "Fauna Grœnlandica" and "Grönland geograph. og statist. beskr." Between the dates of the publication of these two works an interval of upwards of seventy years extends, so that one might suppose that any errors of the first work might have been fully discovered in the interval and corrected in the second. All surprise vanishes, however, when we find that the contrary holds true, and that to-day we know almost as little about the Mammals of Greenland as we did when Fabricius gave us the first systematic account of them. The fact is that naturalists who have visited Greenland have been too much interested in other departments of natural history to pay attention to the larger members of the fauna, or have supposed that there was nothing worth adding to, or (what is just as important) subtracting from it. Accordingly, we find all authors on arctic animals merely contenting themselves with giving a list of Fabricius's species, and at the same time perpetuating the errors which he fell into through ignorance or credulity, independently of the fact that he only wrote of that limited portion of the country then inhabited by the natives over which his authority as a "Grönlandske Missionair" extended. Can we therefore be astonished if we find the fauna of Greenland, in the class Mammalia, burdened with species which have no existence save in the vivid imagination of the Eskimo or the overlearned acuteness of zoologists, and bereft of others which ought to take their place—their history poisoned with fables only worthy of the belief of the last century, and their geographical range in the country over which they are distributed scarcely touched on, or wrongly described. The accounts of the older writers on Greenland (Egede, Saabye, Cranz, &c.) were very unsatisfactory; but a new era in the history of northern zoology dawned when Otho Fabricius, who had passed several years in Greenland as a missionary, published his "Fauna Grœnlandica."* This work, far in advance of its age, and which for the conciseness and accuracy of its descriptions has rarely been surpassed, has most deservedly retained its place as our standard authority on the zoology of Danish Greenland.† Herein are enumerated thirty-one species of Mammalia indigenous to the country, exclusive of man and those which have been introduced by man's agency. Four of these species I have shown in this memoir to have been entered upon imperfect grounds, one was mistaken for another (*Ovibos moschatus* for *Bos grunniens*), and several are now known to be only synonyms of other species. The species of Cetacea are, as might be expected, the most obscurely described of all, and have occasioned much controversy; and the superabundance of literary acumen which has been spent on these descriptions is more than the nature of them will allow of.

Subsequently the elder Reinhardt gave some notes on the Greenland Mammalia in the "Isis" for 1848, which, in the main, are only a reproduction of the earlier account of Fabricius; and in 1857, the present Professor Reinhardt, of Copenhagen, in the Appendix to Rink's "Grönland"‡ furnished a list of the species, also following Fabricius. He has, however, entered the only species then added to the list, viz. *Mus grœnlandicus* of Traill,§ discovered by Scoresby on the east coast in 1822, under the name of *Hypudæus grœnlandicus*,|| and attempts to disentangle the specific history of the *amarok* of the older authors, Fabricius's *Gulo luscus*, the *Phoca ursina*, which Fabricius enters as a member of the Greenland fauna, the *Trichechus manatus*, &c., and with some success, though, not having visited Greenland himself, he is not so successful as he otherwise might have been. This list, as all the others, solely relates to Danish Greenland, extending from Cape Farewell

* Hafniæ et Lipsiæ, 1780.

† In 1867, whilst staying at Claushavn, I occupied as my *study* a little room in the pastor's old house, now deserted and used to accommodate any stray wayfaring men like myself. This was said to be the "dark closet" where Fabricius wrought at his Fauna, Lexicon, and other works. It was afterwards the residence of Saabye the grandson of Egede, who also wrote on Greenland.

‡ Grönland geographisk og statistisk beskrevet, &c., Band ii. Tillaeg Nr. i. (Pattedyr, &c.). This appendix was also published separately, "Naturhistoriske Bidrag til en Beskrivelse af Grönland."

§ Scoresby, "Journal of a Voyage to the Northern Whale Fishery, &c." Appendix, p. 459.

|| Prof. Reinhardt obligingly informs me (March 1868) that he is now quite convinced that this is a *Myodes*, though he only knows it from description.

(lat. 59° 49′ N., long. 43° 54′ W.) to Upernavik (lat. 72° 48′ N., long. 55° 54′ W.), and is valuable as expressing the state of knowledge regarding the Mammalia of Greenland, in Denmark, represented by a naturalist who has paid much attention to the arctic fauna, in the elucidation of some of the marine Mammalia of which he has so highly distinguished himself. This, as far as I am aware, is all that immediately relates to the Mammals of Danish Greenland. Various other writings have thrown much light on their *general* history; but it is with their *special* history and geographical distribution in Greenland that I have to deal. Among these memoirs, I ought not to omit mentioning the excellent paper on the Mammalia of the northern countries by Professor Malmgren,* who accompanied the Swedish expedition of Otto Torell to Spitzbergen.† He has added, incidentally, not a little to our knowledge; but his treatise is mostly a compilation, and, not looking upon the arctic fauna in a comprehensive view, he has fallen into many errors in zoo-geography. For instance, I cannot understand why he has excluded *Balænoptera gigas*, Eschr., and *B. rostrata*, Fab., from the Spitzbergen fauna, nor still less why *Balæna mysticetus*, Linn., is not classed among the Mammals of the seas around. This last is assuredly found there. In Smeerenberg Bay the Dutch used to catch it in abundance, and even erected boiling-houses on shore to "try" out its oil; and the two former are also found there. Indeed nearly all of the Greenland marine Mammalia are also found in Spitzbergen; and certainly Dr. Malmgren's stay was much too short to allow him to come to a decision on the matter.‡

Eschricht and J. T. Reinhardt's memoirs on the Greenland Whale § have added directly to our knowledge; while the numerous papers and catalogues of Gray ‖ and Lilljeborg ¶ on the British and Scandinavian Cetacea (most of which are also found in Green-

---

* "Beobachtungen und Anzeichnungen über die Säugethierfauna Finmarkens und Spitzbergens," in Wiegmann's Archiv für Naturgeschichte (Berlin), 1864, pp. 63–97, translated from Öfversigt af Kong. Svensk. Akad., &c. (1863), ii. pp. 127–155.

† Svenska Expeditionen till Spetsbergen år 1861, under Ledning af Otto Torell: ur detagarnes Anteckningar och andra Handlingar skildrad af K. Cheydenius (Stockholm, 1865). *See* the account of the Walrus in that work, pp. 168–183 (with plate and woodcut), and excellent figures of *Hyperoödon butzkopf*, Lacép., facing p. 480, &c.

‡ It is stated that this Whale has been of late years unknown within many miles of Spitzbergen. The walrus hunters say that the sea is getting too shallow for it. *See* Lamont, Quart. Journ. Geol. Soc., xvi. pp. 152 and 433.

§ Ray Society's Recent Memoirs on the Cetacea, by Professors Eschricht, Reinhardt, and Lilljeborg, edited by W. H. Flower. 4to. London, 1866. With plates. I. On the Greenland Right-Whale (*Balæna mysticetus*). By D. F. Eschricht and J. Reinhardt. 1. The Geographical Range of the Greenland Whale, in former times and at present. 2. The External and Internal Characters of the Greenland Whale; External Conformation; Cavity of the Mouth; Skeleton; Appendix (by the Editor). II. On the Species of *Orca* inhabiting the Northern Seas. By D. F. Eschricht. III. On *Pseudorca crassidens*. By J. Reinhardt. IV. Synopsis of the Cetaceous Mammalia of Scandinavia (Sweden and Norway). By W. Lilljeborg.

‖ Catalogue of Seals and Whales in the British Museum, 1866; and Proceedings of the Zoological Society, and Annals of Nat. Hist., *passim*.

¶ Ray Society's Memoirs on the Cetacea, supra.

land) have helped us to a right understanding of that order. Nilsson has disentangled the northern Pinnipedia in his History of Scandinavian Mammals*; and so has Gray (*libb. citt.*) and, more closely relating to Greenland, Fabricius,† in a supplementary paper to his Fauna, and Dr. Wallace in the short abstract of one read before the Royal Physical Society of Edinburgh,‡ on those killed by the northern seal-hunters. But nearly all of these papers are only local, or relate merely to questions of specific distinctions and synonyms, and touch but lightly upon the Seals either as animals of Greenland, or on their migrations from one part of the arctic regions to another. Our own arctic expeditions halting little, if at all, on the Greenland coast, and many of them unprovided with competent naturalists, have added almost nothing to our knowledge of the arctic or Greenland Mammals; but the American expeditions to Smith's Sound, under Drs. Kane § and Hayes,‖ have supplied us with many interesting notes on the range and habits of species. I wish I could say the same for all the describers of their collections. Professor Cope¶ has attempted to establish two "new" species of *Beluga* from Hayes's collection; but none of them (in my opinion) have the slightest claims to specific distinction,** the supposed differences being merely such as age or the ordinary variations between one individual and another would produce. Lastly, in the Scientific Section of the Narrative of the Second German Expedition will be found some notes by Dr. Peters on the Mammals collected on the East Coast.

Other contributions to arctic mammalogy I shall have occasion to notice as I proceed.

2. *Systematic Distribution of the Greenland Mammalian Fauna.*

As might be expected, the character of the Greenland mammalian fauna partakes of a sarcophagous type, the phytophagous species proper being only three, and the marine species far exceeding in number the terrestrial species. In the nomenclature of the Mammalia, though only a secondary matter, in a paper of this nature, so long as they are correctly named, I have followed some standard authority, without inquiring too strictly into the soundness or priority of the specific names applied, or the value of the tribal or generic divisions under which the writers have classed them.

This subject I may return to more critically at another time; but in this memoir I have allowed convenience of reference to

---

* Skandinavisk Fauna, första Delen, Däggadjuren, pp. 268–326 (1847), also translated in Wiegmann's Archiv für Naturgeschichte, Bd. vii., &c.

† Naturhistorisk Selskabets Skrivter, Bd. i.

‡ Proceedings of the Royal Physical Society of Edinb. 1862–63.

§ Arctic Explorations, 2 vols. 1855.

‖ Voyage towards the open Polar Sea (made in 1860), 1867.

¶ Proceedings of the Philadelphia Academy of Sciences, 1865, p. 278; 1869, pp. 23–31.

** Prof Reinhardt, who, as Inspector of the Zoological Museum of Copenhagen, has every means of arriving at a determination from an examination of a large number of skulls, writes to me that he has arrived at the same opinion.

overrule other considerations, considering that the eminence of the zoologists followed will be a sufficient safeguard that no great error has been committed. Accordingly the nomenclature of Baird's "General Report on the Mammalia of North America" is chiefly followed, as far as relates to the Greenland terrestrial species, and the late Dr. Gray's British-Museum Catalogue (1866) for the marine species, with only a few trifling exceptions, having a view to certain points of the synonymy of Fabricius's species of Cetacea, to be afterwards discussed. I have, however, ventured to differ from Dr. Gray as to the relative rank of the group of Seals, believing, with Illiger,* that they are entitled to ordinal rank, and have accordingly designated them *Pinnipedia* (Illig.)—forming Gray's tribes *Phocina*, *Trichechina*, and *Cystophorina*, for the sake of uniformity, into families under the titles of *Phocidæ*, *Trichechidæ*, *Cystophoridæ*, comprising the same species as the former tribes, without, however, committing myself to an opinion regarding the advisability of so many generic and other subdivisions of so natural a group, or of the good taste displayed by M. Frédéric Cuvier in the formation of some of his genera. Thus, with Professor Nilsson,† I cannot see why, in the formation of the genus Callocéphale ‡ (*Callocephalus*), Linné's *Phoca vitulina* should have been chosen as the type of the genus, while *Phoca barbata*, Fab., should have been retained as the type of the genus *Phoca*.§

Dr. Gray's nomenclature and classification of the Cetacea I have followed almost literally, though some of his species, such as *Lagenorhynchus albirostris*, *L. leucopleurus*, *Delphinus euphrosyne* (*D. Holbœllii*, Eschr.), and *Hyperoödon* (*Lagenocetus*) *latifrons*, are only known from skulls or skeletons. The localities are also very vaguely known; so that in the absence of all details in reference to their habits and distribution, and from the fact, moreover, of their specific (and still more their generic) claims not being in every case universally conceded, the physical geographer or naturalist (strictly speaking) can have little to say regarding them. I have, however, entered them as members of the Greenland fauna, in deference to the opinion of their founder, who, after the death of the lamented Eschricht stood alone in his knowledge of the systematic history of the marine Mammalia. The following table will show the general arrangement, the tribal and numerical distribution of the Mammalia of Greenland, exclusive of all introduced species and others which have been erroneously included in former lists, and of the first with whom Fabricius heads his fauna, "*Homo sapiens : . . . . sine Deo, sine Domino, reguntur consuetudine* :"—

---

* Prodomus, p. 138 (1811).

† Skand. Faun. i., p. 275.

‡ F. Cuvier, Mémoires du Muséum, xi. p. 182.

§ Even Nilsson's genus *Cystophora*, though faultless in aptitude, is liable to the objection that it has also been applied to a genus of Algæ by J. Agardh. This awkward confusion, however, is so common that it is only just to criticise the fault in the abstract.

MAMMALIA GRŒNLANDICA.

Order **Carnivora.**

Family Ursidæ.

Genus Ursus (Thalarctos).

U. maritimus, Linn.

Family Canidæ.

Genus Vulpes.

V. lagopus (Linn.), Rich.

Genus Canis.

C. familiaris, Linn., var. *borealis.*

Genus Mustela.

Mustela erminea, Linn.

Order **Rodentia.**

Family Arvicolinæ.

Genus Myodes.

M. torquatus (Pall.), Keys. & Blas.

Family Leporidæ.

Genus Lepus.

L. glacialis, Leach.

Order **Ruminantia.**

Family Bovidæ.

Genus Ovibos.

O. moschatus (Gm.), Blainv.

Family Cervidæ.

Genus Rangifer.

R. tarandus (L.), Baird.

Order **Pinnipedia.**

Family Phocidæ.

Genus Callocephalus.

C. vitulinus (L.), F. Cuv.

Genus Pagomys.

P. fœtidus (Müll.), Gray.

Genus Pagophilus.

P. grœnlandicus (Müll.), Gray.

Genus Phoca.

P. barbata, O. Fab.

Family Trichechidæ.

Genus Trichechus.

T. rosmarus, Linn.

Genus Halichoerus.

H. gryphus (O. Fab.), Nilss.

Family Cystophoridæ.

Genus Cystophora.

C. cristata (Erxl.), Nilss.

Order **Cetacea.**

Family Balænidæ.

Genus Balæna.

B. mysticetus, Linn.

Family Balænopteridæ.

Genus Physalus.

P. antiquorum, Gray.

Genus Balænoptera.
B. gigas, Eschr.*
B. rostrata (Müll.), Gray.
Genus Megaptera.
M. longimana, Gray.
Family Catodontidæ.
Genus Catodon.
C. macrocephalus (Linn.), Lacép.
Family Delphinidæ.
Genus Delphinus.
D. euphrosyne, Gray.
Genus Lagenorhynchus.
L. albirostris, Gray.
L. leucopleurus (Rasch), Gray.
Genus Orca.
O. gladiator (Bonn.), Sund.
Genus Phocæna.
P. communis, Brookes.
Genus Beluga.
B. catodon (Linn.), Gray.
Genus Monodon.
M. monoceros, Linn.
Genus Globiocephalus.
G. svineval (Lacép.), Gray.†
Family Ziphiidæ.
Genus Hyperoödon.
H. butzkopf (Bonn.), Lacép.
H. latifrons, Gray.

3. *Geographical Distribution of Greenlandic Mammalia.*

Similarity of physical contour, and a general uniformity of climate, varying no doubt in degree, but still sufficiently inhospitable throughout, with an abundance of the food on which all of them subsist throughout the habitable tracks and in the sea washing the shores of Greenland, have failed, contrary to what might have been expected, to produce a geographical distribution of the Mammalia in a like universal manner, or at all corresponding to the physical uniformity hinted at. It is only in the sea and on a narrow strip of land skirting the shores of Greenland that animal life has yet been found. The whole interior of the country appears to be merely a frozen waste, overlain to a depth of many feet by a huge *mer de glace*, extending, so far as yet known, over its entire extent (with the exception of the strip named) from north to south —a sea of freshwater ice whereon no creature lives, a death-like desert with nought to relieve the eye, its silence enlivened by the sound or sight of no breathing thing. This is the *Inlands Iis* of the Danish colonists; the outer strip, with its mossy valleys and ice-planed hills, is the well-remembered *Fastland*. Dreary, doubtless

* *Sibbaldius borealis* (Less.), Gray, Proc. Zool. Soc. 1864, p. 223.
† *Delphinus tursio*, O. Fab. (*Tursio truncatus*, Gray); Greenl. *Nesernak*.

Name of Species.	General distribution.						Nature of distribution in Greenland.		
	Circumpolar.	Circumarctic America.	Circumarctic Europe.	Circumarctic Asia.	Temperate Europe.	Temperate America.	Introduced.	Migratory.	Indigenous all the year round.
Ursus maritimus	*	*	*	*	–	–	–	–	*
Vulpes lagopus	*	*	*	*	–	–	–	–	*
Canis familiaris, *var.* borealis	*	*	–	*	–	–	–	–	*
Mustela erminea	*	*	*	*	*	*	–	–	*
[Felis domestica]	–	–	–	–	–	–	*	–	–
Myodes torquatus	–	*	–	*	–	–	–	–	?
[Mus decumanus]	–	–	–	–	–	–	*	–	–
[—— musculus]	–	–	–	–	–	–	*	–	–
Lepus glacialis	*	*	*	*	–	–	–	–	*
[Sus scrofa]	–	–	–	–	–	–	*	–	–
Ovibos moschatus	*	*	–	–	–	–	–	–	*
Rangifer tarandus	*	*	*	*	–	–	–	–	*
[Ovis aries]	–	–	–	–	–	–	*	–	–
[Bos taurus]	–	–	–	–	–	–	*	–	–
[Capra hircus]	–	–	–	–	–	–	*	–	–
Callocephalus vitulinus	–	*	*	–	*	*	–	–	*
Pagomys fœtidus	*	*	*	–	*	*	–	–	*
Pagophilus grœnlandicus	*	*	*	–	*	*	–	*	–
Phoca barbata	*	*	*	–	*	*	–	*	–
Trichechus rosmarus	*	*	*	*	–	–	–	*	*
Halichœrus gryphus	–	*	*	–	*	–	–	*	–
Cystophora cristata	*	*	*	–	*	*	–	*	–
Balæna mysticetus	?	*	*	–	–	–	–	*	–
Physalus antiquorum	–	*	*	–	*	*	–	*	–
Balænoptera gigas	–	*	*	–	*	*	–	*	–
—— rostrata	–	*	*	–	*	*	–	*	–
Megaptera longimana	–	*	*	–	*	*	–	*	–
Catodon microcephalus	–	*	*	–	*	*	–	*	–
Delphinus euphrosyne	–	–	–	–	–	–	–	*	–
Lagenorhynchus albirostris	–	–	–	–	–	–	–	*	–
—— leucopleurus	–	–	–	–	–	–	–	*	–
Orca gladiator	–	*	*	–	*	*	–	*	–
Phocæna communis	–	*	*	?	*	*	–	*	–
Beluga catodon	*	*	*	*	–	*	–	–	*
Monodon monoceros	*	*	*	*	–	–	–	–	*
Globiocephalus svineval	–	*	*	–	*	–	–	*	–
Hyperoödon butzkopf	–	*	*	–	*	*	–	*	–
—— latifrons	–	–	–	–	–	–	–	*	–
Numerical summary of distribution	14	27	24	10	16	15	7	19	12

NOTE.—This Table, manifestly imperfect, gives the approximate or

Local distribution in Greenland according to latitude and coast.											Remarks.
From Cape Farewell to northern limits.	From Cape Farewell to opening of Smith's Sound.	From Cape Farewell to Melville Bay.	From 67° N. lat. to northerly limits.	From 63° N. lat. to opening of Smith's Sound.	Cape Farewell to 72° N. lat.	Cape Farewell to 69° N. lat.	Cape Farewell to 67° N. lat.	Not north of 61° N. lat.	Not south of 76° N. lat.	Found on the east coast only.	
*	–	–	–	–	–	–	–	–	?		
*											
*											
–	–	–	–	–	–	–	–	–	?	*	
–	–	*	–	–	–	–	–	–	–	–	Only to most northerly outpost of Upernavik.
–	–	–	–	–	–	–	–	–	?		
–	–	*	–	–	–	–	–	–	–	–	Limits of Danish Greenland ?
–	–	*	–	–	–	–	–	–	–	–	Do. do.
*											
–	–	–	–	–	–	–	*				
–	–	–	–	–	–	–	–	–	*	–	Not south of Wolstenholme Sound. Also on the East Coast, N. of Scoresby's Sound.
*											
–	–	–	–	–	–	–	–	*			
–	–	–	–	–	–	–	–	*			
–	–	–	–	–	–	–	*	–	–	–	Not north of Holsteensborg.
*											
*											
*	–	–	–	–	–	–	–	–	–	–	Possibly not north of Melville Bay.
*											
–	–	–	*								
–	–	–	–	–	–	*	–	–	–	–	Range rather doubtful.
*	–	–	–	–	–	–	–	–	–	–	Rare north of 70°.
–	*	–	–	–	–	–	–	–	–	–	Very rarely seen N. of 73° and S. of 65°.
–	–	*									
–	–	*									
–	–	*									
–	–	*									
–	–	–	–	–	*	–	–	–	–	–	Range doubtful.
–	–	–	–	–	–	–	–	–	–	–	Range unknown.
–	–	–	–	–	–	–	–	–	–	–	
–	–	–	–	–	–	–	–	–	–	–	
–	–	*	–	–	–	–	–	–	–	–	Those species of Cetacea marked as extending north only to Melville Bay, probably occasionally reach a higher latitude ; but this bay is the usual limit, and north of this the species is rarely seen.
–	–	–	–	–	–	*	–	–	–	–	
–	–	–	–	*	–	–	–	–	–	–	
–	–	–	*	–	–	–	–	–	–	–	
–	–	*	–	–	–	–	–	–	–	–	
–	–	–	–	–	–	–	*	–	–	–	
–	–	–	–	–	–	–	–	–	–	–	Range unknown, but probably the same as *H. butzkopf.*
10	1	9	2	1	1	2	3	2	1	1	

provisional limits of species. *Canis lupus*, var. *alba*, may be added.

it is to eyes only schooled in the scenery of more southern lands; but, with its covies of ptarmigans flying up at your feet, with their *whir!*, the arctic fox barking its *huc, huc*, on the rocks, and the reindeer browsing in the glens covered with the creeping birch (*Betula nana*, L.), the arctic willows (*Salix herbacea*, L., *S. arctica* Pall., *S. glauca*, L., &c.), the crow-berry (*Empetrum*), the Vacciniums, and the yellow poppies (*Papaver nudicaule*, L.), it is a place of life compared with the cheerless waste lying beyond. It is with it, therefore, and the sea circling around, that we have to deal.*

Many of the animals constituting the mammalian fauna, influenced by no apparent physical cause, have but a limited geographical distribution, not extending south of a certain latitude, or north of another, while other species have a range over the shores of the frozen sea skirting three-quarters of the world. Some species of Seals are migratory, while others are not; and the same is true of various species of Cetacea. All of the terrestrial species proper are indigenous all the year round, confined to the country by its insularity. I have drawn up a table (pp. 8, 9) expressing at a glance the degree and nature of their geographical distribution, local and general. In this table I have divided the distribution under three main heads:—(1) general distribution over the range of the species, (2) nature of its distribution in Greenland, and (3) its local distribution in Greenland. I have, for the sake of convenience, divided the general range of Greenland species into six subdivisions, viz.:—(α) Circumpolar, comprehending the regions around the most northern limits yet reached by man, the particular locality within that region for each species being limited by the nature of its habitat; thus the Bear occupies the shores or frequents the ice-fields and the sea, the Seals the sea and the shore, or the ice-fields, the Dog the vicinity of man's dwellings, and the Hare the land generally, while the Fox keeps more by the shore, but not in the sea, and rarely ventures out on the ice fields; (β) Circumarctic America and (γ) Circumarctic Europe comprehend all the region about Greenland and south of the head of Baffin's Bay, down Davis's Strait, and other places south of the former limits, Hudson's Bay, Labrador, &c., on the one hand, and on the other the Icelandic seas and shores, the regions of Europe generally within or about the arctic circle. It may be called also subpolar, and has been formed to take in the distribution of some species of Seals and Cetacea. The two regions are about the same in zoo-geography.

(δ) Circumarctic Asia comprehends similar limits on the Asiatic continent, and is made to take in the range of the Fox, Lemming, and a few other animals, which extend their range so far east and west. I have not thought fit to create in this table an *Arctic*

* For a further description of the character of the inland ice, &c., the reader is referred to the following papers by the writer of these notes:—

"Das Innere von Grönland," Petermann's Geographische Mittheilungen, 1871; "The Physics of Arctic Ice," Quart. Journ. Geol. Soc., 1871; "Geology of the Noursoak Peninsula, etc.," Trans. Geol. Soc. Glasg., vol. v.; "Disco Bay," The Geographical Magazine, Feb. 1875, and in my section of The Arctic Manual of The Royal Geographical Society, now in preparation.

division proper, limiting it by the arbitrary divisions of geography, divisions which, though necessary enough for the astronomical description of the earth, yet serve no purpose to the physical geographer in tracing the distribution of plants and animals over it. This division is comprehended under my *circumpolar* range, which ends on the seas adjoining Greenland about the head of Baffin's Bay. I have given its general limits there, as many species do not go beyond that barrier, and others do not come south of it. I am well aware that this may appear a somewhat loose way of expressing the limits of regions; but at the same time the species the range of which these divisions are made to express are most wonderfully careless of the degrees, minutes, and seconds which the geographer may erect as their limits, and we can therefore only express their divisional boundaries in an equally elastic manner. I trust, however, that they are sufficiently intelligible.

(ε) To give the southern range of certain species of Seals and Cetacea, I have erected a division for *temperate Europe*, comprehending the British and Scandinavian seas; and in the range of the same latitudes on the shores of the British provinces and the United States of America a (ζ) *temperate American division.* I have not, as in the circumarctic range, erected a division for temperate Asia, as I do not think there is a single species of Seal or Cetacea, found in the seas (and certainly no Mammals on the land) of temperate Asia, found in the corresponding seas of Europe and America, though, as several of the species are common to the circumarctic and circumpolar divisions of all three, some may yet be found. In preparing this table I have endeavoured to give the *natural* range of the species, and have not entered a species in any division because it has been, as an evident *straggler*, seen within that division. For instance, *Balæna mysticetus, Beluga catodon, Monodon monoceros,* and *Trichechus rosmarus* have all of them more than once found their way to the British seas, yet no zoo-geographer would ever think of representing the Right Whale, the White Whale, the Narwhal, or the Walrus as regular members of the British fauna. On the other hand, I need scarcely say that when I put an animal into any division I do not thereby say that it is limited to that division (for, as shown on the table, many extend through several of these divisions), nor that they are found over all that division or series of divisions or regions. I have already explained that the range of each is limited according to its habitat and habits.

I have made these explanations because, as all rules are liable to exceptions, so are systems and systematic divisions. Nature abhors being confined between parallel lines.

Under the division of "Nature of its Distribution in Greenland" I have divided them into (α) Introduced species, (β) Migratory species, and (γ) Species indigenous all the year round.

(α) In Fabricius's day the following Mammals had been introduced into the country, but chiefly into South Greenland:—*Canis familiaris* (European breeds), *Felis domestica, Ovis aries, Capra hircus, Bos taurus, Sus scrofa, Mus decumanus,* and *Mus musculus.* All of these species are yet at times living in the country, but none of them can be said to be acclimatized. The Horse (*Nersasoak*) was once introduced into Greenland, but only remained for a short time. As far as I can discover, its importation was for the purpose of Major Oscean and Capt. Landorff, who in 1728 proposed the mad-cap scheme of "riding across Greenland!"

(β) As the winter approaches, most of the Birds leave the country and do not return again until spring. The terrestrial Mammals are prohibited, by the insularity of the country, from resorting to this method of escaping the rigours of the climate, or the scarcity of food. The Bear to some extent hybernates, though, as I shall afterwards show, this hybernation is not so complete as is usually supposed. The migratory Mammals are therefore limited to the marine species.

All of the Seals, with the exception of *Trichechus rosmarus*, *Callocephalus vitulinus*, and *Pagomys fœtidus*, leave the coast during a portion of the winter, and even of the summer. The migration of the Seals is too complicated a subject to be discussed in a general review; under my notes on each species I shall have occasion to recur to it. In like manner all the Cetacea leave the seas in the winter, with the exception of *Monodon monoceros* and *Beluga catodon*, which can be seen at open places in the ice all the winter through. Why these species should be winter denizens in preference to the others it is difficult to decide. Several species have what may be called a *local migration*, moving from one portion of the coast to another, north and south, during the summer, according to the state of the ice, &c.,—all of which will be noticed in another place.

(γ) The species indigenous all the year round are therefore the terrestrial Mammals and the remainder of the marine species not already mentioned as migratory, viz., *Ursus maritimus*, *Canis familiaris*, var. *borealis*, *Vulpes lagopus*, *Mustela erminea*, *Lepus glacialis*, *Myodes torquatus*, *Ovibos moschatus*, *Rangifer tarandus*, *Trichechus rosmarus*, *Callocephalus vitulinus*, *Pagomys fœtidus*, *Monodon monoceros*, and *Beluga catodon*.

In addition to these well established species there are others frequently entered among the Greenland Mammalia, some of which have but scant right to a place, and others are entirely mythical, as I will show in a section on these animals. Among these I class *Gulo borealis* (*Ursus luscus*), *Phoca ursina* (*Callorhinus ursinus*), and *Trichechus manatus* (*Rhytina gigas*) as animals with little or no claim to be admitted members of the arctic fauna.

The columns for the "Local distribution in Greenland" are arranged solely with reference to our present knowledge of the range of the species in the country, and, being only temporary and to a great extent artificial, are subject to changes as our knowledge of the species extends. At the same time I think it only right to say that they have been very carefully compiled, after considerable study of the *natural range* of the species, and upon principles akin to those for the general distribution of the species.

The column headed "East coast only" I have erected for the reception of *Mustela erminea* solely, all the species of the east coast, so far as we know, being, with this exception, also common to the west. The east coast has, however, been very little explored, and no doubt something remains to be added to our knowledge of the range of species on that coast.

On a comparison of the Greenland fauna with that of other portions of the arctic regions, we can see no reason for looking upon it, in common with the flora and the avi- and ichthy-faunas, as other than essentially Arctic-European, all of the species of Mammalia, with the exception of *Ovibos moschatus*, being found in either Spitzbergen or Nova Zembla, while many of the Arctic-American species are not found in Greenland. The only true American mammal found in Greenland is the Musk-ox, which might have crossed from the western shores of Smith's Sound (where Eskimo tradition describes it as once abundant) on the ice to the eastern shore, where alone in West Greenland it seems to be now found, the great glaciers and ice-floes about Melville Bay seeming to act as a barrier to the southern and northern migrations of the animals on either side of them, and of Man equally with the lower animals.

Looking at the fauna of Spitzbergen,* if we take exception to the very dubious omission which Malmgren has made, we find that there is no species of mammal found in these islands not found in Greenland; and the same is true of the mammals of Nova Zembla, if we take Von Baer's list † as representing the present state of our knowledge, though published more than thirty years ago. In this the exception is a doubtful one ("a little white animal, species uncertain"), but probably an Ermine. I therefore think that we are justified in looking upon the mammalian fauna of Greenland as Arctic-European, and not Arctic-American, though I am aware that opposite views are entertained by naturalists of high eminence.

The mammalian fauna of Iceland has no connexion with that of Greenland, that island possessing only a single species of Mammal indigenous to it (*Mus sylvaticus*); all others have been introduced by man, or, like the *Ursus maritimus* and *Vulpes lagopus*, have drifted from Greenland on ice-floes.

My friend Mr Andrew Murray‡ seems to take exception to the Mouse which is said to be found in Iceland, and regarding which wonderful tales are told;§ and, contrary to the opinion of Povelsen, who considers it *Mus sylvaticus*, L., and of the intelligent Icelanders, who, as represented by Sir W. J. Hooker, do not believe in its existence, thinks that it is *Myodes torquatus* (*hudsonius*, Forst. = *grœnlandicus*, Tr.). If such is the case, it might have been brought over on ice from the east coast of Greenland; but the probability is, according to Steenstrup, who has carefully

---

* Malmgren, *loc. cit.*; Scoresby, "Arctic Regions;" Phipps's "Voyage;" Parry's "Attempt;" Laing's "Voyage to Spitzbergen," &c., &c.

† K. E. von Baer, Wiegmann's Archiv für Naturgeschichte (1839), pt. vii. (*fide* Murray, "Geogr. Distrib. Mamm.," p. 365).

‡ Geographical Distribution of Mammals (1866), p. 267.

§ Pennant, "Arctic Zoology," Introduction, p. lxx.; Hooker's "Tour in Iceland," i. pp. 51, 52.

examined the question, that no Lemming exists in Iceland, and that the only indigenous Mammal is the ***Mus sylvaticus***, showing that the fauna is essentially European, and not American as Murray seemed to suppose.*

From these facts I believe that the island of Iceland is of a newer date than any portion of Scandinavia or Greenland, and, being of a volcanic nature, was formed posterior to the date of the present distribution of land and water in the North Sea; if indeed it, and other detached islands in the North Sea, are not fragments of a more or less continuous land communication, which, when the Miocene flora flourished in the Arctic regions, united Greenland with Europe.†

4. *Notes on the Habits, Distribution, and Synonomy of the Terrestrial Mammalia of Greenland.*

The following notes on certain of the terrestrial species of Mammalia are not intended as either a complete or systematic history of the species, but merely as stray notes on some points in their history hitherto passed over, and on the *species as a Greenland animal.* I have delayed entering upon the history of the marine Mammalia until another time, my observations on these species being too extensive to be included within the limits of one paper; and, as I shall treat of them on a more comprehensive plan than as mere Greenland species, they do not properly come within the scope of a paper on Greenland Mammals.

These notes comprehend my own observations during voyages to the Spitzbergen, Iceland, and Jan Mayen seas, and along the eastern and western shores of Davis's Strait and Baffin's Bay, to near the mouth of Smith's Sound, in 1861. During the past summer (1867) I have again (in company with Messrs. E. Whymper and Tegner) visited Danish Greenland for scientific purposes, but have added little or nothing to my former notes, having seen few Mammalia, except some of the species of Pinnipedia and a Cetacean or two in the sea; and, our travels extending over but a limited portion of the vicinity of Disco Bay, we had but few opportunities of adding to our knowledge of their habits.

I was fortunate enough, however, to obtain the assistance of my friends Dhrr. Knud Gelmeyden Fleischer, Carl Bolbroe, and Octavius Neilsen, whose long acquaintance with the Eskimo language enabled me to discover some of the errors which Fabricius fell into in deciphering the mythical species; and our intelligent travelling companion Hr. Anthon P. Tegner kindly

---

* Steenstrup, "Den opræenda islandske Landpattedyr-faunas Karakter," &c. Vidensk. Meddel. Naturhist. Forening i Kjöbenh., 1867, p. 51; Annals Nat. Hist., ser. 4, vol. iii., p. 445.

† *See* J. D. Hooker, Linn. Trans., vol. xxiii., p. 251; Asa Gray, "Amer. Journ. Science," 1862; J. W. Dawson, "Canadian Naturalist and Geologist," 1862, pp. 334–344; and Murray, "Geogr. Dist. Mamm.," p. 37, for the phytogeographical views of the origin of the Greenland flora and fauna at present received.

gave me the benefit of his experience. These notes I have incorporated in the body of this paper at the proper place.

I have also examined, through the kindness of the curators, the Greenland Mammals in the Copenhagen Museums, and those in the Museum of Science and Art in Edinburgh, comprising many of the typical specimens of Scoresby, Richardson, &c. For this latter favour my thanks are especially due to Professors Archer and Allman, and to the late Mr. J. B. Davies, then Zoological Assistant in the Museum.

1. Ursus Maritimus, Linn.
Greenl. Nennok (*o* guttural).

The well-known "Polar" or "Ice Bear" is found along the whole coast of Greenland from north to south, but not nearly so numerous as in former times, or as is popularly supposed. There are more in the northern than in the southern portion of the country; and it is very seldom seen in mid-Greenland, *i.e.*, between about 69° and 66° N. lat. There are yearly killed from thirty to sixty of them. The Royal Board of Trade in Greenland give the natives about five rigsdaler (11*s.* 3*d.*) for a skin. Occasionally there are a number killed near Cape Farewell which have come round on the Spitzbergen ice-stream. Here a curious custom prevails, viz., that whosoever sights the Bear first, man, woman, or child, is entitled to the skin, and the person who has shot it only to the blubber and flesh.* It is of light creamy colour, rarely pure white, except when young; hence the Scotch whalers call it the "brounie" or "brownie," and sometimes the "farmer," from its very agricultural appearance as it stalks leisurely over the furrowed fields of ice. Its principal food consists of Seals, which it persecutes most indefatigably; but it is somewhat omniverous in its diet, and will often clear an islet of Eider-duck eggs in the course of a few hours. I have seen it watch a Seal for half a day, the Seal continually escaping just as the Bear was about putting its paw on it, at the "*atluk*" (or escape hole) in the ice. Finally, it tried to circumvent its prey in another manner. It swam off to a distance, and when the Seal was again half asleep at its *atluk*, the Bear swam under the ice, with a view to cut off its retreat. It failed, however, and the Seal finally escaped. The rage of the animal was boundless; it roared hideously, tossing the snow in the air, and trotted off in a most indignant state of mind!

During the sealing-season, both in Greenland and in the Spitzbergen seas, the Bear is a constant attendant on the sealer for the sake of the carcasses, in the pursuit of which it is sometimes "more free than welcome." I have often also seen it feeding on Whales of different species, which are found floating dead. In 1861 I saw upwards of twenty all busily devouring the huge inflated carcass of a *Balæna mysticetus* in Pond's Bay, on the western shores of Davis's Strait. We were foolish enough to

* The flesh, and especially the liver, is said to often prove poisonous when eaten. The Eskimo on the western shores of Davis's Strait carefully prohibit their dogs from devouring any portion of it.

fire a few shots among them, when the Bears sprang furiously from the carcass and made for our boat. One succeeded in getting its paws on to the gunwale; and it was only by the vigorous application of an axe that we succeeded in relieving ourselves of so unwelcome an addition to our crew.

On the whole, I do not think that the Polar Bear is a very fierce animal, when not enraged; and I cannot help thinking that a great deal of the impressions which we have imbibed regarding its ferocity are more due to old notions of what it *ought to be*, rather than what *it is*, and that the tales related by Barentz, Edward Pellham, and other old navigators regarding its bloodthirstiness during the time they wintered in Spitzbergen were a good deal exaggerated. When enraged, or emboldened by hunger, I can, however, quite well understand that, like all wild and even domesticated animals, it may be dangerous to man. On the East Coast of Greenland, where they know little of man, they are very bold. The members of the German Expedition, when making out-door observations, had to be continually on their guard against them. I have chased it over the floes of Pond's Bay, and the Bear's only thought seemed to be how best to escape from its pursuers. I should have hesitated a good deal before making so free with the Grizzly Bear of the Californian wilds (*Ursus ferox*), which is, perhaps, the most ferocious animal on the American continent. Though seemingly so unwieldy, the *nennok* runs with great speed; and being almost marine in its habits, it swims well. I have chased it with a picked crew of eight whalemen, and yet the Bear has managed to distance us in the race for the ice-fields. It would every now and again, when its two cubs were getting left in the rear, stop and (literally) push them up behind; and on reaching the steep edge of the ice-floe, finding that we were fast reaching them, it lifted each of them up on the ice with its teeth, seizing the loose skin at the back of the neck. Once on the ice, they were safe.

It is often found swimming at great distances from land (*vide* the statements in arctic voyages, and the works of Richardson, Parry, &c., *passim*). The stories of its making ice-houses, and of their gambols therein, as related by Fabricius, as well as of its combats with the Walrus, are still prevalent in Greenland.

It is curious that the old Eskimo stories about the Polar Bear having no evacuations during the season of hybernation, and being itself the means of preventing them by stopping all the natural passages with moss, grass, or earth (Richardson's "Fauna Bor.-Am." i. 34), prevail also among the North-western American Indians on the other side of the continent, in reference to the Brown Bear (*Ursus americanus*), the substance used in stopping the passages varying according to the tribe among whom the myth is prevalent, from a ball of clay to one of pine-resin!

I do not think that it hybernates during the whole winter, as usually supposed; at all events they are often seen during the winter, though these are probably old males. It is probable that the females, when not pregnant, roam all winter like the males. Unlike its congeners, it does not *hug*, but *bites;* and it will not eat its prey until it is dead, playing with it like a cat with a mouse. I have known several men who, while sitting watching or skinning Seals, have had its rough hand laid on their shoulder. Their only chance has been then to feign being dead, and manage to shoot it while the Bear was sitting at a distance watching its intended victim. Though Eskimo are often seen who have been scarred by it, yet I repeat that, unless attacked, or rendered fierce by hunger, it rarely attacks man. During our last trip to Greenland none of our party saw one; indeed they are only killed in the vicinity of Disco Day during the winter or spring, when they have either come or drifted south on the ice-floes. Six were killed in the vicinity of Omenak during the winter of 1866–67.

2. VULPES LAGOPUS (Linn.); Rich. F. B. A. i. 83.
Greenl. Terienniak, Kaka.

The Arctic Fox is very numerous in south- and mid-Greenland, rarer in the northern parts of the Danish possessions, but quite plentiful again north of Upernavik to high up in Smith's Sound.*

There are two varieties, the blue and the white. This colour is not dependent on the season. The white variety is also more numerous and much less valued than the blue; but again the blue and the white varieties interbreed, and often, the Eskimo say, there is a white mother with blue young, and *vice versâ.* The blue Fox is very valuable, the price for the best kind of skin being from six to seven times as much as for that of the white. Some have been sold at the annual auction of the Greenland furs in Copenhagen at over twenty rigsdaler (nine rigsdaler = 1*l.* sterling). There are yearly killed from 1,000 to 3,000 of the white and blue Foxes, two-thirds being blue and one-third white. In Greenland the white is traded for three marks (1*s.* 1½*d.*), and the blue for two rigsdaler (4*s.* 6*d.*). It is not killed by the Greenlanders in summer, as its summer coat is not valuable. At this time it is found in the mountains preying on the young Ptarmigan (*Tetrao reinhardti*, Brehm). In winter it comes down to prey on shellfish or other marine produce, at the open places near the shore when the tide breaks the ice. About this period it can often be seen barking most impudently at the solitary hunter.

3. CANIS FAMILIARIS, Linn.
a. var. *borealis.*
Greenl. Kemmek or Kremmek.

(*a*) The Dog of the Eskimo is the same species all over the American continent; at least I have seen Dogs from Kamschatka,

* The Fox is often seen hundreds of miles from land during the sealing season in the Greenland Sea, when it feeds on the dead Seals. In pursuit of the wandering Lemming it sometimes loses its way, and has been taken far from its natural haunt. Kalm mentions one being taken in West Gothland, and Pennant (Suppl. Arct. Zool., p. 52) one killed near Lund, Sweden, lat. 55° 42′ N., on Oct. 27, 1786. *See also* Von Baer on the Distribution of the Arctic Fox, Bull. Acad., St.-Petersb., t. ix. p. 89.

Sitka, the western shores of Davis's Strait, and from Greenland which it was impossible to deny were of one species.

(β) Besides this there is, in Danish Greenland, another breed of Dogs of mixed native and European descent, the latter being imported by the whites. These are called by the natives "Meĕkĕ." I have not the slightest doubt that the original breed of the Arctic Dog was the Wolf (*Canis occidentalis*, var. *griseo-alba*, Baird). In its every disposition it agrees with that animal, and there is no point which has been supposed to separate the one from the other which is not common to both of them. I have seen skins of the Wolf which have hair for hair agreed with the typical Arctic Dog. The Wolf is not, however, found in Greenland, unless, as I shall afterwards discuss, the "amorok," which Fabricius erroneously described in his fauna as *Ursus luscus*, be merely a Dog run wild and returned to its original type. The Dog is found as far north as man lives, but is not used by the Eskimo south of Holsteensborg, the sea not being sufficiently frozen over during the winter to permit of sledging. The use of the Dog as a sledge-animal has been so often described * that I may pass it over here without further reference. Being only required during the winter, they lead during the summer and autumn months an idle life, hanging round the settlements, sleeping on the top of the flat earth-huts of their masters, snarling at every one's heels, but running at the first appearance of a stick or stone, snatching up every bit of edible garbage round a village, and, in fact, becoming such a pest to the women when dressing a Seal on the rocks, or when drying meat for winter use, that they are often left to look out for themselves on some barren uninhabited islet. During the summer they are never fed; and often you may pass old Eskimo encampments where the only inhabitants are a few hungry dogs howling from the rock, disconsolate until their lords return. The *appearance* of a stone is enough to send them howling far and near. It is rarely that they bark, generally preferring, with their wolfish instinct, to sit and howl monotonously on some elevated point, and regularly "making night horrible" with their "long cry." The ringing of the workmen's morning and evening bell at the Danish settlements used to be the signal for the commencement of this hyperborean music. This dog can only be kept in subjection by the most unmerciful lashing; for its savage nature will out. When at Clyde River in 1861 I heard of a most horrible tragedy which had been enacted there a few years before. A man, a boy, and a little girl landed from an *omiak* (or open skin boat) on an island where, as is usual, some dogs were confined. Before the poor people could escape to their boat, the animals, infuriated by hunger, sprang upon them. The man and the boy, though much lacerated, managed to regain the omiak; but the poor girl was torn to pieces.

When the Greenland dogs die off, the Greenlander must become extinct, more certainly even than must the "Plain" Indian when

* *Vide* particularly Kane, "Arctic Explorations;" and Hayes, "Voyage towards the open Polar Sea."

the last buffalo is shot. It is impossible for him to drag home the seals, sharks, white whales, or narwhals which he may have shot in the winter at the "strom-holes" in the ice without his dogs —or for the wild native in the far north to make his long migrations, with his family and household goods, from one hunting-ground to another without these domestic animals of his. Yet that sad event seems to be not far distant. About fifteen years ago, a curious disease, the nature of which has puzzled veterinarians, appeared among the Arctic dogs, from high up in Smith's Sound down the whole coast of Greenland to Jakobshavn (69° 13′ N. lat.), where the ice-fjord stops it from going further south; and the government uses every endeavour to stop its spread beyond that barrier, by preventing the native dogs north and south from commingling. Kane and Hayes lost most of their dogs through this disease;* and at every settlement in Danish Greenland the native are impoverished through the death of their teams. It is noticed that whenever a native loses his dogs he goes very rapidly downhill in the sliding scale of Arctic respectability, becoming a sort of hanger-on of the fortunate possessor of a sledge-team.

During the latter portion of our stay in Jakobshavn, scarcely a day elapsed during which some of the dogs were not ordered to be killed, on account of their having caught this fatal epidemic.

The dog is seized with madness, bites at all other dogs, and even at human beings. It is soon unable to swallow its food, and constipation ensues. It howls loudly during the continuance of the disease, but generally dies in the course of a day, with its teeth firmly transfixing its tongue. It has thus something of the nature of hydrophobia, but differs from that disease in not being communicable by bite, though otherwise contagious among dogs. The government sent out a veterinary surgeon to investigate the nature of the distemper; but he failed to suggest any remedy, and it is now being "stamped out" by killing the dogs whenever seized—an heroic mode of treatment, which will only be successful when the last dog becomes extinct in Greenland.

Strange to say, the dogs in Kamschatka are also being decimated by a very similar disease;† and, in a recent communication received from that region, it is said that so scarce have dogs become, that the natives do not care to sell them, and that 100 roubles have been refused for a team of six. Fortunately for the Kamschatkans, they have the reindeer as an ulterior beast of draught and burden. Prof. Otto Torell brought several dogs from Greenland for the use of his expedition to Spitzbergen in 1861; but finding them useless (on account of open water) he set them free, I was informed, on Spitzbergen, where they are now rapidly increasing, and will, doubtless, soon return to the original wolf type.

---

* Kane's "Arctic Explorations," vol. i. p. 157.

† For all that is known about the Dog-disease in Greenland, see Fleming, "Geograph. Mag.," Feb. 1875. [*See also* notes by Dr. W. L. Lindsay, in the Brit. and Foreign Medico-Chirurg. Review, January 1870, pp. 212, 216; and July 1871, pp. 10, 15, 28.—EDITOR.]

Their use in Greenland is almost wholly as sledge-animals. Among the Eskimo on the western shores of Davis's Strait, a loose dog usually precedes the sledge, and, by carefully avoiding broken places in the ice, acts as a guide to the sledge-team, which carefully follows his lead. *En passant* I may remark that dog-driving is by no means an easily acquired or a light labour. In North Greenland and among the wild Arctic highlanders of Cape York and Smith's Sound, dogs are also valuable assistants, by attacking the polar bear while the hunter plants his spears in the animal.* They are also used a little in seal-hunting. Their flesh is also highly appreciated, but rather too valuable for anything except an occasional dainty. The skin is highly valued for socks, and that of the pups for winter clothing; but so scarce have they become, that it is now very hard to raise enough for an *anarak* (jumper), and one of our party paid 18 rigsdaler (2*l.*) for enough to make an overcoat. No longer, as in Giesecke's day,† is it rejected as an article of trade on account of its disagreeable odour.

[4. FELIS DOMESTICA, Briss.
Greenl. Kitsungoak.

The domestic Cat has been kept in Greenland ever since the Danish women came, and it follows them in all their sojournings north and south. In Fabricius's day it was already not uncommon. At present there are many in Julianeshaab district, where mice are quite abundant and troublesome.]

5. MYODES TORQUATUS (Pall.), Keys. & Blas.

This Lemming was found by Capt. Scoresby, in the year 1822, near Scoresby's Sound, on the east coast of Greenland, lat. 69°, and was described by the late Professor Traill, in the appendix to Scoresby's "Journal of a Voyage to the Northern Whale-"fishery, &c.," p. 417, as a new species under the name of *Mus grœnlandicus.* From a careful examination of the original and only specimen, now in the Edinburgh Museum of Science and Art, I am inclined to believe, with Middendorff,‡ that it is not distinct from those already described, and that the *Myodes hudsonius* of Forster (*Mus hudsonius,* Forster in Phil. Trans. lxii. p. 379; *Lemmus hudsonius,* Sab., Parry's Voyage, p. clxxxv) and the *Mus grœnlandicus,* Tr. (*Myodes grœnlandicus,* Wag. and J. E. Gray§, Proc. Zool. Soc. London, xvi. 1848, p. 43, and *Id.* in Rae's Narrative, 1850), are identical with the Siberian *Myodes torquatus* (Pall.), Keys. & Blas.

It can only be classed as a very rare and local (possibly accidental) member of the fauna of Greenland, as it has never since been found

* *See* an interesting account in Kane's "Arctic Explorations."

† Giesecke, article "Greenland," in Brewster's "Edinburgh Encyclopædia (1830)," vol. x., p. 481.

‡ Sib. Reise, II. ii. 1853, p. 87, pls. 4–7 and 10.

§ *Arvicola grœnlandiæ,* Rich. *l. c.* 134; *vide* also Schreber, "Säugethiere," iii., p. 604; Giebel, "Die Säugethiere," &c. (1859), p. 605.

in the country; Graah* did not see it in his two years' journey, nor even hear of its existence. No doubt the east coast of Greenland is almost unapproachable for ice, and has never been visited since Graah's day, except for a little way round Cape Farewell. Whalers, however, have been known to have landed near Scoresby's Sound; but they saw nothing of it, and it may be safely said not to be an inhabitant of the west coast, either within or outside of the Danish possessions.

From Upernavik southward, the Danes have been on the coast, either settled or trading, for at least 120 years, and during that time not a few collectors have visited the country; but, notwithstanding all their exertions and those of the stationary officers of the government there, no specimen of this Mouse has as yet been obtained, nor do the Eskimo know of the existence of such. Murray has therefore taken too wide a generalization, when he portrays, on map lxxxv. of his laborious and generally accurate work the "Geographical Distribution of Mammals" (1866), p. 267, the distribution of the Lemming as extending right along the east and western shores of Greenland to the head of Baffin's Bay, on the supposition that it is a regular member of the Greenland fauna. I am inclined to look upon it as representing the extreme *eastern* limit of the *Myodes torquatus*, as the *Myodes hudsonius* is a climatic species representing the extreme *western* range of the former species. It is almost unnecessary to note, after what I have said, that Fabricius makes no mention of it in his "Fauna Grœnlandica;" and if it had been found, he, ever anxious as he was to add anything to the Greenland Mammals, would have been sure to have heard of it from the natives, credence in whose mythical zoology forms one of the few disfigurations of his work. Neither did Inglefield, Sutherland, Kane, or Hayes see anything of it in Smith's Sound, or southward to the northern limits of the Danish possessions.†

In 1861, the natives at Pond's Bay, on the western shore of Davis's Strait, brought me many skins of this species, which I ascertained to belong to the *hudsonius* form. For the sake of reference, the Arctic species may be classed as follows:—

Myodes torquatus, Pall.
Var. *hudsonius*, Forst.
Var. *grœnlandicus*, Tr.

6. [Mus decumanus, Pall. (1778).
Mus norvegicus, Erxleben (1776).
Greenl. Teriak.

The Brown Rat was introduced as far back as the days of Fabricius by the Danish ships in the summer, and seemed likely to

* Narrative of an expedition to the East Coast of Greenland, Engl. transl. (1837); the original Danish edition is in 4to. Undersögelses-Reise til Ostkysten af Grönland, 1832.

† These remarks (written in 1868) now (1875) require considerable modification. The German Expedition got it on the east coast, on Sabine's Island, in 1869-70. The American Expedition under Hall met with it in Smith's Sound.

prove dangerous in houses; but they gradually and periodically died out, as they could not stand the cold of the winter. Some years ago they were again introduced, and still occasionally one is seen in the summer months in some of the warehouses from Upernavik to near Cape Farewell.]

7. [MUS MUSCULUS, Linn.
Greenl. Teriangoak ("the small rat").

Its history as a colonist animal in Greenland is about the same as the Rat's. At some of the more southern settlements they can occasionally survive the winter and beget abundantly. Both the Mouse and Rat were introduced as far north as Kane's, Hayes's, and Hall's ships wintered, but I cannot learn that they got naturalized.]

8. LEPUS GLACIALIS, Leach.
L. arcticus, ibid.
Greenl. Ukalek.

The Hare is a common animal over the whole coast, from north to south, east and west. It is, however, seen more seldom in the north of the Danish trading limits, and there are only a few hundreds shot annually. They are said to be rather rare on the east coast. I cannot see why its beautiful white skin is not more used. At one time the Danes used to send quantities home, but they could get no market for it. From the Hare the natives spin a kind of yarn which they occasionally knit into caps, for a summer head-dress, for the men and children. It is difficult (indeed, almost impossible) to give characters whereby this species can be separated from the *Lepus variabilis* of Europe when the former is in its summer dress; and the skull presents equal difficulties.

I have, however, preferred to look upon it as nominally distinct, though I really believe that it is only a climatic variety of *L. variabilis*, Pallas.

9. [SUS SCROFA, Linn.
Greenl. Poliké.

It is kept at some of the southern settlements.]

10. OVIBOS MOSCHATUS (Gmel.), Blainv.
Greenl. and Eskimo generally. Umimak.

In the "Fauna Grœnlandica," p. 28. No. 17, Fabricius has classed *Bos grunniens*, L., as one of the animals of Greenland, because he thought that he had found (on a piece of drift ice) some remains of it, consisting of the greater portion of the skull of an animal "very like an ox." He was of opinion that this was a portion of the Yak. He did not, however, consider it to be a native of Greenland, but rather to have been drifted from northern Asia on the ice, the flesh having been eaten by polar bears. Any one can see, by examining the figure which Fabricius afterwards gave of this specimen (Bid. Selsk. Skriv. N. Saml. iii. 82), that it was the Musk-ox; and indeed, he afterwards acknowledged so himself (Bid. Selsk. Skr. 3. N., vi.). It is therefore, after this, somewhat surprising to find a zoologist so well acquainted with the Greenland fauna as the elder Reinhardt stating that the Musk-ox, which, like Fabricius, he called *Bos grunniens*, rarely comes from Melville Island to Greenland.* Mr. Murray seems to doubt on which side of Greenland Fabricius met with his specimen; but there need be no doubt on that matter, as it must have been on the west side. The east was even more unknown in his day than now, and he was certainly never round Cape Farewell. The Musk-ox has, therefore, no right to a place in the fauna of Danish Greenland, nor do I believe that at any time it was an inhabitant of that portion of the continent.

Recent discoveries have, however, shown it to be, with the strongest probability, an inhabitant of the shores of Greenland north of the glaciers of Melville Bay. Dr. Kane met with numerous traces of it in Smith's Sound; and his successor, Dr. Hayes, found at Chester valley in the same inlet, among Eskimo kjokkenmöddings, the skull of a Musk-ox. Eskimo tradition describes the animal as at one time common along the whole coast, and they affirm that it is yet occasionally to be met with. No longer ago than in the winter of 1859 a hunter of Wolstenholme Sound, near a place called Oomiak, came upon two animals, and killed one of them.†

I think, therefore, that we may with some authority assume that the Musk-ox is not yet extinct in Greenland.‡

11. RANGIFER TARANDUS (Linn.), Baird.
Var. *grœnlandicus*, Kerr (Linn. 1792, p. 297).
Greenl. Tukto (tootoo); ♂, Pangnek; ♀, Kollauak.

I will not here enter into any discussion of the vexed question of the identity of the European and American Reindeers, or whether the Greenland Reindeer is specifically distinct from the American species; suffice it to say that the heading of this note sufficiently expresses my views on the subject, after very excellent opportunities of comparison and study, and that I consider the Greenland Reindeer only a climatic variety of the European species. I have, moreover, seen specimens of reindeer horns from Greenland which could not be distinguished from European, and *vice versâ*. On the whole, however, there is a slight variation, which may be expressed by the trivial name to which I have referred at the commencement of these remarks.§

It is found over the whole country, from north to south,‖ but not nearly so plentiful as it used to be. Indeed it is fast on the de-

* "Isis," 1848, p. 248; Schmarda's "Geograph. Verbreitung" (1853), p. 370; *fide* Murray's "Geogr. Dist. of the Mammals," p. 140.

† Hayes's Voyage towards the North Pole (1866), p. 390.

‡ The German Polar Expedition, 1869–70, found it in abundance on the east coast, on Sabine Island; and Hall's Expedition found it in numbers on the shores of the northern reaches of Smith's Sound.

§ *Vide* Murray, Edinb. New Philosophical Journal, Jan. and April, 1859; Newton in Proc. Zool. Soc., 1864; Murray, Geog. Distrib. of Mammals, p. 150 *et seq.*; Baird, North Am. Mammals; *id.* U. S. Pat. Office Rep. (Agric.) 1851 (1852), p. 105.

‖ Rarer on the east coast (apparently).

crease, on account of the unmerciful way in which it is slaughtered by the natives for the skin alone, as is the buffalo in America. The skins are a great article of commerce; sometimes they sell in Copenhagen at from 3 to 7 rigsdaler (6*s.* 9*d.* to 15*s.* 9*d.*) each, according to the quality. (The natives get in Greenland only 72 skillings (1*s.* 6*d.*) for them). The yearly production used to be in the summer time from 10,000 to 20,000, but it is now on the decrease.

Dr. Hayes fed his party luxuriously on them all winter at Port Foulke in Smith's Sound, not many miles from where Kane's party starved a few years before. Behind Holsteensborg are valleys full of Reindeer; and I have heard tales of people climbing the hills in that vicinity, and looking down into glens where the Reindeer were so numerous that they might be supposed to be the herds of a wealthy Laplander. Ten thousand skins were shipped from that post some years ago. They are slaughtered indiscriminately by the natives—these improvident people, in nine cases out of ten, leaving the hides and flesh, and only taking the tongues. They are bad enough shots; and the Danish traders supply them with powder at less than prime cost (viz. 36 skillings, or 9*d.* per lb.), with a view to increase the produce of the hunt; but this ammunition is wasted in a most reckless manner.

On the way to and from these hunts up the fjords ("the interior country," though really the natives know of no place off the coast more than the Europeans do), with that savage desire to ***kill*** every living thing, ducks are shot and left lying, or, if they feel hungry, they will tear off the titbits; a ptarmigan will be shot sitting on its eggs, and the ball cut out of its body to be again used in this murderous sport. There is no necessity for it, for at this time they are abundantly supplied with food, even to excess. It is, however, the season of sport and fun, looked forward to by the natives much in the same light as we do to our grouse-shooting or deer-stalking, and is about as profitable to all parties concerned. In order to pursue this they leave the more lucrative seal-fishery, and neglect to lay in a winter's supply of food; so that when the "banyan" days come they bitterly regret their folly, and weary for the bleached carcases up the frozen fjords. Notwithstanding this, regularly as the season comes round, they are off again to the shooting from far and near, and repeat the same improvident course; nor, if they like it, has anybody a right to complain. In all verity, enjoyments few enough fall to the lot of these hyperborean hunters.

However, the effect of this indiscriminate slaughter is now being felt in the decrease of the Reindeer in many parts where they were once common. They are no longer found on Disco Island, as in the days of Cranz and Fabricius. Indeed there are now very few shot in mid-Greenland, and many of the natives are giving up the hunt for them altogether. During the summer of 1867 only five Reindeers were killed in the district of Ritenbenk (lat. 69° 45′ N.). The yearly average had been about 20 or 30; but the Governor informs me that in his opinion reindeer-hunting days are nearly over in that section of the country.

In the districts of Jakobshavn, Clavshavn, and Christianshaab I did not learn that one had been killed. At Clavshavn a few natives went out hunting, but met with bad weather, and returned for good, having only seen two animals altogether, and shot nothing.

In the southern portion of the country more are seen, not so much on the coast as up the valleys by the fjords. It is in May or June that most of the natives leave their winter houses and go reindeer-hunting. When they do dry any meat, they cover it up in *caches*. The dogs are not taken along with them. In old times, even making every allowance for exaggeration, the Reindeer seems to have been very numerous. In the Icelandic "Sagas" they are spoken of as having been very numerous in the Œster Bygd.

Four hundred years ago the natives seem by these accounts to have hunted the Reindeer much in that section generally supposed to be the site of the Œster Bygd (viz., Julianeshaab district). At the present day they have left that district and it is now nearly sixty years since any have been shot there. Latterly the hunting has been better in Greenland (south). From 1840 to 1845 many were got; and within the last few years they seem (if we might judge by the produce of the hunt) to be on the increase. This, however, is, doubtless, owing a good deal to the use of the rifle; but it is very questionable whether this will not again decrease their numbers as it seems to have done elsewhere. Necessarily we have no better data to go upon than that so many skins have been traded; but, if this is to be received as evidence, more have been traded of late years.

When the hunting was at its best it was at the positions where the country was broadest, or where the great *mer de glace* of the interior was most distant from the coast viz. Holsteensborg, Sukkertoppen, Godthaab, and Fiskernaæsset. Now there are very few killed at the last-named place. Godthaab also yields few; but the Holsteensborg and Sukkertoppen natives have taken a good many of late. At Holsteensborg (formerly mentioned as a favourite lacality) the hunting-ground is behind the large inlets, where the ice lies far back, and where land most free from ice has been found. The Reindeer, living in very large herds, require always to be on the look-out for an extensive feeding-range; and it has been observed that they are going south, in the direction of Julianeshaab, and individuals have been annually shot not far from Fredrikshaab. In order to hunt the Reindeer, the natives go every year, in the month of June, from the southern districts to the two northern districts in the Southern Inspectorate, and return in September. A good number are also shot in the winter time; and not unfrequently, in very snowy winters, they have been known to come down close to the settlements, and the natives have shot them standing in their doorways. The story of the Reindeer going into the *interior* in the winter is founded on erroneous notions of what the interior is. They no doubt go a little way into the valleys; but as for going into the interior, that is a physical impossibility, for the interior is merely one wide frozen waste, surrounded by a circlet of islands. It is to the valleys of these islands that the Reindeer undoubtedly retire; but nobody travels very far afield in Greenland during the winter season, so that we have no means of arriving at a very accurate confirmation of this supposition. Dr. Hayes's people finding them in such abundance at their winter-quarters goes further to prove this. One of his men described to me the party as going over a little ridge, and finding the deer as if in a preserve, like the cattle in the pastures of his native Jutland; "we just shot them as we wanted them." (See also Hayes's "Open Polar Sea," *passim.*)

Their food in Greenland consists chiefly of various species of *Empetrum*, *Vaccinium*, *Betula*, &c.; and I can hardly think that the traditional "reindeer-moss" (*Cladonia* of various species) forms any great portion of its subsistence, as that Lichen is nowhere found in Greenland in such quantity as to afford food for any animal.*

The Greenlanders have no idea of taming the animal; indeed its use to them would be trifling, as it cannot travel well on ice, and the difficulties of transporting supplies of food for it on their long ice-journeys would be great. The Eskimo's sledge-travelling is almost wholly confined to the frozen surface of the sea in winter; and for this purpose dogs answers much better. The meat is very good; and the natives eat the half-digested vegetable contents of the stomach along with blubber as a choice delicacy. They prefer to eat the flesh in a putrid state. It is, with the exception of the breast, for the most part lean. Clothes and thread are made from the skin and sinews. The latter is much sought after in districts where there are no reindeer. From the horn are made all sorts of native implements; but commercially it is of no value in Copenhagen. However, I think its importation ought to answer, if brought to England, though to Denmark it will not pay the freight.

A calculation has been made that from 1840–45 there were about 2,500 persons living in the principal reindeer district. Every family of five persons, it was calculated, would use two skins, &c., which would make 5,000 for themselves; and they sent away 11,500; the total hunt was therefore calculated to be about 16,000 annually. This sum has been taken for a minimum; for every hunter, besides using the skins for clothes, not only for himself and family, also used them for tents, partitions in houses, and for socks, &c., so that the number killed was in all likelihood much greater. Of late years the skins traded by the natives have decreased one half. Between 1851 and 1855 there were annually shot 8,500 deer. It is difficult to say how much meat has been consumed in that period; but every deer may be put down at 80 lbs. of meat alone. This makes the meat, beween 1840 and 1845, amount to 1,280,000 lbs. annually, and between 1851 and 1855 to 680,000 lbs.

---

* On the western shores of Davis's Strait I have known them to come down to feed upon the *Fuci* exposed at low water, as do the cattle and red deer in some places in the north of Scotland.

The Reindeer is often shot in situations where it is impossible for the hunter to carry the meat down, when it becomes a prey to wild beasts and birds. The quantity of meat thus lost is enormous, independently of much more wastefully destroyed, as described in the first portion of these notes. It is so great that during the period first referred to, fully one half was thrown away, and during the last period a quarter. The tallow in a large deer will weigh from 8 to 12 lbs. The tongues are first cut out, after the reindeer is killed. About 3,000 to 4,000 lbs. of reindeer-horn must be used by the natives in South Greenland. The trader at Holsteensborg has (or at least had a few years ago) more than 60,000 lbs. of it lying on the ground in a heap.*

I have gone into the history of the Reindeer in Greenland at some length, because I found that though the Reindeer in Lapland is familiar to many, yet the animal in its wild state is much less known, and I have seen most erroneous statements regarding its distribution in Greenland.

12. [OVIS ARIES, Linn.
Greenl. Saua.

At present it is only known in the district of Julianeshaab, to the number of between 20 and 30. It was already introduced in Fabricius's day. In the summer they feed in the valleys, and in the winter are kept under shelter. They cannot, therefore (nor, indeed, can any of the colonist fauna), be said to be acclimatized.]

13. [BOS TAURUS, Linn.
Greenl. Umimak.

There are 30 or 40 Cattle grazing about in the southern valleys during the summer, and kept at stall in the winter. Some of the more enterprising natives also keep a few cows. I was told by the Danish residents that, though there was quite enough grass occasionally found round the settlements in the summer, even further north, they could not be kept on account of the dogs. The old Icelandic sagas describe the Norsemen as keeping herds of cattle in the valleys of Greenland up to the middle ages; and that the dairy produce was so highly valued that it was sent to Norway for the use of the Royal table. The place where they prosper best now is just on the site of one of these ancient colonies. If any were behind when the colonies were exterminated by the Eskimo, who about this period make their appearance in South Greenland, they must have died out, or, more likely, were slaughtered by the natives (if a people who, to all appearance, were only wandering hordes who had now for the first time crossed Melville Bay from the north, can be so styled); for when Greenland was again visited by the Europeans

* For many of the foregoing statements I am indebted to my friend Dr. Rink, formerly Royal Inspector of South Greenland, and at present Director of the Greenland Board of Trade, and whose work (Grönland geographisk og statistisk, &c.) is the standard on all subjects connected with that country.

no cattle were found. It is somewhat curious that the Greenlanders apply the Eskimo name of the *Musk-ox* to the domestic Ox, showing a recollection of the existence of the former in the land they came from, though it is no longer a native of Greenland to the south of Cape York.]

14. [CAPRA HIRCUS, Linn.
Greenl. Sauarsuk.

As far back as the days of Fabricius, the Goat had been introduced into the southern settlements of Greenland, and was found profitable; they feed on the grass which springs about the old Eskimo camping-places in the summer, and are housed in the winter. I am told that they will eat dried Arctic salmon, if nothing better is forthcoming. It is not kept north of Holsteensborg, as it is found impossible to keep it where there are troops of savage dogs; and it is accordingly only found about the settlements south of that, to the number of about 100.]

15. [*Addit.*] MUSTELA ERMINEA, Linn. The Ermine was found by the Germans on the east coast; see Peters in "Die zweite deutsche Nordpolarfahrt," vol. ii. p. 157. It is entirely unknown in West Greenland.

5. *On some of the doubtful or mythical Animals of Greenland.*

Otto Fabricius used to spend his summers roaming about with the Eskimo, until he had learned to manage a kayak and strike a Seal with a skill which few Europeans can ever acquire. On one of these excursions he found in "Sildefjord, north of the colony of Fredrikshaab," a piece of a skull, about which the native told him something; and from what they related to him, and what he thought himself, he entered no less than two species in the Greenland fauna, "*Trichechus manatus*" (*Rhytina gigas*) and "*Phoca ursina*" (*Callorhinus ursinus,*) being, apparently, not certain to which it belonged. The Greenlanders called this animal *Auvekæjak*, or *Auikæjak*, and said it was like a Walrus and broke things easily to pieces. He was sure that the piece of skull belonged to the first of these animals; and again he repeats the same under the head of *Phoca ursina;* so that it is now difficult to arrive at any conclusion regarding the species of animal to which it belonged. However, I think there can be but one opinion, that neither the Sea-bear nor the *Rhytina* can be entered in the Greenland fauna on such fragmentary evidence. The confused stories of the Greenlanders can give the critic no great hold.

This piece of cranium is not now to be found in Fabricius's Museum. In a posthumous zoological manuscript, entitled "Zoologiske Samlinger," written in Copenhagen during the period between 1808 and 1814, and now preserved in the Royal Library, he has again spoken about the *Auvekæjak* (Bd. ii. p. 298, No. 286), and has thus written about the skull he found in Greenland:—

"The head which I found was full of holes, and looked like that of a Walrus (No. 82), without tusks."

There were many long small teeth in the head (Reinhardt, *op. cit.*, p. 6.); and if such was the case, we cannot be wrong in saying that the animal was not a Mammal. We have, however, no right, when we remember the clear comprehensive style in which Frabicius wrote regarding the Greenland fauna, however much we may be inclined, to say that the whole was erroneous.

It is unfortunate that when Fabricius referred his *Auvekæjak* to the Sea-cow of Steller, he was not acquainted with that animal, and did not know of the horn-plates ; for, if he had, it is impossible that he could have found a resemblance to it in the Auvekæjak. His words regarding it are clear enough, so far as they go—"Rarissimum animal in mari Grœnlandico, cujus solum cranium ex parte conservatum commune cum sequenti specie ab incolis dictum nomine Auvekæjak, vidi, inque hoc dentes spurios tales confertim congestos quales Steller" (*vid. op. cit.* Adel.* § 189). Again, immediately under the head of "*Phoca ursina*," he says :—"Grœnl. AUVEKÆJAK.—Illam esse animal quod sub nomine hoc memorant incolæ non est dubitandum. Dicunt illud in Australiori Grœnlandia, licet raro, dari quadrupes pilosum, ferociter omne occurrens dilacerare, et si visum consumere : ursi maritimi more terra marique degere, impetuosissime natare, venatores valide infestare. Dentes ut amuleta contra ulcera, nec non quodammodo ad instrumenta venatoria adhibentur." There is an evident uncertainty in Fabricius's mind; and he has listened too much to the idle fables of the natives (who have, as I shall presently show, many of that nature); whatever it is, there can, I think, be scarcely a doubt as to the exclusion of *Trichechus manatus* and *Phoca ursina* from the Greenland fauna; nor can their place as yet be supplied by any other species. Prof. Steenstrup thinks that it was a portion of the skull of a Sea-wolf (*Anarrhichas*). The situation of the teeth and the nature of this fish's cellular skull well agree with his description of the skull as "full of holes" ("forhulret," Reinhardt, *op. cit.*, p. 8). Hr. Bolbroe, who understands the Eskimo language intimately, tells me that the word means a "little walrus," and that in all probability it was only the skull of a young walrus, an animal not at all familiar to Fabricius, as they are chiefly confined to one spot, and the natives fear to go near that locality. Fabricius may have only written the description from recollection; and memory, assisted by preconceived notions, may have led him into error in the description of the long teeth, which after all might, without great trouble, be made to refer to the dentition of the young walrus as described by Macgillivray † and Rüppell.‡

This opinion is strengthened by a passage in Fabricius's account of the Walrus, when he again is in doubt whether a certain animal is the young of the walrus or the dugong, "De varietate

* Adelung: "Geschichte der Schifffahrten und Versuche zur Entdeckung des nordöstlichen Weges nach Japan und China" (Halle, 1768) is the book Fabricius refers to. There is a wrong reference in the F. G. to Adelung, viz., 189 for 148.

† Naturalists' Library (Mammalia), vol. vii. (vol. xiii. of series), p. 220. M'Gillivray's Edin. Journ. of Nat. Hist. and Physical Sciences, Aug. 1838, p. 153; Hamilton in Nat. Lib., vol. viii. p. 102.

‡ Bulletin Scien. Nat., vol. xvii. p. 280.

dentibus exertis brevioribus loquuntur incolæ, quam minus recte (ut videtur) ad *Phocas* referunt, si non pullus rosmari, an animal Dugong" (Buff. 205, 245. tab. lvi). So that, after all, perhaps the *Auvekæjak* was only the young of the Walrus; and this opinion I am on the whole inclined to acquiesce in.

Fabricius enters in his "Fauna Grœnlandica," under the name of "*Mustela gulo*, L." (*Gulo borealis*, Retz.), an animal which the natives talked about under the name of *Kappik*. It was said to be found in south Greenland, among high mountains, particularly beside streams, and was especially fond of the hearts of reindeer. He considered it to be the well-known Wolverine, the *Jerf* of Scandanavia (Norse *Arv*, *Erv*, and *Jærv*; Swedish *Jerf*, *Gerf*; Finnish *Kamppi* and *Kamppi-Karhu*). If so, it must be exceedingly rare, for since his time no one has been able to obtain or hear of a specimen. We more than suspect, however, that here, as elsewhere, he was only reproducing in a zoological dress the stories of the natives. So little was then known of the zoology of the Arctic regions, that he might well be excused for entering such animals in his fauna, there existing no reason why they should not be found in Greenland. If Fabricius could have lived to this day, he would have been the first to erase these from his list. The reason why I think so is this:—Under the head of "*Ursus luscus*" he has inserted a very doubtful and preblematical animal, talked of long before his day, and equally so now, under the name of "*Amarok*" ("*Ursus luscus*, Eg., 'Description of Greenland,' Eng. transl., 33, Cr., 'History of Greenland,' Eng. transl., 99, ex descriptione pellis ejus. Cf. 'Continuation,' 287, ubi dicitur subfusca, forsitan etiam veterum Hyæna, Torf., 'Grœnlandia Antiqua, 82"). This animal seems the same as that which he indicated in his fauna under the name of "*Mustela gulo*." He describes it as very fierce, corresponding in this respect with the character of the Wolverine. Depending upon the natives being in the habit of distinguishing animals by different names very clearly, he considered that *Amarok* and *Kappik* were different animals. Neither of them he appears to know anything about. I found the Greenlanders talking to this day about the Amarok all over Greenland; and wonderful stories they tell of its ferocity. It is the terror of the Greenlanders, as Fabricius truly enough remarks; everybody knew about it; but I could find nobody who had ever seen it.* Graah (*Lib. cit.* p. 90.) found the natives of the east coast equally familiar with the name of the *Amarok*; the name *Kappik*, however, was unknown in north Greenland.

Finally, I discovered a man in Claushavn who declared he had seen the *Amarok*; it hunted in packs, he said; and this man made no secret of his belief that it was only native dogs which had escaped and returned to their wild state. In proof of this he told me that, as frequently happens during the annual reindeer-hunting-season,

* Mr. Tegner informs me that one of the natives declares that in July 1867 he saw the marks of the foot of an Amarok at the head of the Tessiursak, an inlet near Claushavn.

one of his dogs escaped and could not be captured again. Three years after, one severe winter, when "looking" his fox-traps, he found the identical dog captured, much subdued by hunger, but still very fierce after living for so long a period out of the reach of the merciless lash. It served its master for many a day after in harness. This man described the "Amarok" as all grey. It has been supposed to be the Wolf (*Canis lupus*, var. *alba*) * and to have crossed over the ice in Smith's Sound; but from what I have said about the Eskimo Dog, it will be apparent that to distinguish between a wild dog and a wolf is a matter of some difficulty. I think, therefore, that the Wolverine has no place in the Greenland fauna, and that the Kappik † and Amarok must be regarded as synonyms of *Canis familiaris*, var. *borealis*, tinctured with a deep hue of fable: Murray portrays the distribution of the Glutton (*Gulo borealis*) on both the east and west coasts of Greenland up to nearly 67° N. lat. (*op. cit.* Map xxiv.); but if I am right in excluding this animal from the Greenland fauna, this distribution is erroneous.

Here I may remark, what must by this time be self-evident, that the Greenlanders cannot be relied upon (independently of the principle in the abstract) for the names of animals. They are not the excellent cetologists we have always been led to suppose, confounding as they do several animals under one name, as I shall have occasion to notice in a future page when discussing the errors which Fabricius was led into by trusting too much to their nomenclature, and which to this time have entangled the history of the northern Cetacea in an almost inextricable knot. Fabricius has notified in his Fauna many species of supposed Seals, &c. under various Eskimo names, but which he was unable to decipher.‡ Hr. Fleischer, Colonibestyrer of Jacobshavn, has aided me in resolving these :—

1. *Siguktok*, "having a long snout and a body similar to *Phoca grœnlandica* perhaps *P. ursina*." This is apparently some Eskimo perversion, if interpreted properly; for I am assured that it is only the name of the Eider Duck (*Somateria mollissima*).

2. *Imab-ukullia*, a Seal with a snow-white coat, "the eye presenting a red iris, probably *P. leporina*," is a rare albino of the Netsik (*Pagomys fœtidus*). The meaning of the word is the Sea-hare.

3. *Atarpiak* or *atarpek*, "the smallest species of Seal, not exceeding the size of the hand, of a whitish colour, and a blackish

* In the winter of 1868–9 a true Wolf was killed at Omenak, and was supposed to have crossed from the western shores of Davis' Strait, where during the same winter they were very abundant.

† Jansen in his "Elementarbog i Eskimoernes Sprog til Brug for Europærne ved Colonierne i Grönland" (Kjöbenhavn, 1862), p. 55, translates "Kappik" as "en Grævling," a badger.

‡ *See* also Giesecke in his "Greenland," in Brewster's Edinburgh Encyclopædia. This article, which is the only original one, as far as I know, ever written upon Greenland in the English language, is a most trustworthy account, for the time it was written. The author, however, copies Fabricius in his errors as well as excellencies.

spot of the form of a half-moon on each side of the body." This description does not correspond to the meaning of the word, which is "the Brown Seal." Hr. Fleischer thinks that it is only a myth, as is—

4. *Kongesteriak*, which has, "according to the description given by the natives, some resemblance to the Sea-ape described by Mr. Heller."* This is one of the northern myths. The natives say it is a Bear which is so covered with an ice-coat that it never comes on land, but is always in the water, &c. These myths, both in the pseudo-Mammalia and in other groups, are endless; but I have given enough to show that no dependence can be placed on their idle superstitious tales.

I may as well close these notes on supposititious or non-existent animals by some remarks on other species, which though not Mammals, yet come fairly under the headings I have given to this section of my paper. The Great Auk (*Alca impennis*, Linn.) once so common in Greenland, in the days of Egede, Cranz, and Fabricius, as, indeed, it was in many other parts of the northern portion of Europe and America, there can be little doubt is now quite extinct in Greenland. I made every inquiry regarding it, but could learn little or nothing about it. The natives about Disco Bay do not now even recollect it by name, though when the old Eskimo name of it (*Isarokitsoc*) was mentioned they immediately repeated it, and said, "Ah! that means little wings!" Though the Royal Museum in Copenhagen has offered large rewards for a specimen, hitherto the efforts have been in vain. One of the stories I was told at Godhavn, on Disco Island, if true, would afford some hope of its yet being found:—Eight years ago (1859), on one of the little islets just outside of the harbour, in the winter time, a half-breed named Johannes Propert (a nephew, by the way, of the well-known interpreter Carl Petersen) shot a bird which he had never seen before, but which, from description, could be no other than the Great Auk. He and his companions ate it, and the dogs in his sledge got the refuse; so that only one feather could afterwards be found. I know the man well. He is rather an intelligent fellow, and was not likely to destroy a bird of such rarity that he had never seen it before, when he knew that it would command a price from the Governor. Moreover Johannes bears the reputation of telling wonderful tales now and then. He says that he saw two, but that one escaped among the rocks. Mr. Frederick Hansen, then Colonibestyrer (Governor) of Godhavn, has offered a reward for it, and is very sanguine that he will yet obtain a specimen of the *Geirfugl*.†

Depending on the native stories of a jumping animal found in the southern part of Greenland, on Grassy meadows, and called

---

* I suppose Giesecke means *Steller's* account of the "Sea-ape," *vide* Pennant, Quadr., ii. p. 301 (*Trichechus hydropithecus*, Shaw, Zool., i. p. 247; *Manatus simia*, Illig.; *M.? hydropithecus*, Fischer, &c.).

† Swedish *Garfogel*, Norse and Icelandic *Geirfugl* and *Goiful*. It is also called in Norse *Stor-Ommer*.

by them *Piglertok* ("the springer"), Frabricius thought that he recognized the Common Frog, and has accordingly entered the *Rana temporaria* as a member of the Greenland fauna. He, however, saw no specimens, nor is such an animal known in Greenland, where there are no species of Reptiles or Batrachians found. About the southern portion of Disco Bay, the natives use the name as a sort of *slang* title to the *Nisa* (*Phocæna communis*, Brookes), the *Marsviin* of the Danes in Greenland,* from its tumbling or springing movements while disporting itself. Jansen (*lib. cit.* p. 59) gives the word in the south Greenland dialect as *pisigsartut* or *pigdlertut*, and translates it "grasshopper" (*græshopper*).

I will not stop to inquire into their grosser myths, which, though relating to animals, are yet only remotely connected with zoological science, and wander away into the domains of mythology, interesting enough, no doubt, but with which we, as zoologists, have but little to do. For instance, as far back as the days of Fabricius, they used to talk about men living away in the glens off from the coast. "They tell tales" (fabulantur), he says, "of other people living away among the mountains, rarely seen by them, never by the Europeans, whom they call *Torngit* (sing. *Tunnek*) or *Tunnersoit*, and even say that they have the appearance, stature, and clothing of Europeans. . . . . . . If they speak truly, which I am not in a position to deny, perhaps they are the remnants of the former Icelandic colonists, who have fled in among the mountains."† About Jakobshavn they still talk of these people, and I collected many such stories. Some of these superstitions describe the *Torngit* as little men; and I know a man who says he saw one of these little men "pop out of a hole and in again" most agilely, and he tells a long story about it. Others describe them as tall men; so that these are undoubtedly only traditions of the old Norsemen. During the Norse possession of the country, the population appears to have got much amalgamated (as indeed we know, because when Hans Egede came, there were many traces of the white stock; and to this day there come from the east coast natives with blue eyes, and fairer hair than is usual in Greenland‡) with the Icelandic adventurers who came with red-haired Erik, and subsequently imbibed much of their superstition. Indeed most of the best Eskimo traditions (as related by Rink in his "Eskimoiske Eventyr og Sagn") are of Scandinavian parentage. Accordingly we find the old Norse tale of that fearful

* Called in Sweden *Marsvin* and *Tumlare*, in Finnish *Merisika*, and in Norse *Ise* and *Nise*, from which, apparently, the Eskimo name *Nisa* is derived, as are not a few of the Greenland words, from their intercourse with the old Norsemen prior to the Middle Ages. I suspect *Piglertok*, now the vulgar term, was originally the native one.

† Fauna Grœnl., p. 4.

‡ A Moravian Missionary at Pamiadluk, near Cape Farewell, told Captain Carl W. Neilsen (who told me) that, in 1850, a party of natives came to that settlement from the east coast, and declared that it was two years since they had left their homes. They were described as tall and fair-haired. Almost every year some come and permanently settle in the Danish colonies.

Kraken * which drew stout ships down to the bottom of the sea, in a Greenlandic version, still terrifying the squat seal-hunters who gather round the blazing *Kotlup* during the long winter nights; but I need say nothing further about it. It is one of the old *trols* of Scandinavia, familiar enough to all of us.

Still more would it be an idle task to inquire regarding that "sea monster" which good Hans Egede saw, and Pastor Bing sketched "off our colony in 64° north latitude."†

I have said enough to show that, though there is yet much to be done to the legitimate zoology of Greenland proper, there is still more to be done in what may be called the illegitimate zoology—the history of zoological myths and errors.

II.—NOTE on additional MAMMALS of GREENLAND. In a letter to Prof. A. NEWTON, from Prof. J. REINHARDT, University Museum, Copenhagen, dated Feb. 2, 1875.

(*See* List of Mammals in the Appendix to H. Rink's "Grönland," by J. Reinhardt, 1857. *See also* the foregoing Article.)

"As to what relates to the Mammals there are two very interesting additions made by the German Arctic Expedition, viz., the *Ovibos moschatus* and *Putorius ermineus*. I believe, however, that both are restricted to the east coast. An animal of the size of the *Ovibos*, at least, could scarcely have escaped being observed during the long period of the Danish colonization, if it really lived along the western coast, where the Danish settlements are. One must also keep in mind that the interior of Greenland is most likely all covered with ice, in fact, one immense glacier. It would be very strange if any Mammal should be able to cross it from east to west. Even the *Myodes torquatus* (*Mus grœnlandicus*) has never been observed at the Danish Settlements.

"But even the list of the Mammals of West Greenland has received an addition during late years. A magnificent white Wolf (*Canis lupus*, var. *alba*) was killed at Omenak during the winter 1868–69. The skin is beautifully stuffed in the Museum. I have been told that there were two individuals in company, but one of them escaped; and afterwards footmarks of an old Wolf and of its whelps seem to have been observed." *See* above, p. 31.

---

* *Kraken*, *Kraxen*, *Krabben*, and *Horven*, *see* Pontopiddan, Nat. Hist. of Norway, vol. ii. p. 211; Ancker-Trold, Olaus, Wormius, Torfæus, &c.

† *Lib. cit.*, p. 86.

III.—On the History and Geographical Relations of the Pinnipedia frequenting the Spitzbergen and, Greenland Seas. By Dr. Robert Brown, F.L.S. F.R.G.S.

[Reprinted by Permission from the Proc. Zool. Soc., 1868, No. XXVII., pp. 405–440, with corrections and annotations by the Author, March 1875.]

Contents.

1. *Introduction.*

In the introduction to the preceding paper I had occasion to refer to the uncertainty which surrounds the history of many of the Arctic Mammalia ; pre-eminently is this true of the Cetacea, but scarcely less so of the order Pinnipedia. Though the specific determination of the species in this group is more easily managed, and has to a great extent been accomplished, yet the end to which these determinations are made,—the history of the life and geographical distribution and migrations of the animals themselves, are yet almost unknown, or accepted on the authority of the old Greenland naturalists, many of whose observations, made in a day when the specific characters were less known, and but a limited portion of the Arctic Ocean explored, have been proved to be far beside the truth. Again, these observations were made on the coast of Greenland where none of our sealers go, while in the Spitzbergen and Jan Mayen seas (the "Old Greenland" or "Greenland Sea" of the whalers) the vast portion of the sealing of commerce is carried on for a few weeks each spring; but regarding the history of the Seals which form the prey of these hunters, the extent, commercial importance of the trade, and the migrations of these animals from one portion of the Arctic Sea to another we absolutely know nothing. . . . .

In the spring of 1861, with a view to acquire a knowledge of the northern Seals of commerce, I accompanied a sealer into the seas between Spitzbergen and Jan Mayen ; that year, however, proved a partial failure, and we returned to England by the end of April, leaving immediately for Baffin's Bay. Dr. John Wallace also made a similar voyage, and was fortunate enough to enjoy better opportunities of observing the habits of Seals than I did, for at the period when I left for Davis's Strait he remained behind, and passed the whole summer in the sea between Spitzbergen, Jan Mayen, and the east coast of Greenland. On my arrival in England he put into my hands an excellent series of notes on these animals, part of which I communicated to the Royal Physical Society of Edinburgh in 1862; and an abstract was published in their "Proceedings" for that year (p. 312). Having some intention of preparing a more extensive work, I reserved my own observations and a great portion of Dr. Wallace's until such time as this might be matured; besides, there were innumerable points in the history of the Seals which I was desirous of investigating before putting any of our observations before the world. However, shortly after this I left on a very long scientific journey, far from the scene of our former studies, and for more than four years the whole subject was laid aside. In the summer of 1867 I again found myself a sojourner as far north as 70° N. lat., in Danish Greenland. During this time I made a very extensive collection of the skeletons, skulls, &c. of these and other animals, besides adding to and correcting some of my former observations. That osteological collection has not yet been examined; but this is the less important, because, so far as I was able to judge during the hasty examination it was possible to give them during the process of preservation, there are no new species among them. Moreover the craniological characteristics of the northern Pinnipedia, thanks to the labours of Nilsson, George and Frederick Cuvier, Blainville, Gray, Gaimard, Lilljeborg, Steenstrup, Murie, and others, are now very satisfactorily determined; and what points are still *sub judice* can easily be settled by an appeal to the collections already in our museums, and to the one formed by me when it is made accessible to science.

. . . . In the following notes are combined most of my own observations with selections from those of Dr. Wallace (distinguished by his name within parentheses when I have been unable to confirm the observation). . . . . The remarks on the species are prefaced by some general observations on the group. For the reasons already stated, I have purposely omitted giving any osteological distinctions, except in a few cases, limiting what descriptive remarks I may have to make to some disputed points regarding the very fallacious distinctive marks derived from the skin. Gray's "Catalogue of Seals in the British Museum," or Bell's "British Quadrupeds" (ed. 2, by Tomes and Alston), will supply all that is necessary on these points. As in the previous paper, I have not attempted a complete history of their habits, geographical distribution, &c., chiefly limiting my remarks to what has fallen within my own observation or knowledge. . . . . The list of popular names attached to each species is the result of not a little work and extensive acquaintance among the seal-hunters and fishermen of the northern coasts. The scientific synonyms are only given when no doubt existed of their applicability, and are not intended to be a complete list.

2. *Physiological Remarks on the Habits of Seals.*

The Seal is, to a considerable extent, fitted for terrestrial progression, which it performs chiefly by the muscles of the trunk, aided by those of the extremities. The result is a rolling, waddling, or shuffling kind of motion—the animal leaning over on one anterior extremity, and then rolling back on the other to make a similar use of it, using them thus alternately and the muscles of the spine continuously, chiefly those of the lumbar region and *erectores spinæ*.* In carnivorous animals the intestinal canal is shorter than in graminivorous species, yet there are exceptions, for the Sloth has a very short intestine, and the Seal a very long one. I have measured the length of the intestine of *Pagophilus grœnlandicus*, and found it to vary between 50 and 56 feet in length.

It is said that the livers of the Seals at Novai Semläj and in the southern seas possess poisonous properties; this is not the case with the livers of any of the Greenland Seals, for they are often eaten, and I never knew of any bad effect ensuing. The lymphatic glands are well developed, these glands being of great size, though not numerous, it being common to find only one in each axilla and groin. In the young Seals the lymphatics of the neck are subject to disease, which appears to be analogous to, if not indeed true scrofula: the glands swell, suppurate, and pour out a purulent discharge, and the animals subject to this disease do not increase in size.

Many theories have been adduced to account for the Seal's capability of remaining with impunity so long below water. That of Buffon and the physiologists of his time was long celebrated: from their finding the *foramen ovale* open in a few instances, they twisted an exception into a rule, and accounted for it by this fœtal peculiarity. Dr. Wallace considers that this theory is erroneous, and from numerous observations he is satisfied that the open foramen must be very rare, for in only one of the Seals which he examined did he find the foramen ovale unenclosed to within a line of the aorta. That of Blumenbach and Houston has been also brought forward, viz., that venous sinuses are to be found in the liver and surrounding parts, and that the large veins have been observed to be enlarged and tortuous; these have been supposed to act as reservoirs for the returning venous blood while the animal is diving under the water. But this theory carries inconsistency in itself. The venous system on the whole, and not in any particular part, unless in the vena cava, from the pressure excited on its walls, is greatly enlarged; but this arises from the great quantity of blood these animals possess. But, even supposing the existence of these venous sinuses, and that the animal will remain below the surface for twenty or twenty-five minutes (though I never saw them remain longer below the surface than fifteen minutes, and from five to eight is the common time), are these sinuses large enough to contain the full quantity of blood that may return in that period from the capillary system? The reply is certainly in the negative. Does the heart's action diminish in rapidity, or come to a full stop? In that case there would be no need of these sinuses. What, then, are their uses?

* For a fuller account of the mechanism of motion in Seals, see Pettigrew, Trans. Linn. Soc., vol. xxvi.; and Murie, Proc. Zool. Soc., 1870.

After a very careful examination, Dr. Wallace informs me that he could not find them in all the Seals which he examined. He certainly remarked the dilated condition of the veins, but referred this to a physiological cause, viz., the pressure of the superincumbent column of blood. He believes that their power of remaining so long below the surface of the water is to be referred to a cause physiological, and not structural. Their expertness in swimming is not possessed from birth, but only developed from an innate instinct. We have often watched young Seals taking the water at first in smooth pools among the ice, and then swimming slowly and quietly about in the still floe-water, —then gradually taking the water, staying below the water at first but a short time, gradually lengthening their stay until they had acquired the faculty of remaining the usual time beneath the surface. Dr. Wallace, then, thinks that this faculty is owing to a cause more physiological than anatomical, and that the explanation he has given, coupled with the enormous quantity of blood which the Seal contains, will account for their power of remaining beneath the water. As I have not examined the anatomy of the Pinnipedia with this object in view, I cannot presume to give an opinion on the matter; in the Narwhal and other Cetacea which I examined, the extensive venous plexus about the vertebral column seemed to explain the possession of this power of temporary subaquatic existence. The flesh of the Seal is quite black, from the enormous quantity of venous blood it is impregnated with; but if exposed to the air or steeped in water, it acquires the usual arterial rosy hue. The flesh of young Seals which have not yet taken the water is, on the contrary, quite red.

3. *Habits and Instincts of Seals in general.*

They spend a considerable part of their time in feeding, but they pass by far the greater part in basking in the sunshine and sleeping on the ice.* It has been remarked that the Seal sleeps and wakes alternately about every 180 seconds. Seals are, however, often killed in considerable numbers when asleep on the ice; and this happens most commonly on a day of warm sunshine. We had a Seal on board about a month old, which I watched attentively for some time, and it certainly seemed to wake and sleep alternately, with the interval mentioned (*Wallace*): when disturbed it made attempts to defend itself; and if left alone for a few seconds, it drew its flippers close to its sides, and gradually it began to look drowsy, then closed its eyes, and, from the long deep breathing, it was evidently asleep for a minute or two (the time varied); and then, without being disturbed in any way, it would suddenly open its large black glassy eyes, stretch out its head, and look about, and, as if satisfied that all was right, would again relapse to sleep, and so on. When asleep they always leave several sentinels on the watch, which, strange to

* "Sternunt se somno diversæ in littore Phocæ" (Virgil, Georgics, lib. 4).

say, are, for the most part, female Seals. These sentinels, however, conduct themselves in the same manner as I have described the individual Seal we had on shipboard. I have been assured by old seal-hunters that Seals can sleep on their back while floating in the sea; and this statement corroborates that of Fabricius and other naturalists. In 1861, in Davis's Strait, the steamer on which I was aboard of ran against a Seal sleeping in this manner. The *blow-holes*, or escape-holes, of the Seals are evidently formed by them when the ice is making, the animal always rising to breathe again at the same place, thus preventing the coagulation of the ice, or breaking it as soon as formed. It has been supposed that the Seal could make such an opening by force or by *keeping its warm nose for a time at one place for the purpose of melting the ice;* but these conjectures are not founded on truth, the following reasons being my grounds for that statement:—It could not break the ice by force, and, moreover, it could not even dare to run its nose against such an obstacle; for the nose of the Seal is a tender point; this was known even to the ancients, and is referred to by Oppian.* This is taken advantage of by the sealers, who secure as many as possible when they are hastening to the water from the ice, by striking them on the nose, and then killing them at their leisure when the others have escaped. Even suppose the muzzle capable of melting the ice (which it certainly is not), where could the animal rise to breathe during the process? The preceding explanation of the formation of the breathing or *blow*-holes was derived from independent observation of the habits of the Seal, but is identical with that given me by the natives of the Arctic regions. It is at such holes that the Eskimo and the Bear watch patiently for their prey.

The *voice* of the Seal is a peculiar cry, somewhat midway between that of a young child and the bleating of a lamb or kid.

They are very fond of music, which was well known to the ancients; and this fondness is often taken advantage of by the hunters at the present day.† I have often seen them raising their heads inquiringly out of the water listening to the sea-songs of the sailors as they wrought at the pumps or tracked the ship to the ice-floe; therefore it seems as if the fabled spell of Orpheus, which was powerless on the Dolphin, takes effect upon the Seals. In moving from one place to another they swim rapidly, sometimes on their backs and often on their sides, occasionally whirling about as if to amuse themselves, and sometimes leaping out of the water altogether.

---

* "Non hami penetrant phocas, sævique tridentes
In caput incutiunt, et circum tempora pulsant.
Nam subito pereunt capitis per vulnera morte."

† It is often alluded to by the ancient poets (thus, "gaudebant carmine phocæ," Apol. Rhod., lib. 1; Val. Flacc., lib. 5, lin. 440, &c.); and all ancient historians especially note that the Seal is "perstudiosa musicæ." The well-known passage in Sir Walter Scott's "Lord of the Isles" (p. 140) also refers to this,—

"Rude Heiskars seals through surges dark
Will long pursue the minstrel's bark."

Their parental love is so great that they will sometimes remain and share the fate of their hapless young. Their instinctive knowledge of danger is very keen; they have been known to seize their young with their flippers and carry them into the water with them when they saw the hunter approaching! I did not see this myself.

Seals are very tenacious of life, and difficult to kill, unless by a bullet through the brain or heart. They are so quickly *flensed*,* that after having been deprived of their skin they have been seen to strike out in the water; so that the sympathies of the rough hunters have been so excited that they will pierce the heart several times with their knives before throwing away the carcass. These movements, however, are apparently reflex or diastaltic, as I have often seen a Seal lying skinned on the deck for an hour, exposed to a temperature of 12° below zero (Fahr.), and yet the muscles of the loins and back retain their contractility to such an extent as to be able to rotate the pelvis on the spine, on those on each side being alternately irritated.

With the exception of the Bladdernose, the other Seals in the Greenland seas appear to have little or no combativeness in their nature, but are a harmless, persecuted, sportive race of graceful athletes making merry the solitary waters of polar lands.

On the other hand, the male Bladdernose is, in truth, the lion of the sea, dividing the empire of the polar waters with its huge ally the Walrus. Instead of flying at the approach of the hunter, he will quite calmly await the approach of danger, preparing for defence by betaking himself to the centre of the piece of ice he is on, and blowing up the air-bladder on his forehead, while he rears his head and snuffs the air like an enraged bull, and often gives battle successfully, making the clubs fly from the hands of his assailants with his flippers, his head being protected as with a helmet by the air-bladder. He will then in turn act on the offensive, and put his opponents to flight, pursuing them with a shuffling, serpent-like motion over the ice, the result often proving somewhat dangerous to the panic-stricken hunter if the boat has left that piece of ice, as the Seal will use his tusks rather ferociously when thus enraged. However, he is not inclined to give battle unless provoked, and looks a dull stupid-looking sort of epicurean as he lolls on the surface of the ice and gazes about with his large black eyes, in an apparently meaningless stare. The "Ground-Seal" and "the Floe-Rat" (*Pagomys hispidus*) in the far north are quite harmless and inoffensive; they apparently delight to swim about in the calm smooth floe-waters, or bask asleep in the sunshine on the surface of the ice. Their greatest enemy is the Polar Bear, who is continually on the alert to take them by surprise, forming, as they do, his chief prey.

Nearly all of the Seals live on the same description of food, varying this at different times of the year and according to the

* A convenient whaler's word (of Dutch origin) to express the operation of taking off the blubber and skin. It is generally pronounced *flinched* by the sealers and whalers.

relative abundance or otherwise of that article in different portions of the Arctic seas. The great staple of food, however, consists of various species of Crustacea which swarm in the northern seas. During the sealing-season in the Spitzbergen sea I have invariably taken out of their stomachs various species of *Gammarus* (*G. sabini*, Leach, *G. loricatus*, Sab., *G. pinguis*, Kr., *G. dentatus*, Kr., *G. mutatus*, Lilljeb., &c.), collectively known to the whalers under the name of "Mountebank Shrimps," deriving the name from their peculiar agility in the water. This "seals' food" is found more plentiful in some latitudes than in others, but in all parts of the Greenland sea from Iceland to Spitzbergen; I have seen the sea at some places literally swarming with them. Again, in the summer in Davis's Strait I have found in their stomach remains of whatever species of small Fish happened to be just then abundant on the coast, such as the *Mallotus arcticus*,* *Salmo* (various species), &c. I have even known them to draw down small Birds swimming on the surface; but their chief food is Crustacea and Fish. They also feed on Medusæ and Cuttlefish (Squids).

4. *Notes on the Species of Pinnipedia.*

(1.) CALLOCEPHALUS VITULINUS (Linn.), F. Cuv.

Phoca vitulina, Linn.
Phoca communis, Linn. (Mus. Ad. Frid., i., 5).
Phoca canina, Pall. (ad partem).
Phoca variegata, Nilss.
Phoca linnæi, Less.
Phoca littorea, Thienem.

Popular names.—*Sea-dog*, *Sea-calf*, *Sea-cat* (English sailors and fishermen generally); *Selkie*, *Selach*,† and *Tangfish* (north of Scotland); *Rawn* (western islands of Scotland); *Sprählig Skäl* (Swedish); in other parts of Scandinavia, and according to age, &c., it is variously designated *Wilkare Skäl*, *Kubbsæl*, *Fjordnacke*, *den spättede Säl* (the Spotted Seal), *Algar*, *Laggar*, *Kutar*, and *Skältokar; Kobbe*, *Stenkobbe* (Norse); *Hylje* (Finnish); *Nuorjo* (Lapp.); *Seehund* (German); *Veau Marin* and *Phoque* (French); *Kassigiak* (Greenland); *Spraglet Sælhund* (Danes in Greenland).

The Eskimo in Pond's Bay, on being shown a good figure of this Seal, called it *Supalo'*; but whether this is their name for the animal and is to be received for a proof that the *C. vitulinus* is found there, I cannot take upon myself to decide. The Greenlanders also call it, according to age, *Kassigiectsak* and *Kassiginak;* but when it attains the age of three years, it is called *Kassiarsoak* ("the big Kassigiak"). It is also, though more rarely, called, according to its age, *Ermik*, *Ermitsak*, *Akutheenak*, and *Akunneklok*. Prof. Newton ("Notes on the Zoology of Spitsbergen," Proc. Zool. Soc. 1864, and Ann. Nat. Hist., Vol. XVI., 3rd Series, p. 423), says that *Pagomys fœtidus* is called *Steen Kobbe* (Stone-Seal) by the Spitzbergen hunters. I

---

* For some account of the Capelin, *see* Pennant's Arctic Zool. Suppl., p. 141.

† The word "Seal" is from the Anglo-Saxon *Selc*, *Seolc*.

suspect that he has erred through his informants mistaking this for *Callocephalus vitulinus.* No doubt Dr. Malmgren seems to think that the latter species is not got in Spitzbergen—an opinion I have ventured to contest in a former paper.

It is also sometimes called "the Freshwater Seal," on account of its following the Salmon high up rivers.*

Remarks, &c.—Any laboured account of a Seal so long and so familiarly known would obviously be out of place in these short notes; I question, however, if all the accounts we posses regarding the Seal under the designation of "*Phoca vitulina*" really refer to this species, and not to *Pagomys fœtidus* and others.† It will, I think, be found that in the western and northern islands of Scotland several species, not hitherto supposed to be regular members of the British fauna, exist, known under the popular names of *Selkie, Selach, Sea-cat,* &c.‡ I do not think I can say anything in regard to its habits further than what is already contained in various works on Mammalia, &c., viz.:—Bingley, British Quadrupeds, p. 57; Bell, History of British Quadrupeds, 2nd ed., p. 240; Hamilton, Amphibious Carnivora (Nat. Lib.), p. 127; James Wilson, in Mag. Zool. and Bot., Vol. I., p. 239; Edmonston, View of Zetland, Vol. II., p. 293, and Mem. Wern. Soc., Vol. VII.; Martin, Western Islands, p. 62; M'Gillivray, British Quadrupeds (Nat. Lib.), Vol. XIII., p. 199; Nilsson, Skandinaviske Fauna, I., p. 276; Fabricius, Naturhistoriske Selskabets Skrifter, I., Band II., p. 98; Œdmann, Vet. Akad. Handl., 1784, p. 84; Rosted, Norske Vidensk. Nye Skrivter, II., p. 185 (good description); Cneiff, "Berättelse om Skälfänget i Œsterbotten," in Vet. Akad. Handl., 1759, p. 179, r. 8 (on the hunt); Holmers, Anteckningar om sättet att skjuta och fänga Skälar, &c. (Stockholm, 1828), (hunt, &c., *fide* Nilss.); Ball, Transactions of the Royal Irish Academy, VIII., and Sketches of British Seals; Bartlett, Proc. Zool. Soc., 1871, p. 701 (where, through an error of identification, it is called *Ph. fœtida*); Gaimard, Voyage en Islande, &c.

Procreation and Young.—On the coast of Greenland it is said to produce its young in the month of June, but the time seems to vary according to season and place. In the Western Isles of Scotland at least it is born pure white, with curly hair, like the young of *Pagomys fœtidus*, but within three days of its birth it begins to take dark colour on the snout and the tips of the flippers (*fide* Capt. McDonald).

---

* I have known a Seal (probably *Halicyon richardsi*, Gray) to be killed at the Dalles of the Columbia River in Oregon, upwards of 200 miles from the Pacific. It was doubtless in pursuit of Salmon. Dog River, a tributary of the Columbia, takes its name from a dog-like animal, probably a Seal, being seen in the lake whence the stream rises.

† In the Appendix to Parry's "Voyage," is a notice of a Seal said to be "*Phoca vitulina.*" It is the young (in second coat) of *Pagophilus grœnlandicus*, which has often been mistaken for this Seal. It can be known by its having the second toe of the fore flippers the longest, while, independently of other characters, *C. vitulina* has the first toe the longest.

‡ "On the Seals of the Outer Hebrides," Fergusson, in McGillivray's "Edin. Journ. Nat. and Phys. Sc.," ii., p. 58.

Geographical Distribution.—This is a Seal peculiar to the coasts of the regions which it affects, but has also a wide range, being found over nearly all the northern coasts of Europe and the colder portions of America. It is even said to be found in the Caspian Sea and Lake Baikal. It does not seem, from its littoral habits, to be found in the Spitzbergen sea, or to form a portion of the commerce of the sealer; it is however found on the coasts of Spitzbergen, tolerably abundant on the eastern shores of Greenland, and in Davis's Strait. It is to be found all the year round all along the coast of Greenland up inlets,* but not to any such extent as *Pagomys fœtidus* and *Pagophilus grœnlandicus.* In Scandinavia it is sometimes called the Fjardskäl on account of its frequenting inlets or fjords.

Economic Value and Hunting.—We have no data to decide as to what extent it is killed in Danish Greenland, its record being united with that of *Pagomys fœtidus.* The skins are highly valued as articles of dress, more especially as material for the women's breeches; and no more acceptable present can be given to a Greenland damsel than a skin of the *Kassigiak.* . . . . In the Danish Settlements they are valued at from three to four rigsdaler. The principal reason which induced the late Admiral Graah's boat-women to accompany him on his memorable voyage along the east coast of Greenland was the hope of obtaining some *Kassigiak* skins from that region, the natives of which value them at even less than the more serviceable hides of the other species, which are sold by the west coast natives for a mere trifle. According to Hr. Cneiff (*l. c.*) a *C. vitulinus* will yield about $6\frac{2}{3}$ Swedish lispunds of blubber, and according to Holmers, even 8 lispunds. Professor Nilsson says that a Seal of this species killed on the coast near Malmö in Sweden yielded over 90 Swedish "potts" of oil, each "pott" being worth 36 skillings, = 67 rigsdaler 24 skillings Rigsmont (Swedish) for the oil of one *C. vitulinus.* In August, when the Seals are poorer, another yielded 75 potts, equal in value to 56 rigsdaler 12 skillings (Swedish). In some of the northern and western islands of Scotland, and at the estuary of the Tay, &c., they are still occasionally hunted for their skins and oil. The skin makes excellent leather; and waistcoats made of it are much valued by fishermen.

No separate returns of the catch of this Seal have been kept; but it is estimated that of *Pagomys fœtidus* and *Callocephalus vitulinus* the yearly capture in Danish Greenland must amount to 70,000 (Rink, *op. cit.*) or more. The flesh is looked upon in Greenland as the most palatable of all "seal-beef."

(2.) PAGOMYS FŒTIDUS (Fab.), Gray.

Phoca fœtida, Fab. (Müller's Prod. Zool. Dan., p. 8.)
Phoca hispida, O. Fab. Nat. Selskab. Skrifter, vol. i. 2., p. 74.
Phoca bothnica, Gm.
Phoca fasciata, Shaw.
Phoca annellata, Nilss.
Phoca discolor, Gray.
Phoca frederici, Less.

* The "Colonie" of Christianshaab in Disco Bay is called *Kassigian-witchz*, or the place of the *Kassigiak.*

Popular names.—*Ringlad Skäl* (Nilsson); *Morunge* (Œdmann Vot. Akad. Handl., 1784, p. 84); *Hringanor* (Mohr, Isl. Naturhistoriske, p. 5); *Kuma* (Tungúses near Baikal); ? *Nerpa* (Russ.); *Neitsiak* (young) and *Neitsik* (old, pronounced *Nesik*, Greenlanders and Danes in Greenland, *Neiturk*, North Greenland), also called the "Fjord Seal," because mostly found up fjords or inlets; *Floe-rat* or *Flaar rat** (of Northern English and Scotch sealers). It has been so often confounded with other Seals that, even on the coasts where it is not uncommon, it has not received many popular names; however, in different parts of the Scandinavian seaboard it is variously called *Inskärsskäl* or *Skärfving*, and *Svart nolled-säl*, or simply the *Nollede*. This is, in all probability, the Seal known in the Hebrides as the *bodach* or old man.

It is doubtful if this is the *Phoca equestris* of Pallas; but I cannot think that there is any serious room for doubt that it is identical with Dekay's *Phoca concolor*. I do not think that anyone now entertains any doubt about its being identical with the *Phoca fœtida* of Fabricius (*l. c.* and in F. Grœnl., p. 13, no. 8), or the *Phoca hispida* described by the same author in the "Naturhistoriske Selskabets Skrifter," *l. c.*, though Nilsson seemed in 1847 to have been doubtful (Skand. Fauna, i., p. 283).

Descriptive Remarks, &c.—This is the smallest of the Greenland Seals; it is chiefly looked upon and taken as a curiosity by the whalers, who consider it of very little commercial importance, and call it the "Floe-rat," as it is always either found on floes or quietly swimming about in the smooth floe-waters.

The young is white, of the yellowish tint of the Polar Bear. The hair is curly.

Habit, &c.—They delight to live in retired bays in the neighbourhood of the ice of the coasts, and seldom frequent the open sea. In the Greenland and Spitzbergen seas they chiefly live upon the floes in retired situations at a considerable distance from the margin of the ice. Dr. Wallace observed them for a considerable time in the months of June and July, between N. lat. 76° and 77°,† in possession of a large floe, part of which was formed of bay ice, where they had their "blow-holes" (the *atluk* of the Danes); his ship lay ice-bound for nearly three weeks, at about three miles from this large floe, and hence he had considerable opportunity of observing them. They passed the greater portion of their time apparently asleep beside their holes; and he never saw them all at one time off the ice, unless alarmed by parties from the ship or by the Polar Bear. When the ice slackened away and the sheets of open water formed around the ships, the Seals used to swim near them; and occasionally at these times a few were killed. In the water they are very cautious, swimming near the hunter, gazing on him as if

---

* I have heard the English sailors call them *Dorrities*, but this term is also used for the Bluebacks (*P. grœnlandicus*).

† Parry met with it in lat. 82° N.

with feelings of curiosity and wonder; but on the ice beside their blow-hole it is almost impossible for the hunter to approach them, so much are they on the alert and so easily alarmed. In Davis's Strait they especially feed about the base of icebergs and up the ice-fjords. The great ice-fjord at Jakobshavn is a favourite haunt of theirs; the reason for this predilection is apparently that their food is found in such localities in greater abundance. The bergs, even when aground, have a slight motion, stirring up from the bottom the Crustacea and other animals on which the Seals feed;* the native, knowing this, frequently endangers his life by venturing too near the icebergs, which not unfrequently topple over upon the eager Seal-hunter.

The old males have a most disgusting smell, which has suggested the name *fœtida*. Even the callous Eskimo is not insensible to it.†

Geographical Distribution, &c.—In the Spitzbergen sea they appear to be confined to high latitudes, and especially to the parallels of 76° and 77° N.; and it is in these latitudes that the whalers chiefly find them. In Davis's Strait they are to be found all the year round, but particularly up the ice-fjords. Their capture constitutes the most important feature of the Seal-hunt in North Greenland; but many are also killed in South Greenland, the Neitsik figuring largely in the trade-returns of that Inspectorate. In Jakobshavn Bay, I am told, they are quite numerous about the middle of August.

Economic Value.—They are extensively captured for food and clothing. Notwithstanding the nauseous smell of the old ones, the flesh of all of them (but especially the younger individuals) is sufficiently palatable to an *educated* taste. During the latter end of summer and autumn it forms the principal article of food in the Danish settlements, and on it the writer of these notes and his companions dined many a time; we even learned to like it and to become quite epicurean connoisseurs in all the qualities, titbits, and dishes of the well-beloved Neitsik! The skin forms the chief material of clothing in North Greenland. All of the οἱ πολλοι dress in Neitsik breeches and jumpers; and we sojourners from a far country soon encased ourselves in the somewhat *hispid* but most comfortable Neitsik nether garments. It is only high dignitaries like "Herr Inspektor" that can afford such extravagance as a Kassigiak (*Callocephalus vitulinus*) wardrobe! The Arctic *belles* monopolise them all.

(3.) PAGOPHILUS GRŒNLANDICUS (Müll.), Gray.

Phoca grœnlandica, Müll. *Callocephalus oceanicus*, Less.
P. oceanica, Lepech. *Phoca semilunaris*, Bodd.

---

* Hr. Distrikts-Læge Pfaff, who has resided at Jakobshavn for many years as district medical officer of North Greenland, suggests this to me; and the idea recommends itself as being that of a very intelligent naturalist.

† "Mares veteres fœtidissimis ad nauseam usque etiam Grœnlandis" (Fab. F. G.). Homer refers to this in another species (probably *Monachus albiventer*):

"Web-footed seals forsake the stormy swell,
And sleep in herbs exhaling nauseous smell."

P. dorsata, Pallas.	*Callocephalus lagurus*, F. Cuv.
P. muelleri, Less.	*Phoca albicauda*, Desm.
Callocephalus grœnlandicus, F. Cuv.	*P. desmarestii*, Less.
Young. *Phoca lagura*, Cuv.	*P. pilayi*, Less.

Popular names.—*Saddleback* (English northern sealers); *Whitecoats* and *Bed Lampiers* (Newfoundland sealers) (young); *Harp Seal* (English authors); *Svartsida* (Norse); *Dælja, Dævok, Aine* (Lapp.); *Svartsiden* (Danish, hence Egede, Green., p. 62); *Blaudruselur* (Icelandic); *Karoleek* and *Neitke* (Eskimo at Pond's Bay, Davis's Strait); *Atak* (Greenlanders). The same people, according to the age of the Seal, call it *Atarak, Aglektok*, or *Uklektok*, and *Atarsoak* (hence Cranz, Greenl., i., p. 163), meaning respectively the little Seal (white), the blueside, and the large Seal, while *Atak* means merely *the* Seal (blackside) without reference to age. A variety having the belly dark also is called by the Danes in Greenland *Svart-svart-siden.* The *Uklektok* of the natives is also called by the whites *Blaa-siden* (the blueside). I shall afterwards refer to some of its other names.

There seems little doubt that the *Phoca oceanica*, Lepech., is identical with this species; indeed Lepechin's description is one of the best we have of the *Pagophilus grœnlandicus.* Lepechin seems to have confounded with this the young of another species, and to have erred by trusting wholly to the deceptive characters of colouring, instead of relying for its distinctive character on the more stable distinction of teeth and skull. What he says about the changes of coat in *P. oceanica* exactly agrees with what I have said regarding the present species.

Remarks.—It seems to be almost unknown to most writers on this group that the male and female of the Saddleback are of different colours; this, however, has long been known to the Seal-hunters. *Male.*—The length of the male Saddleback rarely reaches 6 feet, and the most common length is 5 feet; while the female in general rarely attains that length. The colour of the male is of a tawny grey, of a lighter or darker shade in different individuals, on a slightly straw-coloured or tawny-yellowish ground, having sometimes a tendency to a reddish-brown tint, which latter colour is often seen in both males and females, but especially in the latter, in oval spots on the dorsal aspect. The pectoral and abdominal regions have a dingy or tarnished silvery hue, and are not white, as generally described. But the chief characteristic, at least that which has attracted the most notice, so much as to have been the reason for giving it several names, from the peculiar appearance it was thought to present (*e.g.* "Harp Seal," "Saddleback," &c.), is the dark marking or band on its dorsal and lateral aspects. This "saddle-shaped" band commences at the root of the neck posteriorly, and curves downwards and backwards at each side superior to the anterior flippers,* reaches

* I use this very convenient sealers' vernacular term to express the "paws," "hands," &c., of systematic authors.

downwards to the abdominal region, whence it curves backwards anteriorly to the posterior flippers, where it gradually disappears, reaching further in some individuals than in others. In some this band is broader than in others and more clearly impressed, while in many the markings only present an approximation, in the form of an aggregation of spots more or less isolated. The grey colour verges into a dark hue, almost a black tint, on the muzzle and flippers; but I have never seen it white on the forehead as mentioned by Fabricius. The muzzle is more prominent than in any other northern Seal.

Female.—The female is very different in appearance from the male; she is not nearly so large, rarely reaching 5 feet in length, and when fully mature her colour is a dull white or yellowish straw-colour, of a tawny hue on the back, but similar to the male on the pectoral and abdominal regions, only perhaps somewhat lighter. In some females I have seen the colour totally different; it presented a bluish or dark grey appearance on the back, with peculiar oval markings of a dark colour apparently impressed on a yellowish or reddish-brown ground. These spots are more or less numerous in different individuals. Some Seal-hunters are inclined to think this is a different species of Seal from the Saddleback, because the appearance of the skin is often so very different and so extremely beautiful when taken out of the water, yet as the females are always found among the immense flocks of the Saddleback, and as hardly two of the latter females are alike, but varying in all stages to the mature female, and on account of their being no males to mate with them, I am inclined to believe with Dr. Wallace that these are only *younger female Saddlebacks*. The muzzle and flippers of the female present the same dark-chestnut appearance as in the male.

Procreation, and Changes of Coats in the Young.—I have already spoken of the young as being different from the male; and in my remarks upon their geographical distribution and migrations reference will be made generally to their period and place of procreation, more theoretically, however, than from actual knowledge or observation. I now supply this from a study of this subject in the Spitzbergen sea. The period at which the Saddlebacks take to the ice to bring forth their young may be stated generally at between the middle of March and the middle of April according to the state of the season, &c., the most common time being about the end of March. At this time they may be seen literally covering the frozen waste as far as the eye can reach with the aid of a telescope from the "crow's nest" at the main-royal masthead, and have on such occasions been calculated to number upwards of half a million of males and females. After the females have procured suitable ice on which they may bring forth their young, the males leave them and pursue their course to the margin of the ice; there the Seal-hunters lose them, and are at a loss as to what course they take, the common opinion being that they leave for feeding-banks; but where is unknown. They most probably direct their course along the "cant" of the ice, or among the ice where it is loose and scattered; for in the month of May sealers fall in with the old Seals (male and female) in about from N. lat. 73° to 75°, and in the following month still further north, by which period the young ones have also joined them. The females commonly produce one at a birth, frequently two; and there is good reason for supposing that there are occasionally three, as most sealers can tell that they have often seen three young ones on a piece of ice floating about which were apparently attended by only one female. Yet it is only proper to remark that, of the several ships I have heard of finding the Seals when taking the ice, none of the hunters have been able to tell me that they took more than two from the uterus of the mother.* In contradiction to the opinion of some experienced sealers, I think that it is more than probable that they produce but once a year.

(*a*) The colour after birth is a pure woolly white, which gradually assumes a beautiful yellowish tint when contrasted with the stainless purity of the Arctic snow; they are then called by the sealers "white-coats" or "whitey-coats"†; and they retain this colour until they are able to take the water (when about 14 or 20 days old). They sleep most of this time on the surface of the snow-covered pack-ice and grow remarkably fast. At this stage they can hardly be distinguished among the icy hummocks and the snow—their colour thus acting as a protection to them; for in this state they are perfectly helpless, and the sealer kills them with a blow of the sharp-pointed club, or a kick over the nose with his heavy boot. The mother will hold by her young until the last moment, and will even defend it to her own destruction. I have known them seize the hunter when flaying the young one, and inflict severe wounds upon him. In 1862, during a severe gale of wind many of the young Seals were blown off the ice and drowned. Sometimes the sealing-ships have accidentally fallen among them during the long dark nights of the end of March or beginning of April, and were aware of their good luck only from hearing the cries of the young Seals. The white-coat changes very quickly. In 1862 the late Capt. George Deuchars, to whom science is indebted for so many specimens, brought me two alive from near Jan Mayen; they were white when brought on board, but they changed this coat to a dark one completely on the passage of a week or ten days. They ate fresh beef, and recognised different persons quite readily. The young "white-coat" represented on the plate of *Phoca barbata* by Dr. Hamilton ("Amphibious Carnivora," Naturalist's Library,

* Fabricius says that two at a birth is an exceedingly rare occurrence. Perhaps, after all, Pliny has struck the truth in regard to the order when he says, "*Parit nunquam geminis plures*" (Hist. Nat., lib. 9, sec. 13).

† These are rarely seen in Danish Greenland, and then are called "Isblink" by the Danes from their colour, at least so Fabricius says. He, moreover, informs us that the third year they are called *Aglektok* (as mentioned above), the fourth *Millaktok,* and after a winter *Kinaglit*, when they are beginning to assume the harp-shaped markings of the male (Nat. Selsk. Skrift., i., p. 92). I never heard these names in North Greenland.

vol. viii., pl. 5), from a specimen in the Edinburgh Museum, is not the young of that species, but of *Pagophilus grœnlandicus.* The young white-coat, however, is much plumper than the specimen figured; indeed, in proportion to its size, it has much more blubber between the skin and the flesh than the adult animal.

(β) They take the water under the guidance of the old females. At the same time the colour of the skin begins to change to that of a dark speckled and then spotted hue; these are denominated "hares" by the sealers.*

(γ) This colour gradually changes to a dark bluish colour on the back, while on the breast and belly it is of a dark silvery hue. Young Seals retain this appearance throughout the summer and are termed "Bluebacks" by the sealers of Spitzbergen, "Aglektok" by the Greenlanders, Blaa-siden by the Danes.†

(δ) The next stage is called Millaktok by the Greenlanders. The Seal is then approaching to its mature coat, getting more spotted &c., and the saddle-shaped band begins to form.

(ε) The last stage (in the male to which these changes refer) is the assumption of the halfmoon-shaped mark on either side, or the "saddle" as it is called by the northern sealers.

I consider that about three years are sufficient to complete these changes. This is also the opinion held in Newfoundland, though the Greenland people consider that five years are necessary. I wish, however, to say that these changes do not proceed so regularly as is usually described, some of them not lasting a year, others longer, while, again, several of the changes are gone through in one year; in fact, the coats are always gradually changing, though some of the more prominent ones may be retained a longer and others a shorter time. It would require a very careful and extended study of this animal to decide on this point, which, owing to their migrations, it is impossible to give. After all, these changes and their rapidity vary according to the season and the individual, and really will not admit of other than a general description.

Habits.—It has few other characteristic habits beyond what is mentioned regarding the order generally, or in other sections of this paper on its migrations, &c. It is looked upon by the Greenlanders as rather a careless, stupid Seal, easily caught by a very ordinary *kayaker.* Its food consists of any small Fish (*Mallotus arcticus, Cottus scorpius,* &c.), Crustacea, and even Mollusca. In this its habits agree with those of other species.

Geographical Range and Migrations.—The Saddleback has a wide range, being found at certain seasons of the year in almost all parts of the Arctic Ocean, from the American coast to Nova Zembla, and perhaps even further; it appears that the *Phoca oceanica* (Lepechin, Acta Petropolitana, 1775, t. i., pp. 1, 259,

---

* In this state it is not unlike *Halichœrus gryphus,* but can be distinguished by the characters given by Nilsson, Skand. Fauna, i., p. 301.

† The dental formula of a Seal in this stage killed by me in Davis's Strait, September 1861, was,—incisors $\frac{6}{4}$; canines $\frac{1-1}{1-1}$; molars $\frac{5-5}{5-5}$.

t. 6, 7) is identical with it. Stragglers even find their way into temperate regions; and this is so frequently the case that this Seal may now be classed in the fauna of nearly all of the northern shores of Europe and America. The period of the year influences its position in the Spitzbergen sea (the "Greenland Sea" of the Dutch, the "Old Greenland" of the English whalers). Early in March it is found by the sealing-ships in immense numbers in the proximity of the dreary island of Jan Mayen,* off the east coast of Greenland, not far from the 72nd parallel of north latitude; but, of course, the longitude varies with the extent which the ice stretches out to the eastward, though the common meridian is between 6° and 8° west of Greenwich. They are never found far inwards on the fixed ice, but on the margin of the icebelt which extends along the whole of the eastern shores of Greenland, stretching as far as the longitude of Iceland, and sometimes even for a hundred miles to the eastward of that island and of Jan Mayen island into the ocean. The general direction of its sea-margin is towards the north-east, stretching most commonly as far as Spitzbergen, to N. lat. 80°, but occasionally only to about 75° N. lat., where it joins at an angle another belt of ice which lies in a southern and eastern direction along the coast of Spitzbergen to Cherrie Island. This easterly belt of ice is what the whalers call a "south-east pack"; and at the angle where the two belts join, a passage can generally be accomplished through to the Spitzbergen waters. The nature of the ice, which can easily be perceived by the experienced sealer, determines whether the Seals will be found far from the margin of the ice. Thus, if there is much new light ice, it is probable that the Seals will have taken the ice at a considerable distance from the seaboard margin of the pack, as it is well known that instinctively they select ice of a strong consistence for the safety of their young when in that helpless condition in which they are unable to take to the water. Again, they often take the ice where it stretches out to sea in the form of a long, broad promontory, with apparently this end in view, that their young may easily get to sea when able to do so; this is the great clue which guides the sealer in the choice of the ice where he may find his prey. This was very well exhibited in 1859. Dr. Wallace tells me that there was very little ice that year, and the island of Jan Mayen was altogether free from it; indeed the nearest ice lay away nearly 70 miles or more to the north-west of it. The "Victor," the "Intrepid," and a fleet of other ships met with indications of Seals in 72° N. lat., about eighty miles in a north-westerly direction from Jan Mayen, in the early part of the month of April; they had sailed in an easterly direction through a very loose pack of very heavy ice. The prospects were so good that Capt. Martin, Sen., of the "Intrepid," perhaps the most successful sealer who ever sailed in the

* Hence the Norse sealers often call it the *Jan Mayen Kobbe* (the Jan Mayen Seal), but more often the *Springer*, from its gambolling motions in the whale (Newton, *l. c.*).

Greenland sea, and Capt. Anderson, of the "Victor" (my old fellow voyageur both in the North Atlantic and North Pacific Oceans), were congratulating each other on the almost certain prospect of filling their ships (for, indeed, the old Seals had taken the ice, and some had already brought forth their young), when suddenly there was a change of wind to the eastward, and before many hours it blew a hard gale from that direction. The results were that the ice was driven together into a firm pack and frozen into solid floes, and the "Victor" and many of the best ships of the fleet got ice-bound. The Seals shifted their position towards the edge of the ice to be nearer the sea, and for seven weeks the "Victor" was beset among ice and drifted southwards as far as N. lat. 67° 15′, having described a course of nearly 400 miles. Though I have stated the parallel of 72° N. lat. as being the peculiar whereabouts of the Seals in March, yet they have often been found at a considerable distance from it, as well from Jan Mayen. Thus in 1859 they were found in considerable numbers not far from Iceland, the most northerly point of which is in N. lat. 66° 44′; this leads me to remark that the Seals are often divided into several bodies or flocks, and may be at a considerable distance from each other, although it is most common to find these smaller flocks on the skirts or at no great distance from the main body. After the young have begun to take the water in the Spitzbergen sea, they gradually direct their course to the outside streams, where they are often taken in considerable numbers on warm sunny days. When able to provide for themselves, the females gradually leave them and join the males in the north, where they are hunted by the sealers in the months of May and June; and it is especially during the latter month that the females are seen to have joined the males; for at the "old-sealing" (as this is called) in May, it has often been remarked that few or no males are seen in company with the females. Later in the year, in July, there are seen, between the parallels of 76° and 77° N., these flocks of Seals, termed by Scoresby "Seals' weddings"; and I have found that they were composed of the old males and females and the ***bluebacks***, which must have followed the old ones in the north and formed a junction with them some time in June. There is another opinion, that the old females remain and bring their young with them north; but all our facts are against such a theory (*Wallace*).

These migrations may vary with the temperature of the season, and are influenced by it; it is possible that in the Spitzbergen sea as the winter approaches they keep in advance of it and retreat southward to the limit of perpetual ice, off the coast of Greenland, somewhere near Iceland, where they spend the winter. We are, however, at a loss regarding the winter habits of these Seals in that region; here no one winters, and there are no inhabitants to note their migrations and ways of life. Different is it, however, on the Greenland shores of Davis's Strait, where in the Danish settlements the Seals form, both with the whites and Eskimo, the staple article of food and commerce, and accordingly their habits and arrival are well known and eagerly watched. The *Atarsoak*, as it is commonly called by the Eskimo, the "Svartsidede Sælhund" (Black-sided Sealhound) of the Danes, is the most common Seal in all South Greenland. It is equally by this Seal that the Eskimo lives, and the "Kongl. Grönlandske Handel" makes its commerce. In South Greenland when the Seal generally is talked of, or a good or bad year spoken about, everybody thinks of this Seal; on the other hand, in North Greenland *Pagomys fœtidus* and *Callocephalus vitulinus** are the most common. These last two species are the only Seals which can be properly said to have their home in Greenland, affecting ice-fjords and rarely going far from the coast. This is not the case with *P. grœnlandicus*; at certain times of the year they completely leave the coast; therefore the Seal-hunting in South Greenland is more dependent upon contingencies than in North Greenland. This Seal arrives regularly in September in companies travelling from the south to north, keeping among the islands; occasionally at this time individuals detach themselves from the drove and go up the inlets. The Seal at this period is fatter, and continues so until the winter time. In October and November is the great catching, lessening in December. Very few are seen in January, and in February almost none; but regularly towards the end of May they return to the south of Greenland, and in June further north. The Seal is at this time in very poor condition, and remains for the most part in the fjords. For the second time they disappear in July, again to return regularly in September.† It is therefore seen that this Seal regularly comes and goes twice a year. Every one knows when it commences its migration from the south to the north, but nobody knows where the Seal goes to when it disappears off the coast. Between the time they leave the coast in the spring and return in the summer they beget their young; and this seems to be accomplished on the pack-ice a great distance from land,‡ viz., in the Spitzbergen sea. It is at this period that the seal-ships come after them, as referred to already. Of course a few stragglers occasionally do not leave the coast, and produce their young close to the land; but such exceptions do not at all affect the rule laid down. It is a very familiar fact that round the Spitzbergen seas in April the sealers get the best catch. At this season the Seals accumulate in immense numbers on the pack and can be killed *en masse*; but Dr. Rink cannot believe that in this time they could migrate from the west coast of Greenland to Spitzbergen, the distance being too great. In support of this argument, it is pointed out that in the winter the Seal goes in the

* I was always under the impression that this Seal was rather rare; but, as the return of its capture is not given separately from the former, it is impossible to say accurately.

† This varies a little with latitude, &c.; *e.g.*, this Seal leaves the vicinity of Jakobshavn ice-fjord about the middle of July or beginning of August, and comes back in October very fat. In August and September there are none on that part of the coast.

‡ Rink, *lib. cit.*, et O. Fabricius in Nat. Selsk. Skrift., *l. c.*

opposite direction to that of Spitzbergen, and cannot be seen in the northern parts of Davis's Strait or Baffin's Bay; it is possible therefore, he thinks, that the Seals of Baffin's Bay go in the spring down the west side of Davis's Strait to Newfoundland and Labrador, and supply the bulk of those killed there at that season, that in the winter they cross Davis's Strait and beget their young in that region, and after this cross again to the southern portion of Greenland. One would think that if the Seals came from Spitzbergen there would at this season be great numbers met on the passage round Cape Farewell. At other seasons of the year it is certainly the abundance or otherwise of their food which determines which way the Seal will take. In June the Seals go to feed on Fish up the fjords; but what way they go in July, and where they may be in August, is still a matter of doubt. It is often argued in Greenland that in the "old times" Seals were more numerous than now, and that the great slaughter by the European sealers in Spitzbergen and Newfoundland has lessened their numbers on the shores of Greenland. The worthy Director of Greenland Commerce therefore rejoices that the recent failures of the Seal-hunting in the former localities will have a tendency to again increase their numbers in Davis's Strait and Baffin's Bay, and thereby bring an increase of prosperity to his hyperborean subjects.

Economic Value and Hunting.—To the Greenlander this Seal is of vast importance for its oil, flesh, and hide. One full-grown animal will weigh on an average about 230 lbs., of which the skin and blubber weigh 100 lbs., and the meat 93 lbs., the remainder being the head, blood, and entrails. The edible parts may therefore be said to reach the amount of 100 lbs.; but this weight also includes the bones. The blubber of one at the latter part of the year would probably fill about one-third of a cask, but would not yield over a fourth part of that quantity when the animals return in the spring after procreating. The yearly catch in the Danish settlements is estimated at 36,000. (*Vide* Rink, *l. c.*).

(4.) PHOCA BARBATA, O. Fab.
Callocephalus barbatus, F. Cuv.
Phoca leporina, Lepech.?
Callocephalus leporinus, F. Cuv.

Popular names.—*Hafert skäl* (Swedish)*; *Ajne* (Lapp); *Ursuk* (so written by Fab., but in North Greenland always pronounced *oo-sook*)† (Greenland). It is also called *Takamugak*, and the young *Terkigluk;* but I never heard these terms applied, so they must be rarely used.

What the "great Seals" of Pennant and other authors are has yet to be investigated; they were originally all set down to be this species, but are now generally supposed to belong to the Grey

* Newton (*l. c.*) says that this is the Seal known to the Norse hunters about Spitzbergen as the *Stor-kobbe* (Great Seal), and more frequently as *Blaa-kobbe* (the Blue Seal).

† *Oo-sook* also means blubber. The name may possibly refer to the size or fatness of the animal, and mean "the *big, fat Seal.*"

Seal (*Halichœrus gryphus*). The *Tapvaist* of the western islands of Scotland appears also to belong to that species, *H. gryphus* being a common Seal among the Hebrides.

Descriptive Remarks, &c.—Next to the Walrus this is the largest species of the order found in the northern seas. Perhaps, however, *H. gryphus* may occasionally be found to equal it in size.

Geographical Distribution, &c.—This species has been so often confounded with the Grey Seal (*H. gryphus*) and the Saddleback (*P. grœnlandicus*) in different stages and coats, that it is really very difficult to arrive at anything like a true knowledge of its distribution. At the end of the notice of this species I shall have something to say regarding the probability of its identity with the Ground-Seal of the English Seal-hunters of the Spitzbergen sea. On the coast of Danish Greenland it is principally caught in the district of Julianshaab a little time before the Klapmyds. It is not, however, confined to South Greenland, but is found at the very head of Baffin's Bay, and up the sounds of Lancaster, Eclipse, &c. branching off from the latter sea. The Seals seen by the earlier navigators being nearly always referred in their accounts to either *Phoca vitulina* or *P. grœnlandicus*, it is at present almost impossible to trace its western range; it is, however, much rarer in the north than in the south of Davis's Strait. Accordingly the natives of the former region are obliged to buy the skin from the natives of the more south of settlements, as it is of the utmost value to them. This Seal comes with the pack-ice round Cape Farewell, and is only found on the coast in the spring. Unlike the other Seals, it has no *atluk*, but depends on broken places in the ice; it is generally found among loose broken ice and breaking-up floes.

Economic Value, &c.—This animal is of great importance to the Eskimo; they cut the skin into long strips for harpoon lines—a *sine quâ non* of every *kayak*. Out of every hide can be got four or five lines, and these are cut in a circular form off the animal before it is skinned; after this the lines are dried. These *allunaks* are very strong, and are applied to all sorts of purposes in Greenland travelling. The blubber is more delicate in taste than any other, and is accordingly more prized as a culinary dainty, when such can be afforded. There are only from 400 to 600 caught annually (*Rink, l. c.*).

For long I was puzzled as to what was the "Ground Seal" of the Spitzbergen sealers, but skulls brought me from Spitzbergen in 1869 by Mr. Chas. Edward Smith, surgeon of Mr. Lamont's Expedition, leave no doubt as to their being *Phoca barbata*.

(5.) Halichœrus gryphus, O. Fab.

Phoca gryphus (den krumsnudede sæl), O. Fab.
Halichœrus griseus, Nilss.
Halichœrus gryphus, Nilss.
Phoca gryphus, Licht.
Phoca halichœrus, Thienem.
Phoca thienemanni, Less. (young).
Phoca scopulicola, Thienem. (young, *fide* Gray).

Popular names.—*Grey Seal* (English naturalists); *Graskäl* (or Grey Seal of the Scandinavian naturalists); *Ståtskäl* (Œdm. *l. c.*); *Graskäl* (Swedish); *Sjöskäl*, *Utskärsskäl*, and *Krumnos* (various Scandanavian local names); *Grönfalg?* (Lapps); *Tapvaist?* (western islands of Scotland); *Haaf-fish* (northern islands of Scotland).

General Remarks.—The Grey Seal has no doubt been frequently confounded with other species, particularly *Phoca barbata* and the female of *Pagophilus grœnlandicus.*

It does not seem to frequent the high seas, though possibly this species may be confounded with the "Ground Seal" and some forms of the "Saddleback." It is said to produce on the coast of Sweden in February, and to have one pup at a birth, of a white colour, which attains the dark grey colour of the adult species in about fourteen days.* In 1861, a little south of Disco Island, we killed a Seal the skull of which proved it to be of this species; and again this summer I saw a number of skins in Egedesminde and other settlements about Disco Bay which appeared to be of this species. Though the natives do not seem to have any name for it, the Danish traders with whom I talked were of opinion that the *Graskäl*, with which they were acquainted as an inhabitant of the Cattegat, occasionally visited south and the more southerly northern portions of Greenland with the herds of *Atak* (*P. grœnlandicus*).

The skull to which I refer, though carefully examined at the time, was afterwards accidentally destroyed by a young Polar Bear which formed one of our ship's company on that northern voyage; therefore, though perfectly convinced of its being entitled to be classed as a member of the Greenland fauna, I am not in a position to assert this with more confidence than as being a very strong probability. It should be carefully looked for among the herds of *P. grœnlandicus* when they arrive on the coast. Its hunting forms nowhere an important branch of industry; it is, however, killed on the Scandanavian coasts, at various places, where it is most abundant. A large Grey Seal about eight feet in length will yield (the Swedes say) about 12 lispunds of blubber, equal in value to 36 rigsdaler banco (Swedish); and the hide, which is as large as an ox-hide, will bring the value of such a Seal up to the sum of 60 rigsdaler banco (Swedish).† I have seen and examined this Seal in various collections, and have seen it alive on the coasts of the Cattegat, &c., and among the northern islands of Scotland, but can add nothing of value to the excellent account of Nilsson‡ in his "Skandinavisk Fauna" (Forsta Delen, Däggdjuren, 1847), pp. 298–310.

---

* Capt. McDonald has specimens of a beautiful yellowish-white. It begins to get dark on the snout and flippers within a day or two of birth. It is so abundant in the Hebrides that in one voyage he has killed 70. It is rather rarer on the mainland. (Turner, Journ. Anat. & Phys., 1870; Elwes, Ibis, 1869, p. 25, &c.)

† In the kjökkenmödding of Denmark, in company with remains of *Castor fiber* and *Bos primigenius* are found those of *Halichœrus gryphus*, showing it to have been at one time sufficiently abundant to form part of the food of the primitive inhabitants of Scandinavia.

(6.) Trichechus rosmarus, Linn.

Trichechus rosmarus, Linn.
Rosmarus arcticus, Pall.
Trichechus obesus (et *T. divergens*, Ill., *fide* Gray).
Odobænus rosmarus (L.), Sundeval, Uebers. der Verhandl. der Akad. der Wiss., 1859, p. 441.

Popular names.—*Sea-horse* (English sailors); *Walrus* and *Morse* (Russ., English naturalists and authors); *Hvalross* (Swedish and Danish); *Aavhest* (Sea-horse) and *Rosmar* (Norse); *Morsk* (Lapp); *Awŭk* or *Buvek* (Greenlanders and Eskimo generally): this word is pronounced *āŏŏk* and (like many savage names of animals) is derived from the peculiar sound it utters, a guttural *āŏŏk! āŏŏk!*

General descriptive Remarks.—The general form of the Walrus is familiar enough. However, specimens in museums and the miserably woebegone cubs which have been already twice brought to this country but poorly represent the Walrus in its native haunts. The skin of the forehead (in stuffed specimens) is generally dried to the skull; while in the live animal it is full, and the cheeks tumid. The skin of old animals is generally wrinkled and gnarled. I have seen an old Walrus quite spotted with leprous-looking marks consisting of irregular tubercular-looking white cartilaginous hairless blotches; they appeared to be the cicatrices of wounds inflicted at different times by ice, the claws of the Polar Bear, or met with in the wear and tear of the rough-and-tumble life a Sea-horse must lead in N. lat. 74°. The very circumstantial account of the number of mystacial bristles given in some accounts is most erroneous; they vary in the number of rows and in the number in each row in almost every specimen. They are elevated on minute tubercles, and the spaces between these bristles are covered with downy whitish hair. I have seen several young Walruses in all stages, from birth until approaching the adult stage, and never yet saw them of a black colour, and should have been inclined to look upon the statement that they are so as unfounded, had it not been for the high authority of its author.* All I saw were of the ordinary brown colour, though, like most animals, they get lighter as they grow old. Neither are the muffle, palm, and soles "hairy when young"; in one which I examined before it was able to take the water I saw no difference between it and its mother in this respect. The Walrus appears to cast its nails; for in several which I examined about the same time (viz., in August) most of the nails which had been developed were gone, and young ones beginning to appear. The dentition has been examined by McGillivray (*op. cit.*)†, Rapp, Owen,‡ Flower, Peters,§ and others. In an aged male which I examined at Scott's Inlet, Davis's Strait, August 3, 1861, the small fifth molar on the

---

* Gray, Cat. Seals and Whales in Brit. Mus., 2nd ed., p. 36.

† Bull. Sc. Nat., xvii., p. 280.

‡ Proc. Zool. Soc, 1853, p. 103.

§ Monatsber. der Akad. der Wiss. zu Berlin, Dec. 1864, p. 685; transl. Annals Nat. Hist., xv. (3rd series), p. 355.

right side of the upper jaw still remained, but loose; on the other side the corresponding alveolus was not yet absorbed.*

Shaw (Gen. Zool., i., p. 234) has figured two species of this animal, and inferred their existence principally from the differences in the representations given by Johnston and Cook. Curiously enough, Pontopiddan tells us that the Norwegian fishermen in his day had an idea that there were two species. The whalers declare that the female Walrus is without tusks; I have certainly seen females without them, but again, others with both well developed. In this respect it may be similar to the female Narwhal, which has occasionally no "horn" developed; I do not think, however, that there is more than one species of Walrus in the Arctic regions or elsewhere.

Habits and Food.—On the floes, lying over soundings and shoals, the Walruses often accumulate in immense numbers, and lie huddled upon the ice. More frequently, in Davis's Strait and Baffin's Bay, they are found floating about on pieces of drift ice, in small family parties of six or seven; and I have even seen only one lying asleep on the ice. Whether in large or small parties, one is always on the watch, as was long ago observed by the sagacious Cook; the watch, on the approach of danger, will rouse those next to them; and the alarm being spread, presently the whole herd will be on the *qui vive.* When attacked, unlike the other Seals (unless it be the *Cystophora*), it will not retreat, but boldly meet its enemies. I was one of a party in a boat which harpooned a solitary Walrus asleep on a piece of ice. It immediately dived, but presently arose, and, notwithstanding all our exertions with lance, axe, and rifle, stove in the bows of the boat; indeed, we were only too glad to cut the line adrift and save ourselves on the floe which the Walrus had left, until assistance could reach us. Luckily for us the enraged Morse was magnanimous enough not to attack its chopfallen enemies, but made off grunting indignantly, with a gun-harpoon and new whale-line dangling from its bleeding flanks. Its *atluk* or breathing-hole is cleanly finished, like that of the Seals, but in much thicker ice, and the radiating lines of fracture much more marked.† The food of the Walrus has long been a matter of dispute, some writers, such as Schreber, Fisher, and others, going so far as to deny its being carnivorous at all, because Fisher saw in the stomach of one "long branches of seaweed, *Fucus digitatus*"; and Prof. Bell seems even to doubt whether the small number of grinding-teeth, and more especially their extreme shortness and rounded form, are not rather calculated to bruise the half-pulpy mass of marine

* The anatomy of the Walrus has been described in a beautiful and exhaustive memoir (Trans. Zool. Soc., 1870) by Dr. Murie, F.L.S., F.G.S., an eminent anatomist and zoologist, who has added much to our knowledge of the marine Mammalia.

† There are many interesting details of the habits of the Walrus in Kane's "Arctic Explorations" and "First Grinnel Expedition," in Hayes's "Boat Journey" and "Open Polar Sea," and in Belcher's "Last of the Arctic Voyages."

vegetables than to hold and pierce the Fish's scaly cuirass. I have generally found in its stomach various species of Crustacea, shelled Mollusca, chiefly *Mya truncata* and *Saxicava rugosa*, bivalves very common in the Arctic regions on banks and shoals, and a quantity of green slimy matter which I took to be decomposed Algæ which had accidentally found their way into its stomach through being attached to the shells of the Mollusca of which the food of the Walrus chiefly consists.* I cannot say that I ever saw any vegetable matter in its stomach which could be decided to have been taken in as food, or which could be distinguished as such. As for its not being carnivorous, if further proof were necessary I have only to add that whenever it was killed near where a Whale's carcass had been let adrift its stomach was invariably found *crammed* full of the *krang* or flesh of that Cetacean. As for its not being able to hold the slippery cuirass of a Fish, I fear the distinguished author of "The British "Quadrupeds" (1st ed., p. 287) is in error. The Narwhal, which is even less fitted in its want of dentition for an ichthyophagous existence, lives almost entirely upon Fishes and Cephalopoda. Finally, the *experimentum crucis* has been performed, in the fact that Fish have been taken out of its stomach; and a most trustworthy man, the captain of a Norwegian sealer, has assured me (without possessing any theory on the subject) that he has seen one rise out of the water with a Fish in its mouth. In its stomach I have often seen small stones or gravel; and round its *atluk* considerable quantities are always seen; this is a habit which it possesses in common with *Phoca barbata* and even *Beluga catodon*. These stones may be taken in accidentally, but still they may serve some purpose in its digestive economy.

Next to man, its chief enemy is the Polar Bear. The Eskimo used to tell many tales of their battles; and though I have never been fortunate enough to see any of these scenes, yet I have heard the whalers give most circumstantial accounts of the Walrus drowning the Bear, &c. These accounts may be taken merely for what they are worth; but still this shows that they are not wholly confined to Eskimo fable, and ought therefore not to be hastily thrown aside. There is no doubt, however, that the Bear and the Walrus (like all the Pinnepedia) are but indifferent friends. Another pest I believe I discovered upon this animal for the first time, in 1861, in the shape of two undescribed species of *Hæmatopinus*, one invariably infesting the base of the mystachial bristles, and the other its body. I also found the Seals of Davis's Strait much troubled with another species (*Hæmatopinus phocæ*, Lucas).† I have seen the Walrus *awuking* loudly on the ice, tumbling about, and rushing back from the water to the ice, and from the ice to the water, and then swimming off to another

* In Spitzbergen *Crenella lævigata* constitutes a great portion of its food; the tusks being used to dig it out of the clayey bottom. Torell, in R. S. E. Trans., xxiv., p. 629; and in "Spitzberg. Mollusk.," i., p. 19; also Malmgren, in Wiegmann's Archiv, 1864.

† Proc. Roy. Phys. Soc. Edin., 1863.

piece, and repeating the same operation as if in pain. A few hours afterwards I saw a flock of *Saxicola œnanthe* (it was on a land-floe, close to the Fru Islands) alight on the spot. On going over, I found the ice speckled with one of these species of *Hæmatopinus*, on which the birds had been feeding; and the unfortunate Walrus seems to have been in the throes of clearing itself of these troublesome friends after the approved fashion. Subsequently I have seen these and other small birds alight on the back of the Walrus to peck at these insects, just as crows may be seen sitting on the backs of cattle in our fields. Its tusks it apparently uses to dig up the molluscous food on which it chiefly subsists; and I have seen it also use them to drag up its huge body on to the ice. In moving on shore it aids its clumsy progression by their means.

The Walrus, being an animal of considerable cerebral development, is capable of being readily domesticated. For many years past the Norwegians have frequently brought specimens to different Scandinavian ports; and two have reached England, and survived a short time. More than a century ago one of these animals reached England. De Laet,* quoting from Edward Worst, who saw one of them alive in England which was three months old and had been brought from Novai Semläj, says:—"Every day it was put into water for a short time, but it always seemed happy to return to dry ground. It was about the size of a calf, and could open and shut its nostrils at pleasure. It grunted like a wild Boar, and sometimes cried with a strong deep voice. It was fed with oats and millet, which it rather sucked in than masticated. It was not without difficulty that it approached its master; but it attempted to follow him, especially when it had the prospect of receiving nourishment at his hand." Its naturalisation in our Zoological Gardens having therefore become a subject of considerable interest, I cannot better conclude these notes on the habits of the Walrus than by describing a young one I saw on board a ship in Davis's Strait, in 1861, and which, had it survived, was intended for the Zoological Society.

It was caught near the Duck Islands off the coast of North Greenland, and at the same time its mother was killed; it was then sucking, and too young to take the water, so that it fell an easy prey to its captors. It could only have been pupped a very few hours. It was then 3 feet in length, but already the canine tusks were beginning to cut the gums. When I first saw it, it was grunting about the deck, sucking a piece of its mother's blubber, or sucking the skin, which lay on deck, at the place where the teats were. It was subsequently fed on oatmeal and water and pea-soup, and seemed to thrive upon this *outré* nourishment. No fish could be got for it; and the only animal food which it obtained was a little freshened beef or pork, or Bear's flesh, which it readily ate. It had its likes and dislikes, and its favourites on board, whom it instantly recognised. It became exceedingly

* "Description des Indes Occidentales," *apud* Buffon.

irritated if a newspaper was shaken in its face, when it would run open-mouthed all over the deck after the perpetrator of this literary outrage. When a "fall"* was called it would immediately run at a clumsy rate (about one and a half or two miles an hour), first into the surgeon's cabin, then into the captain's (being on a level with the quarterdeck), apparently to see if they were up, and then out again, grunting all about the deck in a most excited manner "*awuk! awuk!*" When the men were "sallying,"† it would imitate the operation, though clumsily, rarely managing to get more than its own length before it required to turn again. It lay during the day basking in the sun, lazily tossing its flippers in the air, and appeared perfectly at home and not at all inclined to change its condition. One day the captain tried it in the water for the first time; but it was quite awkward and got under the floe, whence it was unable to extricate itself, until, guided by its piteous "*awuking*," its master went out on the ice and called it by name, when it immediately came out from under the ice, and was, to its great joy, safely assisted on board again, apparently heartily sick of its mother element. After surviving for more than three months, it died, just before the vessel left for England. As I was not near at the time, I was unable to make a dissection in order to learn the cause of death.

Regarding the debated subject of the attitude of the Walrus‡ I am not in a position to say more than my own notes taken at the time will allow of; I saw none last summer, and I am afraid to trust to a treacherous memory on such a matter. The entries in my diary, however, are explicit enough on the point so far as relates to this young individual; and I presume that its habits are to be taken as a criterion of those of the old one. When asleep in the cask which served it for a kennel, it lay with both fore and hind flippers extended. When *walking* it moved like any other quadruped, but with its *hind flippers heel first*, the *fore flippers* moving in the ordinary way, *toes first.* I am aware that this is in contradiction to the observations of an eminent zoologist; I, however, merely copy what was expressly noted down at the time. It ought also to be mentioned that, in the excellent figures of the Walrus taken by the artist of the Swedish Expedition to Spitzbergen,§ under the direction of such well-informed naturalists as Torell, Malmgren, Smitt, Goës, Blomstrand, &c., the fore flippers are represented as rather doubled back, and the hind flippers extended.

Geographical Distribution.—The Walrus is an animal essentially of the coast, and not of the high seas. Whenever it is found at

* When a boat gets "fast" to a Whale, all the rest of the crew run shouting about the decks, as they get the other boats out, "a fall! a fall!" It is apparently derived from the Dutch word "Val," a Whale.

† When a ship gets impeded by loose ice gathering around it, the crew rush in a body from side to side so as to loosen it, by swaying the vessel from beam to beam. This is called "sallying the ship."

‡ Gray, Proc. Zool. Soc., 1853, p. 112.

§ *Lib. cit.*, facing p. 169 (chromo-lithograph), and head, p. 308, both drawn by Herr von Yhlen.

any distance from land it is almost always on shoals, where it can obtain the Mollusca which form the bulk of its food. The Seal-hunters never see it, nor is it found among the flocks of Seals on the Spitzbergen and Jan Mayen pack-ice. It is found all along the circumpolar shores of Asia, America, and Europe, sometimes extending into the subpolar, and even stragglers find their way into the temperate, regions of America, Asia, and Europe. It is not unlikely that it may even be found in the Antartic regions. On the north-west coast of America I have known it to come as far south as 50° N. lat. The Indians along the shores of Alaska (lately Russian America) carve the teeth into many fanciful ornaments; * but we should be liable to fall into an error from seeing these teeth among the natives so far south, if we did not know that they are bartered from the more northern tribes. On the American Atlantic seaboard the Walrus comes as far south as the Gulf of St. Lawrence, and stragglers even further. In Lord Shuldham's day they assembled on the Magdalene Islands in that gulf to the number of 7,000 or 8,000; and sometimes as many as 1,600 were killed (or rather slaughtered) at one onset by the hunters who pursued them.† It has been killed several times on the British coast; and I suspect that it is not an unfrequent visitor to our less-frequented shores. Perhaps not a few of the "Sea-horses" and 'Sea-cows" which every now and again terrify the fishermen on the shores of the wild western Scottish lochs, and get embalmed among their folklore, may be the Walrus. In addition to those already recorded I know of one which was seen in Orkney in 1857, and another the Shetland fishermen told me had been seen in the Nor' Isles about the same time. One was killed on East Heiskar, Hebrides, by Capt McDonald, R.N., in April 1841; and another in the River Severn in 1839 ("Edin. Journ. Nat. Phys. "Sciences," 1839-40). There is, however, some ground for believing that at one time it was, if not a regular member of our fauna, at least a very frequent visitor. Hector Boece (or Boethius, as his name has been Latinised), in his quaint "Cronikles "of Scotland," mentions it towards the end of the fifteenth century as one of the regular inhabitants of our shores; and old Roman historians describe the horse-gear and arms of the ancient Britons as ornamented with bright polished ivory. It is difficult to suppose that this could have been anything else but the carved tusks of the Walrus. It is not, however, without the bounds of possibility that this might have been some of the African Elephants' ivory which the Phœnician traders bartered for tin with the natives of the Cassiterides. Except for its occasional movements from one portion of its feeding-ground to the other, the Walrus cannot be classed among the migratory animals. In

* My friend Mr. A. G. Dallas, late Governor-General of the Hudson's Bay Company's territories, has a bust of himself beautifully carved out of a Walrus-tooth, by a Tsimpshean Indian at Fort Simpson, B.C.

† Phil. Trans., lxv., pl. 1, p. 249, &c. *Apud* Pennant, "Arctic Zoology," p. 148-50.

Greenland it is found all the year round, but not south of Rifkol, in lat. 65°. In an inlet called Irsortok it collects in considerable numbers, to the terror of the natives who have to pass that way; and not unfrequently kayakers who have gone "express" have to return again, being afraid of the threatening aspect of "Awuk." A voyager has well remarked that "*Awuk*" is the lion of the Danish Eskimo; they always speak of him with the most profound respect! It has been found as far north as the Eskimo live or explorers have gone. On the western shores of Davis's Strait, it is not uncommon about Pond's, Scott's, and Home Bays, and is killed in considerable numbers by the natives. It is not now found in such numbers as it once was; and no reasonable man who sees the slaughter to which it is subject in Spitzbergen and elsewhere can doubt that its days are numbered. It has already become extinct in several places where it was once common. Its utter extinction is a foregone conclusion. Von Baer has studied its distribution in the Arctic sea; and, so far as they go, his memoir and map may be relied on; both, however, require considerable modifications.*

Economic Value and Hunting.—The ivory tusks of the Walrus always command a good price in the market; and the hides are held in high value as an article of commerce; they are used as material for defending the yards and rigging of ships from chafing. It is also occasionally used for strong bands in various machinery, carriage-making, &c. The flesh tastes something like coarse beef. The whalers rarely or ever use it, having a strong prejudice against it in common with that of Seals and Whales. The Walrus-hunters in Spitzbergen almost exist upon it; and the Eskimo high up in Smith's Sound look upon it as their staple article of food. The American explorers who wintered there soon acquired a liking for it. Accordingly the "Morsk" has been hunted in northern regions from a very early period. The Icelandic Sagas (such as the *Speculum regale*, &c.) speak of it as *Rostungur*; and there is said to be a letter in the library of the Vatican proving that the old Icelandic colonists in Greenland paid their "Peter's Pence" in the shape of Walrus-tusks and hides. However, in 890, as far back as the days of King Alfred of England, Œthere, "the old sea-captain who dwelt "in Helegoland," gave a most circumstantial account to that monarch (who wrote it down in his edition of the *Hormista* of Paulus Orosius†) of slaying, he and his six companions, no less than "three score Horse-whales" in three days.‡ At the present period it is principally captured in Spitzbergen by Russian and

* Mémoires de l'Académie de St.-Pétersbourg, t. iv., p. 97, t. 4 (1836).

† *See* Daines Barrington's Translation (1775), p. 9; and other editions.

‡ This statement need not be doubted when we read how, in 1852, 16 men with lances killed in a few hours 900 out of a herd of 3,000 or 4,000 lying on an island off Spitzbergen: Lamont, Quart. Journ. Geol. Soc., xvi., p. 483. Martens' "Spitzbergen," p. 182, tells us that in 1608 Walruses were huddled together in such numbers on Cherry Island, south of Spitzbergen, that a ship's crew killed above 900 in seven hours.

Norwegian hunters, who visit that island for the purpose. In Danish Greenland, though it was once so abundant that the principal article of trade with Europe, in the days of Erik Raude's colonists, was the tusks of this animal, it may be said nowadays, so far as its hunting or commercial value is concerned, to be extinct. There are never more than a few killed yearly, and it frequently happens that a year passes without any at all being killed within the limits of the Danish trading-posts. It is more than probable that they never were abundant in South Greenland, but that the old colonists went north in pursuit of them. From the Runic column found on the island of Kingatarsoak in 73° N. lat., we know that these enterprising rovers did sail far north; and it is more than reasonable to suppose that it was on one of these Walrus-hunting expeditions that this monument was erected. Indeed so few are now killed in Danish Greenland (whether through degeneracy of the hunters or scarcity of the Walrus it is scarcely worth inquiring too closely) that as, notwithstanding all the appliances of European civilisation now accessible to the natives, ivory cannot be dispensed with in the manufacture of Eskimo implements of the chase, its tusks have sometimes to be reimported from Europe into Greenland. North of the glaciers of Melville Bay, the hardy Arctic highlanders, aided by no *kayak* or rifle, but with a manly self-reliance, enfeebled by no bastard civilisation engrafted upon their pristine savagedom, with their harpoon and *allunaks* still boldly attack the Walrus as he lies huddled upon the ice-foot; and thereby the native supplies to his family the food and light which make tolerable the darkness of the long Arctic night of Smith's Sound. The whalers kill a few annually, striking them, as they do the Whale, with the gun-harpoon, and killing them with steel lances*; but even then it is dangerous work, and not unfrequently brings the hunter to grief. I have been one of a party who have killed several in this manner, and have also seen them captured by the wild Eskimo at Pond's Bay, on the western shores of Davis's Strait, after the aboriginal fashion; but as this has been excellently described by Kane† and Hayes‡ in their different narratives, I will not trouble you with any details. The Swedish expedition to Spitzbergen,§ and Lord Dufferin|| and Mr. Lamont,¶ have given many particulars of its capture by the Spitzbergen hunters. Baron Wrangell** has supplied an account of its chase on different portions of the Siberian coasts; and Nilsson,†† Keilhau,‡‡ and Malmgren §§

* The ordinary rifle is of comparatively little use in hunting this monster Seal. Musket balls will scarcely affect their pachydermatous sides; and I have often seen leaden balls flattened on their skulls. I have more than once seen it snap a steel lance in two with its powerful molars.

† "Arctic Explorations."

‡ "The Open Polar Sea," and "An Arctic Boat Voyage."

§ "Svenska Expeditionen til Spetsbergen år 1861," &c., pp. 168–182.

|| "Letters from High Latitudes."

¶ "Seasons with the Sea-horses."

** Nordküste von Sibirien, ii., pp. 319, 320.

†† *Lib. cit.*, i., pp, 320–325.

‡‡ "Reise i Ost-og Vest-Finnmarken, &c., pp. 146–149."

§§ Wiegmann's Archiv, v., 1864.

complete the list of the principal writers regarding its hunting and commercial importance generally.

(7.) CYSTOPHORA CRISTATA (Erxleb.), Nilss.

Phoca cristata, Erxleb.
Phoca leonina, O. Fab. (non Linn.).
Phoca mitrata, Milbert (Cuv.).
Phoca leucopla, Thienem.
Phoca cucullata, Bodd.
Phoca dimidiata, Cretzsch. (*fide* Rüpp.).
Phoca isidorei, Less.
Mirounga cristata, Gray.
Cystophora cristata, Nilss.
Cystophora borealis, Nilss.
Stemmatopus cristatus, F. Cuv.
Stemmatopus mitratus, Gray.

Popular names.—"*Bladdernose*" or, shortly, "Bladder" (of northern sealers, Spitzbergen sea); *Klappmysta* (Swedish); *Klakkekal*, *Kabbutskobbe* (Northern Norse); *Kiknebb* (Finnish); *Avjor*, *Fatte-Nuorjo*, and *Oaado* (Lapp); *Klapmyds* (Danish; hence Egede, Greenl., p. 46: the word *Klapmyssen*, used by him on page 62 of the same work, Engl. trans., and supposed by some commentators to be another name, means only *the* Klapmyds, according to the Danish orthography): *Klapmütze* (German; hence Cranz, Greenl., i., p. 125: I have also occasionally heard the English sealers call it by this name, apparently learnt from the Dutch and German sailors). All of these words mean the "Seal with a *cap* on," and are derived from the Dutch, who style the frontal appendage of this species a *mutz* or cap, hence the Scotch *mutch*. This prominent characteristic of the Seal is also commemorated in various popular names certain writers have applied to it, such as *Blas-Skäl* (Bladder-Seal) by Nilsson (Skand. Faun., i., p. 312), *Hooded Seal* by Pennant (Synopsis, p. 342), *Seal with a caul* by Ellis (Hudson Bay, p. 134), in the French vernacular *Phoque à capuchon*, and in the sealers' name of *Bladdernose*, ♂ *Neitersoak*, ♀ *Nesaursalik* (Greenland), and *Kakortak* (when two years old).

Descriptive Remarks.—This is one of the largest Seals in Greenland, and in its adult state is at once distinguished by the curious bladder-like appendage to its forehead, which is connected with the nostrils and can be blown up at will.* This has been well described by Dr. Dekay in the "Annals of the Lyceum of Natural History of New York," vol. i.; and with his observations I perfectly agree. The eye of this Seal is large, and of a glassy black colour with a dark-brown iris. It has, like all the family, no external auricle; and the orifice of the ear is very small. The body is long and robust; its colour on the upper or dorsal aspect is dark chestnut or black, with a greater or less number of round or oval markings of a still deeper hue. The hair is long and somewhat erect, and the thick fur-like coating next the skin is often tinged with a reddish coppery colour. The head and flippers are of the same dark chestnut-colour. The pectoral and ventral

* It is often asserted by the sealers that this "bladder" is a sexual mark, and is not found on the female. I do not think there is any just ground for this belief.

regions are of the same dark-grey or tarnished-silvery hue which has been described in the *P. grœnlandicus.*

Habits, &c.—The Bladdernose is not only one of the largest, but the fiercest of the northern Seals; and, as its capture requires some skill, it is only the most expert *kayaker* that can procure any. It will chase a man and bite him, besides making a great commotion in the water. Therefore the hunt is very dangerous to a man in such a frail craft as the Greenland *kayak;* but as long as the memory of the oldest inhabitant of South Greenland extends, only one man in the district of Julianshaab (where they are chiefly captured) has been killed by the bite of the Klapmyds, though not unfrequently the harpoon and line have been broken. The hunting is not so dangerous, however, within late years, as it has been effected by the rifle from the ice; but when the Seal has not been killed outright, the hunter goes out in his kayak and despatches it with the lance. Like all Seals, during the rutting-time, there are great battles on the ice between the males; and the roaring is said to be sometimes so loud that it can be heard four miles off. The skin is often full of scratches from these fights.

With regard to the favourite localities of this species of Seal, Cranz and the much more accurate Fabricius disagree—the former affirming that they are found mostly on great ice islands where they sleep in an unguarded manner, while the latter states that they delight in the high seas, visiting the land in April, May, and June. This appears contradictory and confusing; but in reality both authors are right, though not in an exclusive sense. The hood appears to be an organ of defence from any stunning blow on the nose, the most vulnerable part in a Seal. It only inflates this bladder when irritated. The sailors look upon it as a reservoir of air when under the water.* The story which Fabricius relates about its "shedding tears abundantly" when surprised by the hunter is, I suspect, only an Eskimo tale of wonder. I could find no one credulous enough to believe it; nor during the whole time I passed among the Seal-hunters of the far north did I find anyone esteem my credulity great enough to venture any such story on me.

It is affirmed, curiously enough, that the *Bladdernose* and the *Saddleback* are rarely or ever found together; they are said to disagree. At all events the latter is generally found on the inside of the pack, while the former is on the outside. The latter is also much more common than the Bladdernose,

Procreation and Young.—At first the young Bladdernose is pure white: during the first year, as it grows older and increases in size, a grey tinge appears, and gradually it assumes a deeper and

* Mr. J. Walker, then master of the screw-steamer "Wildfire" (now of the "Erik"), and one of the most intelligent of the whaling captains, assured me (June 1861), from his own observation, that this Seal lies frequently on the top of elevated pieces of ice, and that the use of this hood, or "bladder," appears to be to raise it up with sufficient momentum to the surface (by filling it with air) so as to spring again on to the ice.

deeper hue of the same colour. . . . .
The majority of the "Bladdernoses" which I have seen were about two or three years old, and were apparently, by a slow and gradual change, becoming similar to the old and mature Seals, by turning darker and darker in their colours, and assuming the roundish oval markings, while at the same time they were increasing in size. This species seems to produce its young earlier than *P. grœnlandicus.*

Geographical Distribution and Migrations.—The Bladdernose is found all over the Greenland seas, from Iceland to Greenland and Spitzbergen, but chiefly in the more southern parts. The first Seals which we saw and killed on making the ice early in March 1861, were chiefly young "Bladders" which had not yet got the hood-like appendage. It even finds its way to the temperate shores of Europe and America, and rare stragglers now and then land on the shores of Britain, though it is by no means a member of our fauna proper. This Seal is not common anywhere. On the shores of Greenland it is chiefly found beside large fields of ice, and comes to the coast, as was remarked by Fabricius long ago, at certain times of the year. They are chiefly found in South Greenland, though it is erroneous to say that they are exclusively confined to that section. I have seen them not uncommonly about Disco Bay, and have killed them in Melville Bay, in the most northerly portion of Baffin's Bay. They are principally killed in the district of Julianshaab, and then almost solely in the most southern part, on the outermost islands, from about the 20th of May to the last of June; but in this short time they supply a great portion of the food of the natives and form a third of the colony's yearly production. In the beginning of July the Klapmyds leaves, but returns in August, when it is much emaciated. Then begins what the Danes in Greenland call the *maigre Klapmydse fangst*, or the "lean-Klapmyds-catching," which lasts from three to four weeks. Very seldom is a Klapmyds to be got at other places, and especially at other times. The natives call a Klapmyds found single up a fjord by the name of *Nerimartont*, the meaning of which is "gone after food." They regularly frequent some small islands not far from Julianshaab, when a good number are caught. After this they go further north, but are lost sight of, and it is not known where they go to (Rink, *l. c.*). Those seen in North Greenland are mere stragglers, wandering from the herd, and are not a continuation of the migrating flocks. Johannes (a very knowing man of Jakobshavn) informed me that generally about the 12th of July a few are killed in Jakobshavn Bay (lat. 69° 13′ N.) It is more pelagic in its habits than the other Seals, with the exception of the Saddleback.

Economic Value and Hunt.—The Klapmyds yields, on the average, half a cask of blubber, and the dried meat of every Seal weighs about 24 Danish lbs.; but this is not the whole Seal, which weighs about 200 lbs. The yearly catch in Greenland (Danish) is about 2,000 or 3,000 (Rink, *l. c.*).

5. *Commercial Importance of the "Seal Fisheries."*

The Greenland (*i.e.* Spitzbergen) sealing fleet from the British ports meets about the end of February in Bressa Sound off Lerwick, in Zetland; it leaves for the north about the first week in March, and generally arrives at the ice in the early part of that month. The vessels then begin to make observations for the purpose of finding the locus of the Seals, and this they do by crawling along the edge of the ice, and occasionally penetrating as far as possible between 70° and 73° N. lat.; then continue sailing about until they find them, which they generally do about the first week of April. If they do not get access to them, they remain until early in May, when, if they intend to pursue the whaling in the Spitzbergen sea that summer, they go north to about 74° N. lat. to the "old sealing," or further still (even to 81° N.) to the whaling. Most of them however, if not successful by the middle of April, leave for home to complete their supplies in order to be off by the first of May to the Davis's Strait Whale fishery. During the month of March and the early part of April, the sealers are subject to all vicissitudes of weather, calm and storm suddenly alternating, while the thermometer will stand for weeks at zero, or even many degrees below it.

The number of Seals taken yearly by the British and Continental ships (principally Norse, Dutch and German) in the Greenland sea, when they get among them, will average upwards of 200,000, the great bulk of which are young "Saddlebacks," or, in the language of the sealer, "white-coats." When they have arrived at their maximum quality, 80 generally yield a tun of oil, otherwise the general average is about 100 to the tun. In 1859 good oil sold for about 33*l.* per tun; add to this the value of 100 skins at 5*s.* each, and the whole will amount to 58*l.* sterling. From this simple calculation a very good estimate may be formed of the annual commercial value of the Greenland "Seal fishery," for, supposing 2,000 tuns of oil to be about the annual produce, and assuming 58*l.* as the value per tun, inclusive of the skins, the whole produce of the fishing will amount to the yearly value of 116,000*l.* sterling (*Wallace*). This, of course, does not take into calculation the produce the Danish Government derive from their colonies on the west coast of Greenland (which I notice under the head of each Seal), nor what the Russians derive from the coast of Spitzbergen and from the White Sea. The "fishery," however, is very precarious. Some years little or nothing is got, the ice being too thick for the ships to "get in to them." In one year it may happen that the fishery in the Spitzbergen sea proves a failure, while the Newfoundland one is successful. For some years past it has proved in the former sea almost a failure.*

* It has been rather more successful in Newfoundland. This year (1868), up to the 28th of April, 25,000 Seals had arrived at St. John and Harbour Grace. *See* a good account of the sealing by the continental vessels in Petermann's "Geograph. Mittheil.," Feb. 1868. In 1866 the steamer "Camperdown" obtained the enormous number of 22,000 Seals in nine days. It is nothing uncommon for a ship's crew to club or shoot in one day as many as from 500 to 800 old Seals, with 2,000 young ones.

There seems indeed little doubt that the fishery must fail in course of time, as have the Seal and Whale fisheries in some other parts of the world, and if Seal-hunting is pursued with the energy it is at present, that day cannot be far distant. Some of the sealers laugh at this idea; but where is the enormous produce the South Seas used to yield, superior to anything ever heard of in the North? No doubt the South Sea hunters said the same thing, and doubtless when the inhabitants of Smeerenberg, that strangest of all strange villages, saw the Whales sporting by thousands in their bays, and the oil-boilers steaming above the peaks of Spitzbergen, they laughed at the idea of their ever becoming scarce! Yet how true that idea has proved! For in our day the waters of those high northern seas are rarely troubled, even by a wandering Mysticete, that perchance may have missed its way in making a passage from one secure retreat to another. So will it ultimately be with the Seals. Indeed, some are even now of opinion that they are diminishing in numbers; at least, they have evidently reached their zenith, as shown by statistics; and, taking into consideration the appearance the young Seals presented on the ice in 1861, they did not approach the numbers reported to have been seen by sealers in many previous years. The South Sea "fisheries" became extinct in 15 years, and, making all allowance for the protection afforded to the Greenland Seals by the ice, and supposing the sealing prosecuted with the same vigour as at present, I have little hesitation in stating my opinion that before 30 years shall have passed away the Seal fishery, as a source of commercial revenue, will have come to a close, and the progeny of the immense number of Seals now swimming about in the Greenland waters will number but comparatively few. This event will then form another era in the northern fisheries.*

* History of the Dundee sealers:—

1865		4 vessels	-	63,000 Seals.	
1866	-	7 „	-	58,000 „	
1867	-	11 „	-	56,000 „	
1868	-	12 „	-	16,670 „	
1869	-	11 „	-	45,600 „	
1870	-	9 „	-	90,450 „	
1871	-	9 „	-	62,000 „	up to the 11th of April.

The "Arctic" had 15,000, and the "Esquimaux" 14,330. The St. John's Newfoundland sealers had at about the same time 231,000 Seals, making an average of 21,000 each, the largest for many seasons. Most of them made two trips. From the first trip the average profit of each man was 300*l*.

See also Mr. Yeaman's Notes on the Dundee Seal and Whale Fishery, Report Brit. Assoc., 1867, Trans. of Sect., p. 148.

IV.—On the History and Geographical Relations of the Cetacea frequenting Davis Strait and Baffin's Bay. By Dr. Robert Brown, F.L.S., F.R.G.S.

[Reprinted, by Permission, from the Proc. Zool. Soc., 1868, No. XXXV., pp. 533–556, with corrections and annotations by the Author, March 1875.]

I conclude these papers on the Mammalia of Greenland and adjoining seas by a few notes on the order more intimately associated in popular imagination with the Arctic regions than any other, viz. the Cetacea. Though much more imperfectly known than any other group, yet my observations on them will be more brief than on the other Mammals, and for the same reason which has conduced to the present state of Cetology, viz. the want of opportunities of examining the species. These remarks will therefore necessarily consist of a statement of the geographical range and migrations, and a description of the habits of the better known, and a list of the species, and whatever information can be collected on these points regarding the others only known by skeletons or remains in museums. These I have examined carefully; and the synonymy given is the result of that study, coupled with investigations made in Greenland. With the exception of a few of the more common, such as *Phocæna communis*, *Beluga catodon*, &c., I have not had an opportunity of examining, otherwise than in the manner indicated, most of the species. I have, however, examined at different times above thirty specimens of *Balæna mysticetus*, and many of *Monodon monoceros*; and to these descriptions I have appended various observations derived from my own examination and without reference to other published descriptions, which have in nearly every case been derived from the examination of fœtal specimens or isolated individuals, conveying but an imperfect idea of the species. What I said in another memoir equally applies here, viz. that the descriptions are not given as complete, but merely as fragments of a *mémoire pour servir*. Those who have attempted the examination of any member of the group Cetacea, and still more those whose lot has been to examine with frozen fingers (plunged every now and again into the warm blood of his *subject*) such an unwieldy object on a swaying ice-floe, will appreciate the difficulty of drawing up such descriptions; and to them no apology is necessary for their imperfection. The absolute necessity of recording every description of the members of this order however, apparently well known, must be my excuse for presenting these notes in such a disjointed state. In the original draft I had mentioned various particulars now omitted—the recent reproduction by the Ray Society of the admirable memoirs of Professors Eschricht, Reinhardt, and Lilljeborg rendering their publication unnecessary.

1. BALÆNA MYSTICETUS, Linn.

(α) Popular names.—*Greenland Whale, Right Whale, Common Whale* (English authors); *Whale, Whale-fish*, and "*Fish*" (English whalers). The young are denominated *suckers*, and are also sometimes known by the following names :—*Shortheads* (as long as they continue suckling); *Stunts* (two years); *Skull-fish* (after this stage or until they become *Size-fish*, when the longest splint of whalebone reaches the length of six feet); *Tue qval?* (Norse); *Rhetval* (Danish); *Arbek, Argvek, Arbavik, Sokalik* (Greenland); *Akbek, Akbeelik* pl. (Eskimo of western shores of Davis Strait); I have also heard both the Greenlanders and western Eskimo call it *puma*, but I cannot learn what is the origin of this word, and suspect it to be *whaler*,—a corrupted jargon of Scotch, English, Danish, and Eskimo, joined with some words which seem to belong to no language at all, but to have originated in a misconception on either side, and to have retained their place under the notion that each party was speaking the other's language, something of the nature of the *Lingua Franca* of the Mediterranean, the *Pigeon English* of China, and the *Chinook jargon* of North-west America.

(β) *Descriptive Remarks.*—The lower surface of the head is of a cream-colour, with about half a foot of blackish or ash-colour at the tip (or what corresponds in the higher orders of Mammals to the symphysis) of the lower jaw; further back the colour shades into the general dark blue colour of the body. This colour is generally almost black in adults, but in young ones (or "suckers") it is lightish blue; hence the whalers sometimes call these "blue-skins." The whiskers consist of nine or ten short rows of bristles, the longest bristles anteriorly. There are also a few bristles on the apices of both jaws, and a few hairs stretching all along the side of the head for a few feet backwards. On the tip of the nose are two or three rows of very short white hairs, with fewer hairs in the anterior rows, more in the posterior. I have reason to believe that some of these hairs are deciduous, as I have often found them wanting in old individuals. In older Whales the darker colour of the body impinges on the under surface of the head, leaving the ordinary white of the suckers merely in the form of several irregular blotches, but with two (regular ?) spots, one on each side of the jaw immediately posterior to the eye, composed of a hard cartilaginous material. There is also a little white on the eyelids, and some irregular white markings on the root of the tail. There is likewise a white colour all around the vulva and mammæ. Some individuals may be found quite white on the belly, others piebald, and others with white spots on various portions of the body not mentioned. The presence or absence of a particular white marking on a specimen of a Cetacean under examination ought by no means to be received (as has been done) as a proof that the species is different, or that because such is mentioned in a former description such description is erroneous, because this is one of the most varying characters possessed by the order.* The inside of the mouth inferiorly, where the tongue is not attached, is of a pale blue colour. The tongue is broader posteriorly, and narrowed anteriorly, paler blue than the rest of the mouth, and pale blue all round the edges and where not carnation, which colour prevails in the form of a streak down the mouth of a deep sulcus on the middle and anterior portion of the tongue, terminating irregularly about two feet from the root of the tongue. The contour of the tongue is entire throughout. The substance of the tongue is a fibrous blubber containing very little oil. There are numerous small linear muscles interspersed through the lower part. The roof of the mouth, on each side of the gum, is a continuous curve, broadest anteriorly, pale blue, sides pale blue and carnation mixed. The upper lip is very much smaller than the under. The lips are furrowed immediately behind the edge and bevilled, and are all deep black and speckled. No traces exist of either eyebrows or eyelashes. The eye is very small and hollow, measuring from canthus to canthus 3½ inches (in adult), and 1½ inch deep, with a deep furrow superiorly and inferiorly immediately above and below the eye. The inside of the eyelid is red. The aperture of the auricular canal is difficult to find, and is not larger than the diameter of a goose-quill. The laminæ or "splits" of whalebone are longest in the middle, but grow much shorter posteriorly to this "size-split." The number of laminæ is about 360 on each side. The whalers have a notion that there is a lamina for each day in the year; but this, like the idea that Jonah's face can be seen on the nose of the Whale, is, I am afraid, a rather hasty generalization. Each lamina ends in a tuft of hair, this tuft being continuous with the hair on the inside of the bone, this "hair" again being composed of identically the same substance as the whalebone itself. The outside of the bone is smooth, pale blue-coloured, with the edges overlapping, the free edges pointing posteriorly, but with an interval (varying according to the age of the animal) between the laminæ of so very regular a character that each lamina can be seen and even counted from the outside. Where the bone is placed in the gum it is of a greyish-white colour, and on exposure to air becomes black; all of the portions of the bone most exposed are of a blackish colour. On the outside of the laminæ, a few inches from the end, is a transverse wave or ridge, continuous in a slightly elevated ridge across the whole of the laminæ; and in old Whales there are several of these wavy transverse ridges, which are apparently in some way connected with its growth. The best whalebone has several of these ridges. Interiorly in front of the place where each lamina is inserted into the gum, are several rows of short stumps of whalebone terminated by a tuft, and before these again short white hair laminæ graduating into

* The colour also varies with the age, the back of some being black, of others black and white, and some are all white. Some old Whales are said to have a broad white stripe over their back down to the belly (Laing's "Voyage to Spitzbergen," p. 126; 1815). I cannot confirm this from my own observation.

a velvet-like substance in the mouth. It is said that the laminæ, after once being produced, do not increase in number, but that the interspaces of the laminæ increase in width. This interspace in adult Whales is from about half an inch to one inch in width. Occasionally two splits are found growing together in the gum, but separate below. The length of the whalebone depends, it is said, on the size on the head, and bears no ratio to the length of the body. Occasionally a long Whale has small and short whalebone, whilst a short dumpy individual (for there are individual differences in these as in all other animals, not referable to specific difference) may have much longer. The longest lamina of whalebone which I have heard of being obtained was 14 feet. I have personally known of another 13 feet 3 inches long; but the average length is 12 feet and under. This is the middle split already spoken of, known to the whalers as the "size-split;" but in the measurement of this the tuft of "hair," which sometimes reaches six or seven inches in length, is not included—a very important matter, as much depends upon the size of this split. The breadth and thickness of the laminæ depend upon the age of the animal. It is a common belief that the laminæ of whalebone in the female Whale are broader but shorter than in the male. The colour of the whalebone likewise varies; in the young the laminæ are frequently striped green and black, but in the old animal they are frequently altogether black; often some of the laminæ are striped in alternate streaks of black and white, whilst others want this variegation. Whalebone is said to be occasionally found white, without the animal differing in the slightest degree. That bought from the western Eskimo in the spring is often whitish, because they have kept it lying about or steeped in water all the winter. It also does not necessarily follow that because one whalebone brings a different price from another, the animals that produce them are of different species. For instance, the whalebone brought by the American whalers from Kemisoak (Cumberland Sound, or Hogarth's Sound of Penny) used to bring a less price in the market than that of the English whalers from Davis Strait, Baffin's Bay, and Spitzbergen, because it had lain exposed during the winter and was accordingly worse prepared; therefore, without at all underrating the importance of pressing every point into our service in discriminating the different species of *Balænidæ*, as the whalebone is subject to so much variation, and undergoes so many artificial changes before coming into the hands of the zoologist, I think that we must proceed with the utmost caution in forming species on the mere differences presented by isolated laminæ of whalebone.*

---

* Of late years whalebone has been bringing a better price than formerly, new uses for it having been discovered. A large amount is now used to stiffen silks by being woven into the fabric. By an old feudal law the *tail* of all Whales belonged to the Queen, as a perquisite to furnish Her Majesty's wardrobe with whalebone (Blackstone's Commentaries, vol. i. p. 233, ed. 1783). In commercial parlance whalebone is called "whale-fins."

The *pectoral fins* (or, more properly, swimming-paws*) are of a darkish grey at the axilla, rounded superiorly and bevelled off inferiorily. The upper edge is arcuate in form, with a slight angularity medially; the inferior edge with the outline in a gentle sigmoid curve, with the greater convexity of the curve anteriorly. The *caudal extremity*, if not the homologue, is undoubtedly the analogue of the posterior extremities in other Mammals. It is almost unnecessary to say that the substance of the tail is non-muscular, though it has been described as such in various publications, the only power which it possesses being derived from the attachment of some of the lumbar and other muscles in the extremity of the vertebral column. A transverse section of the root of the tail shows:—1, the epidermis; 2, the soft skin; 3, the blubber, or a cellular substance containing fat-cells; 4, cartilage enveloping the tendinous cells; 5, strong muscular fasciæ, through which the tendons play; 6, spinal canal and vessels; 7, spinal cartilages; 8, blood-vessels; and, 9, synovial glands. A transverse section of the tail shows skin, blubber, tendinous envelope, blood-vessels, and a central cartilaginous mass†. Though, *per se*, the tail has no power, yet as the instrument through which the lumbar muscles (the tendinous attachments of which seem to be prolonged into the cartilaginous substance of the tail) work it exerts enormous force. The figure usually engraved in boys' books of sea adventures, and copied from Scoresby's "Account of the Arctic Regions," of a Whale tossing a boat and its crew up into the air, is generally looked upon by all the whalers to whom I have shown it as an artistic exaggeration. Accidents of this nature are very rare, and never proceed to such an extent; and I have no doubt that Dr. Scoresby's artist has taken liberties with his description, that worthy navigator being himself above any suspicion of exaggeration for the sake of effect. Capt. Alexander Deuchars, who has now made upwards of fifty voyages into the Arctic regions, informed me that he had known a Whale toss a boat nearly 3 feet into the air, and itself rise so high out of the water that you could see beneath it, but that, if Scoresby's figure was correct, the Whale must have tossed the boat very many feet into the air—a feat which he did not think was within the bounds of, if not possibly, yet of probability.

The *teats* are hardly the size of a cow's, are placed about the middle, and one inch from the edge of the sulcus, but in the dead animal are almost universally retracted within the white-coloured or spotted sulcus, in the middle of which they are situated. The *milk* is thick, rich, and rather sweet tasted. The fæcal evacuations of the Whale are red-coloured, most probably due to the red *Cetochili* and other animals which form the bulk of its food. The *skin* (including the cuticle) is about 1½ inch in thickness all over

* Fleming, "Philosophy of Zoology."

† A tolerably good account of these and other points in the economy of the Cetacea, mixed up with a heterogeneous mass of errors, is to be found in the (deservedly?) neglected "Natural History of the Cetacea," &c., by H. W. Dewhurst (1834).

the body, but is rather thicker on the tail, on which organ, however, it is of a uniform thickness. The *blubber* varies from about a foot to eighteen inches in thickness, tolerably uniformly throughout, except on the head, &c.; the colour is like lard or pork fat in young animals, but in the older ones rosy-coloured, from the quantity of nutrient blood-vessels in it. The *flesh* is dark and coarse-fibred, but when properly cooked tastes not unlike tough beef. When the French had whalers in Davis Strait, the sailors, with the usual aptitude of their nation for *cuisine*, made dainty dishes of it; but our seamen, imbued with the virulent dietetic conservatism of the Saxon, prefer to grow scurvy-riddled rather than partake of this coarse though perfectly wholesome food.

The best figure of the Right Whale is that of Scoresby; but in Harris's "Collection of Voyages" there is a very good figure of the animal (almost as good as Scoresby's), accompanied by a very tolerable description. I think Scoresby's figure is erroneous, in so far as I have never been able to see the prominence behind the head which he figures; and the notch shown in the outline figure of the genus in the first edition of the "British Museum Catalogue of Whales" does not exist in nature; but as Dr. Gray does not mention it in his description, I presume that it is placed there through an error of the draughtsman or lithographer.

The size of the Greenland Whale has, I think, been a little under-rated. The late Dr. Scoresby, from abundant data, considered that we have no record of the Whale to be relied upon which gives a greater length than 60 feet. While agreeing with him so far that I believe that to be generally the extreme, I am very doubtful whether they did not at one time, before they were so ruthlessly slaughtered, attain a greater size, or that individuals are not even now found of a greater size. The position in which a Whale is measured alongside the ship, when slightly doubled, is apt to introduce an error into the measurement and make it smaller than it really is. The late Chevalier Charles Louis Giesecke mentions one which was killed at Godhavn in Greenland in 1813 which measured 67 feet, and I shall presently give the measurements of one equally large. The largest one, however, which is known to have been killed in the Arctic seas was one which the late Capt. Alexander Deuchars (whom I have already had occasion to mention as a most trustworthy and experienced whaler, and personally acquainted with the killing of upwards of 500 Whales) obtained in Davis Strait in the year 1849. It measured 80 feet in length: the breadth of the tail, from tip to tip, was 29 feet; the longest lamina of whalebone measured 14 feet; the amount of whalebone in its mouth was large; but the blubber was only about 6 inches in thickness, and only yielded 27 tuns of oil.* The Whales killed in the Spitzbergen sea are said, as a rule, to be generally less and "lighter-boned" (*i. e.* with less whalebone) than those of Davis Strait, which may possibly

* The tun of oil is 252 gallons wine-measure; at a temperature of 60° Fahr. it weighs 1,933 lbs. 12 oz. 14 dr. avoirdupois.

account for the less size of those seen by Scoresby, whose whaling-experience was almost wholly confined to the former region. The females are larger and fatter than the males. I append the measurements of one of the largest Whales recently killed in Davis Strait, for which we are indebted to Dr. Robert Goodsir.

Measurements of a specimen of Balæna mysticetus *killed in Pond's Bay, Davis Strait* (♀).

	ft.	in.
Length from the fork of the tail, along the abdomen, to tip of lower jaw	65	0
Girth behind swimming-paws	30	0
Breadth of tail, from tip to tip	24	0
Greatest breadth between lower jaws	10	0
Length of head, measured in a line from articulation of lower jaw	21	0
Length of vulva	1	2
From posterior end of vulva to anus	0	6
From anterior end of vulva to umbilicus	8	0
Mammæ placed opposite the anterior third of vulva, and 6 inches from tip of it.		
Length of sulcus of mammæ	0	3
Breadth of sulcus, on each side of it	0	2
From tuberosity of humerus to point of pectoral fin	8	0
Greatest breadth of fin	3	11
Depth of lip (interior of lower)	4	7
From the inner canthus of eye to extreme angle of fold of mouth	1	5
From inner to outer canthus	0	6
Length of block of laminæ of baleen, measuring round the curve of the gum, after being removed from the head	16	6
Length of longest lamina on each side	10	6
Distance between the lamina at the gum	0	$0\frac{7}{8}$
Breadth of pulp-cavity of largest lamina	1	0
Average length of pulp when extracted from some of the largest laminæ	0	5
Number of laminæ on either side, about 360.		

The length along the curve of the back and other measurements desirable to have been taken could not be made out, owing to the position of the Whale, as it was suspended in the water alongside.

(γ) *Habits, &c.*—The Right Whale is a gregarious animal, being generally found in small "schools" of three and four, but when travelling from one part of the ocean to another they will sometimes collect in large parties. I am informed by Dr. James M'Bain, R.N., that about thirty years ago he witnessed an extraordinary migration of this nature a little to the south of Pond's Bay. The Whales to the number of several hundreds passed north in a continuous flock, and a few days afterwards were succeeded by an even still more numerous herd of Walruses. The numbers of the latter were beyond all computation; hour after hour did they travel to the northward, never pausing to feed, but all seemingly intent on reaching the opening of Lancaster Sound. A few days subsequently not one was to be seen, as previously there had been no signs of their presence. This was undoubtedly a very rare scene; and the question which must suggest itself is, where could such a number of these huge animals have come from? The Whale is capable of travelling at a very fast rate when irritated by wounds or impelled by fear of its enemies. I was told by the late Capt. Graville, of the screw whaler "Diana," a proverbially experienced and truthful man, that a Whale was struck near the entrance of Scoresby's Sound, on the east coast of Greenland, by the father of the late Dr. Scoresby (with whom Mr. Graville was a fellow apprentice); but, being lost, it was killed next day near the entrance of Omenak Fjord, on the west coast, with the harpoons freshly imbedded in its body. This was adduced in proof of the existence of an inlet in former times (as, indeed, represented on the old maps) across Greenland between these two points. Unless the whole story was founded on a misconception (an event less likely from the searching investigation which took place at the time), we can scarcely believe that the Whale could have reached the west coast by any other means; for, even allowing the greatest credible speed, it comes scarcely within the limits of possibility that it could have doubled Cape Farewell and reached 70° N. latitude within the interval mentioned. The rate at which a Whale travels from place to place whilst feeding, or under other ordinary circumstances, may be stated as being about four miles an hour. Like most of the Cetacea, it generally travels in a course contrary to that of the wind. Its *food* consists, for the most part, of Entomostraca and Pteropoda, but chiefly of the former, and especially of *Cetochilus arcticus*, Baird, and *Cetochilus septentrionalis*, H. Goodsir, *Arpacticus kronii*, Kröy., &c., which are chiefly found in those portions of the sea of the olive-green colour described by Scoresby. This appearance had seen shown* to be produced by vast quantities of *Diatomaceæ*, chiefly *Melosira arctica*, on which the "Whales' food" subsists. It is not, I am of opinion, compatible with facts to suppose that the Right Whale's food is composed in any part of Fishes proper, except, perhaps, a minute individual which may now and then accidentally find its way into its stomach with the mass of *maidre* (as the Whale's food is called). Many of the old whalers contend otherwise, and will adduce measurements of the diameter of the gullet in proof that much larger animals than Acalephæ, Pteropoda, or Entomostraca could be received in the stomach. I have never measured the orifice of any œsophagus which exceeded 2½ inches in diameter, though as these observations were generally made on

* On the Nature of the Discoloration of the Arctic Seas, *see* Seemann's Journ. Botany, Feb. 1868; Trans. Bot. Soc. Edinburgh, vol. ix.; Quart. Journ. Micr. Sci., Oct. 1868; Das Ausland, Feb. 27th, 1868; Petermann's Geogr. Mittheil., 1869; and a reprint in this "Manual."

young Whales, it is not improbable that this size may be exceeded in some individuals. Most of the slimy looking substances found floating in the Arctic seas are generally masses of Diatomaceæ combined with Protozoa, &c.; but in some cases it is the mucous lining of the bronchial passages which has been discharged when the animal was "blowing." This "blowing," so familiar a feature of the Cetacea, but especially of the Mysticete, is quite analogous to the breathing of the higher Mammals, and the "blow-holes" are the perfect homologues of the nostrils. It is most erroneously stated that the Whale ejects water from the "blow-holes." I have been many times only a few feet from the Whale when "blowing," and, though purposely observing it, could never see that it ejected from its nostrils anything but the ordinary breath—a fact which might almost have been deduced from analogy. In the cold Arctic air this breath is generally condensed, and falls upon those close at hand in the form of a dense spray, which may have led seamen to suppose that this vapour was originally ejected in the form of water. Occasionally when the Whale blows just as it is rising out of or sinking in the sea, a little of the superincumbent water may be ejected upward by the column of breath. When the Whale is wounded in the lungs, or in any of the blood-vessels immediately supplying them, blood, as might be expected, is ejected in the death-throes along with the breath. When the whaleman sees his prey "spouting red," he concludes that its end is not far distant; it is then mortally wounded. The Whale carries its young nine or ten months, and produces in March or April. In the latter month a Hull ship obtained a *sucker* with the umbilical cord still attached. It rarely produces more than one at a birth, though it is said that in a few instances two have been seen following the female. It couples during the months of June, July, or August, and, as in most, if not all of the Cetacea, this operation is performed in an upright and not in a recumbent position, as stated in some works, the authors of which might be supposed to speak from personal observation.* Equally erroneous, as far as I can learn, is the idea that it only produces once in two years; but on this subject, as on many others concerning the Cetacea, it would be difficult to pronounce an opinion founded on any decided knowledge. In the month of August I have seen them in the position described, with the pectoral fins adpressed against each other's body, and the male lashing the water with his tail. The young suckles to a considerable age (probably one year), and in order to allow of its getting convenient access to the mammæ the mother lies on its side for a time. Their love of their offspring is so strong, that though the cubs are of very little value, yet the whalers often make a point of killing them in order to render the mother more accessible. During the period of procreation the mother is much fiercer and more dangerous to approach than at other seasons, when it is a timid, harmless animal. I once saw a Whale, when the boats were approaching

---

* Dewhurst, "Natural History of the Cetacea," p. 20.

it, take the young under one pectoral and swim off by aid of the other. When the mother was killed, the cub could not be made to leave the dead body of its mother, though lances were continually run into it by the seamen who were flensing the animal. When the carcass was let go, the young one instantly dived down after it, nor did we see it again. The *sight*, *hearing*, and *smell* of the Whale are all very acute in the water, but are very dull out of it. The power of the Cetacea for remaining beneath the surface of the sea seems to bear a direct ratio to their size. Under ordinary circumstances, the Right Whale will generally remain no longer than half an hour without rising again to breathe; the cubs are, however, more stubborn, and will often remain more than three-quarters of an hour. Whalers and Eskimo have many stories of Whales lying torpid at the bottom of shallow inlets and bays for several days at a time; though I have heard these tales repeated by most credible men, yet I am inclined to hesitate at receiving as facts anything so contrary to physiological laws, and so incapable of receiving any explanation of a reasonable nature.* I have frequently known Whales dive and not come up for hours; but, unfortunately for the acceptance of these wonderful tales of subaquatic being, these universally came up dead! In nearly every case it appears that, diving with tremendous impetus under the tortures of the harpoon, they had struck their heads on the bottom with such force as to stun them for the time being, and before they recovered were drowned; the Whale's nose was in nearly every instance covered with the mud of the bottom. This diving to the bottom is a favourite feat of young Whales; and accordingly these frisky individuals are more difficult to capture than the adult ones of a more staid temperament. All species of Cetacea seem to pass a considerable portion of their time asleep on the surface of the water, and in this position they are often struck. The Right Whale always keeps near the land-floes of ice; and its migrations north and west seem to be due to this habit.†

After man, the chief enemy of the Whale is *Orca gladiator*, the most savage of all the Cetacea, and the only one which feeds upon other animals belonging to the order. The Thresher Shark (*Carcharias vulpes*), the very existence of which Scoresby seemed to doubt, but which is now so comparatively well-known to naturalists and seamen, is also an enemy of the Whale. It is doubtful, however, whether it attacks it in life, or only preys upon it after death. The "Advice" (Capt. A. Deuchars) once took a dead Whale alongside which this Shark was attacking in dozens, the belly being perfectly riddled by them.‡

---

* *Vide* also Dewhurst, *l. c.* p. 36.

† Capt. Wells, in the Dundee whaling steamer "Arctic," is reported to have run, in the summer of 1867, up into Smith's Sound in search of Whales. He found *open water* and *no Whales*—a case of cause and effect (Sherard Osborn, Proc. Roy. Geogr. Soc., vol. xii., p. 103, Feb. 10th 1868).

‡ The sailors have a notion that the Shark does not bite out the pieces, but cuts them by means of its curved dorsal fin, and seizes them as they drop into the water. This belief is widely and firmly received.

The Greenland Shark (*Scymnus* [*Læmargus*] *borealis*, Fl.), though it gorges itself with the dead Whale, does not appear to trouble it during life. Martens' most circumstantial account of the fight between the Whale and Swordfish seems to have originated in a misconception, this name being applied by seamen not only to the Scombroid fish (*Xiphias*), but also to the *Orca*, which, as is well known, fights furiously with the Right Whale. The Whale must attain a great age, nor does it seem to be troubled with many diseases. Whales which are found floating dead are almost always found to have been wounded. They are often killed with harpoon-blades imbedded deep in the blubber; and some of the marks on them have been proved to be the remains of fights of a very ancient date in which the Whale has come off victor.

(δ) *Geographical distribution and migrations.* — The geographical distribution and migration of the Whale on the coast of Danish Greenland has been fully discussed by Eschricht and Reinhardt,* and in the Spitzbergen sea by Scoresby;† so that I confine what few remarks I have to make on this subject to its range along the northern shores of Greenland and the western shores of Davis Strait and Baffin's Bay, where the whalers chase it. They appear on the coast of Danish Greenland early in May, but are not nearly so plentiful as formerly, when the Davis-Strait whaler generally pursued his business on this portion of the coast; but they are now so few that they have generally gone north before the arrival of those ships which have first proceeded to the Spitzbergen sealing. It is rarely found on the Greenland coast south of 65°, or north of 73°; indeed I have only heard of one instance in which it has been seen as far north as the Duck Islands near the entrance of Melville Bay, and even for a considerable distance south of that it can only be looked upon as an occasional straggler. However, after crossing to the western shores of Davis Strait, it occasionally wanders as far north as the upper reaches of Baffin's Bay. The great body, however, leave the coast of Greenland in June, crossing by the "middle ice," in the latitude of Svarte Huk (Black Hook), in about lat. 71° 30′ N. The whaler presses with all speed north through Melville Bay to the upper waters of Baffin's Bay, and across to the vicinity of Lancaster Sound. If there is land-ice in Baffin's Bay at the time they arrive (about the end of July), there are generally some Whales up that Sound and Barrow's Inlet; but they accumulate in greatest numbers in the neighbourhood of Pond's Bay, and even up Eclipse Sound, the continuation of the so-called Pond's Bay, which is in reality an extensive unexplored sound opening away into the intricacies of the Arctic archipelago. The Whales continue "running" here until the end of June, and remain until about the end of August or beginning of September. The whalers think that if they can reach Pond's Bay by the beginning

* Ray Soc. Mem. Cet.

† "Arctic Regions," "Voyage to Greenland," and "Memoirs of the Wernerian Society of Edinburgh" (1811), vol. i. p. 578.

of August they are sure of a "full" ship. The Whales now commence going south, and the whalers continue to pursue them on their austral migration, halting for that purpose in Home Bay, Scott's Inlet, Clyde River, &c. As the season gets more tempestuous and the nights dark, most of them towards the end of September, to avoid the icebergs dashing about in this region at that time of the year, anchor in a snug cove, or *cul de sac*, lying off an extensive unexplored sound, not laid down on any map, in the vicinity of Cape Hooper; others go into a place known by the euphonious name of "Hangman's Cove;"* whilst others go south to *Kemisoak* (Hogarth's Sound of Penny), Northumberland Inlet, or other places in the vicinity of Cumberland Sound and the Meta Incognita of Frobisher,—localities intimately known to many of these hardy seamen, but by name only to geographers. Whilst the good ship lies secure in these unsurveyed and unauthorised harbours (each master mariner according to his predilection), the boats go outside to watch for Whales. If they succeed in capturing one, frequently, if possible, the vessel goes out and assists in securing it. Though they are supposed to return to the ship every night, yet at this time the men are often subjected to great hardship and danger. This is known as the "autumn" or "fall fishing," and this method of pursuing it as "rock-nosing."

M. Guérin, the surgeon of a whaler, has described† what he considers a marked variety of the Right Whale under the name of the "Rock-nosed Whale." The characters which he gives (such as the head being considerably more than one-third the size of the animal, or as 16 to 51) vary in almost every individual. The size of the head, for instance, differs a little in almost all individuals; and Scoresby merely gave one-third the size of the body as the average, not as the unvarying proportion. Whales of different ages keep a good deal together; hence young Whales frequent the bays; the old ones roam in the vicinity of the "middle ice" of Davis Strait, and afterwards come into the bays; and those killed early in the year at Pond's Bay are chiefly young animals. Hence the whaler uses the terms "middle-icers," "rock-nosers," and "Pond's-Bay fish," to designate not a separate species or even variety, but to express a geographical fact and a zoological habit. According to the state of their cargo, the industry of the captain, or the state of the weather, the whalers leave for home from the 1st to the 20th of October, but rarely delay their departure beyond the latter date.

Where the Whale goes to in the winter is still unknown. It is said that it leaves Davis Strait about the month of November, and produces young in the St. Lawrence River, between Quebec and Camaroa, returning again in the spring to Davis Strait. At all events, early in the year they are found on the coast of Labrador, where the English whalers occasionally attack them; but the

* From an Eskimo being found here hung by an *allunak* over a cliff.

† Edinb. New. Phil. Journ., 1845, p. 267.

ships arrive generally too late, and the weather at that season is too tempestuous to render the "South-west Fishing" very attractive. Later in the year the ships enter Cumberland Sound in great numbers; and many of them (especially American and Peterhead vessels) now make a regular practice of wintering there in order to attack the Whales in early spring. It is said that early in September they enter Cumberland (Hogarth's) Sound in great numbers and remain until it is completely frozen up, which, according to Eskimo account, is not until the month of January. It is also affirmed by the natives that when they undertake long journeys over the ice in spring, when hunting for young Seals, they see Whales in great numbers at the edge of the ice-floe. They enter the Sound again in the spring and remain until the heat of summer has entirely melted off the land-floes in these comparatively southern latitudes. It thus appears that they winter (and produce their young) all along the broken water off the coast of the southern portions of Davis Strait, Hudson's Strait, and Labrador. The ice remaining longer on the western than on the eastern shore of Davis Strait, and thus impeding their northern progress, they cross to the Greenland coast; but, as at that season there is little land-ice south of 65°, they are rarely found south of that latitude. They then remain here until the land-floes have broken up, when they cross to the western shores of the Strait, where we find them in July. I am strongly of belief that the Whales of the Spitzbergen sea never, as a body, visit Davis Strait, but winter somewhere in the open water at the southern edge of the northern ice-fields. The Whales are being gradually driven further north, and are now rarely found, even by their traces,* so far south as the Island of Jan Mayen (71° N. lat.), round which they were so numerous in the palmy days of the Dutch whaling trade. I am not quite sure, after all that has been said on this subject, that the Whale is getting extinct, and am beginning to entertain convictions that its supposed scarcity in recent times is a great deal owing to its escaping to remote, less known, and less visited localities. It is said to be coming back again to the coast of Greenland, now that the hot pursuit of it has slackened in that portion of Davis Strait. The varying success of the trade is owing not so much to the want of Whales as to the ill luck of the vessels in coming across their haunts. Every now and again cargoes equal to anything that was obtained in the best days of the trade are obtained. Fourteen years ago I came home to England "shipmates" (as the phrase goes) with no less than thirty Right Whales, in addition to a miscellaneous menagerie of Arctic animals dead and alive, and a motley human crew—a company so *outré* that I question if ever naturalist, or even whaler, sailed with the like before.

* The recent visit of Whales to a particular locality can frequently be known by a peculiar oiliness floating on the water, and (the whalers say, though I confess I was never sensible of it) an unmistakeable odour characteristic of this Cetacean.

(ε) *Economic value.*—After the very excellent account of Scoresby, it would be mere pleonasm on my part to say one word regarding the commercial importance of the Whale. The introduction of steam, the almost universal use of the gun-harpoon, and the discoveries of Ross and Parry on the western shores of Davis Strait have greatly altered the nature of the "Strait fishery" since Scoresby's time. For this reason I have given the outline of a whaler's summer cruise, more especially as it illustrates, according to my observation, the range and migrations of the Right Whale.*

(ζ) *Varieties of* Balæna mysticetus.—The whalers do not recognize any varieties of the Right Whale by specific names, nor do I of my own knowledge know of any entitled to that rank. Professors Eschricht and Reinhardt† consider that there is a second species of Right Whale found in the Greenland and northern seas, the "Nordcaper" (*Balæna nordcaper*, Bonnat.; *Balæna islandica*, Briss., &c.), the "Sletbag" of the Icelanders, and that the following facts have been ascertained regarding it:—1st, that it is much more active than the Greenland Whale, much quicker and more violent in its movements, and accordingly both more difficult and dangerous to capture; 2nd, that it is smaller (it being, however, impossible to give an exact statement of its length) and has much less blubber; 3rd, that its head is shorter, and that its whalebone is comparatively small and scarcely more than half the length of that of the *B. mysticetus;* 4th, that it is regularly infested with a Cirriped belonging to the genus *Coronula;* and (5th) that it belongs to the Temperate North Atlantic as exclusively as the *B. mysticetus* belongs to the icy sea, so that it must be considered exceptional when either of them strays into the range of the other. Moreover they consider that in its native seas it is to be found further towards the south in the winter (viz. in the Bay of Biscay, and near the coast of North America down to Cape Cod), while in the summer it roves about in the sea around Iceland and between this island and the most northerly part of Norway. Dr. Eschricht considers that this was the Whale captured by the Basque whalers in the seventeenth century; hence he has called it *Balæna biscayensis.* A considerable portion of this description corresponds with what I have said regarding the Spitzbergen Whales as a race. I have heard that "barnacles" have been got on Whales; but these were looked upon as a sign of age in the Whale.

It is now a question to what species the Right Whales now and then stranded on the European coasts are to be referred. What the "Scrag Whale" of Dudley‡ (*Balæna gibbosa*, Erxl.) is I cannot

---

* For an elaborate analysis of the German Arctic whale-fishery see Lindemann in the Appendix to Petermann's "Geograph. Mittheil.", 1867; and for that by the Dundee fleet see Yeatman, Rep. Brit. Assoc., 1867. I have given a fuller outline of a Baffin's Bay Whaler's Cruise in "Ocean Highways," 1871. A still better account will be found in Capt. A. H. Markham's "Whaling Cruise," 1872.

† *Loc. cit.*

‡ Phil. Trans. vol. xxxiii., p. 259.

imagine. It is not now known to the whalers; and as neither of the species referred to have as yet been found in Davis Strait or Baffin's Bay,* they do not come within the limits which I have assigned to myself.

2. PHYSALUS ANTIQUORUM, Gray.

Balænoptera musculus, Flem. Brit. An. p. 30.
Rorqualus musculus, F. Cuv. Cétacés, p. 334.
Balæna physalus, Fab. Fauna. Grœnl. p. 35.

Popular names.—*Big Finner*, *Razorback*, (English whalers); *Sillhval* (Swedish); *Sildrör*, *Rören* (Norse); *Sildreki* (Icelandic); *Tunnolik*, *Tekkirsok* (Greenlanders).

This species, in common with most of the family *Balænopteridæ*, does not go far north as a rule, but keeps about the Cod-banks of Rifkol, Holsteensborg, and other localities in South Greenland.† They feed upon Cod and other fish, which they devour in immense quantities. Desmoulins‡ mentions 600 being taken out of the stomach of one; I know an instance in which 800 were found. They often, in common with *Balænoptera Sibbaldii* and *B. rostrata*, wander into the European seas in pursuit of Cod and Herrings, and are quite abundant in the vicinity of Rockal. A few years ago much excitement was got up about the number of "Whales" found in that locality, and companies were started to kill them, supposing them to be the Right Whale of commerce. As might have been expected, they proved only to be "Finners," which prey on the immense quantities of Cod which are found there. This Whale is accounted almost worthless by the whalers; and, on account of the small quantity of oil which it yields and the difficulty of capture, it is never attacked unless by mistake or through ignorance. I remember seeing one floating dead in Davis Strait, to which the men rowed, taking it for a Right Whale; but on discovering their mistake they immediately abandoned it. They had not been the first; for on its sides were cut the names of several vessels which had paid it a visit and did not consider it worth the carriage and fire to try out the oil. The blubber is hard and cartilaginous, not unlike soft glue. Its "blowing" can be distinguished at a distance, by being whiter and lower than that of *Balæna mysticetus*.

3. BALÆNOPTERA SIBBALDII, Gray.

Sibbaldius borealis, Gray, Cat. Seals and Whales, p. 175.

Popular names.—This is popularly confounded with the last, and the same names are applied to it by the whalers and Eskimo. It

* Cranz's description of the *Knotenfisch*, or *Knobbelfisch* (Greenland, vol. i. p. 146), is not derived from his own knowledge, but, like most of his descriptions, is copied from previous authors.

† I am aware that this statement is somewhat at variance with Dr. Eschricht's, as contained in his paper on the "Geographical Distribution of some of the Northern Whales" (Forh. Skand. Naturf. Kjöb., 1847, p. 103); nevertheless I think that it will be found to be substantially correct.

‡ Hamilton on Whales (Jardine's Naturalist's Library).

is probably also the *Kepokarnak* of the Greenlanders, and the *Steypireythr* of the Icelanders.*

It visits the coast of Greenland only in the summer months, from March to November; and its range may be given as the same as the last. Like the former, it is rarely killed by the natives.

4. BALÆNOPTERA ROSTRATA, O. Fab.

Popular names.—*Little Finner*, *Pike Whale* (English whalers and authors); *Waagehval* (Norse); *Tikagulik* (Greenlanders); *Tschikagleuch* (Kamschatkdales); *Seigval* or *Seival* (Finns).

This Whale only comes in the summer months to Davis Strait and Baffin's Bay, or very seldom during the winter to the southern portion of Greenland. It is not killed by the natives; and its range is that of its congeners. The natives of the western shores of Davis Strait seldom recognize the figure of this and allied species of Whales, though the Greenlanders instantly did so.†

5. MEGAPTERA LONGIMANA, Gray.

Balænoptera boöps, O. Fab. Faun. Grœnl. p. 36 (non Linn.?).

Popular names.—*Humpback* (English whalers); *Rörqval*, *Stor Rörhval* (Norse); *Puckelhval* (Swedes); *Keporkak* (Greenlanders and Danes in Greenland).

This Whale is only found on the Greenland coast in the summer months. For many years it has been regularly caught at the Settlement of Frederikshaab, in South Greenland. In North Greenland it is not much troubled. Whilst dredging in the harbour of Egedesminde one snowy June day, I saw a large *Keporkak* swim into the bay; but though there were plenty of boats at the Settlement, and the natives were very short of food, yet they stood on the shore staring at it without attempting to kill it. The natives of this Settlement are no doubt the poorest hunters and fishers in all North Greenland (if we except Godhavn, the next most civilized place); but there were at that time at the Settlement natives from outlying places. Capt. John Walker, in the "Jane" of Bo'ness, one year in default of better game, killed fifteen Humpbacks in Disco Bay. He got blubber from them sufficient, according to ordinary calculation, to yield seventy tuns of oil; but on coming home it only yielded eighteen. The "bone" is short and of little value. Though one of the most common Whales on the Greenland coast, yet, on this account and being difficult to capture, it is rarely troubled.

6. CATODON MACROCEPHALUS, Lacép.

Physeter macrocephalus, Linn. Syst. N. i. p. 107; O. Fab. Fauna Grœnl. p. 41.

---

* Flower, Proc. Zool. Soc., 1804; Turner, Trans. Roy. Soc. Edin., vol. xxvi.

† In a Greenland skeleton at Copenhagen, the lateral processes of the fifth and sixth cervical vertebræ are united, which is not the case with one from Norway. We cannot be too cautious in separating species on such distinctions.

Popular names.—*Sperm-Whale* (English); *Kegutilik* or *Kigutelirksoak* (Greenlanders). It is probably also the *Potvisch* (Norse), and *Tweld-Hval* (Icelandic).

Though currently reported in all compilations as one of the most common animals of the Arctic seas, and especially of Davis Strait and Baffin's Bay, it can only be ranked as a very rare, and possibly accidental, straggler. Whatever it was formerly, it is now only known to Davis Strait whalers by name; many will even ridicule the notion of its being an inhabitant of those seas. I found very few Eskimo who knew it even by tradition; and I could only hear of one recent instance of its being killed on the coast of Greenland, viz. near Proven (72° N. lat.) in 1857. According to Fabricius, however, it is generally found in the more southern parts of Davis Strait.

7. Delphinus euphrosyne, Gray.

Delphinus holbœllii, Eschricht, Skand. Naturf. Möde i Kjöbenhavn, 1847, p. 611.

This species is only known as a member of the Greenland fauna by a skeleton from South Greenland. It is apparently unknown to the natives, for they have no popular names for it.

8. Lagenorhynchus albirostris, Gray.

Delphinus ibsenii, Eschricht, Unders. over Hvald. 5te Afh. i Vid. Selsk. Nat. Math. Afh. xii. 297.

This is only known as a Cetacean of Davis Strait by a skeleton from Greenland in the Copenhagen Museum. It is found also in the Faroe Islands, and in various portions of the North Sea.

9. Lagenorhynchus leucopleurus (Rasch), Gray.

Dr. Gray* has referred a skeleton from Greenland in Mr. Brandt's collection to this species, and on his authority solely I claim it as a member of the Greenland fauna. We possess no particulars of its history as an Arctic animal. The Norwegians know it as the *Qwitskjæving*.

10. Orca gladiator (Bonn.), Sund.

Delphinus orca (L.); O. Fab. Fauna Grœnl. p. 46; Reinhardt, Naturh. Tillæg til Rink's geog. og stat. Breskrev. af Grönl. p. 12.

Physeter microps, Fab. F. G. no. 27.

Popular names.—*Grampus*, *Killer*, *Swordfish* (English seamen); *Späckhuggare*, *Svärdfisk*, (Swedes); *Stourvagn*, *Staurhyning* (Norse); *Ardluik* or *Ardluk* ♀, *Ardlurksoak* ♂, (Greenlanders). In all probability the "*Pernak*," or *Parnak* (*Physeter catodon*, O. Fab.), is also to be referred to *Orca gladiator*. Hr. Fleischer assurred me that it was an *Orca*, but only known to him by name. Curiously enough, the Kamschatdales and Aleutians

* Zool. Erebus and Terror, p. 34, t. 3; Cat. Seals and Whales (1866), p. 273.

have very similar names (*Agluck*, fide Pallas, Zool. Rosso-Asiat. p. 305; and *Aguluck*, fide Chamisso, Nov. Act. Acad. Nat. Cur. vol. xii. p. 262) for animals closely allied to, if not identical with, this species.

The *Ardluk* is only seen in the summer time along the whole coast of Greenland. Wherever the White Whale, the Right Whale, or the Seals are found, there is also their ruthless enemy the Killer. The White Whale and Seals often run ashore in terror of this Cetacean; and I have seen Seals spring out of the water when pursued by it. The whalers hate to see it, for its arrival is the signal for every Whale to leave that portion of the sea. It is said that it will not go among ice, and that the Right Whale, when attacked by it, keeps among ice to escape its persecution. Occasionally the ends of the laminæ of whalebone are found bitten off, apparently by the Killer; and probably this is the origin of the story that it preys on the tongue of the whale. Linné* very happily styles it "Balænarum phocarumque tyrannus† "quas turmatim aggreditur." Though subsisting chiefly on large fishes, they will not hesitate to attack the largest Whalebone Whales, and are able to swallow whole large Porpoises and Seals. Dr. Eschricht took out of the stomach of one thirteen Porpoises and fourteen Seals, the voracious animal having been choked by the skin of a fifteenth. It has been known to swallow four Seals at least immediately one after the other, and in the course of a few days as many as twenty-seven individuals.‡ I know of a case in which they attacked a white-painted herring boat in the Western Islands, probably mistaking it for a *Beluga!* Holböll once witnessed a herd of White Whales, driven into a bay near Godhavn, literally torn to pieces by these voracious sea-wolves.

11. PHOCÆNA COMMUNIS, Brookes.

Popular names.—*Purpess*, *Sea-pig* (English seamen); *Marsuin*,§ *Herring-hogs*, *Pelloch*, *Bucker*, *Puffy-dunter*, *Neesock*§ (fishermen of Northern Islands and coasts of Scotland); *Nesa* or *Nisa* and, more rarely, *Piglertok* (Greenlanders).

The Porpoise arrives in the spring in Davis Strait, and stops there until November, but does not go further north than from

* Mant. Plant., vol. ii., p. 523.

† Gunnerus (Trondh. Selsk. Skriv. iv. p. 99) styles it *Kobbeherre*—Lord of the seals.

‡ Nilsson, Skand. Fauna (Däggdjuren), p. 607.

§ The old Norsemen as they poured forth from Scandinavia on their predatory or colonizing expeditions leavened not only the habits but the language of the conquered. *Marsvin* is the Swedish word for the Porpoise, hence the French *Marsouin* and the same Shetland word. *Nise* (meaning sprite or goblin) is the Norse term for it, hence we have *Nisa* in Greenland and *Neesock* in Shetland (the *ock* being used there, as in many other words, as a diminutive). *Porpoise* is only a corruption of the French *porc poisson*, which we have almost literally translated into *Sea-pig*. So is the German *Meerschwein* identical in origin with the Norse *Marsouin*, also meaning "Sea-pig."

lat. 67° to lat. 69° N. They are now and then caught off the coast during this period. Through the kindness of Hr. Bolbroe, Colonibestyrer of Egedesminde, we obtained the skeleton of a *Nisa*, which had been procured in this vicinity some years ago by his predecessor Hr. Zimmer; but I could see no difference in it, so far as it could be examined in the roughly prepared state, from the one usually found on the British coast. That the *Phocæna tuberculifera*, Gray,* is different from the ordinary Porpoise, I am inclined to doubt. I have examined several Porpoises caught on the British coast, and have invariably found these tubercles on the anterior edge of the dorsal fin more or less developed. Independently of this, it is questionable whether such variable characters (and we know that there are many such characters in Cetacea which give no specific distinction) warrant the separation of *Phocæna tuberculifera* from *P. communis*. The flesh of the Porpoise is far from contemptible as an article of food, and is much relished by sailors.†

Nowhere in the Arctic regions is it hunted, but in Pennant's day, at least, vast numbers were taken in the River St. Lawrence, near Petite Revière, from the end of September to the beginning of November, when they were in quest of eels. Pennant, *Suppl. Arctic Zool.*, p. 62.

12. Beluga catodon (L.), Gray.

Beluga rhinodon, Cope, Proc. Acad. Nat. Sci., 1865, p. 278; 1869, p. 23.

B. declivis, Cope, *op. cit.*, 1865, p. 278; 1869, p. 27.

Popular names.—*White Whale* (English whalers); ***Hvitfisk***, ***Hvidfisk*** (white fish) (Scandinavian seamen, and Danish colonists in Greenland); ***Kelelluak*** (Greenlanders and Eskimo generally). To distinguish it from the Narwhal, it is called also ***Kelelluak-Kakortak***, or simply ***Kakortak***. The young is known as ***Uiak*** (Fabricius).

This is, beyond all comparison, so far as its importance to the Greenlanders and Eskimo is concerned, *the* Whale of Greenland. Like the Narwhal it is indigenous, but is only seen on the coast of Danish Greenland during the winter months, leaving the coast south of 72° N. lat. in June, and roaming about at the head of Baffin's Bay and the western shores of Davis Strait during the summer. In October it is seen to go west, not south, but in winter can be seen, in company with the Narwhal, at the broken

* Proc. Zool. Soc., 1865, p. 320.

† The flesh of the Porpoise and Grampus was eaten in the 14th century in Lent time as fish; and it is lamentable to think how much sin was committed until they were discovered to be Mammals. I have heard of the monks of a Carthusian convent roasting an Otter under a similar zoologico-theological error. A MS. in the British Museum (Harl. MSS., No. 279) contains a receipt for making "puddynge of porpoise;" and we find it served at table as late as the time of Henry VIII., and in Scotland even still later. In the accounts of Holyrood Palace we find frequent entries of moneys paid for "Porpess" for the royal table.

places in the ice. Its range may be said to be the same as the Narwhal's, and during the summer months corresponds with that of the Right Whale, of which it is looked upon as the precursor. It wanders, however, further south than the Narwhal, being found as a regular denizen as far south as 63° N. lat., though on the opposite coast it reaches much further south, being quite common in the St. Lawrence River. The Greenlanders during the summer kill great numbers of them, and preserve their oil, and dry their flesh for winter use. Of this animal and the Narwhal, about 500 are yearly caught; but the majority of this number consists of the White Whale. It feeds on Crustacea, Fish, and Cephalopoda; but in the stomach is generally found some sand. The Greenlanders often jocularly remark, in reference to this, that the *Kelelluak* takes in ballast. Great numbers are caught by means of nets at the entrance of fjords and inlets, or in the sounds between islands. The young are darker-coloured than the adult, and can at once be distinguished among the herds of the ordinary waxy white colour. It is said to be rarely seen far from land. The males and females are together in the drove, and not separate, as has been stated. Their blast is not unmusical; and when under the water they emit a peculiar whistling sound which might be mistaken for the whistle of a bird, and on this account the seamen often call them sea-canaries! It is rarely that the whalers kill a White Whale, their swiftness and activity giving them more trouble than the oil is worth.* They are sometimes also called "Sea-pigs," from their resemblance to that animal when tumbling about in the water.

13. MONODON† MONOCEROS, Linn.

(α) Popular names. — *Narwhal*, *Unicorn*, *Unie* (English whalers); *Narhval* (Scandinavians); *Tugalik*, *Kelelluak-Kernektok*, or *Kernektak* (Greenlanders); *Kelelluak-tuak* (Eskimo at Pond's Bay). The word Narwhal is derived from the Gothic, and means the "beaked whale," the prefix *nar* signifying beak or snout.

(β) *Descriptive remarks.*—The female Narwhal is more spotted than the male. The young is again much darker; and I have seen individuals which were almost white, like the one Anderson describes as having come ashore at the mouth of the Elbe. In a female killed in Pond's Bay, in August 1861, the stomach was corrugated in complicated folds, as were also the small intestines. It contained Crustaceans, bones of Fish, and an immense quantity of the horny mandibles of some species of Cephalopod (probably *Sepia loligo*) firmly packed one within the other. In its stomach was a long *Lumbricus*-like worm; and the cavities behind the palate were filled with froth and an innumerable number of little worms, such as Scoresby describes in his account of the animal.

---

* One of the whalers, a few summers ago, killed several hundreds, but this is an almost isolated case.

† Lamarck subsequently usurped this name for a genus of Pectinobranchiate Mollusca.

In some animals which I examined the bone was quite eaten away by them, and that portion of the lining membrane which remained was red or inflamed. There is a curious anastomosis of reticulating venous blood-vessels inside the lining membrane of the thorax and abdomen and around the spinal cord, which has doubtless a relation to its amphibious life. The blow-holes are placed directly on the top of the head, large, semilunar, opening on either side into two sacs lined with a dark mucous membrane; these openings, again, leading to the bronchiæ and the lungs. The blow-hole has but one opening externally, but about an inch down is divided into two by a cartilaginous septum, continuous a little further down with the bony partition seen in the skull. The rima glottidis is exactly described by the late Prof. Fleming, in the "Wernerian Trans." (vol. i. p. 146). The female (except in very exceptional cases) has no "horns"; but inside the inter-maxillary bone are two undeveloped tusks, each about 10 inches long, rough, and with no inclination to a spiral. On the other hand, the undeveloped tusk (the right) in the male is smooth and tapering, and "wrinkled" longitudinally. Double-tusked ones are not uncommon; I have seen them swimming about among the herd, and several such skulls have been preserved. Among others, there is a specimen presented by Capt. Graville, in the Trinity House, Hull,* another in the University Museum, Cambridge; and, according to Mr. Clarke, nine others in Continental museums. Of course there is no whalebone in its jaw; but it is interesting to notice the laws of homology of structure (as I think) kept up. On the sides of each gum are transverse markings, either corresponding to the alveoli of the teeth or to the position of the laminæ of the whalebone in the *Balænidæ*. The under jaws are very light and quite hollow posteriorly for half their length, as in most species of Cetacea; this cavity is filled with a very fine blubber. The *tongue* is regularly concentrically grooved and attached its whole length, so as scarcely to be recognized as it lies flat on the base of the mouth; the roof of the mouth is correspondingly marked. The *lungs* are each about 1½ foot long; the *kidney* 9 inches long and about 4½ inches broad; the *lacteals* were very distinct and distended; the large intestine at broadest about 4 inches in diameter, at thinnest about 1½ inch, and about 60 feet in length.

The *pectoral fin* is not notched below (as would seem from the plate in Hamilton's book on Whales), but smooth and entire; curved below, the greatest curve pointing posteriorly, but with the thickest part of the fin anteriorly. The animal was greyish or velvet-black, with white spots, sometimes roundish, but more frequently irregular blotches of no certain outline running into one another. There were no spots on the tail or fin; waxy-looking streaks shaded off on each side of the indentation of the tail, which is white at the line of indentation. The *ridge* along its back corresponding to the dorsal fin is of a uniform height of 1 inch

* One of the tusks is 3 ft. long, and the other 4 ft.

throughout, irregularly notched on the top, like the embrasures of a castle-wall, and is formed of blubber covered with the common integument of the body, of which it is merely a raised fold.

(γ) *Habits, &c.*—The Narwhal is gregarious, generally travelling in great herds. I have seen a herd of many thousands travelling north on their summer migrations, tusk to tusk and tail to tail, like a regiment of cavalry, so regularly did they seem to rise and sink into the water in their undulatory movements in swimming. It is very active and will often dive with the rapidity of the *B. mysticetus*, taking out 30 or 40 fathoms of line. These "schools" are not all of one sex, as stated by Scoresby, but males and females mixed. It couples in an upright position; and seems to produce at about the same time as the Right Whale. Usually only one young one is produced, but cases in which a female contained or produced two are known. The use of the tusk has long been a matter of dispute; it has been supposed to use it to stir up its food from the bottom; but in such a case the female would be sadly at a loss. They seem to fight with them; for it is rarely that an unbroken one is got, and occasionally one may be found with the point of another jammed into the broken place where the tusk is young enough to be hollow or is broken near enough to the skull. Fabricius thought that it was to keep the holes open in the ice during the winter; and the following occurrence seems to support this view. In April 1860, a Greenlander was travelling along the ice in the vicinity of Christianshaab, and discovered one of these open spaces in the ice, which, even in the most severe winters, remain open. In this hole hundreds of Narwhals and White Whales were protruding their heads to breathe, no other place presenting itself for miles around. It was described to me as akin to an "Arctic Black Hole "of Calcutta," in the eagerness of the animals to keep at the place. Hundreds of Eskimo and Danes resorted thither with their dogs and sledges, and while one shot the animal, another harpooned it to prevent its being pushed aside by the anxious crowd of breathers. Dozens of both Narwhals and White Whales were killed, but many were lost before they were got home, the ice breaking up soon after. In the ensuing summer the natives found many washed up in the bays and inlets around. Fabricus describes a similar scene. Neither the Narwhal nor the White Whale are timid animals, but will approach close to, and gambol for hours in the immediate vicinity of the ship.

(δ) *Geographical distribution.*—The range and migration of the Narwhal is much the same as that of the White Whale. It is only found on the coast of Danish Greenland during the spring and winter, migrating northward and westward in the summer. It is rarely seen south of 65° N. lat.

(ε) *Economic value.*—In early times the tusk of the Narwhal was highly valued as a medicine; and Master Pomet, in his "Compleat Historie of Drugges," gives special directions regarding the selection of them. The scrapings were esteemed alexopharmic, and used of old in malignant fevers, and against the bite of serpents. Cups made of it were believed to possess the power of detecting and neutralizing any poison contained in them. From this "horn" also was distilled a strong "sal volatile." To this day the Chinese esteem them for their medicinal properties. In old times it was imposed upon the world as the horn of the "unicorn," and sold at a very high price. The heirs of the Chancellor to Christian Frisius of Denmark valued one at 8,000 imperials. (*Mus. Reg. Hafniæ.*) In 1861 the price of Narwhal's ivory was 1*s*. 6*d*. per lb., but of late years it has risen prodigiously in value owing to the repair of the Chinese palaces, but is again falling. In the Palace of Rosenborg is a throne of the kings of Denmark manufactured of this ivory; and Capt. Scoresby (the father of the Doctor) had a bedstead made of it. The oil is highly esteemed, and the flesh is very palatable.* The skin of the Narwhal boiled to a jelly is looked upon, and justly so, as one of the prime dainties of a Greenlander. The hospitable Danish ladies resident in that country always make a point of presenting a dish of *mattak* to their foreign visitors, who soon begin to like it. *See also* Pennant, Supplement to Arctic Zool. p. 100.

14. GLOBIOCEPHALUS SVINEVAL (Lacép.), Gray.

Delphinus melas, Traill, Nicholson's Journal, vol. xxii. (1809) p. 21.

Delphinus deductor, Traill, MS. and Scoresby's Arctic Regions, vol. i. p. 496, t. 13. fig. 1.

Delphinus globiceps, Cuv. Ann. Mus. xix. t. 1. fig. 2.

Delphinus tursio, O. Fabr. Faun. Grœnl. p. 49. no. 31.

Popular names.—*Bottle-nose*, *Caaing Whale* (fishermen and seamen); *Grindaquealur* (Faroe Islands); *Grinde-Hval* (Swedish and Danish); *Nesernak* or *Nisarnak* (Greenland). The term Bottle-nose is applied by sailors to several species of Whales. In fact any Whale which is not a "Right Whale," "finner," "parmacity" (spermaceti), "purpess," "unicorn" (Narwhal) or "White Whale" is with them included under the vague term of "Bottle-nose." The common and most characteristic name for this Whale is that used in the north of Scotland, viz. caaing or driving Whale—a term translated into *deductor*.†

There seems little doubt that this is the *Delphinus tursio* of Fabricius, as the Eskimo name *Nesernak* is applied to the present animal. If so, Fabricius's name has the priority; but, as it has been confounded with another species, it is better to keep Lacépède's most barbarous trivial name. Gray and other authors look upon Fabricius's *Nesernak* as the type of a distinct species, and have described it as *Tursio truncatus*. The *Delphinus truncatus* of Montagu (Wernerian Society's Trans. vol. iii. t. 5. fig. 3) is a totally different animal. Fabricius's description ("Frons rotunda,

* Though indeed the learned Wormius warns us that it is a deadly poison.

† It has no connexion with *calling*, as it has sometimes been translated, even in works written by Scotchmen. It is derived from the Scotch word *caa*, signifying to drive, relating to their ordinary method of capture, viz., by driving them ashore.

declivis s. sursum repanda, desinens rostra attenuatiore; sic fronti anatis mollissimæ non absimilis,"), though seemingly contradictory of the identity of the *Globiocephalus svineval* and *Delphinus tursio* of O. Fabr., must in reality be received for no more than it is worth. Cetological critics have received the descriptions of Fabricius as if they were infallible, or superior to those of any other author who has succeeded him. We know that many of his descriptions of other animals, which are well known, were erroneous, and that few of those regarding which there could be no mistake were altogether free from error; therefore I cannot see why we should receive the others otherwise than as approximately correct. Fabricius enjoyed during the few years he passed in Greenland no better opportunities than any other naturalist in that country at the present day. Many of the animals which he describes are very rarely killed or seen by the natives; and many of his descriptions bear on the face of them the marks of having been derived from the natives' narration, and not from actual specimens. Any one who has examined such unwieldy animals as the Cetacea must know how difficult it is, even under the most favourable circumstances, to arrive at anything like an accurate idea of the animal the external appearance of which we may be desirous of describing. Therefore, as the Greenlanders call this animal *Nesernak*, as the description does not widely differ from the appearance of the Caaing Whale, and as Montagu's *Delphinus truncatus*, with which it has been supposed to be synonymous, has never been found in Davis Strait, while the present species has, we are warranted in concluding, with Dr. Reinhardt, that the synonymy given under this species is correct.

This Whale is not a regular visitor of Davis Strait or Baffin's Bay, but is occasionally to be seen in droves in the summer time along the whole coast of Danish Greenland. An excellent account of this species is given by Turner and M'Bain, derived from the examination of some individuals of a drove which came into the Frith of Forth in the spring of 1867 (Journ. Anat. and Phys. 1867, and Proc. Roy. Phys. Soc. Edin. 1866–67 ined.)*

15. Hyperoodon butzkof, Lacép.

Monodon spurius, O. Fab. Faun. Grœnl. p. 31. no. 19.

Chænocetus rostratus (Müll.), Eschr. Undersög. over Hvaldyr. 4de Afh. 1845; Reinhardt, Tillæg til en Beskrev. af Grönland (Rink), p. 11.

---

* For the anatomy of this species, *see* Murie, Trans. Zool. Soc. vol. viii., p. 235. In the Zoological Society's "Proceedings" for 1853, p. 103, there is a notice of a paper "On the capture of *Delphinus orca* in South Greenland," by M. Rehüller, in which it is said that the number taken at Westmanshavn since 1843 was 2,200, whereas between 1819 and 1843 there were only 280. This additional capture, amounting in the aggregate to the value of 4,000*l.* sterling, was described as being due to the introduction of nets. Now there is no such place as "Westmanshavn" in Greenland, and I question if 2,200 *Orcas* have ever been killed in Greenland since the beginning of time. Apparently the notice refers to the capture of *Globiocephalus* in the Faroe Islands.

Popular names. — *Bottle-nose* or *Bottlie* (English whalers); *Nabbhväl* (Scandinavians); *Andarnefia* (Icelanders); *Dögling* (Faroe-islanders); *Anarnak* (Greenlanders).

This is undoubtedly the *Monodon spurius* of Fabricius, that author having made the not uncommon mistake of describing the upper for the lower jaw. As it is a rare animal on the Greenland coast, Fabricius could have been but little acquainted with it. This Whale is only seen about the mouth of Davis Strait, swimming in threes or fours; it is occasionally captured, as one will yield as much oil as a Narwhal. One ship's crew some years ago killed fifteen, and the oil was represented to me as mixing well with spermaceti, and selling for the same price, viz. 10*s.* 6*d.* per gallon.

16. HYPEROODON LATIFRONS, Gray.

Lagenocetus latifrons, Gray, Proc. Zool. Soc. 1864, p. 241.

This species is known from skulls and skeletons in various museums; and as an Arctic animal from a skull brought from "Greenland" by Capt. Wareham, and now in the Newcastle Museum, and by a skeleton from the same region in the Copenhagen University Museum. Greenland, however, is a loose term; but from what I have said as to the range and habits of *H. butzkof*, we may safely conclude that this has been obtained in Davis Strait. I am not aware that we have any external characters to separate it from the preceding, but yet the apparently constant distinction presented by the skull would lead us to believe in its distinctness. Therefore, though we may not go so far as Eschricht in believing it to be the male of *H. butzkof*, yet we must hesitate before joining in the opinion of even such an experienced zoologist as Dr. Gray as to its claim to generic rank.

V.—On some CETACEA of GREENLAND. By DR. E. D. COPE and DR. I. I. HAYES. (From the Proceedings Acad. Nat. Sci. Philadelphia, Dec. 1865, p. 274.)

"He also alluded to the existence of several species of White Whales, probably confounded hitherto, owing to their uniform coloration. Similar uniformity exists in various genera, as *Corvus*, *Chasmarhynchus*, etc. A species brought by Dr. I. I. Hayes, from Upernavik, was called *Beluga rhinodon*, and a large one presented by Dr. E. K. Kane was characterized under the name *B. concreta*."

"Dr. I. I. Hayes stated that the two skulls, mentioned by Prof. Cope as belonging to the genus *Beluga*, brought by him from Greenland, were obtained from the Governor of Upernavik, as those of the 'White Whale.' He also observed, that during his voyage he had seen the White Whale abundantly as far north as 78° N. lat."

VI.—Notes on Birds which have been found in Greenland. By Alfred Newton, M.A., F.R.S., Professor of Zoology and Comparative Anatomy in the University of Cambridge.

Though many authorities have been consulted in making the following compilation, it is founded mainly on the excellent "List of the Birds hitherto observed in Greenland", by Professor Reinhardt, which was printed in 'The Ibis' for 1861 (pp. 1-19) and gives the most complete catalogue of the species of that country as yet published. Some additions to it have since been communicated by him to the Natural-History Union of Copenhagen*, and these I have here incorporated. I have further to acknowledge, with sincere thanks, his great kindness in sending me the proof-sheets of his latest contribution to the subject, made during the present year and as yet unpublished (*op. cit.* 1875, p. 127), that I might avail myself of its valuable contents. On the other hand, it must be confessed that Prof. Reinhardt's "List", though all one could desire as regards the stray visitors to Greenland, gives few or no particulars of the habitat of some of the species which regularly frequent that country, and this information I have had to supply from the work of the ill-fated Holböll†, whose long residence there as an officer of the Danish Government, and taste for Ornithology rendered him a most trustworthy authority on this head. The works of the naturalists of the last century, Bruennich ‡ and Otho Fabricius§, have not been neglected by me, and as evidence of the completeness of the latter I may repeat Prof. Reinhardt's remark, that since its publication the number of birds known to breed in Greenland has been only increased by eleven. I have of course examined also the 'Memoir on the Birds of Greenland'‖, published in 1819, by the venerable Sir Edward Sabine, and the far too meagre Natural-History Supplements to the several 'Voyages' of Parry and of Ross—works which excite regret at the glorious opportunities so ingloriously missed through the absence of special naturalists, and only redeemed from utter opprobrium by the zeal of volunteers.¶ The long series of expeditions in search of Franklin's ships from the same cause was still more barren of results in respect to Arctic Ornithology, so that a single discovery of Sir Leopold McClintock's,**

---

* Videnskabelige Meddelelser, 1864, p. 246; 1865, p. 241; 1872, p. 131.

† 'Ornithologiske Bidrag til den grönlandske Fauna.' Naturhistoriske Tidsskrift, 1843, pp. 361–457. A German translation of this memoir by Dr. Paulsen was published at Leipzig in 1846, and again reissued in 1854.

‡ Ornithologia Borealis. Hafniæ: 1764. 8vo. 80 pp.

§ Fauna Groenlandica. Hafniæ et Lipsiæ: 1780, 8vo. pp. 53–124.

‖ Transactions of the Linnean Society, xii. pp. 527–559.

¶ The result of nearly all that was then ascertained about Birds is embodied in the second volume of the well-known 'Fauna Boreali-Americana' by Swainson and Richardson. (London: 1831, 4to., 523 pp.)

** Journal of the Royal Dublin Society, 1856, pp. 57–60.

and the notes of Dr. David Walker*, who did not possess any special proficiency in the study, furnish almost the only increase to our knowledge of the subject gained during that period.† The different American expeditions, judging from what has been published about them, added absolutely nothing—a fact particularly to be regretted when we regard the high latitudes they successively reached. More in this respect was achieved by the Germans, and to the observations of Dr. Pansch, contained in the elaborate work of Dr. Finsch‡, we owe information of some value. To various works not especially treating of Arctic Ornithology, or of the Ornithology of Davis Strait at least, there is no need for me here to refer more in detail.

It is now beginning to be recognized by ornithologists that to draw any sound conclusions from the avifauna of a country we must strictly limit our basis to the species of Birds which either breed in or annually, for a longer or shorter period, frequent it, and consequently to obtain a true notion of its peculiarities all accidental stragglers should be dismissed from consideration. They are indeed eminently worthy of regard from another point of view, throwing light as they do on the general question of the wanderings of Birds, but they are of little account in the aid they give to elucidating the great subject of Geographical Distribution. It has, therefore, seemed to me expedient to distinguish between these two categories by using a different series of numbers to indicate them, and also by indenting the paragraphs in which the stragglers are noticed. Without some such precaution the interspersal of stragglers among true denizens only leads to confusion, and especially would it do so in the present case when the two categories are almost equal in number, while most of the stragglers have occurred outside of the Arctic Circle, and in places lying many degrees of latitude to the southward of the tracts which the new Expedition is to explore. Still further to direct attention to these last tracts, the names of those species which, so far as one can judge, may be not unreasonably looked for in Smith Sound, and some of them thence to the northward, are printed in thick type, while the names of those which are known to breed in Greenland and yet may not be expected to occur beyond the Danish Settlements are in small capitals. The native (Esquimaux) names when given by Fabricius or others are marked by inverted commas. I have further to premise that the Danish Settlements are divided into two Inspectorates, roughly speaking, separated by the 68th parallel, as well as to observe that when a species is said to be

---

* Ibis, 1860, pp. 165–168; Journal of the Royal Dublin Society, 1860 pp. 61–67.

† The majority of such ornithological specimens as were collected during the Franklin search passed into the possession of Mr. Barrow, who subsequently gave his collection to the Museum of the University of Oxford, and a catalogue of it has been published by Mr. Harting (Proceedings of the Zoological Society, 1871, pp. 110–123).

‡ Die zweite deutsche Nordpolarfahrt. Leipzig: 1874. 2 vols. 8vo. vol. ii. pp. 178–239.

found generally or throughout Greenland the words "in suitable "localities" must be understood to follow, even though not inserted.

1. Haliaetus albicilla. White-tailed Eagle. "Nektoralik," "Tertersoak."

Inhabits generally and breeds in the whole of Danish Greenland, including the eastern coast. Its northern range not as yet determined. Being the only Eagle found in the country, there seems no need to give here its diagnostic characters.

(1.) *Pandion haliaetus.* Osprey.

A single specimen obtained (25 Sept.) at Godhavn, by Mr. E. Whymper, and sent to the Museum of Copenhagen. Must be regarded as a straggler (most likely from America), since it is not found in Iceland, and has only once been known to occur in the Færoes (1848).

2. **Falco candicans.** Greenland Falcon. "Kirksoviarsuk-kakortuinak."

The white form of Great Northern Falcon. In summer more common in the Northern Inspectorate than in the Southern, but occurring, according to Dr. Finsch, also on the Eastern Coast. The limits of its breeding-range in either direction have not been determined.

3. Falco islandus. Iceland Falcon. "Kirksoviarsuk-kernek-tok."

The darker form of Great Northern Falcon, by some held to be distinct both from *F. candicans* and *F. gyrfalco.* The northern limits of its breeding-range have not yet been determined. A young male Falcon, killed 24th September 1872, on the Fiskenæs, referred by Dr. Finsch to *F. gyrfalco*, probably belonged to this form.

4. Falco peregrinus. Peregrine Falcon. "Kirksoviarsuk-millekulartok."

Said to breed generally throughout Greenland, certainly up to lat. 69° N., and in many of the lands to the westward of Baffin's Sea. Examples obtained by Dr. Walker, of the 'Fox,' R.Y.S., at Port Kennedy (lat. 72° N.), are specifically indistinguishable from European specimens.

(2.) *Falco æsalon.* Merlin.

A specimen caught at sea (lat. 57° 41′ N., long. 35° 23′ W.) in May 1867, by Mr. E. Whymper, and by him presented to the Norfolk and Norwich Museum, seems to have reached the most western limit of the species known. A common species in Iceland; in North America replaced by the nearly allied *F. columbarius.*

(3.) *Tinnunculus alaudarius.* Kestrel.

One said to have flown on-board ship off Cape Farewell, on Parry's first return voyage, and killed. (Sabine, Suppl. App. p. ccx.)

5. **Nyctea scandiaca.** Snowy Owl. "Opik," "Opirksoak."

Very common; in summer more numerous in the Northern Inspectorate than in the Southern. Found also on the Eastern Coast, and extends westward to Liddon Island and Melville Island (75° N.). A thoroughly circumpolar species, migrating in winter to lower latitudes, and, from its white colour and large size, incapable of being confounded with any other species.

6. *Asio accipitrinus.* Short-eared Owl. "Siutitok."

A scarce species in Greenland, but perhaps breeds there, though not further to the southward than 65°. Its northern range altogether unknown, but it has been shot on the Green Islands in Disco Bay, lat. 68° 50′ N.

(*4.*) *Sphyropicus varius.* Yellow-bellied Woodpecker.

One found dead near Julianehaab, July 1845; another sent from Greenland about 1858.

(*5.*) *Colaptes auratus.* Flicker or Golden-winged Woodpecker.

Herr Möschler has recorded the receipt of a specimen from Greenland in 1852 (Journ. für Orn. 1856, p. 335).*

(*6.*) *Chætura pelasgia.* Chimney-Swift.

One shot in 1863 near the Sukkertop (Reinhardt, Vid. Medd. 1865, p. 241).

(*7.*) *Hirundo horreorum.* Barn-Swallow.

Two known to have been obtained, one at the Fiskenæs about 1830, the other at Nenortalik.†

(*8.*) *Vireosylvia olivacea.* Red-eyed Flycatcher.

One received from Greenland in 1844, and most likely from the Southern Inspectorate. Sir Oswald Mosley has recorded the occurrence of this American species in England (Nat. Hist. Tutbury, p. 385, pl. 6).

(*9.*) *Empidonax pusillus.* Little Flycatcher.

Two received from Godthaab in 1853.

(*10.*) *Contopus borealis.* Olive-sided Flycatcher.

One shot at Nenortalik, 29 August, 1840, and sent to the Royal Museum at Copenhagen.

(*11.*) *Dendrœca virens.* Black-throated Green Warbler.

One sent from Julianehaab in 1853.

(*12.*) *Dendrœca coronata.* Yellow-rumped Warbler.

Three examples prior to 1860.

(*13.*) *Dendrœca striata.* Black-polled Warbler.

One sent from Godthaab in 1853.

---

* *Chordediles popetue.* American Night-Hawk. One found dead on Melville Island.

† *Cotyle riparia.* Sand-Martin. A pair said to have been seen on Melville Island, 9 June, 1820 (Parry, Journal, &c. p. 195).

(14.) *Dendrœca blackburniæ?* Orange-throated Warbler.
A young bird shot at Frederikshaab, 16 October 1845, has been referred to this species with hesitation owing to the bad state of the specimen.

(15.) *Parula americana.* Particoloured Warbler.
One sent from the Southern Inspectorate in 1857, in a very bad state, but quite recognizable.

(16.) *Helminthophaga ruficapilla.* Nashville Warbler.
Obtained twice:—once at Godthaab about 1835, and again at the Fiskenæs, 31 August, 1840.

(17.) *Geothlypis philadelphia.* Mourning Warbler.
One obtained at the Fiskenæs in 1846, another at Julianehaab in 1853.

(18.) *Troglodytes palustris.* Long-billed Marsh-Wren.
One procured at Godthaab in May 1823.

(19.) *Regulus calendula.* Ruby-crowned Wren.
One sent from Nenortalik in 1859.

7. SAXICOLA ŒNANTHE. Wheatear. "Kyssektak."
Known to breed in Greenland from the time of Otho Fabricius, and, according to Holböll, extending its range to lat. 73° N. and even further. Strays also to the westward, and observed by James Ross, 2 May, 1830, in Felix Harbour (lat. 70° N., long. 91° 53′ W.). Obtained on Shannon Island by the German Expedition (Finsch). The peculiar distribution of this species in the northern part of the Nearctic Region has yet to be explained (*cf.* Yarrell, Br. B. ed. 4, i. pp. 352, 353).

(20.) *Turdus migratorius.* American Robin.
An adult male shot near Kornuk in the Godthaab Fjord (Reinhardt, Vid. Medd. 1865, p. 241).

(21.) "*Turdus minor.*"
One specimen, so named by Prof. Reinhardt, obtained in June 1845, at Amaraglik, near Godthaab. Prof. Baird says it is difficult to say which of the three North-American species is thereby meant (Am. Journ. Sc., ser. 2, xli. p. 339).

(22.) *Turdus iliacus.* Redwing.
One sent to Dr. Paulsen in 1845, another shot at Frederikshaab, 20 October, 1845.

(23.) *Motacilla alba.* White Wagtail.
One sent from the Southern Inspectorate in 1849, another, obtained by Dr. Walker, at Godhavn, in August 1857.

8. ANTHUS LUDOVICIANUS. Pennsylvanian Pipit.
Supposed to breed in Greenland not further south than lat. 67° N., but unquestionably does so in the northern parts of the North-American continent.

(*24.*) *Anthus pratensis.* Meadow-Pipit.
Received by Dr. Paulsen from Greenland in 1845.

(*25.*) *Otocorys alpestris.* Shore-Lark.
One shot at Godthaab in October 1835, but known before to occur on the other side of Davis Strait: *e.g.*, at Cape Wilson, 10 July, 1822.

9. **Plectrophanes nivalis.** Snow-Bunting. "Kopanauarsuk."
Breeds generally throughout the country, and said to be the commonest land-bird on the Eastern Coast (Pansch). Breeds also on Melville Peninsula, and is very numerous on the Parry Islands. Seen by Kane at Rensselaer Harbour in June 1854.

10. PLECTROPHANES LAPPONICUS. Lapland Bunting. "Narksarmiutak."
Also breeds generally throughout the country, as well as on Melville Peninsula and other lands to the westward of Davis Strait.

11. ZONOTRICHIA LEUCOPHRYS. White-crowned Bunting.
Seems to be confined to Southern Greenland: not numerous, but certainly a breeding bird, though its nest has not as yet been found in the country.

12. LINOTA LINARIA. Mealy Redpoll. "Orpingmiutak," "Anarak."
Said to breed generally throughout Greenland, suitable localities being, of course, understood, but is *migratory* there. Seems to be indistinguishable from the *Fringilla linaria* of Linnæus, the *F. borealis* of most English authors, but not their *F. linaria*, which is a much smaller and more rufescent form.

13. LINOTA CANESCENS. Greenland Redpoll.
Said to be constantly *resident*, and a regular breeder, but not further south than lat. 70° N. Occurred also in Kaiser Franz-Josef's Fjord, 1 August, 1870 (Finsch). The *Linota hornemanni* of Holböll, and possibly the *Ægiothus rostratus* of Dr. Coues.

(*26.*) *Loxia leucoptera.* American White-winged Crossbill.
An adult specimen procured about 1831 from the east coast by an Esquimaux. Subsequently another adult and three young were obtained in South Greenland.

(*27.*) *Xanthocephalus icterocephalus.* Yellow-headed Maize-bird.
One obtained, 2 September 1820, at Nenortalik.

(*28.*) *Sturnus vulgaris.* Starling.
A single specimen sent by Holböll. (Qu. *S. færoensis*, Feilden, if that be a distinct species?)

14. **Corvus corax.** Raven. "Tullugak," "Kernektok."
Breeds more in South than in North Greenland, and also observed on the East Coast. Several pairs seen on Melville Island. A

specimen from Beechey Island in the Barrow Collection. Noticed several times on Parry's Second Voyage.

15. **Lagopus rupestris.** Rock-Ptarmigan. "Akeiksek," "Kauio."

The only species of the genus which inhabits Greenland, where it occurs equally on the East as on the West Coast. Found by the German Expedition on Sabine and Clavering Islands. In great abundance on the Parry Islands, and thence southward throughout Melville Peninsula, but its southern range west of Davis Strait still undetermined. Its specific distinctness from *L. mutus* is questioned by several authorities, but the males of *L. rupestris* (including under that name *L. reinhardti* and *L. islandorum*) seem never to acquire entirely black feathers on the breast as do the males of *L. mutus*—the Ptarmigan of Scotland and the European continent. The females and the males in winter of the different forms can hardly be distinguished.

(29.) *Crex pratensis.* Corncrake.

One obtained at Godthaab and sent to the Museum of Copenhagen in 1851.

(30.) *Crex porzana.* Spotted Rail.

One obtained at Godthaab, 28 September, 1841 ; a second taken at Nenortalik was sent to Copenhagen in 1856.

(31.) *Crex carolina.* Carolina Rail.

One killed at the Sukkertop, 3 October, 1822.

(32.) *Fulica americana.* American Coot.

Twice obtained in Greenland, and in the same year (1854) —once at Godthaab, and once in Disco Bay. The latter example is in the Barrow Collection.

(33.) *Ardea cinerea.* Heron.

Said by Crantz to have been seen in South Greenland, 27 August, 1765. A young bird found dead near Nenortalik in 1856, and sent to Copenhagen.

(34.) *Botaurus lentiginosus.* American Bittern.

One caught by dogs during a storm at Egedesminde in 1869, and identified by its remains.*

(35.) *Hæmatopus ostralegus.* Oystercatcher.

One sent from Julianehaab in 1847, another in 1871 from Godthaab, and a third from Nenortalik in 1859.

16. **Strepsilas interpres.** Turnstone. "Telligvak."

Not common according to Holböll, but breeds generally along the coast. Found by the German Expedition in Sabine Island and at Cape Broer-Ruys. Recorded from Winter Island in June, and breeds

* *Grus canadensis.* Brown Crane. One obtained near Igloolik 25 June.

on the Parry Islands. Its quaintly marked black and white head, deep black breast, chestnut and black back, and white belly, render this one of the most easily recognized of shore-birds.

(36.) *Vanellus cristatus.* Lapwing.

One obtained, 7 January, 1820, near the Fiskenæs; a second received from Julianehaab in 1847.

17. **Squatarola helvetica.** Grey Plover.

Rare, but found in both Inspectorates, and, according to Holböll, increasing in numbers—an assertion which Prof. Reinhardt doubts. Said to breed on Melville Peninsula, where, according to Richardson, its eggs were obtained. Specimens of these, however, exist in very few collections, and apparently only from Siberia and Alaska. The bird is to be distinguished from the Golden Plover by its larger size, its deep black axillary feathers (which are very apparent in flight) and its rudimentary hind-toes.

18. **Charadrius virginicus.** American Golden Plover. "Kajorrovek," "Kajordlek."

Somewhat rare in Greenland, but possibly breeds there, as it does in considerable abundance on swampy places in the Parry Islands. Seen in plenty on Parry's Second Voyage. Not distinguished by the older writers (including Richardson) from the following species, but is always recognizable by its smoky-grey axillary feathers, and more slender form.

(37.) *Charadrius pluvialis.* Golden Plover.

One, in summer plumage, shot in the spring of 1871 on the Noursoak Peninsula. Believed by Dr. Finsch to breed in East Greenland. To be distinguished from the foregoing species by its pure white axillaries and somewhat stouter build.

19. **Ægialitis hiaticula.** Ringed Plover. "Tukagvajok."

Breeds generally in Greenland and found on Sabine and Clavering Islands. Said to be abundant on the shores of Possession Bay and Regent's Inlet, but was perhaps mistaken for a nearly-allied species. Was found by Professors Torell and Nordenskjöld on the Seven Islands (lat. 80° 45′ N.), and therefore has possibly the highest northern range of any known shore-bird.

The "*Charadrius hiaticula*" of Richardson (App. Parry's Second Voyage, p. 351), apparently brought from Mount Sabine, was subsequently identified by him with the North American *Ægialitis semipalmata* (Faun. Bor.-Am. ii. p. 367)—a species believed to have been obtained in Boothia Felix on Ross's Second Voyage, but not hitherto recognized from Greenland, where it may however not unreasonably be expected to occur. This differs from *Æ. hiaticula* in being smaller and slenderer, in wanting the white patch above and behind the eye, and in having a much narrower pectoral band. On closer examination also the middle and outer toes of *Æ. semipalmata* will be seen to be united at their base by a very distinct web.

(38.) *Totanus flavipes.* Yellowshank.

One sent from Greenland in 1854 to Herr Möschler (Journ. f. Orn. 1856, p. 335).*

20. **Calidris arenaria.** Sanderling.

Scarce, and said not to breed further south than lat. 68°, but the young have been obtained at Godthaab. Found on the East Coast by Graah, and by the German Expedition on Sabine Island where it was breeding. Said to have been found breeding in considerable numbers on the Parry Islands; but authentic eggs have been only recently made known to naturalists (Proc. Zool. Soc. 1871, pp. 56, 546, pl. iv., fig. 2.; Zweite deutsche Nordpolarfahrt, ii. p. 240), and are very rare in collections. About the size of a Skylark. May be distinguished from other Sandpipers by wanting the hind toe, and from the small Plovers, which have only three toes, by the mottled colouring (grey, rufous and black) of its upper plumage. The abundance of this bird during many months of the year on the coasts of the British Islands, and many other countries both of the Old and New World, together with the absolute want of any positive and trustworthy information as to the peculiarities which would seem to accompany its habits during the breeding-season, and the selection of its places of nidification, render these matters deserving of close attention.

21. **Phalaropus fulicarius.** Grey [or Red] Phalarope. "Kajok?"

Said to be the latest summer-bird to arrive, to be very rare in the south and not to breed below lat. 68° N., but thence northward to be common. Its common English name of "Grey" Phalarope is exceedingly inapplicable when in its summer plumage, for then the whole of the lower parts are of a bright orange-red colour, the upper parts being diversified with dark brown and tawny-yellow. The breeding-habits of this bird are little known, and it would seem to be often mistaken for the next species, which is far more common, and readily distinguished by the white plumage of its lower parts—even in summer, and its more slender bill.

22. **Phalaropus hyperboreus.** Red-necked Phalarope. "Nelloumirsortok."

Seems to be the commonest species of Phalarope throughout the country, and possibly occurs very far to the northward, though in the Arctic Regions of the Old World it does not go anything like so far as the preceding. The difference between the two birds has been given above.†

* *Catoptrophorus semipalmatus.* Willet. A bird seen by the late Prof. Goodsir in Exeter Sound was ascribed by him to this species (Arctic Voyage, p. 145); but the matter must be regarded as doubtful in the highest degree.

† *Phalaropus wilsoni*, though never yet met with far to the northward, may be not unreasonably expected to occur, if only as a straggler, within the Arctic Circle. It can be readily distinguished from either of the foregoing by its longer and more slender bill and legs.

(39.) *Tringa minutilla.* American Stint.
One shot in the spring of 1867 on Noursoak Peninsula.*

(40.) *Tringa maculata.* Pectoral Sandpiper.
One was received from Greenland in 1851 by the Copenhagen Museum, and two more examples were sent thither from Nenortalik in 1859.

23. TRINGA BONAPARTII. Bonaparte's Sandpiper.
Believed by Holböll (according to Dr. Paulsen) to breed near Julianehaab, where small flocks of both old and young birds have been observed in August. A very young bird was obtained at Nenortalik in 1835, one undergoing the change to winter-plumage in 1840, and three were procured there in 1841.

24. *Tringa alpina.* Dunlin. "Tojuk."
Dr. Paulsen has more than once received this species from Greenland both in young and autumn plumage. It probably breeds there, as it certainly does on Melville Peninsula, and elsewhere on the coast of Davis Strait. The Dunlin of the American continent seems to be constantly larger than that of Europe, and has been described as distinct by the name of *Tringa americana.* No appreciable difference in plumage is, however, perceptible.†

25. **Tringa striata.** Purple Sandpiper. "Sarbarsuk," "Sirksariarsungoak."
Occurs in winter even so far as the sea is open, and is of general distribution. Though not mentioned by Graah as met with on the East Coast, some twenty or thirty were seen on Sabine Island by Dr. Pansch.

26. **Tringa canutus.** Knot. "Kajok? ‡" "Kajordlik?"
Rare in the South, but often met with in the North: believed not to breed below lat. 68° N. Is thought to have its nest in the bays of Greenland, but authentic eggs seem never to have been obtained in that country, nor are such known to exist in collections. After the breeding-season resorts to the outer islands. Is reported to have been found breeding on Melville Peninsula, and in great abundance on the Parry Islands. The large flocks of this bird which in autumn and spring throng our own coasts, as well as those of Europe and temperate North America, to say

---

* "*T. minuta.*" A single specimen brought home by Mr. Edwards (Richardson, App. Parry's Second Voyage, p. 354). The "*T. minuta*" of Dr. Walker was *T. striata.*

† There are several other species of Shore-Sandpipers which may be not unreasonably looked for (perhaps as stragglers) in high latitudes. Little, if anything, is known of their breeding-habits, and therefore the occurrence of such birds is especially worthy of attention.

‡ This name is also common to *Phalaropus fulicarius,* doubtless from the similarity in the colour of the summer-plumage of the two species. The Knot, however, is at least twice as large as the Phalarope. In Iceland, where both birds occur, they are equally confounded by the natives.

nothing of countries lying much further to the southward, while its breeding-habits are not known with any certainty, render it especially an object of interest; and any light that can be thrown on its place and mode of nidification will be most valuable, for there is no common bird respecting the summer-haunts of which ornithologists are at present more ignorant. About the size of a large Snipe, but with much shorter bill and legs, it is in summer of a bright orange-red on all the lower parts, and above mottled with black, reddish-brown and white, the rump being white or white tinged with red. In its chief breeding-quarters, wherever they may be situated, it must be numerous, judging from its abundance at other times of the year. Large flocks are known to occur in Iceland, but these do not stay there many days and pass on—obviously to the northward. It has not been met with on the east coast of Greenland nor in Spitsbergen; the presumption, therefore, is that the countries to the west or north of Greenland are the goal of its vernal migration.

(*41.*) *Macrorhamphus griseus.* Brown Snipe.
One sent from the Fiskenæs in 1824.

27. *Gallinago media.* Common Snipe.
One received by Dr. Paulsen in 1845, but the species has been so often observed in Greenland, that it may very likely breed there, though positive information as to the fact is not forthcoming.*

(*42.*) *Limosa ægocephala.* Black-tailed Godwit. "Sargvarsurksoak."
Fabricius seems to have seen a single specimen, and one is said to have been obtained at Godthaab prior to 1820.

(*43.*) *Numenius borealis.* Esquimaux Curlew.
Two specimens supposed to have been of Greenland origin have been received at Copenhagen; one was brought in 1858 and was said to have been shot at Julianehaab; about the other Prof. Reinhardt knows nothing.†

(*44.*) *Numenius hudsonicus.* Hudsonian Curlew.
One sent from Godthaab many years since by Holböll, who says he had seen two others from Julianehaab and the Fiskenæs respectively.

28. *Numenius phæopus.* Whimbrel.
Nearly a dozen examples, sent from all parts of the country, have been received, and, though Holböll doubts its doing so, Prof. Reinhardt thinks that this species may breed in Greenland.

---

* The American-Snipe (*Gallinago wilsoni*) which very closely resembles our own bird, but differs in possessing sixteen instead of fourteen tail-feathers, may perhaps be looked for to occur in Greenland.

† Three individuals of a species of *Numenius* flew past the ships' boats in Regent Inlet. (Sabine, Suppl. App. p. ccx.)

29. **Sterna hirundo.** Arctic Tern. "Imerkoteilàk."

Breeds in various suitable localities on both coasts of Greenland, as well as on the western shores of Baffin's Sea.

30. **Xema sabinii.** Sabine's Gull.

Said not to breed further south than lat. 75° N. and appears not to be common in Danish Greenland, but was found by Sir E. Sabine breeding in great numbers on three small islands in lat. 75° 30′ [*qu.* Sabine Islands in Melville Bay ?] associated with the Arctic Tern. Many specimens were obtained in June and July at Winter Island and Aulitiwick, where subsequently flocks were seen flying high, as if migrating to the southward. Has been found breeding in North-western America, but nothing has yet been recorded of its habits in that quarter. Sir E. Sabine informed Richardson that he killed two in Spitsbergen, and the latter says that the specimen brought thence was in full summer-plumage, but it has not since been observed by others in that country. Dr. von Middendorff found it breeding abundantly at the mouth of the Taimyr, again in company with the Arctic Tern. The fact of these two species resorting to the same spot in localities so far apart should put observers on their guard against the possibility of confounding the nests and eggs of each. The eggs of this Gull are extremely rare in collections, and such as have been seen do not so much differ from those of the Tern (which are common enough) as to obviate the need of the most careful identification. This Gull is of small size and may be distinguished from others by its grey head, black collar and forked tail. From the Arctic Tern it may be known by its stouter build, less pointed wings and tail, and black bill and feet, the former having a yellow tip—the Tern having the bill and feet red, while in it the dark colour of the head is confined to a cap and does not extend below the eyes.

31. **Rhodostethia rosea.** Cuneate-tailed or Ross's Gull.

One of the rarest of birds, to be distinguished from other Gulls by its small (almost Dove-like) black bill, white head and neck, with a black ring round the latter, and wedge-shaped tail—the plumage, especially of the lower parts, deeply tinged with rose-colour. Four specimens have been received from Greenland by the Museum of Copenhagen, of which three were shot in Disco Bay, and the fourth near the Sukkertop, while a fifth is believed to have been obtained by Holböll. Originally discovered at Alágnak, in Melville Peninsula, where two examples were killed. Nothing whatever is known of the breeding-habits of this species, and only three examples are believed to exist in this country, one of which is said to have been killed in Yorkshire. It has occurred once in Heligoland, and once in the Færoes. The only specimen known on the continent of Europe is in the Museum of Mainz, and there appear to be none in America.

32. **Pagophila eburnea.** Ivory-Gull. "Nayauarsuk."

The well-known circumpolar "Ice-bird" needs no description, but long as Arctic navigators have been acquainted with it, its nest seems to have been undiscovered until 1853, when Sir L. M'Clintock found one on Cape Krabbé (lat. 77° 25′ N.), containing a single egg (Journ. R. Dubl. Soc., i. p. 57, pl. 1). Subsequently two eggs were obtained by one of the Swedish Expeditions in Spitsbergen, and these seem to be the only authenticated specimens that have been brought to the notice of naturalists. The bird itself is far from being uncommon in collections, and in some parts of the Arctic Regions is pretty plentiful. It is subject to some variation in size, and especially in the relative dimensions of some of its parts, but there is no good reason to suppose that there is more than one species of the genus.

33. **Rissa tridactyla.** Kittiwake. "Tattarak."

Breeds in both Inspectorates, but more commonly in the Southern. Recorded by Graah from the Eastern Coast of Greenland, though not observed there by the German Expedition. Its limits to the northward have not been laid down. The black quill-feathers of its wings are an unfailing distinction between this Gull and any other of its size likely to be met with far north.

(*45.*) *Larus argentatus*. Herring-Gull.

An accidental and extremely rare bird in Greenland, where it can only be a straggler, and is not known to have occurred further north than Godthaab. Dr. Walker says he saw it at Frederikshaab. A pair observed at Winter Island, 29 June, 1822. Larger than the preceding species, but like it has black primary quills. A doubtful species (*L. affinis*, Reinhardt), with a darker back, is said to have been obtained in Greenland, while on the other hand a form, with a paler back (*L. chalcopterus*, Licht.)—of which only three specimens have been procured,—seems to indicate a transition to the next.

34. **Larus leucopterus.** Iceland or Lesser White-winged Gull. "Nayangoak."

Breeds in both Inspectorates, but more commonly in the Southern. Also observed on the East Coast, and said to breed on the Parry Islands. In Greenland it is reported to be the most common Gull after the Kittiwake. Its comparatively small size, pale blue mantle (which, however, is subject to some variations of shade), and white primaries distinguish this species from any other. Immature birds vary greatly in the intensity of the brown clouding of the plumage.

35. **Larus glaucus.** Glaucous Gull or Burgomaster. "Naya." "Nayavek," "Nayainak."

The most common large Gull in Greenland. At Najartut, south of Godthaab, said to breed by itself, but most generally in company with *Rissa tridactyla* and *L. leucopterus*. Subject to the same variation of shade as the latter, but the existence of species called *L. arcticus* and *L. glacialis* has not been confirmed. Found also on the west side of Davis Strait and the East Coast of

Greenland, and said to be as numerous in the Polar Sea as it is in Davis Strait.

36. LARUS MARINUS. Great Black-backed Gull. "Nayardluk," "Nayardlurksoak."

Breeds generally throughout Danish Greenland, but most commonly between lat. 63° and lat. 68°. As large as the preceding species, or larger, but easily distinguished therefrom by its black back and primaries.

(*46.*) *Stercorarius catarrhactes.* Great Skua.
Seen twice on the south coast by Holböll.

37. **Stercorarius pomatorhinus.** Pomatorhine Skua.

Said to be the commonest species of Skua in the north. Breeds in societies from Bjornenæs, north of Egedesminde, to the northward. Several were killed in Regent Inlet, and it was also seen on the Parry Islands, but more rarely than the next species. Authenticated eggs of this bird are rare in collections. It is easily distinguished in flight by the peculiar formation of the two middle tail-feathers, which are twisted near the tip, so as to take a vertical direction, and give the appearance of a disc or ball attached to the bird's tail.

38. **Stercorarius parasiticus.** Common Skua. "Isingak," "Meriarsairsok."

Breeds in both Inspectorates, but most commonly in the Southern. Found on the East Coast by Graah, but not by the German Expedition. Obtained also on the west coast of Davis Strait. Equally abundant in the Polar Sea as in the latter. To be distinguished from the preceding species by its smaller size and perfectly straight tail. This and the next species appear to be "dimorphic," a wholecoloured * and a particoloured bird being often found paired, and the difference in plumage seems to be irrespective of sex or age; but on this point further information is desired.

39. **Stercorarius longicaudatus.** Buffon's Skua.

Said not to breed further south than lat. 70° N. One example obtained by the Germans. To be distinguished from the last species by its smaller size, more slender bill, and, even on the wing, by its exceedingly long tail. Would seem to be rather less "dimorphic" than *S. parasiticus.*

40. **Procellaria glacialis.** Fulmar or Mallemoke. "Kakordluk," "Kakordluvek;" dark variety, "Igarsok."

Said not to breed further to the south than lat. 69° N. Occurs also in East Greenland (Pansch). A very unmistakeable bird, but worthy of attention since individuals vary a good deal in the

---

* It is to this wholecoloured form that the name *S. richardsoni* properly applies.

shade of colouring. The young are supposed to be darkest in hue, but some seem to keep this sign of immaturity all their life.

41. PUFFINUS MAJOR. Greater Shearwater. "Kakordlungnak."

Marked by Prof. Reinhardt as breeding in Greenland, and said by Holböll to be found in great numbers from the southern point of the country to lat. 65° 30′ N.; the eggs of this bird are utterly unknown. Shearwaters of some species have many times been noticed in abundance off Cape Farewell.

(47.) *Puffinus kuhli.* Grey Shearwater.

Only known from Greenland by a specimen received thence by Herr Möschler and now in the Leyden Museum (Schlegel, Mus. Pays-Bas, *Procellariæ*, p. 24).

(48.) *Puffinus anglorum.* Manks Shearwater.

Once received from Greenland. The changes of plumage undergone by Shearwaters seem to be somewhat analogous to those of the Skuas, and no ornithologist at present has been able to give a rational explanation of them.

42. THALASSIDROMA LEACHI. Fork-tailed Petrel.

Constantly observed near the coast to lat. 64° or 65° N., and most frequently about the entrance of Godthaab Fjord, on the islands in which it is said to breed.*

(49.) *Thalassidroma bulweri.* Bulwer's Petrel.

Only known from Greenland by a specimen received thence at the Museum of Leyden (Schlegel, Mus. Pays-Bas, *Procellariæ*, p. 9), from the Moravian missionaries.

43. **Fratercula arctica** (?) Puffin. "Killangak."

Puffins seem to be nowhere common in Greenland, and are said by Holböll not to breed further south than lat. 63° 30′ N., which seems a questionable assertion. Whether two species are found there is also a doubtful matter.† The Puffin of Spitsbergen appears to the compiler to be justifiably separable from that which inhabits more southern stations in Europe on account of its much larger size, and to it should probably be assigned the name of *F. glacialis* (Leach), but the type of that supposed species is said to have been received from Greenland, whence Cassin also says he has seen it. On the other hand Prof. Reinhardt says that all the Puffins he has examined from Greenland belong to the common species *F. arctica.* The difference between the two is admittedly only one of size, though that difference is

* Two examples of *Procellaria pelagica*, the common Stormy Petrel, with the locality "Groënland" are contained in the Museum of Leyden, having been received direct from Holböll, who doubtless obtained them on one of his voyages, but whether in the Greenland seas is another matter.

† *Fratercula cirrhata*, the Tufted Puffin, a bird of the north-west coast of North America, is said to have been received from Greenland (Möschler, Journ. f. Orn. 1856, p. 335); but there is most likely some mistake about it.

great. A series of specimens which would help to clear up this matter is something to be desired.

44. **Uria grylle.** Black Guillemot or Greenland Dove. "Serbak," "Sergvak"; (in summer) "Kernekungojuk," "Kernektarsuk"; (in winter) "Kakortungojuk."

Very numerous on both coasts of Greenland, and said to remain longer than any other bird. Plentiful also on Melville Peninsula but more rarely seen in the Polar Sea. The distribution of the various species of Black Guillemot (which it may be observed is, except in the breeding plumage, anything but "black") is matter deserving of the fullest attention. The ordinary form from Spitsbergen is of slender build, and has the wing-spot in the adult purely and entirely white. That of the Norwegian and British coasts (*U. grylle*, vera) is stouter, and has the white feathers of the wing-spot with black at the base, but this colour does not shew outwardly. That of the North Pacific (*U. columba*) has a distinct black bar across the wing-spot, while another form (*U. carbo*) is altogether black. Now a specimen not to be distinguished from the typical *U. columba* was obtained in the Spitsbergen seas by Dr. von Heuglin, and Holböll says he has seen in Greenland an entirely black example, which, therefore, may perhaps be regarded as *U. carbo*. Whether these were exceptional varieties of the normal form, or examples which had accidentally wandered from their proper habitats is a question which cannot be decided—but in the latter case the question has an important geographical aspect, as tending to show the occasional means of water communication between opposite parts of the circumpolar region.

45. **Mergulus alle.** Rotge or Little Auk. "Akpalliarsuk," "Kaerrak."

Said not to breed further south than lat. 68° N., but, though its great stations are in the northern parts of Baffin's Sea, not to be common in the Polar Sea. Found also in East Greenland.

46. ALCA TROILE. Willock or Common Guillenot.

Two examples sent by Holböll from Godthaab, where, and perhaps in other places on the coast, it breeds, but still, to all appearance, very rarely. Its variety, *A. lacrymans*, seems to be still more rare in Greenland.

47. **Alca arra.** Bruennich's Guillemot. "Akpa."

Doubtless the commonest bird on the Greenland coasts, but said not to breed south of lat. 64° N. Occurred on Parry's Second Voyage. Holböll met with three specimens entirely black, two near Godthaab and one at the Sukkertop, but all in winter! Some recent writers have most unreasonably questioned or even denied the specific distinction of this and the foregoing.

48. ALCA TORDA. Razor-bill. "Akparnak," "Akpartluk."

Not rare either in the Northern or Southern Inspectorate, but not hitherto observed on the East Coast.

(*50.*) *Alca impennis.* Gare-fowl or Great Auk. "Isarokitsok."

The earliest discovery of this remarkable and interesting species in Greenland was in or about the year 1574, when an Icelander, by name Clemens, visited certain islands on the east coast, then called Gunnbjarnareyjar, and since identified with Danell's or Graah's Islands, lying in lat. 65° 20′ N., whereon he found it so plentiful that he loaded his boat with the birds. It has not since been known to occur on that coast. Bruennich, in 1764, did not mention Greenland as a locality for it. Fabricius, in 1780, while giving its Esquimaux name, says that it was rarely seen on the outer islands, and that in winter; he had, however, examined a young bird, only a few days old, taken in August. Old birds, he adds, were very rare. The Museum of Copenhagen possesses a specimen, said to have been killed on Disco in 1821, but this is very possibly that which is known to have been procured by Heilmann at the Fiskenæs in 1815. The last examples with certainty known to have existed were killed on Eldey, off the south-west point of Iceland in 1844.

(*51.*) *Podiceps auritus.* Horned Grebe.

A few immature specimens have been obtained in the southern part of Greenland.

(*52.*) *Podiceps holbœlli.* American Red-necked Grebe.

This New-World representative of the Old-World *P. griseigena*, was first described as a distinct species from specimens obtained in Greenland, but its specific validity is questioned by many ornithologists. It seems to have occurred three times in that country.

49. **Colymbus septentrionalis.** Red-throated Diver. "Karksauk."

Found on the East Coast and breeds in both Inspectorates, as also on the western coast of Davis Strait.

50. Colymbus glacialis. Great Northern Diver. "Tudlik."

Observed by Graah on the East Coast, on the West breeds generally, but more in the South than the North, where indeed it seems to be rare. Examples of this bird from the Fur Countries and west of North America, with a pale-coloured bill have been described as forming a distinct species, under the name of *C. adamsi*, but the like are to be met with in Europe.*

(*53.*) *Sula bassana.* Gannet. "Kuksuk."†

Accidental and rare.

51. Phalacrocorax carbo. Cormorant. "Okaitsok."

Said by Holböll to breed from the Godthaab Fjord northward so far as he had been. Observed also on the East Coast.

---

* *Colymbus arcticus*, the Black-throated Diver, was found in considerable numbers in Parry's Second Voyage.

† This name seems to be also applied to the Swan.

52. **Mergus serrator.** Red-breasted Merganser. "Pajk," "Nyaliksak."

53. CLANGULA ISLANDICA. Barrow's Goldeneye. "Kærtlutorpiarsuk," more properly "Niakortok."

Breeds in South Greenland only, and apparently not further north than Godthaab.

(54.) *Clangula albeola.* Buffel-headed Duck.

One obtained at Godthaab about the year 1830.

54. HISTRIONICUS TORQUATUS. Harlequin-Duck. "Tornauiarsuk."

Observed on the East Coast: most common between lat. 62° and 65° N., rarer to the northward. The male of this species, from its singularly marked plumage, cannot be confounded with any other species; the female is known by its dusky head and the white spot on either side.

55. **Harelda glacialis.** Long-tailed Duck. "Aglek."

Common on the whole coast, and breeds also on the Parry Islands and on the land westward of Davis Strait. The long tail of the male sufficiently distinguishes it from that of any other Duck; the female has a white or dirty-white head with dusky spots.

(55.) *Fuligula marila.* Scaup-Duck.

Dr. Walker, of the "Fox," R.Y.S., obtained one at Godhavn, in August 1857. Three specimens were sent from Nenortalik in 1859.*

(56.) *Fuligula affinis.* American Scaup-Duck.

A pair was shot in June on Innusulik, an islet some ten miles from Egedesminde. It may possibly breed in Greenland.

56. *Œdemia perspicillata.* Surf-Scoter.

A few specimens have been obtained from the Danish settlements. It was observed by Graah on the East Coast.

57. **Somateria mollissima** (?). Eider. ♂ "Amaulik," ♀ "Arnauiak," "Mittek."

Common along all the coasts, northern limit unknown. In the Eider of the New World (*S. dresseri*), regarded by Mr. Sharpe as distinct from that of the Old, the bill is more gibbous, and the bare space behind the nostril more extended than in the European bird. The Eider of Davis Strait, and thence northward, will probably be found to belong to the American form, but the Eider of the east coast of Greenland is very likely to be the European. The Eider of Spitsbergen has also been separated from *S. mollissima* by Dr. Malmgren under the name of *S. thulensis*, but the asserted difference between them, if it can be maintained at all, is but

* The Tufted Duck, *Fuligula cristata*, is said to have been obtained, at Godhavn, by Dr. Walker, but this was probably a mistake.

slight. In Western Arctic America occurs a very good species, the *S. v-nigrum*, larger than *S. mollissima* or *S. dresseri*, and the male having a black chevron under the chin, as in that of the following.

58. **Somateria spectabilis.** King-Duck. "Siorakitsok," ♂ "Kingalik," ♀ "Kaiortok," "Arnauiartak."

Said not to breed further south than lat. 67° N., but in some numbers at lat. 73°. Also on the East Coast of Greenland and on the western shores of Davis Strait. Breeds abundantly on the Parry Islands. The male easily distinguished from other species of the genus by its grey head and protuberant nasal disc. The female much resembles that of *S. mollissima* or *S. dresseri*, but is smaller and more ruddy, and the sides of the bill are not feathered up to the nostrils, while the central nasal ridge extends as far as the nasal openings. Identified eggs of the King-Duck are scarce.

59. ANAS BOSCHAS. Wild Duck. "Kærtlutok."

Breeds in both Inspectorates, and is not rare.

(*57.*) *Anas acuta.* Pintail. "Kærtlutorpiarsuk."

Of accidental but not very rare occurrence.

(*58.*) *Anas crecca.* Teal. "Kærtlutorpiarsuk."

A few examples have been killed at different places among the Danish settlements.

(*59.*) *Anas carolinensis.* American Teal.

Four specimens are known to have been obtained in South Greenland prior to 1860.

(*60.*) *Anas penelope.* Widgeon.

A young drake sent by Holböll in 1851. Prof. Reinhardt has seen two others also killed in South Greenland.

60. **Bernicla brenta.** Brent-Goose. "Nerdlek."

Said not to breed in Greenland lower than lat. 70° N., but does so in great numbers in the Polar Sea. Is the smallest species of Goose found in the Arctic Regions, and easily distinguished by its black head and neck, each side of the latter having only a small semilunar patch of white. In the form called *B. nigricans*, which, though most common on the Pacific coast of North America, also occurs on the Atlantic, the black of the throat extends lower down and over part of the breast, and the white patches of the neck almost or quite meet in front.

61. *Bernicla leucopsis.* Bernacle-Goose.

A regular autumnal visitor at Julianehaab, and may perhaps breed in Greenland. Recorded also by Graah from the East Coast. The breeding of this species in a wild state seems only to have been observed by Dr. von Middendorff in Siberia, though eggs laid by tame birds are common enough. Two or more forms intermediate between this and the next species have been described. It may possibly happen that the Bernacle-Goose of the New World hitherto attributed to *B. leucopsis* is distinct.

(61.) *Bernicla canadensis*? Canada Goose.

A specimen, supposed to be from Greenland, in the Museum of Copenhagen, has been doubtfully assigned to this species, which is perhaps the biggest Goose known. It may possibly, however, be the *B. hutchinsi*, which is said to be distinguishable from the true *B. canadensis* by the possession of *sixteen* instead of *eighteen* tail-feathers. But the American Geese of this form have not as yet been clearly differentiated, and it seems impossible to furnish a true diagnosis of the supposed species which have received the name of *B. leucopareia* and *B. leucolæma*.

62. **Chen hyperboreus.** Snow-Goose.

A few young birds only have been seen, and these more frequently in the Northern Inspectorate than in the Southern. Is found also on the west coast of Davis Strait. Probably breeds in the far north, but a doubt may perhaps be entertained whether the examples killed in Greenland belong to the true *C. hyperboreus* or to *C. albatus* (if these be really distinct), which is said to have occurred in Ireland.

63. Anser gambeli. American White-fronted Goose. "Nerdlernak."

Not rare in fresh water between lat. 66° and 68° 30′, and also observed by the German Expedition on the east coast. Though the White-fronted Goose of Greenland has been generally assigned to the European form, *A. albifrons*, it would seem to belong rather to the larger American *A. gambeli*; but the difference between the two appears to be that of size only. The true *A. albifrons* is a regular visitant to Iceland, and therefore the specimen obtained by Dr. Copeland on the East Coast may well belong to that form, though it does not follow that the birds which frequent the west coast are of the same form.

(62.) *Cygnus ferus*? Wild Swan. "Kuksuk."

The Swan which occurs occasionally in Greenland has been generally referred to the European species (*C. ferus*), but that which was observed at Igloolik, on Parry's Second Voyage, and is said to breed on the Parry Islands (though not numerously), seems more likely to be one of the American species, *C. buccinator* or *C. americanus*. Hence a reasonable doubt may exist as to which of the three the Greenland examples are.

From the foregoing list it will be seen that, while *sixty-two* of the birds therein enumerated are nothing but stragglers to Greenland, the number of those which may be called denizens of the country cannot be raised above *sixty-three*, to reach which we must even, in some cases, stretch a point. That Greenland, so far as its birds are concerned, belongs to the Nearctic Region has long been known, and the fact in respect of the *species* can be most conveniently shown thus :—

	Species belonging to the Old World.	Species belonging to the New World.	Species common to both Worlds.	Doubtful.
Stragglers	19	34	8	1
Inhabitants	5	11	45	2

The result with regard to the *genera* under which the species are named is not very different :—

	Genera belonging to the Old World.	Genera belonging to the New World.	Genera common to both Worlds.
Stragglers -	2	12	28
Inhabitants -	0	2	45

Turning to the range of the species in Greenland itself, we find that of the 62 stragglers only 9 are known to have penetrated to North Greenland, while the localities whence 13 were procured are not named. Supposing that the same proportion of northern stragglers exists among the 13 of which no particulars have been given as among the 49 of which we know the locality, the number of stragglers to North Greenland may be raised to 12, all of which may reasonably be supposed to have passed through the limits of South Greenland. Four-fifths of the stragglers named in this list may accordingly be safely dismissed from our mind, when considering even the casual visitors to that part of Greenland which lies nearest to the scene of the new Expedition's labours. The remainder are not Arctic Birds in any sense, since they have not crossed the Polar Circle, and indeed many of them have hardly been within 400 miles of it.

Then of the regular denizens, which, taking the highest estimate, cannot be put at more than 63, we find that 16—or nearly one-fourth—do not occur within the Polar Circle, and are therefore not entitled to the name of Arctic Birds. The remaining 47 are recorded as inhabiting North Greenland, but their northward extension is uncertain. Considering, however, what is known of them in other parts of the world, and various facts which seem to bear on their geographical range, we may arrive at something like an approximation of the number which may not unreasonably be looked for in Smith Sound. Yet, making the most liberal allowance, this number cannot be raised above 36,* and to these

* I am quite aware that this allowance is too great, but I think it best to err on the safe side. If the Expedition meets with 30 species in Smith Sound it will surpass expectation. The number of species, including stragglers, at present known to have occurred in Spitsbergen does not exceed 30.

36 species should attention be particularly directed—how much further in the direction of the Pole any of them may go it is of course impossible to forecast. The principal features by which each may be distinguished have been briefly noticed by me, and, I trust, in a way that may lead to an easy and correct determination even by those observers who are not professed naturalists.

Magdalene College, Cambridge,
20 March, 1875.

VII.—A REVISED CATALOGUE of the FISHES of GREENLAND. By Dr. CHR. LÜTKEN, University Museum, Copenhagen. 1875.*

Obs.—The only general account of the Fishes of Greenland published since the time of Fabricius is that of the late Professor J. Reinhardt (K. D. Vid. Selsk. Skr. VII., 1838); and the following list must still be regarded as provisional and open to corrections. Several species have only been indicated, not described, and must be regarded as doubtful until the revision of the Ichthyology of Greenland, in preparation by the writer, is finished. For further particulars the papers of the late Professor Kröyer in "Naturhistorisk Tidsskrift," Series I. and II., and Ser. III., Vol. I., and the Altases of the "Voyage en Islande et au Groenland," and the "Voyage en Scandinavie, en Laponie, &c.," par Gaimard, should especially be consulted.

[The species marked with an asterisk (*) are well represented in British collections, or can be obtained from dealers. With respect to them see the "Instructions for collecting Fishes," issued with this "Manual" for the Expedition.—EDITOR.]

Gasterostei.

*1. *Gasterosteus aculeatus*, L. (Greenl. *Kakilisak.*)
(Var. trachurus.)
F. Gr. 122; G. loricatus, Rhdt. (var. dimidiatus et gymnurus, Rhdt.).
G. noveboracensis, C. V. (pp); Gthr. Cat., I., p. 2, &c.

Scomberoidei.

2. *Lampris guttatus* (Retz.).
C. V., t. X., p 39, t. 282; Gthr. Cat., II., p. 416.
Gaimard, Voyage en Islande et au Groenland, t. 10.
(The skeleton of a specimen caught at Arsuk is in the Museum of Copenhagen.)

---

* *See also* Reinhardt's Catalogue of Fishes in Rink's Naturhist. Bidrag til en Beskrivelse af Grönland, af J. REINHARDT, J. C. SCHIÖDTE, O. A. L. MÖRCH, C. F. LÜTKEN, J. LANGE, H. RINK. Tillæggene til "Grönland, "geograph. og statist. beskrev." af H. RINK. 8vo. Copenhagen, 1857.

Notacanthini.

3. *Notacanthus Fabricii* (Rhdt.).
Campylodon, O. Fabr., Nat. Hist. Selsk. Skr. IV., 2, p. 21, t. 10, f. 1.
N. nasus, Bl.; Voyage en Islande, &c., t. 11.
C. V., t. VIII., p. 467, t. 241; Gthr. Cat., III., p. 544.
(Fabricius's original specimen is lost.)

Cottoidei.

*4. *Cottus scorpius*, L. (Greenl. *Kaniok.*)
F. Gr. 113.
C. grœnlandicus and C. porosus, C. V., IV., p. 185; VIII., p. 498; Gthr. Cat., II., p. 161, 159.; Voyage en Islande, t. 9, f. 2.
C. ocellatus and labradoricus, Stor.; C. glacialis, Rich.; Acanthoc. variabilis, Gir.
†C. quadricornis, Ross (non Sab.).

5. *C. scorpioides*, Fabr. (Gr. *Pokudlek, Igarsok, Akullikitsok*).
F. Gr. 114; C. V., IV., p. 187.

6. *Phobetor ventralis* (Val.). (Greenl. *Itekivedlek, Kanikitsok, Ujarangmio.*)
F. Gr. 115 (C. gobio); C. V., IV., p. 194, t. 79, f. 1; Gthr. Cat., II., p. 168.
Cottus tricuspis, Rhdt.; Gymnacanthus tric., Gill.
Acanthocottus patris, Stor.
Phobetor tricuspis, Voy. en Scandinavie, t. 4, f. 1; Nat. Tidsskr., II. 1., p. 263.

7. *Centridermichthys uncinatus* (Rhdt.).
K. D. Vid. Selsk. Skr., VI., p. liii.
Gthr. Cat., II., p. 172.

8. *Centridermichthys bicornis* (Rhdt.).
K. D. Vid. Selsk. Skr., VIII., p. lxxv.
Gthr. Cat., II., p. 172.

9. *Icelus hamatus*, Kr.
Nat. Tidsskr., II., 2, p. 253; Voy. en Scandin., t. I., f. 2; Gthr. Cat., II., p. 172.
? Cottus polaris, Sab.

10. *Triglops Pingelii* (Rhdt.).
Nat. Tidsskr., II. 2, p. 260; Voy. en Scand., t. 1, f. 1.
? T. pleurostictus, Cope.
Gthr. Cat., II., p. 173.

11. *Aspidophorus decagonus* (Schn.). (Greenl. *Kaniordluk, Kaniornak.*)
F. Gr. 112; C. V., IV., p. 223; Gthr. Cat., II., p. 251.
Nat. Tidsskr., II., I., p. 243; Voyage en Scand., t. 5, f. 1. Archagonus decagonus, Gill.

---

† The true *C. quadricornis*, L. (*hexacornis*, Rich.), (*Oncocottus quadricornis*, Gill), according to Sabine and Richardson an inhabitant of Arctic America, has never been sent from the Danish Settlements in Greenland. Peters determined a *Cottus* from the East Coast of Greenland as *C. hexacornis*, Rich., but is inclined to regard it as identical with *C. scorpius*, Fab. (Zte deutsche Nordpolarfahrt, II., p. 169).

Young : *A. spinosissimus*, Kr.
Nat. Tidsskr., l. c., p. 250; Voyage en Scand., t. 5, f. 2; Gthr. Cat., II., p. 214.

12. *Aspidophoroides monopterygius* (Bl.).
C. V., IV., p. 224; VI., p. 554, t. 169; Gthr. Cat., II., p. 216.

13. *A. sp.*
(An undescribed species in the Museum at Copenhagen.)

*14. *Sebastes norvegicus* (Müll.). (Greenl. *Sullupaugak.*)
F. Gr. 121; C. V., IV., p. 327, t. 87; Günth. Cat., II., p. 95; Nat. Tidsskr., II. 1, p. 270; Voyage en Islande, &c., t. 9, f. 1.
(*S. viviparus*, Kr. Nat. Tidsskr., II. 1, p. 275, Voy. en Scand., t. 6, Gthr., p. 96, is now commonly regarded only as a variety of *S. norveg.*)

Lophioidei.

15. *Himantolophus grœnlandicus*, Rhdt.
K. D. Vid. Selsk. Skr., VII., p. 132, t. 4.
(Only a single mutilated specimen, which could not be preserved, with the exception of the frontal tuft, has been received.)

16. *Ceratias Holbœlli*, Kr.
Nat. Tidsskr., II. 1, p. 639; Voyage en Scand., t. 9.
Gthr. Cat., III., p. 205.
(Only a few specimens have been received; one only could be preserved.)

17. *Oneirodes Eschrichtii*, Ltk.
Overs. K. D. Vid. Selsk. Forh., 1871, p. 56, t. 2; Ann. Nat. Hist., 1872, Vol. IX., p. 329, t. 9.
(Only one specimen, not quite complete, is known.)

Discoboli.

*18. *Cyclopterus lumpus*, L. Greenl. *Nepisa*, *Angusedlok*, (♂), *Arnardlok* (♀).
F. Gr. 92 & 94 (C. minutus, the young).
Gthr. Cat., III., p. 155.
Voyage en Islande, &c., t. 8.

19. *C. spinosus*, Müll. (Greenl. *Nepisardluk.*)
F. Gr. 93; Nat. Tidsskr., II. 2, p. 262; Voy. en Scand., t. 4., f. 2; Gthr. Cat., III., p. 157.
Eumicrotremus spinosus Gill.

20. *Liparis Fabricii*, Kr.
Nat. Tidsskr., II. 2, p. 274; III. 1, p. 235.
Voy. en Scandin., t. 8, f. 2.
? L. communis, Sab. (sec. Gthr.); Gthr. Cat., III., p. 161.

21. *L. arctica*, Gill.
Proc. Acad. Phil. 1864, p. 191 (Port Foulke).

21*a*. *L. Montagui*, Don.
Nat. Tidsskr., III. 1, p. 243; Voy. en Scand., t. 13, f. 1.

22. *L. lineata* (Lep.), Kr.
Nat. Tidsskr., II. 2, p. 284; III. 1, p. 244.
Voy. en Scandinavie, t. 13, f. 2.
23. *L. tunicata*, Rhdt. (Greenl. *Abapokitsok, Amersulak.*)
F. Gr. 95 (pp.); Nat. Tidsskr., III. 1, p. 236.
L. barbatus, Ekstr. (sec. Malmgren).
L. (Actinochis) major, Gill.
24. *L.* (*Careproctus*) *Reinhardti*, Kr.
Nat. Tidsskr., III. 1, p. 252.
L. gelatinosus (Pall.), Rhdt.

Obs.—Prof. Peters (Zte deutsche Nordpolarfahrt, p. 172) refers several of these to one species, but is evidently not acquainted with Kröyer's last paper on the subject.

Blennioidei (**et Lycodini**).

25. *Stichæus præcisus* (Kr.).
Nat. Tidsskr., I. i., p. 25 et p. 372, ; III. 1., p. 295; Voy. en Scand., t. 20., f. 1.
Clinus (Stich.) unimaculatus, Rhdt.; Gthr. Cat., III., p. 283.
Eumesogrammus præcisus, Gill.
26. *St. punctatus* (Fabr.). (Greenl. *Akulliakitsok.*)
F. Gr. 110; Nat. Hist. Selsk. Skr., II. 2, p. 84, t. 10, f. 3; Nat. Tidsskr., III. 1, p. 303; Voy. en Scandinavie, t. 20, f. 2; Gthr. Cat., III., p. 283.
27. *Lumpenus aculeatus*, Rhdt.
Nat. Tidsskr., III. 1, p. 268; Voy. en Scandinavie, t. 14, f. 2; Gthr. Cat., III., p. 282.
Leptoclinus aculeatus, Gill. ? Lump. maculatus, Fries.
28. *L. Fabricii*, Rhdt. (Greenl. *Tejarnak.*)
F. Gr. 109; Nat. Tidsskr., III. 1, p. 274; Voy. en Scand., t. 14, f. 1.
Stichæus lumpenus, Gthr. Cat., III., p. 280.
29. *L. medius*, Rhdt.
Nat. Tidsskr., III. 1, p. 280; Stichæus medius, Gthr. Cat., III., p. 281. Anisarchus medius, Gill.
30. *L. gracilis*, Rhdt.
Nat. Tidsskr., III. 1, p. 282.
? Leptoblennius gracilis, Gill.

Obs.—*L. nubilus*, Rich. (Centroblennius nubilus, Gill), from Northumberland Sound, is not known from Greenland.

31. *Centronotus fasciatus*, Schn. (Greenl. *Kurksaurak.*)
F. Gr. 108 (Bl. gunellus); Gthr. Cat., III., p. 287.
Gunellus grœnlandicus, Rhdt.; C. V., XI., p. 441–42, t. 340. Murænoides fasciatus, Gill.
? Asterropteryx gunelliformis, Rüpp. (Gthr. Cat., III., p. 288.)
32. *C. affinis*, Rhdt.
K. D. Vid. Selsk. Skr., VII., p. 123.
33. *Lycodes Vahlii*, Rhdt. (Greenl. *Misarkornak.*)
K. D. Vid. Selsk. Skr. VII., p. 153, t. 5; Gthr. Cat., IV., p. 319.

34. *L. reticulatus*, Rhdt. (Greenl. *Akulliakitsok, Kussaunak.*)
K. D. Vid. Selsk. Skr., VII., p. 167, t. 6; Gthr. Cat., IV., p. 320.

35. *L. seminudus*, Rhdt.
K. D. Vid. Selsk. Skr., VII., p. 223; Gthr. Cat., IV., p. 320.

36. *L. perspicillum*, Kr.
Nat. Tidsskr., III. 1, p. 289; Voy. en Scandin., t. 7 Gthr. Cat., IV., p. 320.

37. *L. nebulosus*, Kr.
Nat. Tidsskr. III., 1, p. 293.

Obs.—L. polaris (Sab.) (Melville Island), and *L. mucosus* (Rich.) (Northumberland Sound), have not been received from the Danish Settlements.

*38. *Anarrichas lupus*, L. (Greenl. *Kigutilik.*)
F. Gr. 97; C. V., XI., p. 349; Gthr. Cat., III., p. 208; Voy. en Scand., t. 12, f. 2; Voy. en Islande, &c., t. 4.
A. vomerinus, Gill.

39. *A. denticulatus*, Kr.
K. D. Vid. Selsk. Overs., 1844, p. 140; Voy. en Scand., t. 12, f. 1.
(A single specimen in the Museum at Copenhagen.)

*40. *A. pantherinus*, Zouiew? (Greenl. *Kærrak.*)
F. Gr. 97, b (A. minor).
A. Steenstrupii, Gill.
(Günther, l. c., regards it only as a variety of *A. lupus.*)

41. *Bythites fuscus*, Rhdt. (Greenl. *Amersulak.*)
K. D. Vid. Selsk. Skr., VII., 175, t. 7; Gthr., IV., p. 375.
(Only known from a single specimen.)

42. *Gymnelis viridis* (Fabr.). (Greenl. *Unernak.*)
F. Gr. 99; Nat. Tidsskr., III. 1, p. 258; Voy. en Scand., t. 15.
G. punctulatus et lineolatus, Rhdt.; G. pictus, Gthr. Cat., IV., p. 323 and 324.
Cepolophis viridis, Kaup. Ophidium stigma, Rich.

Obs.—Uronectes Parryi (Ross), Gthr. Cat., IV., p. 326, found in Baffin's Bay during Parry's Third Voyage, has not yet been sent from the Danish Settlements in Greenland.

Gadoidei.

*43. *Gadus morrhua*, L. (Greenl. *Saraudlik, Saraudlirksoak.*)
F. Gr. 101 et 102 (G. callarias); Gthr. Cat., IV., p. 328; Voy. en Islande, &c., t. 16.
Morrhua americana, Stor.

*44. *G. ovak*, Rhdt. (Greenl. *Ogak* or *Ovak.*)
F. Gr. 103 (G. barbatus); G. ojac, Rich.
Voy. en Scandinavie, &c., t. 19. (According to Günther a variety of No. 43.)

*45. *G. agilis*, Rhdt. (Greenl. *Misarkornak.*)
F. Gr., No. 100 (G. æglefinus); G. saida, Lep., Gthr., IV., p. 337.
Merlangus polaris, Sab., Ross, Richardson.
G. Fabricii, Richards. Gthr. Cat., IV., p. 336.
Boreogadus polaris, Gill.

Obs.—Prof. Peters has established a new species, *G. glacialis*, on a specimen of Cod from Sabine Island (Zte deutsche Nordpolarfahrt, II., p. 172).

*46. *Merlangus carbonarius* (L.). (Greenl. *Ordlit.*)
F. Gr. 104 (G. virens); Gthr. Cat., IV., p. 339.; Voyage en Islande, &c., t. 6, f. 2.
Pollachius carb., Gill.

*47. *Merluccius vulgaris*, Cuv. (Greenl. *Akulliakitsok.*)
F. Gr. 105; Gthr. Cat., IV., p. 344.

*48. *Lota molva* (L.). (Greenl. *Ivirksoak.*)
F. Gr. 106; Gthr. Cat., IV., p. 461 (Molva vulgaris, Fl.).

*49. *Brosmius vulgaris*, Cuv. (Greenl. *Nejorpallugak*).
F. Gr. 107; Gthr. Cat., IV., p. 369 (B. brosme).
Voy. en Islande, &c., t. 5.

50. *Motella Reinhardti*, Kr.
K. D. Vid. Selsk. Skr., VII., p. 115 (M. mustela).
Onos Reinhardti, Gill.

51. *M. ensis*, Rhdt.
K. D. Vid. Selsk. Skr., VII., p. 128; Gthr. Cat., IV., p. 366. Onos ensis, Gill.

*52. *M. argentata*, Rhdt.
K. D. Vid. Selsk. Skr., VII., p. 128; Couchia argentata, Gthr. Cat., IV., p. 363. Ciliata argentata, Gill.
(A young form?)

Macruridæ.

53. *Coryphænoides Strœmii* (Rhdt.).
Macrurus Strœmii, Rhdt., Sundev.; Voy. en Scandin., t. 11.
Lepidoleprus norvegicus, Nilss.; Gthr. Cat., IV., p. 396.

54. *Macrurus rupestris* (Fabr.). (Greenl. *Ingmingoak.*)
F. Gr. 111.; M. Fabricii, Sundev.; Gthr. Cat., IV., p. 390.

55. *M. trachyrhynchus* (Risso).
Lepidoleprus trachyrh., R.; Gthr. Cat., IV., p. 395.
(A head alone was found on the ice at Kangek (Godthaabsfjord.)

Pleuronectidæ.

*56. *Hippoglossus vulgaris*, Fl. (Greenl. *Netarnak.*)
F. Gr. 117; H. maximus, Mind.; Gthr. Cat., IV., p. 403; Voy. en Islande, &c., t. 14.
? H. americanus, Gill.

57. *H. pinguis* (Fabr.). (Greenl. *Netarnarak, Kalleraglek.*)
F. Gr. 118 (Pleur. cynoglossus); K. D. Vid. Selsk. Skr., I., p. 43, t. 2, f. 1; Voy. en Scandinavie, t. 22.
H. grœnlandicus, Gthr., IV., p. 404 (cfr. p. 450).

58. *Drepanopsetta (Hippoglossoides) platessoides* (Fabr.).
F. Gr. 119; K. D. V. Selsk. Skr., I., p. 50, t. 2, f. 2; Voyage en Scandinavie, t. 21 (Greenl. *Okotak, Kollevsak*). Citharus platessoides, Rhdt. sen.

Ammodytidæ.

59. *Ammodytes dubius*, Rhdt. (Greenl. *Putsrotok.*)
F. Gr. 98 (A. tobianus).
K. D. Vid. Selsk. Skr., VII., p 131; Gthr., IV., p. 381.

Obs.—The name will perhaps be dropped. Kröyer distinguished two Greenland species which he regarded as new.

Anguillini.

60. *Anguilla*, sp. (Greenl. *Nimeriak.*)
F. Gr. 96.

Clupeacei.

*61. *Clupea harengus*, L. (Greenl. *Kapiselik.*)
F. Gr. 129; Gthr. Cat., VII., p. 415.; Cuv. & Val. xx., p. 30, t. 591–93.

Salmones.

62–66. *Salmo*, sp.
The Salmons of Greenland (F. Gr. 123–27), "Kapisarliksoak" (*S. salar*, Fabr.), "Ekalluk" (*S. carpio et alpinus*, Fabr.), "Ekallukak" (*S. stagnalis*), and "Aunardlek" (*S. rivalis*) are doubtful species, which have received no revision since the time of Fabricius. Cfr. Günther Cat., VI., p. 124. Peters (Zte deutsche Nordpolarfahrt, II., p. 174) determined two fresh-water specimens from Sabine Isl. (East Greenland) as *S. Hoodii*, Rich.?

67. *Microstoma* (?) *grœnlandica*, Rhdt.
K. D. Vid. Selsk. Skr., VIII., p. lxxiv.; Gthr. Cat., VI., p. 205.

*68. *Mallotus villosus* (Müll.). (Greenl. *Angmaksak.*)
F. Gr. 128 (*S. arcticus*); S. grœnlandicus, Bl., Rich.; Günth. Cat., VI., p. 170; Cuv. & Val. xxi., p. 392, t. 620–23; Voy. en Scand., t. 16, f. 1; Voy. en Islande, &c. t. 18, f. 1.

Scopelini.

69. *Scopelus glacialis*, Rhdt. (Greenl. *Keblernak.*)
F. Gr. 120 (Labrus exoletus); Gthr. Cat., V., p. 407.
Nat. Tidsskr., II. 2, p. 230; Voy. en Scandin., t. 16, f. 2.

70. *Stomias ferox*, Rhdt.
Nat. Tidsskr., II. 2, p. 253; Voy. en Scandin., t. 16 B., f. 2; Gthr. Cat., V., p. 426.

71. *Paralepis borealis*, Rhdt. (Greenl. *Saviliursak.*)
F. Gr. 130 (Clupea encrassicholus).
Nat. Tidsskr., II. 2, p. 241; Voy. en Scandinavie, t. 16 B., f. 1; Gthr. Cat., V., p. 419.
Arctozenus borealis, Gill.

Squali.

72. *Selachus maximus* (Gunn.). (Greenl. *Kaksibkannioa.*)
F. G. 90; Sq. peregrinus, Blv.; Sq. elephas, Les.
Cetorhinus maximus, Gill.
Gthr. Cat., VIII., p. 394.

73. *Centroscyllium Fabricii*, Rhdt. (Greenl. *Kukilik.*)
F. G. 88 (Squalus acanthias); Gthr. Cat., VIII., p. 425.

74. *Somniosus microcephalus* (Schn.). (Greenl. *Ekallurksoak.*)
F. Gr. 89 (Squalus carcharias); Somniosus brevipinna, Les.
Squal. borealis, Scor.; Squ. glacialis, Faber; Scymnus micropterus, Val.; Sc. Gunneri, Rich.
Læmargus borealis, M. H.†; Gthr. Cat., VIII., p. 426.
Voyage en Islande, &c., t. 22 & t. 1.

Rajæ.

75. *Raja radiata*, Don. (Greenl. *Taralikisak.*)
F. G. 87 (R. fullonica); Gthr. Cat., VIII., p. 460.

76. *R. sp.*
The eggs of a larger species have been received from Greenland.

Cyclostomi.

77. *Petromyzon fluviatilis*, L.
Gthr. Cat., VIII., p. 502.
Two specimens have been sent from the southern part of Greenland.

*78. *Myxine glutinosa*, L. (Greenl. *Ivik.*)
F. Gr. 334; Gthr. Cat., VIII., p. 510.

Obs.—The Fishes of Greenland which are of economical importance are especially *Cottus scorpius*, *Sebastes norvegicus*, *Cyclopterus lumpus*, *Hippoglossus vulgaris*, *Gadus agilis* and *ovak*, *Salmo carpio*, *Mallotus arcticus*, and *Somniosus microcephalus.*

Many of the rarer Fishes of Greenland are inhabitants of *great depths*, and owe their rarity in collections to this circumstance.

The short time allowed for the compilation of this list has prohibited every attempt to solve the doubts of identity, &c. with which the history of certain species is perplexed.

---

† The name *Læmargus* (M. H.) is preoccupied by Kröyer for a genus of parasitic Crustacea.

VIII.—On the STRAINING APPENDAGES or BRANCHIAL FRINGES of the BASKING SHARK (*Selachus maximus*, Gunn.). By Prof. Dr. JAPETUS STEENSTRUP.

[Abstract of the Memoir in the Overs. over d. K. D. Vidensk. Selsk. Forhandl., 1873.]

Dr. Steenstrup offers an explanation of certain appendages many feet in length, consisting of long, hornlike rays resembling beard or comb-like fringes, which have long been the object of research.

Professor Hannover showed, in his work on the dermal spines of Rays and Sharks, that these rays have the same structure as the spines, being formed of bony matter and identical with true teeth. Not admitting with Hannover that these rays are situated on the outer skin, like the spines of certain Rays, Steenstrup has always supposed, from their form and disposition, that they filled an office similar to that of the beard-like gums of the Whale.

Such straining appendages, composed of a series of distinct teeth set upon the branchial arches, occur in a great number of fishes,* notably those living on animalcules only. Having been led to suppose that such appendages belong to certain great Sharks, Dr. Steenstrup has been fortunate enough to find a remark made by Gunnerus relative to the Pelerin, dated more than a century ago, and so exactly descriptive of this organ that there can be no doubt of its identity. Thus he has been able to show that other authors also have observed that there was such an apparatus in the Pelerin, although the indications are so incomplete that without Gunnerus' description it would be impossible to understand them.

With this description by Gunnerus, and corresponding indications by other authors—for example Low, Pennant, Mitchell, and R. Foulis—Steenstrup has arrived at the following conclusions:—

1. The Pelerin (*Selachus maximus*, Gunn.) or Basking Shark has the interior of the mouth furnished with a fringe or branchial strainer of a special character, as a little beard-like apparatus, with rays 5 or 6 inches long, and resembling that of *Balæna*. This strainer is situated along the enormous branchial openings of the animal, and takes the part of a sieve to strain the food.

2. From this branchial fringe come (as Gunnerus' description has enabled us to see) the beard-like apparatuses which have been long preserved in the Museums of Copenhagen, Kiel, Christiania, and Frondhjem, and which Professor Hannover has studied and described in this above-mentioned memoir, in the K. D. V. Selsk. Skrifter, 5 ser., vol. vii., 1867.

---

* Dr. Andrew Smith found the branchial openings in the mouth of *Rhinodon typicus* guarded with a cartilaginous, sieve-like apparatus for straining animalcules from water ejected through the branchial canals.—*Illust. Zool. S. Africa, Pisces*. London, 1849. 4to.

3. The existence of such a fringe places beyond a doubt that the manner of living of the Pelerin is similar to that of the "whalebone" Whales; so that this colossal Shark obtains its food from the small animals that it sifts from the mass of water ejected through the fringe.

4. The rays of the branchial fringe or the elements of the branchial "beard," as the microscopic researches of Hannover go to prove, must be considered *as very long and attenuated teeth*, an arrangement which gives to the genus *Selachus* a generic character at present unique.

5. Characterised by these branchial appendages, Selachians existed in the seas of Europe during the Tertiary period, being represented by *Hannovera aurata* of P. J. Van Beneden,† found in the Belgian Crag near Antwerp.

IX.—Prodromus Faunæ Molluscorum Grœnlandiæ (in Rink's "Grönland," &c., 1857, pp. 75–100). By Dr. O. A. L. Mörch. Revised and augmented by Dr. O. A. L. Mörch, University Museum, Copenhagen. April 1875.

[The species marked with an * are doubtful inhabitants of Greenland.]

MOLLUSCA GRŒNLANDICA.

Classis I.—Androgyna, Mörch.

Ordo I.—**Geophila,** Fér.

*1. Arion fuscus, Müll. Probably introduced.
L. agrestis, L. According to Wormskiold.

2. Vitrina angelicæ, Bk. & Möll. Pfr.
"The neat little Snail," Olaf. & Paulsen's Reise i Island. Helix pellucida, Fabr., Faun. Grœnl. Found alive all the winter, chiefly at the hot springs where *Angelica* grows.

3. Conulus Fabricii (Helix), Bk. & Möll. Pfr.

*4. Helix alliaria, Miller.
Helicella Steenstrupii, Mörch. Helicella sp., Stp., Conch. von Island. Helix nitida, Fabr., F. G., n. 385. H. alliaria, Forbes, Brit. Assoc., 1839, p. 142.

*5. Helicogena (Tachea) hortensis, Müll. (Igaliko; Wormskiold.)

6. Pupa (Vertigo) Hoppii, Möll. Pfr.

7. Succinea grœnlandica, Bk. & Möll. Pfr. (Gr. *Kuksuk.*)

Ordo II.—**Hygrophila,** Fér.

8. Planorbis (Nautilina) arcticus, Bk. & Möll. Dkr.

9. Limnæa (Limnophysa, subg.) Vahlii, Bk. & Möll.

(10.) var. α. nitens. L. Pingelii, Bk. & Möll. (In a pond at Nepisetsundet.)

---

† Bullet. Acad. R. Belg., ser. 2, vol. xxxi., 1871, p. 504.

(11.) var. β. leucostoma. L. Mœlleri, Bk., 1847. L. grœnlandica, Jay, Cat. (Bk. olim.)
(12.) var. γ. malleata.
(13.) var. δ. parva: peristomate sæpe soluto, linea elevata parietali.
14. L. Wormskioldii, Bk., 1847. Species intermedia. Testa umbilicata, solidissima; spira elongata, acuminata; sutura profunda; apertura semilunari, interdum soluta. Taken with the dredge near Arsut, outside a river, by L. Barrett and C. P. Möller.
15. L. Holbœlli, Bk. & Möll.

Ordo III.—**Ptenoglossata,** Troschel.

16. Menestho albula, (Turbo) Fabr. Möll. Stimps.
Pyramis striatula, Couthuoy, Bost. Journ.
17. Scalaria grœnlàndica, Perry, Conch., 1811.
Turbo clathrus grœnlandicus, Ch., XI., f. 1878–79. Scalaria communis, var., Lam. S. planicostata, Kiener.
18. Scalaria (Acirsa, subg.) borealis, Bk. Proc. Geol. Soc., 1841; Bronn, Index Pal.
Scalaria, Lyell "On the Rising of Sweden," t. 2, f. 11, 12. S. Eschrichtii, Holb. & Möll. S. undata, Sow., Thes.
19. Philine quadrata, S. Wood. Bullæa granulosa, "Sars," Möll.?
20. Ph. punctata, (Bullæa) Möll. non Adams.
21. Cylichna alba, (Bullæa) Brown. Lovèn.
Bulla corticata, Bk., Möll. B. triticea, Couthouy.
22. C. Reinhardti, Holb. & Möll. (Bulla oryza, Totten ?)
23. C. (Roxania, subg.) insculpta, Totten. Gould. (Bulla Reinhardti, Möll., p.p.)
24. Utriculus turritus, (Bulla) Möll. Bulla obstricta, Gould ?
25. Diaphana debilis, Gould. Phil.
Bulla subangulata, Möll. Amphisphyra globosa, Lovèn?
26. D. expansa, Jeffr.
27. Physema hiemalis, Couth.
28. Dolabrifera Holbœllii, Bergh. (Belonging to a genus of *Aplysiadæ* not found nearer than the West Indies. It is about 2 inches long, with an internal shell.)

Ordo IV.—**Gymnobranchia,** Cuvier.

29. Dendronotus Reynoldsii, Couthouy, Boston Journ.
Doris arborescens, Fabr. Tritonia, Möll.
(30.) junior ? Tritonia Ascanii, Möll.
31. Lamellidoris liturata, Bk. Möll.
Doris muricata (Müll.), Sars. D. bilamellata, Ald. et Hanc.
32. D. acutiuscula, Stp. Möll.
33. Doris (Acanthochila) repanda, Ald. et Hanc.
34. Polycera Holbœllii, (Euplocamus) Möll.
35. P., another species, drawn by Holböll.
*36. Proctaporia fusca, (Doris) Fabr., F. Gr., fig. 10.
37. Æolis salmonacea, Couthouy, Boston Journ.
Doris papillosa, Fab. Æolis papilligera, Bk., 1847. Æ. bodoensis, Möll. non Gun.
38. Æ. bostoniensis, Couth. (Omenak; Olrik.)

39. Æ. Olrikii, Mörch. Aff. *Æ. gymnotæ*, Couthouy, sed utrinque fasciculis c. xii. fasciculorum papillis confertis, in linea recta transversa digestis; dorso nudo.
40. Galvina rupium, Möll.
41. Campaspe pusilla, Bergh.
42. Cratena hirsuta, Bergh.
43. Limapontia ? caudata, Möll. According to Fabricius.

Ordo V.—**Pteropoda,** Cuvier.

44. Clione limacina, Phipps. (Gr. *Augursak, Ataursak.*)
   Clione papilionacea, Pallas. Clio retusa, Müll., Fabr. Clio borealis, Brug.
45. Limacina helicina, Phipps. (Gr. *Tullukaursak.*)
   L. helicialis, Lam. Argonauta argo ?, Müll., Prod. Arg. arctica, Fabr.
46. Heterofusus balea, (Limacina) Möll.
   Spirialis Gouldii, Stimpson, Shells of New Engl., p. 27.
47. Clio pyramidata, L. (Mouth of Davis's Strait; Holböll & Rink.)

CLASSIS II.—DIOICA, Latr.

Ordo I.—**Tænioglossata,** Troschel.

48. Onchidiopsis grœnlandica, Bergh.
49. Marsenina grœnlandica, (Sigaretus) Möll.
   Marsenia grœnlandica, Bergh. (Bk.), t. v., f. 12 ? Oxynoe glabra, Couth., probably.
50. M. (Oithonella, subgen.) micromphala, Bergh. (Probably the same as No. 49.)
51. Velutina (Velutella, subg.) flexilis, Mont. Lask.
   V. plicatilis, Lovèn, non Müll. quæ *Akera bullata* (Fabr.) est.
52. V. (Velutina, Flem., subgen) lanigera, Möll. Sars.
   Helix haliotoides, Fabr., *teste MS. auctoris.*
53. V. haliotoides, (Helix) Müll. Möll. Lovèn.
(54.) var. grandis. Nerita bullata, Ch., X., p. 307, f. 1598–9.
   Bulla neritoidea, Ch., l. c., lin. 20. (Julianehaab).
55. V. (Morvillia, subg.) zonata, Gould. (Godthaab, Holb.)
   Galericulum undatum, Brown, *teste* Gould.
(56.) var. grandis. Velutina canaliculata, Bk., 1847. (Arsut, Barrett.)
57. Lacuna (Epheria, subg.) vincta, Mont.
   Turbo divaricatus, Fabr. non L. Lacuna, Möll. Lovèn.
58. L. glacialis, Möll. Middendorf.
59. L. (Temana, subg.) pallidula, Da Costa. Möll.
60. Littorina grœnlandica, Ch. Mke.
   Nerita littorea, Fabr. T. Davidii, Bolt. L. castanea, Desh.
(61.) var. lævior. Nerita littoralis, Müll., Prod., 2953. Fabr.
   Lit. palliata, Sars. Lit. arctica, Möll.
*62. L. obtusata, L. An. monstrum præcedentis ?
63. Natica affinis, (Nerita) Gm.
   Natica clausa, Sow. N. septentrionalis, Bk. Möll. N. consolidata, Couthouy.
64. Mamma (Lunatia, subg.) grœnlandica, Bk. Möll. Lovèn.
   Natica pallida, Brod. & Sow. ?

65. M. (Mamma, subg., Kl. ?) borealis, Gray, 1839, Beechy's Voy., t. 37, f. 2.
N. nana, Möll.
66. M. (Amauropsis, subg., Mörch) islandica, (Nerita) Gm.
Natica helicoides, Johnst.
(67.) var. fragilis. Natica cornea, Möll.
68. Amaura candida, Möll. H. & A. Adams, Gen.
69. Aclis Walleri, Jeffr. (Hamilton Inlet, Labrador; Wallich.)
70. Rissoa (Onoba, subgen.) saxatilis, Möll.
Trochus striatellus, Fab. non L. Rissoa arctica, Lovên.
71. R. (Paludinella) globulus, Möll.
72. Rissoa (Onoba, subg.) castanea, Möll.
R. exarata, Stimps. Sh. New Engl., p. 34, t. 1, f. 3, probably.
73. R. scrobiculata, Möll.
74. Rissoella eburnea, (Rissoa) Stimpson.
Boston Journ., Proc., IV., 14. Shells of New Engl., p. 34, t. 1, f. 1, sed spm. grœnlandica differunt: labro medio coarctato et columella torta. Möller, Index, Addende, No. 4, Gasteropus. (Godthaab, 60 fathoms, Holb.)
75. Skenea planorbis, (Turbo) Fabr. Möll. Lovèn.
76. Homalogyra rota, Forbes & Hanley. (1,622 faths., Wallich.)
77. Mœlleria costulata, (Margarita) Möll. F. & H. (Godthaab, 60 fathoms, Holb.)
78. Turritella (Tachyrhynchus, subg.) erosa, Couth., Bost. Journ.
T. polaris, Bk. & Möll.
79. T. (Tachyrhynchus, subg.) reticulata, Mighels & Adams, Bost. Journ., iv., 50.
T. lactea, Möll.
80. Cerithium (Bittium, subg.) arcticum, Mörch. (Sukkertoppen, 65 fathoms, Holb.)
Turritella ? costulata, Möll. (Nec Lam. nec Risso.)
81. Trichotopis borealis, Brod. & Sow.
Tr. atlantica, Bk.
82. Tr. conica, Möll. (Near Fladöerne, 30 fathoms; Söndre Strömfjord, 60 fathoms.)
83. Aporrhais occidentalis, Bk. (A fragment from Dr. Vahl.)
84. Cancellaria (Admete, subg.) viridula, (Tritonium) Fabr. Adams, Gen., t. 29, f. 5.
Admete crispa, Möll. C. buccinoides, Couth. C. Couthouyii, Jay.

Ordo II.—**Toxoglossata,** Troschel.

85. Pleurotoma (Ischnula, Clark; Pleurotomina, Bk., subg.) turricula, Mont., var.
Subsp. 1. Murex angulatus, Don.
Defrancia nobilis, Möll. Pleurotoma Mœlleri, Reeve, f. 324. Defr. lactea, Möll., *teste* Reeve.
86. Subsp. 2. Defrancia scalaris, Möll.
Fusus turriculus, Gould.
87. Subsp. 3. Defrancia exarata, Möll.
Pleurotoma rugulata, "Möll." Reeve, f. 345.

88. Pl. Woodiana, Möll.
Pleurotoma leucostoma, Reeve, f. 278. D. reticulata, Vahl., *teste* Reeve. D. harpularia, Couth., *teste* Lovèn. Tritonium roseum, Sars. Lovèn, No. 89.

89. Pl. elegans, Möll., non Scachi.

90. Pl. pyramidalis, Ström.
Subsp. 1. Fusus pleurotomarius, Couth. Defrancia Vahlii, Bk. Möll. Reeve.

91. Pl. cancellata, (Fusus) Mighels & Adams, 1841.
Defrancia cinerea, Möller. Tritonium Pingelii, Sars.

(92.) var. purpurea. D. Pingelii, Bk. Möll.

93. Pl. violacea, M. & Adams.
Defrancia cylindracea, Möll. Pleurotoma grœnlandica, Reeve f. 343. D. suturalis, Möll., *teste* Reeve.

(94.) var. spira breviori. Pl. livida, Reeve, f. 316, non Möll.

(95.) var. ventricosa. Defrancia Beckii, Möll.

96. Pl. borealis, Reeve, f. 277 (Errata).
Defrancia scalaris, "Vahl.," Reeve *olim*. D. livida, Möll. non L.

(97.) var. ventricosa, pallida.
D. viridula, Möller, non Fabr. Pl. decussata, Couthouy, non Brown.

Ordo III.—**Rhachiglossata,** Troschel.

98. Tritonium glaciale, L. Ch. Lam.
Buccinum carinatum, Phipps, Voyage.

*99. Tr. Hancockii, Mörch. B. grœnlandicum, Hancock. Reeve, non Ch.

100. Tr. scalariforme, Bk. & Möll.
Buccinum tortuosum, Reeve, f. 11. B. tenue, Gray; Reeve, f. 27?

101. Tr. undatum, L., Midd., Beitr., p. 482, pl. 4, f. 3. (Olrik, Holb.)

102. Tr. terræ-novæ, Beck. B. Donovani, "Gray," Reeve, f. 2. (Olrik.)

103. Tr. grœnlandicum, Ch., X., p. 177.
Tr. undatum, Fabr. Buccinum cyaneum, Brug. Bk. Möll. Lovèn. Sars. B. boreale, Leach, Ross's Voyage, ii., 1819, p. 173.

(104.) var. B. tenebrosum, Hanc. An., vol. xviii., t. 5, f. 12.

(105.) var. B. Humphreysianum, Möll. non Bennett.

106. Tr. hydrophanum, (Buccinum) Hancock (Olrik, 200–300 fathoms.)

107. Tr. undulatum, Möll.
Buccinum glaciale, Don. B. labradorense, Reeve. B. undatum, Midd., Beitr., 482; non L.

108. Tr. Humphreysianum, Bennet?
Buccinum ciliatum, Reeve, fig. 1?, non Fabr.

109. Tr. ciliatum, Fabr. Moll.
B. Mœlleri, Reeve, f. 29. B. tenebrosum, var. borealis, Midd., Beitr., t. 3, f. 7, 8.

(110.) var. lævior.

111. Fusus (Neptunea, subg.) despectus, L. Fabr.

(112.) var. Tritonium fornicatum, Fabr. Voy. de la Recherche, t. 2, Fabricius' original specimen.

(113.) var. Fusus carinatus, Pennant. Lam.

114. F. tornatus, Gould, var. (Holböll collected one specimen.)
Fusus borealis, Phil., Abbild.
115. F. (Tritonofusus, subg.) Krœyeri, Möll. (Arsut, Lucas Barrett & Holböll.)
Fusus arcticus, Phil., Abbild.
116. F. latericeus, Möll.
Tritonium incarnatum, Sars, "Reise til Lofoten."
117. F. (Sipho, subg.) islandicus, Ch., 4, f. 1312–13. (Dr. Pingel.) Dr. Pfaff, 1 spcm.
Tritonium antiquum, Fabr. non L.
118. F. Holbœllii, Möll.
119. F. propinquus, Alder.
F. Sabinii, Hanck., An., xviii., pl. 5, f. 10.
120. F. ebur, Mörch.
121. F. togatus, Mörch.
122. F. Lachesis, Mörch. Unique. 1½ inch long, with turretted spire.
123. F. (Volutopsius, subg.) norvegicus, Ch., var. (Collected by Pastor Jörgensen.)
F. Largillierti, Petit, Journ. de Conch.
124. Murex (Trophon, subg.) clathratus, L.
Tritonium Rossii, Leach. Found in a Fish's stomach.
(125.) var. Tr. Bamffii, Don.
126. Tr. Gunneri, Lovèn. Trophon Bamffii, Möll. pp.
127. Tr. craticulatus, (Tritonium) Fabr.
Trophon Fabricii, Bk. Hancock. Murex borealis, Reeve.
128. Purpura lapillus, L. Fabr. (Neritiksokfjorden.)
var. P. imbricata, Lam., No. 31.
129. Columbella (Astyris, subg.) rosacea, Gould. Möll. Mangelia Holbœllii, Beck.
Pleurotoma viridula, "Möll." Reeve.
130. Mitra (Volutomitra, Gray, subg.) grœnlandica, Bk. Möll.

Classis III.—Exocephala, Latr.

Ordo I.—**Rhipidoglossata,** Troschel.

131. Trochus occidentalis, Mighels & Adams. Stimpson.
Trochus formosus, Forbes, Ann. & Mag. Margarita alabastrum, Bk. Lovèn. F. & H.
132. Margarita grœnlandica, Ch., V., f. 1781, p. 108. Gm. Wood.
M. umbilicalis, Brod. & Sow. M. undulata, var. lævior, Möll.
(133.) var. α. M. sulcata, Sow. Midd., t. 8, f. 45, 46.
(134.) var. β. M. costellata, Sow. & Brod.
(135.) var. γ. M. undulata, S. & B.
Trochus cinereus, Fabr. p.p. Margarita striata, Leach, *teste* Forb. & Hanl.
136. M. cinerea, Couth., Bost. Journ.
Trochus cinerarius, var. Fabr., F. Gr. Margarita striata, var. grœnlandica, Möll. M. sordida, Hancock, Ann. & Mag. M. grœnlandicæ (Ch.) forma despecta, Mörch.
(137.) var. grandis. M. striata, B. & S. (Olrik.)

138. M. helicina, Phipps. Fabr.
Trochus neritoides, Gm. Helix margarita, Laskey. Margarita arctica, Leach.
139. M. argentata, Gould. Middend.
M. glauca, Möll., Index. M. Harrisoni, Hancock ?
140. M. Vahlii, Möller.
M. pusilla, Jeffr. F. & H.
141. Cemoria noachina, (Patella) L., Mantiss. plant.
Patella fissurella, Müll., Prod. 2865. Gm.
142. Scissurella crispata, Fleming, var. Möll., Append. No. 3.

Ordo II.—**Heteroglossata,** Trosch.

143. Pilidium rubellum, (Patella) Fabr. Gm. Möll. Lovèn. Sars.
144. Lepeta cæca, (Patella) Müll. Lovèn.
var. Patella candida, Couthouy. P. cerea, Möll.
145. Tectura testudinalis, (Patella) Müll. Möll. Lovèn.
Patella testudinaria grœnlandica, Ch., X., f. 1814–15.
146. Chiton (Tonicia, subg.) marmoreus, Fab.
junior? Ch. cinereus, "L." Fabr.
147. Ch. (Leptochiton, subg.) albus, L. Fabr. Möll. Lovèn.
148. Chiton ruber, L.
149. Ch. cinereus, L.
Ch. asellus, Ch.
150. Siphonodentalium, sp. According to O. Torell.

Classis IV.—Cephalopoda, Cuv.

151. Octopus grœnlandicus, (Sepia) Dewhurst, 1834. (Gr. *Imab-puirsà.*)
Sepia octopodia, Fabr. Octopus granulatus, Möll. non Lam. O. arcticus, Prosch.
152. Cirroteuthis Muelleri, Esch. Sciadephorus, Reinh. & Prosch. (Jacobshavn.)
153. Rossia palpebrosa, Owen, Ross's Second Voyage.
154. R. Mœlleri, Stp., 1856, Act. Hafn.
155. Leachia hyperborea, Stp., 1856, Act. Hafn.
156. Gonatus Fabricii, (Onychoteuthis) Lichenst. (Gr. *Amikok.*)
Sepia loligo, Fabr., F. G. Onychoteuthis Kamschatica, Midd.
(157.) junior (*teste* Stp.), Onychoteuthis ? amœna, Möll. (Gr. *Amikungoak.*)
*158. Sepiola atlantica, d'Orb. ? *teste* Stp. (Holb.)

Classis V.—Acephala, Cuv.

159. Teredo denticulata, Gray, Ann., 1850, VIII. Differt a *T. nana*, Turt. (*T. megotara*, F. & H.) alæ parte dorsali antice rotundata, nec acuminata.
Pholas teredo, Fabr. (Gr. *Kerksuk-Kuma.*) Teredo navalis, Möll. non L. T. dilatata, Stimps. Bost. Proceed., Oct. 1851, probably.
160. Mya truncata, L. Fabr. Möll.
var. M. Uddevalensis, Forb., Geol. Surv., I., p. 407. junior M. arenaria, Fabr. *teste MS. auctoris.*
161. M. arenaria, L. Möll. Fabr. MS.

162. Cyrtodaria siliqua, Spgl. (Fossil, Dr. Rink.)
Glycimeris incrassata, Lam.

163. C. Kurriana, Dkr. About 2 inches long; pale-brown in colour. (Found at low water near Jacobshavn; Dr. Pfaff.)

164. Saxicava arctica, (Mya) L. Fabr. (Gr. *Imennek.*)
Mya byssifera, Fabr. Saxicava grœnlandica, Pot. & Mich. S. pholadis, L. Gray, Parry's Voy.

165. Panomya norvegica, (Mya) Spgl. Woodward. (Fossil, Dr. Rink.)
Panopæa Spengleri, Valenc.

166. Lyonsia arenosa, (Pandorina) Möll.
Lyonsia gibbosa, Hancock, Ann., XVIII. t. 5, f. 11, 12.

167. Thracia myopsis, Bk. Möll.
Thracia Couthouyi, Stimpson.

168. Th. septentrionalis, Jeffr. Th. truncata, Mighels & Adams, Boston Journ., 1842, t. 4, f. 1 (sed margine dorsali magis declivi.) Long. 27 mm.; height 19 mm. (Eight Danish miles off land outside the Söndre Strömfjord, at 60 fathoms, Möller.)

169. Neæra cuspidata, Olivi. (Wallich.)

170. Tellina (Macoma, subg.) calcarea, Ch.
Tellina proxima, Smith.

171. T. mœsta, Desh. Perhaps only a variety of *T. calcarea*, influenced by fresh water. (Arsut, L. Barrett.)

172. T. crassula, Desh. Small, triangular. Known from Greenland by a few specimens only.

173. T. tenera (Macoma, Leach), Ross's Voyage, ii., 1819, p. 175; Thomson's Annals Philos., xiv., 1819, p. 204.
T. grœnlandica, Bk. Gould, p. 66. Lovèn, No. 299. T. Fabricii, Hanley. Sow. Thes., f. 112. Venus fragilis, Fabr., No. 414.

174. Venus fluctuosa, Gould, p. 57, f. 50. (Narsalik; Holb.)

175. Pisidium Steenbuchii, (Cyclas) Möll. (In a pond near Baals River.)

176. Thyasira Gouldii, Phil., M. Z., 1845, p. 74.
Cryptodon flexuosum, Möll. Gould, f. 52.

177. Montacuta elevata, Stimpson.
M. bidentata, Gould, p. 59, non Mont.

178. M. Mœlleri, Holb.

179. M. Davvsoni, Jeffr., Brit. Conch., ii., 216.

180. Kellia planulata, Stimpson.
K. rubra, Gould, p. 60, f. 33, non Mont.

181. Turtonia minuta, (Venus) Fabr. (Gr. *Ipiksaunatak.*)
Chione minuta, Desh., Cat. Ven., No. 121.

182. Astarte compressa, (Venus) L., Mantis. plant. non Mtg.
Venus borealis, Ch., VII., f. 413, 414. Astarte elliptica, Brown. A. semisulcata, Gray. Möll. Phil., Abbild., non Leach.

183. A. crebricostata, McAndrew & Forbes, Ann., 1847, XIX., p. 98, t. 9, f. 4.
A. crenata, Gray (?), Parry's Voy., 1824

184. A. (Tridonta, subg.) semiculcata, Leach, Ross's Voy., App., 175; non Möll.
Crassina arctica, Gray, Parry's Voy. Möll. Venus borealis, Ch., VII., f. 412. Crassina corrugata, Brown.
185. A. (Nicania, subg.) striata, Leach, 1819, Ross's Voy., App., 170. Gray, Beechy's Voy., t. 44, f. 9. Moll.
A. multicostata, Macgill. Phil.
(186.) var. A. globosa, Möll.
187. A. Banksii, Leach. Möll. Beechy's Voy., t. 44, f. 10.
*188. A. pulchella, Jonas. Phil., Abbild., X., p. 60, t. 1, f. 12.
A. Warhami. Hancock?
189. Cyprina islandica, (Venus) L. non Fabr. (Dr. Rudolph.)
190. Cardium ciliatum, Fabr.
C. islandicum, Ch. C. arcticum, Sow., Ill., f. 26.
191. C. elegantulum, Beck. Möller. Sars.
192. C. (Serripes, subg.) grœnlandicum, Ch.
Venus islandica, Fabr. non L. (Gr. *Ipiksaunak.*)
(193.) var. borealis, Reeve; striated anteriorly and posteriorly.
194. Arca pectunculoides, Scacchi. (Wallich.)
195. Nucula inflata, Hancock, Ann.
N. tenuis, Gray. Möll.
196. N. nitida, Sow., Illust. Hanley.
197. N. delphinodonta, Mighels & Adams, Boston Journ. iv., 40. Gould.
N. corticata, Holb. & Möll.
198. Nuculana buccata, Stp. & Möll. Sars.
N. Jacksonii (Gould) differt umbonum sculptura.
199. N. pernula, Müll.
Arca Martinii, Bolt., Mus. Bolt.
Leda macilenta, Stp. & Möll.
Nucula rostrata, Mart. Lam.
N. fluviatilis, Sow., Gen.
200. N. minuta, (Arca) Müll. Fabr. Ch. Möll. (Gr. *Imenningoak.*)
N. parva, Sow., Illustr.
(201.) var. grandis. Leda complanata, Möll.
202. N. (Portlandia, subg.) arctica, Gray, Parry's Voy. Wood, Suppl. t. 6.,
Yoldia portlandica, Woodward, non Hitch.
Nucula truncata, Brown, Ill., xxv., f. 19.
203. N. pygmæa, Münster. Lovèn. Forbes & Hanley.
var. Nucula lenticula, Möll.
204. Yoldia limatula, Say. Torell.
205. Y. hyperborea, Lovèn.
206. Y. thraciæformis, Storer. (Fiskernæsset and Sukkertoppen, at 60 to 70 fathoms.)
Yoldia angularis, Möll.
207. Modiolaria nigra, Gray, Parry's Voy., 1824. Wood. Ch., f. 767.
Mytilus discors, var. suevica, Fabr., Vid. Selsk, 1788.
M. discrepans, Mont., Suppl. Leach. Möller.
Modiola compressa, Menke.
Modiolaria striatula, Beck, Voy. de la Recherche, non L., Mantiss.

208. M. lævigata, Gray, Parry's Voy. Wood. (Gr. *Bibibiarsuk.*)
Mytilus discors, Fabr. (Modiolaria) Bk. non L.
209. M. corrugata, Stimpson, Conch. N. Engl., 1851. (Godthaab, 50 to 60 fathoms., Holb.)
Modiola discors, Gould, p. 130, f. 184, non L.
210. M. (Dacrydium) vitrea, Holb. & Möll. (Sukkertoppen, at 100 fathoms, Holböll.)
211. Crenella decussata, Mont., var.
Modiola ? cicercula, Möll.
212. Mytilus edulis, L. Fabr., var. (Gr. *Uilok.*)
Mytilus edulis exornatus, Ch., XI., f. 2024.
M. borealis, Lam., No. 25 ?
M. exornatus, Pfr., Register.
213. Limatula sulculus, Leach. Lovèn.
Lima sulcata, Möll.
L. conclusa, Möll. Beck, 1847.
214. Pecten islandicus, Möll. Ch. (Gr. *Kirksoaursak.*)
junior P. Fabricii, Phil.
(215.) var. costis elevatis latis.
216. P. (Pseudamussium, subgen.) grœnlandicus, Sow., Thes. Lovèn.
P. vitreus, Gray, Parry's Voy., non Ch.

The ACALEPHÆ and HYDROZOA, which follow here in the *Appendix* No. 4 of RINK's "Grönland," &c., 1857, pp. 95–98, were then enumerated by DR. MÖRCH, with their synonymy mostly after O. Fabricius' MSS.: they are now given in a Revised Catalogue by DR. LÜTKEN further on, pp. 187–190.

BRACHIONOPODA.

217. Rhynchonella psittacea, Ch. Gm. Terebratula, Möll.
218. Terebratella Spitzbergiensis, Davidson. (Inside dead *Rhynchonellæ* and *Balani.*)
219. Terebratulina septentrionalis, Couthouy. T. caput-serpentis ?
220. Terebratula cranium, Müll. (108–228 fathoms, Wallich.)

Greenland Names for Mollusca, &c.

Ajuaursak ; from *Ajuck*, a blain or boil ; *Cemoria, Tectura,* &c.
Akoperursak, from *Akopiut,* an animal which sits on its back ; *Velutina.*
Amikok, from *Amek,* a skin or hide ; *Gonatus Fabricii,* Licht.
Amikorsoak, the same.
Amikoungoak ; Gonatus amœnus, Möll.
Ataursak ; Clio limacina.
Augursak, the same. (A ram.)
Bibibiarsuk ; Modiolaria lævigata.
Daitsimetit, see *Tessermetut.*
Imab Imata ɔ, heart of the sea ; *Chrysaora pontocardia.*

Imab-puirsă ɔ, who shoot themselves out of the sea; *Octopus grœnlandicus.*

Imeningoak; Saxicava arctica ?, L.

Imennek, because they squirt so much water (*Imek*). *Tellina tenera*, Leach. *Saxicava arctica.*

Imennigoak, the little *Imennek*, because they resemble *Saxicava arctica; Nuculana minuta.*

Ipiksaunak, like a grindstone; *Cardium grœnlandicum.*

Ipiksaunatak, the young of *Ipiksaunak; Turtonia minuta.*

Ipiarsursak, like a tent-bag; the *Beroïdæ.*

Kallaliassut, because they seem to jump in the water. The Sepioid kind of Molluscs, found in North Greenland according to Mdm. Lytzen. Undetermined Cuttle-fish, Möll.

Kemiarsursak, like a dog's pup; *Æolidia*, Tritonia.

Kerksuk-Kumă ɔ, Wood-worm; *Teredo denticulata.*

Killiortout, from *Killiorpok*, scraping tool; *Mytilus edulis.*

Kirksoaursak, because they jump out of the pots in which they are to be cooked (*Kirkserpok*, to jump down); or a likeness to *Kirksoak; Pecten Islandicus.*

Korotungoak; the little crenulated; *Trophon.*

Korsoak; see *Amikorsoak.*

Nakkarsursak, a bladder; *Boltenia Bolteni*, L.

Nuertlek, from *Nuak*, thick spittle; *Medusæ.*

Nuertleksoak; Medusa capillata, Fabr., Fn. Gr. No. 203.

Nyaursæt ɔ, hair-like; *Sertularia.*

Puirsarsoakasik ɔ, the big bad ones which jump up; an *Octopus*, Möll.

Sàrpangaursæt; Sertularia.

Siuterursak; Vitrina angelicæ, Margarita helicina, Menestho, Skenea, Littorina groenlandica, &c.

Siuterok, from *siut*, ear; all Snails; *Margarita, Littorina*, &c.

Siuterungoak, a little Snail; *Lacuna, Trophon*, &c.

Siuterursoak, the big Snails; *Tritonium, Fusus.*

Terkeingak, like a shade for the eyes; *Chiton marmoreus.*

Terkungoak, in common with Oniscus; a declivity or slope on a rock; *Chiton.*

Tessermètut ɔ (*Daisimetit*), which live in freshwater lakes (*Tessek*); *Limnæa.*

Tullukaurtak, from *Tulluk*, sea-raven, like a raven; *Limacina arctica.*

Tupilek, having a tent over it; *Cirroteuthis.*

Uilok, from *Uivok* (?), to increase or swell out, or rather from *Uiunge*, nymphæ muliebres; *Mytilus edulis*, L.

Umataursak; Chrysaora pontocardia; Julianehaab.

Usursak, from *Usuk*, penis; *Mya truncata.*

APPENDIX.

Notes on Shells, by Dr. Mörch.

1. The stomach of the Shark (*Squalus*) often contains very rare animals; often uninjured Cephalopods, *Buccinum hydro*-

phanum, &c., also a large *Anthipathes* (a black sponge-like Coral).

The Sharks are fished for by the Greenlanders in the winter, through a hole in the ice by a fine line of at least 100 faths., formerly made of whalebone fibre. The Shark gently follows the hook to the surface, where it is killed.

2. In the stomach of *Anas mollissima*, and especially of *Anas spectabilis*, rare shells are often found. Large specimens from Greenland are nearly all found in this way.

3. The shell of *Limacina* is always broken by the animal when captured. In the stomach of a species of *Cottus*, found in North Greenland, these shells, however, are found *entire* and *empty*.

4. The Whelks (*Buccinum* and *Fusus*) are taken in sunken baskets, baited with dead fish, but the basket must not lie too long as the bait is eaten away by *Gammarus*, &c., in an incredibly short time.

5. The land and freshwater shells are particularly interesting. On the west coast of Greenland are found one *Vitrina*, two small *Helices* (*Helix fulva* and *H. alliaria* of England), one *Pupa*, and one *Succinea*.

Of freshwater shells there are one small *Planorbis*, several *Limnææ*, and one Bivalve (*Pisidium*).

None have yet been found in Spitzbergen and Nova Zembla.

In a memoir "On the Land and Freshwater Mollusca of Green-"land," ("American Journal of Conchology,") Dr. Mörch offers the following remarks (p. 27) :—

"The land shells of Greenland are nearly allied to those of Iceland. The *Vitrina*, the *Succinea*, and *Hyalina alliaria* accord better with the species of Iceland than with American species. The freshwater species of Greenland, however, are entirely different from those of Iceland. The *Limnææ* of Iceland belong to *Radix*, while those of Greenland all belong to *Limnophysa*; both genera common to Europe and America. The *Planorbis* of Greenland is perhaps *Pl. parvus*, Say, of America, or *Pl. spirorbis*, Rm. f. 63, (*Pl. Dazuri*, nob.). *Pisidium Steenbuchii* of Greenland is most nearly allied to *P. pulchellum*, found in Iceland and Europe, but it is much larger. They have been all found in Labrador by Packard. Although the Arctic species are small, they are the largest in the group of species to which they belong. Thus *Succinea Grœnlandica* is larger than *S. arenaria*; *Vitrina* larger than any European species, except, perhaps, *V. major*; the *Pupa* is the largest of the Vertigos."

X.—MARINE INVERTEBRATA collected by the ARCTIC EXPEDITION under DR. I. I. HAYES. By W. STIMPSON, M.D. (May 1862). From the Proceed. Acad. Nat. Sciences Philadelphia, 1862–3, p. 138–142. 1863.

[In the following list those species are enumerated which were brought home from *Port Foulke* and *Littleton Island*, on the

Eastern or Greenland shore of Smith Sound, lat. 78½°, from *Cape Faraday* on the opposite shore, lat. 79° 45′, and *Gale Point* (lat.?). Others are mentioned as coming from Godhavn and from the W. and N.W. Coast of Greenland, which are included, by name or synonym, in the Catalogues given elsewhere in this "Manual."]

1. **Mollusca.**

Clione limacina, Phipps. Port Foulke.

Buccinum cyaneum, Beck. Port Foulke.

Mya truncata, Lin. Mostly var. Uddevallensis. The siphons were found in great numbers in the stomach of a Walrus. Port Foulke.

Saxicava arctica, Desh. Do. Do.

Astarte plana, J. Sow. Port Foulke.

Cardium Hayesii, n. sp. Disco Island. Also found in Nova Scotia.

Crenella faba (O. Fab.), Stimp. N.W. Coast of Greenland.

2. **Crustacea.**

Crangon boreas, J. C. Fabr. Godhavn; Port Foulke; Littleton Isl.

Hippolyte Gaimardii, M.-Edw. Port Foulke.

H. gibba, Kröyer. Port Foulke.

H. turgida, Kr. Godhavn; Port Foulke.

H. Phippsii, Kr. Port Foulke.

H. polaris, Owen. Port Foulke; Littleton Isl.

H. borealis, Owen. Littleton Isl.

Mysis oculata, Kr. Port Foulke.

Anonyx ampulla, Kr. (var.?). Gale Point.

Pherusa tricuspis, n. sp. Littleton Isl.

Gammarus locusta, J. C. Fab. Port Foulke.

Themisto arctica, Kr. In the stomach of a Seal at Cape Faraday.

Bopyrus hippolytes, Kr. Port Foulke.

Lernæopoda elongata, Grant. Port Foulke.

Hæmobathes cyclopterina, Steenstrup & Lütken. On the gills of *Gymnelis viridis* at Littleton Isl.

Balanus balanoides, Darwin. Port Foulke.

3. **Annelida.**

Lepidonote cirrata, Œrst. Port Foulke.

L. punctata, Œrst. Port Foulke and Littleton Isl.

Phyllodoce grœnlandica, Œrst. Port Foulke.

Cirratulus borealis, Lamk. Godhavn; Littleton Isl.

Siphonostomum plumosum, Rathke. Port Foulke.

Tecturella flaccida, Stimpson. Port Foulke.

Brada inhabilis, Stimpson. Gale Point.

Pectinaria Eschrichtii, Rathke. Godhavn; Port Foulke.

Spirorbis nautiloides, Lam. Port Foulke.

Priapulus caudatus, Lam. From the stomach of a Walrus Port Foulke.

4. **Echinodermata.**

Myriotrochus Rinkii, Steenstr. Port Foulke.

Asterias grœnlandica, Stimps. Port Foulke.
A. albula, Stimps. Godhavn; Port Foulke.
Ophioglypha squamosa, Stimps. Godhavn; Port Foulke.

XI.—SHELLS, &c., from the HUNDE ISLANDS, DAVIS' STRAIT; dredged by DR. P. C. SUTHERLAND, October 1852; named by DR. S. P. WOODWARD, 1865. (Phil. Trans. clv. 1865, p. 328.)

No. I.—28–30 fathoms. (See page 192, for the material of these dredgings.)

Balanus porcatus, DC.? Water-worn fragments.
B. crenatus, Brug.? Water-worn fragments.
Mya truncata. Fragment.
Saxicava arctica. Small valve.
Tellina calcarea (=proxima=lata). Fragment.
Echinus, sp. Fragments of plates and spines.

No. II.—30–40 fathoms.

Leda minuta. Odd valve (large) and fry.
Crenella decussata. Small.
Limatula sulcata.
Astarte striata. Young.
A. semisulcata. Young.
Saxicava. Fry.
Rissoa castanea.
R. scrobiculata.
Scissurella crispata.
Turritella lactea. Young.
Margarita undulata.
M. cinerea. Young.
Echinus. Small spine.
Spirorbis. Whorls furrowed.

No. III.—25–50 fathoms.

Saxicava arctica. Adult.
Lyonsia striata. Fry.
Astarte striata. Adult and fry.
Leda truncata. Fragments.
L. pygmæa. Fry.
Crenella decussata.
C. faba.
Nucula tenuis. Fry.
Cardium elegantulum.
Natica pusilla (grœnlandica). Fry.
Cylichna Gouldii. Young.
Rissoa scrobiculata.
Spirorbis.
Echinus. Spine.

No. IV.—50–70 fathoms.

Pilidium fulvum.
Acmæa. Fragment.
Chiton albus? Two valves.
Astarte striata. Fry.
Spirorbis nautilus?
S. sulcata.

No. V.—60–70 fathoms.

Pecten islandicus. Fragments.
Mya truncata.
Astarte borealis, *var.* semisulcata. Young.
A. striata.
Saxicava. Fry.
Crenella decussata.
Limatula sulcata.
Turritella lactea. Fragment.
Rissoa castanea.
R. scrobiculata.
Margarita helicina.

Margarita undulata. Fragment and fry.
M. cinerea. Fry.
Scissurella crispata.
Littorina obtusata. Fry.
Cemoria noachina. Fry.
Pilidium fulvum.
Serpula.
Spirorbis.
Balanus porcatus. Tergum and fragments of parietes.
Echinus. Fragments of spines.

XII.—A Revised Catalogue of the Tunicata of Greenland. By Dr. Chr. Lütken, University Museum, Copenhagen. 1875.

TUNICATA.

Ascidiæ simplices (Greenl. Nakasursak).

1. *Boltenia Bolteni*, L.
   F. Gr. 323 (A. clavata, Fabr.).
   B. reniformis et ciliata, Möller, Ind. Moll. Gr., p. 95.
2. *Cynthia chrystallina* (Möll.).
   Clavellina chryst., Möll. l. c., p. 95.
3. *C. rustica* (L.).
   F. Gr. 316 and 317 (?). (Jun.?) (A. quadridentata, Fabr.)
   A. monoceros, Möll. l. c., p. 95.
4. *C. pyriformis* (Rthk.).
   Jun.? F. Gr. 322 (A. villosa).

Obs.—A species from East Greenland is described by Kupffer as *C. villosa*, Fabr. (Zte deutsche Nordpolarfahrt, II., p. 244).

5. *C. echinata* (L.).
   F. Gr. 318.
6. *C. conchilega* (Möll.).
   Möller, l. c.
7. *C.* (*Molgula*) *glutinans*, Möll.
   Möller, l. c., p. 94.
8. *C. tuberculum* (Fabr.).
   F. Gr. 321.
9. *C. Adolphi*, Kupf.
   Zte deutsche Nordpolarfahrt, II., p. 245.
10. *Phallusia lurida* (Möll.).
    Möller, l. c., p. 95.
11. *Ph. complanata* (Fabr.).
    F. Gr. n. 320.
12. *Chelyosoma Macleayanum*, Sow. Brod.
    Cp. Eschricht's paper in K. D. Vid. Selsk. Skr. V. (1841).
13. *Pelonaia*, sp.

(An undetermined specimen from Jacobshavn in the Museum of Copenhagen.)

Obs.—The *Ascidiæ* of Greenland require revision, with examination of fresh specimens; several of the above-named species are only imperfectly known, and others remain undescribed in collections. The *Compound Ascidiæ* are rather numerous, but no attempt has been made to identify them. "*Alcyonium rubrum* and *digitatum*," F. Gr. 462 and 463, probably belong to this division. "*Synoicum turgens*" owes its introduction into the "Fauna of Greenland" to a mistake.

XIII.—The POLYZOA of GREENLAND. By Dr. CHR. LÜTKEN, University Museum, Copenhagen. 1875.

This List is an abstract from Dr. Smitt's Monograph of the Scandinavian and Arctic Polyzoa in the "Proceedings of the Swedish Academy" for 1864-8, to which work the reader is referred for further particulars. References to Kirchenpauer's list of Polyzoa from East Greenland, in "Die zweite deutsche Nordpolarfahrt," are added.*

1. Cyclostomata.

1. *Crisia eburnea* (L.).
2. *Diastopora simplex* (Busk).
3. *Diastopora hyalina* (Flemg.).
   Kirchp., p. 426.
4. *Mesenteripora mæandrina* (Wood).
5. *Tubulipora atlantica* (Forb.).
   Idmonea atl. (Forb.), Kirchp., p. 427.
6. *T. fimbria* (Lmk.).
   T. serpens, Fabr. F. Gr. 428.
7. *T. flabellaris* (Fabr.).
   Fabr. F. Gr. 431.
   Phalangella fl., Kirchp., p. 427.
8. *T. incrassata* (D'Orb.).
9. *T. fungia* (Couch).
10. *T. penicillata* (Fabr.).
    Fabr. F. Gr. 430.
11. *Hornera lichenoides* (Fabr.).
    Fabr. F. Gr. 436; Kirchp., p. 425.
12. *Discoporella verrucaria* (L.).
    Fabr. F. Gr. 432; Kirchp., p. 427.

12*a*. *D. hispida*, (Fl.).
Kirchp., p. 427.

13. *Defrancia lucernaria* (Sars).

2. Ctenostomata.

14. *Alcyonidium hirsutum* (Flemg.).
    Kirchp., p. 428.
15. *A. gelatinosum* (L.).
    Kirchp., p. 428.

---

* The "*Isis*," "*Tubipora*," "*Madrepora*," and "*Millepora*" of Fabricius are apparently all Polyzoa, with the exception of Nos. 438 (*Nullipora*, sp.) and 434 (*Corallina officinalis*?).

16. *A. hispidum* (Fabr.).
    Fabr. F. Gr. 448.

3. Chilostomata.

17. *Cellularia ternata* (Sol.).
    Sertularia reptans, Fabr. F. Gr. 459.
    Menipea arctica (Busk) and *M. Smittii*, Nordm., Kirchp., l. c., p. 417–8.
18. *C. scabra* (v. Ben.).
    Sertularia halecina, Fabr. F. Gr. 455.
    Scrupocellaria inermis (Norm.), Kirch. l. c., p. 418.
19. *Gemellaria loricata* (L.).
    Fistulana ramosa, Fabr. F. Gr. 451.
20. *Caberea Ellisii* (Flemg.).
21. *Bugula Murrayana* (Bean).
    Flustra foliacea, Fabr. F. Gr. 445.
22. *Flustra chartacea* (Gm.).
23. *Fl. membranacea* (L.).
24. *Fl. papyracea* (Pall.)
25. *Fl. foliacea* (L.).
26. *Cellaria articulata* (Fabr.).
    Isis hippuris, Fabr. F. Gr. 427.
27. *Membranipora lineata* (L.).
    ? *Flustra membranacea*, F. Gr. 446.
    Kirchenpauer, l. c., p. 419.
28. *M. spinifera* (Johnst.).
29. *M. Flemingii*, Busk.
    M. Flemingii, B., and *M. minax*, B., Kirchp., p. 419.
30. *M. pilosa* (Linn.).
31. *Escharipora annulata* (Fabr.).
    Fabr. F. Gr. 444.
32. *Porina Malusii* (Aud.).
33. *P. ciliata*, Pall.
34. *Anarthropora monodon* (Busk).
35. *Escharella porifera* (Smitt).
    ? *Hemeschara* (?) *contorta*, Kirchp., l. c., p. 422.
36. *Esch. palmata* (Sars).
37. *E. Legentilii* (Aud.).
    Lepralia Smittii, Kirchp., p. 420.
38. *E. Jacotini* (Aud.).
39. *E. auriculata* (Hass.).
40. *E. Landsborovii* (Johnst.).
    Lepralia Landsborovii, Kirchp., p. 421.
41. *E. linearis* (Hass.).
42. *Mollia hyalina* (L.).
    Cellepora nitida, Fabr. F. Gr. 443.
    Lepralia nitida, Kirchp., p. 420.
43. *Myriozoon crustaceum* (Sm).
44. *M. subgracile* (D'Orb.).
    Millepora truncata, Fabr. F. Gr. 435 (p.p.).
45. *M. coarctatum* (Sars).
    Mill. truncata, F. Gr. (p.p.).

46. *Lepralia spathulifera* (Sm.).
47. *L. hippopus* (Sm.).
48. *Porella acutirostris* (Sm.).
49. *P. lævis* (Flemg.).
50. *Eschara verrucosa* (Busk).
51. *E. cervicornis* (Pall.).
    Kirchp., p. 424.
52. *E. elegantula* (D'Orb.).
53. *Escharoides Sarsii* (Sm.).
    ? *Cellepora spongites*, Fabr. F. Gr. 439.
54. *E. rosacea* (Busk).
55. *Discopora coccinea* (Abdg.).
    Lepralia Peachii and *L. sinuosa* (Busk), Kirchp., p. 421.
56. *D. appensa* (Hass.).
57. *D. sincera* (Sm.).
58. *D. Skenei* (Sol.).
    Cellepora Skenei, Kirchp., p. 424.
59. *Cellepora scabra*, Fabr.
    Millepora reticulata, Fabr. F. Gr. 437.
    Kirchp., p. 423.
60. *C. ramulosa* (L.).
    C. verrucosa, Fabr. F. Gr. 440.
61. *Celleporaria incrassata* (Lam.).
    Kirchp., p. 423.
62. *Retepora cellulosa* (Linn.).
63. *Loxosoma*, sp.*

XIV.—Insects and Spiders of Greenland: an Abstract of the Sketch of the Insect-fauna, Arachnida, &c., of Greenland (Freshwater, Land, and Littoral Arthropoda), by J. C. Schiödte, in the "Naturh. Bidrag til Rink's Beskrivelse af Grönland," 1857, pp. 50–74.

I.—INSECTA.

Eleutherata (Coleoptera). *Carabi.*

Nebria nivalis, Payk., Mon. Car. 52, xxxi.
Patrobus hyperboreus, Dej., Sp. Col. III. 30, 3; O. Fab., F. Gr. 190, 139. *Gr.* Siutisortak.
Bradycellus cognatus, Gyll., Ins. Suec. IV., App. 455.
Bembidium Grapei, Gyll., *ibid.* IV., App. 403.

Dytisci.

Hydroporus, sp.
Colymbetes dolabratus, Payk., Fn. I. 204, 13; O. Fab., F. Gr. 189, 138.

---

* Kirchenpauer, l. c., adds to the list of Greenland Polyzoa, *Celleporella pepralioides* (Norm.) and *Lepralia pertusa*, Busk.

Gyrini. Gyrinus, sp.
Staphylini.
Quedius fulgidus, Fabr., Mant. Ins. I. 220, 14.
Quedius, sp.
Micralymma brevilingua, Schiödte, Naturh. Tidskr., ser. 2, I., 337, 2, t. 4, f. 2.
Anthobium Sorbi, Gyll. *op. c.* II. 206, 8.
Staphylinus maxillosus, L., O. Fab., F. Gr. 140.
S. fuscipes, O. Fab., F. Gr. 141.
S. lignorum, O. Fab., F. Gr. 142.
Byrrhi.
Cistela (Byrrhus) stoica, O. Fab., F. Gr. 131, var. of *Byrrhus fasciatus*, F.
Simplocaria metallica, Sturm, Deutschl. Ins. II. 111, 18, t. 34, f. B.
Curculiones.
Rhytidosomus scobina, Schiödte.
Phytonomus, sp.
Otiorhynchus maurus, Gyll., *op. c.* III. 293, 24; O. Fab., F. Gr. 136.
O. arcticus, Fab., F. Gr. 137.
Coccinellæ. Coccinella trifasciata, O. Fab., F. Gr. 133.

Ulonata (**Neuroptera,** *parte*).
Ephemera culiciformis, Linn., Faun. Suec. 1475.

Synistata (**Neuroptera,** *parte*, et **Trichoptera**).
Hemerobius obscurus, Zett., Ins. Lapp. 1049, 7.
Phryganea grisea, Lin., Faun. Suec. 1484.
P. interrogationis, Zett., Ins. Lapp. 1063, 12.

Piezata (**Hymenoptera**).
Nematus ventralis, Dahlb., Consp. Tenthred. 9, 91.
Bombus hyperboreus, Schönh., Vet. Ak. Handl. 1809, I. 57, t. 3, f. 2 (fœm.); *Apis alpina*, O. Fab., F. Gr. 155; *Bombus arcticus*, Kirby, Suppl. App. Parry's Voy. ccxvi. (fœm.).*
B. balteatus, Dahlb. Bombi Scand. 36, 8 (fœm.); *Bom. Kirbiellus*, Curtis, App. Ross's Second Voy. lxii. (masc.); an etiam fœm. ?); *Bomb. arcticus*, Kirby, *l. c.* (masc.).†
Cryptus arcticus, Schiödte.
C. Fabricii, Schiödte. O. Fab., F. Gr. 198, 154.

Glossata (**Lepidoptera**). *Papiliones.*
Argynnis chariclea, Herbst., Pap. 10, 125, 47, t. 272, f. 5, 6; *Papilio Tullia*, O. Fab., F. Gr. 143.
Chionobas Balder, Boisd., Icon. Lep. 19, 189, 4, t. 39, f. 1–3.
C. Bore, Hübn., Pap. t. 29, f. 134–136.
Colias Boothii, Curtis, App. &c. lxv. 10, t. A., f. 3–5.
Noctuæ.
Agrotis quadrangula, Zett., Ins. Lapp. 935, 4.

---

* *See* Curtis's List of Insects from the Parry Islands, &c., further on.

† *See* Kirby's List of Insects from the Parry Islands, &c., further on. The Insects and Arachnids of the East Coast of Greenland are also enumerated further on.

Agrotis rava, Herr.
A. islandica, Staudinger.
A. Drewsenii, Staudinger.
Noctua Westermanni, Staudinger.
Hadena exulis, Lefeb., Ann. Soc. Entom. France, V. 392, t. 10, f. 2.
H. Sommeri, Lefeb., *op. cit.*, 391, f. 1.
H. grœnlandica, Zett., Ins. Lapp., 939, 9.
H. picticollis, Zett., Ins. Lapp., 939, 8.
Aplecta occulta, Rossi, var. *implicata*, (Hadena) Lefeb., *op. cit.*, 394, t. 10, f. 5.
Plusia gamma, L., Fn. Suec., 1171.
P. interrogationis, L., Fn. Suec., 1172.
P. parilis, Hübn., Noct., t. 90, f. 422.
P. diasema, Dalm., Boisd., Index, 93.
Anarta algida, Lefeb., *op. cit.*, 395, f. 5, probably *Phalæna Myrtilli*, O. Fab., Fn. Gr., 147.
A. amissa, Lefeb., *op. cit.*, 397, f. 6, 7.
A. leucocycla, Staudinger.
A. vidua, Hübn., var. *lapponica*, Thunb., Diss. Ins. Sv., 2, 42.

Phalænæ.

Phæsyle polaria, Boisd., Duponch., var. *Brullei*, Lefeb., *l. c.*, 399, f. 8.
Cidaria brumata, Lin., Fn. Sv., 1293.

Pyralidæ.

Botys hybridalis, Hübn., Pyral., t. 17, f. 114.

Tortrices.

Teras indecorana, Zett., *op. cit.*, 989, 3.

Tineæ.

Eudorea centuriella, Schifferm., Syst. Verz.
Pempelia carbonariella, Fischer von Roeslerst. Abb. 30.
Plutella senilella, Zetterstedt, *op. cit.*, 1001, 2.

Antliata (Diptera).

Chironomus polaris, Kirby, Suppl. App. Parry's Voy., ccxviii., Curtis, App., &c. lxxvii., 27, t. A., f. 14.
C. turpis, Zett., *op. cit.*, 811, 82.
C. frigidus, Zett., *ibid.*, 812, 14.
C. variabilis, Stæger, Naturh. Tidssk., ser. 2, I., 351, 4; II., 571, 44.
C. basalis, Stæg., *ibid.*, 351, 6.
C. byssinus, Meigen, Zweifl. Ins., I., 46, 56.
C. aterrimus, Meig., I., 47, 59.
C. picipes, Meig., I., 52, 74.
Diamesa Waltlii, Meig., Stæg., 353, 10.
Tanypus crassinervis, Zett., Ins. Lap., 817, 1.
T. tectipennis, Zett., *ibid.*, 815, 5.
T. tibialis, Stæg., *op. cit.*, 354, 13.
Ceratopogon sordidellus, Zett., *op. cit.*, 820, 6. *Culex pulicans* (*pulicaris*), O. Fabr., F. Gr., 211, 173.
Tipula arctica, Curtis, *op. cit.*, lxxviii, 29, t. A. f. 15. *T. rivosa*, O. Fabr., F. Gr., 156.

Erioptera fascipennis, Zett., *op. cit.*, 831, 9.
Trichocera maculipennis, Meig., I., 214, 4? *Tipula regelationis*, Fabr., F. Gr., 202, 157.
Boletina grœnlandica, Stæg., 356, 18.
Sciara iridipennis, Zett., 827, 9.
Simulia vittata, Zett., 803, 3. *Culex reptans*, O. Fabr., F. Gr., 210, 172.
Rhamphomyia nigrita, Zett., Ins. Lapp., 567. *Empis borealis*, O. Fabr., F. Gr., 211, 174.
Dolichopus grœnlandicus, Zett., Dipt. Scand., II., 528; *D. tibialis*, var. *b.*, Zett., Ins. Lapp., 711.
Helophilus grœnlandicus (Tabanus), O. Fab., Fn. Gr., 208, 170; *H. bilineatus*, Curtis, *op. cit.*, lxxviii., 30.
Syrphus topiarius, Meig. III,, 305, 47.
Sphærophoria strigata, Stæg., 362, 31.
Sarcophaga mortuorum, Lin., Fn. Suec., 1830; *Volucella mort.*, Fabr., F. Gr., 206, 166.
Musca erythrocephala, Meig. V. 62, 22; *Volucella vomitoria*, Fabr., F. Gr., 207, 167?
M. grœnlandica, Zett., Ins. Lapp., 657, 16; *Vol. cæsar*, O. Fab., F. Gr. 207, 168?
Anthomyia dentipes, Fabr., Syst. Antliat, 393, 95.
A. irritans, Fallen, Musc., 62, 58.
A. frontata, Zett., Ins. Lapp., 669, 35.
A. trigonifera, Zett., *ibid.*, 669, 36.
A. arctica, Zett., *ibid.*, 669, 34.
A. triangulifera, Zett., *ibid.*, 680, 83.
A. scatophagina, Zett., *ibid.*, 677, 69?
A. striolata, Fall., Musc., 71, 77.
A. ruficeps, Meig. V., 177, 62?
A. ciliata, Fabr., Ent. Syst. IV., 333, 87.
Scatophaga squalida, Meig. V., 252, 10.
S. littorea, Fall. Scatom., 4, 4.
S. fucorum, Fall. *ibid.*, 5, 5.
Cordylura hæmorrhoidalis, Meig. V., 237, 17.
Helomyza tibialis, Zett., Ins. Lap., 767, 12.
H. geniculata, Zett. *ibid.*, 767, 13.
Piophila casei, Lin., F. Suec., 1850.
P. pilosa, Stæg., *op. cit.*, 368, 52.
Ephydra stagnalis, Fall., Hydromyz., 5, 5.
Notiphila vittipennis, Zett., Ins. Lap, 718, 6?
Phytomyza obscurella, Fall., Phytomyz., 4, 8.

Suctoria.

Pulex irritans [?], L., O. Fabr., F. Gr., 221, 193; on the Hare only. *Gr.* Ukalib-Koma. Piksiksak.

Rhyncota.

Heterogaster grœnlandicus, Zett., Ins. Lap., 262, 3.
Cicada lividella, Zett., *ibid.*, 290, 5.
Aphis punctipennis, Zett., *ibid.*, 311, 7.
Dorthesia chiton, Zett., *ibid.*, 314, 1.

Siphunculata.

Pediculus humanus, L., O. Fab., Fn. Grœnl., 215, 182. *Gr.* Komak. (Egg) Erkek.

[The following *Bird-lice* are enumerated by O. Fabricius in "Fn. Grœn." under "Pediculus" (No. 184–192):—

184. strigis. *Gr.* Opib-Koma.
185. corvi. *Gr.* Tullukab-Koma.
186. clangulæ. *Gr.* Kærtlutorpiabsab-Koma.
187. gryllæ. *Gr.* Serbab-Koma.
188. bassani. *Gr.* Kubsab-Koma.
189. lari. *Gr.* Najab-Koma.
190. tringæ. *Gr.* Sargvarsub-Koma.
191. hiaticulæ. *Gr.* Tukagvajub-Koma.
192. lagopi. *Gr.* Akeisib-Koma.]

Mallophaga.

Trichodectes (?) canis, De Geer, Mém., VII., t. 4, f. 16; Fn. Gr., 215, 183. *Gr.* Kemmik-Koma.

Thysanura.

Podura, spp.

II.—ARACHNIDA.

Araneæ.

Lycosa saccata (Fabr.), and Attus, spp., Fn. Gr., 204–208.

Opiliones.

Phalangium opilio ?, L. O. F., Fn. Gr., p. 225, No. 203. *Gr.* Niutôk.

Acari.

Bdella, &c., spp., Fn. Gr., 194–202.

[Under "Acarus" O. Fabricius enumerates (No. 194–202):—

194. siro. Itch-mite. *Gr.* Okok. Killib-Innua.
195. cadaverum. In dried Fish especially. *Gr.* Okok.
196. holosericeus. *Gr.* Okok.
197. aquaticus. *Gr.* Imak-Koma.
198. muscorum. *Gr.* Merkub-Koma.
199. gymnopterorum. *Gr.* Anarirsab-Koma.
200. coleoptratorum. *Gr.* Egyptsab-Koma.
201. longicoruis. *Gr.* Ujarkab-Koma.
202. littoralis. *Gr.* Sirksat-Koma].

Pycnogona (p. 71).

(See Dr. Lütken's Catalogue at p. 163.)

III.

Isopoda, Amphipoda, Entomostraca (pp. 72, 73).

(See Dr. Lütken's Revised Catalogue of the Crustacea, p. 146.)

XV.—The CRUSTACEA of GREENLAND. By Dr. CHR. LÜTKEN, University Museum, Copenhagen. 1875.

This list is chiefly a revised copy of that given by Prof. Reinhardt in Rink's "Greenland," containing the corrections and additions published of late years.*

Decapoda.

1. *Chionocoetes phalangium* (Fabr.). *Gr.* Arksegiarsuk, &c.
   Cancer phalangium, Fabr. Fauna Grœnl., n. 214.
   Cancer opilio, Fabr., Vid. Selsk. Skr., N. S., III. p. 180.
   Chionocoetes opilio, Kröyer, Naturh. Tidsskr., II., p. 249.
   Kröyer, Voyage en Scandinavie, &c., Crustac., t. I.
2. *Hyas aranea* (Linn.). *Greenl.* Arksegiak, &c.
   Cancer araneus, Linn., Fauna Suec., II., 2030.
   Cancer araneus, Fabr., F. Gr. 213.
3. *Hyas coarctata* (Leach).
   Leach, Malac. podophthalm. britt., t. 21, C.†
4. *Pagurus pubescens*, Kr.
   Kröyer, Naturh. Tidsskr., II., p. 251 ; Voyage, &c., t. 2, f. 1.
   Eupagurus pubescens, Stimpson, Proc. Philad. Acad., 1858, p. 75.
5. *Crangon boreas* (Phipps). *Greenl.* Umiktak.
   Cancer homaroides, Fabr., F. Gr., p. 218 ; Mohr, Islands Naturh., n. 245, t. 5.
   Cancer boreas, Phipps, Voyage, p. 190, t. 12, f. 1.
   Sabine, Suppl. App., p. 235 ; Beechey's Voy. Zool., p. 87 ; Zool. Dan., t. 132, f. 1.
   Kröyer, Naturh. Tidsskr., IV., p. 218, t. 4, f. 1–14.
   Bell, Belcher's Voy., p. 402.
   Buchholz, Zte deutsche Nordpolarf., p. 271.
6. *Sabinea septemcarinata* (Sab.).
   Crangon septemcarinata, Sabine, App. Voy. Parry, p. 58, t. II., f. 11–13.
   Owen, App. Voy. Ross, p. 82 ; Kröyer, Naturh. T., IV., p. 244, t. 4–5, f. 34–44.
7. *Argis lar* (Owen).
   Crangon lar, Owen, Zool. Beechey's Voyage, p. 88.
   Kröyer, Naturh. Tidsskr., IV., p. 255, t. 5, f. 45–62.

---

* The synonyms given are principally taken from authors on Arctic or Scandinavian Zoology.

† The occurrence of *Lithodes maja* and *Nephrops norvegicus* in Greenland needs confirmation. *Cancer gammarus*, F. Gr. 215 (*Homarus vulgaris*), must be omitted ; also, 220 (*Cancer arctus ; Gr.* Tillektoutelik), &c.

8. *Hippolyte Fabricii*, Kr.
Kröyer, Naturh. Tidsskr., III., p. 571; Vid. Selsk. Skr., IX., p. 277, t. 1, f. 12–20.

9*a*. *Hippolyte Gaimardii*, M. Edw.
Milne Edwards, Hist. Natur. d. Crust., II., p. 378.
Kröyer, Nat. T., IV., p. 572; Vid. Selsk. Skr., l. c., p. 282, t. I., f. 21–29.

b. *Hippolyte gibba*, Kr.
Kröyer, Nat. T., III., p. 572; Vid. Selsk. Skr., l. c., p. 288, t. I.–II., f. 30–37.

Obs.—Auct. cl. Goësii (Öfvers. Vetensk. Akad. Förhandl., Stokholm, 1863) a præcedente sexu (masculo) modo diversa; huc quoque accedit *Hip. Belcheri*, Bell (Belcher's Voy., p. 402, t. 24, f. 1).

10. *H. incerta*, Buchh. Zte deutsche Nordpolarfahrt, p. 272.

11. *Hippolyte spinus* (Sow.).
Cancer spinus, Sowerby, Brit. Miscell., t. 21.
Alpheus spinus, Leach, Trans. Linn. Soc., XI., p. 247; Owen, Append. Ross, p. 83, t. B., f. 2.
Hippolyte Sowerbei, Leach, Malac. podophthalm. britt., t. 30.
Hippolyte Sowerbei, Kr. N. T., III., p. 573; Vid. S. Skr. l. c., p. 298, t. II., f. 45–54.
Bell, Brit. Crust., p. 284.

12. *Hippolyte macilenta*, Kr.
Kröyer, N. T., III., p. 574; Vidensk. Selsk. Skr., IX., p. 305, t. II., f. 55–56.

13*a*. *Hippolyte Phippsii*, Kr.
Kröyer, N. T., III., p. 575; Vid. S. Skr. l. c., p. 314, t. III., f. 64–68.

b. *Hippolyte turgida*, Kr.
Kröyer, N. T., III., p. 575 Vid. S. Skr., l. c., p. 308, t. II., III., f. 57–63.

Obs.—Auctorit. cl. Goësii (l. c.) fœmina præcedentis. According to Buchholz (l. c., p. 274) the difference is not of a sexual character, but still he regards them only as varieties of the same species.

14*a*. *Hippolyte polaris* (Sab.). *Gr.* Pikkutak.
Cancer squilla, Fabr., var. β., Fauna Gr., n. 216.
Alpheus polaris, Sabine, Suppl. App. Parry, p. 238, t. 2, f. 5–8.
Owen, App. Voy. Ross, p. 85.
Kröyer, N. T., IV., p. 577; Vidensk. Selsk. Skr., l. c., p. 324, t. III., IV., f. 78–82.
Bell, Belcher's Last Arctic Voy., p. 407.

b. *Hippolyte borealis*, Owen.
Owen, App. Voy. Ross, p. 89.
Kröyer, N. T., IV., p. 577; Vid. S. Skr., l. c., p. 330, t. 3, f. 74–77.
Bell, l. c., p. 400.

Obs.—Auct. cl. Goësii (l. c.) a *H. polari* haud distincta. Also Buchholz (l. c., p. 275) is inclined to regard them as one species.

15. *Hippolyte aculeata* (Fabr.). *Greenl.* Naularnak.
Astacus grœnlandicus, J. C. Fabricius, Systema Entomol., p. 416.
Cancer aculeatus, O. Fabr., F. Gr., n. 217.
Alpheus aculeatus, Sabine, Suppl. App. Parry's Voy., p. 237, t. II., f. 9–10.
Hippolyte aculeata, cornuta, armata, Owen, Zool. Beechey's Voy., p. 86–89.
Kröyer, Nat. Tidsskr., III., p. 578; Vid. Selsk. Skr., p. 334, t. 4–5, f. 83–104.
Bell, l. c., p. 401; Buchholz, l. c., p. 276.

16. *Hippolyte microceras*, Kr.
Kröyer, Nat. T., III., p. 578; Vid. S. Skr., p. 341, t. 5, f. 105–9.

17. *H. Panschii*, Buchh., l. c., p. 277, t. 1, f. 1.

18. *Pandalus borealis*, Kr.
Kröyer, N. T., II., p. 254; II. R., I., p. 461; Voyage, &c., t. 6, f. 2.

19. *Pandalus annulicornis* (Leach).
Leach, Malac. podophth. britt., f. 40.
Kröyer, N. T., II. R., I., p. 469; Voyage, t. 6, f. 3.

20. *Pasiphaë tarda*, Kr.
Kröyer, N. T., II. R., I., p. 453; Voyage, t. 6, f. 1.

21. *P. glacialis*, Buch.
Buchholz, l. c., p. 279, t. 1, f. 2 (70° lat. N.).

22. *Sergestes arcticus*, Kr.
Kröyer, Vid. Selsk. Skr., V. R., IV., p. 24, t. 3, f. 7, et t. 5, f. 16.*

23. *Thysanopoda inermis*, Kr.†
Kröyer, Voy., t. 7, f. 2.

24. *Th. norvegica*, Sars.
Buchholz, l. c., p. 285.

25. *Thysanopoda longicaudata*, Kr.†
Kröyer, Voy., t. 8, f. 1.

26. *Th. Raschii*, Sars.
Buchholz, l. c., p. 285.

27. *Mysis oculata*, Fabr. *Greenl.* Irsitugak.
Cancer oculatus, Fabr., F. Grœnl., n. 222, f. 1; Vid. Selsk. Skr., N. S., I., 563.
C. pedatus, Fabr. F. Gr. 221?
Mysis Fabricii, Leach, Trans. Linn. Soc., XI., 350.
Kröyer, Voyage, &c., t. 8, f. 23; Nat. Tidsskr., III., 1 R. I., p. 13.
Buchholz, l. c., p. 284.

28. *Mysis latitans*, Kr.
Kröyer, N. T., III. R., I., p. 30, t. I., f. 4.

---

* I have omitted *Sergestes Rinkii*, because this species was not taken exactly in Greenland, but in the Northern Atlantic, between Greenland and Scotland.

† The exact habitat of these two species is unknown; they are inserted here on the authority of Prof. Reinhardt, who, I believe, consulted Prof. Kröyer on the subject.

29. *Mysis arctica*, Kr.
Kröyer, N. T., III. R., I., p. 34, t. 1, f. 5.*

Cumacea.

30*a*. *Diastylis Edwardsii* (Kr.).
Cuma Edwardsii, Kr. N. T., III., p. 504, t. 5 f. 1–16; II. R., II., p. 128, t. I., f. 1–3, 5, 9, 14; Voyage, t. 4.

b. *Diastylis brevirostris* (Kr.).
Cuma brevirostris, K. N. T., II. R., II., p. 174, t. 2, f. 6; Voyage, t. 5 A., f. 1.

Obs.—Auct. cl. Sarsii a *D. Edwardsii* sexu (masculo) modo distincta.

31*a*. *Diastylis Rathkii* (K.).
Cuma Rathkii, Kr. N. T., III., p. 513, t. 5–6, f. 17–30; II. R., II., p. 144, t. 1, f. 4–6; Voyage, t. 5, f. 1.

b. *Diastylis angulata* (Kr.)
Cuma angulata, Kr. N. T., II. R., II., p. 156, t. I., f. 2, t. 2, f. 1; Voy., t. 5, f. 2.

Obs.—According to Sars, the male of 31*a* (Christiania Vidensk. Selsk. Forh., 1864).

32. *Diastylis resima* (Kr.).
Cuma resima, Kr. N. T., II. R., II., p. 165, t. 2, f. 2; Voy., t. 3, f. 1.

33. *Leucon nasica*, Kr.
Cuma nasica, Kr. N. T., III., p. 524, t. 6, f. 31–33.
Kröyer, N. T., II. R., II., p. 189, t. 2, f. 5; Voyage, t. 3, f. 2.

34. *Eudorella deformis* (Kr.).
Leucon deformis, Kr. l. c., p. 194, t. 2, f. 4; Voyage, t. 5 A., f. 3.†

Isopoda.

35. *Arcturus Baffini* (Sab.).‡
Idothea Baffini, Sabine, App. Parry's Voy., p. 59, t. 1, f. 4–6.
Milne Edwards, Hist. d. Crust., II., p. 123, t. 31, f. 1.
Bell, Belcher's Arct. Voy., p. 408.

36. *Idothea Sabini*, Kr.
Idothea entomon, Sabine, Suppl. App. Parry's Voy., p. 227; Bell, Belcher's Arct. Voy., p. 408.
Kröyer, Naturh. Tidsskr., II. R., II., p. 395; Voyage, t. 27, f. 1.

---

* *Dymas typicus* (Kr. Naturh. Tidsskr., III. R., I., p. 63) is omitted, because I believe it to be, with *Myto Gaimardii* of the same author, only a larval form of some typical long-tailed Decapodous Crustacean.

† *Alauna Goodsiri*, Bell, l. c., p. 403, t. 34, f. 2, should be compared with the Greenland *Cumacea* enumerated above.

‡ *Arcturus Baffini*, Sab., has not, as far as I know, been found in Greenland; but of late years the Museum at Copenhagen has received several specimens from the Färö Islands, and from North-eastern Iceland.

37. *Idothea nodulosa*, Kr.
Kröyer, Naturh. Tidsskr., II. R., II., p. 100; Voyage, t. 26, f. 2.*

38. *Oniscus*, sp.? *Gr.* Kerksub-Koma.
Oniscus asellus, Fabr., F. Gr. 228.

39. *Asellus grœnlandicus*, Kr. *Greenl.* Teitsib-Terkeinga.
Oniscus aquaticus, Fabr., F. Gr. 227.

40. *Henopomus tricornis*, Kr.
Kröyer, Naturh. Tidsskr., II. R., II., p. 372; Voyage, t. 30, f. 2.

41. *Jæra nivalis*, Kr. *Greenl.* Sirksab-Koma.
Oniscus marinus, Fabr., F. Gr., n. 229.
Kröyer, Vid. Selsk. Skr., VII., p. 303, t. 4, f. 21.

42. *Munna Fabricii*, Kr.
Kröyer, Nat. T., II. R., II. B., p. 380; Voyage, t. 31, f. 1.

43*a*. *Anceus elongatus*, Kr.
Kröyer, Nat. T., II. R., II. B., p. 388; Voyage, t. 30, f. 3.

b. *Praniza Reinhardti*, Kr.
Kröyer, Vid. Selsk. Skr., VII., p. 301, t. 4, f. 20.†

44. *Æga psora* (L.). *Greenl.* Saraulib-Koma, &c.
Oniscus psora, Linn., Syst. Nat. (X.), I., p. 636; Fabr. F. G., n. 226.
Æga marginata, Leach; Milne-Edwards; Cuvier Regn. An., t. 67, f. 1.
Lütken, Vidensk. Medd. N. For., 1858, p. 66, t. I. A., f. 9–11.

45. *Æga arctica*, Ltk.
Lütken, l. c., p. 71, t. I. A., f. 1–3.

46. *Æga crenulata*, Ltk.
Lütken, l. c., p. 70, t. I. A., f. 4–5.

Obs.—The Greenland *Ægæ* are especially found on the Shark, *Somniosus microcephalus*; also probably on the large Cod-fishes.

47. *Bopyrus hippolytes*, Kr.
Kröyer, Vidensk. Selsk. Skr., VII., p. 306, t. 4, f. 22.
Gyge hippolytes, Spence Bate, Brit. Cr., II., p. 230; Buchholz, l. c., p. 286. (On *Hip. polaris.*)

48. *Bopyrus abdominalis*, Kr.
Kröyer, Naturh. Tidsskr., III., p. 102 and 289, t. 1–2; Voyage, t. 29, f. 1.
Phryxus hippolytes, Rathke, Nov. Act. Ac. Nat. Cur., XX., p. 40. (On *Hip. Gaimardii, turgida*, Kr.).

49. *Dajus mysidis*, Kr.
Kröyer, Voyage, t. 28, f. 1.

---

* *I. robusta*, Kr., is omitted, because I am not aware that this widely diffused *pelagic* Crustacean really inhabits the shores of Greenland. Kröyer's specimens were captured between Iceland and Greenland, in 60° lat. N.

† Considering the known relations between *Anceus* and *Praniza*, it might be presumed that these (43*a* and *b*) are but the two sexes of one species.

Leptophryxus mysidis, Buchholz, l. c., p. 288, t. 2, f. 2. (On *Mysis oculata*, Buchholz.)

Amphipoda (**et Læmidopoda**).*

50. *Pontoporeia femorata*, Kr.
Kröyer, Naturh. Tidsskr., IV., p. 153; II. R., I., p. 530; Voyage, t. 23, f. 2.
Boeck, Crust. Amphip., p. 123.

51. *Opis typica*, Kr.
Kröyer, Nat. T., II. R., II., p. 46; Voyage, t. 17, f. 1.
Boeck, Crust. Amphip., p. 120.
Opis Eschrichtii, Kr., N. T., IV., p. 149.

52. *Lysianassa gryllus* (Mandt).
Gammarus gryllus, Mandt, Observ. in itinere ad Grœnland. facto, p. 34.
Lysianassa magellanica, Milne-Edwards, Ann. Sc. Nat., 3 s., t. 9, p. 398; Voyage de Castelnau.
Eurytenes magellanicus, Lilljeborg, Acta Upsal., 3 s., 1865, p. 11, t. 1–3, f. 1–22.
Lysianassa magellanica, Sp. Bate, Cat. Amp., t. 10, f. 5.
Goës, Öfv. Vet. Ak. Förh., 1865, p. 1 (sep.), t. 36, f. 1; Boeck, Crust. Amphip., p. 105; Skand. Arkt. Amphip., p. 144.

53. *Socarnes Vahlii* (Rhdt.).
Kröyer, Vid. Selsk. Skr., VII., p. 233.
Anonyx Vahlii, Kr., N. T., II., p. 256; II. R., I., p. 599; Voyage, t. 14, f. 1.
Anonyx Vahlii, Bruzelius, Skand. Amphipod., Vet. Akad. Handl., n.s., III., p. 43.
Gammarus nugax, Owen, App. Ross. Voy., p. 87.
Socarnes Vahlii, Boeck, Crust. Amphip., p. 100; Skand. Arkt. Amph., p. 129, t. 6, f. 8.

54. *Anonyx lagena* (Rhdt.).
Kröyer, Vid. Selsk. Skr., VII., p. 237, t. 1, f. 1; Bell, l. c., p. 406.
Cancer nugax, Phipps, Voy., t. 12, f. 2, p. 192.
Lysianassa appendiculosa, Kr. l. c., p. 240, t. 1, f. 2 Nat. T., II., p. 257.
Anonyx lagena, Kr., N. T., II., p. 256; Sp. Bate, Cat. Amph., p. 17, t. 12, f. 7.
Anonyx ampulla, Kr., l. c., II. R., I., p. 578; Voyage, t. 13, f. 2; Bruzelius, l. c., p. 39; Stimpson, Proc. Philad., 1863; non Sp. Bate, Cat., p. 79, t. 13, f. 5.
Lysianassa appendiculata, Sp. Bate, l. c., p. 67, t. 10, f. 8.
Anonyx lagena, Boeck, Crust. Amphip., p. 108; Skand. Arkt. Amph., p. 152; Buchholz, l. c., p. 300.

---

* Species dubiæ: *Oniscus arenarius*, F. Gr. 234; *O. Strœmianus*, F. Gr. 235 (*Gr.* Kingupek); et *O. abyssinus*, F. Gr. 236.

55. *Anonyx gulosus* (Kr.). *Gr.* Kingungoak-aukpilartok.
? *Oniscus cicada*, Fabr., F. Gr., 233.
Anonyx gulosus, Kr. l. c., II. R., I., p. 611; Voyage, t. 14, f. 2; Bruzel., p. 44; Boeck, Crust. Amphip., p. 110; Skand. Arkt. Amph., p. 157, t. 5, f. 4.
A. norvegicus, Lillj., Öfv. Vet. Ak. Förh., 1851, p. 22; *A. Holbœlli*, Sp. Bate, Cat., p. 75, t. 12, f. 4.

56. *Aristias tumidus* (Kr.).
Anonyx tumidus, Kr. N. T., II. R., II., p. 16; Voyage, t. 16, f. 2; Bruzelius, l. c., p. 41.
Lysianassa Audouiniana, Sp. Bate, Cat., p. 69, t. 11, f. 1 (fide Boeck).
Aristias tumidus, Boeck, Crust. Amphip., p. 107; Skand. Arkt. Amphip., p. 148, t. 3, f. 4.

57. *Hippomedon abyssi* (Goës).
Anonyx abyssi, Goës, l. c., p. 3, t. 371, f. 5; *Hippomedon abyssi*, Boeck, Crust. Amphip., p. 103; Skand. Arkt. Amphip., p. 138.

58. *Hippomedon Holbœlli* (Kr.).
Anonyx Holbœlli, Kr., N. T., II. R., II., p. 8; Voyage, t. 15, f. 1; Bruzel., l. c., p. 43.
A. denticulatus, Sp. B., Cat., p. 74, t. 12, f. 2; *Hippomedon Holbœlli*, Boeck, Crust. Amphip., p. 102; Skand. Arkt. Amph., p. 136, t. 5, f. 6, et t. 6, f. 7.

59. *Orchomene minuta* (Kr.).
Anonyx minutus, Kr., l. c., p. 23; Voy., t. 18, f. 2.
Orchomene minuta, Boeck, Crust. Amphip., p. 116; Skand. Arkt. Amph., t. 5, f. 3.

60. *Onisimus Edwardsii* (Kr.).
Anonyx Edwardsii, Kr., l. c., II. R., II., p. 1; Voyage, t. 16, f. 1 (non Sp. Bate).
Onisimus Edwardsii, Boeck, Crust. Amph., p. 113; Skand. Arkt. Amph., t. 6, f. 4.

61. *O. plautus* (Kr.)
Anonyx plautus, Kr., l. c., II. R., I., p. 629; Voy., t. 15, f. 2.
Sp. Bate Cat., p. 78, t. 13, f. 1; Buchholz, l. c., p. 303.
Onisimus plautus, Boeck, Crust. Amph., p. 112; Skand. Arkt. Amph., t. 4, f. 2.

62. *Onisimus littoralis* (Kr.).
Anonyx littoralis, Kr., l. c., II. R., I., p. 621; Voy., t. 13, f. 1; Bruzelius, l. c., p. 46; Buchholz, l. c., p. 302.
Alibotrus littoralis, Sp. Bate, Cat., p. 86.
Onisimus littoralis, Boeck, Crust. Amph., p. 112; Skand. Arkt. Amph., t. 5, f. 7.

63. *Cyphocaris anonyx*, Ltk.
Boeck, Crust. Amphip., p. 104; Skand. Arkt. Amph., p. 141, t. 6, f. 1.

64. *Stegocephalus ampulla* (Phipps).
Cancer ampulla, Phipps, Voy., p. 191, t. 12, f. 3; Herbst., Naturg. Kr., p. 117, t. 35, f. 2.
Gammarus ampulla, Ross, App. Parry's Voy., p. 20.
Stegocephalus inflatus, Kr., N. T., IV., p. 150; II. R., I., p. 522, t. 7, f. 3; Voyage, t. 20, f. 2; Bruzelius, p. 38.
Bell, Belcher's Voy., p. 406, t. 35, f. 1; Sp. Bate, Cat., t. 10, f. 2; Goës, l. c., t. 38, f. 8–9.
Stegocephalus ampulla, Boeck, Crust. Amphip., p. 128.

65. *Metopa Bruzelii* (Goës).
Leucothoë clypeata, Bruzel. l. c., p. 96.
Montagua clypeata et Bruzelii, Goës, l. c., p. 6, t. 38, f. 10; Boeck, Crust. Amph., p. 192.

66. *Metopa clypeata* (Kr.).
Leucothoë clypeata, Kr., N. T., IV., p. 157; II. R., I., p. 545, t. 6, f. 2; Voy., t. 22, f. 2; Boeck, Crust. Amphip., p. 140.

67. *Metopa glacialis* (Kr.).
Leucothoë glacialis, Kr., N. T., IV., p. 159; II. R., I., p. 539, t. 6, f. 3; Voy., t. 22, f. 3.
Metopa glacialis, Boeck, Crust. Amphip., p. 141.

68. *Syrrhoë crenulata*, Goës.
Goës, Crust. Amph., p. 11, f. 25; Boeck, Crust. Amph., p. 147; Buchholz, l. c., p. 304.

69. *Odius carinatus*, (Sp. Bate).
Otus c., Sp. Bate, Cat. Amphip, p. 126, t. 23, f. 2; Goës, l. c., p. 6.

70. *Vertumnus cristatus* (Owen).
Acanthonotus cristatus, Owen, App. Ross 2nd Voy., p. 90, t. B, f. 8.
Boeck, Crust. Amphip., p. 179.

71. *Vertumnus serratus* (Fabr.). *Greenl.* Kingungoak-Kappinartolik.
Oniscus serratus, Fabr., F. Gr. 237.
Amphithoë serra, Kr., Vid. Selsk. Skr., VII., p. 266, t. 2, f. 8; Nat. T., II., p. 260.
Acanthonotus serra, Bruzelius, l. c., p. 78.
Vert. serra, Boeck, Crust. Amph., p. 180; Buchholz, l. c., p. 342.

72. *Vertumnus inflatus* (Kr.).
Acanthonotus inflatus, Kr., N. T., IV., p. 161.
Goës, l. c., p. 7, t. 38, f. 11; Boeck, Crust. Amph., p. 180.

73. *Paramphithoë glabra*, Boeck.
P. exigua, Goës, l. c., p. 7., t. 38, f. 12; Boeck, Crust. Amph., p. 175.
? *Parapleustes glacialis*, Buchholz, l. c., p. 337, t. 7, f. 1.

74. *Paramphithoë panopla* (Kr.).
Amphithoë panopla, Kr., Vid. S. Skr., VII., p. 270, t. 2, f. 9; Voyage, t. 11, f. 2.

Bruzelius, l. c., p. 69; *Paramph. p.*, Boeck, Crust. Amph., p. 176.

Pleustes tuberculatus, Sp. Bate, Cat., p. 62, t. 9, f. 8; *Pl. panoplus*, Buchholz, l. c., p. 334, t. 7.

75. *Paramphithoë bicuspis* (Rhdt.).

Amphithoë bicuspis, Kr., Vid. Selsk. Skr., VII., p. 273, t. 2, f. 10.

Paramphithoë bicuspis, Bruzel., l. c., p. 73.

Pherusa bicuspis, Sp. Bate, Brit. Crust., p. 253; Cat., p. 144, t. 27, f. 7.

Ph. cirrus, Sp. Bate, Cat., p. 143, t. 27, f. 6.

76. *Paramphithoë pulchella* (Kr.).

Amphithoë pulchella, Kr., Voyage, t. 10, f. 2; Bruzelius, l. c., p. 70.

Pherusa p., Sp. Bate, Cat., p. 143, t. 20–7, f. 5; Boeck, Crust. Amph., p. 177.

77. *Atylus carinatus* (Fabr.).

Gammarus carinatus, Fabr., Ent. Syst., II., p. 515.

Atylus carinatus, Leach, Linn. Trans., XI., 357; Zool. Misc., III., p. 22, t. 69.

Amphithoë carin., Kr., Vid. S. Skr., VII., p. 256, t. 2, f. 6; N. T., II., p. 259; Voy., t. 11, f. 1.

Buchholz, l. c., p. 357, t. 40; Boeck, Crust. Amphip., p. 190.

78. *Atylus Smitti* (Goës).

Goës, l. c., p. 8, t. 38, f. 14; Boeck, Crust. Amphip., p. 191; Buchholz, l. c., p. 361.

79. *Pontogeneia crenulata*, (Rhdt.).

Amphithoë crenulata, Kr., Vid. S. Skr., VII., p. 278, t. 3, f. 12; N. T., IV., p. 165.

Amph. inermis, Kr., l. c., p. 275, t. 3, f. 11; *Pont. inermis*, Boeck, Crust. Amphip., p. 194; Buchholz, l. c., p. 366.

80. *Tritropis fragilis* (Goës).

Paramphithoë fragilis, Goës, l. c., p. 8, t. 39, f. 16.

Tritropis fr., Boeck, Crust. Amph., p. 160; Buchholz, l. c., p. 320.

81. *Tritropis aculeata* (Lepechin).

Oniscus aculeatus, Lepech., Acta Petrop., 1778, I., p. 247, t. 8, f. 1.

Talitrus Edwardsii, Sabine, Suppl. App. Parry, p. 233, t. 2, f. 1–4; Ross, App. Parry's Voy., p. 205.

Amphithoë Edwardsii, Owen, App. Ross Voy., p. 90; Kröyer, N. T., II., p. 76; Voyage, t. 10, f. 1.

Tritropis aculeata, Boeck, Crust. Amph., p. 158; Buchholz, l. c., p. 316, t. 4.

82. *Calliopius læviusculus* (Kr.).

Amphithoë læviuscula, Kr., Vid. S. Skr., VII., p. 281, t. 3, f. 13; Bell, l. c., p. 406.

Amph. serraticornis, Sars, Christiania Vid. Selsk. Forh., 1858, p. 140.

Paramphithoë læviuscula, Bruzel., p. 76.

Calliope læviuscula et grandoculis, Sp. Bate, Cat. Amph., p. 148–9, t. 28, f. 2 et 4.
Boeck, Crust. Amph., p. 197.

83. *Amphithopsis longimana*, Bk.
Boeck, Crust. Amph., p. 199.

84. *Cleïppides tricuspis* (Kr.).
Acanthonotus tricuspis, Kr., N. T., II. R., II., p. 115; Voyage, t. 18, f. 1.
Boeck, Crust. Amphip., p. 201.

85. *Halirages fulvocinctus* (Sars).
Amphithoë fulvocincta, Sars, l. c., p. 141.
Pherusa tricuspis, Stimpson, Proc. Ac. Phil., 1863, p. 138.
Paramphithoë fulvoc., Goës., l. c., t. 38, f. 15; Boeck, Crust. Amph., p. 116 (*Halirages fulvoc.*); Buchholz, l. c., p. 367.

86. *Paramphithoë ? megalops* (Buchh.).
Buchholz, *op. cit.*, p. 369, t. 12.

87. *Acanthozone cuspidata* (Lep.).
Oniscus cuspidatus, Lep., Act. Petr., 1778, t. 8, f. 3.
Acanthosoma hystrix, Owen, App. Ross. Voy., p. 91, t. B., f. 4; Bell, l. c., p. 406.
Amphithoë hystrix, Kr., Vid. S. Skr., VII., p. 259, t. 2, f. 7; Nat. T., II., p. 259.
Bruzelius, l. c., p. 71; *Acanth. cuspid.*, Boeck, Crust. Amphip., p. 184.
A. hystrix, Buchholz, l. c., p. 362, t. 11.

Obs.—"*Amphithoë Jurinii* ?, Kröy.," Bell, l. c., p. 406. I am not aware that Prof. Kröyer ever described a species of that name.

88. *Œdicerus saginatus*, Kröyer, Nat. T., IV., p. 156; Bruzelius, l. c., p. 94; Goës, l. c., t. 39, f. 18; Boeck, Crust. Amphip., p. 162.

89. *Œdicerus lynceus*, Sars.
Sars, l. c., p. 144; Boeck, Crust. Amphip., p. 162.
Œdicerus propinquus, Goës, l. c., p. 10, t. 39, f. 19.
Buchholz, l. c., p. 331, t. 7, f. 2.

90. *Œ. borealis*, Bk.
Crust. Amphip., p. 162.
Buchholz, l. c.

91. *Monoculodes affinis* (Bruz.).
Œdic. aff., Bruzelius, l. c., p. 93, f. 18; Goës, l. c., p. 11, t. 39, f. 21.

92. *Monoculodes norvegicus*, Boeck.
Crust. Amphip., p. 164.

93. *Monoculodes latimanus* (Goës).
Œdic. l., Goës, l. c., p. 11, t. 39, f. 23; Boeck, Crust. Amph., p. 168.

94. *M. borealis*, Bk.
Œd. affinis, Goës, l. c., p. 11, f. 21.
Boeck, Crust. Amph., p. 168.
Buchholz, l. c., p. 325, t. 5.

95. *Tiron acanthurus*, Lillj.
Tessarops hastata, Norman, Annals, 1868, p. 412, t. 22, f. 4, 7.
Syrrhoë bicuspis, Goës, l. c., t. 40, f. 26.

96. *Harpina plumosa* (Kr.).
Phoxus plumosus, Kröyer, Nat. T., IV., p. 152, II. R., I., p. 563; Bruzelius, l. c., p. 66; Sp. Bate, Br. Cr., p. 146.
Harpina plumosa, Boeck, Crust. Amph., p. 135.

97. *Phoxus Holbœlli*, Kr.
Kröyer, l. c., IV., p. 151; II. R., I., p. 551; Bruzelius, l. c., p. 68; Sp. Bate, l. c., p. 143.
Boeck, Crust. Amphip., p. 135; Skand. Arkt. Amph., t. 7, f. 5.

98. *Haploöps tubicola* (Lilljeborg) (var.).
Lilljeborg, Öfvers. Vet. Ak. Förhandl., 1855, p. 135; Bruzelius, l. c., p. 88; Goës, l. c., p. 12.
Boeck, Crust. Amphip., p. 226.

99. *Ampelisca Eschrichtii*. Kr.
Kröyer, Nat. T. IV., p. 155; Boeck, Crust. Amph., p. 224.
Buchholz, l. c., p. 375, t. 13, f. 1.

100. *Byblis Gaimardi* (Kr.).
Ampelisca Gaimardi, Kröyer, Voyage, &c., Crust., t. 23, f. 1.
Bruzelius, l. c., p. 86; Sp. Bate, l. c., p. 127; *Byblis G.*, Boeck, Crust. Amphip., p. 228.

101. *Pardalisca cuspidata*, Kr.
Kröyer, Nat. T., IV., p. 153; Bruzelius, l. c., p. 101; Boeck, Crust. Amphip., p. 151.
Buchholz, l. c., p. 306, t. 1, f. 3, et t. 2, f. 1.

102. *Eusirus cuspidatus*, Kr.
Kröyer, Nat. T., II. R., II., p. 501, t. 7, f. 1; Voyage, t. 19, f. 2; Bruzelius, l. c., p. 63.
Boeck, Crust. Amphip., p. 156.
Buchholz, l. c., p. 313, t. 3, f. 2.

103. *Melita dentata* (Kr.).
Gammarus dent., Kröyer, Nat. T., IV., p. 159; Bruzelius, l. c., p. 61.
Gammarus Krœyeri, Bell, Belcher's Arctic Voy., p. 405, t. 34, f. 4.
Megamæra dentata, Sp. Bate, Cat., p. 225, t. 39, f. 4.
Boeck, Crust. Amph., p. 211.

104. *Gammarus locusta*, (Linn.). *Greenl.* Kingak.
Oniscus pulex, Fabr., F. Gr. 231.
Cancer locusta, Linn., Faun. Suec., II., p. 497.
Gammarus locusta, Mont., Linn. Soc. Trans., IX., p. 92, t. 4, f. 1.
Gammarus boreus, Sabine, Ross, Owen, Bell (Parry's, Ross's, and Belcher's Voyages).

Gammarus arcticus, Sowerby, Account Arct. Reg., p. 541, t. 16, f. 14.
Gammarus locusta, Kr., Vid. Selsk. Skr., VII., p. 255; Bruzelius, l. c., p. 52; Lilljeborg, l. c., 1853, p. 448.
Gammarus mutatus et Duebenii, Lilljeb., l. c., 1853, p. 448; 1851, p. 22.
Gammarus pulex, Stimps., Mar. Invert. Gr. Man., p. 55.
Boeck, Crust. Amphip., p. 204; Buchholz, l. c., p. 343.

105. *Gammaracanthus loricatus* (Sabine).
Gammarus loric., Sabine, Suppl. App. Parry's Voy., p. 231, t. 1., f. 7; Bell, l. c., p. 405.
Kröyer, Vid. S. Skr., VII., p. 250, t. I., f. 4; Nat. T., II., p. 258.
Lovèn, Öfv. Vetensk. Akad. Förhandl., 1861, p. 287.
Gammaracanthus loricatus, Sp. Bate, Cat. Amph., p. 202, t. 36, f. 2.
Boeck, Crust. Amph., p. 135.

106. *Amathilla Sabini* (Leach).
Gam. S., Sabine, Sup. Parry's Voy., p. 232, t. 1, f. 8–11.
Ross, App. Parry's Voy., p. 204; Owen, App. Ross's Voy., p. 89; Bell, l. c., p. 404.
Kröyer, Vid. Selsk. Skr., VII., p. 244, t. I., f. 3; Nat. Tidsskr., II., p. 257.
Bruzelius, l. c., p. 50.
Amathia Sabini, Sp. Bate, l. c., p. 197, t. 35, f. 9.
Boeck, Crust. Amphip., p. 217; Buchholz, l. c., p. 346, t. 8, f. 1–2, et t. 9, f. 1.

107. *A. pinguis* (Kr.).
Gam. p., Kröyer, Vid. Selsk. Skr., VII., p. 252, t. 1, f. 5; Nat. Tidsskr., II., p. 258.
Boeck, Crust. Amph., p. 218; Buchholz, l. c., p. 353, t. 9, f. 2.

108. *Autonoë macronyx* (Lilljeb.).
Gammarus macronyx, Lilljeb., l. c., 1853, p. 458; 1855, p. 125.
Bruzelius, l. c., p. 29, t. 1, f. 6; Goës, l. c., p. 15, t. 40, f. 31.

109. *Protomedeia fasciata*, Kr.
Kröyer, Nat. Tidsskr., IV., p. 154; Boeck, Crust. Amph., p. 239.
Gam. macronyx, Lilljeb., K. V. A. H., 1854, p. 458.

110. *Photis Reinhardti*, Kr.
Photis Reinhardti, Kr., Nat. T., IV., p. 155.
Amphithoë pygmæa, Lilljeb., l. c., 1852, p. 9; Bruzelius, l. c., p. 32 (*A. Reinhardti*); Boeck, Crust. Amph., p. 233.

111. *Podocerus anguipes* (Kr.).
Ischyrocerus anguipes, Kr., Vid. Selsk., VII., p. 283, t. 3, f. 14; Nat. T., IV., p. 162.

Gammarus zebra, Rathke, Nov. Acta A. C.-L., t. XX., p. 74, t. 3, f. 4.
Bruzelius, l. c., p. 21; Boeck, Crust. Amphip., p. 167; Buchholz, l. c., p. 378, t. 13, f. 2, et t. 14.

112. *Podocerus latipes* (Kr.).
Ischyrocerus latipes, Kr., Nat. T., IV., p. 162; Boeck, Crust. Amphip., p. 167.

113. *Siphonocoetes typicus*, Kr.
Kröyer, Nat. T., II. R., I., p. 481, t. 7, f. 4; Voyage, t. 20, f. 1.
Boeck, Crust. Amphip., p. 177.

114. *Glauconome leucopis*, Kr.
Kröyer, Nat. T., II. R., I., p. 491, t. 7, f. 2; Voyage, t. 19, f. 1.
Unciola gl., Sp. Bate, Cat., p. 279 Boeck, Crust. Amphip., p. 259.
Buchholz, l. c., p. 385.

115. *Themisto libellula* (Mandt).
Themisto Gaudichaudii, Ross, App. (non Guérin).
Gammarus libellula, Mandt, Obs. itin. Gr., p. 32.
Themisto arctica, Kr., Vid. Selsk. Skr., VII., p. 291, t. 4, f. 16; Stimpson, Philad. Proc., 1863.
Themisto crassicornis, Kr., l. c., p. 295. t. 4, f. 17.
Goës, l. c., p. 17, t. 41, f. 33; Boeck, Crust. Amphip., p. 88; Skand. Arkt. Amph., p. 88, t. 1, f. 5; Buchholz, l. c., p. 385, t. 15, f. 1.

116. *Th. bispinosa*, Boeck.
Boeck, Crust. Amphip., p. 88; Skand. Arkt. Amph., p. 87, t. 1, f. 4.

117. *Parathemisto compressa* (Goës).
Themisto compr., Goës, l. c., p. 17, t. 41, f. 34.
Boeck, Crust. Amphip., p. 87; Skand. Arkt. Amphip., p. 86.

118. *Hyperia medusarum* (Müll.).
Lestrigonus exulans et Hyperia oblivia, Kr., Vid. Selsk. Skr., p. 298, t. 4. f. 18,
Lestrigonus exulans et Kinnahani, Sp. Bate, Brit. Sess. Cr., p. 5 et 8.
Hyperia medusarum, Sp. Bate, Cat. Amph., p. 295, t. 49, f. 1; Boeck, Crust. Amph., p. 85; Skand. Arkt. Amph., p. 79, t. 1, f. 1.
Hyperia galba (Mont.), Sp. Bate, Brit. Crust., p. 12.*

119. *Tauria medusarum* (Fabr.). *Greenl.* Urksursak.
Oniscus medusarum, Fabr., F. Gr. 232.
Metoecus medusarum, Kr., l. c., p. 288, t. 3, f. 15;

---

* As *Hyperoödon rostratus* and *Globiocephalus melas* are occasionally seen in Baffin's Bay, their parasites (*Platycyamus Thompsoni, Pennella crassicornis, Xenobalanus gl.*, and *Cyamus globicipitis*) might also be enumerated among the Crustacea of Greenland; but they are omitted here because they have not actually been sent down from Greenland.

Boeck, Crust. Amphip., p. 86; Skand. Arkt. Amph., p. 82, t. 2, f. 2.

120. *Dulichia spinosissima*, Kr.
Kröyer, Nat. T., II. R., I., p. 512, t. 6, f. 1; Voyage, t. 22. f. 1; Boeck, Crust. Amph., p. 262.

121. *Caprella septentrionalis*, Kr. *Greenl.* Napparsariak.
Squilla lobata, Fabr., F. Gr. 225 (non Müll.).
Kröyer, Nat. T., IV., p. 590, t. 8, f. 10–19; Voyage, t. 25, f. 2.
Caprella cercopoides, White, App. Sutherland's Journ., p. 203, f. 1 et 207.
Boeck, Crust. Amphip., p. 276.

122. *Cercops Holbœlli*, Kr.
Kröyer, Nat. Tidsskr., IV., p. 504, t. 6, f. 1–13; Boeck, Crust. Amph., p. 269.

123. *Ægina longicornis*, Kr.
Kröyer, Nat. Tidsskr., IV., p. 509, t. 7, f. 1–12; Boeck, Crust. Amph., p. 270.

124. *Æ. echinata*, Boeck.
? *Caprella spinifera*, Bell, Belcher's Last Arctic Voy., p. 407, t. 35, f. 2; Buchholz, l. c., p. 388.

125. *Cyamus mysticeti*, Ltk. *Greenl.* Arberub-Koma.
Martens, Spitzberg. Reise., p. 85, t. Q., f. D.
Oniscus ceti, Pallas, Spicil. Zool., f. IX., p. 76, t. 4, f. 14.
Squilla balænæ, de Geer, Mémoir. VII., p. 540, t. 42, f. 6–10.
Cyamus ceti, Kröyer, Nat. Tidsskr., IV., p. 476, t. 5, f. 63–70; Sp. Bate, Brit. Crust., p. 85.
C. mysticeti, Lütken, K. D. Vid. Selsk. Skr., 3 R., X., p. 251, t. 1, f. 1.

Obs.—On *Balæna mysticetus.*

126. *Cyamus boopis*, Ltk.
Oniscus ceti, Fabr., F. Gr. 230.
Lütken, l. c., p. 262, t. 3, f. 6.

Obs.—On *Megaptera boops.*

127. *Cyamus monodontis*, Ltk.
Lütken, l. c., p. 256, t. 1, f. 2.

Obs.—On *Monodon monoceros.*

128. *Cyamus nodosus*, Ltk.*
Oniscus ceti, Zoologia Danica, t. 119, f. 113–117.
Lütken, l. c., p. 274, t. 4, f. 8.

Obs.—With the preceding.

Phyllopoda et Cladocera.

129. *Apus glacialis*, Kr.
Kröyer, Nat. T., II. R., II., p. 431; Voy., t. 40, f. 1.

130. *Branchipus paludosus* (Müll.). *Greenl.* Taitsim-illærkei.
Cancer stagnalis, Fabr., F. Gr. 224; Zool. Dan., t. 48.

---

* Quid est *Talitrus cyaneæ*, Sabine, Suppl. App. Parry's Voy., t. I., f. 12–18 ?

131. *Nebalia bipes* (Fabr.).
Cancer bipes, Fabr. F. Gr. 223, f. 2.
Cancer gammarellus bipes, Herbst. Naturg. Krabb. u. Krebse.
Nebalia Herbstii, Leach; Milne-Edwards, Hist. N. de Crust.
Kröyer, Nat. T., II. R., II., p. 436; Voyage, t. 40, f. 2.
Buchholz, l. c., p. 388.

132. *Daphnia rectispina*, Kr. *Greenl.* Taitsim-illærangoa, &c.
Daphne pulex, F. Gr. 238.

133. *Lynceus*, sp.
Lynceus lamellatus, ? Kröyer, Vid. Selsk. Skr., VII., p. 320.

Ostracoda.*

134. *Cypridina*, sp. ?
? *Cypridina excisa*, Stimps. Marine Inv. Gr. M., p. 39, t. 2, f. 28 [*Bradycinetus brenda*, Baird].

Copepoda.

135. *Pontia Pattersonii* (Templ.).†
Anomalocera Pattersonii, Templeton, Trans. Ent. Soc. II., p. 34, t. 5.
Kröyer, Nat. Tidsskr., II. R., II., p. 561, t. 6, f. 1–7; Voyage, t. 42, f. 1.

136. *Diaptomus castor*, Jur. ?
Buchholz, l. c., p. 392.

137. *Harpacticus chelifer* (Müll.).
Cyclops chelifer, Müll. Z. D. Prod. 2413.
Harp. chel., Lilljeborg, Cladocera, t. 22, f. 2–11.
Buchholz, l. c., p. 393.

138. *Tisbe furcata* (Baird).
Canthocamptus f., Bd., Brit. Entom., p. 210.
Tisbe f., Claus, Copepoden, t. 15, f. 1–12.
Buchholz, l. c., p. 393.

139. *Cleta minuticornis* (Müll.).
Cyclops m., Müller, Entom., p. 117, t. 19, f. 14–15.
Canthocamptus m., Baird, l. c.
Buchholz, l. c., p. 393, t. 15, f. 3.

140. *Zaus spinosus*, Claus.
Buchholz, l. c., p. 394.

141. *Zaus ovalis* (Goodsir).
Sterope ovalis et armatus, Goods.
Claus, Copepoden, p. 146, t. 13, f. 11–18.
Buchholz, l. c., p. 394.

142. *Thorellia brunnea*, Boeck.
Vid. Selsk. Forh. Christ., 1864, p. 26.
Buchholz, l. c., p. 395.

---

* See further on, page 166, for the Ostracods from the Hunde Islands, &c.
† Should, perhaps, be omitted for similar reasons as *Idothea robusta*.

143*a*. *Calanus hyperboreus*, Kr.
Kröyer, Vid. Selsk. Skr., VII., p. 310; Nat. T., II. R., II., p. 542; Voyage, t. 11, f. 2.
Cetochilus septentrionalis, Goodsir.
Buchholz, l. c., p. 392, t. 15, f. 2.

Obs.—According to B., the two following species are probably the same as 143*a*.

143*b*. *Calanus quinqueannulatus*, Kr.
Kröyer, N. T., II. R., II., p. 545; Voy., t. 41, f. 3.

143*c*. *Calanus spitzbergensis*, Kr.
Kröyer, l. c., p. 531; Voy., t. 41, f. 1.

144. *Calanus caudatus*, Kr.*
Kröyer, l. c., p. 550; Voy., t. 42, f. 2.

145. *Canthocamptus? hippolytes*, Kr.
Kröyer, Nat. Tidsskr., III. R., II. B., t. 17, f. 10, p. 334. On the gills of *Hipp. aculeata.*

146. *Thersites gasterostei* (Kr.).
Pagenstecher, Archiv f. Naturg., 1861, p. 126, t. 6, f. 1–9.
Ergasilus gasterostei, Kr., l. c., p. 233, t. 12, f. 2.
On *Gast. aculeatus.*

147. *Lernæopoda elongata* (Grant).
Scoresby, Account Arct. Reg., I., 538, t. 15.
Lernæa elongata, Grant, Edinb. Journ. Science, 1827.
Kröyer, Nat. Tidsskr., I., p. 259, t. 2, f. 12, et t. 3, f. 3; Steenstrup et Lütken, Vid. S. Skr., V. R., V., p. 422, t. 15, f. 37.
On the eye of *Somniosus microcephalus.*

148. *Lernæopoda carpionis*, Kr. *Gr.* Ekallub-massimioa.
Lernæa salmonea, Fabr., F. Gr. 327.
Kröyer, Nat. T., I., p. 268, t. 11, f. 6; III. R., II., p. 275, t. 14, f. 4.
On *Salmo carpio*; on *Gasterost. aculeatus?*

149. *Lernæopoda sebastis*, Kr.
Kröyer, Nat. T., III. R., II., p. 279, t. 17, f. 7.
On *Sebastes norvegicus.*

150. *Brachiella rostrata*, Kr.
Kröyer, l. c., I., p. 207, t. 2, f. 1; III. R., II., p. 290, t. 17, f. 8.
On *Hippoglossus maximus* and *pinguis.*

151. *Anchorella uncinata* (Müll.). *Gr.* Saraulib-massimioa.
Lernæa unc., Fabr., F. Gr. 328; Zool. Dan., t. 33, f. 2.
Kröyer, Nat. T., I., p. 290, t. 3, f. 8.
On *Gadus morrhua.*

152. *Anchorella agilis*, Kr.
Kröyer, N. T., III. R., II., p. 300, t. 16, f. 2.
On *Gadus agilis.*

153. *Anchorella stichæi*, Kr.
Kröyer, N. T., III. R., II., p. 298, t. 16, f. 1.
On *Stichæus punctatus.*

---

* *Obs.—Quid Cyclops brevicornis*, Fabr., F. Gr. 240 (*Gr.* Ingnerolanek)?

154. *Lesteira lumpi*, Kr.
Kröyer, Nat. T., III. R., II., p. 325, t. 17, f. 7.
On *Cyclopterus lumpus*.

155. *Diocus gobinus* (Müll.). *Gr.* Itekiudlib-massimioa.
Lernæa gobina, Fabr., F. Gr. 329; Zool. Dan., t. 33, f. 3.
Chondracanthus gobinus, Kr. N. T., I., p. 280, t. 2, f. 8, et t. 3, f. 12.
Stp. et Ltk., Vid. Selsk. Skr. V. R., V., p. 423, t. 15, f. 39; Kröyer, l. c., III. R., II., p. 259.
On *Phobetor ventralis*.

156. *Chondracanthus radiatus* (Müll.). *Greenl.* Ingmingursab-massimioa.
Lernæa rad., Fabr., F. Gr. 330; Zool. D., t. 33, f. 4.
Kröyer, l. c., III. R., II., p. 251, t. 19, f. 1.
On *Macrurus rupestris*.

157. *Chondracanthus nodosus* (Müll.). *Greenl.* Sullupaukak-massimioa.
Lernæa nod., Fabr., F. Gr. 331; Zool. D., t. 33, f. 5.
Kröyer, l. c., II., p. 133, t. 3, f. 2.
On *Sebastes norvegicus*.

158. *Chondracanthus cornutus* (Müll.).
Lernæa cornuta, Zool. Dan., t. 33, f. 6.
On *Pleuronectidæ*.

159. *Tanypleurus alcicornis*, Stp. et Ltk.
Steenstrup et Lütken, Vid. Selsk. Skr., l. c., p. 424, t. 15, f. 38. On *Cyclopterus spinosus*.

160. *Herpyllobius arcticus*, Stp. et Ltk.
Steenstrup et Lütken, l. c., p. 426, t. 15, f. 40.
Silenium polynoës, Kr., l. c., III. R., II., p. 329, t. 18, f. 6. On *Lepidonoti* and other Chætopodous Annulata.

161. *Caligus* (*Lepeophtheirus*) *hippoglossi*, Kr. *Greenl.* Netarnab-Koma.
Binoculus piscinus, Fabr., F. Gr. 239.
Kröyer, N. T., I., p. 625; III. R., II., p. 131, t. 6, f. 5. On *Hippoglossus maximus*.

162. *Caligus* (*Lepeophtheirus*) *robustus*, Kr.
Kröyer, N. T., III. R., II., p. 135, t. 6, f. 6.
On *Raia radiata*.

163. *Dinematura ferox*, Kr.
Kröyer, N. T., II., p. 40, t. 1, f. 5; Stp. et Ltk., l. c., t. 7, f. 14. On *Somniosus microcephalus*.

164. *Peniculus clavatus* (Müll.).
Lernæa clavata, Müll., Zool. Dan., p. 38, t. 33.
Kröyer, Nat. T., III. R., II. B., p. 266, t. 14, f. 8.
On *Sebastes norvegicus*.

165. *Hæmobaphes cyclopterina* (Müll.). *Greenl.* Nepisardlub-massimioa.
Lernæa cyclopterina, Fabr., F. Gr. 326; Kröyer, Nat. T., I., p. 502, t. 5, f. 4.

Steenstrup et Lütken, l. c., p. 705, t. 13, f. 30; Stimpson, Proc. Philad., p. 139.
On *Cyclopterus spinosus*, *Cottus scorpius*, *Cendronotus fasciatus*, and *Sebastes norvegicus.*

166. *Lernæa branchialis*, L.* *Gr.* Okab-massimioa.
Lernæa gadina, Fabr., F. Gr. 325.
Kröyer, Nat. T., I., p. 293, t. 3, f. 10; Stp. et Ltk., l. c., p. 403, t. 13, f. 28–29.
On *Gadus morrhua*, *ovak*, and *agilis.*

(*Incertæ Sedis.*)

167. *Psilomallus hippolytes*, Kr.
Kröyer, N. T., III. R., II., p. 336, t. 17, f. 10.
On *Hippolyte aculeata.*

Cirripedia.

168. *Peltogaster paguri*, Rathke.
Rathke, N. A. Acad. C. L.-C. N. C., XX., p. 245, t. 12, f. 17.
On *Pagurus pubescens.*

169. *Sylon*, sp.
Kröyer, Vid. Selsk. Overs., 1855 p. 128.
On *Hippolyte*, sp.

170. *Balanus porcatus* (Da Costa). *Greenl.* Katungiak.
Lepas balanus, Fabr., F. Gr. 423.
Buchholz, l. c., p. 396.

171. *Balanus balanoides* (Linn.). *Greenl.* Katungiak.
Fabr., F. Gr. 424.

172. *Balanus crenatus*, Brug.
Enc. Méthod. Vers.
Lepas foliacea, var. A., Naturh. Selsk. Skr., I., 1, 174.

173. *Coronula diadema* (Linn.). *Gr.* Keporkab-Katungiarsoa.
Lepas balænaris, Fabr., F. Gr. 425.
On *Megaptera boops.*

174. *Conchoderma auritum* (Linn.).
Lepas aurita, Syst. Nat. (XII.), p. 1110.
Vidensk. Selsk. Skr., 1809–10, p. 94.
Lepas balænaris, jun., Fabr., F. Gr. 425.†

APPENDIX.

Pycnogonida.

175. *Nymphon grossipes* (Linn.). *Greenl.* Niutok.
? *Phalangium grossipes*, Linn., S. N. (XII.), p. 1027.
Pycnogonum grossipes, Fabr., F. Gr. 210 (p.p.).
Sabine, Suppl. App., p. 225; Kröyer, N. T., II. R., I., p. 108; Voyage, t. 36, f. 1.
Buchholz, l. c., p. 336.

---

* On *Pennella crassicornis* from *Hyperoödon rostratus*, and *Xenobalanus globicipitis* from *Globiocephalus melas*, cfr. the note to p. 158.

† *Pegesimallus spiralis*, Kr. (N. T., III. R., II., p. 336, t. 18, f. 7), does not belong to the Crustacea, but to the Hydrozoa (Siphonophora).

176. *Nymphon mixtum*, Kr.
Kröyer, N. T., l. c., p. 110; Voyage, t. 35, f. 2.
Buchholz, l. c., p. 397.

177. *Nymphon longitarse*, Kr.
Kröyer, N. T., l. c., p. 112; Voy., t. 26, f. 2.

178. *Nymphon hirtum*, Fabr.
Fabr., Entomol., IV., p. 417.
Nymphon hirsutum, Sabine, l. c., p. 226.
Kröyer, l. c., p. 113; Voy., t. 36, f. 3; Buchholz, l. c., p. 397.

179. *Nymphon brevitarse*, Kr.
Kröyer, l. c., p. 115; Voy., t. 36, f. 4.

180. *Eurycyde hispida* (Kr.).
Zetes hispidus, Kr., l. c., p. 117; Voyage, t. 38, f. 1.
Euryc. hisp., Schiödte, Rink's Grönland, Nat. Til., p. 71.

181. *Pallene spinipes* (Fabr.).
Pycnogonum spinipes, Fabr. F. Gr., p. 211.
Kröyer, l. c., p. 118; Voy., t. 37, f. 1.

182. *Pallene intermedia*, Kr.
Kröyer, l. c., p. 119; Voy., t. 37, f. 3.

183. *Pallene discoidea*, Kr.
Kröyer l. c., p. 120; Voy., t. 37, f. 3.

184. *Phoxichilidium femoratum* (Rathke).*
Pycnogonum grossipes, var., Fabr., F. Gr. 210.
Nymphon femoratum, Rathke, Nat. Selsk. Skr., V. 1, p. 201.
Phoxichilus proboscideus, Kr., Vid. Selsk. Skr., VII., p. 321.
Orithyia coccinea, Johnst.; *Phoxichilid. coccineum*, Milne-Edw.
Kröyer, Nat. Tidsskr., l. c., p. 122; Voy., t. 38, f. 2.

PRINCIPAL WORKS AND MEMOIRS ON THE CRUSTACEA OF GREENLAND.

Boeck: Crustacea amphipoda borealia et arctica (Vid. Selsk. Forh. Christiania, 1870).

—— De Skandinaviske og arktiske Amphipoder. 1ste Hefte, 1872.

Buchholz: Crustaceen; Die zweite deutsche Nordpolarfahrt, 1874.

---

* I am not aware that *Pycnogonum littorale*, Ström (Fabr., F. Gr. 212), has actually been found on the shores of Greenland. Here also should be mentioned *Phoxichilus proboscideus*, Sab. (Suppl. App. Parry), from North Georgia, and *Nymphon hirtipes* and *Nymphon robustum*, Bell (Belcher's Last of Arctic Voyages, p. 408–9, t. 35, f. 3–4), from Northumberland Sound. These two should especially be compared with the species from Greenland, the descriptions and figures of which were apparently unknown to the English author.

Goës: Crustacea amphipoda maris Spitzbergiam alluentis cum speciebus aliis arcticis enumerat. . . . (Öfvers. Vetensk. Akad. Förh. Stokholm, 1865).

Kröyer: Om Snyltekrebsene især med Hensyn til den danske Fauna (Naturhistorisk Tidsskrift, I., p. 172, 252, 475, 650; II., p. 7, 131. 1837–38).

—— Conspectus Crustaceorum Grœnlandiæ (*ibid.*, II., p. 249, 1838).

—— Grönlands Amphipoder (the Amphipods of Greenland, with descriptions of other Greenland Crustacea, and an enumeration of the known species, remarks on the geographical distribution, &c.). (Kongl. Danske Videnskabernes Selskabs naturvid.-mathemat. Afh., VII., 1839.)

—— *Bopyrus abdominalis* (Naturh. Tidsskr., III., pp. 102 and 289. 1840).

—— Fire nye Arter af Slægten *Cuma* (*ibid.*, III., p. 508, 1841).

—— Udsigt over de nordiske Arter af Slægten *Hippolyte* (*ibid.*, IV., p. 570, 1841).

—— Monographisk Fremstilling af Slægten *Hippolytes* nordiske Arter (Kongl. D. Vid. Selsk. Nat. Math. Afh., IX., 1842).

—— Nye nordiske Slægter af Amphipodernes Orden (Nat. Tidsskr., IV., p. 141. 1842).

—— Beskrivelse af nordiske *Crangon* Arter (*ibid.*, IV., p. 217).

—— Om *Cyamus Ceti*, Linn. (*ibid.*, p. 474, 1843).

—— Beskrivelse af nogle nye Arter og Slægter af *Caprellina* (*ibid.*, pp. 490 and 585, 1843).

—— Bidrag til Kundskab om Pycnogoniderne (*ibid.*, II. R., I., p. 90, 1844).

—— Carcinologiske Bidrag (*ibid.*, II. R., I. B., p. 453, 1845; II., p. 113, 1846; p. 366, 1847; p. 527, 1848; p. 561, 1849).

—— Om Cumaernes Familie (*ibid.*, II. R., II., p. 123, 1846).

—— Forsög til en monographisk Fremstilling af Kræbsdyrslægten *Sergestes* (with an appendix on the auditory organs of Crustacea) (Kongl. Danske Vidensk. Selsk. Skr., V. R., naturv.-math. Afh. IV., 1856).

—— Et Bidrag til Kundskab om Krebsdyrfamilien *Mysidæ* (Nat. Tidsskr., III. R., I., 1861).

—— Bidrag til Kundskab om Snyltekrebsene (on Parasitic Crustacea) (Nat. Tidsskr., III. R., II. Bd., 1863).

—— The Carcinological portion of Gaimard's "Voyage en Scandinavie, en Laponie," &c. (Plates only.)

Lütken: Nogle Bemærkninger om de nordiske *Æga*-Arter (Videnskabelige Meddelleser fra den naturh. For., 1858, p. 65).

—— Bidrag til Kundskab om Arterne af Slægten *Cyamus* Latr. eller Hvallusene (K. D. Vid. Selsk. Skr., X., 1873).

Steenstrup and Lütken: Bidrag til Kundskab om det aabne Havs Snyltekrebs og Lernæer (on Parasitic Crustacea and Lernæidæ), Vid. Selskabs Skr., V. R., V. Bd., 1861).

XVI.—OSTRACODA from GREENLAND, &c. By G. S. BRADY, ESQ., C.M.Z.S.

1.—OSTRACODA from the HUNDE ISLANDS, DISCO BAY, dredged by DR. P. C. SUTHERLAND, and determined by G. S. BRADY, ESQ., C.M.Z.S. (Phil. Trans. 1862, civ., p. 327; Trans. Zool. Soc. 1865, v., p. 360, &c.; Annals Nat. Hist. 1868, ser. 4, vol. ii., p. 30, and Revision, February 1875).

1. Cythere limicola (Norman).	25–30	fathoms.
2. C. angulata? (G. O. Sars).	60–70	„
3. C. tuberculata (Sars).	60–70	„
4. C. abyssicola (Sars).	?	
5. C. septentrionalis, Brady.	60–70	„
6. C. costata, Brady.	60–70	„
7. C. lutea, Müller.	?	
8. C. emarginata, Sars.	60–70	„
9. C. Finmarchica, Sars.	60–70	„
10. Cytheridea papillosa, Bosquet.	25–30	„
11. C. pulchra, Brady.	28–40	„
12. C. oryza, Brady.	?	
13. C. punctillata, Brady.	60–70	„
14. Cytheropteron latissimum. (Norman).	25–30	„
15. Bythocythere simplex (Norman)	?	

2.—OSTRACODA from CUMBERLAND INLET, 15½ fathoms, lat. 66° 10′ N., long. 67° 15′ W. Collected by a Whaler. By G. S. BRADY, ESQ., C.M.Z.S. (Annals Nat. Hist., 1868, ser. 4, vol. ii., p. 31).

1. Cythere Dunelmensis (Norman).
2. Cytheropteron Montrosiense, C. B. & R. (Pl. V., f. 1–5.)
3. C. arcuatum, Brady, non *vespertilio*, Rss. (Pl. V., f. 6, 7.)
4. C. inflatum, C. B. & R. (Pl. V., f. 8–10.)
5. Cytherura undata, G. O. Sars.

3.—OSTRACODA from DAVIS'S STRAIT, lat. 67° 17′ N., long. 62° 21′ W., 6 feet below low-water mark. Collected by a Whaler. By G. S. BRADY, ESQ., C.M.Z.S. (Annals Nat. Hist., ser. 4, vol. ii., 1868, p. 31.)

Cythere lutea, Müller.
C. villosa (G. O. Sars).
C. Finmarchica (G. O. S.).
C. borealis, Brady. (Pl. IV., f. 1–4, 6, 7.)
C. emarginata (Sars).
C. angulata (Sars).
C. pulchella, Brady. (Pl. V., f. 18–20.)
C. tuberculata (Sars).
C. concinna, Jones.
Cytheridea papillosa, Bosq.
Cytherura rudis, Brady. (Pl. V., f. 15–17.)

4.—*Supplement.* FROM ICELAND (IN SHELL-SAND).

Cythera lutea, Müller.
C. borealis, Brady.
C. emarginata (Sars).

XVII.—A REVISED CATALOGUE of the ANNELIDA and other, not Entozoic, WORMS of GREENLAND. By Dr. CHR. LÜTKEN, University Museum, Copenhagen. 1875.

ANNULATA CHÆTOPODA ET DISCOPHORA, GEPHYREA, ETC.

As far as the marine *Chætopoda* are concerned, this List is chiefly based upon Dr. *Malmgren's* memoirs on the Arctic Annulata. To the Greenland species enumerated by this author are added a few from the Museum of the University in Copenhagen, for instance, the *Hirudinidæ*, identified by Mr. *Malm*, of Göteborg; the *Sipunculidæ*, by the late Prof. *Keferstein*, in Göttingen, &c.

The following memoirs should especially be consulted:—

A. S. Œrsted: Grönlands Annulata dorsibranchiata (K. D. Vidensk. Selsk. mathem.-natur. Afh., X. Deel).

Malmgren: Nordiska Hafs-Annulater (Öfvers. K. Vet. Akad. Förh., 1865), 1–3.

——— Annulata polychæta Spitzbergiæ, Grœnlandiæ, &c., 1867.

W. Keferstein: Beiträge zur anatomischen und systematischen Kenntniss der Sipunculiden (Zeitschr. f. wiss. Zool., XV., 1865).

O. A. L. Mörch: Revisio critica Serpulidarum (Naturhist. Tidsskr., 3 R., 1 B., 1863).

In the Catalogue of *Entozoa* (Art. XVIII., p. 172) the species are added which have been identified from Greenland species by Dr. *Krabbe* in the Museum at Copenhagen.

Euphrosynidæ.

1. *Euphrosyne borealis*, Œrstd.

Euphrosyne borealis, Œrsted, Grönland's Annul. dorsibr., p. 170, f. 23–27.

Polynoidæ.

2. *Lepidonotus squamatus* (L.).

Aphrodita squamata, Linn., S. N., Ed. X., p. 655.
Pall. Miscel. Zool., p. 91 (pp.), t. 7, f. 14 *a–d.*
A. punctata, Müll. Pr. Z. D. 2642; v. Würmern, p. 170, t. 13; Abildgd. Zool. Dan., III., p. 25, t. 96, f. 1–4; Fabr. F. Gr. 291.
Polynoë squamata, Aud. et M.-Edw. Rech. Annel., p. 80, t. 1, f. 1–16.

Lepidonote punctata, Œrsted, Annul. Dan. Comp., p. 12, f. 2, 5, 39, 41, 47, 48.
Lepidonotus squamatus, Kinberg, Eugenies Resa, II., p. 13, t. 4, f. 15.
Malmgren, Nord. Hafs.-Annul., p. 56.

(A single specimen in the Museum at Copenhagen, labelled "Greenland.")

3. *Nychia cirrosa* (Pall.).
Aphrodita cirrosa, Pall., Miscell. Zool., p. 95, t. 8, f. 3–6.
Aphrodita scabra, Fabr., Fauna Grœnl. 292.
Nychia cirrosa, Malmgren, Nordiska Hafs-Annulater, p. 58, t. 8, f. 1.

4. *Nychia Amondseni*, Mlgr.
Malmgren, Annulata polychæta, p. 5, t. 1, f. 4.

5. *Eunoa Œrstedii*, Mlgr.
Lepidonote scabra, Œrstd. (non Fabr.), l. c., p. 164, f. 2, 7, 10, 12, 13, 17, 18.
Eunoë Œrstedii, Mlgr., Nordiska Hafs-Annulater, p. 61, t. 8, f. 3.

6. *Eunoa nodosa* (Sars).
Polynoë nodosa, Sars, Christiania Vid. Selsk. Forh., 1860, p. 59.
Eunoë nodosa, Malmgren, l. c., p. 64, t. 8, f. 4.

7. *Lagisca rarispina* (Sars).
Polynoë rarispina, Sars, Christiania Vid. Selsk. Forh., 1860, p. 60.
Lagisca rarispina, Mlgr., l. c., p. 65, t. 8, f. 2.

8. *Harmothoë imbricata* (L.).*
Aphrodita imbricata, Linn., Syst. Nat. (ed. XII.), p. 1804.
Aphrodita cirrata, Fabr., Faun. Grœnl. 290, t. 1, f. 7.
Lepidonote cirrata, Œrsted, l. c., p. 166, f. 1, 5, 6, 11, 14, 15.
Harmothoë imbricata, Mlgr. l. c., p. 66, t. 9, f. 8.

9. *Antinoë Sarsii* (Kinbg., *grœnlandica*, Mlmgr.).
Antinoë (*Sarsii*) *grœnlandica*, Malmgren, Annul. polych., p. 13, et Nordiska Hafs-Annul., p. 75, t. 9, f. 6.

Sigalionidæ.

10. *Pholoë minuta* (Fabr.).
Aphrodita minuta et *A. longa*, Fabr., Faun. Gr. 293 et 294.
Pholoë (?) *minuta*, Œrsted, l. c., p. 169, f. 3, 4, 8, 9, 16.
Pholoë minuta, Malmgren, Nord. Hafs-Annul., p. 89, t. 11, f. 13.

---

* According to Möbius, *Harmothoë imbricata* and *Antinoë Sarsii* are but one species.

Nephthydidæ.

11. *Nephthys ciliata* (Müll.).
    Nereis ciliata, Müll., Zool. Dan., t. 89, f. 1–4.
    Nephthys ciliata, Malmgr., l. c., p. 104, t. 12, f. 17.
12. *Nephthys lactea*, Malmgr.
    Malmgr., Annul. polych., p. 18 (name only). Undescribed specimens, perhaps of this species, are in the Copenhagen Museum.
13. *Nephthys cœca* (Fabr.). (Greenl. *Sengiarsoak.*)
    Nereis cœca, Fabr., F. Gr. 287; Naturh. Selsk. Skr., V., p. 185, t. 4, f. 24–29.
    Nephthys cœca, Œrsted, l. c., p. 193, f. 73, 74, 77–86.
    Nephthys cœca, Malmgr., Nord. Hafs-Annul., p. 104, t. 12, f. 18.
14. *Nephthys longosetosa*, Œrstd.
    Œrstd. l. c., p. 195, f. 75–76.
    Malmgren, l. c., p. 106, t. 13, f. 24.

Phyllodocidæ.

15. *Phyllodoce citrina*, Mlgr.
    Phyllodoce maculata, Œrstd. (non Fabr.), l. c., p. 191, f. 46, 48.
    Ph. citrina, Malmgr., l. c., p. 95, t. 13, f. 24.
16. *Phyllodoce grœnlandica*, Œrstd.
    Œrsted, l. c., p. 192, f. 19, 21, 22, 29–32.
    Malmgren, l. c., p. 36; Annul. polych., t. 2, f. 9.
17. *Phyllodoce Rinki*, Mlgr.
    Annul. polych., p. 23, t. 2, f. 11.
18. *Phyllodoce Luetkeni*, Malmgr.
    Annal. polych., p. 24, t. 2, f. 10.
19. *Phyllodoce incisa*, Œrsted.
    ? *Nereis maculata*, Fabr. (non Müll.), F. Gr. 281.
    Phyllodoce ? incisa, Œrsted, Grönl. Ann. dors., p. 189, f. 44. (Perhaps a doubtful species.)
20. *Eulalia viridis* (Müll.). (Greenl. *Sengiarak.*)
    Die grüne Nereide, Müll., Würm., p. 162, t. 11.
    Nereis viridis, Fabr. F. Gr. 279.
    Œrsted, l. c., p. 188.
    Eulalia viridis, Malmgr., Nord. Hafs-Ann., p. 98, t. 15, f. 39.
21. *Eulalia problema*, Malmgr.
    Nord. Hafs-Ann., p. 99, t. 14, f. 29.
22. *Eteone longa* (Fabr.). (Gr. *Sengiak.*)
    Nereis longa, Fabr., F. Gr. 289; Naturh. Selsk. Skr., V., p. 171, t. 4, f. 11–13.
    Eteone longa, Œrsted, l. c., p. 185, f. 20, 28.
23. *Eteone cylindrica*, Œrsted.
    Œrsted, l. c., p. 187, f. 42, 49, 57.
24. *Eteone flava* (Fabr.).* (Greenl. *Sengiarak.*)

---

* *Nereis cærulea*, Fabr. F. Gr., 280, perhaps a Phyllodocean, not determinable ("Sengiarak" in Greenland, as many other species).

Nereis flava, Fabr., F. Gr. 282; Nat. Selsk. Skr., V., p. 168, t. 4, f. 8–10.
Eteone flava, Œrsted, l. c., p. 186, f. 47.
Malmgren, Nord. Hafs-Ann., p. 102, t. 15, f. 35.

Hesionidæ.

25. *Castalia aphroditoides* (Fabr.). (Greenl. *Sengiarak.*)
Nereis aphroditoides, Fabr., F. G. 278; Nat. Selsk. Skr., V., p. 164, t. 4, f. 4–6.
Castalia Fabricii, Malmgr., Ann. polych., p. 32.

26. *Castalia rosea* (Fabr.). (Greenl. *Sengiarak.*)
Nereis rosea, Fabr. F. Gr. 284; Nat. Selsk. Skr., V., p. 175, t. 4, f. 14–16.

Syllidæ.

27. *Autolytus longisetosus*, Œrstd. (Greenl. *Sengiarak, Iglolualik.*)
? *Nereis prismatica*, Fabr., F. Gr. 285; Nat. S. Skr., V., p. 177, t. 4, f. 17–20.
? *Nereis bifrons*, Fabr., F. Gr. 303; l. c., p. 181, f. 21–23.
Polybostrichus longisetosus, Œrstd., l. c., p. 182, f. 62, 67, 71.
Autolytus longisetosus, Malmgr., Ann. polych., p. 34, t. 7, f. 38.

28. *Autolytus Alexandri* (Malmgr.).*
Malmgr., Ann. polych., p. 37, t. 7, f. 39.

29. *Autolytus incertus*, Mlgr.
Malmgr. *op. cit.*, p. 35, t. 6, f. 40.

30. *Syllis incisa* (Fabr.). (Greenl. *Sengiak.*)
Nereis incisa, Fabr., F. Gr. 277; Nat. Selsk. Skr., V., p. 160, t. 4, f. 1–3.

31. *Syllis Fabricii*, Malmgr. (Greenl. *Sengiarak.*)
Nereis armillaris, Fabr. F. Gr. 276 (non Müll.).

32. *Chætosyllis Œrstedi*, Malmgr. ?
Joida sp., Œrsted, l. c., p. 182.
Malmgren, Annul. polych., p. 45, t. 8, f. 51.

Nereidæ.

33. *Nereis zonata*, Malmgr. (Greenl. *Sengiak.*)
? *Nereis diversicolor*, Fabr., F. Gr. 274 (non Müll.).†
Malmgren, Annul. polych., p. 46, t. 5, f. 34.

34. *Eunereis paradoxa* (Œrsted).
Heteronereis paradoxa, Œrsted, l. c., p. 177, f. 50, 63, 64, 66. (Known from a single specimen.)

35. *Nereis pelagica*, Linn. (Greenl. *Sengiarsoak.*)
Linn. Syst. Nat. (X.), p. 654.
Nereis verrucosa, Fabr., F. Gr. 275.

---

* *Nereis noctiluca*, Fabr. F. Gr. 273 (Greenl. "Ingnerolak"), is a doubtful, undetermined species.

† *N. diversicolor* is cited from East Greenland by Möbius, Zte deutsche Nordpolarfahrt, II., p. 254.

Nereis pelagica, Œrsted, l. c., p. 175, f. 52, 53, 55, 58, 59.
Malmgren, Annul. polych., p. 47, t. 5, f. 35.

(36.) *Heteronereis grandifolia* (Rathke).
Nereis grandifolia, Rathke, Beiträge z. Fauna Norwegens (Nova Acta C. L.-C. N. C., XX.), p. 155, t. 7, f. 13–14.
Heteronereis arctica et assimilis, Œrsted, l. c., p. 179–180, f. 50, 51, 54, 60, 61, 65, 68, 70, 72.
Heter. grandifolia, Malmgr., Nord. Hafs-Ann., p. 108, t. 11, f. 15–16.

Obs.—*Eunereis* and *Heteronereis* are now known to be the natatory, sexually mature state of *Nereis; H. grandifolia* of *Nereis pelagica*.

Lumbrinereidæ.

37. *Lumbrinereis fragilis* (Müll.).
Lumbricus fragilis, Müll. Prodr. Zool. Dan. 2611; Zool. Dan., p. 22, t. 22, f. 1–3.
Malmgren, Annul. polych., p. 63, t. 14, f. 83.

Eunicidæ.

38. *Nothria conchylega* (Sars).
Onuphis conchylega, Sars, Beskr. og Jagttag, p. 61, t. 10, f. 28.
Onuphis Eschrichtii, Œrsted, l. c., p. 172, f. 33–41, 45.

Glyceridæ.

39. *Glycera capitata*, Œrstd. (Greenl. *Pullateriak.*)
Nereis alba, Müll. Prodr. Z. D. 2634; Zool. Dan., II., p. 29, t. 62, f. 6–7.
Glycera capitata, Œrstd. l. c., p. 196, f. 87, 88, 90–94, 96, 99.

40. *Glycera setosa*, Œrd.
Œrsted, l. c., p. 198, f. 89, 95, 97.

Ariciidæ.

41. *Scoloplos armiger* (Müll.). (Greenl. *Pullateriak.*)
Lumbricus armiger, Müll. Zool. Dan., I., p. 22, t. 22.
Scoloplos armiger, Œrsted, l. c., p. 201, f. 113, 117, 118.

42. *Naidonereis quadricuspida* (Fabr.).
Nais quadricuspida, Fabr., F. Gr. 296.
Scoloplos quadricuspida, Œrsted, l. c., p. 200, f. 106–10.

Opheliidæ.

43. *Ammotrypane aulogaster*, Rathke.
Rathke, Beitr. z. F. Norw., l. c., p. 188, t. 10, f. 1–3.
Ophelina acuminata, Œrstd. Archiv f. Naturg., X., p. 111, t. 3, f. 24–26.

44. *Ophelia limacina* (Rathke).
Ammotrypane limacina, Rathke, l. c., p. 190, t. 10, f. 4–8.
Ophelia bicornis, Œrsted, Grönl. Ann., p. 204, f. 104–5, 115, 116, 121.

45. *Travisia Forbesi*, Johnst.
Johnston, Ann. Nat. Hist., IV., p. 373, t. 11, f. 11–18.
Ammotrypane oestroides, Rathke, l. c., p. 192, t. 10, f. 9–12.
Ophelia mamillata, Œrsted, Grönl. Ann., p. 205, f. 103, 112, 114, 119, 120; Archiv f. Naturg., X., p. 110, t. 3, f. 21–23.

Scalibregmidæ.

46. *Scalibregma inflatum*, Rathke.
Rathke, l. c., p. 184, t. 9, f. 15–21.
Oligobranchus roseus et grœnlandicus, Sars, Fauna littor. Norvegiæ, I., p. 91, 92, t. 10, f. 20–27.

Telethusæ.

47. *Arenicola marina* (Linn.). (Greenl. *Inellualuak.*)
Lumbricus marinus, Linn., Syst. Nat. (XII.), p. 1077.
Lumbricus marinus, Fabr., F. Gr. 262, et *L. papillosus*, ibid. 267.
Arenicola piscatorum, Œrstd., Grönl. Ann., p. 207.

Sphærodoridæ.

48. *Ephesia gracilis*, Rathke.
Rathke, l. c., p. 176, t. 7, f. 5–8.
Sphærodorum flavum, Œrsted, Annul. Dan. Comp., p. 43, f. 7, 92, 101.
Pollicita peripatus, Johnst., Ann. Nat. Hist., XVI., p. 5, t. 2, f. 1–6.
Sphærodorum peripatus, Claparède, Beob. Anat. Entw. wirbellos. Thiere, p. 50, t. 11, f. 8–18.
Ephesia gracilis, Malmgr., Annul. polych., p. 79.

Chloræmidæ.

49. *Trophonia plumosa* (Müll.). (Greenl. *Merkolualik.*)
Amphitrite plumosa, Müll., Prodr. Z. D. 2621 (Abildgaard Zool. Dan., III., t. 90, f. 1–2).
Amphitrite plumosa, Fabr., F. Gr. 271.
Siphonostoma plumosa, Rathke, l. c., p. 208, t. 11, f. 1–2.
Trophonia Goodsiri, Johnston, Ann. Nat. Hist., IV., p. 371, t. 11, f. 1–10.

50. *Flabelligera affinis*, Sars.
Sars, Bidrag til Södyrenes Naturh., p. 31, t. 3, f. 16.
Siphonostoma vagininiferum, Rathke, l. c., p. 211, t. 11, f. 3–10.
Tecturella flaccida, Stimpson, Marine Invert. Gr. Manan, p. 32, t. 3, f. 21.

51. *Brada villosa* (Rthk.) ?
Siphonostoma villosum, Rathke, N. Act. Acad. C. L.-C. N. C., XX., p. 215, t. 14, f. 11, 12.

52. *Brada granulata*, Malmgr.
Annul. polychæta, p. 85, t. 12, f. 71.
Brada inhabilis (Rathke ?); Stimpson, Proc. Acad. Philad., 1863.

Sternaspidæ.

53. *Sternaspis fossor*, Stimps.
Stimpson, Marine Invertebrata of Grand Manan, p. 29, t. 2, f. 19.

Chætopteridæ.

54. *Spiochætopterus typicus*, Sars.
Sars, Fauna littor. Norv., II., p. 1, t. 1, f. 8–21.
Malmgren, Annul. polych., p. 98.

Spionidæ.

55. *Scolecolepis* (*Laonice*) *cirrata* (Sars).
Nerine cirrata, Sars, Nyt. Mag. f. Natur., VI., p. 207.
Scol. (*Laon.*) *cirr.*, Malmgr., Annul. pol., p. 91, t. 9, f. 54.

56. *Spio filicornis* (Fabr.). (Greenl. *Igloluatik.*)
Nereis filicornis, Fabr. F. Gr. 289 ; Spio filicornis, Fabr., Schr. Naturf. Freunde, VI., p. 264, t. 5, f. 8–12.
Spio filic., Malmgren, l. c., p. 92, t. 1, f. 1.

57. *Spio seticornis*, Fabr.
Nereis seticornis, Fabr., F. Gr. 288.
Spio seticornis, Fabr., Schr. Naturf. Freunde, VI., p. 260, t. 5., f. 1–7.

58. *Spiophanes Krœyeri*, Grube.
Grube, Archiv f. Naturg., 1860, p. 88, t. 5, f. 1.
Malmgren, Annul. polych., p. 94, t. 9, f. 56.

59. *Leipoceras uviferum*, Möb.
Zte deutsche Nordpolarf., II., p. 254, t. 1, f. 10.

Cirratulidæ.

60. *Cirratulus cirratus* (Müll.). (Greenl. *Nyaurselik.*)
Lumbricus cirratus, Müll., Prodr. Z. D. 2608.
Lumbricus cirratus, Fabr., F. Gr. 266.
Cirratulus borealis, Œrsted, l. c., p. 206, f. 98, 102.
Cirr. bor., Rathke, l. c., p. 180, t. 8, f. 16, 17.

Halelminthidæ.

61. *Notomastus latericeus*, Sars ?
Sars, Nyt. Mag., VI., p. 199 ; Fauna litt. Norv., II., p. 12, t. 2, f. 8–17.
Malmgren, Annul. polych., p. 97.

62. *Capitella capitata* (Fabr.). (Greenl. *Pullateriak.*)
Lumbricus capitatus, Fabr., F. Gr. 262.
Lumbriconais marina, Œrsted, Naturh. Tidsskr., IV., p. 128, t. 3., f. 6, 11, 12.

Maldanidæ.

63. *Nicomache lumbricalis* (Fabr.).
Sabella lumbricalis, Fabr., F. Gr. 369.
Clymene lumbricalis, Sars, F. litt. Norv., II., p. 16, t. 2, f. 23–26.
Malmgren, Nord. Hafs-Ann., p. 190; Ann. polych., t. 10, f. 60.

64. *Axiothea catenata,* Mlgr.
Malmg., N. Hafs-Ann., p. 190 ; Ann. pol., t. 10, f. 59.

Ammocharidæ.

65. *Ammochares assimilis,* Sars.
Sars, Nyt. Mag. f. Natur., VI., p. 201.
Malmgren, Annul. polych., t. 11, f. 65.
66. *Myriochele Heeri,* Mlgr.
Annul. polych., p. 101, t. 7, f. 37.

Amphictenidæ.

67. *Cistenides granulata* (L.). (Greenl. *Imab-polia.*)
Sabella granulata, L., Syst. Nat. (XII.), p. 1268.
Amphitrite auricoma, Fabr., F. Gr. 272.
Amphitrite Eschrichtii, Rathke, l. c., p. 219.
Pectinaria grœnlandica, Grube, Arch. f. Naturg.
Cistenides granulata, Malmgr., Nord. Hafs-Ann., p. 359.
68. *Cistenides hyperborea,* Malmgr.
Malmgr. Nord. Hafs-Ann., p. 360, t. 18, f. 40.

Ampharetidæ.

69. *Ampharete Grubei,* Mlgr.
? *Amphicteis acutifrons,* Grube, Archiv f. Naturg., XXVI., p. 109, t. 5, f. 6.
Amphar. Grubei, Malmg., N. Hafs-Ann., t. 19, f. 44.
69*a*. *Ampharete Goësi,* Mgr.
Nord. Hafs-Ann., p. 364, t. 19, f. 45.
70. *Amphicteis Gunneri* (Sars).
Amphitrite Gunneri, Sars, Beskr. Jagtt., t. 11, f. 30.
Crossostoma midas, Gosse, Ann. Nat. Hist., 1855, XVI., p. 310, t. 8, f. 7–12.
Amphicteis grœnlandica, Grube, Archiv f. N., XXVI., p. 106, t. 5, f. 3.
Amphicteis Gunneri, Malmgr., Nord. Hafs-Ann., p. 365, t. 19, f. 46.
71. *Sabellides borealis,* Sars.
Sars, Fauna litt. Norv., II., p. 22, 23.
Malmgren, Nord. Hafs-Ann., p. 368, t. 20, f. 47.
72. *Melinna cristata* (Sars).
Sabellides cristata, Sars, l. c., p. 19 et 24, t. 2, f. 1–7.
Melinna cristata, Malmg., N. Hafs-Ann., t. 20, f. 50.
72*a*. *Lysippe labiata,* Mgr.
Nord. Hafs-Ann., p. 367, t. 26, f. 78.

Terebellidæ.

73. *Amphitrite cirrata,* Müll. (Greenl. *Iglulualik.*)
O. F. Müller, Prodr. Zool. Dan. 2617.
O. Fabricius, Fauna Grœnl. 269.
Malmgren, Nord. Hafs-Ann, p. 375, t. 21, f. 53.
74. *Amphitrite grœnlandica,* Mlgr.
Malmgren, l. c., p. 376, t. 21, f. 52.
75. *Nicolea arctica,* Malmgren.
Nordiska Hafs-Annul., p. 381, t. 24, f. 66, 67.

76. *Scione lobata*, Mlgr.
l. c., p. 383, t. 23, f. 62.
77. *Axionice flexuosa* (Grube).
Terebella flexuosa, Grube, Archiv f. Naturg, XXVI., p. 102, t. 5, f. 2.
Axionice flexuosa, Malmgren, l. c., p. 384, t. 24, f. 68.
78. *Leæna abranchiata*, Malmgr.
l. c., p. 385, t. 24, f. 64.
79. *Thelepus cincinnatus* (Fabr.). (Greenl. *Iglulualik.*)
Amphitrite cincinnata, Fabr., F. Gr. 270.
Terebella pustulosa, Grube, l. c., p. 100.
Thelepus Bergmanni, Leuck., Archiv f. N., XV., p. 169, t. 3, f. 4.
Lumara flava, Stimpson, Marine Invert. Gr. M., p. 30.
Thelepus circinnatus, Malmgr., l. c., p. 387, t. 22, f. 58.
80. *Leucariste albicans*, Malmgr.
Nord. Hafs-Ann., p. 390, t. 23, f. 61.
Polycirrus arcticus, Sars, Christiania Vid. Selsk. Förh., 1864, p. 14.
81. *Ereutho Smitti*, Malmgr.
Nord. Hafs-Annul., p. 391, t. 23, f. 63.
82. *Artacama proboscidea*, Malmgr.
l. c., p. 394, t. 23, f. 60.
83. *Trichobranchus glacialis*, Malmgr.
l. c., p. 395, t. 24, f. 65.
84. *Terebellides Strœmii*, Sars.
Sars, Beskrv. og Jagttag., p. 48, t. 13, f. 31.
Malmgren, l. c., p. 396, t. 19, f. 48.

Sabellidæ.

85. *Laonome? Fabricii* (Kr.).
Sabella Fabricii, Kr., Bidrag til Sabellerne, Vid. Selsk. Overs., 1856, p. 20.
86. *Potamilla reniformis* (Müll.).
Die nierenförmige Amphitrite, Müller, v. Würmern, p. 194, t. 16.
Sabella reniformis, Leuckart, l. c., p. 183, t. 3, f. 8.
Sabella aspersa et oculata, Kröyer, l.c., p. 19 et 22.
Potamilla reniformis, Malmgren, Annul. polych., p. 114, t. 13, f. 77.
87. *Euchone analis*, (Kr.).
Sabella analis, Kr. l.c., p. 17.
Malmgren, Nord. Hafs-Ann., p. 406, t. 28, f. 88.
88. *Euchone tuberculosa*, Kr.
Sabella tuberculosa et S. rigida, Kr., l. c., p. 18.
Euchone tuberculosa, Malmgr., l. c., p. 401, t. 29, f. 92.
89. *Dasychone infarcta*, Kr.
Sabella infarcta, Kröyer, l. c., p. 21.
Malmgren, Nord. Hafs-Annul., p. 403, t. 28, f. 86.
90. *Chone infundibuliformis*, Kr. (Greenl. *Iglualik.*)

Tubularia penicillus, Fabr., F. Gr., 438 (non Müll. nec Linn.).
Chone infundibuliformis et suspecta (?), Kr., l. c., p. 33.
Sabella paucibranchiata, Kr., l. c., p. 22.
Chone infund., Mlgr., N.Hafs.-Ann., p. 404, t. 28, f. 87.

91. *Amphicora Fabricii* (Müll.).
Tubularia Fabricii, Müll., Prodr. Z. D. 3066.
Tub. Fabricii, Fabricius, Fauna Grœnl., 450, f. 12.
Othonia Fabricii (Johnst.), Gosse, Ann. Nat. Hist., 2 ser., V., p. 33, t. 4, f. 22.
Amphicora sabella (Ehrbg.), O. Schmidt, Neue Beitr. Würmer, p. 21, t. 2.
Fabricia quadripunctata, Leuck. Beitr. wirbell. Th., p. 151, t. 2, f. 3; Claparède Mém. Soc. Phys. Genève, XVI., p. 118, t. 4, f. 11–15.

91*a*. *Sabella crassicornis*, Sars.
S. picta, Kr., l. c., p. 24.
Malmgr., Nord. Hafs-Ann., p. 399, t. 27, f. 83.

Eriographididæ.

92. *Myxicola Steenstrupii*, Kr.
Myx. Steenstrupii et Sarsi, Kr. l. c., p. 9 et 35.
Malmgren, Nord. Hafs-Ann., p. 408, t. 29, f. 90.

Serpulidæ.*

93. *Hydroides norvegica*, Gunn.
Serpula triquetra, Fab. F. Gr. 374.
Hydroides norvegica, var. ε. *grœnlandica*, Mörch, Revisio critica Serpulidarum, p. 31 (sep.).

94. *Spirorbis verruca* (Fabr.), Mörch.
Serpula glomerata, Fabr., F. Gr. (non Linn.), n. 377.
Spirorbis verruca, Mörch, l. c., p. 85.

95. *Spirorbis quadrangularis*, Stimps.
Serpula contortuplicata, Fabr., F. Gr. 376.
Spirorbis quadrangularis, Stimps., Mar. Inv. Gr. Manan, p. 29.
Spir. quadr., var. α. *Fabricii*, Mörch, l. c., p. 89.

96. *Spirorbis borealis*, Daud.
Serpula spiriorbis, Fabr., F. Gr. 372; Mörch, p. 83.

97. *Spirorbis spirillum*, L.
Serpula spirillum, Linn. S. N. (X.), n. 692.
Malmg., Annul. polych., p. 123; Mörch, op. cit., p. 92.

98. *Spirorbis lucidus* (Mont.). (Greenl. *Katungiak.*)
Serpula porrecta et spirillum, Fab., F. Gr. 371, 373.
Serpula lucida, Mtg., Test. Brit., p. 507.
Spirorbis (*Spirillum*) *lucidus*, var. γ. *grœnlandica*, Mörch, l. c., p. 93.

99. *Spirorbis vitreus* (Fabr.).
Serpula vitrea, F. Gr. 378.

---

* *Serpula semilunium* (*seminulum*, L.), Fabr., F. Gr. 370, and *S. stellaris*, Fabr., F. Gr. 380, are Foraminifera (Mörch, Rev. crit. Serpulid., p. 118–119).

Spirorbis (*Spirillum*) *vitreus*, Mörch, l. c., p. 94.

100. *Spirorbis cancellatus* (Fabr.).
Serpula cancellata et granulata, Fabr., F. Gr. 378 et 380.
Spirorbis (*Spirillum*) *cancellatus*, Mörch, l. c., p. 94.

101. *Protula media*, Stmps.
Möbius, Zte deutsche Nordpolarfahrt, II., p. 256, t. I., f. 21–24.

Tomopteridæ.

102. *Tomopteris septentrionalis*, Stp.
Nat. For. Vid. Medd., 1849–50, p. iv.

Lumbricidæ.†

*103. *Lumbricus, sp.* (Gr. *Pullateriak.*)
L. terrestris, Fabr., F. Gr. 258.

*104. *Lumbricus* (?) *rivalis*, Fabr.
Fauna Grœnl. 260.

*105. *Enchytræus vermicularis* (Müll.)? (Greenl. *Kumangoak.*)
Lumbricus vermicularis, Fabr. F. Gr. 259.

*106. *Sænuris lineata* (Müll.)? (Greenl. *Kumak.*)
Lumbricus lineatus, Fabr. Faun. Gr. 261.

*107. *Clitellio arenarius* (Müll.). (Greenl. *Pullateriak.*)
Lumbricus arenarius, Fabr. F. Gr. 264.

*108. *Clitellio minutus* (Müll.). (Greenl. *Sirksab-Kumangoa.*)
Lumbricus minutus, Fabr. F. Gr. 265.

*109. *Opsonais* (?) *marina*, (Fabr.). (Greenl. *Kumak.*)
Nais mar., Fauna Grœnl. 295.
Opsonais marina, Gervais, Bullet. l'Acad. de Belg., V., p. 5 (sep.).

Hirudinidæ.

110. *Platybdella versipellis* (Diesing). (Greenl. *Kaneisib-Kuma.*)
Hirudo piscium, Fabr., F. Gr. 301.
Ichthyobdella versipellis, Diesing.
Platybdella scorpii (Fabr. MS.), Malm., Göteborg Kongl. Vet. och Vitt. Samh. Handl., VIII., 1863, p. 253.
(From *Cottus scorpius* and *scorpioides.*)

111. *Platybdella Fabricii*, Malm.
Malm. l. c., p. 248.

112. *Platybdella Olriki*, Malm.
Förhandl. Skand. Naturf. Stokholm, 1863, p. 414.
(From *Hyas aranea*).

113. *Platybdella affinis*, Malm.
l. c., p. 413 (from *Phobetor ventralis*).

---

* The species marked * are doubtful, and have not been revised since the time of Fabricus.

† Eisen described *Lumbriculus variegatus*, Müll., and *Enchytræus Pagenstecheri*, R., from Greenland (Öfv. Vet. Akad. Förh., 1872).

Obs.—Other species of Fish-leeches, not yet determined, have been found on *Anarrichas*, sp. (*Pl. anarrichæ*, Malm. ? l. c., p. 122), *Liparis tunicatus*, *Hippoglossus vulgaris* (*Pl. hippoglossi*, Malm. ? l. c., p. 257), and *Macrurus rupestris*.

114. *Udonella, sp.* (an hujus loci ?).
(On *Caligus hippoglossi.*)

Echiuridæ.

115. *Echiurus forcipatus* (Fabr.). (Greenl. *Illulualik.*)
Lumbricus echiurus, Fabr., F. Gr. 268, et *Holothuria forcipata*, ejusdem 349.

Priapulidæ.

116. *Priapulus caudatus* (Lmk.). (Greenl. *Tarkiksunak.*)
Holothuria priapus, Linn., Syst. Nat. (XII.), p. 1091.
Holothuria priapus, O. Fabr., F. Gr. 347.
Holothuria priapus, Zool. Dan., III., p. 27, t. 96, f. 1; IV., p. 18, t. 135, f. 2.
Priapulus caudatus, Ehlers, Zeitschr. f. wissensch. Zool., XI., p. 205, t. 20–21.

117. *Priapulus glandifer* (Ehlers).
Ehlers, l. c., p. 209, t. 21, f. 24.

Sipunculidæ.

118. *Phascolosoma Œrstedii*, Keferst.
Keferstein, Zeitschr. f. wissensch. Zool., XV., p. 436, t. 31, f. 8, et 33, f. 39.

119. *Phascolosoma boreale*, Keferst.
Keferstein, l. c., p. 437, t. 31, f. 7, et t. 33, f. 33.

Myzostomidæ (incertæ sedis).

120. *Myzostoma gigas*, Ltk. (MS.).
On *Antedon Eschrichtii*, M. Tr.; Copenhagen Museum.

Chætognatha (ad *Nematodas* ?).

121. *Sagitta*, sp.
Not uncommon in the Arctic seas in the vicinity of Greenland.

TURBELLARIA.

Obs.—The *Planariæ* and *Nemerteæ* of Greenland have not been studied since the time of Fabricius. The following list does little more than show in what manner his species have been partly interpreted, and does no justice to the richness of this branch of the Arctic Fauna.

1. *Monocelis subulata* (Fabr.). (Gr. *Kekkursab-Kuma.*)
F. Gr. 308.

2. *Planaria lactea*, Müll. (Gr. *Kumak.*)
F. Gr. 309.*

---

* Doubtful species:—*Pl. operculata*, Fabr. (F. Gr. 310), and *Pl. caudata*, Müll., F. Gr. 310 (both by the Greenlanders termed "Kekkursab-Kuma," as are other flat Worms). The latter is perhaps a naked Snail (Œrsted, Naturh. Tidsskr., IV., p. 546).

3. *Amphiporus grœnlandicus*, Œrd.
   Nat. Tidsskr., IV., p. 581.
4. *Omatoplea rubra* (Müll.). (Gr. *Kekkursab-Kuma.*)
   F. Gr. 304.
5. *Polystemma roseum* (Müll.).
   Möbius, Zte deutsche Nordpolarfahrt, II., p. 257.
6. *Tetrastemma grœnlandicum*, Dies.
   F. Gr. 311 (Pl. candida).
7. *Notospermum viride* (Müll.) (Gr. *Kekkursab-Kuma.*)
   F. Gr. 305.
8. *Meckelia fusca* (Fabr.). (Gr. *Pullateriak.*)
   F. Gr. 306.
9. *M. angulata* (Müll.). (Gr. *Pullateriak.*)
   F. Gr. 303.

XVIII.—A Revised Catalogue of the Entozoa of Greenland. By Dr. Chr. Lütken, University Museum, Copenhagen. 1875.

Cestoida.*

1. *Tænia pectinata*, Goeze.
   (Lepus glacialis.)
2. *T. expansa*, Rud.
   (Cervus tarandus, Ovibos moschatus.)
3. *T. coenurus*, Küch.
   (Canis lagopus: Möbius, Zte deutsche Nordpolarfahrt, II., p. 258.)

Obs.—*T. canis-lagopodis* will probably also be found in the Arctic Fox in Greenland.

4. *T. armillaris*, Rud. (Greenl. *Akpab-Kuma.*)
   F. Gr. 298 (T. tordæ, Fabr.).
   Krabbe, K. D. Vid. Selsk. Skr., ser. 5, VIII., p. 259, t. 1, f. 4–6.
   (Uria Bruennichii.)
5. *T. sternina*, Kr.
   Krabbe, l. c., p. 259, t. 1, f. 7–9.
   (Sterna macrura.)
6. *T. larina*, Kr.
   Krabbe, l. c., p. 261, t. 1, f. 16, 17.
   (Larus glaucus, tridactylus.)
7. *T. micracantha*, Kr.
   Krabbe, l. c., p. 262, t. 1, f. 18–21.
   (Larus glaucus, tridactylus, eburneus.)

---

* *Obs.*—For the Tapeworms of *Birds* in Greenland Dr. Krabbe's "Bidrag "til Kundskab om Fuglenes Bændelorme," should be consulted, and for those of the Seals the same author's "Helminthologiske Undersögelser i Danmark "og paa Island," in the "Transactions of the R. Danish Academy of Sciences," vols. VII. and VIII. (1868 and 1870).

8. *T. campylacantha*, Kr.
Krabbe, l. c., p. 263, t. 1, f. 22–24.
(Uria grylle.)
9. *T. microrhyncha*, Kr.
Krabbe, l. c., p. 206, t. 2, f. 38–40.
(Charadrius hiaticula.)
10. *T. clavigera*, Kr.
Krabbe, l. c., p. 267, t. 2, f. 41–43.
(Strepsilas interpres.)
11. *T. retirostris*, Cr.
Krabbe, l. c., p. 282, t. 5, f. 97–99.
(Strepsilas interpres.)
12. *T. megalorhyncha*, Kr.
Krabbe, l. c., p. 284, t. 5, f. 104–105.
(Tringa maritima.)
13. *T. teres*, Kr.
Krabbe, l. c., p. 284, t. 5, f. 106–108.
(Somateria mollissima, spectabilis, Larus glaucus.)
(14.) *T.* "*malleus*," Goeze (formæ monstrosæ).
T. fasciolaris, Pall.
Krabbe, l. c., p. 288.
(Somateria mollissima, Mergus serrator.)
15. *T. minuta*, Kr.
Krabbe, l. c., p. 292, t. 6., f. 127–129.
(Phalaropus fulicarius, hyperboreus.)
16. *T. microsoma*, Cr.
Krabbe, l. c., p. 296, t. 6, f. 146–150.
(Somateria mollissima, spectabilis, Larus glaucus.)
17. *T. fusus*, Kr.
Krabbe, l. c., p. 307, t. 7, f. 180, 181.
(Larus glaucus, L. marinus.)
18. *T. brachyphallos*, Kr.
Krabbe, l. c., p. 310, t. 8, f. 193, 194.
(Tringa maritima.)
19. *T. grœnlandica*, Kr.
Krabbe, l. c., p. 316, t. 8, f. 210, 211.
(Harelda glacialis.)
20. *T. fallax*, Kr.
Krabbe, l. c., p. 319, t. 8, f. 221, 222.
(Somateria mollissima.)
21. *T. borealis*, Kr.
Krabbe, l. c., p. 338, t. 10, f. 282, 283.
(Emberiza nivalis.)
22. *T. trigonocephala*, Kr.
Krabbe, l. c., 339, t. 10, f. 284–286.
(Saxicola œnanthe.)
23. *Bothriocephalus cordatus*, Leuckart.
Krabbe, K. D. Vid. Selsk. Skr., VII., p. 377, t. 7, f. 114–116.
(Homo grœnlandicus, Canis fam. grœnl., Phoca barbata, Œdobænus rosmarus.)

24. *B. variabilis*, Kr.
Krabbe, l. c., p. 378.
(Phoca vitulina.)
25. *B. lanceolatus*, Kr.
Krabbe, l. c., p. 378.
(Phoca barbata.)
26. *B. phocarum* (Fabr.). (Greenl. *Urksub-Kuma.*)
F. Gr. 296, b.; Nat. Hist. Selsk. Skr., I., 2, p. 153, t. 10.
Tetrabothrium anthocephalum, Rud.
Krabbe, l. c., p. 379, t. 7, f. 101–105, 117.
(Phoca barb., vitul., Cystophora cristata.)
27. *B. fasciatus*, Kr.
Krabbe, l. c., p. 379.
(Phoca hispida.)
28. *B. elegans*, Kr.
Krabbe, l. c., p. 378.
(Cystophora cristata.)
29. *B. similis*, Kr.
Krabbe, l. c., p. 379.
(Canis lagopus.)
30. *B. ditremus*, Cr.?
(Colymbus septentrionalis.)
31. *B. rugosus*, Rud.?
(Gadus ogak.)
32. *B. punctatus*, Rud.
(Cottus scorpius.)
33. *B. crassiceps*, Rud.?
(Cottus scorpius, Gadus ovak, morrhua, Delphinapterus leucas.)
34. *B. proboscideus*, Rud.
(Salmo carpio.)
35. *B.* (*Tetrab.*) *macrocephalus*, Rud. (et sp. aff.)
F. Gr. 297, b. (T. alcæ). (Greenl. *Akpab-Kuma.*)
(Larus glaucus, marinus, tridactylus, Procellaria glacialis, Uria Bruennichii, grylle, Mergus serrator, Colymbus septentrionalis, Corvus corax, Falco islandicus.)
36. *Octobothrium rostellatum*, Dies. (Greenl. *Sullukpaukab-Kuma.*)
F. Gr. 297 (T. erythrini). (Sebastes norvegicus.)
37. *Fasciola intestinalis*, L. (Greenl. *Kakillisab-Kuma.*)
F. Gr. 300 (T. gasterostei).
Schistocephalus solidus (Müll.); S. dimorphus, Cr.
(Gasterosteus aculeatus, Mergus serrator, Larus glaucus.)
38. *Anthobothrium perfectum*, Rud.
(Somniosus microcephalus.)
39. *Diplocotyle Olrikii*, Kr.
Krabbe, Vid. Medd. Nat. For., 1874, p. 22, t. 3.
(Salmo carpio.)

Obs.—The Nos. 299, of the "Fauna Grœnlandica" (*T. scorpii*, "Kaneisub-Kuma"), from *Cottus scorpius*, *Gadus ovak*, and *Salmo carpio* (probably two different species, cfr. Rudolphi); and 313 (*Fasciola barbata*, "Amikorsub-Kuma"), from *Gonatus Fabricii*, have not been identified by modern helminthologists. It has been suggested that the last-named (313) is only the spermatophore of the Squid.

Several Tapeworms have been found in *Anser brenta*, *Haliætus albicilla*, *Larus Sabini*, *Colymbus glacialis*, *Fringilla lapponica*, *Hippoglossus vulgaris*, &c., but not in such a state that they could be determined satisfactorily.

Trematoda.

1. *Distomum hepaticum*, L. (Gr. *Sauab-Kuma.*)
   F. Gr. 312.
   In Sheep (imported ?).
2. *D. seriale* (Rud.). (Greenl. *Ivisarkub-Kuma.*)
   F. Gr. 314 (Fasciola umblæ).
   (Salmo alpinus.)

Obs.—Undetermined species of Flukes have been found in *Phoca barbata*, *Lumpenus aculeatus*, and *Mergus serrator*.

3. *Onchocotyle borealis*, van Ben.
   (Somniosus microcephalus, on the gills.)
4. *Phylline hippoglossi* (Fabr.). (Greenl. *Netarnab-Kuma.*)
   F. Gr. 302.
   (Hippoglossus vulgaris.)

Nematoda.

Obs.—The Nematoda of Greenland have not been worked out. The following species, with a few exceptions, are enumerated in the "Fauna Grœnlandica," and have been interpreted by later authors in the manner indicated :—

1. *Ascaris mystax*, Zed.
   (Canis lagopus; Möbius in Zte deutsche Nordpolarfahrt, p. 257).
2. *A. vermicularis*, L. (Greenl. *Koartak.*)
   F. Gr. 248.
   (Homo grœnlandicus.)
3. *A. lumbricoides*, L. (Greenl. *Kumarksoak.*)
   F. Gr. 249.
   (Homo grœnlandicus.)
4. *A. osculata*, Rud.
   (Phoca grœnlandica.)
5. *A. gasterostei*, Rud. (Greenl. *Kakillisab-Kuma.*)
   F. Gr. 242 (Gordius lacustris).
   (Gasterosteus aculeatus.)
6. *A. rajæ*, Fabr. (Greenl. *Taralikkisab-Kuma.*)
   F. Gr. 253.
   (Raja radiata.)
7. *Eustrongylus gigas*, Rud.
   (Canis familiaris grœnlandicus.)
8. *Liorynchus gracilescens*, Rud. (Greenl. *Urksub-Kuma.*
   F. Gr. 251 (A. tubifera).
   (Phoca barbata.)

9. *Ophiostomum dispar*, Rud. (Greenl. *Atab-Kuma, Neitsib-Kuma.*)
   F. Gr. 250 et 252 (A. phocæ, ♀, et A. bifida, ♂).
   (Phoca grœnlandica, Ph. hispida.)
10. *Agomonema commune* (Desl.). (Greenl. *Kumak.*)
   F. Gr. (241).
11. "*Nematoideum Alcæ-picæ*," Rud. (Gr. *Akpab-Kuma.*)
   F. Gr. 257 (Ascaris alce).
   (Uria Bruennichii.)
12. "*Dubium gasterostei aculeati*, Rud."
   F. Gr. 243 (Gordius globicola).
   (Gasterosteus aculeatus.)

Obs.—*Fabricius* describes four species of *Gordius* (F. Gr. 244–247 : *G. intestinalis, cinctus, capillaris, lacteolus*—"Kumak, Kumangoak") which apparently have not been interpreted by later authors; the first, at least, must probably be referred to the *Turbellaria* (*Nemerteæ*).

Acanthocephala.

1. *Echinorhynchus strumosus*, Rud.
   (Phoca hispida, vitulina, grœnlandica, Cystophora cristata, Canis familiaris grœnlandicus).
2. *E. acus*, Rud. (Gr. *Okab-Kuma.*)
   F. Gr. 255–56 (Ascaris versipellis, A. gadi).
   (Gadus ovak, morrhua; ? Hippoglossus vulgaris.)
3. *E. polymorphus*, Br.
   (Somateria mollissima; ? Harelda glacialis.)
4. *E. porrigens*, Rud.
   (Balænoptera gigas.)
5. *E. hystrix*, Br.
   (Graculus carbo, Mergus serrator.)
6. *E. inflatus*, Cr.
   (Charadrius hiaticula.)
7. *E. micracanthus*, Rud.
   (Saxicola œnanthe.)
8. *E. pleuronectis-platessoides*, Rud. (Gr. *Okotab-Kuma.*)
   F. Gr. 254 (Ascaris pleuronectis).
   (Drepanopsetta platessoides.)

Obs.—An undetermined species of *Echinorhynchus* has been found in *Salmo carpio;* another in *Squatarola helvetica:* undescribed species of *Ascaris* in *Liparis tunicatus* and *Reinhardti, Ammodytes*, sp., and *Motella Reinhardti.*

XIX.—A Revised Catalogue of the Echinodermata of Greenland. By Dr. Chr. Lütken, University Museum, Copenhagen. 1875.

ECHINODERMATA.

Holothuridæ.

1. *Cucumaria frondosa* (Gunn.). (Greenl. Innellualik, Irksolik.)
   F. G. 343, 344; Ltk., Grönl. Ech., p. 2.
2. *C. calcigera*, Ag.
   C. Koreni, Ltk., Grönl. Ech., p. 4.
3. *C. minuta*, Fabr. (Greenl. Kavmarsungoak.)
   F. Gr. 346; Ltk. l. c., p. 7. Ocnus Ayresii, Stmps.
4. *Orcula Barthii*, Tr.
   Ltk. l. c., p. 9.
5. *Psolus phantapus*, Str.
   Ltk. l. c., p. 12.
6. *Ps. Fabricii* (D. K.).
   F. Gr. 348; Ltk. l. c., p. 13.
   Cuvieria Fabricii, Auct.; Lophothuria Fabricii, Verr.
7. *Chiridota læve* (Fabr.). (Greenl. Kaumarsorsoak.)
   F. G. 345; Ltk. l. c., p. 16.
8. *Myriotrochus Rinkii*, Stp.
   Vid. Medd. Nat. For. 1851, p. 55; Ltk. l. c., p. 22.
9. *Eupyrgus scaber*, Ltk.
   Ltk., Grönl. Echin., p. 22.

Echinida.*

10. *ToxopneustesDroebachiensis* (Müll.). (Greenl. Ekkursak.)
    F. Gr. 368 (E. saxatilis); Ltk. l. c., p. 24.
    Strongylocentrotus Droebach., Agassiz, Revision of the Echin.

Asterida.

11. *Asterias polaris*, M. Tr. (Greenl. Nerpiksoak, Nerpiksout.)
    F. Gr. 362 (pp.) et 365; Ltk. l. c., p. 28.
12. *A. groenlandica*, Stp.
    F. G. 362 (pp.); Ltk. l. c., p. 29.
13. *A. stellionura*, Val.
    Perrier, Recherches s. l. Pedicellaires (1869), p. 48.
    Found in the stomach of the Greenland Shark.
14. *A. albula*, Stmps.
    A. problema, Stp.; Ltk. l. c., p. 30.

Obs.—Spontaneous division! Ltk., Overs. Vid. Selsk. Forh., 1872, p. 117; Ann. Mag. Nat. Hist., 1873, &c.

---

* *Obs.*—*Brissopsis lyrifera*, Forb., was dredged in Davis' Strait, according to Goodsir (Forbes, Nat. Hist. Europ. Seas., p. 51).

15. *A.* (*Stichaster*) *rosea*, M. Tr.
Cribella rosea, Forb. Brit. Starf.
(With No. 13.)
16. *Cribella sanguinolenta* (Müll.). (Greenl. Nerpiksout.)
F. Gr. 363 (Ast. spongiosa) ; Ltk. Grönl. Ech., p. 31.
Cr. oculata, Fabr. ; Echinaster eschrichtii, M. Tr.
17. *Solaster papposus* (L.). (Greenl. Nerpiksout.)
F. G. 364; Ltk. l. c., p. 40.
18. *S. endeca*, L.
Ltk. l. c., p. 35.
19. *Pteraster militaris* (Müll.).
Ltk. l. c., p. 73.
20. *Ctenodiscus crispatus* (Retz.).
Ch. pygmæus et polaris, M. Tr.
Ltk. Grönl. Echin., p. 45.
21. *Archaster tenuispinus* (D. K.).
Vid. Medd. Nat. For., 1871, p. 240.
(With Nos. 13, 15, and 22.)

Ophiurida.

22. *Ophioscolex glacialis*, M. Tr.
(With Nos. 13, 15, and 21.)
23. *Ophioglypha Sarsii* (Ltk.).
Additam. ad hist. Ophiurid., I., p. 42, t. 1, f. 3–4.
24. *O. robusta*, Ayr.
Ophiura squamosa, Ltk. l. c., p. 46, t. 1, f. 7.
25. *O. nodosa* (Ltk.).
Addit., I., p. 48, t. 2, f. 9.
26. *O. Stuwitzii* (Ltk.).
Addit., I., p. 51, t. 1, f. 8.
27. *Ophiocten sericeum* (Forb.).
Ophi. Kroeyeri, Ltk. l. c., p. 51, t. 1, f. 5.
28. *Ophiopus arcticus*, Lgn.
Öf. Vet. Akad. Förh., 1866, p. 309.
29. *Ophiopholis aculeata* (Müll.). (Greenl. Nerpiksoursak.)
F. Gr. 366; Ltk. Addit., I., p. 59, t. 2, f. 15–16.
Ophiolepis scolopendrica, M. Tr. ; Ophiocoma bellis, Forb.
30. *Amphiura Sundevalli*, M. Tr.
A. Holbœlli, Ltk. l. c., p. 55, t. 2, f. 13.
31. *Ophiacantha spinulosa*, M. Tr.
Ltk. Addit., I., p. 65, t. 2, f. 14.
Ophiocoma arctica and Ophiacantha grœnlandica, M. Tr.
Ast. bidentata, Retz. ; Ophiocoma echinulata, Forb.
32. *Asterophyton eucnemis*, M. Tr.
F. Gr., p. 367 (Ast. caput-medusæ).
Ltk., Addit., I., p. 70, t. 2, f. 17–19.
33. *A. Agassizii*, Stmps.
Ltk. Addit., III., p. 66.

Crinoida.

34. Antedon Eschrichtii (M. Tr.).
Ltk., Grönl. Echinod., p. 55.

XX.—A REVISED CATALOGUE of the ANTHOZOA and CALYCOZOA of GREENLAND. By Dr. CHR. LÜTKEN, University Museum, Copenhagen. 1875.

ANTHOZOA.

Polyactinia (Actinida).

1. *Actinia* (*Urticina*) *crassicornis*, Fabr. (Greenl. Kettuperak.) F. Gr. 340; U. Davisii, Ag.?
2. *A. spectabilis*, Fabr. (Greenl. Kettuperarsoak.) F. Gr. 342, *b*.
3. *A.* (*Chrondractinia*) *nodosa*, Fabr. (Greenl. Aitsib-pa.) F. Gr. 341.
4. *A.*(*Acthelmis*)*intestinalis*, Fabr. (Greenl. Kettuperangoak.)

5–6. *Edwardsia*, spp. 2.

7–8. *Peachia*, *sp.* and *P.* (?) sp.

5–8 are preserved in the Museum at Copenhagen, but cannot be identified more accurately from specimens in alcohol.

Antipatharia.

9. *Antipathes arctica*, Ltk.

Overs. K. D. Vid. Selsk. 1871, p. 18; Ann. Mag. Nat. Hist. 4., X., p. 77.

A single specimen, found in the stomach of *Somniosus microcephalus* (the Greenland Shark). (Rödebay.)

Octactinia (Alcyonaria).

10. *Ammothea arctica*, Ltk. (MS.)

? *Briareum grandiflorum*, Möbius, Zte deutsche Nordpolarfahrt, II., p. 260.

(Not uncommon.)

11. *Alcyonium*, sp.

An undetermined specimen in the Museum at Copenhagen.

12. *Umbellula Linddahlii*, Köll.

Umb. miniacea et pallida, Lindd., K. Vet. Akad. Handl., XIII., 3, t. 1–3.

(Baffin's Bay and entrance to Omenak Bay, 400 and 122 faths.). The identity with the type described by Mylius and Ellis, 1753, "*Isis encrinus*," L., is left undecided.

Obs.—The *Actinidæ* want revision, with examination of living specimens. The "*Alcyonia*" of the "Fauna Grœnlandica" are apparently either Compound *Ascidiæ* (462 and 463) or *Spongozoa* (464 and 465).

CALYCOZOA.

1. *Lucernaria* (*Manania*) *auricula*, Fabr.

F. G. 332; Steenstrup Vid. Medd. Nat. For., 1859, p. 108.

2. *L. quadricornis*, Müll.
 L. fascicularis, Flmg.
3. *L.* (*Halicyclus*) *octoradiata* (Lmk).
 L. auricula, Rathke.
 Steenstrup, l. c., p. 108.
4. *L.* (*Craterolophus*) *convolvulus*, Johnst.
 L. campanulata, Lmx.
 A single specimen, collected by the late Governor Olrik ; the other species are not uncommon. (Greenl. Akuilisaursak, Unnerarsuk, Ornigarsuk.)

XXI.—A Revised List of the ACALEPHÆ and HYDROZOA of GREENLAND. By Dr. CHR. LÜTKEN, University Museum, Copenhagen. 1875.

The *Medusæ* and *Hydroidæ* of Greenland have not been satisfactorily worked out. The following list must be regarded only as preliminary, being limited to those species whose occurrence in Greenland can be stated on tolerably good authority.

Ctenophora (**Beroidæ**). (Greenl. *Ikpiarsursak.*)
1. *Mertensia ovum* (Fabr.).
 F. Gr. 355 (*Beroë ovum*).
 Beroë pileus, Scoresby, Arct. Reg. II. t. 16, f. 4.
 Cydippe ovum et C. cucullus, Esch., Syst. d. Akal., p. 25.
 Al. Agassiz, Illust. Cat. North Amer. Acal., p. 26, f. 29–37.
2. *Pleurobrachia rhododactyla*, Ag.
 F. Gr. 354 (*Beroë pileus*).
 L. Agassiz, Mem. Am. Ac., IV., p. 314, t. 1–5; Contr. Nat. Hist. Un. St., III., p. 203–248, t. 2*a*.
 Al. Agassiz, Cat. N. Am. Acal., p. 30, f. 38–51.
3. *Idya cucumis* (Fabr.).
 F. Gr. 353 (*Beroë cucumis*); Esch., Syst. d. Ak., p. 36.
 ? *I. borealis*, Less., Zooph. Acal., p. 134.
Obs.—Quid Beroë infundibulum, Fabr. (F. Gr. 352) ?

Discophora (Greenl. *Nuertlek*) ; **Hydrozoa.**
4. *Aurelia flavidula*, Per. Les.
 F. Gr. 356 (*Medusa aurita*).
 L. Agassiz, Contrib. Nat. Hist. U. St., III., pl. 6–9.
 Al. Agassiz, Cat. N. Am. Ac., p. 42, f. 65–66.
5. *Cyanea arctica*, Per. Les. (Gr. *Nuertlersoak.*)
 F. Gr. 358 (*M. capillata*).
 L. Agassiz, Contrib., III., pl. 3–5*a*., 10, 10*a*.
 Al. Agassiz, Cat., p. 44, f. 67.
Obs.—The difference of Nos. 4 and 5 from their European representatives may still be questioned.

6. *Charybdea hyacinthina* (Faber).
   Medusa (*Melitea*) *hyacinthina*, Faber, Fische Islands, p. 197.

Obs.—*Dodecabostricha dubia*, Brdt., *Quoyia bicolor*, Q. G., and *Charybdea periphylla*, P. L., apparently belong to this or to closely allied species of this genus !

7. *Trachynema digitale* (Fabr.).
   F. Gr. 361 (*Medusa digitalis*).
   Eirene digitalis, Esch.
   Al. Agassiz, Cat., p. 57, f. 81–86.
8. *Hydra*, sp.
   A fresh-water Polype was observed and collected in Greenland by the late Governor Olrik.
9. *Hydractinia echinata* (Fl.).
   F. Gr. 338 (*Hydra squamata*).
   H. polyclina, L. Agassiz, Contrib., IV., p. 227–239, t. 16 et 26, f. 18; Al. Ag., Cat., p. 198, f. 329–330.
   Hincks, British Hydr. Zooph., p. 23, t. 4; Allman, Tubularian Hydroids, p. 345–347, t. 15, 16, f. 10, 11.
10. *Syncoryne* (*Sarsia*) *mirabilis*, Ag.
   Coryne mirab., L. Agassiz, Contrib., IV., p. 185–217, t. 17–19, 20, f. 1–9; Mem. Am. Ac., IV., p. 224, t. 4, 5.
   Al. Agassiz. Cat., p. 175, f. 283–287; Allman, l. c., p. 278.
   (Distinct from the *S. tubulosa* of Northern Europe?)
11. *Coryne*, sp.
   F. Gr. 452 (*Fistulana muscoides*).
12. *Coryne*, sp.
   F. Gr. 339 (*Hydra ramosa*, Fabr.).
13. *Myriothela phrygia* (Fabr.).
   F. Gr. 333 (*Lucernaria phrygia*).
   Candelabrum phrygium, Blainv.
   Allman, Tubul. Hydr., p. 382.
14. *Tubularia indivisa*, L.
   T. Couthouyi, L. Ag., Contrib., p. 266, t. 23*a*, 24, 26.
   Allman, Tubul. Hydr. p. 403 et 400, pl. 20.
15. *Monocaulis grœnlandica*, Allm. (in litt.)
   (An undescribed species in the Museum of Copenhagen, which will be described in a forthcoming paper by Prof. Allman.)
16. *Melicertum campanula* (Fabr.).
   F. Gr. 360 (*Medusa campanula*).
   Campanella Fabricii, Less. Zooph. Ac., p. 281.
   Al. Agassiz, Cat., p. 130, f. 202–214.
17. *Eudendrium*, sp.
   F. Gr. 457 (*Sertularia volubilis*).
18. *Bougainvillia superciliaris*, Ag.
   Hippocrene supercil., L. Ag., Mem. Am. Acad., IV., p. 250, t. 1–3.

Bougainv. supercil., L. Ag., Contrib., IV., p. 289, t. 27, f. 1–7.
Al. Ag., Cat., p. 153, f. 232–240.
Allman, Tubul. Hydr., p. 315.

? 19. *Stomobrachium tentaculatum*, Ag.
F. Gr. 359 (*Medusa bimorpha*).
? Al. Agassiz, Cat. p. 98. f. 140–142.

20. *Tiaropsis diademata*, Ag.
L. Agassiz, Mem. Am. Acad., IV., p. 289, t. 6; Contrib., IV., p. 308, t. 31, f, 9–15.
Al. Ag., Cat. p. 69, f. 91–93.

21. *Campanularia verticillata*, L.
C. olivacea, Lmx.
Hincks, Brit. Hydr. Zooph., p. 167, t. 32, f. 1.

22. *Lafoëa fruticosa*, Sars.
Hincks, l. c., p. 22, t. 41, f. 2.
Kirchenpauer, Zte deutsche Nordpolarfahrt, II., p. 416.

23. *Cuspidella*, sp. In the Museum, Copenhagen.

24. *Salacia abietina*, Sars.
Hincks, Brit. Hydr. Zooph., p. 212, t. 41, f. 3.
Grammaria robusta, Stmps.

25. *Eucope* (*Thaumantias*) *diaphana*, Ag.
L. Agassiz, Mem. Am. Ac., IV., p. 300, f. 1, 2.
Al. Ag., Cat., p. 83, f. 115–125.

26. *Zygodactyla grœnlandica* (Per. Les.).
F. Gr. 357 (*Medusa æquorea*).
Al. Ag., Cat., p. 103, f. 153–156.

27. *Halecium muricatum*, Ell. Sol.
Hincks, Brit. Hydr. Zooph., p. 223, t. 43, f. 1.

Obs.—Sertularia halecina, Fabr., belongs to the Polyzoa !

28. *Sertularia pumila*, L.
F. Gr. 456 (*S. thuja*).
L. Agassiz, Contrib., IV., p. 326, t. 32 (*Dynamena pumila*).
Al. Ag., Cat., p. 141, f. 225–226.
Hincks, Brit. Hydr. Zooph., p. 260, t. 53, f. 1.

29. *Sertularia abietina*, L.
F. Gr. 453.
Hincks, Brit. Hydr. Zooph., p. 266, t. 55.

30. *Sertularia fastigiata*, Fabr.
F. Gr. 458 (*S. fastigiata*).
(Vix *S. argentea*, L., Hincks, Brit. Hydr. Zooph., p. 268, t. 56.)

31. *Sertularella rugosa*, L.
F. Gr. 454 (*Sertularia rugosa*).
Hincks, Brit. Hydr. Zooph., t. 47, f. 2.
Amphitrocha rugosa, Ag.

32. *Sertularella polyzonias*, L. (Greenl. *Nyaursæt.*)
F. Gr. 460 (*Sertularia ciliata*); K. D. Vid. Selsk. Skr., 1824, p. 37.

Cotulina polyzonias, Ag.
Hincks, Brit. Hydr. Zooph., p. 235, t. 46, f. 1.
33. *S. tricuspidata* (Alder).
Hincks, Brit. Hydr. Zooph., p. 239, t. 47, f. 1.
Kirchenpauer, Zte deutsche Nordpolarfahrt, II., p. 416.

Obs.—Several boreal *Siphonophora* (Diphyes, Physophora, &c.) are also, at least occasionally, found in the neighbourhood of South Greenland.

XXII.—A Revised Catalogue of the Spongozoa of Greenland. By Dr. Chr. Lütken, University Museum, Copenhagen. 1875.

SPONGOZOA.

Obs.—The Sponges from Greenland in the Museum of Copenhagen, and those brought home by the German Expedition to East Greenland, were determined by *Oscar Schmidt* in "Grundzüge einer Spongien-Fauna des atlantischen Gebietes" (1870), by *E. Häckel* in his monograph, "Die Kalkschwämme" (1873), and by both authors in "Die zweite deutsche Nordpolarfahrt," zte Abth. (1874). The following species are enumerated :—

1. *Filifera*, sp. (*Hircinia variabilis*).
O. Schmidt. Grundz., p. 31.
2. *Cacospongia*, sp.
Nordpolarf., II., p. 430.
3. *Chalinula ovulum*, O. S.
Grundz., p. 38, t. 5, f. 1.
4. *Reniera*, sp.
Nordpolarf., p. 430.
5. *Amorphina genitrix*, O. S.
Grundz., p. 41, t. 5, f. 9.
6. *Eumastia sitiens*, O. S.
l. c., p. 42, t. 5, f. 12.
7. *Suberites Luetkenii*, O. S.
l. c., p. 47, t. 5, f. 7.
8. *S. arciger*, O. S.
l. c., p. 47, t. 5, f. 6.
9. *Thecophora semisuberites*, O. S.
l. c., p. 50, t. 6, f. 2.
10. *Isodictya fimbriata*, Bbnk.
l. c., p. 56.
11. *I. infundibuliformis*, Bbnk.
Nordpolarf., II., p. 430.
12. *Desmacidon anceps*, O. S.
l. c., p. 430.
13. *Esperia intermedia*, O. S.
l. c., p. 433.

14. *E. fabricans*, O. S.
   Ibid.
15. *Geodia simplex*, O. S.
   Grundzüge, p. 70.
16. *Halisarca Dujardinii*, Johnst.
   Nordpolarf., II., p. 435.
17. *Ascaltis Lamarckii*, H.
   Kalkschw, p. 60, t. 9, f. 5, t. 10, f. 4.
18. *Ascortis Fabricii* (O. S.).
   Leucosolenia Fabricii, O. S., Grundz., p. 73; Kalkschw., p. 71, t. 11, f. 3, t. 12, f. 3.
19. *A. corallorhiza*, H.
   Kalkschw., p. 73, t. 11, f. 4, t. 12, f. 4.
20. *Ascandra reticulum* (O. S.).
   Nardoa retic., O. S., Grundz., p. 73 ; Kalkschw., p. 87, t. 14, f. 4, t. 20.
21. *Leucandra Egedii* (O. S.).
   Sycinula Eg., O.S., Grundz., p. 74 ; Kalkschw., p. 173, t. 32, f. 1.
22. *L. ananas* (Mont.).
   Sycinula penicillata, O. S., Grundz., p. 73, t. 2, f. 25 ; Kalkschw., p. 200, t. 32, f. 5, t. 40, f. 1–8.
23. *L. stilifera* (O. S.).
   Leucandra stilif., O. S., Grundz., p. 73, t. 2, f. 24; Kalkschw., p. 225, t. 33, f. 4, t. 40, f. 11.
24. *Sycaltis glacialis*, H.
   Kalkschw., p. 269, t. 45, f. 4–7.
25. *Sycandra ciliata* (Fabr.).
   F. Gr. 466 ; O. Schm., Grundz., p. 74; Kalkschw., p. 296, t. 51, f. 1, t. 58, f. 9.
26. *S. arctica*, H.
   Sycon raphanus, O. S., l. c., p. 74 ; Kalkschw., p. 353, t. 55, f. 1, t. 60, f. 15.
27. *S. compressa* (Fabr.).
   F. Gr. 464 ; Sycinula clavigera, O. Schm., l. c., p. 74, t. 2, f. 26.
   Kalkschw., p. 360, t. 55, f. 2, t. 57.
28. *S. utriculus* (O. S.).
   Ute utric., O. S., l. c., p. 74, t. 2, f. 27.
   Kalkschw., p. 370, t. 55, f. 3, t. 58, f. 3.

XXIII.—FORAMINIFERA from the HUNDE ISLANDS in SOUTH-EAST or DISCO BAY, DAVIS' STRAIT, on the WEST COAST of GREENLAND (lat. 68° 50′ W., long. 53° N.), from soundings taken by Dr. P. C. SUTHERLAND in 1850. By Professors W. K. PARKER, F.R.S., and T. RUPERT JONES, F.R.S. From a Memoir on some Foraminifera from the North-Atlantic and Arctic Oceans, including Davis' Strait and Baffin's Bay: Phil. Trans. clv., 1865, pp. 326, &c.

I. Hunde Islands, 25 to 30 fathoms. Pale-grey micaceous clay; more than half small mica-flakes. With vegetable matter (Fucal); Hydrozoa (*Sertularia*); Polyzoa (*Berenicea*, &c.); Entomostraca (*Cythere*, &c.); bivalve and univalve Mollusca. Foraminifera:—*Polymorphina*, *Truncatulina*, *Pulvinulina*, *Polystomella*, *Nonionina*, *Nummulina*, *Cassidulina*, *Bulimina*, *Textularia* and *Verneuilina*, *Cornuspira*, *Quinqueloculina*, *Triloculina*, *Lituola*.

II. 28 to 30 fathoms. Gravel of hornblende-schist and syenite. Sea-weed (*Fucus*); Nullipores; fragments of *Balanus* (predominant); Crustacea (*Talitrus*, *Cythere*, &c.); spines and plates of *Echinus*; Polyzoa; Univalves and Bivalves. Foraminifera:—*Globigerina*, *Truncatulina*, *Pulvinulina*, *Discorbina*, *Polystomella*, *Nonionina*, *Cassidulina*, *Quinqueloculina*, *Lituola*.

III. 30 to 40 fathoms. Shelly sandy mud, with syenitic fragments. Fragments of *Balani*; *Cythere*; *Serpula*; spines of *Echinus*; Bivalves and Univalves. Foraminifera:—*Nodosaria*, *Cristellaria*, *Lagena*, *Polymorphina*, *Uvigerina*, *Globigerina*, *Truncatulina*, *Pulvinulina*, *Discorbina*, *Polystomella*, *Nonionina*, *Cassidulina*, *Bulimina*, *Virgulina*, *Bolivina*, *Textularia*, *Verneuilina*, *Patellina*, *Trochammina*, *Quinqueloculina*, *Lituola*.

IV. 50 to 70 fathoms. Shelly, fine, syenitic sand. *Serpula*. Bivalves and Univalves. Foraminifera:—*Lagena*, *Polymorphina*, *Uvigerina*, *Truncatulina*, *Pulvinulina*, *Discorbina*, *Polystomella*, *Nonionina*, *Cassidulina*, *Patellina*, *Quinqueloculina*, *Lituola*.

V. 60 to 70 fathoms. Shelly, sandy, syenitic mud. *Serpula*, *Balanus* (predominant); Univalves and Bivalves. Forminifera:—*Dentalina*, *Cristellaria*, *Lagena*, *Polymorphina*, *Uvigerina*, *Globigerina*, *Truncatulina*, *Pulvinulina*, *Discorbina*, *Polystomella*, *Nonionina*, *Cassidulina*, *Bulimina*, *Virgulina*, *Bolivina*, *Textularia*, *Bigenerina*, *Verneuilina*, *Spirillina*, *Patellina*, *Trochammina*, *Cornuspira*, *Quinqueloculina*, *Triloculina*, *Lituola*.

The five specimens of sea-bottom above-mentioned, taken at depths of from 25 to 70 fathoms, and consisting mainly of shelly muddy sands, afford a good local example of the Foraminiferal fauna of the "Arctic Province"* of naturalists, at the "Coralline

---

* *See* "The Natural History of the European Seas," by E. Forbes and R. Godwin-Austen, pp. 28, &c. 8vo. Van Voorst, London, 1853.

zone" (15–50 fathoms) and the "Coral zone" (50–100 fathoms) of Davis' Strait. *Lagenæ* abound in these dredgings at from 30 to 70 fathoms; *Polymorphina* is small here and rather common; *Uvigerina* common at from 30 to 70 fathoms, but small. *Globigerinæ* are not rare at the same depths, but are very small. *Truncatulina* (*Planorbulina*) flourishes at all the depths (25–70 fathoms). *Pulvinulina* is freely represented by the small *P. Karsteni*. *Discorbina* gets more abundant with the greater depth. The simple forms of *Polystomella*, including the feeble *Nonioninæ*, have their home evidently in this region. *Cassidulina* abounds, but is not large. A small *Nummulina*, the feeble representative of a once highly potent species, still abounding in some warm seas, is not wanting in the "Coralline zone." The essentially Arctic form of *Bulimina* (*B. elegantissima*) flourishes at from 30 to 70 fathoms at the Hunde Islands, and other varieties are not wanting, though not abundant. The *Textulariæ* are represented by some small specimens of the type, and by three of its modifications in small but numerous individuals. *Spirillina* is very rare and small. *Patellina* is small and common from 30 to 70 fathoms. *Trochammina* is common, though small, in the deepest soundings. *Cornuspira* is common at the least and the greatest depths. *Quinqueloculina* is common, but not large, throughout. *Triloculina* occurs freely at 25 to 30 fathoms. *Lituola* abounds from 25 to 70 fathoms.

XXIV.—FORAMINIFERA from BAFFIN'S BAY. By Prof. W. K. PARKER, F.R.S.; and Prof. T. RUPERT JONES, F.R.S. From a Memoir on some Foraminifera from the North Atlantic Oceans, including Davis' Strait and Baffin's Bay: Phil. Trans. clv. 1865, pp. 325, &c.

Soundings from Baffin's Bay, between 76° 30′ and 74° 45′ N. lat., derived from seven deep-sea soundings and some iceberg mud taken during one of the Arctic Expeditions under Sir Edward Parry.

This material from the "Arctic Province" of naturalists* is but scanty. None of the Foraminifera here obtained are numerous except *Polystomella striatopunctata*, *Nonionina scapha*, *Truncatulina lobatula*, and *Cassidulina lævigata*, the first two of which are at home in Arctic waters; and none have attained here a large size except *Lituolæ*. The material from 150 fathoms yielded these relatively large and numerous specimens.

I.—From lat. 75° 10′, long. 60° 12′; ? fathoms; fine grey syenitic sand, with syenitic fragments: *Dentalina, Lagena, Truncatulina, Polystomella* and *Nonionina, Cassidulina, Quinqueloculina, Lituola.*

---

* *See* E. Forbes and R. Godwin-Austen's "Natural History of the European Seas." 8vo. London, Van Voorst, 1853.

II.—Lat. 76° 30′, long. 77° 52′; 150 fathoms; greyish muddy micaceous sand, with syenitic fragments: *Globigerina*, *Truncatulina*, *Pulvinulina*, *Polystomella* and *Nonionina*, *Cassidulina*, *Lituola*.

III.—Lat. 74° 45′, long. 59° 17′; 250 fathoms, grey sandy mud; quartzose sand, angular and rounded; no Foraminifera.

IV.—Lat. 75° 25′, long. 60°; 314 fathoms; syenitic sand, with fragments of syenite: *Triloculina*, *Lituola*.

V.—Lat. 76° 20′, long. 76° 27′: no Foraminifera.

VI.—Lat. 75°, long. 59° 40′; 230 fathoms; *Truncatulina*, *Polystomella* and *Nonionina*, *Quinqueloculina*, *Lituola*.

VII.—Lat. 76° 10′, long. 76°; sand from an iceberg; grey, fine, micaceous syenitic, with fragments of syenite: no Foraminifera.

XXV.—Foraminifera, Polycystina, &c., from Davis' Strait. Extracted from Dr. Ch. G. Ehrenberg's "Microgeological Studies on the Microscopic Life of "the Sea-bottom of all Zones, and its Geological "Influences," in the "Abhandlungen k. Akad. Wiss. "zu Berlin," for 1872, pp. 131-399, with 12 plates and a map. 4to. 1873.

Soundings in Davis' Strait from 6,000 to 10,998 feet, and in one instance 12,540 feet, are noticed by Ehrenberg in the "Monatsbericht k. Akad. Wiss. Berlin" for 1861, p. 275, &c., and several Foraminifera, containing their animal matter, are figured in the "Abhandl. für 1872," Pl. I., and partly in Pl. II. Thus in Pl. I.:—

Fig. 1. Aristerospira Liopentas, 6,000′.
„ 2. A. Microtretas, 6,000′.
„ 3. A. porosa, 6,000′.
„ 4. A. Pachyderma, 6,000′.
„ 5. A. glomerata, 6,000′.
„ 6. Miliola Dactylus, 10,998′.
„ 7. Quinqueloculina oblonga, 10,998′.
„ 8. Grammostomum ? euryleptum, 6,000′.
„ 9. Phaneroptomum microporum, 6,000′.
„ 10. Planulina lævigata, 6,000′.
„ 11. Phanerostomum Micromega, 6,000′.
„ 12. P. Alloderma, 10,988′.
„ 13. P. scutellatum, 6,000′.
„ 14. P. Globulus, 12,540′.
„ 15. Planulina abyssicola, 9,240′.
„ 16. P. Globigerina, 6,000′.
„ 17. P. grœnlandica, 10,988′.
„ 18. Nonionina borealis, 6,000.
„ 19. Rosalina Hexas, 6,000′.
„ 20. Planulina depressa, 6,000

Fig. 21. Rotalia profunda, 10,998′.
„ 22. R. globulosa, 10,998′.
„ 23. Pylodexia glomerulus, 6,000′.
And in Pl. II., figs. 24, 25, Pylodexia Uvula, 6,000′.

According to the nomenclature of English rhizopodists (*see* "Annals Mag. Nat. Hist.," ser. 4, vol. ix., p. 211, 280; x., p. 184, 253, 453, and for March 1, 1873), the study of Dr. Ehrenberg's genera and species would lead us to regard figs. 1–5, 9–16, and 22 as various individuals of *Globigerina bulloides*, D'Orb., with some modifications of spirality and size of chambers, whilst fig. 23 and figs. 24 and 25 of Pl. II. are *Globigerinæ* within the specific limits of *Gl. bulloides*, but varietal, near *elevata* and *helicina* (D'Orb.). Fig. 6 is *Lagena globosa*, var.; 7, *Quinqueloculina;* 8. *Textularia sagittula.* Figs. 17 and 19-21, are *Planorbulinæ*, small growths and varieties of *Pl. vulgaris* probably; and fig. 18, *Nonionina.* Looked at in this light, this group of Foraminifera, excepting its far great richness in *Globigerinæ*, is to a great extent such as has been described from Davis' Strait and Baffin's Bay by Parker and Jones in the "Phil. "Trans.," clv., 1865. *See above*, pp. 192, 193.

Besides Davis' Strait, Hingston Bay, Greenland, has also yielded Microzoa to Dr. Ehrenberg ("Monatsber.," 1853, p. 523, &c.). All of the foregoing, together with the materials he obtained from the First and Second* German North Polar Expeditions ("Monatsb.," 1869, p. 253, &c., and 1872, p. 282, &c.) and one sounding from Behring's Strait ("Abhandl.," 1872, p. 195), Dr. Ehrenberg has grouped in the "Abhandl." for 1872 (Table, pp. 220–329), with the following results for the "North-Polar Zone":—

Polythalamia (Foraminifera), 36.
Polygastrica (Diatomaceæ, &c.), 92.
Polycystina, 6.
Phytolitharia (51), Geolithia (3), and Zoolitharia (4), (spicules of Sponges, &c.), 51.

XXVI.—On the EXISTENCE of MARINE ANIMALS at various DEPTHS in SEAS abounding in FLOATING ICE, in ARCTIC REGIONS near GREENLAND and SPITZBERGEN. From the Appendix B. of SIR C. LYELL'S "Antiquity "of Man," p. 508. 1863.

"Dr. Torell, after he had examined, between the years 1856 and 1860, the glaciers of Switzerland, Norway, Iceland, Greenland, and Spitzbergen, was appointed to command in 1861 a scientific expedition fitted out at the joint expense of the Swedish Government and Prince Oscar of Sweden. It consisted of two

* For notes on the Foraminifera of the N.E. coast of Greenland, *see* further on.

ships, and a survey was made of the coast of Spitzbergen and the adjoining seas.

"So far from finding any scarcity of Mollusca, these explorers collected no less than 150 living species, chiefly on the west and north coasts of Spitzbergen, in lat 79° and 80° N., and the number of individuals, as well as the variety of species, was often great, especially where the bottom consisted of fine mud derived from moraines of glaciers, and from the grinding action of the land-ice on the rocks below.

"Between Spitzbergen and the north of Norway, but nearer the former country, Dr. Torell and his fellow-labourer Mr. Chydenius obtained, at the enormous depths of 1,000 and 1,500 *fathoms* (September 1861), Mollusca (a *Dentalium* and *Bulla* or *Cylichna*), a Crustacean, Polythalamian Shells, a Coral three inches long, with several red Actinias attached to it, and a few Annelids. These occurred to the west of Beeren's Island, in latitude 76° 17′ N., and longitude 13° 53′ E., in a sea where floating ice is common for ten months in the year. The temperature of the mud at the bottom was between 32° and 33° Fahrenheit, and that of the water at the surface 41°, and of the air 33° Fahrenheit.

"In Greenland, north of Disco Island, between latitude 70° and 71° N., in a deep channel of the sea, separating the peninsula of Noursoak from the island of Omenak, a region where the largest icebergs come down into Baffin's Bay, Dr. Torell dredged up, besides more than twenty other Mollusks, *Terebratella Spitzbergensis*, living at a depth of 250 fathoms. This shell I found fossil in 1835, at Uddevalla, in the ancient glacial beds, far south of its present range. The bottom of the sea in the Omenak channel consisted of impalpable mud, and on the surface of some of the floating bergs was similar mud, on which they who trod sank knee-deep; also numerous blocks of granitic and other rocks of all sizes, most of them striated on one, two or more sides. Here, therefore, a deposit must be going on of mud containing marine shells, with intermingled glaciated pebbles and boulders.

"A species of *Nucula* (*Leda truncata* or *Yoldia truncata*, Brown), now living in the seas of Spitzbergen, North Greenland, and Wellington Channel, Parry Islands, was found by Dr. Torell to be one of the most characteristic species in the mud of those icy regions. Of old, in the Glacial Period, the same shell ranged much farther south than at present, being found embedded in the Boulder-clay in the south of Norway and Sweden as well as of Scotland. It has been observed by the Rev. Thomas Brown, together with several other exclusively Arctic species, at Elie, in the south of Fife, in glacial clay, at the level of highwater-mark. I have myself collected it in a fossil state in the glacial clay of Portland and other localities in Maine in North America. It is the shell well known as *Leda Portlandica* of Hitchcock.

"In ponds and lakes in 'the outskirts' of North Greenland, in Disco Island for example, no freshwater Mollusca were met with by Dr. Torell, though some species of Crustacea of the genera *Apus* and *Branchipus* inhabit such waters. This may help us to explain the want of fossils in all glacial deposits of fluviatile or lacustrine origin. The discoveries above referred to show that the marine glacial beds of the Clyde and those of Elie in Fife, with their Arctic shells, are precisely such formations as might be looked for as belonging to a period when Scotland was undergoing glaciation as intense as that to which Spitzbergen and North Greenland are now subjected."*

XXVII.—The First Part of the "Outlines of the Distribution of Arctic Plants." By J. D. Hooker, M.D., F.R.S., &c. &c. Trans. Linn. Soc., vol. 23, 1861, pages 251–348. Read June 21, 1860. Reprinted by permission.

[Owing to want of time, this Memoir is reprinted as it appeared in 1861, without reference to slight modifications in the calculations which subsequent discoveries necessitate. The most important of these discoveries, however, are given in foot-notes, &c.—J. D. Hooker, April 1875.]

Contents.

I.—Introduction.

I shall endeavour in the following pages to comply, as far as I can, with a desire expressed by several distinguished Arctic voyagers, that I should draw up an account of the affinities and distribution of the flowering plants of the North-Polar regions. The method I have followed has been, first, to ascertain the names and localities of all Plants which appear on good evidence to have

---

* See also the notes on the Continental Ice of Greenland, former Climate of Greenland, and range of fossil *Sequoia* over Arctic Regions, on Mackenzie River, and in Iceland, "Antiquity of Man," 4th edit., 1873, pp. 274–281.

been found north of the Arctic Circle in each continent; then to divide the Polar Zone longitudinally into areas characterised by differences in their vegetation; then to trace the distribution of the Arctic Plants, and of their varieties and very closely allied forms, into the Temperate and Alpine regions of both hemispheres. Having tabulated these data, I have endeavoured to show how far their present distribution may be accounted for by slow changes of climate during and since the Glacial Period.

The Arctic flora forms a circumpolar belt of 10° to 14° latitude north of the Arctic Circle. There is no abrupt break or change in the vegetation anywhere along the belt, except in the meridian of Baffin's Bay, whose opposite shores present a sudden change from an almost purely European flora on its east coast, to one with a large admixture of American plants on its west.

The number of flowering plants which have been collected within the Arctic Circle is 762 (Monocot. 214, Dicot. 548). In the present state of cryptogamic botany it is impossible to estimate accurately the number of flowerless plants found within the same area, or to define their geographical limits; but the following figures give the best approximate idea I have obtained:—

Filices -	- 28	Characeæ	- 2	Fungi	- 200?
Lycopodiacea	- 7	Musci	- 250	Algæ	- 100
Equisetaceæ	- 8	Hepaticæ	- 80	Lichenes	- 250

Total Cryptogams	- 925
„ Phænogams	- 762
	1,687

Regarded as a whole, the Arctic flora is decidedly Scandinavian, for Arctic Scandinavia, or Lapland, though a very small tract of land, contains by far the richest Arctic flora, amounting to three-fourths of the whole; moreover, upwards of three-fifths of the species, and almost all the genera of Arctic Asia and America, are likewise Lapponian, leaving far too small a percentage of other forms to admit of the Arctic, Asiatic, and American floras being ranked as anything more than subdivisions, which I shall here call "Districts," of one general Arctic flora.

Proceeding eastwards from Baffin's Bay, there is, first, the Greenland District, whose flora is almost exclusively Lapponian, having an extremely slight admixture of American or Asiatic types; this forms the western boundary of the purely European flora. Secondly, the Arctic-European District, extending eastward to the Obi River, beyond the Ural range, including Nova Zembla and Spitzbergen; Greenland would also be included in it, were it not for its large area and geographical position. Thirdly, the transition from the comparatively rich European District to the extremely poor Asiatic one is very gradual; as is that from the Asiatic to the richer fourth or West-American District, which extends from Behring's Straits to the Mackenzie River. Fifthly, the transition from the West to the East-American district is even less marked, for the lapse of European and West American species is trifling, and the appearance of East-American one is equally so; the transition in vegetation from this district again to that of Greenland is, as I have stated above, comparatively very abrupt.

The general uniformity of the Arctic flora, and the special difference between its subdivisions may be thus estimated: the Arctic Phænogamic flora consists of 762 species; of these, 616 are Arctic-European, many of which prevail throughout the polar area, being distributed in the following proportions through its different longitudes:—

	Scandinavian forms.	Asiatic and American.	
Arctic Europe -	616	- 586	- 30=1 : 19·57
„ Asia -	233	- 189	- 44=1 : 4·2
„ W. America	364	- 254	- 110=1 : 2·3
„ E. America	379	- 269	- 110=1 : 2·4
„ Greenland -	207	- 195	- 12=1 : 16·2

This table places in a most striking point of view the anomalous condition of Greenland, which, though so favourably situated for harbouring an Arctic-American vegetation, and so unfavourably for an Arctic-European one, presents little trace of the botanical features of the great continent to which it geographically belongs, and an almost absolute identity with those of Europe.

Moreover, the peculiarities of the Greenland flora are not confined to these; for a detailed examination shows that it differs from all other parts of the Arctic regions in wanting many extremely common Scandinavian plants, which advance far north in all the other polar districts, and that the general poverty of its flora in species is more due to an abstraction of Arctic types than to a deficiency of temperature. This is proved by an examination of the Temperate portion of the Greenland peninsula, which adds very few plants to the entire flora as compared with a similar area south of any other Arctic region, and these few are chiefly Arctic plants and almost without exception Arctic-Scandinavian species.

There is nothing in the physical features of the Arctic regions, their oceanic or aërial currents, their geographical relations, nor their temperature, which, in my opinion, at all accounts for the exceptional character of the Greenland flora; nor do I see how it can be explained except by assuming that extensive changes of climate, and of land and sea, have exerted great influence, first in directing migration of the Scandinavian species over the whole polar zone, and afterwards in introducing the Asiatic and American species with which the Scandinavian are so largely associated in all the Arctic Districts except those of Europe and Greenland. It is inconceivable to me that so many Scandinavian plants should, under existing conditions of sea, land, and temperature, have not only found their way westward to Greenland, by migration across the Atlantic, but should have stopped short on the east shore of Baffin's Bay, and not crossed to America; or that so many

American types should terminate so abruptly on the west coast of Baffin's Bay, and not cross to Greenland and Europe; or that Greenland should contain actually much fewer species of European plants than have found their way eastwards from Lapland by Asia into Western and Eastern Arctic America; or that the Scandinavian vegetation should in every longitude have migrated across the tropics of Asia and America, whilst those typical plants of these continents which have found their way into the Arctic regions have there remained restricted to their own meridians.

It appears to me difficult to account for these facts, unless we admit Mr. Darwin's hypotheses,* first, that the existing Scandinavian flora is of great antiquity, and that previous to the Glacial Epoch it was more uniformly distributed over the polar zone than it is now; secondly, that during the advent of the Glacial Period this Scandinavian vegetation was driven southward in every longitude, and even across the Tropics into the South Temperate zone; and that on the succeeding warmth of the present epoch, those species that survived both ascended the mountains of the warmer zones, and also returned northwards, accompanied by aborigines of the countries they had invaded during their southern migration. Mr. Darwin shows how aptly such an explanation meets the difficulty of accounting for the restriction of so many American- and Asiatic-Arctic types to their own peculiar longitudinal zones, and for what is a far greater difficulty, the representation of the same Arctic genera by most closely allied species in different longitudes.

To this representation and the complexity of its character, I shall have to allude when indicating the sources of difficulties I have encountered, whether in limiting the polar species, or in determining to what southern forms many are most directly referable. Mr. Darwin's hypothesis accounts for many varieties of one plant being found in various Alpine and Arctic regions of the globe, by the competition into which their common ancestor was brought with the aborigines of the countries it invaded; different races survived the struggle for life in different longitudes; and these races again afterwards converging on the zone from which their ancestor started, present there a plexus of closely allied but more or less distinct varieties or even species, whose geographical limits overlap, and whose members very probably occasionally breed together.

Nor is the application of this hypothesis limited to this inquiry; for it offers a possible explanation of a general conclusion at which I had previously arrived,† and which I shall have again to discuss here, viz., that the Scandinavian flora is present is every latitude of the globe, and is the only one that is so; and it also helps to explain another class of most interesting and anomalous facts in Arctic dis-

---

* This theory of a southern migration of northern types being due to the cold epochs preceding and during the Glacial originated, I believe, with the late Edward Forbes; the extended one of their transtropical migration is Mr. Darwin's, and is discussed by him in his "Origin of Species," chap. xi.

† Introd. essay to the "Flora of Tasmania," p. ciii.

tribution, at which I have now arrived from an examination of the several polar districts, and especially of that of Greenland.

A glance at the appended chart [not reproduced here] shows how this theory bears upon the Greenland flora, explaining the identity of its existing vegetation with that of Lapland, and accounting for its paucity of species, for the rarity of American species, of peculiar species, and of marked varieties of European species. If it be granted that the polar area was once occupied by the Scandinavian flora, and that the cold of the Glacial Epoch did drive this vegetation southwards, it is evident that the Greenland individuals, from being confined to a peninsula, would be exposed to very different conditions to those of the great continents.

In Greenland many species would, as it were, be driven into the sea, that is, exterminated, and the survivors would be confined to the southern portion of the peninsula; and not being there brought into competition with other types, there could be no struggle for life amongst their progeny, and, consequently, no selection of better adapted varieties. On the return of heat these survivors would simply travel northwards, unaccompanied by the plants of any other conntry.

In Arctic America and Asia, on the other hand, where there was a free southern extension and dilatation of land for the same Scandinavian plants to occupy, these would multiply enormously in individuals, branching off into varieties and sub-species, and occupy a larger area the further south they were driven; and none need be altogether lost in the southern migration over plains, though many would in the struggle that ensued, when they reached the mountains of those continents and were brought into competition with the Alpine plants which the same cold had caused to descend to the plains. Hence, on the return of warmth, many more Scandinavian species would return to Arctic America and Arctic Asia than survived in Greenland; some would be changed in form, because only the favoured varieties could have survived the struggle; some of the Alpine, Siberian, and Rocky-Mountain species would accompany them to the Arctic Zone, while many Arctic species would ascend those mountains, accompanying the Alpine species in their re-ascent.

Again, as the same species may have been destroyed in most longitudes, or at most elevations, but not at all, we should expect to find some of those Arctic Scandinavian plants of Greenland which have not returned to Arctic America still lurking in remote Alpine corners of that great continent; and we may account for *Draba aurea* being confined to Greenland and the Rocky Mountains, *Potentilla tridentata* to Greenland and Labrador, and *Arenaria Grœnlandica* to Greenland and the White Mountains of New Hampshire, by supposing that these were originally Scandinavian plants, which, on the return of warmth, were exterminated on the plains of the American continent, but found a refuge on its mountains, where they now exist.

It appears, therefore, to be no slight confirmation of the

general truth of Mr. Darwin's hypothesis, that, besides harmonizing with the distribution of Arctic plants within and beyond the Polar Zone, it can also be made, without straining, to account for that distribution and for many anomalies of the Greenland flora, viz., 1, its identity with the Lapponian ; 2, its paucity of species ; 3, the fewness of temperate plants in temperate Greenland, and the still fewer plants that area adds to the entire flora of Greenland ; 4, the rarity of both Asiatic and American species or types in Greenland ; and 5, the presence of a few of the rarest Greenland and Scandinavian species in enormously remote Alpine localities of West America and the United States.

II.—On the Local Distribution of Plants within the Arctic Circle.

The greatest number of plants occurring in any given Arctic District is found in the European, where 616 flowering plants have been collected from the verge of the Circle to Spitzbergen. From this region vegetation rapidly diminishes in proceeding eastwards and westwards, especially the latter. Thus, in Arctic Asia only 233 flowering plants have been collected ; in Arctic Greenland, 207 species ; in the American continent east of the Mackenzie River, 379 species ; and in the area westwards from that river to Behring's Straits, 364 species.

A glance at the annual and monthly Isothermal Lines shows that there is little relation between the temperature and vegetation of the areas they intersect beyond the general feature of the scantiness of the Siberian flora being accompanied by a great southern bend of the annual isotherm of 32° in Asia, and the greatest northern bend of the same isotherm occurring in the longitude of West Lapland, which contains the richest flora. On the other hand, the same isotherm bends northwards in passing from Eastern America to Greenland, the vegetation of which is the scantier of the two, and passes to the northwards of Iceland, which is much poorer in species than those parts of Lapland to the southward of which it passes. The June isothermals, as indicating the most effective temperatures in the Arctic regions (where all vegetation is torpid for nine months, and excessively stimulated during the three others), might have been expected to indicate better the positions of the most luxuriant vegetation ; but neither is this the case, for the June isothermal of 41°, which lies within the Arctic zone in Asia, where the vegetation is scanty in the extreme, descends to 54° N. lat. in the meridian of Behring's Straits, where the flora is comparatively luxuriant, and the June isothermal of 32°, which traverses Greenland north of Disco, passes to the north both of Spitzbergen and the Parry Islands. In fact, it is neither the mean annual, nor the summer (flowering), nor the autumn (fruiting) temperature that determines the abundance or scarcity of the vegetation in each district, but these combined with the ocean-temperature and consequent prevalence of humidity, its geographical position, and its former conditions, both climatal and geographical. The relations between the isothermals and floras in each longitude being, therefore, special and not general, I shall consider them further when defining the different Arctic floras.

The northern limits to which vegetation extends varies in every longitude, and its extreme limits are still unknown; it may, indeed, reach to the pole itself. Phænogamic plants, however, are probably nowhere found far north of lat. 81°. 70 flowering plants are found in Spitzbergen; and Sabine and Ross collected nine on Walden Island, towards its northern extreme, but none on Ross's Islet, 15 miles further to the north. Sutherland, a very careful and intelligent collector, found 23 at Melville Bay and Wolstenholme and Whale Sounds, in the extreme north of Baffin's Bay, lat. 76° 77° N. Parry, James Ross, Sabine, Beechey, and others together found 60 species on Melville Island, and Lyall 50 on the islands north of Barrow Straits and Lancaster Sound.

About 80 have been detected on the west shores of Baffin's Bay and Davis's Straits, between Ponds Bay and Home Bay. To the north of Eastern Asia, again, Seemann collected only 4 species on Herald Island, lat. 71½° N., the northernmost point attained in that longitude. On the east coast of Greenland, Scoresby and Sabine found only 50 between the parallels of 70° and 75° N.; whilst 150 inhabit the west coast between the same parallels.

The differences between the vegetations of the various polar areas seem to be to a considerable extent constant up to the extreme limits of vegetation in each. Thus *Ranunculus glacialis* and *Saxifraga flagellaris*, which are all but absent in West Greenland,* advance to the extreme north in East Greenland and Spitzbergen. *Caltha palustris, Astragalus alpinus, Oxytropis Uralensis* and *nigrescens, Parrya arctica, Sieversia Rossii, Nardosmia corymbosa, Senecio palustris, Deschampsia cæspitosa, Saxifraga hieraciifolia* and *Hirculus*, all of which are absent in West Greenland, advance to Lancaster Sound and the polar American islands, a very few miles to the westward of Greenland.

On the other hand *Lychnis alpina, Arabis alpina, Stellaria cerastioides, Potentilla tridentata, Cassiopeia hypnoides, Phyllodoce taxifolia, Veronica alpina, Thymus Serpyllum, Luzula spicata*, and *Phleum alpinum*, all advance north of 70° in West Greenland, but are wholly unknown in any part of Arctic Eastern America or the polar islands.

The most Arctic plants of general distribution that are found far north in all the Arctic areas are the following; all inhabit the Parry Islands, or Spitzbergen, or both:—

Ranunculus nivalis.
R. auricomus.
R. pygmæus.
Papaver nudicaule.
Braya alpina.
Cardamine bellidifolia.
C. pratensis.
Draba alpina.

* Both were found by Kane's Expedition, but by no previous one.

Draba androsacea.
D. hirta.
D. muricella.
D. incana.
D. rupestris.
Cochlearia anglica.
C. officinalis.
Silene acaulis.
Lychnis apetala.
Arenaria verna.
A. arctica.
Stellaria longipes.
Cerastium alpinum.
Potentilla nivea.
P. frigida.
Dryas octopetala.
Epilobium latifolium.
Sedum Rhodiola.
Chrysosplenium alternifolium.
Saxifraga oppositifolia.
S. cæspitosa.
S. cernua.
S. rivularis.
S. nivalis.
S. stellaris.
S. flagellaris.
S. Hirculus (E. Greenland only).
Antennaria alpina.
Erigeron alpinus.
Taraxacum Dens-leonis.
Cassiopeia tetragona.
Pedicularis hirsuta.
P. sudetica.
Oxyria reniformis.
Polygonum viviparum.
Empetrum nigrum.
Salix herbacea.
S. reticulata.
Luzula arcuata.
Juncus biglumis.
Carex fuliginosa (not yet found in Arctic Asia, but no doubt there).
C. aquatilis (not yet found in Arctic Asia, but no doubt there).
Eriophorum capitatum.
E. polystachyum.
Alopecurus alpinus.
Deyeuxia Lapponica.
Deschampsia cæspitosa (East Greenland only).
Phippsia algida.
Colpodium latifolium.
Poa flexuosa.
P. pratensis.
P. nemoralis.
Festuca ovina.

Of the above, *Saxifraga oppositifolia* is probably the most ubiquitous, and may be considered the commonest and most Arctic flowering plant.

The following are also inhabitants of all the five Arctic areas, but do not usually attain such high latitudes as the foregoing :—

Ranunculus Lapponicus.
Draba rupestris.
Viola palustris.
Honkeneya peploides.
Epilobium augustifolium.
E. alpinum.
Hippuris vulgaris.
Artemisia borealis.
Vaccinium uliginosum.
V. Vitis-idæa.
Ledum palustre.
Pyrola rotundifolia.
Polemonium cæruleum, and vars. (East Greenland only).
Pedicularis Lapponica.
Armeria vulgaris.
Betula nana.
Salix lanata.
S. glauca.
S. alpestris.
Luzula campestris.
Carex vesicaria.
Eriophorum vaginatum.
Atropis maritima.

The absence of *Gentiana* and *Primula* in these lists is very unaccountable, seeing how abundant and very Alpine they are on the Alps and Himalaya, and *Gentiana* on the South-American Cordilleras also. The few remaining plants, which are all very northern, and almost or wholly confined to the Arctic zone, are the following. † indicates those species absolutely peculiar; †† the only peculiar genus.

Ranunculus Pallasii.
R. hyperboreus.
Trollius Asiaticus.
Corydalis glauca.
Cardamine purpurea.
Turritis mollis.
Cochlearia sisymbrioides.
Hesperis Pallasii.
† Braya pilosa.
Eutrenia Edwardsii.
Parrya arctica.
† P. arenicola.
Odontarrhena Fischeriana.
Sagina nivalis.
Stellaria dicranoides.
Oxytropis nigrescens.
Sieversia Rossii.
S. glacialis.
Rubus arcticus.
Parnassia Kotzebuei.
Saxifraga Eschscholtzii.
S. serpyllifolia.
† S. Richardsoni.
Cœnolophium Fischeri.
† Nardosmia glacialis.
Artemisia Richardsoniana.
A. glomerata.
† A. androsacea.
Erigeron compositus.
Chrysanthemum arcticum.
Pyrethrum bipinnatum.
† Saussurea subsinuata.
Campanula uniflora.
Gentiana arctophila.
G. aurea.
Eutoca Franklinii.
Pedicularis flammea.
† Douglasia arctica.
† Monolepis Asiatica.
Betula fruticosa.
Salix speciosa.
† S. glacialis.
S. phlebophylla.
S. arctica.
Orchis cruenta.
Plantanthera hyperborea.
Carex nardina.
C. glareosa.
C. rariflora.
Hierochloe pauciflora.
Deschampsia atropurpurea.
Phippsia algida.
Dupontia Fisheri.
Colpodium pendulinum.
C. fulvum.
C. latifolium.
†† Pleuropogon Sabini.
† Festuca Richardsoni.

III.—Distribution of the Arctic Flowering Plants in various Regions of the Globe.

There is but one distinct genus confined to the Arctic regions, the monotypic and local *Pleuropogon Sabini;* and there are but seven other peculiarly Arctic species, together with one with which I am wholly unacquainted, viz., *Monolepis Asiatica.* The remaining 762 species are all found south of the Circle, and of these all but 150 advance south of the parallel of 40° N. lat., either in the Mediterranean basin, Northern India, the United States, Oregon, or California; about 50 are natives of the mountainous regions of the Tropics, and just 105 inherit the South Temperate zone.

The proportion of species which have migrated southwards in the Old and the New World also bear a fair relation to the facilities for migration presented by the different continents. Thus

Of 616 Arctic-European species—
- 496 inhabit the Alps, and
- 450 cross them;
- 126 cross the Mediterranean;
- 26 inhabit Africa.

Of 379 Arctic-East-American—
- 203 inhabit the United States (of which 21 are confined to the mountains);
- 34 inhabit Tropical American mountains;
- 50 inhabit Temperate South America.

Of 233 Arctic-Asiatic species—
- 210 reach the Altai, Soongaria, &c.;
- 106 reach the Himalaya;
- 0 are found on the Tropical mountains of Asia;
- 5 inhabit Australia and New Zealand.

Of 346 Arctic-West-American species—
- 274 are North-temperate;
- 24 on Tropical mountains;
- 37 in South-temperate zone.

These tables present in a very striking point of view the fact of the Scandinavian Flora being the most widely distributed over the globe. The Mediterranean, South-African, Malayan, Australian, and all the floras of the New World have narrow ranges compared with the Scandinavian, and none of them form a prominent feature in any other continent than their own; but the Scandinavian not only girdles the globe in the Arctic Circle, and dominates over all the others in the North-temperate zone of the Old World, but intrudes conspicuously into every other Temperate flora, whether in the northern or southern hemisphere, or on the Alps of Tropical countries.

The severest test to which this observation could be put is that supplied by the Arctic-Scandinavian forms, for these belong to the remotest corner of the Scandinavian area, and should of all plants be the most impatient of temperate, warm, and tropical climates. The following will, approximately, express the result:—

Total Arctic-Scandinavian forms	586
In North-United-States and Canada, &c.	360
In Tropical America	40
In Temperate South America	70
In Alps of Middle Europe, Pyrenees, &c.	490
Cross Alps, &c.	480
Reach South Africa	20
Himalaya, &c.	300
Tropical Asia	20
Australia, &c.	60

In one respect this migration is most direct in the American meridian, where more Arctic species reach the highest southern latitudes. This I have accounted for ("Flora Antarctica," p. 230)

by the continuous chain of the Andes having favoured their southern dispersion.

But the greatest number of Arctic plants are located in Central Europe, no fewer than 530 out of 762 inhabiting the Alps and Central and Southern Europe, of which 480 cross the Alps to the Mediterranean basin. Here, however, their further spread is apparently suddenly arrested; for, though many doubtless are to be found in the Alps of Abyssinia and the Western Atlas, there are few compared with what we found further east in Asia, and fewer still have found their way to South Africa.

The most continuous extension of Scandinavian forms is in the direction of the greatest continental extension, namely, that from the North Cape in Lapland to Tasmania,* for no less than 350 Scandinavian plants have been found in the Himalayas, and 53 in Australia and New Zealand, whereas there are scarcely any Himalayan and no Australian or Antarctic forms in Arctic Europe. Now that Mr. Darwin's hypotheses are so far accepted by many botanists, in that these concede many species of each genus to have had in most cases a common origin, it may be well to tabulate the Generic distribution of the Arctic plants as I have done the Specific; and this places the prevalence of the Scandinavian types of vegetation in a much stronger light:—

Scandinavian Arctic *Genera* in Europe - - -		280
Found in North-United-States - -	(approximately)	270
„ Tropical American mountains	„	100
„ Temperate South America -	„	120
„ Alps - - - - - -	„	280
Cross Alps - - - - - -	„	260
Found in South Africa - - -	„	110
„ Himalaya, &c. - - -	„	270
„ Tropical Asia - - -	„	80
„ Australia, &c. - - -	„	100

The most remarkable anomaly is the absence of *Primula* in Tropical America, that genus being found in Extra-tropical South America, and its absence in the whole Southern Temperate zone of the Old World, except the Alps of Java.

Thalictrum, *Delphinium*, *Impatiens*, *Prunus*, *Circæa*, *Chrysosplenium*, *Parnassia*, *Bupleurum*, *Hieracleum*, *Viburnum*, *Valeriana*, *Artemisia*, *Vaccinium*, *Rhododendron*, *Pedicularis*, and *Salix* are all Arctic Genera found on the Tropical mountains of Asia (Nilghiri, Ceylon, Java, &c.), but not yet in the South-tem-

* The line which joins these points passes through Siberia, Eastern China, the Celebes Islands, and Australia; but the glacial migration has been due south from the Arctic and North-temperate regions in various longitudes to the Pyrenees, Alps, Carpathians, Caucasus, Asia Minor, and the Persian and North-Indian mountains. The further migration south to the distant and scattered alpine heights of the Tropics, and thence to South Australia, Tasmania, and New Zealand, is, in the present state of our knowledge, to me quite unaccounted for. Mr. Darwin assumes for this purpose a cooled condition of the globe that must have been fatal to all such purely tropical vegetation as we are now familiar with.

perate zone of Asia, and very few of them in Temperate South Africa.

There are, however, a considerable number of Scandinavian plants which are not found in the Alps of Middle Europe, though found in the Caucasus, Himalaya, &c., and conversely there are several Arctic-Asiatic and -American plants found in the Alps of Central Europe, but nowhere in Arctic Europe. In other words, certain species extend from Arctic America westward to Arctic Europe, and there are certain other species which extend from Arctic Europe to the Caucasus and Central Asia, which neither exist on the Alps of Central Europe nor extend eastward to Arctic America:—

[Here follows a list of 103 species of plants common to Arctic Europe and Temperate Asia, &c., but not to the Alps of Europe, pp. 260-1.]

It is curious to remark how many of these Boreal European plants, which are absentees in the Alps, have a very wide range, not only extending to the Himalaya and North China, but many of them all over Temperate North China; only one is found in the South-temperate zone. In the present state of our knowledge we cannot account for the absence of these in the Alps; either they were not natives of Arctic Europe immediately previous to the Glacial Period, or, if so, and they were then driven south to the Alps, they were afterwards there exterminated; or, lastly, they still inhabit the Alps under disguised forms, which pass for different species. Probably some belong to each of these categories. I need hardly remark that none inhabit Europe south of the Alps, or any part of the African continent.

The list of Arctic-American and -Asiatic species which inhabit the Alps of Europe, but not Arctic Europe, is much smaller. Those marked † are Scandinavian, but do not enter the Arctic Circle:—

Anemone patens.
A. alpina.
A. narcissiflora.
† Ranunculus sceleratus.
† Aconitum Napellus.
† Arabis petræa.
† Cardamine hirsuta.
Draba stellata.
† Thlaspi montanum.
† Lepidium ruderale.
† Sagina nodosa.
† Linum perenne.
Phaca alpina.
† Astragalus hypoglottis.
† Spiræa salicifolia.
† Potentilla fruticosa.
P. sericea.
† Ceratophyllum demersum.
Bupleurum ranunculoides.
† Viburnum Opulus.
Galium rubioides.
† G. saxatile.
Ptarmica alpina.
Aster alpinus.
Gentiana prostrata.
Polygonum polymorphum.
Corispermum hyssopifolium.
Alnus viridis.
Pinus cembra.
† Sparganium simplex.
† Typha latifolia.
Carex ferruginea.
C. supina.
C. stricta.
† C. pilulifera.
Scirpus triqueter.
Deyeuxia varia.
Spartina cynosuroides.
† Glyceria fluitans.
Hordeum jubatum.

IV.—Botanical Districts within the Arctic Circle.

The following are the prominent features, botanical, geographical, and climatal, of the five Districts of the Arctic zone:—

1. **Arctic Europe.**—The majority of its plants are included in the Lapland and Finland Floras; and, owing to the temperature of the Gulf Stream, which washes its coasts, Lapland is by far the richest province in the Arctic regions. The mean annual temperature at the Polar Circle, where it cuts the coast-line, is about 37°, and the June and September temperatures throughout Lapland are 40° and 57° respectively; thus rendering the climate favourable both to flowering and fruiting. Spitzbergen belongs to this flora, as do Nova Zembla and the Arctic countries west of the river Obi, which forms its eastern boundary; for the Ural mountains do not limit the vegetation any more than do the Rocky Mountains in America. Gmelin observed, more than a century ago, that the River Obi in lower latitudes indicates the transition longitude from the European to the Asiatic flora.

Even in this small area, however, there are two floras, corresponding to the Arctic-Norwegian and Arctic-Russian. The latter, commencing at the White Sea, though comparatively excessively poor in species, contains nearly 20 that are not Lapponian, including *Braya rosea, Dianthus alpinus, D. Seguieri, Spiræa chamædrifolia, Saxifraga hieracifolia, Hieracleum Sibiricum, Liguria Sibirica, Ptarmica alpina, Gentiana verna, Pleurogyne rotata*, and *Larix Sibirica.*

There are, further, several Scandinavian plants which cross the Arctic Circle or the east shores of the White Sea, but do not do so in Lapland, as *Athamanta Libanotis, Chrysanthemum Leucanthemum, Bidens tripartita*, and others.

Iceland and Greenland also botanically belong to the Arctic-Lapland province, but I have here excluded both: the former because it lies to the south of the Arctic Circle; the latter because both its magnitude, position, and other circumstances require that it should be treated of separately.

As far as I can ascertain, 616 species (Monocotyledons 183, Dicotyledons 433, = 1 : 2·3) enter the Arctic Circle in this region, of which 70 advance into Spitzbergen; but no phænogamic plant is found in Ross's Islet beyond its northern extremity. The proportion of genera to species is 266 : 616 (1 : 2·3). Of these Arctic-European plants, 453 cross the Alps or Pyrenees to the Mediterranean basin; a few occur on the mountains of Tropical Africa (including *Luzula campestris* and *Deschampsia cæspitosa*); and 23 are found in South Africa.

No fewer than 264 species do not enter the Arctic Circle in any other longitude; and 184 are almost exclusively natives of the Old World, or of this and of Greenland, not being found in any part of North America; 24 are confined to Arctic Europe and Greenland.

The following Arctic-European plants are of sporadic occurrence in North America :—

Ranunculus acris, Rocky Mountains.

Arabis alpina, Greenland and Labrador.

Lychnis alpina, Greenland and Labrador.

Arenaria arctica, Greenland and Rocky Mountains.

A. verna, Greenland, Arctic Islands, and Rocky Mountains.

Alchemilla vulgaris, Greenland and Labrador.

Gnaphalium sylvaticum, Greenland and Labrador.

G. supinum, Greenland, Labrador, and United States Mountains.

Vaccinium myrtillus, Rocky Mountains only.

Cassiopeia hypnoides, Greenland, United States Mountains, and Labrador.

Phyllodoce taxifolia, Greenland, United States Mountains and Labrador.

Gentiana nivalis, Greenland and Labrador.

Veronica alpina, Greenland and United States Mountains.

Bartsia alpina, Greenland and Labrador.

Pedicularis palustris, Labrador.

Primula farinosa, Labrador.

Salix phylicifolia, United States Mountains.

S. arbuscula, Greenland and United States Mountains.

Juncus trifidus, Greenland and United States Mountains.

Carex capitata, Greenland and United States Mountains.

Phleum alpinum, Greenland, United States Mountains, and Labrador.

Calamagrostis lanceolata, Labrador.

There are, besides, a considerable number of Arctic-European plants which, in the New World, are confined to Greenland, being nowhere found in East America; these will be enumerated when treating of the Greenland Flora. [The plants (29 species), which are widely distributed in temperate America and Asia, but almost exclusively Arctic in Europe, are enumerated, p. 263.]

The works upon which I have mainly depended for the habitats of the Arctic-European plants are Wahlenberg's "Flora Lapponica," Ledebour's "Flora Rossica," Fries' "Summa Vegetabilium Scandinaviæ" and "Mantissæ," and various admirable treatises by Andersson, Nylander, Hartmann, Lindblöm, Wahlberg, Blytt, C. Martins, Ruprecht, and Schrenk.

For Spitzbergen plants I have depended on Hooker's enumeration of the Spitzbergen collections made during Parry's attempt to reach the North Pole, Captain Sabine's collection made in the same island, and on Lindblöm and Beilschmied's "Flora von Spitzbergen" (Regensburg, Flora, 1842).

For the southern distribution of the Arctic-European plants, I have further consulted Nyman's excellent "Sylloge," Ledebour's "Flora Rossica," Grisebach's "Flora Rumelica," Grenier and Godron's "Flore de France," Parlatore's "Flora Italiana," Koch's "Synopsis Floræ Germaniæ," Munby's "Catalogue of Algerian Plants," A. Richard's of those of Abyssinia, Visiani's "Flora Dalmatica," Delile's "Flora Ægyptiaca," Boissier's noble "Voyage botanique dans l'Espagne," and Tchihatcheff's "Asia Minor," besides numerous local floras of the Mediterranean regions, Madeira, the Azores, and Canaries.

2. **Arctic Asia.**—This District, which, for its extent, contains by far the poorest flora of any on the globe, reaches from the Gulf of Obi eastward to Behring's Straits, where it merges into the West-American. The climate is marked by excessive mean cold; at the Obi the isotherm of 18° cuts the Arctic Circle in its S.E. course, and at the eastern extremity of the province the isotherm of 20° cuts the same circle; while the central part of the district is all north of the isotherm of 9°. The whole of the district is hence far north of the isotherm of 32°, which descends to 52° N.L. in its middle longitude. The extremes of temperature are also very great; the June isotherm of 41° ascending eastward through its western half to the Polar Sea, whilst the September isotherm of 41° descends nearly to 60° N.L., whence the low autumn temperature must present an almost insuperable obstacle to the ripening of seeds within this segment of the Polar Circle.

The warming influence of the Atlantic currents being felt no further east than the Obi, and the summer desiccation of the vast Asiatic continent, combine to render the climate of this region one of excessive drought as well as cold, whence it is in every way most unfavourable to vegetation of all kinds.

The total number of species hitherto recorded from this area is 233:—

Monocotyledons - -	42	= 1 : 4·5.
Dicotyledons - - -	191	

The proportion of genera to species is 1 : 2. Of the 233 species, 217 inhabit Siberia, as far south as the Altai or Japan, &c.; 104 extend southwards to the Himalaya or mountains of Persia; none are found on the mountains of the two Indian peninsulas, and 85 on those of Australia and New Zealand. All but 37 are European, and nine of these are almost exclusively Arctic. [The table follows, page 264.]

Thus out of 37 non-European species only 12 are confined to Asia, the remaining 25 being American. On the other hand, there are only 22 European species in Arctic Asia which are not also American, which scarcely establishes a nearer relationship between Arctic Asia with Europe than with America. [The table follows, page 264.]

In other words, of the 233 Asiatic species 196 are common to Asia and Europe, 22 are confined to Asia and Europe, 25 are confined to Asia and America only, and 12 are confined to Asia, of which three are peculiar to the Arctic Circle.

The rarity of Gramineæ, and especially of Cyperaceæ, in this region is its most exceptional feature, only 21 of the 138 Arctic species of these orders having hitherto been detected in it. Cryptogamic plants seem to be even more rare; *Woodsia Ilvensis* and *Lastrea fragrans* being the only Filices hitherto enumerated. Further researches along the edge of the Arctic Circle would doubtless add more Siberian species to this flora, as the examination of the north-east extreme would add American species, and possibly lead to the flora of the country of the Tchutchis being ranked with that of West America.

The works which have yielded me the most information regarding this flora are Ledebour's "Flora Rossica," and the valuable memoirs of Bunge, C. A. Meyer, and Trautvetter on the vegetation of the Taimyr and Boganida Rivers, and on the plants of Jenissei River, in Von Middendorff's Siberian "Travels." For their southern extension, Trautvetter and Meyer's "Flora Ochotensis," also in Middendorf's "Travels"; Bunge's enumeration of North-China and Mongolian plants; Maximovicz's "Flora Amurensis"; Asa Gray's paper on the botany of Japan (Mem. Amer. Acad. N.S., vi.); Karelin and Kiriloff's enumeration of Soongarian plants; Regel, Bach, and Herder on the East-Siberian and Jakutsk collections of Paullowsky and Von Stubendorff. For the Persian and Indian distribution, I have almost entirely depended on the herbarium at Kew, and on Boissier's and Bunge's numerous works.

3. **Arctic West America.**—The District thus designated is analogous in position, and to a considerable extent in climate, to the Arctic-European, but is much colder, as is indicated both by the mean temperature and by the position of the June isotherm of 41°, which makes an extraordinary bend to the south, nearly to 52° N.L., in the longitude of Behring's Straits.

It extends from Cape Prince-of-Wales, on the east shore of Behring's Straits, to the estuary of the Mackenzie River; and, as a whole, it differs from the flora of the province to the eastward of it by its far greater number both of European and Asiatic species, by containing various Altai and Siberian plants which do not reach so high a latitude in more western meridians, and by some Temperate plants peculiar to West America. This eastern boundary is, however, quite an artificial one; for a good many eastern plants cross the Mackenzie and advance westwards to Point Barrow, but which do not extend to Kotzebue's Sound; and a small colony of Rocky-Mountain plants also spreads eastward and westwards along the shores of the Arctic Sea, which further tends to connect the floras; such are *Aquilegia brevistylis, Sisymbrium humile, Hutchinsia calycina, Heuchera Richardsonii, Crepis nana, Gentiana arctophila, Salix speciosa,* none of which are generally diffused Arctic plants, or natives of any other parts of Temperate America but the Rocky Mountains.

The Arctic Circle at Kotzebue's Sound is crossed by the isotherm of 23°, and at the longitude of the Mackenzie by that of 12° 5′; whilst the June isotherm of 41° ascends obliquely from S.W. to N.E., from the Aleutian Islands to the mouth of the Mackenzie, and passes south of this province; the June and the September isotherms of 41° and 32° both traverse it obliquely, ascending to the N.E.

The vast extent of the Pacific Ocean and its warm northerly currents greatly modify the climate of West Arctic America, causing dense fogs to prevail, especially throughout the summer months, whilst the currents keep the ice to the north of Behring's

Straits. The shallowness of the ocean between America and Asia north of lat. 60°, together with the identity of the vegetation in the higher latitudes of these continents, suggests the probability of the land having been continuous at no remote epoch.

The number of phænogamic plants hitherto found in Arctic West America is 364:—

$$\left.\begin{array}{lr}\text{Monocotyledons} & 76\\ \text{Dicotyledons} & 288\end{array}\right\} = 1:3{\cdot}7.$$

The proportion of genera to species is 1 : 1·7. Of these 364 species, almost all but the littoral and purely Arctic species are found in West-temperate North America or in the Rocky Mountains, 26 in the Andes of Tropical or Sub-tropical America, and 37 in Temperate or Antarctic South America. Comparing this flora with that of Temperate and Arctic Asia, I find that no less than 320 species are found on the north-western shores and islands of that continent, or in Siberia, many extending to the Altai and the Himalaya. A comparison with Eastern Arctic America shows that 281 are common to it, and 38 are found in Temperate but not Arctic East America. [The list follows, page 266.]

These, it will be seen, are for the most part North-temperate plants, common in many parts of the globe, and which are only excluded from Eastern Arctic America by the greater rigour of its climate.

The best marked European and Asiatic species that are not found further east in Temperate or Arctic America are 18 in number. [The list follows, page 267.]

Hence it appears that of the 364 species found in Arctic West America, 319 inhabit East America (Arctic or Temperate, or both), and 320 are natives of the Old World—a difference hardly sufficient to establish a closer affinity of this flora with one continent rather than with the other.

The species peculiar to this tract of land (Arctic West America) are:—

Braya pilosa.
Saxifraga Richardsonii.
Artemisia androsacea.
Saussurea subsinuata.
Salix glacialis.

The rarity of Monocotyledons, and especially of the glumaceous orders, is almost as marked a feature of this as of the Asiatic flora; of the 138 Arctic species of *Glumaceæ*, only 54 are natives of West Arctic America.

The materials for this flora are principally the plants of Chamisso, collected during Kotzebue's voyage, and described by himself and Schlechtendahl; Lay and Collie's collections, described in Beecher's voyage; the "Flora Boreali-Americana"; and Seemann's plants, described in the "Botany of the Herald." Most of the above collections are from Behring's Straits. For the Arctic coast flora I am mainly indebted to Richardson's researches, and to Pullen's and other collections enumerated by Seemann in his account of the flora of Western Eskimo Land. For the southern extension of the flora I have had recourse to the "Flora

"Boreali-Americana," Ledebour's "Flora Rossica," which includes the Sitcha plants; the American floras of Nuttall, Pursh, Torrey, Gray, &c., and to the collections of Drs. Lyall and Wood formed in Vancouver's Island and British Columbia: for the Californian, Mexican, and Cordillera floras generally, to the Herbarium at Kew, the works above mentioned, and the various memoirs of Torrey and of Gray on the plants of the American Surveying Expedition.

4. **Arctic East America (exclusive of Greenland).**— This tract of land is analogous to the Arctic-Asiatic in many respects, of position and climate, but is very much richer in species. It extends from the estuary of the Mackenzie River to Baffin's Bay, and its flora differs from that of the western part of the continent, both in the characters mentioned in the notice of that province, and in possessing more East-American species. The western boundary of this province is an artificial one; the eastern is very natural, both botanically and geographically, for Baffin's Bay and Davis' Strait (unlike Behring Strait) have very deep water and different floras on their opposite shores. The Arctic Circle is crossed in the longitude of the Mackenzie River by the isotherm of 12°, which thence trends south-eastward to the middle of Hudson's Bay; and in longitude of Davis' Strait it is crossed by the isotherm of 18½°. The June isotherm of 41° descends obliquely from the shores of the Arctic Sea, near the mouths of the Mackenzie, to the northern parts of Hudson's Bay, south of the Arctic Circle, and the September isotherm of 41° is everywhere south of the circle. Hence the western parts of this province are very much warmer than the eastern, so much so, that the whole west coast and islands of Baffin's Bay lie north of a southern inflection of the June isotherm of 32°, which passes north of all the other polar islands. The Parry Islands have an analogous temperature of 40°. The warmth of the western portion of this tract is no doubt mainly due to the influence of the Pacific Ocean being felt across the continent of West America, though possibly also to the presence of a comparatively warm polar ocean, or to Atlantic currents crossing the Pole between Nova Zembla and Spitzbergen, of which nothing certain is known.* Be this as it may, the comparative luxuriance of the flora of Melville Island is a well-known fact, and one inexplicable by considerations of temperature, if unaccompanied by a humid atmosphere. The whole region is of course far north of the isotherm of 32°, which, in the longitude of its middle district, descends to Lake Winnipeg in lat. 52°.

That portion of this province which is richest in plants is the tract which intervenes between the Copper mine and Mackenzie River; east of this vegetation rapidly diminishes, as also to the northwards. The flora of the Boothian Peninsula, surrounded as it is with glacial straits, and placed centrically among the

* It is a well-known fact that the temperature always rises rapidly with the north (as well as other) winds over all this Arctic-American area.

Arctic islands, is perhaps the poorest of any part of the area, those of Banks' Land and Melville Island to the N.W. being considerably richer, as are those of the shores of Lancaster's Sound and Barrow's Strait, and the shores of Baffin's Bay to the north and east.*

The phænogamic flora of Arctic East America contains 379 species.

$$\left.\begin{array}{lr}\text{Monocotyledons - - -} & 92\\ \text{Dicotyledons - - -} & 287\end{array}\right\} = 1:3\cdot 1.$$

The proportion of genera to species is 1:2·0; of these 379 species, 323 inhabit temperate North America, east of the Rocky Mountains; 35 the Cordillera; and 49 Temperate or Antarctic South America. Comparing this flora with that of Europe, I find that 239 species (or two thirds) are common to the Arctic regions of both continents, whilst but little more than one third of the Arctic European species are Arctic-East-American; of 105 non-European species in Arctic East America, 32 are Asiatic; leaving 73 species confined to America, of which the following are furthermore confined to the eastward of the Rocky Mountains and Mackenzie River:—

Corydalis glauca.
Sarracenia purpurea.
Viola cucullata.
Silene Pennsylvanica.
Arenaria Michauxii.
Polygala Senega.
Lathyrus ochroleucus.
Rubus triflorus.
Prunus Virginiana.
Heuchera Richardsonii.
Cornus stolonifera.
Grindelia squarrosa.
Vaccinium Canadense.
Dracocephalum parviflorum.
Douglasia arctica.
Elæagnus argentea.
Urtica dioica.
Salix cordata.
Populus tremuloides.
Picea nigra.
Spiranthes gracilis.
Crypripedium acaule.
Carex oligosperma.
Pleuropogon Sabini.

Of these, *Douglasia* and *Pleuropogon* are the only ones absolutely peculiar to Arctic East America. It is a noticeable fact that not one of them is found in any part of Greenland. Compared with Greenland, the Arctic-East-American flora is rich, containing, besides those just enumerated, no less than 165 other species not found in Greenland. The following are found on the Arctic islands, and many on the west coast of Baffin's Bay, but not in West Greenland:—

Caltha palustris.
Parrya Arctica.
Merkia physodes.
Stellaria crassifolia.
Astralagus alpinus.
Oxytropis campestris.
O. Uralensis.
Oxytropis nigrescens.
Sieversia Rossii.
Saxifraga hieracifolia.
S. Virginiensis.
S. Hirculus (East Greenland only).
Valeriana capitata.

* Details of these florulas will be found in the 5th volume of the "Linnean "Journal," under the notice of Dr. Walker's Collections made during the Voyage of the "Fox."

Nardosmia corymbosa.
Ptarmica vulgaris.
Crysanthemum arcticum.
Artemisia vulgaris.
Senecio frigidus.
S. palustris.
S. pulchellus.
Solidago Virga-aurea.
Aster salsuginosus.
Crepis nana.
Saussurea alpina.
Andromeda polifolia.
Arctostaphylos alpina.
Kalmia glauca.
Phlox Sibirica.
Castilleja pallida.
Pedicularis capitata.
P. versicolor.
Androsace septentrionalis.
A. Chamæjasme.
Salix phlebophylla.
Lloydia serotina.
Hierochloe pauciflora.
Deschampsia cæspitosa (East Greenland only).
Glyceria fluitans.
Pleuropogon Sabini.
Bromus purgans.
Elymus mollis.

There are thus no fewer than 184 of the 379 Arctic-East-American species (fully half) which are absent in West Greenland, whilst only 105 (much less that one third) are absent in Europe. This alone would make the limitation of species in the meridian of Baffin's Bay more decided than in any other Arctic longitude; and I shall show that it is rendered still more decisive by the number of Arctic-Greenland plants that do not cross to Arctic East America.

Of the 379 Arctic-East-American species, only 56 are not found in Temperate East America, of which two are absolutely confined to this area; two others (*Parrya arenicola* and *Festuca Richardsoni*) to Arctic East and West America; 25 are found in Temperate West America, and about 20 are Rocky-Mountain species, and not found elsewhere in Temperate America.

For our knowledge of this flora I am principally indebted to the "Flora Boreali-Americana," and to Richardson's * botanical appendix to Franklin's First Voyage and his "Boat Journey "through Rupert's Land."

I have also examined the materials upon which the above works were founded, and the collections of almost every subsequent journey and voyage, up to those of Dr. Walker in the "Fox." To enumerate the numerous botanical appendices to Voyages, and separate opuscules to which these have given rise, from Ross' First Voyage to the present time, would be out of place here. I have endeavoured to embody in the essay the information gleaned from all of them. For the southern distribution of these plants in the United States, &c., I have had recourse primarily to Asa Gray's excellent "Manual of the Botany of the Northern United "States," to Chapman's "Flora of the S. E. States," and to the Reports on the Botany of various Exploring Expeditions.

5. **Arctic Greenland.**—In area Arctic Greenland exceeds any other Arctic District except the Asiatic, but ranks lowest of all

* I am indebted to Sir John Richardson for some corrections to this list, which account for a few discrepancies between his lists of Arctic-American plants and my own; these refer chiefly to genera and species introduced into his lists, but here excluded.

in number of contained species. In many respects it is the most remarkable of all the provinces, containing no peculiar species whatever, scarcely any peculiarly American ones, and but a scanty selection of European.

A further peculiarity is that the flora of its Temperate regions is extremely poor, and adds very few species to the whole flora, and with few exceptions, only such as are Arctic in Europe also. Being the only Arctic land that contracts to the southward, forming a peninsula, which terminates in the ocean in a high northern latitude, Greenland offers the key to the explanation of most of the phenomena of Arctic vegetation ; and as I have already made use of it for this purpose, I shall be more full in my description of its flora than of any other.

The east and west coasts of Greenland differ in many important features ; the eastern is the largest in extent, the least indented by deep bays, is perennially encumbered throughout its entire length by ice-fields and icebergs, which are carried south by a branch of the Arctic current that sets between Iceland and Greenland, and is hence excessively cold, barren, and almost inaccessible. The west coast again is generally more or less free from pack-ice from Cape Farewell (lat. 60°) to north of Upernavik in lat. 73°. It is washed by a southerly current, which is said to carry drift timber from the Siberian rivers into its fiords, and enjoys a far milder climate, and consequently has a more luxuriant vegetation.

A somewhat similar contrast is exhibited between West Greenland and the opposite shores of Baffin's Bay and Davis' Strait, because they may in some degree explain their differences of vegetation. There is also another difference between the polar islands and Greenland, inasmuch as the former are for the most part low, without mountains or extensive glaciers ; while the latter is exceedingly mountainous, with valleys along the shore terminating in glacier-headed fiords, and the coast is bound by glaciers of prodigious extent from Melville Bay to Smith's Sound.

The isothermal lines in Greenland all follow one course, from S.W. to N.E., running more parallel to one another in this meridian than in any other. The isotherm of 32° passes through the southern extremity of the peninsula, and that of 5° through its north extreme at Smith's Sound. The June isotherm of 41° skirts its east coast, and that of 32° passes north of Disco. The June temperature of Disco is hence as low as that of the north of Spitzbergen, of middle Nova Zembla, and of the extreme north of Asia ; and yet Disco contains quadruple their number of plants. The autumn cold is very great, the September isotherm of 32° crossing the Arctic Circle on the west coast ; and to this scantiness of the flora may to some extent be attributed.

The Arctic Greenland flora contains 206 species according to Lange's catalogue (in Rink's "Grönland"); or 207, according to my materials (Monocot. 67, Dicot. 140, = 1 : 2·1), the proportion of genera to species being 1 : 2. Of these 207 species, the following 11 alone are not European :—

Anemone Richardsonii (Asiatic).
Turritis mollis (Asiatic).
Vesicaria arctica (American only).
Draba aurea (Rocky Mountains and Labrador only).
Arenaria Grœnlandica (Mountains of U.S.).
Potentilla tridentata (Labrador only).
Saxifraga triscuspidata (Labrador only).
Erigeron compositus (American only).
Pedicularis euphrasioides (Asia).

On the other hand, no less than 57 Arctic-Greenland species are absent in Arctic East America, and the following 36 Arctic-Europe and Greenland species are either absent in all parts of Eastern Temperate America, or are extremely local there :—

Arabis alpina (Labrador only).
Lychnis alpina (Labrador only).
L. dioica (absent).
Spergula nivalis (absent).
Arenaria uliginosa (absent).
A. ciliata (absent).
Stellaria cerastioides (absent).
Alchemilla alpina (absent).
A. vulgaris (Labrador only).
Sibbaldia procumbens (United States only).
Rubus saxatilis (absent).
Potentilla verna (Labrador only).
Sedum villosum (absent).
Saxifraga Cotyledon (Labrador and Rocky Mountains only).
Galium saxatile (absent).
Gnaphalium sylvaticum (Labrador only).
G. supinum L. (Labrador and White Mountains only).
Cassiopeia hypnoides (Labrador only).
Phyllodoce taxifolia (Labrador and White Mountains).
Gentiana nivalis (Labrador only).
Thymus serpyllum (absent).
Veronica alpina (White Mountains only).
V. saxatilis (absent).
Euphrasia officinalis (N. U. States).
Bartsia alpina (Labrador only).
Rumex acetosella (absent).
Salix Arbuscula (absent).
Peristylus albidus (absent).
Carex capitata (White Mountains only).
C. microglochin (absent).
C. microstachya (absent).
C. pedata (absent).
Elyna caricina (Rocky Mountains only).
Phleum alpinum (Labrador and White Mountains only).
Calamagrostis lanceolata (Labrador only).
Deschampsia alpina (absent).

When it is considered how extremely common most of these plants are throughout Europe and Northern Asia, and that some of them inhabit also N.W. America, their absence in Eastern America is even more remarkable than their presence in Greenland.

Another singular feature of both Arctic and Temperate Greenland is its wanting a vast number of Arctic plants which are European, and are found also in America. The following is a list of most of these, excluding about 15, which are water-plants, or species whose range is limited. The letter "I." placed before a species signifies that it is *Icelandic*, and I have introduced it to show how many are absent from this island also, but how many are present. The letter "S." indicates that the species is found in the South Temperate or Antarctic Circle. The asterisk (*) indicates that the species is Arctic both in East America and Europe.

Anemone alpina.
A. nemorosa.
A. narcissiflora.
* Ranunculus Purshii.
*I. Caltha palustris.
* Aconitum Napellus.
Actæa spicata.
Nuphar luteum.
Nasturtium amphibium.
S. Barbarea præcox.
S. Turritis glabra.
Thlaspi montanum.
Sisymbrium Sophia.
*I. Erysimum lanceolatum.
Arabis hirsuta.
I.S. Cardamine hirsuta.
* Parrya artica.
I. Draba muralis.
I. Subularia aquatica.
*I. Drosera rotundifolia.
I. D. longifolia.
I. Viola tricolor.
*I. Arenaria laterifolia.
* Stellaria longifolia.
I. S. crassifolia.
Linum perenne.
Geranium Robertianum.
Hypericum 4-angulum.
Oxalis acetosella.
* Phaca frigida.
* Astragalus alpinus.
* A. hypoglottis.
* Oxytropis campestris.
O. Uralensis.
Lathyrus palustris.
Spiræa salicifolia.
S. Geum urbanum.
I. G. rivale.
* Rubus àrcticus.
Potentilla fructicosa.
P. Pennsylvanica.
P. argentea.
*I.S. Fragaria vesca.
I. Sanguisorba officinalis.
Rosa cinnamomea.
R. blanda.
* Circæa alpina.
*I.S. Epilobium tetragonum.
*I.S. E. alsinæfolium.
S. Lythrum salicaria.
* Ribes rubrum.
* R. alpinum.
*I. Parnassia palustris.
Saxifraga Sibirica.
* S. hieraciifolia.
S. bronchialis.
* Bupleurum ranunculoides.
Conioselinum Fischeri.
Cicuta virosa.
*I. Carum carui.
Adoxa moschatellina.
Viburnum Opulus.
Lonicera cærulea.
*I. Linnæa borealis.
*I. Galium boreale.
G. rubioides.
I. G. trifidum.
S. G. aparine.
* Valeriana capitata.
* Nardosmia frigida.
* Crysanthemum arcticum.
I. Pyrethrum nodosum.
P. bipinnatum.
* Artemisia vulgaris.
S. Bidens bipartita.
Tanacetum vulgare.
Antennaria Carpatica.
* Senecio resedæfolius.
* S. frigidus.
* S. palustris.
* S. campestris.
S. aurantiacus.
* Solidago Virga-aurea.
* Aster Sibiricus.
* A. alpinus.

S. Erigeron acris.
S. Sonchus arvensis.
I. Hieracium boreale.
* Saussurea alpina.
I. Vaccinium myrtillus.
* Andromeda polifolia.
Cassandra calyculata.
*I. Arctostaphylos alpina.
*I. Pyrola secunda.
* Gentiana amarella.
I. G. tenella.
* Myosotis sylvatica.
M. palustris.
I. M. arvensis.
* Sentellaria galericulata.
I.S. Prunella vulgaris.
Glechoma hederaceum.
S. Stachys palustris.
* Gymnandra Pallasii.
* Castilleja pallida.
I.S. Veronica officinalis.
S. V. scutellata.
I.S. V. serpyllifolia.
Melampyrum pratense.
M. sylvaticum.
*I. Pedicularis palustris.
* P. versicolor.
Scrophularia nodosa.
Utricularia vulgaris.
* Pinguicula villosa.
Glaux maritima.
Trientalis Europæa.
* Androsace septentrionalis.
* A. Chamæjasme.
Naumbergia thyrsiflora.
I.S. Primula farinosa.
I. Plantago major.
P. lanceolata.
S. Chenopodium album.
I.S. Atriplex patula.
Corispermum hyssopifolium.
* Pologonium Bistorta.
I. P. amphibium.
* Myrica Gale.
I. Betula alba.
I. B. pumila.
I. Alnus incana.
I. Salix pentandra.
I. S. myrtilloides.
I. Triglochin maritimum.
Scheuzeria palustris.
Veratrum album.
* Lloydia serotina.
* Allium schænoprasum.
* Smilacina bifolia.
* Platanthera obtusata.
* Calypso borealis.
Godyera repens.
Cypripedium guttatum.
Calla palustris.
Typha latifolia.
Narthecium ossifragum.
Luzula maxima.
S. Juncus communis.
I. J. articulatus.
I. J. bulbosus.
J. stygius.
Carex pauciflora.
C. tenuiflora.
S. C. stellulata.
I. C. chordorrhiza.
C. teretiuscula.
C. paradoxa.
S. C. Buxbaumii.
I. C. limosa.
S. C. Magellanica.
C. ustulata.
C. livida.
I. C. pallescens.
C. maritima.
I. C. cæspitosa.
I. C. acuta.
C. stricta.
C. filiformis.
I.S. Eleocharis palustris.
S. E. acicularis.
S. Scirpus triqueter.
S. S. lacustris.
Eriophorum alpinum.
Rhynchospora alba.
Alopecurus pratensis.
I. Milium effusum.
S. Phalaris arundinacea.
I.S. Phragmites communis.
*I. Hierochloe borealis.
* H. pauciflora.
*I. Catabrosa aquatica.
*I.S. Glyceria fluitans.
*I. Atropis distans.
I. Festuca elatior.
S. Bromus ciliaris.
I.S. Triticum caninum.
S. Hordeum jubatum.

Altogether there are absent in Greenland upwards of 230 Arctic-European species, which are all of them American plants. The most curious feature of this list is the absence throughout Greenland of the genera *Spiræa, Senecio, Astragalus, Trifolium, Phaca, Oxytropis, Androsace, Aster, Myosotis, Rosa, Ribes, Thlaspi, Sisymbrium, Geranium*, &c., and of such ubiquitous Arctic species as *Fragaria vesca, Caltha palustris*,* *Barbarea præcox.* It is remarkable that *Astragalineæ* are also absent from Spitzbergen and Iceland.

Iceland possesses 432 species (Monocot. 157, Dicot. 275), amongst which I find about 120 Arctic-European plants that do not enter Greenland; whereas only 50 of the European plants that inhabit Greenland are absent in Iceland. The more remarkable desiderata of Iceland are *Astragalineæ, Anemone, Aconitum, Braya, Turritis, Artemisia*, and *Androsace; Alopecurus alpinus, Luzula arcuata, Hierochloe alpina, Rubus chamæomorus, Cassiopeia tetragona, Arnica montana, Antennaria dioica*, and *Chrysoplenium alternifolium.* On the other hand Iceland contains of Arctic genera absent in Greenland, *Caltha* (one of the most common plants about Icelandic dwellings), *Cakile, Geranium, Trifolium, Spiræa, Senecio*, and *Orchis.*

But perhaps the most remarkable fact of all connected with the Greenland flora is that its Southern and Temperate districts, which present a coast of 400 miles extending south to lat. 60° N., do not add more than 74 species to its flora, and these are almost unexceptionally Arctic-European plants; and, inasmuch as these additional species increase the proportion of Monocotyledons to Dicotyledons of the whole flora, Greenland as a whole is botanically more Arctic in vegetation than Arctic Greenland alone is!

The only American forms which Temperate Greenland adds to its flora are, *Ranunculus Cymbalariæ, Pyrus Americana*, a very trifling variety of the European *Aucuparia, Viola Muhlenbergii* (a mere variety of *V. canina*), *Arenaria Grœnlandica* (a plant elsewhere found only on the White Mountains of New Hampshire), and *Parnassia Kotzebuei* (a species which is scarcely different from *palustris*).

The only plants which are not members of the Arctic flora elsewhere, and which are confined in Greenland to the Temperate zone, besides the above American plants, are *Blitum glaucum, Polamogeton marinus, Sparganium minimum*, and *Streptopus amplexifolius;* the rest will all be found in the column of the Arctic Plant Catalogue devoted to Greenland, where S. signifies that the species is found south only of the Arctic Circle in that country. On the other hand, Temperate Greenland adds very materially to the number of European-Arctic species that do not enter Eastern America (Arctic or Temperate), amongst which the most remarkable are—

Cerastium viscosum.	Sedum annuum.
Vicia cracca.	Galium uliginosum.
Rubus saxatillis.	G. palustre.

* This is the more remarkable because it forms a conspicuous feature in Iceland, and is a frequent native of all the Arctic-American coasts and islands.

Leontodon autumnale.
Hieracium murorum.
H. alpinum.
Gentiana aurea.
Betula alpestris.
Juncus trifidus.
J. squarrosus.
Anthoxanthum odoratum.
Nardus stricta.

Another anomalous feature in Greenland flora is the presence, on the East-Arctic coast, of some species not found on the West, nor in the Temperate southern end of the peninsula. These are *Lychnis dioica* (Arctic Europe), *Saxifraga Hirculus* (abundant in all extreme Arctic latitudes but West Greenland), *Polemonium cæruleum* (all Arctic longitudes but West Greenland), *Deschampsia cæspitosa* (all Arctic longitudes, but also absent in Spitzbergen).

For data connected with the Greenland flora, I am mainly indebted to the collections of the various polar voyagers in search of a North-west Passage, especially to Drs. Lyall's and Sutherland's; to Lange's catalogue in Rink's "Grönland"; and to the notices of Vahl, Greville, Sir William Hooker, &c.; to Sutherland's Appendix to Penny's Voyage, and Durand's to Kane's Voyage.

There is a curious affinity between Greenland and certain localities in America, which concerns chiefly a few of the European plants common to these countries. First, there are in Labrador, or on the Rocky Mountains, or White Mountains of New Hampshire, a certain number of European plants found nowhere else in the American continent. They are—

Ranunculus acris (Rocky Mountains).
Arabis alpina (Labrador).
Lychnis alpina (Labrador).
Sibbaldia procumbens (Rocky Mountains).
Potentilla verna (Labrador).
Montia fontana (Labrador).
Gnaphalium sylvaticum (Labrador).
G. supinum (Labrador and White Mountains).
Cassiopeia hypnoides (Labrador and White Mountains).
Phyllodoce taxifolia (Labrador and White Mountains).
Gentiana nivalis (Labrador).
Veronica alpina (White Mts.).
Bartsia alpina (Labrador).
Salix Arbuscula (White Mts.).
Luzula spicata (White Mts.).
Juncus trifidus (White Mts.).
Carex capitata (White Mts.).
Kobresia scirpina (Rocky Mts.).
Phleum alpinum (White Mountains and Labrador).
Calamagrostis lanceolata (Labrador).

There are also three plants, peculiar to Greenland and Labrador, or the White or Rocky Mountains, which have not hitherto been found elsewhere in America. They are—

Draba aurea (Rocky Mountains).
Arenaria Grœnlandica (White Mountains and Labrador).
Potentilla tridentata (Labrador).

V.—On the Arctic Proportions of Species to Genera, Orders, and Classes.

The observations which have hitherto been made on this subject are almost exclusively based on data collected on areas too small to yield general results. Especially in determining the influence of temperature in regulating the proportions of the great group of flowering plants, it is of the highest importance to take comprehensive areas, both because of the wider longitudinal dispersion of some orders, especially the Monocotyledons, and the effects of local conditions, such as bog land, which determine the overwhelming preponderance of Cyperaceæ in some Arctic provinces compared with others. The proportion of genera to species in the whole Arctic phænogamic flora is 323 : 762 or $1:2\cdot3\left(\begin{matrix}\text{Monocot.} & \frac{1\,:\,2\cdot8}{1\,:\,2\cdot2}\\ \text{Dicot.} & \end{matrix}\right)$, and that of orders to species 1 : 10·8 : in the several provinces as follow:—

——	Gen.	Gen. to Sp.	Orders.	Ord to Sp.
Arctic Europe - -	277	1:2·3	64	1:9·6
Asia - - -	117	1:2·0	38	1:6·1
West America - -	172	1:2·1	48	1:7·6
East America - -	193	1:2·5	56	1:6·8
Greenland - - -	104	1:2·0	38	1:5·5

Thus Europe presents the most continental character in its Arctic flora and West America the most insular; which may be attributable to the same cause in both, namely, the uniformity of variety of type. In West America we have, as in an oceanic island, a great mixture of types (Asiatic, European, East and West American) and paucity of species; in Europe the contrary. The proportions of species to orders are still more various; but here, again, Europe takes the lead decidedly. The proportions of genera and orders to species of all Greenland differ but little from those of its Arctic regions; whereas the contrast between Arctic Europe and this, together with Norway as far south as 60° N. lat., is very much greater. This is in accordance with the observation I have elsewhere made, that the whole of Greenland is comparatively poorer in species than Arctic Greenland is:—

	Gen. Sp.	Ord. Sp.
Arctic Scandinavia - -	1 : 2·3	1 : 9·6
All Scandinavia - - -	1 : 2·8	1 : 11·6
Arctic Greenland - -	1 : 2·0	1 : 5·5
All Greenland - - -	1 : 2·3	1 : 6·6

The proportions of Monocotyledons to Dicotyledons are—

Arctic Flora - - -	1 : 2·6
„ Europe - - -	1 : 2·3
„ Asia - - -	1 : 4·5
„ West America - -	1 : 3·8
„ East America - -	1 · 3·1
„ Greenland - -	1 : 2·1
All Greenland - - -	1 : 2·0

[A Table of the proportion of the largest Orders to the whole Flora (p. 276) is here omitted.]

The great differences between the proportions of largest Orders to the whole Flora show how little confidence can be placed in conclusions drawn from local floras. *Ericeæ* is the only order which is more numerous proportionally to other plants in every province than in the entire Arctic flora, and *Cruciferæ* is the only one that approaches it in this respect; and *Leguminosæ* is the only one which is less numerous proportionally in them all. East and West America agree most closely of any two provinces; then (excluding *Leguminosæ*) all Greenland and Europe; next Arctic Greenland and all Greenland. The greatest differences are between Arctic Europe and Asia, and Arctic Asia and West America; they are less between Arctic Greenland and Asia (excluding *Leguminosæ*); they are great between Arctic Greenland and East America; and as great between all Greenland and Arctic America.

The proportion formerly deducted by Brown, and others for the high Arctic regions was a much smaller one; the Monocotyledons being in comparison with the Dicotyledons 1 : 5; and this still holds for some isolated, very Arctic localities, as North-east Greenland; whereas Spitzbergen presents the same proportion as all the Arctic regions, 1 : 2·7; the Parry Islands 1 : 2·3; the west coast of Baffin's Bay, from Pond's Bay to Home Bay, 1 : 3·3; and the extreme Arctic plants mentioned at p. 205, 1 : 3. Of the prevalent Arctic plants mentioned at p. 203–4 the proportion is 1 : 3·4. I have dwelt more at length on these numerical proportions than their slight importance seems to require; my object being to show how little mutual dependence there is amongst the Arctic florulas. Each has profited but little through contiguity with its co-terminous districts, though all bear the impress of being members of one northern flora.

VI.—On the Grouping of Forms, Varieties, and Species of Arctic Plants for the Purposes of Comparative Study. Pages 276–281. [Not reprinted.]

VII.—Tabulated View of Arctic Flowering Plants, and Ferns, with their Distribution. Pages 281–309. [Abstract of two columns here given, with some additions.]

[Two columns only of the original table are here abstracted; one for "Arctic-Eastern America," and another for "Greenland." A few additions to these have been made by the Author from various Expeditions and other sources of information since 1860.]

The Arctic Flowering Plants and Ferns indicated by the following table are—1. Those from *East Arctic America* ("**EA** or **M**"), from Mackenzie River to Baffin's Bay: the "**M**" signifies that the plant extends to the islands north of Lancaster Sound, and to the Parry Islands, including Melville Island, the best explored of them. 2. Those from *Arctic Greenland* ("**G**," "**S**," "**E**," "**Ss**," or "**NE**"): the "**S**" indicates that the species has been found south only of the Arctic Circle in Greenland; the "**E**" refers to those found on the east coast only, the explored portions of which lie to the north of lat. 70°; the "**Ss**" stands for Smith's Sound; and the plants marked "**NE**," together with four of the eight marked "**E**," have been noted on the east coast by the "Second "German Polar Expedition."* *Alp.* means "Alpine in Europe."

DICOTYLEDONES.

Ranunculaceæ.

Thalictrum dioicum, L. **EA.**
alpinum, L. *Alp.* **G.**
Anemone patens, L. **EA.**
Richardsoni, Hk. **EA. G.**
parviflora, Mich. **EA.**
decapetala, L. **EA.**
Pennsylvanica, L. **EA.**
Ranunculus aquatilis, L. **EA. G.**
confervoides, Fr. **G.**
glacialis, L. *Alp.* **G. NE.**
Flammula, L. **EA. S.**
reptans, L. **G.**
Cymbalaria, Psh. *Alp.* **EA. S.**
auricomus, L. **M. G. NE.**
sceleratus, L. **EA.**
Purshii, Rich. **EA.**
nivalis, L. *Alp.* **M. G. Ss. NE.**
sulphureus, Sol. **G.**
acris, L. **S.**
Lapponicus, L. *Alp.* **EA. G.**
hyperboreus, Rottb. *Alp.* **EA. G.**
pygmæus, Wahl. **M. G. NE.**
hispidus, Mich. **EA.**
Pennsylvanicus, L. **EA.**
Caltha palustris, L. **M.**
Coptis trifolia, Sal. *Alp.* **S.**
Aquilegia Canadensis, L. **EA.**
brevistylis, Hook. **EA.**

Papaveraceæ.

Papaver alpinum, L. **M. G.**
nudicaule, auct. **Ss. NE.**

---

* Flowering Plants occurring on the East Coast only. Enumerated in Dr. Hooker's list:—

Lychnis dioica.	Polemonium cæruleum.
Saxifraga Hirculus.	Deschampsia cæspitosa.

Mentioned in the Appendix to the "Zweite deutsche Nordpolarfahrt":—

Cochlearia fenestrata.	Polemonium humile.
Saxifraga Hirculus, var. alpina.	Deschampsia brevifolia.

All the plants yielded by the more lately explored portion of the coast from 75° to 78° N. lat. are enumerated, with localities, further on.—Editor.

Corydalis glauca, Psh. **EA.**
pauciflora, Pers. **EA.**

Saraceniaceæ.
Saracenia purpurea, L. **EA.**

Cruciferæ.
Nasturtium palustre, DC. **EA. S.**
Barbarea vulgaris, Br. **EA.**
Turritis mollis, Hook. **EA. G.**
(Arabis) Holboellii, Horn. **G.**
Arabis hirsuta, L. **EA.**
alpina, L. *Alp.* **EA. G.**
petræa, Lamk. *Alp.* **EA. G. NE.**
Cardamine bellidifolia, L. *Alp.* **M. G. NE.**
hirsuta, L. **EA.**
pratensis, L. **EA. G.**
Parrya arctica, Br. *Alp.* **M.**
arenicola. **EA.**
Vesicaria arctica, Rich. **EA. G. Ss.*** **NE.**
Draba alpina, L. *Alp.* **EA. G. Ss. NE.**
var. *glabra* and *hispida.* **Ss.**
androsacea, Wahl. *Alp.* **M. G.**
Wahlenbergii, Hartm. **G. NE.**
corymbosa, Br. **G. S.**
muricella, Wahl. *Alp.* **EA. G. NE.**
nivalis, Lilj. non DC. **G.**
stellata, Jacq. non. DC. **EA.**
hirta, L. *Alp.* **EA. G.**
arctica, Vahl. **G. NE.**
incana, L. **EA. G.**
rupestris, Br. *Alp.* **EA. G. Ss. NE.**
aurea, Vahl. **G.**
Cochlearia Danica, L. **EA. G.**
Anglica, L. **EA. G.**
fenestrata, Br. **E.**
officinalis, L. **M. G. Ss.**
Hesperis Pallasii, T. & G. *Alp.* **EA. G. Ss.**
Sisymbrium Sophia, L. **EA.**
canescens, Nutt. **EA.**
humile, C. A. M. *Alp.* **EA.**
salsugineum, Pall. **EA.**
Erysimum hieraciifolium, L. **EA.**
cheiranthoides, L. **EA.**
Braya alpina, Sternb. *Alp.* **EA. G.**
(Platypetalum) purpurascens, Br. **G.**
Eutrema Edwardsii, Br. **M. G.**
Thlaspi montanum, L. **EA.**
Capsella bursa-pastoris, L. **S.**
Lepidium ruderale, L. **EA.**

---

* *Vesicaria arctica* and *Hesperis Pallasii* were found also in Washington Land, beyond Smith's Sound.

Droseraceæ.

Drosera rotundifolia, L. **EA.**

Violarieæ.

Viola palustris, L.
canina, L. **S.**
cucullata, Ait. **EA.**

Caryophylleæ.

Silene acaulis, L. *Alp.* **EA. G. Ss. NE.**
Pennsylvanica, Mich. **EA.**
Lychnis apetala, L. *Alp.* **M. G. Ss. NE.**
affinis, Vahl. **G. NE.**
triflora, Br. **G. NE.**
dioica, L. **E.**
alpina, L. *Alp.* **G.**
Sagina procumbens, L. **S.**
nodosa, E. M. **EA. S.**
nivalis, Fr. **G.**
Linnæi, Presl. **EA. G.**
saxatilis, Wimm. **G.**
Arenaria lateriflora, DC. **EA.**
formosa, Fisch. *Alp.* **EA.**
uliginosa, Schl. **G.**
(Alsine) stricta, Wahl. **G.**
Rossii, Br. **EA. G.**
Michauxii, Fenzl. **EA.**
verna, L. **M. G.**
rubella, Br. **G. NE.**
arctica, Stev. *Alp.* **EA. G.**
biflora, Wahl. **G. NE.**
ciliata, L. *Alp.* **G. NE.**
Grœnlandica, Spr. **G.**
Honkeneja peploides, Ehr. **EA. G.**
Merkia physodes, Fisch. **EA.**
Lepigonium salinum, Fr. **EA. G.**
Stellaria borealis, Big. *Alp.* **EA. G.**
humifusa, Rottb. *Alp.* **EA. G. Ss. NE.**
longipes, Goldie. *Alp.* **M. G. NE.**
Edwardsii, Br. **G. Ss.**
uliginosa, Murr. **EA. G.**
media, L. **EA. G.**
longifolia, Fries. *Alp.* **EA.**
crassifolia, Ehr. **EA.**
cerastioides, L. *Alp.* **G.**
Cerastium alpinum, L. *Alp.* **M. G. NE.**
viscosum, L. **S.**
vulgatum, L. **S.**
Fischerianum, Ser. **Ss.**

Balsamineæ.

Impatiens fulva, DC. **EA.**

Lineæ.
Linum perenne, L. **EA.**
Polygaleæ.
Polygala Senega, Willd. **EA.**
Leguminosæ.
Phaca frigida, L. *Alp.* **EA.**
Astralagus alpinus, L. *Alp.* **M.**
hypoglottis, L. **EA.**
Oxytropis campestris, DC. *Alp.* **EA.**
Uralensis, DC. *Alp.* **M.**
nigrescens, Fisch. *Alp.* **EA.**
deflexa, DC. *Alp.* **EA.**
Hedysarum boreale, Nutt. **EA.**
Mackenziei, Rich. **EA.**
Lathyrus maritimus, L. **EA. S.**
ochroleucus, Hook. **EA.**
Vicia Americana, Mühl. **EA.**
Cracca, L. **S.**
Lupinus perennis, L. **EA.**
Rosaceæ.
Alchemilla alpina, L. *Alp.* **G.**
vulgaris, L. **G. Ss. NE.**
Dryas octopetala, L. *Alp.* **M. G. Ss. NE.**
integrifolia, Vahl. **G. Ss.**
Drummondii, Rich. *Alp.* **EA.**
Geum urbanum, L. **EA.**
Sieversia Rossii, Br. **M.**
Sibbaldia procumbens, L. *Alp.* **G.**
Rubus arcticus, L. *Alp.* **EA.**
Chamæmorus, L. *Alp.* **EA. S.**
saxatilis, L. *Alp.* **S.**
Potentilla fruticosa, L. *Alp.* **G.**
anserina, L. **EA. G.**
nivea, L. *Alp.* **M. G. NE.**
Vahliana, L. **G.**
pulchella, Br. **G. Ss. NE.**
hirsuta, Vahl. **Ss.**
biflora, Lehm. **EA.**
frigida, Vill. *Alp.* **M. G.**
emarginata, Psh. **G. NE.**
verna, L. *Alp.* **G.**
maculata, Lehm. **G.**
tridentata, L. **EA. G.**
Comarum palustre, L. **S.**
Fragaria vesca, L. **EA.**
Sanguisorba officinalis, L. **EA.**
Rosa cinnamomea, L. **EA.**
blanda, Ait. **EA.**
Pyrus aucuparia, L. **S.**
Prunus Virginiana, DC. **EA.**
Amelancheir Canadensis, Torr. & Gray. **EA.**

Onagrarieæ.

Epilobium angustifolium, L. **EA. G.**
latifolium, L. *Alp.* **M. G. NE.**
alpinum, L. **EA. G.**
origanifolium, Lam. **G.**
palustre, L. **EA. S.**

Halorageæ.

Callitriche verna, L. **EA. S.**
Myriophyllum spicatum, L. **EA.**
alterniflorum, DC. **EA. S.**
Hippuris vulgaris, L. **EA. G.**
Ceratophyllum demersum, L. **EA.**

Portulaceæ.

Montia fontana, L. **G.**

Crassulaceæ.

Sedum Rhodiola, DC. *Alp.* **EA. G. NE.**
villosum, L. *Alp.* **G.**
annuum, L. **S.**

Grossularieæ.

Ribes lacustre, Pursh. **EA.**
rubrum, L. **EA.**
Hudsonianum, Rich. **EA.**

Saxifrageæ.

Mitella nuda, L. **EA.**
Chrysosplenium alternifolium, L. **M. G.**
Parnassia palustris, L. **EA.**
Kotzebuei, C. & S. **EA. S.**
Saxifraga cotyledon, L. *Alp.* **G.**
Aizoon, Jacq. **G.**
oppositifolia, L. *Alp.* **M. G. Ss. NE.**
cæspitosa, L. *Alp.* **M. G. NE.**
uniflora, Br. **Ss.**
cernua, L. *Alp.* **M. G. Ss. NE.**
rivularis, L. **M. G. Ss. NE.**
nivalis, L. *Alp.* **M. G. Ss. NE.**
Virginiensis, Mich. **EA.**
hieraciifolia, W. & R. *Alp.* **EA. NE.**
stellaris, L. *Alp.* **M. G.**
Hirculus, L. **M. E.**
alpina, Engler. **E.**
flagellaris, Willd. *Alp.* **M. G. Ss. NE.**
tricuspidata, Retz. **M. G. Ss.**
aizoides, L. *Alp.* **EA. G. NE.**
punctata, L. *Alp.* **EA.**
Heuchera Richardsonii, Br. *Alp.* **EA.**

Umbelliferæ.

Bupleurum ranunculoides, L. *Alp.* **EA.**
Conioselinum Fischeri, Wimm. *Alp.* **EA.**

Archangelica officinalis, DC. **G.**
Ligusticum Scoticum, L. **S.**
Ciccuta virosa, L. **EA.**
maculata, DC. **EA.**
Seseli divaricatum, Pursh. **EA.**

Corneæ.
Adoxa moschatellina, L. **EA.**
Cornus stolonifera, Mich. **EA.**
Canadensis, L. **EA.**
Cornus Suecica, L. *Alp.* **G.**

Caprifoliaceæ.
Viburnum Opulus, L. **EA.**
Lonicera cærulea, L. **EA.**
Linnæa borealis, L. **EA.**

Rubiaceæ.
Galium boreale, L. **EA.**
uliginosum, L. **S.**
triflorum, Muhl. **S.**
trifidum, L. **EA.**
palustre, L. **S.**
saxatile, L. **G.**

Valerianeæ.
Valeriana capitata, Willd. **EA.**

Compositæ.
Nardosmia frigida, Hk. *Alp.* **M.**
palmata, Hk. *Alp.* **EA.**
Achillea millefolium, L. **S.**
Ptarmica alpina, L. **EA.**
Chrysanthemum arcticum, L. **EA.**
integrifolium, Richd. **EA.**
Pyrethrum inodorum, Sm. **EA.**
Artemesia vulgaris, L. **EA.**
biennis, Willd. **EA.**
desertorum, Spr. **EA.**
borealis, Pall. *Alp.* **EA. G.**
Helenium autumnale, Hk. **EA.**
Antennaria alpina, L. *Alp.* **M. G.**
dioica, Br. **EA. G.**
Gnaphalium sylvaticum, L. **EA. G.**
Norvegicum, Gunn. **G.**
supinum, L. *Alp.* **G.**
uliginosum, L. **S.**
Arnica montana, L. *Alp.* **M. G.**
alpina, Læst. **S. NE.**
Senecio aureus, L. **EA.**
frigidus, Less. **EA.**
palustris, L. **M.**
campestris, L. **EA.**

Solidago Virga-aurea, L. **EA.**
Aster Sibiricus, L. **EA.**
salsuginosus, Rich. *Alp.* **EA.**
alpinus, L. *Alp.* **EA.**
multiflorus, Ait. **EA.**
Erigeron compositus, Pursh. *Alp.* **EA. G.**
Eriocephalus, Vahl. **NE.**
alpinus, L. *Alp.* **EA. G.**
Philadelphicus, L. **EA.**
Grindelia squarrosa, Duval. **EA.**
Taraxacum Dens-leonis, Desf. **M. G.**
ceratophorum, DC. **G.**
palustre, DC. **Ss.**
phymatocarpum, J. Vahl. **G. NE.**
Troximon glaucum, Nutt. **EA.**
Crepis nana, Rich. *Alp.* **EA.**
Sonchus arvensis, L. **EA.**
Leontodon autumnalis, L. **S.**
Mulgedium pulchellum, Nutt. **EA.**
Hieracium murorum, L. **G.**
alpinum, L. *Alp.* **S.**
umbellatum, L. **EA. S.**
Saussurea alpina, L. *Alp.* **EA.**

Campanulaceæ.

Campanula rotundifolia, L. **EA. G.**
arctica, Lange. **NE.**
linifolia, Henk. **Ss.**
uniflora, L. *Alp.* **M. G. NE.**

Vaccinieæ.

Vaccinium uliginosum, L. **EA. G. Ss. NE.**
oxycoccos, L. *Alp.* **EA. S.**
vitis-Idæa, L. *Alp.* **EA. G.**
Canadense, Kalm. **EA.**

Ericeæ.

Cassiopeia hypnoides, L. *Alp.* **G.**
tetragona, L. *Alp.* **EA. G. Ss. NE.**
Andromeda polifolia, L. **EA. G.**
Arctostaphylos Uva-ursi, Spr. **EA. G.**
alpina, Spr. *Alp.* **EA. G. NE.**
Diapensia Lapponica, L. *Alp.* **EA. G.**
Loiseleuria procumbens, L. *Alp.* **EA. G.**
Rhododendron Lapponicum, L. *Alp.* **EA. G. NE.**
Kalmia glauca, L. **EA.**
Ledum palustre, L. **EA. G.**
Grœnlandicum, Retz. **G.**
Phyllodoce taxifolia, Sol. *Alp.* **EA. G.**
Pyrola minor, L. **EA. G.**
secunda, L. **EA. G.**
rotundifolia, L. **EA. G. NE.**
grandiflora, Rad. **G. Ss.**
Grœnlandica, Horn. **G.**

Gentianeæ.

Gentiana amarella, L. **EA.**
aurea, L. **S.**
propinqua, Rich. **EA.**
detonsa, Fr. *Alp.* **EA. S.**
nivalis, L. *Alp.*
Pleurogyne rotata, Gr. *Alp.* **EA. G.**
Menyanthes trifoliata, L. **G.**

Hydrophylleæ.

Eutoca Franklinii, Br. **EA.**

Polemoniaceæ.

Polemonium cæruleum, L. **EA. E.**
humile, Willd. **E.**
Phlox Sibirica, L. *Alp.* **EA.**

Boragineæ.

Myosotis sylvatica, Hoffm. **EA.**
Mertensia maritima, Don. **EA. G.**
denticulata, Don. **EA.**
Virginica, DC. **EA.**

Labiatæ.

Thymus serpyllum, L. **G.**
Dracocephalum parviflorum, Nutt. **EA.**
Stachys palustris, L. **EA.**

Orobancheæ.

Boschniakia glabra, C. A. M. **EA.**

Scrophularineæ.

Limosella aquatica, L. **S.**
Gymnandra borealis, Pall. *Alp.* **EA.**
Castilleja pallida, Kth. *Alp.* **EA.**
Veronica alpina, L. *Alp.* **G.**
serpyllifolia, L. **EA.**
saxatilis, L. *Alp.* **G.**
Euphrasia officinalis, L. **EA. G. NE.**
Rhinanthus Crista-galli, L. **EA. S.**
Bartsia alpina, L. *Alp.* **G. Ss.**
Pedicularis capitata, Ad. **EA.**
Lapponica, L. *Alp.* **EA. G.**
euphrasioides, Ster. *Alp.* **EA. G.**
hirsuta, L. *Alp.* **EA. G. Ss. NE.**
Sudetica, L. *Alp.* **M. G.**
Langsdorffii, Fisch. **G.**
flammea, L. *Alp.* **EA. G.**
versicolor, Wahl. *Alp.* **EA.**

Lentibularineæ.

Utricularia vulgaris, L. **EA.**
minor, L. **EA. G.**
Pinguicula vulgaris, L. **EA. G.**
villosa, L. **EA.**

Primulaceæ.

Dodecatheon Meadia, L. **EA.**
Androsace septentrionalis, L. **EA.**
Chamæjasme, L. *Alp.* **EA.**
Douglasia arctica, Hk. *Alp.* **EA.**
Primula stricta, Horn. *Alp.* **EA. G.**
Sibirica, Jacq. *Alp.* **EA. S.**

Plumbagineæ.

Armeria vulgaris, Willd. **EA. G.**
Labradorica, Wallr. **Ss.**
Sibirica, Turc. **G. NE.**

Plantagineæ.

Plantago major, L. **EA.**
lanceolata, L. **EA.**
maritima, L. **G.**

Polygoneæ.

Kœnigia Islandica, L. *Alp.* **EA. G.**
Oxyria reniformis, Hk. *Alp.* **M. G. Ss.**
digyna, L. **NE.**
Rumex acetosa, L. **S.**
Acetosella, L. **G.**
aquaticus, L. **EA. S.**
salicifolius, Weinm. **EA.**
Polygonum Bistorta, L. **EA.**
viviparum, L. *Alp.* **M. G. Ss.**
aviculare, L. **S.**

Chenopodieæ.

Chenopodium album, L. **EA.**
maritimum, L. **EA.**

Elæagneæ.

Elæagnus argentea, L. **EA.**
Shepherdia Canadensis, Nutt. **EA.**

Santalaceæ.

Comandra livida, Rich. **EA.**

Empetreæ.

Empetrum nigrum, L. **EA. G. Ss. NE.**
rubrum, L.

Urticeæ.

Urtica dioica, L. **EA.**

Betulaceæ.

Betula papyracea, Ait. **EA.**
nana, L. **EA. G. Ss. NE.**
pumila, L. **EA.**
fruticosa, Pall. **S.**
Alnus viridis, DC. *Alp.* **EA. G.**
incana, Willd. **EA.**

Salicineæ.

Salix lanata, L. *Alp.* **EA. G.**
speciosa, H. & A. *Alp.* **EA.**
myrtilloides, L. **EA.**
cordata, Muhl. **EA.**
Arbuscula, L. *Alp.* **G.**
glauca, L. *Alp.* **EA. G.**
arctica, Br. *Alp.* **M. G. Ss. NE.**
alpestris, And. **EA. G.**
myrsinites, L. *Alp.* **EA. S.**
phlebophylla, And. **EA.**
reticulata, L. *Alp.* **EA. G.**
herbacea, L. *Alp.* **EA. G. Ss.**
polaris, L. *Alp.* **M.**
Populus tremuloides, Mich. **EA.**
balsamifera, L. **EA.**

Coniferæ.

Pinus Banksiana, Lamb. **EA.**
Abies alba, L. **EA.**
Picea nigra, L. **EA.**
Larix Americana, Mich. **EA.**
Juniperus communis, L.
Virginiana, L. **EA.**

Fluviales.

Triglochin maritimum, L. **EA.**
palustre, L. **S.**
Potamogeton rufescens, Schr. **S.**
pusillus, L. **S.**
gramineus, L. **S.**
Zostera marina, L. **S.**

Melanthaceæ.

Tofieldia palustris, L. *Alp.* **EA. G. Ss.**
borealis, Wahl. **G.**
coccinea, Richards. **EA.**
Zigadenus chloranthus, Rich. **EA.**

Liliaceæ.

Lloydia serotina, L. *Alp.* **EA.**
Allium Schœnoprasum, L. **EA.**

Smilaceæ.

Smilacina bifolia, Desf. **EA.**

Orchideæ.

Peristylus albidus, L. **G.**
Platanthera hyperborea, Lindl. **EA. G.**
Kœnigii, Lindl. **G.**
obtusata, L. **EA.**
Calypso borealis, L. **EA.**
Listera cordata, Br. **S.**
Corallorrhiza innata, L. **G.**
Spiranthes gracilis, Br. **EA.**

Cypripedium guttatum, Sw. **EA.**
humile, Salisb. **EA.**

Irideæ.
Sisyrinchium Bermudianum, L. **EA. S.**

Aroideæ.
Sparganium natans, L. **EA. S.**
simplex, Sm. **EA.**
Typha latifolia, L. **EA.**

Junceæ.
Luzula spadicea, DC. **G.**
parviflora, Desv. **G.**
campestris, Sm. **EA. G.**
congesta. **Ss.**
multiflora, Ehr. **G.**
spicata, Desv. *Alp.* **EA. G.**
arcuata, Hook. *Alp.* **M. G.**
hyperborea, Br. **G. NE.**
pilosa, Willd. **G.**
Juncus biglumis, L. *Alp.* **M. G. NE.**
triglumis, L. *Alp.* **EA. G. NE.**
castaneus, L. *Alp.* **EA. G. NE.**
arcticus, Willd. *Alp.* **EA. G.**
filiformis, L. **EA. S.**
trifidus, L. *Alp.* **S.**
squarrosus, L. **S.**
bufonius, L. **EA. S.**
polycephalus, Mich. **EA.**
articulatus, L. **EA. S.**

Cyperaceæ.
Carex dioica, L. **EA. G.**
gynocrates, Wimm. **EA.**
rupestris, All. *Alp.* **G. NE.**
nardina, Fr. **EA. G. NE.**
capitata, L. **G.**
microstachya, Ehr. **G.**
scirpoidea, Mx. **EA. G.**
Wormskioldiana, Horn. **G.**
canescens, L. **EA. G.**
curta, Good.; vitilis, Fr. **G.**
glareosa, Wahl. **EA. G.**
ursina, Dewey. **G.**
Heleonastes, Ehr. **EA. G.**
lagopina, Wahl. *Alp.* **G.**
festina, Dew. *Alp.* **EA. G.**
leporina, L. **EA.**
incurva, Light. **EA. G.**
stenophylla, Light. **S.**
alpina, Sw. *Alp.* **EA. G.**
holostoma, Drej. **G.**
atrata, L. *Alp.* **EA. S.**

Carex fuliginosa, St. E. Hpe. *Alp.* **M. G. NE.**
misandra, Br. **G.**
rariflora, Sm. *Alp.* **EA. G.**
Magellanica, Lam. **EA.**
ustulata, Wahl. *Alp.* **EA.**
podocarpo, Br. **EA.**
livida, Wahl. **EA.**
panicea, L. **EA. S.**
supina, Wahl. **EA. G.**
flava, L. **S.**
pedata, Warl. *Alp.* **G.**
capillaris L. **EA. G.**
salina, Wahl. **G.**
subspathacea, Wormsk. **NE.**
vulgaris, Fr. **EA. S.**
cæspitosa, L. **EA.**
rigida, Good. *Alp.* **EA. G. Ss.**
hyperborea, Drej. **G.**
aquatilis, Wahl. **M. G.**
pilulifera, L. **EA. S.**
vesicaria, L. **EA. G.**
pulla, Good. **G.**
ampullacea, Good. **EA. G.**
oligosperma, Mich. **EA.**
Kobresia scirpina, Willd. *Alp.* **EA. G.**
(Elyne) spicata, Schrad. **NE.**
caricina, Willd. *Alp.* **EA. G. NE.**
Elæocharis palustris, Br. **EA. S.**
Scirpus triqueter, L. **EA.**
cæspitosus, L. **M. G.**
Eriophorum capitatum, Host. *Alp.* **M. G.**
Schenchzeri, Hpe. **G. NE.**
vaginatum, L. **EA. G.**
polystachyum, L. **M. G. NE.**
angustifolium, Rth. **G.**

Gramineæ.

Alopecurus alpinus, L. *Alp.* **M. G. NE.**
geniculatus, L. **EA. G.**
Phleum alpinum, L. *Alp.* **G.**
Phalaris arundinacea, L. **EA.**
Agrostis rubra, L. *Alp.* **G.**
vulgaris, L. **EA. S.**
canina, L. **G.**
Deyeuxia Canadensis, P.B. **EA.**
lapponica, Vahl. **EA. G.**
neglecta, Rupr. **EA.**
varia, P.B. **EA. G.**
strigosa, Wahl. **G.**
Calamogrostis lanceolata, Roth. **G.**
purpurascens, Br. **G. NE.**
phragmitoides, Hartm. **G.**

Spartina cynosuroides, W. **EA.**
Anthoxanthum odoratum, L. **S.**
Hierochloe borealis, L. **Ss.**
alpina, L. **M. G. NE.**
pauciflora, Br. **M.**
Deschampsia cæspitosa, P. B. **M. E.**
brevifolia, Br. **E.**
atropurpurea, Wahl. **EA. S.**
alpina, L. *Alp.* **G.**
flexuosa, L. **S.**
Trisetum subspicatum, P. B. *Alp.* **M. G. NE.**
Phippsia algida, Br. **M. G. NE.**
Catabrosa aquatica, P. B. **EA. S.**
vilfoidea, And. **G.**
Colpodium latifolium, Br. **M. G. NE.**
pendulinum, Læstd. **S.**
Dupontia Fisheri, Br. **M. G.**
Glyceria fluitans, Br. **EA. G.**
arctica, Hk. **Ss.**
Vahliana, Th. Fr. **G.**
Pleuropogon Sabini, Br. **M.**
Atropis maritima, L. **EA. G.**
(Poa) angustata, Br. **G.**
Poa annua, L. **S. NE.**
alpina, L. *Alp.* **EA. G.**
pratensis, L. **EA. G.**
nemoralis, L. **M. G.**
cæsia, Sm. **G. NE.**
flexuosa, Wahl. *Alp.* **M. G.**
Cenisea, All. **G.**
arctica, Br. **Ss. NE.**
abbreviata, Br. **G. NE.**
Festuca Richardsoni, Hk. **EA.**
ovina, L. **M. G.**
rubra, L. **G.**
brevifolia, Br. **G. NE.**
Bromus ciliatus, L. **EA. S.**
Triticum repens, L. **EA. G.**
violaceum, Horn. **G.**
Elymus arenarius, L. **EA. G.**
mollis, Trin. **EA.**
Hordeum jubatum, **EA.**
Nardus stricta, L. **S.**

ACOTYLEDONES.

Filices.

Polypodium Dryopteris, L. **S.**
Rhæticum, L. **S.**
Phegopteris, L. **S.**

Woodsia Ilvensis, Br. *Alp.* **EA. G. NE.**
hyperborea, Br. *Alp.* **EA. G.**
glabella, Br. *Alp.* **EA. G.**
Cistopteris fragilis, Bernh. **EA. G. NE.**
Lastrea fragrans, Sw. **EA. G.**
filix-mas, Sw. **S.**
Polystichum lonchitis, L. *Alp.* **EA. G.**
Cryptogramma acrostichoides, R. *Alp.* **EA.**
Botrychium lunaria, Sw. **G.**
virginianum, Sw. **S.**

Lycopodiaceæ.

Lycopodium selago, L. **EA. G.**
annotinum, L. **EA. G.**
clavatum, L. **S.**
selaginoides, L. **S.**
alpinum, L. **EA. G.**
Isoetes lacustris, L. **S.**

Equisetaceæ.

Equisetum palustre, L. **EA.**
variegatum, L. **EA. G.**
arvense, L. **EA. G. NE.**
sylvaticum, L. **G.**
scirpoides, Mich. **EA. G. NE.**

VIII.—OBSERVATIONS ON THE SPECIES. Pages 310–348. [Not reprinted.]

XXVIII.—CRYPTOGAMIC PLANTS from BAFFIN'S BAY (Lat. 70° 31′ to 76° 12′ on the East Side, and at Possession Bay, Lat. 73° on the West Side). By ROBERT BROWN. 1819.

[From CAPTAIN JOHN ROSS'S "Voyage of Discovery," &c., 2nd edit., 2 vols. 8vo. London, 1819. Vol. ii., Appendix, pp. 194–5.]

Lycopodium Selago, L.
Polytrichum juniperinum, Hooker & Taylor.
Orthotrichum cupulatum, H. & T.
Trichostomum lanuginosum, H. & T.
Dicranum scoparium, H. & T.
Mnium turgidum, Wahl.
Bryum, sp.
Hypnum aduncum, L.
Jungermannia, sp.
Gyrophora hirsuta, Achar.
G. erosa, Ach.
Cetraria Islandica, Ach.
C. nivalis, Ach.
Cenomyce rangiferina, Ach.
C. fimbriata, Ach.
Dufurea ? rugosa, n. sp.
Cornicularia bicolor, Ach.
Usnea ? sp. nov.
Ulva crispa, Lightf.
"Red Snow," N. lat. 76° 25′, W. long. 65°.

[*See further on*, DR. W. L. LINDSAY'S Catalogue of Greenland Lichens.]

XXIX.—Flowering Plants and Algæ of Greenland, Davis' Strait, and Baffin's Bay, collected by Dr. P. C. Sutherland, and determined by Sir W. J. Hooker and G. Dickie, M.D., Professor of Natural History, Queen's College. Belfast. 1853.

[From Commander E. A. Inglefield's "Summer Search (in 1852) " for Sir John Franklin," &c. 8vo. London, 1853. Appendix, pp. 133–144.]

I.—The Flowering Plants and a Fern, named by Sir W. J. Hooker, have been incorporated in the foregoing List of Arctic Plants by Dr. J. D. Hooker, C.B., Pres. R.S., &c. (*See* pages 225–238.)

II.—The Algæ, named by Dr. Dickie.

1. **Melanospermeæ.**

Fucaceæ.

Fucus vesiculosus, L. Hunde Islands, 40–50 fathoms; floating near Beechy Island (Barrow Strait); on the beach, Whale Sound. The specimens nearly all destitute of vesicles.

F. nodosus, L. Fiskernaes and Whale [Whale-fish ?] Island; and floating in 70° 50′ N.

Sporochnaceæ.

Desmarestia viridis, Lam. Hunde Isl., 50–100 fathoms.

D. aculeata, Lam. Fiskernaes; Hunde Isl., 80–100 fathoms; Whale Isl.; floating in 73° 20′ N.

Laminariaceæ.

Alaria esculenta, Grev. [Pylaii ?]. On the beach, Whale Sound. Large; some of the fronds upwards of 6 inches broad.

Laminaria fascia, Ag. Hunde Isl., 40–50 fathoms.

L. saccharina, Lam. Hunde Isl., 50–100 fathoms.

L. longicruris, De la Pyl. Melville Bay; Whale Sound; Cape Saumarez; floating off Dark Head, Greenland (lat. 72° 15′ N.), upwards of 10 feet in length, and their roots abounding with animal forms peculiar to deep water.

L. digitata, Lam. Whale Sound.

Agarum Turneri, Post. & Rupr. Hunde Isl., 10–100 fathoms; Whale Isl., 40–50 fathoms; Melville Bay.

Dictyotyaceæ.

Dictyota fasciola, Lam. Hunde Isl., 40–50 fathoms; Whale Isl., 20–40 fathoms.

Dictyosiphon fœniculaceus, Grev. Hunde Isl., 50–70 fathoms; and floating in lat. 73° 20′ N.

Asperococcus Turneri, Hook. Fiskernaes.

Chordariaceæ.

Chordaria flagelliformis, Ag. Fiskernaes; Hunde Is., 40–100 fathoms; Whale Isl.; Melville Bay.

Elachista fucicola, Fries. Fiskernaes; Whale Isl.

Elachista flaccida, Aresch (?). On *Desmarestia aculeata*, Whale Isl.

Myrionema strangulans, Grev. A minute plant, probably identical with this species, was found infesting *Callithamnion Rothii*, at low-water-mark, Hunde Isl.

Ectocarpaceæ.

Chætopteris plumosa, Kutz. Hunde Isl., 25–30 fathoms; on the beach, Whale Sound.

Ectocarpus littoralis, Lyngb. Fiskernaes; Hunde Isl., 50–100 fathoms; and floating in lat. 73° 20′ N.

E. Durkeei, Harv. (?). Fragments apparently of this species, mixed with the following.

E. Landsburgii, Harv. Hunde Isl., 70–80 fathoms.

2. **Rhodospermeæ.**

Rhodomelaceæ.

Polysiphonia nigrescens, Grev. Fragments apparently of this variable species were found at Hunde Isl., 40–50 fathoms; and cast up in Whale Sound.

Corallinaceæ.

Melobesia polymorpha, Linn. Erebus-and-Terror Bay, in 15 fathoms.

M. fasciculata, Harv. Erebus-and-Terror Bay, 8–10 fathoms.

M. lichenoides, Borl. At low-water-mark, Fiskernaes; Hunde Isl., 7 fathoms; Cape Adair, 12–18 fathoms.

Sphærococcoideæ.

Delesseria sinuosa, Lam. Dark Head.

D. angustissime, Griff. Whale Isl.

Calliblepharis ciliata, Kutz. On the beach, Whale Isl.

Squamarieæ.

Peyssonnelia Dubyi, Crouan. Cape Adair, 12–15 fathoms, on stones.

Rhodymeniaceæ.

Euthora cristata, J. Ag. Hunde Isl., 90–100 fathoms.

Cryptonemiaceæ.

Callophyllis laciniata, Kutz. Whale Isl., floating and on beach.

Halosaccion ramentaceum, J. Ag. Whale Isl., cast up.

Ceramiaceæ.

Ptilota serrata, Kutz. Whale Isl., 30–40 fathoms; Whale Sound, floating.

Callithamnion Rothii, Lyngb. Hunde Isl., low-water-mark; Cape Adair, on stones dredged in 12–18 fathoms.

3. **Chlorospermeæ.**

Confervaceæ.

Cladophora Inglefieldii, n. s. Low-water-mark, Fiskernaes.

C. rupestris, Kg. Low-water-mark, Fiskernaes.

C. arcta, Kg. Low-water-mark, Fiskernaes.

C. uncialis, Harv. Omenak, and Whale Sound.

Conferva melagonium, Web. & Mohr. Cape Bowen; Whale Sound; and Beechy Island (Barrow Strait).

C. sp., probably near *C. youngana*; fragments. Cape Bowen; Hunde Isl., 25–30 fathoms.

Conferva capillaris, L. Freshwater pools, Hunde Isl.
C. bombycina, Ag. Pools, Hunde Isl.

Ulvaceæ.

Enteromorpha intestinalis, Link. Hunde Isl.; Cape Bowen.
E. percursa, Hook. Hunde Isl., beach.
Ulva latissima, Linn. Low-water-mark, Omenak.
U. crispa, Lightf. [? *Prasiola fluviatilis*, Somm.] * Whale Isl.
Porphyra vulgaris, Ag. Whale Sound.

Nostochineæ.

Nostoc sphæricum, Vauch. In freshwater pools, Hunde Isl.

Diatomaceæ.

Fragments of a minute species of *Schizonema*, too imperfect for recognition, were found on Drift Wood in lat. 62° N., long. 51° W.; also on stones at Cape Bowen and Whale Sound.

XXX.—Plants from West Greenland and Smith Sound, collected by Dr. E. K. Kane, U.S.N., and determined by Elias Durand and T. B. James. (From E. K. Kane's "Arctic Explorations in the Years 1853–1855," vol. ii., Appendix XVIII., pp. 442–467; 1856.)

[For the Phænogamous Plants and Equiseta, *see* Dr. Joseph D. Hooker's List, above, p. 225.]

Filices (p. 464).

Polypodium phegopteris? lat. 65° N. lat.
Woodsia Ilvensis, R. Br. 64° and 72° N. lat.
Cystopteris fragilis, Bernh. As high up as N. lat. 76°.
C. fragilis, var. dentata, Hooker (?). N. lat. 80°.

Lycopodiaceæ.

Lycopodium selago, L.
L. annotinum, L.
L. alpinum, L.

Musci. (T. B James.)

Sphagnum squarrosum, Pers.
S. acutifolium, Ehrt.
S. recurvum, Brid.
Tetraplodon mnioides, Bruch & Schimper.
Splachnum vasculosum, L.
S. Wormskioldii, Horn.
Bryum lucidum, James, n. s.
B. Muhlenbeckii, B. & Sch.
Aulacomnion turgidum, Schwæg.
Polytrichum juniperinum, Hedw.
Dicranum scoparium, β. orthophyllum, B. & Sch.
Dicranum elongatum, Schw.
D. virens, Hedw.
D. virens, β. Wahlenbergii, B. & Sch., and another var.
D. Richardsoni, Hook.
D. Muhlenbeckii, B. & Sch.
D. 2 spp.
Racomitrum lanuginosum, Bird.
Weissia crispula, Hedw.
Hypnum riparium, L.
H. uncinatum, Hedw.
H. cordifolium, Hedw. and var.
H. stramineum, Dickson.
H. sarmentosum, Vahl.
H. Schreberi, Willd.

---

* *See* Trans. Bot. Soc. Edinb., vol. ix., p. 426; and *further on*, p. 281.

HEPATICEÆ. (T. P. James.)

Ptilidium ciliare, Nees.
Sarcocyphus Ehrharti, Cord.
Jungermannia divaricata, E. B.
J. squarrosa, Hook.

THALLOPHYTES. (T. P. James.)

Cetraria Islandica, Ach.
Peltigera canina, Hoffm.
Cladonia pyxidata, Fries.
Cladonia rangiferina, Hoffm.
C. furcata, Floerk.
C. sp.

XXXI.—NOTICE of FLOWERING PLANTS and FERNS collected on both Sides of DAVIS' STRAIT and BAFFIN'S BAY. By MR. JAMES TAYLOR, ABERDEEN.

[Reprinted, by Permission, from the Transactions of the Botanical Society of Edinburgh, vol. vii., 1862, pp. 323–334. Read 13 March 1862.]

The Plants named in the following list were collected by me in the course of five voyages made to Davis Strait, &c., as surgeon on board whaling vessels in the years 1856–61. Some of these years were more favourable for making such collections than others. My time was often very limited, and the ground I could explore much circumscribed by the short stay of the vessels at particular localities. The vessels usually remained longest in Cumberland Gulf, and accordingly the districts round about it have been the most completely investigated. But with longer time and more means at my disposal for making protracted excursions into the interior, I have no doubt that the subjoined list could have been very greatly increased. I have here given only the Flowering Plants and Ferns. I collected a great many Mosses and Lichens, but they have not yet been thoroughly examined. On the east side of Davis Strait and Baffin's Bay, I have had opportunity of exploring parts of the country from Disco Island to Wilcox Point; and on the west side my observations have extended with some intervals, from Cumberland Inlet to Cape Adair, a little north of Scott's Bay. It may be explained, that in the list, E., or E. side, means Danish Greenland, and W., or W. side, the islands lying to the west of Davis Strait and Baffin's Bay, forming part of the Arctic islands of North America; also, when any particular place is named, it is to be understood as including the district surrounding it. To obviate the necessity of giving latitudes and longitudes in the list, it seems advisable to give here the latitude and longitude of the principal places named:—

East side.	N. Lat.	W. Long.
Disco island, - - -	69° 10′	54° 30′
Hasen or Hare Island, - -	70° 30′	54° 15′
Dark Head or Svarthuk, -	71° 40′	56° 0′
Upernavik, - - - -	73° 25′	57° 26′
Wilcox Point, - - -	74° 18′	58° 8′

West side.	N. Lat.	W. Long.
Cape Enderby, - - -	63° 45′ -	64° 30′
Cape Mercy (of Davis), -	65° 10′ -	64° 40′

According to the maps, the latitude and longitude of these two places are—Cape Enderby, lat. 63° 45′, long. 67°; Cape Mercy, lat. 65°, long. 63° 20′.

Niatolik (Nawaktolik), -	65° 50′ -	65° (68′?)
Cape Searle, - - -	67° 20′ -	62° 30′
Scott's Inlet, - - -	71° 10′ -	71° 0′
Cape Adair, - - -	71° 20′ -	72° 0′

The Kickertine Islands, and the islands called Midliattwack, are in the middle of Cumberland Gulf. They are composed of metamorphic rocks, which rise in Midliattwack to the height of 557 feet, and in the Kickertines to that of 450 feet. These measurements, as well as those given in the list, were all made by means of the aneroid. To give an idea of the temperature of an ordinary fine day in these latitudes, and show the conditions under which Arctic vegetation makes so rapid a growth, I subjoin the temperature of one of the Kickertine Islands at various altitudes, on the 20th August 1861, when there was a clear sky, a bright sun, and little wind:—At 50 feet, exposed thermometer 69° Fahr.; in shade, 48°·5; sunk 1½ foot in soil, 45°; water of a small lake, 58°. At 100 feet in a valley, exposed thermometer, 70°; in shade (a little more wind), 46°; sunk 1½ foot, in somewhat moist soil, 44°. At 200 feet, exposed thermometer, 58°; in shade, 51°·5; sunk 1½ foot in sandy soil, 45°. At 450 feet, exposed thermometer, 62°; in shade, 41°; sunk 9 inches in sandy soil, 48°. Also, at one of these islands a thermometer was sunk 22 inches in a gravelly soil, and examined every two hours for twenty-four hours. The mean of all the observations was 42°·38.

The following is a list of the Flowering Plants and Ferns collected:—

Ranunculaceæ.

Ranunculus affinis, Br.—Flowers in August. Coast to 500 feet. Soil granitic. W. side, at Kingnite, Cumberland Gulf. Grows to about the height of 18 inches.

R. nivalis, L.—Fl. in June. Sea to snow-line. Perennial; on any soil, but most luxuriant on volcanic. E. and W.

R. hyperboreus, Rottb.—Fl. June and August. Alt. 200 feet. E., Disco, Dark Head, and Wilcox Point. W., Cumberland Gulf, Cape Searle, Scott's Bay, Cape Adair. Grows in small pools of water, the depth of which modifies its appearance.

R. pygmæus, Wahl.—Fl. June to August. Range same on both sides. Alt., Sea to 1000 feet. Any soil, and grows in small tufts of from 6 to 12 plants. E., Disco and Hassen Islands, Dark Head, Danish Head, and Wilcox Point. W., Cumberland Gulf, Cape Searle, Scott's Bay, &c.

R. sulphureus, DC.—Fl. June to Sept. Range as in last species. Alt., Sea to 200 feet. Any soil. Flower often white, and the whole plant is often under the snow, except the flower stalk; petals very deciduous, and the seeds are often not shed till next spring. E., Dark Head and Danish Head. W., Kingnite, Cumberland Gulf, Cape Searle, Scott's Bay, Cape Adair.

Papaveraceæ.

Papaver nudicaule, L.—Fl. June to Sept. Alt., Sea to 1,500 feet. Any soil, but chiefly on glacier-drift of a clayey nature; and amongst animal refuse, where Esquimaux huts have been. Flower sometimes of an inky blue colour, with yellow flowers often on the same plant. The natives appear to make no use of this species. Common on both sides.

Cruciferæ.

Arabis alpina, L.—Fl. June to Sept. Alt., Sea to 500 feet. Range limited to the following localities; on trap in loose soil, and associated with few other plants. E., Disco, Dark Head. W., Cape Searle.

Cardamine bellidifolia, L.—Fl. June to Aug. Range, 64° to 74°. Alt., Sea to 1,500 feet. Mossy soil, amongst *Sphagna*, *Cyperaceæ*, &c. E., Disco and Hassen Islands, Dark Head, Horse Head, Wilcox Point. W., Cumberland Gulf, Cape Searle, Scott's Bay, &c.

C. pratensis, L.—Fl. June and July. Alt., 200 feet. E., Disco.

Draba glacialis, Adams.—Fl. June. Alt., 500 to snow-line. On trap soil, growing singly, with long filiform roots running deep into the soil; appearance much modified by elevation and exposure. E., Dark Head. W., Cape Searle.

D. hirta, L.—Fl. June to Aug. Alt., Sea to 2,000 feet. In common with some other Drabas, it sends up two sets of flowering stems, in the earlier and later parts of the season respectively; the siliculæ of the latter generally retaining their seed during the winter. E., Dark Head, Upernavik, Horse Head, Wilcox Point. W., Cumberland Gulf, Capes Searle and Adair.

D. rupestris, Br.—Fl. June to Aug. Alt., Sea to 1,000 feet. On any rocky and granitoid soil. E., Dark Head, Danish Head, Wilcox Point. W., Kingnite, Cumberland Gulf, Capes Searle and Adair.

D. muricella, Wahl.—Fl. June and July. Alt., Sea to 500 feet. Soil granitoid. Perennial. E., Wilcox Point. W., Cape Searle, Scott's Bay.

D. stellata, Jacq.—Fl. June and July. Soil granitic. E., Wilcox Point.

D. lapponica, DC.—Fl. June, July. Alt., 1,000 feet. Soil granitoid. Very variable. E., Wilcox Point, Dark Head. W., Cape Searle, Scott's Bay.

Vesicaria arctica, Rich.—June. Alt., 500 feet. Soil granitic. Roots sinking deep. E., Wilcox Point.

Cochlearia officinalis, L.—Fl. June to Aug. Alt., Sea to 200 feet. Grows very profusely. The varieties *fenestrata*, *arctica*, and *anglica* are not so common. E., Hassen Island, Dark Head, Upernavik, Horse Head, Duck Islands, Wilcox Point. W., Cumberland Gulf, Cape Searle, Scott's Bay.

Caryophyllaceæ.

Silene acaulis, L.—May to July. Alt., Sea to snow-line. Common on both E. and W. side.

Lychnis apetala, L., and varieties.—Fl. June and July. Alt., Sea to 1,000 feet. Any moist soil. Some specimens are but an inch in height, and covered with long hairs; others nine inches, branching freely, and glabrous; flowers pink or white, the former colour most frequent on trap soils, where also the whole plant had a reddish appearance; the latter on granitic soils, the plant being of a dark green. E., Wilcox Point. W., Cape Searle, Midliattwack, and Niatoling.*

L. alpina, L.—June and July. E., Disco.

Honkeneja peploides, Ehr.—July, Aug. Alt., Sea to 50 feet. On the coast, but was also collected about three miles up a river, at the Winter Harbour, Kingnite, on an old sea beach, now raised about twenty feet above spring tides; while it also grew on the present beach, just below, in plenty. W., Kingnite, Cumberland Gulf, Kickertine Island, Cape Searle.

Arenaria verna, L., var. *rubella*, Br.—Fl. June to Aug. Alt., 500 feet (?). Soil granitic. In crevices of rocks. W., Winter Harbour, Kingnite, Cumberland Gulf.

A. Rossii, Br.—July, Aug. Alt., 200 to 1,000 feet. Most frequent in trap soil, moistened by melting snow. E., Hassen Island, Dark Head. W., Kickertine Island, Cumberland Gulf, and Cape Searle.

A. arctica, Stev.—July, Aug. Alt., Sea to 1,000 feet. On any soil. E., Dark Head, Upernavik, Horse Head, Wilcox Point. W., Cumberland Gulf, Cape Searle, Scott's Bay, &c.

Stellaria Edwardsii, Br.—July, Aug. Sea to 500 feet. Soil granitic. W., Winter Harbour, Kingnite, Kickertine Island, Niatolik, Cumberland Gulf.

S. stricta, Br.—June to Aug. Sea to 500 feet. Found in great profusion about the ruins of Esquimaux settlements. E., Women's Island, Duck Island, and Wilcox Point. W., Cape Searle, Scott's Bay, and along the coast.

S. longipes, Goldie.—July, Aug. W., Niatolik, Cumberland Gulf.

S. humifusa, Rottb.—July, Aug. Plentiful in sandy beaches. E., Dark Head, Women's Island. W., Capes Adair, Searle, Kickertine Island, Cumberland Gulf.

S. læta, Rich.—July, Aug. W., Niatolik, Cumberland Gulf, Cape Searle.

Cerastium alpinum, L.—May to Aug. Sea to snow-line. Varieties not unfrequently occur. Very common on both sides.

C. trigynum, Fries.—July. Alt., 1,000 feet. E., Disco.

Rosaceæ.

Dryas octopetala, L. (*integrifolia*, Vahl).—June, July. Alt., 1,000 feet. E., Disco, &c. W., Cape Searle, Cumberland Gulf, &c.

* "Niatoling" is the name applied to the *district* round the station of Niatolik.

Potentilla tridentata, L.—July, Aug. Alt., 300 feet. In crevices of granite rocks. W., Niatolik, Midliattwack Islands, Cumberland Gulf.

P. emarginata, Psh.—July, Aug. Alt., 500 feet. W., Niatolik, Kickertine Island, and Kingnite, Cumberland Gulf.

P. nivea, L.—June to Aug. Sea to snow-line. The specimens sent to Professor Balfour did not appear to him to be the true *nivea*. Are they nearer var. *pulchella* of Br.? E., Dark Head, Women's Island, Horse Head, Wilcox Point. W., Kickertine, Midliattwack, Niatolik Islands, Cumberland Gulf, Cape Searle, Scott's Bay, Cape Adair.

P. Vahliana, L.—July, Aug. Coast to snow-line. A very variable plant, giving rise to many of the varieties of authors. Thus, *P. sericea* seems a two-flowered form, while *P. hirusta*, Vahl., and *P. Jamesoniana*, Grev., are also varieties of it. E., Hassen Island, Dark Head, Horse Head, Wilcox Point. W., Cumberland Gulf, Cape Searle, Scott's Bay.

Onagraceæ.

Epilobium alpinum, L.—July. E., Disco.

E. latifolium, L.—June to Aug. Alt., 1,000 feet. Any soil; spreads much, but in some places seldom flowers. In warm valleys, in a southern exposure, it grows luxuriantly; in such places, I have several times found the exposed thermometer to indicate 80° to 90°. The highest temperature I ever observed in these regions was 106° Fahr. E., Hassen Island, Dark Head, Women's Island, Wilcox Point. W., Cumberland Gulf, Cape Searle, Scott's Bay, &c.

E. angustifolium, L.—Aug., Sept. Alt., 1,000 feet. Only found in the locality indicated, where it occupied a large space of ground amongst *Salix arctica*. W., North side of Winter Harbour, Kingnite, Cumberland Gulf.

Haloragraceæ.

Hippuris vulgaris, L.—Fl. Aug. Range limited. Alt., 100 feet. In small pools, to a temperature of 56° Fahr. W., Kickertine Islands, Cumberland Gulf.

Saxifragaceæ.

Saxifraga oppositifolia, L.—Fl. May to July. Alt., 1,500 feet. E., Disco, Hassen Islands, Dark Head, Wilcox Point. W., Cumberland Gulf, Cape Searle, Scott's Bay, &c.

S. tricuspidata, Retz.—Fl. June to Sept. Alt., Snow-line. Any soil, and may reach a foot in height. E., Disco, Hassen Islands, Dark Head, Women's Islands, Horse Head, Wilcox Point. W., Cumberland Gulf, Capes Searle and Adair.

S. Aizoon, Jacq.—Aug., Sept. Alt., 300 feet. In clefts of granitic rocks. W., Kingnite, Middliattwack Islands, and Niatoling in Cumberland Gulf.

S. nivalis, L.—Fl. July, Aug. Alt., Coast to 1,000 feet. Best in damp soil, mossy. E., Dark Head and Wilcox Point. W., Cumberland Gulf, Scott's Bay, Cape Searle.

S. cernua, L.—Aug. Coast to 200 feet. By the sides of

rivulets, amongst mosses, &c. W., Kickertine Islands, Kingnite, Cumberland Gulf, Scott's Bay.

S. rivularis, L.—July, Sept. Coast to 2,000 feet. Any soil, but varies in height from 1 to 6 inches, and often flowers twice a year. E., Hassen Island, Dark Head, Women's Island, Wilcox Point. W., Cumberland Gulf, Cape Scarle, Scott's Bay, Cape Adair.

S. cæspitosa, L.—June to Aug. Coast to snow-line.

The following varieties occur :—

1. Leaves variable, the cauline ones entire.
2. Leaves tripartite and cuneate.
3. Leaves of both forms, and in 2 or 3 flowers on the same stem.

Common on both sides.

S. Hirculus, L.—Aug. Alt., 100 feet; on clay soil; it grows singly. W., Scott's Bay.

S. stellaris, L.—Aug. Alt., 200 feet. On granite, and often viviparous. W., Kickertine Island, Cumberland Gulf.

S. foliolosa, Br.—July. E., Disco.

S. hieraciifolia. W. & K.—Aug. Alt., 100 feet; on moist granitoid soils. W., Banks of a river south of Scott's Bay.

Chrysosplenium alternifolium, L.—Aug., on the beach amongst mosses. W., Middliattwack Islands, Cumberland Gulf.

Compositæ.

Gnaphalium sylvaticum, L.—June to Aug. Alt., 1,000 feet. Any soil; very variable. E., Dark Head, Women's Islands, Wilcox Point. W., Cumberland Gulf, Cape Searle, Scott's Bay, Cape Adair.

Antennaria alpina, L.—June to Aug. Alt., 1,000 feet. Professor Balfour has doubts whether this be *alpina*. W., Kingnite, Kickertine, Middliattwack Islands, Cumberland Gulf.

Arnica montana, L. (*angustifolia*, Vahl).—June to Aug. Alt., 500 feet. Varies much; height, 1 inch to 1½ foot; is smaller in trap than in granitic soils. E., Dark Head, Horse Head, Wilcox Point. W., Cumberland Gulf, Cape Searle.

Artemisia borealis, Pallas.—June to July. Alt., 500 feet. In crevices of rocks. W., Kingnite, Cumberland Gulf.

Erigeron uniflorus, L.—July to Aug. Alt., 700 feet. Varies much in size. Largest specimens, 18 inches. E., Hassen Island, Dark Head, Wilcox Point. W., Cumberland Gulf, Capes Searle and Adair, Scott's Bay.

Taraxacum palustre, DC.—July, Aug. Sea to 500 feet. E., Disco, Dark Head, Wilcox Point. W., Cumberland Gulf, Cape Searle.

Campanulaceæ.

Campanula linifolia, Hænk.—Aug., Sept. Alt., 500 feet on granitic soils. W., Cumberland Gulf, Cape Searle, Scott's Bay.

C. uniflora, L.—June to Aug. Alt., 500 feet. There seem to be two varieties of the plant. E., Disco, Hassen, Dark Head, Wilcox Point. W., Cumberland Gulf, Capes Searle and Adair.

Vacciniaceæ.

Vaccinium uliginosum, L.—Fl., May, June. Sea to snow-line; often covers large spaces, singly or associated with *Cladonia rangiferina*. The large and juicy fruit was abundant wherever I saw the plant. Common on both sides.

V. Vitis-Idæa, L.—May, June. Alt., 500 feet. E., Wilcox Point.

Ericaceæ.

Cassiopeia tetragona, Don.—July, Aug. Alt., Snow-line. Occurs everywhere, like the *Calluna* of Scotland, and made a fire for us at night when travelling, besides an excellent couch under the shelter of a boulder—no unnecessary luxuries in my longer journeys inland to the West of Cumberland Gulf, &c. Common on both sides.

C. hypnoides, Don.—June, July. Alt., 50 feet. In sandy flats on the coast, in dense masses. W., Kingnite, Kickertine Islands, Niatoling, and Cape Searle, Scott's Bay.

Azalea procumbens, L.—June, July. Alt., 500 feet. E., Wilcox Point.

Ledum palustre, L.—Aug. Mossy soil. Has a powerful odour, and is hence used by the natives in packing and preparing their Seal skins; also used as tea. E., Dark Head, Women's Islands, Horse Head, Wilcox Point. W., Cumberland Gulf, Capes Searle and Adair, Scott's Bay.

Andromeda polifolia, L.—July. Mossy soil. E., Disco, Wilcox Point.

Menziesia (*Phyllodoce*) *cærulea*, L.—July, Aug. Alt., 300 feet. In granitic and mossy soils, with *Ledum*. E., Dark Head, Horse Head, Wilcox Point. W., Cumberland Gulf, Cape Searle.

Rhododendron Lapponicum, L.—June, July. Alt., 1,000 feet. Often covers large spaces, flowering in great profusion; more frequent on the E. side. E., Disco, Hassen and Women's Islands, Dark Head, Horse Head, Wilcox Point. W., Cumberland Gulf, Cape Searle, Scott's Bay.

Pyrola rotundifolia, L. (*chlorantha*, Sw.).—Aug. Very common on any soil; specimens occur with only one flower. Common on both sides.

Diapensia Lapponica, L.—July, Aug. Sea to 500 feet. The withered leaves of former years remain on the stem, closely packed below the living ones. Often grows on exposed plains, and on dry soil. Common on both sides.

Boraginaceæ.

Mertensia maritima, Don.—July, Aug. Beach, amongst sand. W., Cape Searle.

Scrophulariaceæ.

Pedicularis arctica, Br.—July, Aug. Sea to 500 feet. E., Dark Head, Wilcox Point. W., Cape Searle.

P. Kanei, Durand.—July, Aug. Sea to 1,000 feet. Trap soil. E., Hassen Island, Dark Head, Horse Head. W., Cape Searle.

P. hirsuta, L.—June, July. Alt., 800 feet. The most common of the genus. E., Disco and Hassen, Women's Islands, Dark Head, Wilcox Point. W., Cumberland Gulf, Cape Searle, Scott's Bay.

P. Langsdorfii, Fischer.—June, July. Alt., 500 feet. E., Dark Head, Wilcox Point. W., Kingnite, Cape Searle, Scott's Bay.

P. Nelsoni, Br.—June, July. Alt., 300 feet. W., Kickertine Islands, Middliattwack, Cumberland Gulf.

P. Lapponica, L.—June, July. E., Disco.

Euphrasia officinalis, L.—July, Aug. Sea to 50 feet. The plants were very small. W., Kingnite, Cumberland Gulf.

Plumbaginaceæ.

Armeria vulgaris, Willd.—July, Aug. Sea to 500 feet. Only seen at Cape Searle, on trap soil. W., Cape Searle.

Polygonaceæ.

Polygonum viviparum, L.—June to Aug. Coast to 500 feet. Common on both sides.

Oxyria reniformis, Hook.—June to Aug. Sea to snow. Very common, but not used by the natives as a cure for scurvy, for which they use the stomach of a recently killed Deer. Common on both sides.

Kœnigia Islandica, L.—July, Aug. 50 to 500 feet. On the ground moistened with water from melted snow, and in moist crevices of rocks. W., Cape Searle, Middliattwack Island, and Kingnite, Cumberland Gulf. Annual.

Empetraceæ.

Empetrum nigrum, L.—May, June. Reaches snow-line. Very common in some places, and with fine fruit, which often survives the winter. In autumn its berries, with those of *Vaccinium uliginosum*, are collected and eaten by the Natives. These are also eaten by *Corvus corax*, var. *Americanus*, and *Plectrophanes nivalis* and *P. lapponica*; while the Grouse are fond of the young twigs. Common on both sides.

Betulaceæ.

Betula nana, L.—June. Not seen north of Disco, nor on the west side. E., Disco.

Salicaceæ.

Salix arctica, Br.—May, June. Alt., 1,500 feet. The tallest plant seen was 4 feet in height; it often grows to a considerable size, spreading over the southern face of some boulder. Common on both sides.

S. reticulata, L.—May, June. Alt., 500 feet. E., Dark Head, Wilcox Point. W., Scott's Bay.

S. herbacea, L.—May, June. Coast to snow-line. Covers extensive tracts, and that too where most other plants cease to appear, except *Junci* and *Luzulæ*. The Grouse feed on the leaves in spring. In dry fine weather in September, I have often seen its downy seeds wafted in clouds over land and sea. Common on both sides.

S. vestita, Pursh.—May, June. Alt., 200 feet. W., Niatolik, Cumberland Gulf.

S. desertorum, Rich.—May, June. Alt., 100 feet. W., Kingnite and Scott's Bay.

S. arbutifolia, Sm.—May, June. Alt., 200 feet. Detected by Professor Balfour. W., Kickertine and Middliattwack Islands, Cumberland Gulf. Frequent.

Melanthaceæ.

Tofieldia palustris, L.—June. Alt., 500 feet. Mossy soil. E., Wilcox Point. W., Kingnite, Cumberland Gulf, Cape Searle, Scott's Bay.

Juncaceæ.

Luzula spicata, Desv.—June, July. Alt., 100 feet. W., Niatolik, Kickertine Islands, Kingnite, all in Cumberland Gulf.

L. spadicea, DC.—June, July. Alt., 100 feet. In marshes amongst *Sphagna*. W., Kickertine, Kingnite.

L. arcuata, Hook.—July, Aug. Coast to snow-line. The seeds of this and other *Luzulæ* often survive the winter, being suddenly covered with snow, and thus afford a supply of food in spring to many Birds in their northward migration, while they are equally serviceable in autumn on their return. Frequent on both sides.

L. hyperborea, Br.—July, Aug. Reaches snow-line. The leaves and stems of succeeding seasons often remain on the same plant, and from their size, &c. in a measure indicate the character of each season. Common on both sides.

L. campestris, Br., v. *congesta*.—July, Aug. Alt., 200 feet. W., Cumberland Gulf, in various places.

Juncus biglumis, L.—June, July. Coast to snow-line. Common on both sides.

J. castaneus, Sm.—June, July. Alt., 150 feet. Grows where water has stood in the early part of the year. W., Cumberland Gulf, in various places.

J. arcticus, Willd.—Aug. Alt., 100 to 150 feet. Mossy soil W., Middliattwack Islands, Cumberland Gulf, Scott's Bay.

Cyperaceæ.

Eriophorum capitatum, Host.—June, July. W., Cumberland Gulf, Cape Searle, Scott's Bay, Cape Adair.

E. angustifolium, Roth.—June, July. Coast to 300 feet. Common on both sides.

Carex rigida, Good.—June, July. Coast to snow-line. E., Wilcox Point. W., Cape Searle and Scott's Bay.

C. nardina, Fries.—July, Aug. In dry, stony places, like a dead tuft of grass. W., Cumberland Gulf, var. loc., Cape Searle.

C. misandra, Br.—July, Aug. W., Kingnite, Cumberland Gulf.

C. saxatilis, L.—July, Aug. Alt., 500 feet. Grows in marshes. E., Wilcox Point. W., Scott's Bay.

C. vulgaris, L.—June, July. Coast to 300 feet. W., Cumberland Gulf.

C. glareosa, Wahl.—June, July. At the sea-level on the sandy beach, in large circular tufts. W., Cumberland Gulf, var. loc.

C. stans, Drej.—June, July. Sea to 1,500 feet. On the sandy shore; this plant is so stunted as to seem very different from specimens at higher elevations and more favourable situations. E., Dark Head, Wilcox Point. W., Cumberland Gulf, Cape Searle, Scott's Bay.

C. fuliginosa, Hoppe.—July, Aug. Coast to 500 feet. On exposed plains. E., Wilcox Point. W., Cumberland Gulf.

C. compacta, Br.—June, July. Coast to 200 feet. Very common, and grows amongst mosses in marshes. Specimens nearly 2 feet high seen. E., Dark Head, Wilcox Point. W., Cumberland Gulf, Cape Searle, Scott's Bay.

C. aquatilis, Wahl.—July, Aug. In bogs. E., Disco.

C. leporina, L.—June, July. Alt., 200 feet. In marshes amongst mosses. W., Kickertine Islands, Cumberland Gulf.

C. capillaris, L.—July, Aug. Alt., 1,000 feet. On granite cliffs. W., Cumberland Gulf, var. loc.

C. serpoides, Mich.—June, July. On granite cliffs. E., Women's Islands, Wilcox Point. W., Cumberland Gulf, Cape Searle, Scott's Bay.

C. Vahlii, Schkh.—June, July. Coast, 500 feet. In mossy soil, and about the edges of bogs. W., Cumberland Gulf, var. loc.

C. rariflora, Smith.—June, July. Coast to 500 feet. In marshes. W., Cumberland Gulf, various localities.

Gramineæ.

Alopecurus alpinus, Sm.—June to Aug. Most plentiful about old Esquimaux settlements; greatest height, 2 feet. Common on both sides.

Calamagrostis canadensis, Nutt.—Aug. W., Cape Searle.

Agrostis vulgaris, L.—Aug. Alt., 200 feet. On both dry and moist cliffs. W., Cumberland Gulf.

A. rupestris, Willd.—June, July. Alt., 50 feet. On dry rocky soil. W., Middliattwack Islands, Cumberland Gulf.

Hierochloe alpina, Br.—July, Aug. Reaches the snow-line. A very common Arctic grass; greatest height, 2 feet. Common on both sides.

H. pauciflora, Br.—June, July. Sea to 100 feet. On sandy soil by rivers, &c., and on the coast. W., Cumberland Gulf, var. loc.

Poa alpina, L.—Aug. Alt., 1,000 feet. W., Cumberland Gulf, var. loc., Cape Searle, Scott's Bay.

P. arctica, L.—July, Aug. Coast to snow-line. Perhaps the commonest Arctic grass, and would often form fine pasture. Common on both sides.

P. cenisia, All.—July. Alt., 500 feet. W., Kickertine and Middliattwack Islands, Cumberland Gulf.

P. Balfourii, Parn.—July. Alt., 500 feet. On cliffs of granite. W., Kingnite, Cumberland Gulf, Cape Searle.

P. angustata, Br.—July. Alt., 50 feet. Confined to the coast, on sandy soil. W. Cumberland Gulf, var. loc.

Festuca ovina.—June, July. W., Kingnite, Cumberland Gulf.

F. brevifolia, Br.—July. Coast to 1,500 feet. Very common on any soil. Common on both sides.

F. Richardsoni, Huds.—July. Coast to 500 feet. On granitic soils. W., Middliattwack Islands, in Cumberland Gulf.

Elymus arenarius, L.—July, Aug. E., Disco.

E. mollis, Br.—Aug., Sept. Sea-level, on sandy soil. W., North side of Winter Harbour, Kingnite.

Colpodium latifolium, Br.—July, Aug. Coast to 500 feet. In marshes and in sandy soil. Common on both sides.

Phippsia algida, Br.—June, July. Coast to 100 feet. In marshes. A very common grass. Common on both sides.

Trisetum subspicatum, Beauv.—July, Aug. Coast to snow-line; very common, growing sometimes 2½ feet high. Common on both sides.

Dupontia Fischeri, Br.—July. Sea to 50 feet; soil, sandy. W., Cumberland Gulf, var. loc.

Pleuropogon Sabinei, Br.—July, Aug. Coast to 200 feet. Grows in pools of water, on any kind of soil. It is, perhaps, the finest of Arctic grasses; its leaves float on the surface, the culm rising from 9 inches to 1 foot above the water, bearing its beautiful purple florets. W., Cumberland Gulf, Cape Searle, Scott's Bay, Cape Adair.

Filices.

Woodsia Ilvensis, Br.—July, Aug. Range limited. Alt., 100 to 500 feet, on granitic rocks. I have not found this nor any other of the Ferns descend below 100 feet. W., Niatolik, Kingnite, Cumberland Gulf, Cape Searle.

W. hyperborea, Br.—July, Aug. E., Dark Head, Women's Islands, Wilcox Point. W., Niatolik, Kingnite (very abundant), Cape Searle, Scott's Bay, &c.*

W. glabella, Br.—July, Aug. Range more limited than the last, not abundant in some localities. E., Horse Head, Wilcox Point. W., Cumberland Gulf, Cape Searle, Scott's Bay.

Cystopteris fragilis, Bernh.—July, Aug. A Fern of great beauty and of rapid growth; ascends 100 feet higher than the other species. The most characteristic specimens of the var. *dentata* were found growing among the dead roots of *Salix arctica*, on the N. side of Winter Harbour, Kingnite, in Cumberland Gulf. E., Disco, Women's Islands, Wilcox Point. W., Kickertine Islands, Kingnite, Cumberland Gulf, Scott's Bay.

C. alpina, Desv. (?)—I have generally brought home with me in a living state the roots of such plants as seemed most eligible for cultivation; and with a little care and experience in packing,

---

* None of the *roots* brought home have developed into *W. hyperborea*. They have all turned out to be either *W. Ilvensis* or *W. glabella*, though the fronds brought home corresponded with *W. hyperborea*.

this practice affords very satisfactory results, as in the hands of Mr. John Roy, sen., nurseryman, and the Rev. Mr. Beverly.*

Equisetaceæ.

Equisetum arvense, L.—Only seen barren. Alt., Coast to 500 feet. E., Disco, Dark Head, Wilcox Point. W., Cumberland Gulf, Scott's Bay.

E. variegatum, Schleich.—Only seen barren. Alt., 100 feet. Soil, granitic. W., Inland from Cape Searle.

Lycopodiaceæ.

Lycopodium annotinum, L.—Aug. Only seen once, and at an elevation of 200 ft. W., North side of Winter Harbour, Kingnite, in Cumberland Gulf.

L. alpinum, L.—Coast to snow-line; seen in all parts of these regions visited by me. Common on both sides.

XXXII.—Mr. John Sadler's List of Arctic Cryptogamic and other Plants, collected by Robert Brown, Esq., during the Summer of 1861, on the Islands of Greenland, in Baffin's Bay and Davis' Strait, and presented to the Herbarium of the Botanical Society.

[Reprinted, by Permission, from the Trans. Bot. Soc. Edinburgh, vol. vii., 1862, pp. 374–5.]

No. 1. Collected on the Big Duck Island and Duck Islands, Baffin's Bay, June 8–11, 1861.

Stereocaulon paschale.
Cladonia uncialis.
C. papillaria.
Cetraria Islandica.
C. nivalis.
Parmelia parietina var.
P. saxatilis.
P. omphalodes.
P. conspersa.
Lecanora tartarea.
L. ventosa.
Gyrophora hirsuta.
Cornicularia ochroleuca.
C. bicolor.
Pogonatum alpinum.
Bryum cæspiticium.
Hypnum aduncum.

---

* Among the Arctic Ferns brought home by me, and reared by Mr. Beverly, were found last year several plants of one which he, and several others who examined it, suspected to belong to this species. They were led to this suspicion by observing the form and habit of the fronds, and especially the nature of the rhizome, which spreads more widely, and throws up its small tufts of upright fronds at greater intervals than *C. fragilis*. Just now (June 1862) the plants are in good condition, but the fronds seem not quite so like those of *C. alpina* as they were last year. Though this Fern is evidently different from the common forms of *C. fragilis*, and in several respects approaches the so-called *C. tenuis*, in others *C. alpina*, it may perhaps prove to be only an extreme form of *C. fragilis*. But it shall be carefully watched as it grows, in order to fix its identity. At all events, in its *present* form, if it is not *C. alpina*, it is intermediate between that species and *C. fragilis*, and as worthy of being raised to the rank of a separate species, as many other varieties that have been so treated.

No. 2. Collected on Browne Island, one of the Women's Islands, Baffin's Bay, off North Greenland, lat. 74° 7', long. ——; primary rocks and boggy wet soil almost wholly composing the islands. June 5, 1861.

Several species of Stereocaulon and Cladonia.	Pogonatum alpinum.
Dicranum Richardsoni?	Hypnum uncinatum, var.
Aulacomnion turgidum.	H. sp.
	Urceolaria scruposa.

No. 3. From Hare Island, west coast of Greenland; lat. 70° 43 N.; long. 55° 42' W.; greenstone, gneiss, and other rocks jutting out above the snow. 27th May 1861.

Bryum nutans.	Gyrophora proboscidea.
Cornicularia bicolor.	G. proboscidea, var.
C. pubescens.	Cetraria nivalis.
Lecidea rupestris.	Lecidea geographica.
L. petræa.	L. geo., var. apicula.
Gyrophora arctica.	Parmelia caperata.
G. hyperborea.	P. olivacea.

No. 4. The following were the only Flowering Plants seen and brought home intermixed with the Mosses.

Salix Lapponum. On all the islands.	Papaver nudicaule. On all the islands.
Stellaria humifusa. Duck Island.	Silene acaulis. Duck Islands.
Saxifraga rivularis. Duck Islands.	Empetrum nigrum. Duck Isl.
	Poa alpina. Browne Island.
	P. danica. Browne Island.

Dr. Brown discovered in addition to several Mosses and Lichens rare to the Arctic Flora, *Laminaria longicruris* of De la Pylaie's "Flora of Newfoundland" occurring plentifully within the Arctic Circle; and *Melobesia calcarea*, hitherto only recorded from Spitzbergen, for the first time in Davis' Strait (Hasen Island, four fathoms).

XXXIII.—PLANTS from SMITH'S SOUND.

(From the "Enumeration of the Arctic Plants collected by "Dr. I. I. HAYES in his Exploration of Smith's Sound, "between parallels 78th and 82nd, during the months "of July, August, and beginning of September 1861. "By E. DURAND, THOS. P. JAMES, and SAML. ASHMEAD." Proceed. Acad. Nat. Sciences of Philadelphia, March 1863; vol. for 1863–1864).

I.—PHÆNOGAMOUS PLANTS (52; the localities mentioned are Gale Point, Netlik, Port Foulke, Cape Isabella, and Tessuissak). By E. DURAND.

[These are included in Dr. Hooker's List of Arctic Plants, *see above*, page 225.—EDITOR.]

II.—Cryptogamous Plants.

1. *Lycopodiaceæ.*

Lycopodium annotinum, L. Tessuissak.

2. *Musci* (36). By T. P. James.

Andræa petrophila, Ehrh. ?
Barbula ruralis, Hedw.
Orthotricum affine, Schr.
Grimmia spiralis, Hook. & Tayl.
Racomitrium lanuginosum, Brid.
Polytrichum juniperinum, Hedw.
Aulacomium turgidum, Schw.
Bryum Duvallii, Voit.
B. purpurascens.
B. arcticum, Brid. & Sch.
B. rutilans, Br. & Sch.
B. cyclophyllum, Br. & Sch.
B. crudum, Sch.
B. nutans, Schr.
B. palustre, L.
B. æneum, Blytt.
Mnium affine, var.
Mnium rugicum, Bland.
M. rostratum, Schw.
Meersia Albrotinii.
Bartramia (aff. calcareæ).
Conostomum boreale, Swartz.
Splachnum Wormskioldii, Brid.
S. vasculosum, L.
Hypnum uncinatum, Hedw.
H., aduncum, L.
H. oligorhizon, Br. & Sch.
H., n. sp. ?
Cladonia pyxidata (L.), Fries.
C. furcata, var. racemosa, Hoff.
C. ignota ?
Lecidea geographica ? Hoff.
Umbilicaria hyperborea ! Hoff.
U. ignota ?
Verrucaria popularis, Floerk.
V. maura, var. striatula, Hoff.

3. *Lichenes* (15). By T. P. James.

Alectoria bicolor (Ehrh.), Nylander.
A. sulcata ? (Lev.), Nyl.
A. ochroleuca (Ehrh.), Nyl.
Lecanora ventosa, Ach.
Neuropogon Taylori, Hook., Nyl.
Platysma cucullata, Hoff.
P. nivalis, Ach.
Plocadium elegans (Ach.), Nyl.
Parmelia saxatilis (L.), Ach.
P. Borreri, Turner.
P. Stygia (L), Ach.
P. conspersa ? (Ehrh.), Ach.
Dactylina arctica (Rich), Nyl.
Stereocaulon denudatum, Floerk.
S. condensatum, Hoff.

4. *Algæ* (16). By S. Ashmead.

Fucus vesiculosus, L.
Alaria esculenta, Grev.
Ulva latissima, L.
Laminaria phyllitis, Lam.
L. longicruris, Pylaie.
L. fascia, Ag.
L. saccharina ? Lam.
Rhodymenia interrupta, Grev.
Enteromorpha compressa, Grev.
Solieria chordalis, Ag.
Cladophora arcta, Dill.
Bryopsis plumosa, Ag.
Desmarestia aculeata, Lam.
Chætomorpha littorea, Haw.
Ectocarpus ?
Sp. ignota ?

No new species were determined; living roots brought home ceased to live in the spring at Philadelphia; and the seeds collected, or found in the soil brought home, failed to germinate, though the Arctic soil was apparently very rich, and though every care was taken.

XXXIV.—FLORULA DISCOANA: CONTRIBUTIONS to the PHYTO-GEOGRAPHY of GREENLAND, within the Parallels of 68° and 70° North Latitude. By DR. ROBERT BROWN, F.L.S., F.R.G.S., &c.

[Reprinted, by Permission, from the "Transactions of the Botanical Society of Edinburgh," vol. ix., part 2, 1868, pp. 430–465.] Read July 9, 1868. Slightly abridged; and revised by the Author, March 1875.*

I. *Review of Greenland Botanical Literature.*—The flora of Greenland has been at various times partially examined by different botanists. The early missionaries, Egede, Fabricius, Saabye, and others, made collections of the plants of the districts over which their ministerial functions extended, and some of these are yet in the Herbarium at the Botanic Garden in Copenhagen. In 1826 the Chevalier Charles Louis Giesecke (better known as Sir Charles Giesecke), Professor of Mineralogy to the Royal Dublin Society, who had passed several years in Greenland as a mineral collector, published a list of the plants of that country.† His list comprehends a large number of species, but he is manifestly wrong in regard to many of them. Some, which may possibly be members of the Greenland flora, have never been found since his day. The various explorers in search of Franklin, and the Surgeons of Whalers, have at different times added to our knowledge of the distribution of the plants, by collecting on various portions of the coast.‡ But by far the most important collections which ever came from Greenland were those of Vahl, who botanised with the utmost assiduity over the whole extent of Danish Greenland, and has published various papers on the plants. The most valuable literary contribution, however, to the history of the Greenland flora, is the list in the Appendix to Rink's "Grönland geographisk og statistisk," by my friend Professor

---

* Reprinted materially as in the original publication, without augmentation from the later researches of Berggren, Th. Fries, and others, this paper will serve as a specimen of a Botanist's summer-work in Greenland.

† Article "Greenland," Brewster's Edinburgh Encyclopædia.

‡ Lyall's collections, by Hooker, in Journ. Linn. Soc. Bot. vol. i. pp. 114–124; Notes on Arctic Plants, Dickie, Journ. Linn. Soc. Bot. vol. iii. (1859) pp. 109–112 (plants collected by Clarke, Clark, Maitland, Philips, Craig, and Sutherland); Dickie (Sutherland's Plants) in Appendix to Inglefield's "Summer Search for Sir John Franklin," (1853); Dickie on Philpott's Plants from Lancaster Sound, Linn. Soc. Journ. Bot. vol. xi. p. 92; Sir W. J. Hooker and Dickie in Appendix to Sutherland's Narrative of Penny's Expedition; Account of the Botany of M'Clintock's Expedition (Walker's Plants), Hooker and others, Journ. Linn. Soc. Bot. vol. v. p. 85; Taylor on Davis' Strait Plants, Trans. Bot. Soc. vol. vii. p. 323, or Edin. Phil. Journ. 1862; Sadler's Notice of Cryptogamia collected by R. Brown on islands of Baffin's Bay, Trans. Bot. Soc. vol. vii. p. 374; Sutherland on *Cystopteris alpina*, Trans. Bot. Soc. vol. vii. p. 393; and generally Hooker, Linn. Soc. Trans. 1861.

Johann Martin Lange of Copenhagen, forming a summary of the labours of all former Danish botanists, and a determination of the collections of Egede, Vahl, Rink, Holböll, and others contained in the Herbarium of the University of Copenhagen.* Drs. Kane† and Hayes‡ have added to our knowledge of the plants of the extreme northern shores of Greenland. Professor Lange's list, dealing only with the Danish possessions in that country, does not touch upon these. It is to be hoped, however, that he will yet undertake an extended flora of Greenland, a task for which he is so well qualified, both from his knowledge of the subject and the opportunity which he possesses of consulting Herbaria.

II. *The present Collections.*—During the summer of 1867, from June until September, I passed the season in Danish Greenland, collecting specimens in all departments of natural history, and pursuing scientific investigations. The summer was very favourable for botanical research. Accordingly, though my time was very limited, and greatly occupied with other pursuits, I made a large collection of the plants, of all orders, found in the country between Egedesminde and Kudlesæt. As the country was chiefly in the vicinity of Disco Bay, I have denominated the account of these collections the *Florula Discoana.* These plants are here enumerated by the assistance of various botanical friends, whose reputation is a sufficient guarantee for the accuracy of the lists under their names. Though containing few plants really new to science, the list is interesting as being the most complete one of the plants of that section of country, and as adding to our knowledge of the phyto-geography of the coast,—the earlier collections being to a great extent useless for that purpose, as the labels merely afforded the information that they were collected in "Greenland." . . . . .

III. *Climate.*—During the winter the country is covered with snow, and the plants protected under its warm covering. Darkness then covers the whole face of the country for about four months. About May and the beginning of June, according to the state of the season, the earth again begins to appear. By July the snow has generally cleared off all the lower grounds, and only lies in hollows, on the hills, or in places shaded from the sun.

---

* Oversigt over Grönlands Planter af Joh. Lange (Bibliothekar og Assistent ved den botaniske Have) Tillæg Nr. 6 til Rink in lib. cit.; Vahl om Stellaria Grœnlandica og Dryas integrifolia (Nat. Selsk. Skriv. 4 Band. 2 H. Ss. 169–172); *see* also Rink, "Om den geographiske Beskaffenhed af de danske "Handels-distrikter i Nordgrönland, &c." (Det Kongl. danske Vidensk. Selskab. Sk. 5 Række, 3 Bind, 1853, p. 71). Drejer's Revisio critica Caricum borealium (Kröyer's Tidsskr. iii. p. 423). Hornemann in Graah's Journey to East Coast of Greenland (Transl.) Appendix; Greville on Jameson's West-Greenland Plants, Mem. Wern. Soc. vol. iii. p. 426; Hooker on Sabine's Plants, Trans. Linn. Soc. vol. xiv., and on Scoresby Plants in App. to Scoreby's "Greenland"; Flora Danica; Retzius' Floræ Scandinaviæ Prodromus, &c.

† Durand, in Appendix to Kane's "Arctic Explorations," vol. ii. *Above*, p. 241.

‡ Hayes' Open Polar Sea; and Durand in Proc. Phil. Acad. Nat. Sciences, March 1863; and partially in "Das nordlichste Land der Erde," Petermann's Geographische Mittheil. (1867), p. 176 *et seq.* *Above*, p. 254.

From this period until the middle of September, very little snow ever falls, and the climate is mild, and even warm and sunny, as during the summer of 1867. A little rain also falls during most seasons. Vegetation springs up apace, and during the long summer day, of four months, soon comes to maturity. By the beginning of August the flowers are on the wane, and by the end of that month have wholly disappeared. The weather in September is uncertain, showers of snow falling, and the nights being dark and cold. By October "bay ice" begins to form in quiet harbours or inlets, and the ground gets its winter mantle of snow. The soil freezes hard to the depth of several feet (where it is so thick), and all nature slumbers. Meteorological observations have been taken at various royal trading posts throughout Greenland.* At Jakobshavn, one of these settlements, Dr. Rudolph, now Governor of Upernavik, kept for upwards of three years a careful register of the thermometer. Jakobshavn was our headquarters, and the locality for the chief portion of the species here enumerated, and it may be taken as typical of the climate of Disco Bay. I therefore present the means of temperature there, as a mean of the climate over the region embraced in the title of this paper.

THERMOMETRICAL MEANS OF THE CLIMATE OF JAKOBSHAVN.

Lat. 69° 13′ 26″ N.

January,	− 2·4 Fahr.	July,	45·4 Fahr.
February,	0·3	August,	42·4
March,	8·2	September,	34·6
April,	+ 18·8	October,	25·1
May,	32·5	November,	12·5
June	41·5	December	− 7·5
Winter (Mean Temp.)	− 3·4	Summer,	+ 43·1
Spring, ,,	19·9	Autumn,	24·1

Whole year, 22·5.

IV. *Character of the Country in which the Plants were collected.*—The country in which the specimens were collected consists chiefly of bare rounded gneissose hills, planed by old ice-action, and covered with boulders and travelled blocks of stone. In the hollows, where the melting of the snow collects, are peaty bogs, and in other places dry heath-looking tracts, covered with *Empetrum nigrum*, *Cassiopeia* (*Andromeda*) *tetragona*, *Betula nana*, and such like plants. The eastern side of these glens is richest in plants, and the vicinity of streams and dripping springs yields a considerable variety. In the Waigat Strait, about Kudlesæt, Ounartok, and Atanakerdluk, the geology changes, and bold trap cliffs and dykes burst through sedimentary rocks of Miocene age. Here is the limited district containing the now celebrated fossil beds of Greenland.

I may shortly describe each individual district, taking the

* *Collectanea Meteorologica*, Fasc. iv. Hauniæ, 1856. Rink, Tillæg Nr. 8, "Meteorologie" til *Grönland geographisk og statistisk beskrevet*, Andet Bind, 1857.

Danish Trading Divisions as guides, and looking upon the chief post in each district as the centre and type of the division. It was also in the immediate vicinity of these posts that the greater number of the plants here enumerated were collected.

(1.) *Egedesminde.*—Lat. 68° 42′ 39″ N., long. 52° 43′ 48″ W.* The island on which the settlement is built is low-lying, bare, and bleak. The vegetation is very stunted, and is affected by the cold wind,—no high mountains being in that vicinity to shield the low-lying ground, and few cliffs which can radiate the sun on the soil. The climate here is more foggy than in other places further to the east and nearer the mainland. The Cranberries and Whortleberries on the small hills in general bear no ripe fruit; the Arctic Willows and the Birch do not grow in any great luxuriance; and the greater part of the country is covered with swampy Moss, only allowing a little green to appear now and then. Warm springs . . . . . are found on the island of Sakartloek, lying at the head of Tessiursak Bay, about eight miles from Egedesminde, and near the mouth of a little river flowing over a level tract scattered with boulders. One of these springs issues in a large stream out of a very solid granite wall and over a smooth mossy ground, out of which other two or three springs run between the stones and moss with about the same force. The temperature, according to Dr. Rink, is 42°·1 Fahr., or 20°·2 Fahr. higher than the mean temperature of the island. A little basin, a few hundred feet in length, which the spring forms, is never frozen; and at the bottom of the bay, where the stream debouches, no ice lies in the winter. Large banks of *Bartramia fontana*, &c., form round the springs, which keep these Moss banks always in a tremulous motion. On the island of Aito, and the surrounding islands, the same characteristics prevail as in the vicinity of Egedesminde. The vegetation is exceedingly scanty, and but little can be seen but brown rust-coloured rocks and stunted vegetation. Here is found *Sedum Rhodiola*, DC.—found nowhere farther north than South-East Bay. It is said to be here very abundant on the top of the small sterile islands, tipped by turf and the excrement of birds. We arrived at Egedesminde on the 6th of June, and left on the 14th of the same month. During most of this time the weather was snowy, and little or nothing except a few Lichens and Mosses rewarded my search. I am, however, under obligations to Fröken Julie Levesen for most kindly presenting to us a small collection of Egedesminde plants, made by her in the preceding

* In most cases, and in reference to the latitudes invariably, I follow my own observations made during the past summer. In reference to the longitudes, my own observations not being all yet (owing to the arrangements of the expedition) accessible, I have followed either Graah's observations (in "Tabel over adskillege Punkters observerede Brede og Laengde paa Vest-"kysten af Grönland," in "Beskrivelse til det vissende Situations Kaart over "den vestilige Kyst af Grönland," &c., &c. Kjöbenhavn, 1825), or others given to me through the politeness of Premier-Lieutenant H. L. M. Holm of the Kongl. Kaart-Archiv in Copenhagen. The position is that of the chief "colonie" or trading-post.

year, which has enabled us to add some localities to the Disco flora, and a few additions to the scanty list. The general character of the country at this season of the year may be gathered from the following jotting in my journal: and as it is equally characteristic of other portions of Disco Bay, I may be excused quoting it:—

"*June* 6.—To-day we took an excursion over the island on which the settlement of Egedesminde ('the memory of Egede') is built. The Eskimo name of it is *Arsiat*, and means the summer place; and they remark, not inaptly, that it lies in its little archipelago of islands, like a spider in its web. Nothing was to be seen but bare granite rocks, worn by ice, or covered with poor Franklin's *tripe de roche*—the *tudluak* of the natives—with snowy drifts in every shady place, and bogs in the hollows, or lakes with the surface ice yet unmelted. Few living things were out: a Bee, a Spider or two, and a *Dyticus* in the pools, with a Snow-bunting (*Emberiza nivalis*) looking out for a nesting-place, were the only specimens of animal life we came across in our rambles. No flowers were as yet above the ground to any extent. The Willows were shooting up, and the Empetrum was green above the half-thawed soil. Eriophorums were coming into flower, but the only plant in bloom was *Cassiopeia tetragona*. Masses of woolly-looking matter, apparently bleached Confervaceæ, mantled some of the stagnant pools near the village, which were half choked up with rotting fragments of Seals and other animals. Near the top of the island were found larvæ and cocoons of Lepidoptera, pieces of the shell of *Echinus drobachiensis*, Müll., and the shell of a Decapodous Crustacean, apparently carried up there by sea-birds, or perhaps by the wind. If we are to credit the Eskimo tales of *Asaminah*, the south-east wind, it has force enough to carry for some distance much heavier bodies than shells. In some of the little valleys we met Greenland women laden with the Dwarf Birch, Empetrum, and Willows —collectively the *Brændsel* of the Danes—for fuel in their houses."

(2.) *Christianshaab*.—Lat. 68° 49′ 19″ N., long. 51° 8′ 14″ (Nordenskiöld) W.—I visited this locality in the first week of August, and added several plants to my collection. My notes describe it as possessing "more varied scenery than any of the " other settlements I have yet seen, lying in a long 'hope' with " green slopes to the water's edge, and fells of syenite 1,600 feet " in height in front of the 'colonie,' and beyond, — the way " leading through a green grassy valley,—a lake alive with wild " Geese (*Anas brenta*, Pall.). Behind and all around are sunny " 'braes,' green with the moisture of rushing rivulets, and many " flowers as yet strangers to my collection." The coast between Christianshaab and Claushavn is low and easily landed on, with green slopes, and streams running down from the hills and bursting through the boulder-clay. On one of the islands (particularly *Krikertasasuk*, "the long big island") about six miles from Christianshaab, I added several plants to my collection, particularly *Potentilla anserina*, L., which, though found further north, is yet only entered in Lange's list on Vahl's authority, and was not found by me elsewhere in the vicinity of Disco Bay.

(3.) *Claushavn*—Lat. 69° 7′ 31″ N., long. 50° 55′ 30″ W.—This commercial establishment is built on a flat, backed by hills of considerable height. On this flat is a small lake, round the marshy borders of which plants grow luxuriantly. This flat is divided off into little glens by *roches moutonnées* like knolls of rocks, each glen ending in a terminal moraine at the lower edge, and exhibiting the same evidences of ancient glaciers. Many plants are found here on this sunny flat which I did not observe at Jakobshavn, only seven miles north of it across the Icefjord. For Greenland, Claushavn is a sunny spot, and not unpleasant. Here *Epilobium latifolium*, L., luxuriates, and *Lychnis apetala*, L., is found growing in considerable quantity among the rocks behind the Colonibestyrers house. *Armeria vulgaris*, Willd., *Trisetum subspicatum*, P. B., and *Juncus triglumis*, L., were found by me only in this locality. From Il-ŭl-iă-mĭn-ĕr-sŭăk ("the big mountain overlooking the Icefjord"), rising to the height of 1,400 feet, can be seen the Icefjord, and little lakes lying in rugged valleys, with the commencement of the Tessiusak just peering out, and away beyond to the eastward the dreary stretch of the inland ice. *Rhododendron lapponicum*, Stellarias, and Drabas were the plants most prominent. *Papaver nudicaule*, the hardiest of all Arctic plants, was found here long after *R. lapponicum* had disappeared. I visited Claushavn first on the 24th June, and subsequently at various times in July, and afterwards while travelling to Christianshaab in the beginning of August.

(4.) *Jakobshavn*—Lat. 69° 13′ 26″ N., long. 50° 55′ W.—This was our head-quarters for the whole of our residence in the country, and the greater number of the plants were collected here. The settlement is built on rounded knolls of rocks, with boggy little valleys between, where the vegetation springs; further back are various boulder-clay valleys, where considerable vegetation appears, though very little exposed to the sun. The flora is not nearly so profuse as at Claushavn. The whole country in this region is composed of rounded syenitic hills of various heights up to 1,200 feet, bare or polished with ice-action, or covered with black, horny Lichens, and with scattered boulders and angular blocks of stone lying in all kinds of positions over their summits and faces wherever it is possible for them to lie. Between these fells and rocks lie flat valleys, composed of boulder-clay beneath, but capped with a boggy covering of turfy Peat, which the natives cut and dry in stacks for winter fuel. Early in the summer these are mere bogs of marshes, into which you sink over the knees. Here the meltings of the winter's snows accumulate, forming miniature lakes in the hollow places, permanent all the year round, bordered by a thicket of Cyperaceæ and bright with the yellow Ranunculus and other Arctic marsh-plants, and the overflow goes off by streams which pour in mimic cascades over the sea-cliffs. In some of these lakes or boggy places I found *Hippuris vulgaris,* which I did not observe anywhere else in the district. A fruitful habitat for plants was the dripping rocks, where a little stream flowed in through a valley at the head of the harbour. Outside of the little harbour a few Algæ were found, the continual grinding of ice-bergs off the shore hardly allowing of their growth. However, just below the "kirke" where we lived, the rocks yielded not a few species, and the scum of pools furnished some interesting freshwater species. North of Jakobshavn the coast is very similar—low-lying, with glens and valleys, the outlets of former glaciers, scattered with old moraines, but presenting nothing particularly worthy of notice in a botanical point of view. On the site of Eskimo villages (such as *Akatout,* in Rode Bay) a very luxuriant growth of vegetation springs up; and here I gathered some plants, which will be found recorded in their proper places.

(5.) *Illartlek Inlet.*—This inlet breaks the coast in lat. 69° 27′ N. Like all of such fjords or inlets, it is the site of an ancient glacier which here reached the coast. The entrance of it is in Pakitsok Bay, and is marked by an immense terminal moraine, where many plants grow luxuriantly.* I have always noticed that plants grow most luxuriantly near large rocks or boulders, the rock attracting a greater amount of heat to the soil. This is very evident on the broad American prairies, where stones are rare; and was equally apparent here, though on a lesser scale. At the head of this inlet (or at least one of the heads) a muddy glacier stream flows in, silting up the head of the inlet for several miles. On the left hand is a bold bluff of boulder (glacier ?)-clay and boulders, a remnant, as all such are, of the former upheaval of the coast, though at present, in the vicinity of Disco Bay at least, the coast is perceptibly sinking. This clay was very sandy, and was kept together by a turf of Empetrum, Betula, and Grasses; but on the windward side, where it meets the blast from the glacier, it was bare of vegetation, and the fine powdery clay was blown into hillocks around a few Willow tufts. On the less exposed places a few stunted plants grew, particularly *Ledum palustre,* here at least belying its trivial name, for it grows mostly on dry ground. Between the glacier and this place is a flat valley, after ascending the first slope, covered with a spongy turf and permeated by streams, and ornamented with a little lake where the wild Geese breed. On the slope, just before crossing over a little ridge to the glacier, I found the rare Lichen *Dactylina arctica,* Nyl., in considerable profusion, but nowhere else. This valley is plentifully tufted with the fragrant *Hierochloe alpina,* which is used for stuffing the native boots. Crossing the ridge mentioned, we descend a little slope and face the glacier, the overflow of that great *mer de glace* which overspreads the whole interior of Greenland with an icy covering. The slope facing the glacier and the

* In this catalogue "Illartlek" refers to this locality; "Illartlek glacier,' to the immediate vicinity of the glacier and inland ice, &c.

cliffs around are bare of vegetation, and the whole vicinity is very chilly and dreary. The cold blasts have even nipped the usual profusion of Arctic vegetation, and we have to go far afield to gather the Dwarf Birch for our cooking fire. "On the slope, "however, survive nearly all the species of Saxifraga, and on "the sunny spots *Vaccinium uliginosum* is bearing its pleasant-"tasted berries, all of which tell us that autumn (after which "cometh the winter, when no man can work) is travelling on "apace. Stellarias and Oxyria show themselves frequently, as "do also *Epilobium latifolium*, and the Eriophorum with its "tasselled head of cottony down, in the boggy places here "and there, while *Stellaria Edwardsii* is occasionally seen quite "abundant at the head of the inlet. *Papaver nudicaule* is "coming into seed, as well as the species of Pedicularis, which, "with *Lycopodium annotinum*, &c., maintain their ground in "appropriate situations." The glacier face was in lat. 69° 24′ 12″ N. We entered the inlet on the 20th of July, and left on the 29th of the same month.

(6.) *Ritenbenk*—Lat. 69° 45′ 34″ N., long 51° 7′ W.—The island on which this settlement is situated is called *Akpaet*, and presents nothing phytographically remarkable. There is a considerable amount of Dwarf Willow and turf on it. By the time we arrived here (August 20) the Arctic flora was nearly gone, so that Ritenbenk does not figure much in this catalogue. The shore afforded, however, a few Sea-weeds.

(7.) *Sakkak*—Lat. 70° 0′ 28″ N., long. 52° W. (approx.)—At this little outpost there is a broad sunny flat, with the "inland ice" appearing as miniature glaciers down between the cliffs behind. Here I found *Festuca ovina*, L., in great luxuriance, but except a few Algæ from the shallow muddy ice-choked harbour I did not add greatly to my collection.

(8.) *Atanakerdluk*—Lat. 70° 02′ 30″ N., long. 52° W. (approx.)—By the time we arrived here Phanerogamic vegetation was nearly over; and except a few Cryptogamic plants I have little to add from this locality. Here, as I have remarked, the geology entirely changes from the primitive to sedimentary formations; and the few days we spent here (22nd to 24th August) were occupied by me almost entirely in collecting the Miocene plants, and describing and making sections of the strata, the arid slope presenting no recent plants to collect. Though, of course, the limited materials possessed will scarcely admit of deciding what influence the change of soil, consequent on the altered geological conditions, may have in giving an altered character to the flora; yet, so far as I was able to judge from the decayed plants which remained above ground, it seems that they were, to a great extent, different from those gathered on the granitic soil.

(9.) *Ounartok*—Lat. 70° 2′ N., long. 52° 24′ W. (both approx.)—The locality known under this name seems to have been at one time a native "house-place," and traces can yet be seen of former habitations at the mouth of a gurgling creek which flows from the mountains, and it is yet a favourite camping place for the rare visitors and wayfaring-men along this dreary coast. Much débris has been brought down by this creek as it dashes from the mountain and the inland ice of Disco Island (for it is situated on the opposite shore of the Waigatz Strait, as are also the two next localities mentioned), and bursts though the sedimentary strata which lie in its way.

(10.) *Kudlesæt*—Lat. 70° 5′ 35″ N., long. 52° 32′ W. (approx.) —This was the most northern locality reached by us in 1867. Here are green mossy slopes, but as the sun does not reach this spot for several hours in the day, the vegetation, even on the 27th of August, was backward. Here several streams flow down and form a marshy flat at one place before reaching the sea. On this wet ground, and on the sandy "links" which skirt the coast for a few yards in breadth at this place, I found one or two plants, such as *Juncus triglumis*, L., *Equisetum variegatum*, L., &c., which, though not peculiar to the locality, are yet rather uncommon in this region.

(11.) *Godhavn* or *Lievely*—Lat. 69° 14′ 58″ N.,* long. 53° 24′ 40″ W.—This little post, situated at the south-western point of Disco Island, is perhaps the best known botanical locality in all Greenland, having been a regular halting place for whalers and the numerous Arctic Expeditions. Hence we find plants from this locality figuring in all the lists hitherto published, and containing some not in this catalogue, as by the time we arrived (4th Sept.) vegetation had almost entirely disappeared. The settlement itself is built on an off-lying islet of syenite; but on the other side of the harbour on Disco Island, where the syenite meets with that great trap dyke which, either in its main body or in its offshoots, traverses the whole breadth of the island of Disco and the Noursoak peninsula, there is a "warm" stream of the same character as that on the island near Egedesminde, already described. This stream falls into the harbour, flowing through a little green valley called Lyngemarken (or the "heath field"), backed by huge fells of trap. This Lyngemarken is the best botanical locality which I have yet seen in Greenland. Though most of the plants had faded down in this valley, yet, from what I was able to identify, or from other small collections, it appears to be very rich in species. The most characteristic plants are *Salix glauca*, *Betula nana* (seldom over one foot high), *Rhododendron lapponicum*, *Cassiopeia tetragona*, *Empetrum nigrum*, *Saxifraga tricuspidata*, *S. Aizoon*, *S. cæspitosa*, *S. rivularis*, *Azalea procumbens*, *Gnaphalium norvegicum*, *Veronica alpina*, *Arnica alpina*, *Bartsia alpina*, *Campanula uniflora*, *Epilobium angustifolium*, *E. latifolium*, *Dryas octopetala* var. *integrifolia*, *Papaver nudicaule*, *Pedicularis flammea*, *Silene acaulis*, *Armeria maritima*, *Alchemilla vulgaris*, &c.; and among Cryptogamia,

---

* Graah gives the lat. as 69° 14′ 22,″ while the late Lieut. Ulrich (in general a very good observer), according to a meridian altitude given me by the Royal Chart-Office of Denmark, states it as 69° 13′ 30″ N.; but as Capt. Graah's position and mine agree so closely, I believe that we are nearer the truth.

Cetraria islandica, *C. nivalis*, *Cladonia gracilis*, *Peltidea aphthosa*, *Polytrichum juniperinum*, *Racomitrium canescens*, *Sphærophorum coralloides*, &c. (*vide* Dr. Rink, &c.) The valley graduates by a gentle slope to a dark beetling precipice. At between one or two thousand feet from the shore the vegetation seems to be lost, and there is only seen mountain cliffs or débris of rocks rolled from above, through which the stream runs gurgling along. The most remarkable of all the plants, however, which I saw in this valley were remains of the "Qvan" (*Angelica officinalis*, Hoffn.), well known by its native and Norse names (apparently one of the words of the old Norsemen which have got incorporated in the Eskimo language), which grew in patches by the side of the stream, and occasionally in the moist ground. It is one of the most interesting plants of Greenland, and is only found on the island of Disco, in North Greenland. It is, however, abundant in the vicinity of South Greenland fjords, and particularly in the district of Julianshaab, so much so, that the natives say that Disco was once a portion of Julianshaab district, and that a great angekok or wizard towed it north. He would have towed it still further had not a rival cut the rope! That is what may be called a "myth of observation." The Danes and Greenlanders use the leaves much as an antiscorbutic. On the leaves is occasionally found *Vitrina angelicæ*. By the borders of the stream, and at the northern head of the valley, I found *Achemilla vulgaris*, L., growing. I heard much of a place, about twelve miles from Godhavn, called *Qvannersoit*, "the place "of the Qvan," which, if all stories are true, seems to be the most agreeable spot in the district. It is situated between high falls and "jokulls," with numerous waterfalls from them, and green slopes covered with the most luxuriant vegetation in all North Greenland. Angelica has been found at various places on the island of Disco, but nowhere so abundantly as here, as the name indicates. The Willow is here eight feet high* when raised up from the ground. Numerous flowers grow here. *Rhododendron lapponicum*, *Pedicularis flammea*, *Ledum grœnlandicum* (*palustre*), &c., are seen in profusion. Godhavn was the last locality we visited in Greenland, and on the 12th of September we left in the royal trader "Hvalfisk," Capt. Hans Seiftrup, for Denmark, just as the snow was beginning to cover the hills, and the nights were getting cold, dark, and dreary. My time was much occupied in zoological, geological, and astronomical work, besides having a full share of the varied duties of the party, so that my leisure for botany was limited; and when we take into account the time occupied in going from place to place, the period over which the collecting extended did not much exceed two months, the whole extent of our residence in the country being only three months.†

* I have seen a stem of *Betula nana* from Upernavik (72° 48′ N.) two inches in diameter, and another from South-East Bay equally thick.

† It has been necessary to give these dates, in order to show the times of flowering, and to avoid repetitions, though the object of this paper is not to furnish any narrative of the journey.

V. *Economic Botany of Disco Bay.*—(1.) *Gardens.*—Around most of the little trading posts the Danish officers have attempted to cultivate a few garden vegetables, and by bringing soil from old Eskimo houses, and taking the greatest care, a few of the hardier vegetables are raised in small quantities. Potatoes never get bigger than marbles; but spinach, radishes, lettuces, &c. prosper, and are ready for use about the middle or beginning of August. Of Dr. Pfaff's and Hr. Andersen's gardens at Jakobshavn and Ritenbenk we have most pleasant remembrances. The garden at the latter place deserves honourable mention, and as it was, perhaps, one of the most favoured and favourable specimens of such, the description will suffice for all. It is situated on a sunny slope, with a southern exposure, and composed of earth brought from old Greenland houses (and therefore richly manured), heaped up to the depth of two feet. The vegetables were most luxuriant—lettuce, cabbage, turnips (white), carrots, parsley, and onions. This garden parallelogram of 18 by 12 yards, with its luxuriant vegetation, the gravel walk, the miniature summer-house in the centre, the green watering-pot, and the bird nets over the lettuce, had quite a home aspect amid the barren grey syenite and granite, with hundreds of icebergs in sight at any hour. The Danish ladies cultivate in their houses most of our garden flowers, —geraniums, fuchsias, roses, nasturtiums (a great favourite), ivy, &c.; but they are apt to be destroyed if placed out of doors.

(2.) *Fuel.*—It is a great mistake to suppose that the Eskimo burn nothing but blubber for fuel. Their principal fuel is turf, the Birch, Empetrum, Willow, Andromeda, Ledum, Vaccinium, &c., which they collect and store for winter use, or use immediately in the summer. We used this in all our travels, though, indeed, an armful soon blazes up like a bunch of straw. The collection, storing, and cutting of the various descriptions of fuel is interesting; but I must pass it over with this notice.

(3.) *Food Plants.*—Equally erroneous is the notion that they use no vegetable food. Berries form their principal article of vegetable diet, and comprehend Blaeberries (*Vaccinium uliginosum*), Cranberries, *Empetrum*, *Vaccinium Vitis-idæa*, &c. Though the latter is used by the Danish residents as a preserve, yet it is not eaten generally by the natives; and even the Blaeberries are eaten cautiously by them, on account of some supposed noxious quality.

(4.) *Plants used Hygienically.*—There are some plants, of which the flowers, leaves, or roots are eaten raw or boiled, such as *Sedum Rhodiola*, the flowers of *Epilobium*, *Pedicularis hirsuta*, of which the flower tops are boiled and eaten as a sort of cabbage; the Sorrel (*Oxyria*), and the well-known Scurvy-grass (*Cochlearia*), which is used in scurvy by the natives, who are often affected by that disease, though never touching salt. I have already spoken of the use of the Angelica by the Danes and Greenlanders. Iceland-moss (*Cetraria islandica*) is found in various places; but is rarely, if ever, used by the natives. Various species of Algæ are used as food, but only resorted to when hard pressed by hunger. The species chiefly used is called

Aukpadlurtok (*Chorda Filum*, Ag.). *Fucus vesiculosus*, L., *Alaria Pylaii*, Grev. (Sutluitsok), (the ally of which, *Alaria esculenta*, is eaten on our own shores), *Rhodymenia palmata*, Grev., are also used. *Lycoperdon Bovista* is said to be applied to bleeding wounds.

VI. *Introduced Plants.*—In another memoir I propose discussing the origin and nature of the Greenland flora, its geographical range in Greenland, and the hypsometrical distribution of the species; but I believe it will not be out of place to conclude these introductory remarks on the Disco flora, by calling the attention of future collectors to the subject of introduced or colonist species. Species at all tender, if accidentally introduced into Greenland, though they may survive the summer, yet can scarcely be expected to live over the winter. There are, however, some plants found in Greenland, the indigenous character of which is doubtful. On the sides of the fjords, up to 61°, is found, in the form of small shrubs, the well-known *Sorbus Aucuparia*, L., and from its position there seems to be same good reason for supposing it was brought to Greenland by the old Norse and Icelandic colonists. Again, *Xanthium strumarium*, L., was found by Giesecke in the garden of the Moravian Brethren at Lichtenau in the Frith of Agluitsok, near Cape Farewell, in 60° N. lat.; but was probably sent from Europe in seed. These subjects, as well as the means by which plants may be transported from place to place, the hybridising of some of the more variable species, especially the Drabas, are all eminently worthy of being attended to; and as several Arctic expeditions will be in the field next summer, we may hope to obtain some more enlightenment on these matters.

VII. In addition to the gentlemen who have so minutely examined the collections, and regarding whose work I will not say a single word, as it speaks for itself, I have specially to thank Dr. Hooker, Professor Oliver, and Mr. J. G. Baker, of the Herbarium at Kew, for much assistance, and a *carte blanche* in the way of whatever aid the magnificent collections under their charge could afford to me while studying and assorting my collections. These collections comprehend all the species actually brought home as far as flowering plants and ferns are concerned. Several other species. however, were identified, but too far gone to be preserved. It is possible that a further examination of some of the marine Algæ and Lichens may show some of them to be distinct, and during the examination of the zoological collections, a few minute species of Algæ may be found. The Diatomaceous and Desmidious collections are so extensive that it was found impossible to present the result of their examination in this place, and a large portion of them is not yet accessible to science.* Though a large number of the species recorded in this *Florula* were identified by me at the time of collection, yet for the nomenclature as it now

* For the species causing the discoloration of the sea, see *Trans. Bot. Soc. Edin.* for Dec., *Quart. Journ. Science*, and *Seeman's Journ. Bot.*, 1868, and Translations in *Das Ausland*, Feb. 27; 1868, Geogr. Mitt., 1868, &c.

stands, the botanists whose names are placed after each division are responsible. For remarks regarding locality, I am solely answerable.

(I.) *Phanerogamia ; and Vascular Cryptogamia.* By D. OLIVER, F.R.S., F.L.S., Professor of Botany, University College, London, &c.

1. *Thalictrum alpinum*, L. (In leaf only.) Lyngemarken, Disco I.
2. *Ranunculus hyperboreus*, Rottb. Jakobshavn, Akatout.
3. *R. pygmæus*, Wahl. Akatout Jakobshavn, Christianshaab, Illartlek, Claushavn.
4. *R. lapponicus*, L. Jacobshavn.
5. *Papaver nudicaule*, L. Greenl.* *Nasoot.* Claushavn, Jakobshavn, &c.
6. *Cochlearia officinalis*, L., var. *fenestrata.* (R. Br.) Jakobshavn, Egedesminde.
7. *C. officinalis*, L. Illartlek Inlet.
8. *Arabis alpina*, L. Claushavn, Ounartok.
9. *Cardamine bellidifolia*, L. Jakobshavn.
10. *Draba incana*, L. Jakobshavn.
11. *D. incana*, var. Claushavn, Jakobshavn.
12. *D. hirta*, L. Claushavn, Jakobshavn, Ounartok, Egedesminde.
    D. hirta, var. Ounartok, Illartlek.
13. *D. hirta*, var. (?) Siliqua ovato-elliptica v. ovato-oblonga demum parce puberula valvis reticulatis, pedicello æquilonga v. eod. longiore. Godhavn.
14. *D. muricella*, Wahl. (*D. nivalis*, Lilj.) Jakobshavn.
15. *D. rupestris*, R. Br. Jakobshavn, Egedesminde, Illartlek.
16. *D, aff. D. rupestri*, differt : glabrillima, pedicellis inferioribus longioribus, siliquis late ovato-ellipticis v. fere rotundatis. Jakobshavn.
17. *Silene acaulis*, L. Egedesminde, Claushavn.
18. *Lychnis apetala*, L. Claushavn.
    L. apetala, var. *triflora.* (R. Br.) Claushavn.
19. *L. alpina*, L. Claushavn, Jakobshavn.
20. *Cerastium alpinum*, L. Jakobshavn, Claushavn, Egedesminde.
21. *C. alpinum*, var. From the same localities.
22. *Stellaria humifusa*, Rottb. Jakobshavn, Akatout, Godhavn, Island north of Christianshaab (Kritertasasuk).
23. *S. longipes*, Goldie. Christianshaab.
24. *S. longipes*, var. (*S. Edwardsii*, R. Br.) Claushavn, Illartlek, Akatout, Jakobshavn, Christianshaab.
25. *S. media*, L. (Near houses only.) Christianshaab.
26. *S. cerastioides*, L. (*Cerastium trigynum.*) Claushavn, Godhavn, Ounartok, Lyngemarken.

---

* "Greenl."—Greenlanders. The name succeeding is the native one in the North-Greenland dialect.

27. *Arenaria arctica*, Stev. (*A. biflora*, Wahl.) Lyngemarken, Claushavn.
28. *A. verna*, L. Jakobshavn, &c.
29. *Montia fontana*, L. Akatout, Claushavn.
30. *Alchemilla vulgaris*, L. Lyngemarken.
31. *Dryas octopetala*, var. *integrifolia* (V.) (*D. integrifolia*, V.)—foliis speciminibus nonnullis basin versus crenato-dentatis. Egedesminde (Miss Levesen, 1866), Illartlek, Christianshaab, Claushavn, Jakobshavn.
32. *Potentilla nivea*, L. β. (*P. grœnlandica*, R. Br.) Claushavn.
33. *P. nivea*, L. Illartlek, Claushavn, Jakobshavn.
34. *P. tridentata*, L. Christianshaab, Lyngemarken.
35. *P. anserina*, L. Krikertasusuk Island, six miles north of Christianshaab.
36. *Sibbaldia procumbens*, L. Jakobshavn.
37. *Saxifraga oppositifolia*, L. Greenl. *Kakethlanglet*. Jakobshavn, Illartlek, Christianshaab, Egedesminde.
38. *S.* (*Aizoon*, Jacq.) *Cotyledon*, L. Christianshaab.
39. *S. cæspitosa*, L. Illartlek, Christianshaab, Jakobshavn, Egedesminde, Claushavn.
40. *S. stellaris*, L. Jakobshavn.
41. *S. rivularis*, D. Jakobshavn, Illartlek, Egedesminde (Miss Levesen).
42. *S. cernua*, L. Greenl. *Akudleloot*. Illartlek, Christianshaab, Egedesminde (Miss Levesen), Claushavn, Jakobshavn.
43. *S. tricuspidata*, Retz. Greenl. *Nōonēet*. Proven, lat. 72° (Miss Levesen), Egedesminde (Miss Levesen), Claushavn, Jakobshavn.
44. *S. nivalis*, L. Jakobshavn, Sakkak, Egedesminde (Miss Levesen).
45. *Hippuris vulgaris*, L. Jakobshavn.
46. *Epilobium latifolium*, L. Christianshaab, Claushavn, Jakobshavn, Egedesminde (Miss Levesen).
47. *E. angustifolium*, L. Varietas foliis oblongo-lanceolatis basi obtusis sessilibus v. subsessilibus interdum ternatim approximatis, racemis brevibus foliosis, stylo staminibus breviore. Lyngemarken.
48. *Campanula rotundifolia*, L., var. *linifolia* (Haenk). Claushavn, Jakobshavn, Illartlek Inlet.
49. *C. uniflora*, L. Jakobshavn.
50. *Vaccinium uliginosum*, L. Greenl. *Pĕdlōot*. Egedesminde, Illartlek, Christianshaab, Jakobshavn.
51. *Pyrola rotundifolia*, L., var. *grandiflora*, DC., Greenl. *Lapasert*. Jakobshavn, Illartlek, Proven, and Egedesminde, 1866 (Miss Levesen).
52. *P. rotundifolia*, var. Christianshaab, Claushavn.
53. *Diapensia lapponica*, L. Jakobshavn, Claushavn, Egedesminde (Miss Levesen).
54. *Cassiope hypnoides*, D.M. Egedesminde (Miss Levesen).
55. *C. tetragona*, D.M. Greenl. *Isutseet*. Egedesminde (Miss Levesen), Jakobshavn, Claushavn.

56. *Phyllodoce taxifolia*, Salisb. Egedesminde and Proven (Miss Levesen), Claushavn, Christianshaab.
57. *Ledum palustre*, L. Greenl. *Karasatch*. Claushavn, Jakobshavn, Godhavn, Egedesminde (Miss Levesen).
58. *Loiseleuria procumbens*, Desf. Egedesminde (Miss Levesen), Jakobshavn, Claushavn.
59. *Rhododendron lapponicum*, Wahl. Egedesminde, Jakobshavn, Godhavn, Claushavn, Christianshaab.
60. *Erigeron alpinus*, L. Jakobshavn, Claushavn.
61. *E. compositus*, Pursh. Atanakerdluk.
62. *Artemisia borealis*, Pall. Christianshaab.
63. *Gnaphalium norvegicum*, Gunn. Lyngemarken.
64. *Artemisia alpina*, L. Atanakerdluk, Lyngemarken, Jakobshavn, Claushavn.
65. *Arnica montana*, L., var. *angustifolia*. Claushavn, Illartlek, Jakobshavn.
66. *Taraxacum Dens-leonis*, Desf., var. *palustris*. Claushavn.
67. *Pedicularis lapponica*, L. Greenl. *Udenarooset*.* Claushavn, Christianshaab. Jakobshavn.
68. *P. flammea*, L. Jakobshavn, Claushavn, Egedesminde (Miss Levesen).
69. *P. hirsuta*, L. Jakobshavn, Illartlek, Egedesminde (Miss Levesen).
70. *Veronica alpina*, L. Lyngemarken.
71. *Bartsia alpina*, L. Christianshaab.
72. *Pinguicula ; sine flore, verisim. P. vulgari*. Christianshaab.
73. *Armeria vulgaris*, Willd. Claushavn.
74. *Plantago maritima*, L. Claushavn, Illartlek.
75. *P. borealis*, Lange. Flora Danica, t. 2707, Suppl. Jakobshavn. (Rocks near Dr. Pfaff's house, very sparingly.)
76. *Polygonum aviculare*, L. Christianshaab (Colonist ?), Jakobshavn, Claushavn, Christianshaab, Proven, and Egedesminde (Miss Levesen).
77. *Oxyria reniformis*, Hk. Greenl. *Somnit*. Jakobshavn, Illartlek.
78. *Betula nana*, L. Greenl. *Modikoote*. Jakobshavn, Egedesminde (Miss Levesen), Godhavn.
79. *Empetrum nigrum*, L. Greenl. *Panukojet*. Egedesminde, &c. (universally distributed).
80. *Salix glauca*, L. Jakobshavn, Claushavn, Egedesminde.
81. *S. arctica*, R. Br.? Greenl. *Seet*. Egedesminde, Jakobshavn.
82. *S. herbacea*, L. Jakobshavn.
83. *S.*, sp. (♀ fl.)Egedesminde.
84. *S.*, an var. *S. arcticæ ?* Egedesminde (Miss Levesen).
85. *S. glauca*, L., var. foliis latioribus apice rotundatis late acutatisve (poll. latis). Jakobshavn.
86. *Tofieldia palustris*, L. Claushavn, Jakobshavn, Christianshaab.
87. *Juncus biglumis*, L. Kudlesæt.

---

* Probably all the genus has the same name.

88. *Juncus triglumis*, L. Claushavn.
89. *J. castaneus*, Sm. Claushavn, Jakobshavn.
90. *Luzula spadicea*, DC. Lyngemarken.
91. *L. hyperborea*, R. Br. Jakobshavn, Lyngemarken.
92. *L. campestris*, Sm., var. *congesta*. Claushavn, Jakobshavn, Illartlek.
93. *Scirpus cæspitosus*, L. Claushavn.
94. *Eriophorum capitatum*, Hist. Greenl. *Okăliousăk*.* Illartlek, Egedesminde.
95. *E. vaginatum ?* L. Jakobshavn.
96. *E. angustifolium*, Hoppe. Jakobshavn.
97. *Carex rupestris*, All. Claushavn, Jakobshavn.
98. *C. lagopina*, Wahl. Godhavn.
99. *C. rigida*, Good. (*et* varr.). Jakobshavn, Lyngemarken, Egedesminde (Miss Levesen).
100. *C. aquatilis*, Wahl. Jakobshavn.
101. *C. rariflora*, Sm. Akatout, Illartlek, Jakobshavn.
102. *C. alpina*, Sm. (*C. Vahlii*, Sch.) Single specimen, Jakobshavn.
103. *C.*, *aff. C. stenophyllæ*. Jakobshavn.
104. *Alopecurus alpinus*, L. Jakobshavn, Egedesminde.
105. *Hierochloe alpina*, L. Greenl. *Eeweek*. Claushavn, Jakobshavn, Illartlek, &c.
106. *Phippsia algida*, R. Br. Jakobshavn.
107. *Calamagrostis lanceolata*, Robb.(var. *C. phragmitoides*, Hart.) Lyngemarken, Jakobshavn.
108. *Trisetum subspicatum*, P. B. Claushavn.
109. *Elymus arenarius*, L. Illartlek, Akatout, Claushavn.
110. *Agrostis rubra*, L. (*A. alpina*, Wahl.) Jakobshavn, Christianshaab.
111. *Poa annua*, L. Jakobshavn.
112. *P. alpina*, L. Claushavn, Lyngemarken, Jakobshavn.
113. *P. alpina*, *forma elatior*. Akatout.
114. *P. cæsia*, Sm. Claushavn, Christianshaab, Jakobshavn.
115. *P. nemoralis*, L. Claushavn, Jakobshavn.
116. *P. pratensis*, L. Proven (Miss Levesen), Claushavn, Jakobshavn, Christianshaab.
117. *P. flexuosa*, Wahl. Illartlek, Jakobshavn.
118. *P. nemoralis*, L., var. Jakobshavn, Akatout, Illartlek.
119. *P. flexuosa*, Wahl., var. (*P. cenisia*, All.) Illartlek.
120. *Glyceria maritima*, M. & K. Illartlek.
121. *G.* (*Poa*) *angustata* (Br.) Akatout, Christianshaab.
122. *Festuca ovina*, L. Sakkak, Claushavn, Illartlek. Jakobshavn.
123. *Lycopodium Selago*, L. Greenl. *Toterurese*. Jakobshavn, Egedesminde, &c.
124. *L. annotinum*, L. Jakobshavn, Illartlek.
125. *Equisetum arvense*, L. Jakobshavn, Kudlesæt, Lyngemarken, Claushavn.

---

* A generic name.

126. *Equisetum variegatum*, L. Lyngemarken, Kudlesæt.
127. *Cystopteris fragilis*, Bernh. Claushavn, Jakobshavn, Illartlek Inlet.
128. *Woodsia Ilvensis*, R. Br. Claushavn, Jakobshavn, Christianshaab.
129. *W. Ilvensis*, var.? Too young to determine, but possibly this may be *W. glabella*, R. Br. Jakobshavn.

(II.) *Mosses.* By M. A. Lawson, M.A., Professor of Botany in the University of Oxford.

1. *Andreæa rupestris*, Hedw. Jakobshavn.
2. *Sphagnum squarrosum*, Persoon. Egedesminde.
3. *Splachnum sphæricum*, Hedw., var. *luridum.* Jakobshavn.
4. *Sp. Wormskjoldii*, Asch. Jakobshavn.
5. *Aulacomnium palustre*, Schw. Jakobshavn.
6. *Polytrichum juniperinum*, Linn., var. *alpestre.* Jakobshavn
7. *P. sexangulare*, Hoppe. Jakobshavn.
8. *Bryum pallens*, Sw. Jakobshavn.
9. *B. Wahlenbergii*, Br. Lyngemarken, Disco. I.
10. *B. crudum*, Schreb. Jakobshavn.
11. *B. inclinatum*, Dicks. Jakobshavn.
12. *B. cæspiticium*, Schw. Jakobshavn.
13. *B. Zierii* (? no fruit), Dicks. Jakobshavn.
14. *B. carneum* (?) B. Jakobshavn.
15. *B. capillare*, H. & W. Jakobshavn.
16. *Leptobryum pyriforme*, Wils. Jakobshavn.
17. *Psilopilum arcticum*, Brid. Jakobshavn.
18. *Dicranum virens*, Hedw. Jakobshavn.
19. *D. cerviculatum*, Hedw. Jakobshavn.
20. *D. squarrosum*, Starke. Lyngemarken and Egedesminde.
21. *D. palustre*, Brid. Jakobshavn.
22. *D. polycarpum*, H. & T. Jakobshavn.
23. *Grimmia pulvinata*, Hook. et Tayl. Godhavn.
24. *Orthotrichum rupestre*, Schlech. Jakobshavn.
25. *Conostomum boreale*, Sw. Jakobshavn.
26. *Bartramia ithyphylla*, Bred. Jakobshavn.
27. *B. fontana*, Schw. Jakobshavn.
28. *Tortula fallax*, Hdw. Jakobshavn.
29. *Ceratodon purpureus*, Bred. Jakobshavn.
30. *Didynodon rubellus*, Br. Jakobshavn.
31. *Weissia cirrhata*, Hdw. Godhavn.
32. *Distichium capillaceum*, Br. et Sch. Jakobshavn.
33. *Hypnum Schreberi*, Willd. Jakobshavn.
34. *H. uncinatum*, Hdw. Jakobshavn.
35. *H. riparium*, L. Jakobshavn.
36. *H. fluitans*, L. Jakobshavn.
37. *H. pulchellum*, Dicks. Jakobshavn.
38. *H. molle*, Dicks. Jakobshavn.
39. *H. rutabulum*, Linn. Jakobshavn.
40. *H. stramineum*, Dicks. Jakobshavn.

In addition to the above, Mr. Alex. Croall detected among Algæ gathered on the shore, or washed up from the harbour of Godhavn, several species of Mosses, which had been swept down by the mountain-torrents from Lyngemarken Fell, and other of the bold mountains surrounding the "good harbour." They may have possibly some geological interest in reference to the imbedding of land species in marine formations in company with marine plants. They are as follows:—

Dicranum scoparium. Polytrichum urnigerum, sexangulare, et piliferum. Bryum albicans, nutans, et Wahlenbergii. Hypnum fluitans et stramineum.

Some are additions to the muscological flora of Greenland. Mr. J. SADLER, R. Bot. Gardens, Edinburgh, has made some corrections in this list.

(III.) *Hepaticæ.* By BENJAMIN CARRINGTON, M.D., F.L.S., Eccles.

[The Hepaticæ here enumerated were almost solely collected at Jakobshavn along with the Mosses already described. As none of them are of any great rarity, it has not been thought necessary to affix in this summary the exact localities in every case.—R. B.]

1. *Jungermannia barbata*, var. *attenuata*, Mart.
2. *J. barbata*, var. *Floerkii*, N. ab E.
3. *J. barbata*, var. *lycopodiodes.*
4. *J. catenulata*, Hübner.
5. *J. divaricata*, E. B.
6. *J. acuta*, Lindbg.
7. *J. minuta*, Swz.
8. *J. alpestris*, Schleich. A few stems among *J. minuta.*
9. *J. setiformis*, Ehrh.
10. *Ptilidium ciliare*, N. ab E.
11. *Marchantia polymorpha*, Linn.

Dr. Carrington (in letter, Aug. 7, 1868) remarks.—"I have also (from Greenland) *J. grœnlandica*, N. ab E.; *J. cordifolia*, Hook.; *J. albescens*, Hook.; *J. saxicola*, Schradr.; *J. bicuspidata*, L.; *J julacea*, Lightf.; *J. laxifolia*, Hook." The following species have also been recorded from Greenland:—*Sarcoscyphus sphacelatus*, N. ab E.; *Gymnomitrum concinnatum*, Corda; *Alicularia compressa*, Hook.; *Scapania compacta*, Lindbg.; *S. uliginosa*, N. ab E.; *Marpanthus Flobovianus*, N. ab E.; and *Fimbriaria pilosa*, Tayl.

(IV.) *Lichens.** By W. LAUDER LINDSAY, M.D., F.R.S.E., F.L.S., Perth.

1. *Alectoria jubata*, L., var. *chalybeiformis*, L. Jakobshavn, &c.
2. *A. ochroleuca*, Ehrh. Jakobshavn and Godhavn.

---

* *See also* Dr. W. L. Lindsay's "Lichen-flora of Greenland," *further on*, p. 284, comprising the later determinations of these species and varieties.—EDITOR.

3. *Alec. och.*, var. *nigricans*, Ach. Illartlek glacier.
4. *Cetraria Islandica*, L., var. *Delisei*, Bory.
5. *C. Islandica*, var. *leucomeloides*, Linds.
} Lyngemarken, Godhavn, Egedesminde, Jakobshavn, &c.
6. *C. cucullata*, Bell. Jakobshavn, Egedesminde, and Illartlek glacier.
7. *C. nivalis*, L. Jakobshavn, &c.
8. *C. aculeata*, Ehrh. Jakobshavn, &c.
9. *Dactylina arctica*, Nyl.* Near Illartlek glacier.
10. *Nephroma arcticum*, L. Godhavn.
11. *Peltigera aphthosa*, Ach. Egedesminde.
12. *P. venosa*. L. ? Jakobshavn, &c.
13. *P. canina*, Hoffm.
14. var. *rufescens*, Auctt. pr. p. Egedesminde, Lyngemarken, &c.
15. *Solorina crocea*, L. Lyngemarken, Illartlek glacier.
16. *Parmelia saxatilis*, L. Jakobshavn, Illartlek glacier.
17. *P. saxatilis*, var. *panniformis*, Ach. Illartlek, &c.
18. *P. saxatilis*, var. *sphærophoroidea*, Linds. Egedesminde, &c.
19. *P. saxatilis*, var. *omphalodes*, L. Jakobshavn, Egedesminde, Illartlek glacier, &c.
20. *P. olivacea*, L. Egedesminde. The collection contains several varieties of this.
21. *P. Fahlunensis*, L. Jakobshavn, &c.
22. *P. lanata*, L. Jakobshavn, Illartlek glacier.
23. *P. encausta*, Sm. Jakobshavn, &c.
24. *P. stygia*, L. (Several varieties.) Jakobshavn, Illartlek glacier.
25. *Physcia pulverulenta*, Pers. Jakobshavn.
26. *P. cæsia*, Hffm. Kudlesæt.
27. *P. stellaris*, L. Jakobshavn.
28. *Placodium elegans*, Link. Jakobshavn, &c.
29. *P. chrysoleucum*, Sm.
30. *P. chrysoleucum*, var. *opacum*, Ach.
} Kudlesæt.
31. *Pannaria brunnea*, Sw.
32. *P. brunnea*, var. *coronata*, Hffm.
} Ounartok, Godhavn, Lyngemarken, and Illartlek glacier.
33. *Squamaria saxicola*, Poll. Jakobshavn, Godhavn.
34. *Lecanora ventosa*, L. Jakobshavn.
35. *L. tartarea*, L.
36. *L. tartarea*, var. *frigida*, Sw.
37. *L. tartarea*, var. *gonatodes*, Ach.
} Jakobshavn, Lyngemarken, Godhavn, Illartlek glacier.
38. *L. parella*, L. Kudlesæt.
39. var. *Upsalienis*, L. Jakobshavn, &c.
40. *L. oculata*, Dicks. (Various varieties.) Jakobshavn and Illartlek glacier.

---

* This rare fungoid-looking Lichen was found by me in considerable abundance on a dry mossy slope before reaching the Illartlek glacier. It was detected by Mr. W. G. Smith, having been accidentally packed in the Fungi parcels.—R. B.

41. *Lecanora polytropa*, Ehrh. Jakobshavn, Ounartok, Egedesminde, &c.
42. *L. polytropa*, var. *intricata*, Schrad. Ounartok.
43. *L. badia*, Ehrh. Jakobshavn, Ounartok.
44. *L. subfusca*, L. Jakobshavn, &c.
45. *L. subfusca*, var. *epibrya*, Ach. Jakobshavn, &c.
46. *L. bryontha*, Ach. Jakobshavn, &c.
47. *L. turfacea*, Whlb. Jakobshavn, &c.
48. *L. sophodes*, Ach. (Many varieties.) Jakobshavn, &c.
49. *L. calcarea*, L. Kudlesæt and Jakobshavn.
50. *L. cinerea*, L. Kudlesæt (various forms).
51. *L. smaragdula*, Whlb. Jakobshavn.
52. *Stereocaulon paschale*, L.
53. *S. tomentosum*, var. *alpinum*, Laur.
54. var. *deundatum*, Auctt. Egedesminde, &c.
55. *Cladonia pyxidata*, L. (Various vars.) Jakobshavn and Illartlek glacier.
56. *C. verticillata*, Hffm., var. *cervicornis*, Ach. Jakobshavn, &c.
57. *C. gracilis*, L. (Various vars.) Jakobshavn, Illartlek glacier, &c.
58. *C. amaurocræa*, Flk. Egedesminde and Godhavn.
59. *C. furcata*, Schreb. Godhavn (various forms).
60. *C. cornucopioides*, L. (Various vars.) Jakobshavn, &c.
61. *C. fimbriata*, L. Jakobshavn, &c.
62. *C. deformis*, L. Jakobshavn, Egedesminde, &c.
63. *C. rangiferina*, L. Egedesminde.
64. *C. degenerans*, Flk. Jakobshavn, &c.
65. *C. uncialis*, L. Godhavn.
66. *Thamnolia vermicularis*, Sw. Jakobshavn.
67. *Umbilicaria hyperborea*, Ach. (Various vars.) Jakobshavn, &c.
68. *U. arctica*, Ach. Illartlek glacier.
69. *U. cylindrica*. L. (Various vars.) Jakobshavn and Egedesminde.
70. *U. vellea*, L. Jakobshavn.
71. *Lecidea Grœnlandica*, Linds.* Jakobshavn. (Kudlesæt.)
72. *L. vernalis*, L. Jakobshavn, &c.
73. *L. parasema*, Ach. (Vars.) Godhavn, Atanakerdluk.
74. *L. lapicida*, Ach. Jakobshavn. &c.
75. *L. fusco-atra*, L. Jakobshavn.
76. *L. castanea*, Hepp. ? Illartlek glacier.
77. *L. sabuletorum*, Schreb. Jakobshavn, &c.
78. *L. obscurata*, Smrf. ? Jakobshavn, &c.
79. *L. disciformis*, Fr. (Various vars.) Illartlek glacier, &c.
80. *L. atro-alba*, Ach. (Various vars.) Jakobshavn, &c.
81. *L. petræa*, Wulf. (Var.) Atanakerdluk, Jakobshavn, &c.

---

* This species, with *new* forms (species and varieties) not here enumerated, will be described in a separate Memoir in preparation. [Trans. Linn. Soc. xxvii. pp. 305-368, with five 4to. coloured plates.]

82. *Lecidea geographica*. L. (Various vars.) Jakobshavn, Egedesminde, &c.
83. *L. alpicola*, Schr. Jakobshavn, Egedesminde, &c.
84. *L. sulphurella*, Th. Fr.? Atanakerdluk.
85. *L. myriocarpa*, DC. Jakobshavn, Egedesminde, &c.
86. *Sphærophora coralloides*, Ach. (Various vars.) Egedesminde, Illartlek glacier, &c.
87. *Coniocybe furfuracea*, L. Illartlek glacier, Claushavn.
88. *Collema melænum*, Ach. Jakobshavn.
89. *Leptogium lacerum*, Sw. Jakobshavn.
90. *Ephebe pubescens*, L. Egedesminde.
91. *Normandina viridis*, Ach. Lyngemarken.

(V.) *Marine Algæ*. By ALEX. CROALL, Associate B.S., Stirling, and Joint Author of "The Nature-Printed British Seaweeds."

[I did not make the collection of Algæ a special object, and the comparatively large number of species here recorded is due more to the skill and patient industry of Mr. Croall, than to any special acumen or diligence on the part of the collector. Hitherto, exclusive of freshwater forms, there have been found beyond 60° north latitude over the whole Arctic region 63 species of Marine Algæ.* The well-known algologists who have examined this collection have been able to detect, by critically examining every scrap, 41 species of marine and 11 freshwater forms in or around Disco Bay alone.—R. B.]

MELANOSPERMEÆ.

1. *Fucus vesiculosus*, L. Ritenbenk shore, with *Ectocarpus crinitus*, &c. 30th August, v.c. Egedesminde, off Rifkol, &c.

Most of the specimens are rather dwarfish, some of them even less than an inch in length, yet even some of these bear fruit, and among them are specimens both with and without air-vesicles.

2. *Fucus nodosus*, L.

Rather more slender than usual, and the receptacles more globose; but similar forms may be seen on our own shores. Floating in sea out of sight of land, in Davis's Strait, off Lichtenau. June.

3. *Desmarestia aculeata*, L.

Scarcely differing from ordinary specimens, and barren as usual. Jakobshavn harbour, in 4 to 5 fathoms, muddy bottom; very plentiful.

4. *Alaria Pylaii*, Grev.

Common just within low-water mark at Jakobshavn, &c. Eaten by the natives.

---

* Dickie, Journ. Linn. Soc. Botany, vol. ix. pp. 235–243.

5. *Laminaria saccharina*, Lam.
Everywhere in the Laminarian zone; called *kăk-wŏk* by the natives.

6. *Chorda filum*, Ag.
Rockpools, and similar to such on our own shores, also in 2 fathoms; abundantly. Sakkak, Waigatz Strait. August.

7. *Agarum Turneri*, P. & R.
Very abundant in Egedesminde harbour, in from 5 to 6 fathoms; substance of frond very brittle when fresh.

[In both 1861 and 1867 I identified as common along the Arctic shores, *Laminaria longicruris*, De la Pyl., and *L. fascia*, Ag. Among a small collection made at Godhavn by Frue Smith, wife of the Royal Inspector of North Greenland, I detected *L. digitata*, L., in addition.—R. B.]

DICTYOTACEÆ.

8. *Dictyosiphon fœniculaceus*, Grev.
The specimens exhibits both the solid and fistulose state of the plant, but no spores were observed. Most of the specimens had been found floating. Jakobshavn, 3 fathoms, July 5, with many Diatomaceæ; Egedesminde, June 10th, highwater-mark; pools on shore, with *Schiezonema obtusum*; Claushavn, 30th June, floating near shore; Disco Bay generally; Ritenbenk, on shore, August; Sakkak, in 2 fathoms, August; everywhere common.

CHORDARIACEÆ.

9. *Chordaria flagelliformis*, Lam.
Parasitic on *L. longicruris*, floating in Davis's Strait, May 29th; rock-pools at highwater-mark, Egedesminde, 10th June.

10. *Elachista fucicoli*, Fries.
Parasitic on *Fucus vesiculosus*, in company with *E. crinitus*; in rock-pools within highwater-mark, where streams of fresh water flow over it during a portion of the day. Jakobshavn. June.

ECTOCARPACEÆ.

11. *Sphacellaria plumosa*, Ag.

12. *S. cirrhata*, Ag.
The first species appears to be the most common, as small fragments of it were continually occurring entangled with almost every specimen . . . . No separate specimens were observed; but very satisfactory fragments were found among a mass of rubbish washed up at highwater-mark, at Godhavn harbour, with other Algæ and freshwater plants.

13. *Ectocarpus siliculosus*, Lyngb.
Very dwarfish, scarcely an inch in height, and barren; floating in the sea off Holsteensborg, &c. May.

14. *Ectocarpus crinitus*, Carm.

In rock-pools, Jakobshavn, common; also found at Ritenbenk and Godhavn. The specimens referred to this species are sufficiently numerous, but without fruit; if not identical with this species, they are certainly very closely allied, at least in structure. . . . . It cannot be a very uncommon species, as fragments of it were occurring with almost every specimen, often much decomposed, and covered with Diatomaceæ.

RHODOSPERMEÆ.

Rhodomelaceæ.

15. *Polysiphonia urceolata*, Grev.

Most of the sections exhibit five siphons, occasionally four, while *P. arctica* is said to have seven. The specimens were parasitical on the stems of Laminariæ, from a depth of 3 fathoms (Jakobshavn); also in rocky pools at Egedesminde, &c.

16. *Melobesia polymorpha*, L.

A very characteristic specimen of this species, smooth, rounded; the upper surface covered with the dot-like punctures of the Ceramidia; the margin free, and somewhat curved upwards. A fracture shows that only the upper surface, forming a narrow zone of about $\frac{1}{6}$th of an inch in depth is alive, being still filled with the colouring matter of the cells, the rest being pure white.

17. *Delesseria angustissima*, Griff.

A small fragment only of this was detected entangled among the filaments of *Conferva Melagonium*, without fruit, but sufficiently characteristic of the species. Jakobshavn, 5 fathoms; plentiful. July.

18. *Hypnea purpurascens*, Harv.

Small fragments only of this species were observed, mixed up with others, also without fruit. Egedesminde. June.

19. *Euthora cristata*, J. Ag.

A small but very distinct specimen of this was found growing on a fragment of *Tubularia indivisa*, L.

20. *Rhodophyllis veprecula*, J. Ag.

A single small specimen of this was dredged from 12 fathoms. . . . . . Egedesminde. June, with *Flustra avicularis* and *Callithamnion americanum*.

21. *Dumontia filiformis*, Grev.

Just below high-water mark at the "Kirke" of Jakobshavn. June, plentiful; Ritenbenk shore, August.

22. *Kallymenia reniformis*, Turn.

We have some hesitation in referring the specimens to this species, not from any doubts entertained as to their identity, but from the circumstance of the specimens of previous collections having been referred to a different species, or at least designated by a different name (*Kallymenia Pennyi*, Har.) . . . .

23. *Holosaccion ramentaceum*, J. Ag.

There are numerous examples of this, many of them quite simple, and others abundantly branched, and from 3 to 7 inches in length. Jakobshavn, Sakkak (2 fathoms), &c., v.c.

24. *Rhodymenia palmata*, L. Jakobshavn, in 3 fathoms, July.

CERAMEACEÆ.

25. *Ceramium rubrum*, J. Ag.

A rather slender form, the main stems opaque, the branches sub-diaphanous. It may prove a distinct species when this puzzling genus is better understood. Washed up on the beach at Godhavn. September.

26. *Ptilota serrata*, Kutz.

Another intermediate form, and equally puzzling. Only a few fragmentary specimens were observed, mixed with other species; the smallest of these, however, exhibit the doubtful character of the species. Jakobshavn, 3 fathoms; just below highwater-mark at Claushavn, covered with *Cellularia reptans.*

27. *Callithamnion americanum*, Harv.

Only a few fragments of what seems referable to this species; were detected among *Rhodophyllis veprecula.* It is remarkable for the length of the joints, and the patent and attenuated branches.

28. *C. Rothii*, Lyngb.

Scarcely differing even in luxuriance from well-grown specimens on our own shores.

CHLOROSPERMEÆ.

SIPHONACEÆ.

29. *Bryopsis plumosa*, Ag.

Specimens small, but characteristic of the species. The size may depend partly upon the season (June), and partly on the locality (in rock-pools).

CONFERVACEÆ.

30. *Cladophora arcta*, Kutz.

In the collection there are numerous specimens exhibiting the plant both in the spring and summer form, some of them in a beautifully sporiferous condition. Pools on shore, Jakobshavn, June; Egedesminde, sandy shore close to high-water mark, June 10; Sakkak, 2 fathoms, August. Godhavn, washed on beach, September.

31. *Conferva arenosa*, Carm.

Agrees in structure, but the filaments finer than usual in this species. Jakobshavn, rocks on shore. July.

32. *C. melagonium*, Web. et Mohr.

The specimens agree well in structure, but their luxuriance is remarkable; from 18 to 24 inches being the average size. They are thus much less rigid than if they were more dwarfish and stunted in their growth. Jakobshavn harbour, 5 fathoms, July, plentiful; and outside of the harbour, in 3 fathoms, not so common.

33. *Conferva Youngeana*, Dillw.

To this species we refer a few filaments apparently attached to a slice from the stem of *Laminaria longicruris*, being unable satisfactorily to refer them to any other. Found floating in Davis' Strait, off Rifkol, mixed with *Schizonema obtusum*. June.

ULVACEÆ.

34. *Enteromorpha compressa*, Grev.

As plentiful and as polymorphous apparently as on our own shores. All the forms, however, seem easily referable to the present species. Jakobshavn, 3 fathoms, July; Sakkak, August.

35. *Ulva latissima*, L.

Equally abundant with the last. Jakobshavn, 3 fathoms. August, &c.

36. *Porphyra vulgaris*, Ag.

Sakkak, 2 fathoms. August.

37. *Bangia fusco-purpurea*, Lyngb.

. . . . . Floating in the sea, Davis' Strait. May.

OSCILLATORIACEÆ.

38. *Lyngbya Carmichaelii*, Harv.

Filaments of this species are frequently entangled among other specimens, so that the species does not appear to be uncommon, but they are very variable in thickness, and several species may be included. Floating in the sea, parasitic on the stem of *Laminaria longicruris*, in Davis' Strait, off Holsteensborg. May.

39. *L. flacca*, Harv.

Only a few filaments of this were observed mixed with the last and others.

40. *L. speciosa*, Carm.

A fine specimen of this beautiful species was picked up floating in Davis' Strait, north of Holsteensborg. May.

(VI.) *Freshwater Algæ.* By G. DICKIE, M.A., M.D., F.L.S., Professor of Botany in the University of Aberdeen.

ULVACEÆ.

1. *Prasiola fluviatilis*, Sommerfeldt.

This was first described as a native of Norway by Sommerfeldt in 1828. Meneghini has more recently published it as *P. Sauteri*, and Grunow has constituted var. β, *Hausmanni;* it is, however, right to retain the name first given it. Both forms are in this collection, gathered "close to highwater-mark in freshwater "pools, visited by the spray, at Christianshaab, August; and "dried-up pools near the sea, Jakobshavn, June. The large "form, var. β, in freshwater pools, Egedesminde, &c.* The plant varies considerably in size. In places only partially moist, it is small; and when superficially examined in the dry state, it might be mistaken for the large forms of *P. crispa*. This error I committed in a list of species collected by Dr. P. C. Sutherland in the summer of 1852.† It is reported, on the authority of Vahl, as a native of Spitzbergen. Mr. Brown's specimens comprehend various stages. At first it is linear with narrow stripes, subsequently oblong-ovate with undulate and crenate margin, and finally very broad and irregular in outline.

2. *Enteromorpha compressa*, Linn.

The specimens might be referred to the smaller and simpler forms of this variable species. "Jakobshavn (June), in a rock "pool within highwater-mark; but the same pools during slack "water fill by influx of freshwater streams, so that the plants grow "half the time in fresh, and the other half among salt water."—R.B.

3. *E. percursa*, J. Ag.

"Lyngemarken, Disco Island (September); and a few specimens mixed with *Prasiola fluviatilis*, at Jakobshavn, in slightly brackish pools, out of reach of the tide, but visited by sea spray, June 1867."—R. B.

CONFERVACEÆ.

4. *Conferva bombycina*, Ag.

"Stagnant pools of fresh water, among the rocks near the sea, Jakobshavn, Egedesminde, and Lyngemarken (July, June, and September)."—R. B.

5. *C. floccosa*, Ag. (*Microspora floccosa*, Kutz.)

The specimens are not in very good condition, but cannot be referred to in any other genus or species. "Pools of fresh water "near the sea, Jakobshavn (July), and Egedesminde (June)."—R. B.

ULOTHRICACEÆ.

6. *Ulothrix mucosa*, Thunb.

There are only two small specimens, which, however, may be referred to this species. "Moist places near running water, "Egedesminde (June), and in streams, Jakobshavn (June)."

7. *U. minutula*, Kutz. ?

Certainly distinct from the former, judging from the diameter

---

* Many of the marine Algæ were encrusted with various Foraminifera, Polyzoa, and other Zoophytes. Among the latter were observed *Cellepora pumicosa*, *C. ramulosa*, *Cellularia reptans*, *Crisia eburnea*, *Flustra avicularis*, *Tubulipora patina*, *Tubularia indivisa*, &c.

† Appendix to Inglefield's "Summer Search for Sir John Franklin," 1853.

of the filaments; but in a genus where the length of the joints and the diameter vary so much, it is not easy to name from dried specimens. Very much entangled and mixed with *Conferva bombycina*, "as a floating scum on freshwater pools close to the "sea and within reach of the spray. June 1867."—R. B.

PALMELLACEÆ.

8. *Hydrurus penicillatus*, Ag., var., *parvulus*.

"In filaments (attached to small stones) in the current flowing "from the spring at Lyngemarken; Sept. This spring maintains a "uniform temperature all the year round, and remains unfrozen "during the winter."—R. B. The specimens, in the dry state, were not in a condition to show general outline; but there can be no doubt about the genus, a comparison with an authentic specimen having been made. The cells ("gonidia," Kutz.) are in lineal series, ovate or oblong-ovate, and with one end, usually the narrower, colourless. I have little doubt that it is a small variety of the above species, which is widely diffused, and varies much in size and branching. In "Nereis Americana," by the late Professor Harvey, it is reported as attaining a length of one or two feet at Santa Fé, New Mexico; in Northern Europe its size is very much less, and in every respect liable to great variation. In Fries' "Summa Vegetabilium Scandinaviæ," it is reported as found in Norway and Lapland. Mr. Brown's discovery of it in Greenland, in nearly lat. 70° N., is of some interest, when the circumstances are so peculiar.

OSCILLARIACEÆ.

9. *Lyngbya cincinnata*, Kutz.

Abundant in "Lyngemarken Spring. Sept."—R. B. After careful examination, I am constrained to refer this to the above species, which appears to be widely diffused in North Europe.

10. *Oscillaria*.

There are two species (mere fragments), and mixed up with the last, but they are too imperfect for specific recognition.

CHROOESCCACEÆ.

11. *Microcystis*, sp.

Mr. Brown found this on the petrous bone of a Seal lying on damp ground at Egedesminde, in the form of a faint green crust. It may be referred to the above genus, but further I cannot venture to decide.

During examination of portions of the species already enumerated, the following Desmidieæ and Diatomaceæ were incidentally noted :—

DESMIDIEÆ.

Cosmarium undulatum, Corda.
C. connatum, Breb. ?
Staurastrum pygmæum, Breb. ?
Penium truncatum, Breb.
Closterium Cornu, Ehr.

DIATOMACEÆ.

Epithemia ocellata, Kutz.
Eunotia monodon, Ehr.
Synedra ulna, Ehr.
Navicula dicephala, Kutz.
Stauroneis anceps, Ehr.
S. gracilis, Ehr.
Cocconema lanceolatum, Ehr.
Meridion circulare, Ag.
Odontidium mesodon, Kutz.
Tabellaria flocculosa, Kutz.
Himantidium majus, W. Sm.

In conclusion, it may be remarked, that while greater special attention to the freshwater forms would probably have produced a few more species, nevertheless that department of the flora of Danish Greenland cannot be otherwise than limited, on account of the peculiar conditions of the region—the interior being at all seasons a frozen waste, and merely a narrow line along the shores presenting streams and pools of fresh water, with here and there a few springs.

(VII.) *Fungi*. BY WORTHINGTON GEORGE SMITH, F.L.S., London.

[The species of Fungi found in an Arctic country like Greenland can be but few, and it being almost impossible to preserve them in a condition fit for after study and determination caused me to be less anxious than otherwise I might have been to pay much attention to their collection. The extraordinary heat and dryness of the summer of 1867 were not favourable for the growth of this order of plants. Mr. Smith has, however, been able to identify with some certainty nearly all of the species which I brought home.—R. B.]

1. *Agaricus* (*Amanita*) *vaginatus*, Bull. Mossy bank opposite the Settlement of Godhavn. Not uncommon. Sept.
2. *A.* (*Clitocybe*) *infundibuliformis*, Schaeff. Godhavn, &c. Sept.
3. *A.* (*Clitocybe*) *brumalis*, Fr. Jakobshavn, &c.
4. *Hygrophorus virginius*, Fr. This common esculent species I found at Jakobshavn, &c.
5. *Boletus scaber*, Fr. Jakobshavn. July.
6. *Uromyces intrusa*, Lev. Parasitic on leaves of *Alchemilla vulgaris*, L., at Lyngemarken, Disco Island. Sept.

XXXV.—THE LICHEN-FLORA of GREENLAND, with Remarks on the Lichens of other Arctic Regions. By W. LAUDER LINDSAY, M.D., F.R.S.E., F.L.S.

[Reprinted, by Permission, from the Transactions of the Botanical Society of Edinburgh, vol. x., pp. 32–65. Read January 14, 1869.]

I. *Introduction.*

My attention was drawn to the Lichen-flora of Greenland by being requested, in the winter of 1867–8, by my friend Robert Brown, Ph.D., F.R.G.S., to examine and determine the Lichens collected by him in West Greenland in the course of the "West "Greenland Exploring Expedition" of 1867. On studying, in connexion with the determination of the species submitted to me, the literature of Greenland Lichenology, I was surprised to find that there is not on record any separate and modern list of the Lichens of that country. It has occurred to me to endeavour to supply this want in lichenological literature by drawing up the Enumeration, hereto appended, of all Lichens up to this date found, or recorded as having been found, in Greenland, compiled from all the sources of information accessible to me.

The basis of the said list is a catalogue of the Lichens collected by Dr. Brown. But it includes also all those mentioned as occurring in Greenland by Th. M. Fries in his "Lichenes Arctoi "Europæ Grœnlandiæque."* This work includes a record of all the Greenland Lichen-collections of Danish botanists, by whom chiefly contributions to its Lichen-flora have been made. The largest and most important of these collections appears to have been that of J. Vahl, which was made exclusively on the *west* coast, as was also Dr. Brown's. Minor collections were made by Rink and Wormskiold.

The localities of Dr. Brown's collections were chiefly on, or in the vicinity of, Disco Island. They have been elsewhere particuearly specified.† The additional localities mentioned by Fries include the following, which, having no means of ascertaining their exact geographical position, I arrange in alphabetical order:—

Alluk, Amaralik, Amitsuarsuk.
Godhaab.
Holsteinborg and district—including Ikkatok.
Isortok.
Julianehaab.
Kukiarsuk.
Neunese, Njarasurksoit, Nenortalik.
Okivisekan.
Pakitsok (Jakobshavn district).
Sarmalik, Serkunsuk, Sydostbugten (district).
Tiksulik, Tunnudliarbik, Tessarmiut.
Umanak, Upernavik.
Wajgattet [Waigat].

I have also included in my Enumeration all Greenland localities, or Lichens, recorded in the following works or papers:—

1. Crantz: "History of Greenland" (1820). His list of Lichens (p. 318) is unimportant and very meagre, not amounting

* Trans. Royal Society of Sciences, Upsala, series iii. vol iii. 1860.
† "Observations on Greenland Lichens."—Trans. Lin. Soc. xxvii. (1869).

to thirty species, named according to the nomenclature and classification of Dillenius and Hudson. Several names it is impossible now to identify with modern species.

2. Th. M. Fries: "Lichenes Spitsbergenses," 1867; published in the "Kongl. Svenske Vetenskaps-Akademiens Handlingar."

3. Nylander: "Lichenes Scandinaviæ," 1861.

4. Walker and Mitten: Lichens collected by Dr. Walker of the "Fox" Expedition, under Sir Leopold M'Clintock, on the coast at Frederikshaab, Godhaab, Fiskernaes, Uppernavik, and on Disco (Godhavn). Determinations by Mitten. Journal of Linnean Society, Botany, vol. v., p. 87. This list contains some that are not mentioned by Fries. Extra-Greenland localities were—Port Kennedy, 72° N. lat., on the Boothian peninsula, which occupies a central position among the Arctic-American islands; Pond's Bay and Lancaster Sound, on the west side of Baffin's Bay; and Cape Osborne, with whose geographical position I am unacquainted.

5. Hayes and James: Lichens collected by Dr. Hayes; determinations by Professor Thomas P. James; Proceedings of the Academy of Natural Sciences of Philadelphia, 1863, p. 96. These collections were made much more to the north than any of the others, viz., in Smith's Sound, between parallels 78° and 82°. It is not, however, always or clearly stated on which shore they were collected,* though it would appear to have been the eastern or Greenland side.

Professor James remarks, "Not a single *fruited* specimen was "to be found in the entire collection," a circumstance of interest in connexion with a fact I have pointed out elsewhere †—the frequency of barrenness (in apothecia) of the Lichens of Arctic countries. It would almost appear that this sterility, or its frequency, bears a proportion to the northernness of the latitude.

James enumerates the following, which were not found, or are not recorded, by other collectors or lichenologists :—

Alectoria sulcata, *Lév.*‡
Neuropogon Taylori, *Hook.*§
Parmelia Borreri, *Turn.*‖
Stereocaulon condensatum, *Hffm.*¶
Cladonia furcata, *Hffm.*, *var.* racemosa, *Flk.***

---

* Such an omission becomes of more importance where the Strait is much broader. Thus, in the Kew and other Herbaria, I have found specimens labelled "Baffin's Bay" or "Davis Straits." Now, Greenland occupies so decidedly an intermediate position between Europe and America in regard to its general flora, that it is always desirable to know on what side of the bay and straits in question given plant-collections have been made. In the Kew Herb., however, I have frequently met with labels of a much vaguer kind, *e.g.*, "North Pole," "Arctic regions," "Franklin's first journey," or Parry's "first voyage," without specifying any precise locality! *Vide also* p. 300.

† "Observations on Greenland Lichens."

‡ Nylander (Syn., p. 281) gives it only as an *Indian* species.

§ Nylander (Syn., p. 273) gives it only as an *Antarctic* species. Probably it has been confounded with its Arctic representative *N. melaxanthus.*

‖ Nylander (Syn., p. 389) gives its northern limit as Central Norway.

¶ Nylander, (Syn., p. 250) gives its northern limit as Central Sweden and New England, U.S.

** Nylander (Syn., p. 206) records it as a central-European form, and describes the type as becoming rare northwards. *Racemosa* is a not uncommon

Verrucaria popularis, *Flk.**
maura, *Whlnb.*
var. striatula, *Hffm.*†

But these determinations appear to me so little trustworthy, for the reasons assigned in the foot-notes, that I have not included the Lichens in question in my Enumeration.

6. Ross and Brown: Lichens of the East Side of Baffin's Bay, lat. 70°. to 76°, and West Side of Possession Bay, lat. 73°; determined by the late Robert Brown, F.R.S., of the British Museum; published in the "Voyage of Discovery" by Sir John Ross (London, 1819, 2nd. ed., vol. ii, p. 195). The same Lichens are probably what are enumerated as Baffin's Bay Lichens in the collected works of the said Robert Brown (vol. i, 1866, p. 178). This list contains, however, no Lichens not enumerated in my Catalogue on other authority.

There are, probably, other minor papers on Greenland Lichens which I have not seen,‡ *e.g.*, one by Nylander, "Ad Lichenographiam Grœnlandiæ quædam Addenda" (Regensburg, "Flora," 1827), describing certain collections of J. Vahl, including, according to Krempelhuber ("Geschichte," p. 361), a record of three new species.

In general terms, Lichen-collections in Greenland may be said to have been made between lat. 60°, the extreme south, and about 75°, the latitude of Upernavik. Certain exceptional collections have been made as high as lat. 82°, while the majority have come from about the latitude of Disco, 70°.

Geologically Greenland appears to consist, for the most part of—(1), granites; (2), metamorphic schists, especially gneiss and mica-slate; (3), various traps—porphyritic or amygdaloidal; and (4), various superficial Tertiary strata, exhibiting at some points a rich *fossil* flora.

There is in Greenland a great scarcity of arboreal vegetation—a circumstance that, more than any other perhaps, determines the

---

British form. I collected it both in Norway and Faroe ("Northern Cladoniæ," pp. 420–1, Journal of Linnean Society, vol. ix. Botany). It would appear to be a much more northern Lichen than Nylander supposes. I have given its northern distribution in a paper on the "Arctic Cladoniæ" (p. 172, Transactions of Botanical Society of Edinburgh, vol. ix. 1867).

* This is probably a synonym, but I do not find it in any of the lichenological works in my library.

† If this be *V. striatula*, Whlnb., it is recorded by Fries (Arct., p. 267) as occurring in Finmark.

It thus appears, that, while in the case of certain of these Lichens (*e.g.*, the *Alectoria* and *Neuropogon*) it is most unlikely they *can* occur in Greenland, in no case is the determination such that it can be relied upon!

‡ Thus Krempelhuber refers ("Geschichte," p. 361) to—

1. Collections by Breutel. 2. A list of Lichens, determined by Mr. John Sadler, collected by Robert Brown, F.R.G.S., in North Greenland (Browne and Women's Islands), and on its west coast (Hare Island).—Trans. Bot. Soc. of Edin., vol. vii, 1862, p. 374. In a letter to me, Dr. Brown describes the said Lichens as "only a few" collected, in 1860, "on the Duck Islands, off "the north (?) coast of Greenland. . . . There was only a short list. . . . "When I landed that summer, which was rarely, the ground was covered "with snow, and the only things which peeped out were a few Lichens on the "rock-summits, all of which I . . . . collected." (*See* above, p. 253.—Editor.)

peculiarities of its Lichen-flora. There is a total absence of forests, and, consequently, of the shade and moisture which they provide and conserve. Hence there is a comparative absence of the *Graphideæ*,* *Strictæ*, *Collemata*, *Calicia*, *Usneæ*, *Ramalinæ*, *Pertusariæ*, *Endocarpa*, and generally of the *corticolous* Lichens so common in Central Europe and America. The want of forests can scarcely, however, account for the paucity of *Verrucariæ*, many at least of which are saxicolous. Though, as a general rule, trees are absent, they occur, as they do in Iceland, Faroe, Shetland, Orkney, and the Hebrides—exceptionally, of stunted growth; while there seems to be no scarcity, at least at certain points on the western coast, of woody bushes or shrubs. Thus the following trees or shrubs are reported† as occurring in Greenland:—Service-tree, Birch, Alder, Willow, Juniper, Crowberry, Whortleberry, and Black Crakeberry.

While there is absence or scarcity of corticolous, fruticulose, and foliaceous Lichens, there is an abundance of saxicolous forms, referable mostly to the genera *Lecidea*, *Lecanora*, *Squamaria*, *Parmelia*, and *Umbilicaria*; of terricolous species, referable to the genera *Cladonia*, *Alectoria*, *Cetraria*; and of muscicolous species and varieties, belonging to the genera *Lecidea* and *Lecanora*. But the only prominent feature of the Lichen-flora of Greenland recorded by travellers is the abundance of the *Umbilicariæ*, which in many localities give a character to the colouring of the landscape. Thus the author of the "Edinburgh Cabinet Library" volume on Greenland writes of the interior:—"The mountains "are either *entirely bare*, or covered with a mourning veil of "black Lichens" (p. 226). . . . "The dark rocks are clothed "with numerous sombre-coloured Lichens, which *grow with great* "*rapidity* beneath the snow"‡ (p. 380). Near Cape Lister, on the east coast, he describes a "pavement of loose quartz or horn-"blende stones, either naked or covered with black Lichens. "These, with a few tufts of hardy plants, were *all* the vegetation "visible" (p. 247.)

Some parts of the country are described as *absolutely barren*. "Even the Greenlander, accustomed as he is to the horrors of "nature, calls these spots places of desolation" (p. 226). Of certain parts of the coast, Graah says (1837), "*No sign of vege-* "*tation* was observable on these walls of rock . . . at many "places *not even a bit of moss*" (pp. 47, 48)—"moss" being a comprehensive and vague term generally used by travellers to include Lichens, especially of the fruticulose kinds (*e.g.*, *Cladonia*, *Ramalina*, *Alectoria*, *Usnea*, *Cetraria*). In works of travel I not unfrequently find rocky or desert districts of country described as *barren* of vegetation. For instance, Lord Haddo, speaking of the

---

* *Arthonia trabinella* is the sole representative of this large family.

† Edinburgh Cabinet Library volume on "Iceland, Greenland, and the "Faroe Islands," 1840, p. 377, chapter on Botany.

‡ This assertion, if it is founded on fact, is of much interest in connexion with the question of Lichen-growth as a test of age, a subject of which I have treated shortly in the Report of the British Association for 1867, p. 88, and more fully in "The Farmer" of October 23, 1867.

rocky banks of the Nile in Egypt and Nubia, says they were "without a particle of vegetation"* (p. 173). This does not expressly exclude Lichens, as the term "vegetation" is in such cases used in reference to phænogamic vegetation only. But elsewhere, in the same narrative, he specially excludes even Lichens, *e.g.*, where he describes "precipices and cliffs without "the least particle of vegetation, or *even a Lichen* on the surface" (p. 129) . . . "they have *not even a single Lichen*" (p. 134) —as if Lichens were an inferior growth to vegetation!

All such pictures are imaginative or poetical; they are not scientific—not the assertion of naturalists with specially trained, all-observant eyes. They contradict the observations both of geologists and botanists, *e.g.*, Sir Charles Lyell's observations on the vegetation of the young lavas of Vesuvius and Etna, and my own on that of the older lavas of Iceland.† All observation and inquiry lead me to conclude that in no part of the world are rocks of any age—that is, of more than a few months old—absolutely devoid of lichenose vegetation. I have made careful observations on the rapidity of Lichen-growth and development, and have shown elsewhere ‡ that a very few months or years, in different localities, suffice for the appearance on fresh surfaces, whether of rock or wood, of a luxuriant lichenose vegetation.§ Beaumont therefore sings, with more truth than the travellers quoted—

"The bleakest rock upon the loneliest heath
Feels in its barrenness some touch of spring;
And, in the April dew or beam of May,
Its moss and *lichen* freshen and revive."

Even in the desert of loose sand, or equally loose volcanic dust, where there is not sufficient cohesion of particles to permit of higher vegetable growth, Lichens are developed both on the said sand or dust itself, and on all foreign substances of sufficient density to permit of the adhesion of their thallus, or their apothecia; for they are much more frequently athalline than is generally supposed. Thus they coat the bleached bones of the men and animals

---

* "Memoir of Lord Haddo," by Elliott. London, 1867.

† "Northern Lichen-Flora," p. 403, Journal of Linnean Society, vol. ix., Botany.

‡ "To what Extent is Lichen-Growth a Test of Age?"—Report of Brit. Association, 1867, p. 88; "Farmer," October 23, 1867, p. 528.

§ An extract from "M'Clintock's Reminiscences of Arctic Ice-travel," in the "Journal of the Royal Dublin Society," vol. ii., p. 235, bears on the subject.

Sir F. L. M'Clintock, searching for Parry's encampment at Point Nias, recognised it "by the stones arranged for keeping down the sides of his tent," and others used as seats or pillows, and he found also that "the narrow-"rimmed wheels of Sir. E. Parry's cart had left tracks, still wonderfully dis-"tinct, in the soft, wet earth, thinly coated with moss! In one place these "cart-tracks were continuous for 30 yards. . . . No Lichens had grown "upon the upturned stones, and even their deep beds in the soil, whence "Parry's men removed them, were generally distinct. . . . This astonishing "freshness of traces, after a lapse of 33 years, compels us to assign a very con-"siderable antiquity to the circles of stones and other Esquimaux traces, which "we find sparingly strewed along the southern shore of the Parry Group, "since they are always moss-covered, and often indistinct."—EDITOR.

that have fallen victims to the inhospitality of nature; and all manner of articles of metal, leather, or wood left by passing travellers. Tennyson sings, with perfect truthfulness of description—

> "And there they lay till all their *bones* were bleached
> And *lichened* into colour with the crags."

Some deserts on the Pacific coast of South America appear to be as barren as any in Africa or Asia, if not so extensive. One of them is described by Darwin, who was all day riding across it, as "a complete and utter desert." But he goes on to say that "the loose sand was strewed over with a *Lichen*, which lies on "the surface quite *un*attached.* This plant belongs to the "genus *Cladonia*, and somewhat resembles the Reindeer Lichen.† "In some parts it was in sufficient quantity to tinge the sand, as "seen from a distance, of a pale yellowish colour. Further inland, "during the whole ride of fourteen leagues, I saw only one other "vegetable production, and that was a most minute yellow *Lichen* ‡ "growing on the bones of the dead Mules." §

The "Old Bushman," writing of East Finmark, describes "long "stretches of shingle and gravel without the least signs of vege- "tation—*not even Lichens*" (p. 373).‖ Von Baer ¶ says that absence of vegetation is characteristic of the deserts of Nova Zembla. Nevertheless, he goes on to remark, that while foliaceous Lichens are scarce, every block of augitic porphyry is clothed with crustaceous species, which occur also, though less frequently or copiously, on rocks or stones of other mineralogical character. He specialises *Lecidea geographica* and *Stereocaulon paschale* as prominent forms.

My Catalogue enumerates 268 species and varieties (that is, those separately named by systematists) of Lichens in Greenland. In 1840 the Greenland Lichens on record amounted only to 59, and the difference between these figures shows the extent of the contributions that have been made in the interval to its Lichen-flora. I have elsewhere estimated the Lichens of Iceland at about 150.** But it must be borne in mind that Iceland is a very much smaller country than Greenland—occupying only about 3 or 3½ degrees of latitude (from 63½° to 66½°)—while its Lichens have certainly not been collected and studied to the same extent. I believe that the Lichens both of Greenland and Iceland are at present under-estimated. I have no doubt that considerable

---

* A type of an *un*attached desert or steppe Lichen is *Lecanora esculenta*, Pall.; described in my "Hist. Brit. Lichens," pp. 228, 211, 51.

† Probably a form of *Cl. rangiferina*, which is known to occur in Brazil.

‡ Probably a form of *Placodium elegans* or *Pl. murorum*.

§ "Naturalist's Voyage," chap. xvi.

‖ "Spring and Summer in Lapland."

¶ "Voyage to Nova Zembla:" Bulletin Scientifique de l'Acad. Impér. des Sciences de St.-Petersbourg.

** "Northern Lichen-Flora," pp. 393–4.

additions remain to be made to the Lichen-flora of both countries. That of Greenland cannot be set down at *less* than 300, and it will probably considerably exceed this. There are few special collections of Lichens made in Greenland that do not contain new forms. Thus, Th. Fries, Nylander, and myself have detected novelties in the collections respectively submitted to our examination—a circumstance that shows what might be achieved by the visit of an experienced Lichen-collector even to Greenland. As regards the geographical distribution of Greenland Lichens, it is sufficient here to refer to those—

1. That are confined to Greenland.
2. " " the Arctic regions.
3. " common to Britain.
4. " " the European Alps.

Those that are confined to Greenland, or that may meanwhile be held as so restricted in their distribution, are the new species or varieties described by Fries, Nylander, or myself.* I have little doubt, however, that the majority at least of these Lichens will sooner or later be found in other countries — Arctic or more southern.

The purely or generally *Arctic* species are very few, viz. :—

Dactylina arctica.
Usnea melaxantha.
Pyrenopsis hæmatopis.
Alectoria jubata, var. *nitidula.*
Peltidea scabrosa.
Umbilicaria Pennsylvanica.
Pannaria lepidiota, var. *tristis.*
P. Hookeri, var. *macrior.*
Squamaria chrysoleuca, var. *feracissima.*
S. melanaspis, var. *alphoplaca.*
S. geophila.
Lecanora tartarea, vars. *grandinosa* and *thelephoroides.*
L. varia, var. *leucococca.*
L. atro-sulphurea.
L. ferruginea, vars. *cinnamomea* and *hypnophila.*
Lecidea spilota, var. *polaris.*
L. auriculata.
L. armeniaca, var. *melaleuca.*
L. pallida.
L. insignis, var. *geophila.*
L. scabrosa, var. *cinerascens.*
L. urceolata, and var. *deminuta.*
L. coronata.
L. cumulata.
L. castanea.
L. Tornœensis.
L. subfuscula.
Arthonia trabinella.
Verrucaria maura, var. *aractina.*
V. ceuthocarpa.

But some of these Lichens occur in countries or districts south of the Arctic Circle. Thus *U. melaxantha* occurs in Iceland (according to Th. Fries, Arct., p. 25, and Carroll in Seemann's "Journal of Botany," vol. v, p. 109). It occurs also very frequently in the southern hemisphere; in Patagonia and its islands, on the Andes, in New Zealand and Tasmania, and on the Antarctic islands. In the Arctic regions it is invariably sterile, while in the Antarctic it is often fertile. *P. hæmatopis* I have found in

* The *new* species or varieties found by myself in Dr. Brown's collections are described in my "Observations on Greenland Lichens."

Iceland * nearly as far as south as lat. 64°. *U. Pennsylvanica* occurs in the United States as far south, at least, as lat. 40°. *L. Tornœensis* descends below the Arctic Circle, but apparently not many miles. *L. subfuscula* appears to be what Nylander named in my Herbarium *L. bacillifera*, var. *subfuscula*. It occurs in Iceland.† *L. coronata* (*Rhexophiale* of Th. Fries, Arct., p. 705) appears to be the *Lecidea rhexoblephara*, Nyl. (Scand., 240, and Carroll, 290). If so, it occurs in Scotland, on Ben Lawers, according to Jones.

Deducting the species that are mainly or entirely confined in their distribution either to Greenland or to Arctic countries, the majority at least of the remainder occur on the Scandinavian Alps, and many of them on the Alps of Scotland and Switzerland, or generally on those of continental Europe; while a considerable number are common *British* forms. Some Lichens, which are only alpine in Britain (Scotland), occur, as might be expected, at low elevations, and even on or near the sea-level, in Greenland (*e.g.*, *Thamnolia vermicularis*, which is abundant in the Jakobshavn district, on or about the coast).

The *non-British* species are the following:—

Alectoria divergens.
Cladonia carneola.
cyanipes.
Cetraria odontella.
Nephroma arcticum.
papyraceum.
Parmelia centrifuga.
Umbilicaria anthracina.
hirsuta.
spodochroa.
Squamaria chrysoleuca.
straminea.
Lecanora chlorophona.
epanora.
oreina.
peliscypha.
molybdina.
Lecanora Jungermanniæ.
turfacea.
nimbosa.
mniaræa.
Lecidea alpestris.
aglæa.
elata.
geminata.
obscurata.
fuscescens.
cuprea.
cinnabarina.
leucoræa.
squalida.
Endocarpon dædaleum.
Verrucaria mucosa.
clopima.

A comparison of the Greenland Lichen-flora with that of Arctic America (assuming Leighton's Catalogue of Sir John Richardson's collections in 1826 ‡ to be representative of the Arctic-American Lichen-flora), shows that there exists a considerable difference in the elements of which they are respectively made up. There are certain genera and species in Arctic America that do not occur in Greenland, while there are in Greenland at least many species that do not occur in Arctic America. The

* "Northern Lichen-Flora," p. 370.

† Ibid., p. 372.

‡ Apparently during Franklin's Second Land Expedition, 1825–6.—Journal of Linnean Society, vol. ix, Botany, p. 184.

first category includes Lichens that are more peculiarly *American*; the second those which are characteristically *Scandinavian*.*

In Arctic America there are four genera that are unrepresented in Greenland, viz., *Odontotrema*, *Gyrostomum*, *Xylographa*, and *Graphis*; and the following species:—

Nostoc commune.†
Cladonia cornuta; cariosa; pityrea.
Usnea barbata; lacunosa.
Ramalina calicaris.
Evernia Prunastri.
Nephromium tomentosum.
Platysma Richardsoni; lacunosum.
Sticta herbacea.
Physcia ciliaris; candelaria.
Umbilicaria pustulata; Muhlenbergii.
Squamaria ambigua.
Lecanora fulvo-lutea.
Pertusaria leioplaca.
Lecidea coarctata; vesicularis; tessellata; chalybeia; protrusa; parasitica.
Verrucaria Frankliniana; nigrescens; glabrata.
Odontotrema Richardsoni.
Gyrostomum urceolatum.
Xylographa flexella.
Graphis serpentina.
Arthonia intumescens.

There is, however, a much larger number of Lichens belonging to the Greenland flora, and for the most part to that of Scandinavia, that do not occur in Arctic America. These include, besides the new forms found in Brown's Collection, the following species enumerated in the present Catalogue of Greenland Lichens:—

Lecidea—all except geographica; sanguineo-atra; disciformis; atro-brunnea; turgidula; sabuletorum.
Lecanora—all save vitellina; cerina; cinerea; parella; subfusca; glaucoma; oreina; turfacea; varia; bryontha; smaragdula; verrucosa.
Parmelia—all save physodes; saxatilis; olivacea; lanata; stygia.
Umbilicaria—all except vellea; cylindrica; erosa.
Squamaria—all save elegans; chrysoleuca; saxicola.
Stereocaulon—all save tomentosum.
Verrucaria—all except epidermidis.
Ephebe—all.
Endocarpon—all.
Sticta—all.
Urceolaria—all.
Sphærophoron—all.
Pyrenopsis—all.
Arthonia trabinella.
Pannaria lepidiota; Hookeri.
Cetraria aculeata; odontella.
Cladonia bellidiflora; digitata; carneola; cyanipes; verticillata; cervicornis.
Collema flaccidum; melænum; lacerum.
Usnea melaxantha.
Alectoria Thulensis.
Solorina crocea.
Peltidea scabrosa.
Nephroma arcticum.

* So far as we may venture to judge from the present imperfect data, it would appear that the affinities of the Greenland Lichen-flora are greater towards that of Scandinavia, and generally of *Europe*, than to that of *America*.

† This may be included in lists of Greenland *Algæ*. Berkeley (in Treasury of Botany") places *Nostoc*, and "Falling stars," among *Algæ*.

The difference between the Greenland and Arctic-American Lichen-floras is obvious from a comparison of the summary appended to my Catalogue of the former with Leighton's similar summary of the latter (p. 185). But comparisons based on such tables alone are most fallacious, inasmuch as they are drawn up on very different principles. Leighton, for instance, gives 35 genera, while I give only 28 ; the number of species and varieties in Greenland being 268, and in Arctic America 203. My genera are fewer, however, mainly because I do not split up such genera as *Collema*, *Cladonia*, *Usnea*, *Alectoria*, *Cetraria*, *Nephroma*, *Parmelia*, *Physcia*, *Squamaria*, *Lecanora*, *Lecidea*, *Endocarpon*, and *Verrucaria* into the host of sub-genera into which they have been divided of late years by Continental Lichenologists. Another class of discrepancies necessarily arises from the different position given to certain anomalous species, such as *Lecanora* or *Pertusaria bryontha*, *Thamnolia* or *Cladonia vermicularis*, *Endocarpon viride* or *Normandina viridis*. Richardson's Lichens, according to Leighton, amount to 163 (p. 184)—a number which does not correspond with the total given in his table or summary (p. 185). But he names or numbers a series of trivial or inconstant forms or conditions, which in the hands of some other lichenologists—certainly in mine—would not receive separate nomenclature or enumeration.* Treated in a similar way, the number of species and varieties given in the following Catalogue of Greenland Lichens would be largely increased, for I have neither named nor numbered the "various forms" or "several forms" of many variable species (*e.g.*) of the genera *Cladonia*, *Parmelia*, *Physcia*, *Lecanora*, and *Lecidea*.

Leighton's enumeration, however, is very far from giving an adequate idea of the Lichen-flora of the vast area known as Arctic America. Its deficiencies may be judged of by the number of species mentioned in other works or herbaria as Arctic-American Lichens that are not enumerated by Leighton. Thus Tuckerman in his "Synopsis," published in 1848, records the following, which do not occur in Leighton's list:—

Trachylia tigillaris.

Calicium lenticulare, *Ach.*; subtile, *Pers.*; phæocephalum, *T. & B.*

Cladonia alcicornis; carneola; turgida, *Hffm.*

Sphærophoron fragile; globiferum, *L.*; compressum, *Ach.*

Alectoria ochroleuca and *var.* rigida; jubata, *var.* bicolor; implexa, *Fr.*

Ramalina polymorpha.

Dufourea ramulosa, *Hook.*

Cetraria odontella ; aculeata.

Solorina crocea.

Nephromium arcticum ; resupinatum, *Ach.*

Parmelia tristis ; Fahlunensis ; caperata ; conspersa ; diversicolor, *Ach.*

Physcia parietina, *vars.* polycarpa, *Fr.*, and laciniosa, *Duf.*

---

* His elaboration of the genus *Cladonia* may be taken as an illustration, and compared with my remarks on that genus in my "Arctic Cladoniæ," Trans. Bot. Soc. of Edin., vol. ix, p. 179.

Pannaria hypnorum; triptophylla, *Fr.*
Squamaria straminea; muscorum.
Lecanora atra; tartarea; pallescens, *Fr.*; oculata; badia; ventosa; exigua; aurantiaca; fusco-lutea, *Hook. & Dicks.*
Urceolaria scruposa.
Thelotrema lepadinum.
Lecidea parasema; fusco-atra; confluens; lapicida; variegata, *Fr.*; rivulosa, *Fr.*; galbula; candida, *Ach.*; vernalis; decolorans, *Fr.*; lucida, *Fr.*
Umbilicaria proboscidea; hyperborea; hirsuta; anthracina, *var.* reticulata, *Sch.*
Pertusaria faginea.
Endocarpon miniatum; lætevirens.
Verrucaria punctiformis, *Pers.* (= *var.* of epidermidis).
Pyrenothea leucocephala; *var.* lecidina, *Fr.* (= Lecidea abietina, *Ach.*)

Nylander, in his "Enumération générale des Lichens,"* published in 1858, mentions in addition,—

Siphula ceratites.
Platysma septentrionale, *Nyl.*
Physcia aquila, *Fr.*, *var.* compacta, *Nyl.*
Umbilicaria arctica.
Lecanora frustulosa, *var.* subventosa, *Nyl.*; subsophodes, *Nyl.*
Pertusaria concreta, *Nyl.*
Lecidea cinnabarina.

In the Menziesian Herbarium (Edinburgh) I found the following—labelled as collected by Richardson, but not included in Leighton's list:—

Alectoria ochroleuca (type).
Parmelia conspersa.
Peltidea scutata.
Umbilicaria hyperborea.
Lecidea fusco-lutea, *Dicks.*

While that of Kew contains the following—also collected by Sir John—but in 1848-9, and not mentioned by Leighton:—

Parmelia saxatilis, *var.* furfuracea.
Peltidea scutata.
Physcia aquila, *var.* compacta; parietina (type).
Umbilicaria proboscidea.
Placodium murorum.
Lecanora polytropa.
Lecidea anomala.

These omissions amount to 84 species and varieties, which, added to Leighton's totals—163 or 203—give an aggregate of 247 or 287; the mean of the two estimates being 267—a number that may be said to equal the aggregate Greenland Lichen-flora—268. Both Nylander and Leighton found new species in the Arctic-American Lichen-collections which they examined. What is known as Arctic America comprises a very large area of country, much of it wooded; and there can be no doubt, I think, that if its Lichens had been collected and studied with the same care as those of Greenland or Spitzbergen, its Lichen-flora would have

* "Mémoires de la Société Impériale des Sciences Naturelles de Cherbourg," vol. v, 1857, p. 85.

attained a much higher numerical position than that of either of these countries.*

No proper comparison, however, can be made between Greenland and Arctic America as regards their lichenose vegetation. The area of Richardson's collections—catalogued by Leighton—lies between 47° and 67° N. lat., while that of the various Greenland collections reaches from 60°, the southmost point of Greenland, as high as 82° N. lat. Though called "Arctic," no part of the so-called Arctic America † of Leighton's catalogue lies within the Arctic Circle ; ‡ while of equal importance, with mere latitudinal difference, is the abundance of forests in America, and their absence in Greenland,—a circumstance that has a similar influence in determining the difference between the Lichen-floras of Iceland and Scandinavia.§ In other words, America and Scandinavia possess a large and varied Lichen-flora of *corticolous* ‖ species, which cannot be looked for in Greenland or Iceland. This marked difference in the arboreal vegetation of the two countries renders it unnecessary to contrast the Lichen-flora of Greenland with that of Scandinavia.

It is, however, both legitimate and interesting to institute a comparison between the Lichen-floras of Greenland and Spitzbergen. The latter island is equally devoid of wood ; it extends nearly as far to the north as Greenland (76° to 80° N. lat.) ; and its Lichens have been examined and catalogued by the same distinguished Swedish botanist, Fries the younger, so that uniformity of nomenclature and classification is secured. The Lichens of Spitzbergen and its islets amount, according to Fries' "Lichenes Spitsbergenses," to 266,—that is, about the same as those of Greenland. Considering the very much smaller area of Spitzbergen, this is a large total; but, on the other hand, that Arctic island is so easily accessible from Norway that it has been repeatedly visited by Scandinavian botanists specially with a view to plant-collection. Its Lichen-flora has thus been much more fully studied than that of Greenland.

Prior to the publication of the "Lichenes Spitsbergenses," our knowledge of the Lichen-flora of Spitzbergen consisted mainly of the determinations by Sir William Hooker¶ and Robert Brown ** (of the British Museum) of the few Lichens collected by Sir

---

* The additions that yet remain to be made will occur, probably, in the group of microscopic saxicolous *Lecideæ* and *Lecanoræ*, which require for their collection, as well as examination and description, the eyes and the special knowledge of a skilled Lichenologist.

† *See* definition of the term *arctic*, in the author's "Arctic Cladoniæ."

‡ Arctic America includes also what was, till lately, known as Russian America; and its Lichen-flora ought to embrace the species collected by Dr. Seemann during Beechey's Voyage in 1848 between Norton and Kotzbue Sounds.

§ "Northern Lichen-Flora," p. 402.

‖ It will be observed that Richardson's collections were, in great measure, of *corticolous* forms.

¶ In the Appendix to "Parry's Fourth Voyage," 1827.

** In the Appendix to Scoresby's "Arctic Regions," vol. i, p. 75; and also in Robert Brown's collected works, edited by Bennett, 1866, vol. i, p. 181.

Edward Parry and Dr. Scoresby. Parry's collection was made, apparently, chiefly on the Spitzbergen islets, viz. :—

	N. Lat.		N. Lat.
Low Island - -	80° 20′	Little Table Island -	80° 48′
Walden Island -	80° 38′	Ross' Islet - -	80° 49′

as well as in Hecla Cove, which is, I presume, on the main island. *En route* he also made collections * at Hammerfest, near the North Cape of Norway, about 71° N. lat. Scoresby's collection, again, appears to have been made on the main island, in King's Bay or about Mitre Cape. The determinations of Hooker and Brown were, doubtless, made without microscopical examination; hence their lists of Spitzbergen Lichens are no exception to the rule, that all determinations founded exclusively on external non-microscopical characters include many forms that cannot be identified with modern species. The following illustrations will show the difficulty connected with synonymy in the catalogues of Hooker and Brown.†

Gyrophora deusta, *Ach.*, may be either Umbilicaria flocculosa, *Hffm.;* or U. arctica; or U. proboscidea,—to which both Leighton and Th. Fries refer it, and which is a Greenland species.

G. tessellata, *Ach.*, is U. anthracina, *Sch.*, var. reticulata, *Sch.*, according to Fries.

G. hirsuta does not occur in Spitzbergen, according to Th. Fries (p. 53), and is therefore an error in determination. He suggests that the plant may be a form of U. vellea.

Cladonia alcicornis is a similar error for similar reasons (Th. Fries, p. 53). Fries suggests that the plant was perhaps Cl. macrophylla.

Stereocaulon paschale, Fries suggests, that was perhaps really form of *S. tomentosum* (pp. 27 and 53).‡

Sphærophoron fragile. He expresses a similar doubt here, suggesting that it may belong to *coralloides.* I hold such doubts and distinctions, however, to be unnecessarily nice, inasmuch as I see no good ground for separating the different forms of *Sphærophoron* or *Stereocaulon* under separate *species.*

Parmelia stygia, he suggests (p. 12), may belong rather to his *alpicola.* I have seen no authentic specimen of his *alpicola;* but from the circumstance of his recording its occurrence in Scotland, it appears to me that Nylander is probably correct in considering *alpicola* a mere form of *stygia.*

Nephroma polaris, *Ach.*; without fruit: Hammerfest; is probably *N. arcticum*, L.

Alectoria ochroleuca occurs only as var. *rigida*, and A. jubata only as *var.* chalybeiformis (*Th. Fries*).§

---

* Mentioned in his First, Second, and Third, as well as Fourth Voyages.

† Compare also "L. Spitsberg.," p. 53, "Species ab auctoribus allatæ, "verisimiliter omnino excludendæ."

‡ *S. denudatum, Flk.*, occurs in Kew Herb., labelled "Ross' Islet (Parry)."

§ *Usnea melaxantha*, labelled "Spitzbergen, Sabine and Scoresby," occurs in Kew Herb.

Isidium oculatum * is doubtless Lecanora oculata.
Parmelia recurva, *Ach.*=P. incurva, *Pers.*
Endocarpon sinopicum, *Whlnb.*=Lecanora smaragdula, *var.*
Lecidea atro-virens, *Ach.*=L. geographica.
Cornicularia spadicea, *Ach.*=Cetraria aculeata.

There is less or no difficulty as to the identification of the remainder of the Lichens enumerated by Hooker or Brown, which are the following†:—

Cladonia rangiferina; pyxidata; gracilis; cornucopioides; furcata; bellidiflora.
Cetraria nivalis; Islandica; cucullata. ‡
Sphærophoron coralloides.
Peltidea canina; aphthosa.
Thamnolia vermicularis.
Parmelia saxatilis, *var.* omphalodes; lanata.§
Umbilicaria proboscidea; || erosa.¶
Placodium elegans; murorum.
Lecanora tartarea.**
Solorina crocea.
Alectoria divergens.

The whole number of species and varieties enumerated by Hooker and Brown is only 37, while the Spitzbergen Lichens catalogued by Fries amount to 266,—the very great difference being a measure of the progress that has been made in the collection and study of the Lichens of that island since the voyages of Parry and Scoresby. All English lists of Spitzbergen Lichens are included in the "Lich. Spitsbergenses" of Fries, which is—and is likely long to remain—a standard work on the Lichens of that island.

Contrasting his list with the Catalogue which follows of Greenland Lichens, it is at once obvious that—as in the case of

* The specimen in the Kew Herb., labelled "Walden Island (Parry)," appears to me, however, to be referable to *Lecanora tartarea.*

† I have had the opportunity of examining several of them for myself in the Kew Herb.

‡ *C. juniperina* occurs in the Kew Herb., labelled "Arctic Islets (Parry)."

§ *P. caperata,* in abundant fruit, labelled "Spitzbergen, Ross," occurs in the Kew Herb. But I am not aware of Ross having visited Spitzbergen. *Physcia parietina* (sub nom., *candelaria*), labelled "Ross' Islet (Parry)," occurs in Kew Herb.

|| *U. hyperborea* occurs in Kew Herb., labelled "Walden Island (Parry);" and *U. vellea,* labelled "Ross' Islet (Parry)." What is labelled in the same Herb. *U. proboscidea,* "Parry's Voyage to the North Pole," appears to me to belong to *cylindrica.* In the same Herb. *U. hyperborea* occurs, labelled "Spitzbergen, Parry, Voyage to the North Pole" (partly sub nom., *Gyrophora tessellata,* partly *G. vellea*).

¶ It does not appear which species of "very large Tripe de Roche" it is that is or are described by Parry as very abundant on rocks on the south side of Walden Island, and on the sides of Little Table Island. In his "Narrative" (1829, pp. 65, 67, and 175) he refers to the abundance on Walden Island of *Umbilicaria proboscidea, Cladonia rangiferina,* and *Alectoria divergens,* while the "Tripe de Roche" was more luxuriant than he had ever seen it elsewhere. Scoresby, too, describes the rocks of Spitzbergen as "covered "with a mourning veil of black Lichens" (consisting apparently of three species of *Umbilicaria; Parmelia stygia;* and *Alectoria chalybeiformis*).

** Var. *frigida* (sub nom. *Upsaliensis*) occurs in the Kew Herb., labelled Spitzbergen, 1773, C. J. Phipps."

Arctic America—there is a large proportion of species in the one country that does not occur in the other.* It must suffice, as an illustration, to enumerate those which—occurring in Spitzbergen—have not hitherto been found in Greenland. This category includes, in the first place, no less than 25 new species or varieties found for the first time in Spitzbergen, and described by Fries, viz.:—

Toninia conjungens.
Bilimbia microcarpa.
Biatorina globula, *Flk.*, *var.* polytrichina; tuberculosa; Stereocaulorum.
Lecidea polycarpa, *Körb.*, *var.* clavigera; ramulosa; pullulans; sulphurella; impavida; associata.
Sporastatia tenuirimata; Spitsbergensis; cinerea, *Sch.*, *var.* haplocarpa.
Buellia vilis; urceolata; convexa.
Arthonia excentrica.
Polyblastia Gothica.
Verrucaria rejecta; extrema.
Arthopyrenia conspurcans.
Sticta linita, *Ach.*, *var.* complicata.
Lecanora coriacea; erysibe, *Ach.*, *var.* personata.

Deducting these new species and varieties, there still remains the large number of 67 species and varieties not as yet found in Greenland, viz.:—

Dufourea muricata, *Laur.*
Sticta linita, *Ach.*
Physcia parietina, *var.* aureola, *Ach.*
Peltidea malacea, *Ach.*; polydactyla, *Hffm.*
Pannaria microphylla, *Sw.*; arctophila, *Th. Fr.*
Lecothecium asperellum, *Whlnb.*
Arctomia delicatula, *Th. Fr.*
Placodium albescens, *Hffm.*
Lecanora glaucocarpa, *Whlnb.*; subsimilis (*sub* Gyalolechia), *Th. Fr.*; aipospila, *Whlnb.*; gibbosa, *Ach.*, and *var.* squamata, *Ach.*; mastrucata, *Whlnb.*; cinereo-rufescens, *Ach.*, *var.* alpina, *Smrf.*; rhodopis, *Smrf.*, *var.* melanopis, *Smrf.*; flavida, *Hepp*; Dicksoni, *Ach.*; pyracea, *Ach.*; oligospora, *Rehm.*
Hymenelia Prevostii, *Fr.*
Lecidea (*sub* Toninia) fusispora, *Hepp*; (*sub* Bacidia) viridescens, *Mass.*; venusta, *Hepp*; (*sub* Bilimbia) syncomista, *Flk.*; (*sub* Biatorina) fraudans, *Hellb.*; (*sub* Biatora) miscella, *Smrf.*; collodea, *Th. Fr.*; Lulensis, *Hellb.*; curvescens, *Mudd.*; rupestris, *Scop.*; terricola, *Anzi.*; (*sub* Lecidea) rhætica, *Hepp*; confluens, *Web.*; tenebrosa, *Fw.*; (*sub* Sporastatia) Morio, *Ram.*, *v.* coracina, *Smrf.*; cinerea, *Sch.*; privigna, *Ach.*; (*sub* Buellia) punctata, *Flk.*; cæruleo-alba,

* Some of the deficiencies of the Spitzbergen flora are remarkable, *e.g.*, *Parmelia saxatilis, physodes, olivacea, Fahlunensis*; *Cladonia furcata*; *Lecanora glaucoma, cinerea*; *Urceolaria scruposa*; *Lecidea fusco-atra, icmadophila*; *Endocarpon miniatum.*

Kremp.; Rittokensis, *Hellb.;* coracina, *Hffm.;* coniops, *Whlnb.*

Arthonia fusca, *Mass.;* clemens, *Tul.*

Endocarpon (*sub* Dermatocarpon) cinereum, *Pers.;* pulvinatum, *Th. Fr.*

Verrucaria (*sub* Microglæna) sphinctrinoides, *Nyl.;* (*sub* Polyblastia) theleodes, *Smrf.*, and *v.* Schæreriana, *Mass.;* Helvetica, *Th. Fr.;* hyperborea, *Th. Fr.;* bryophila, *Lönnr.:* gelatinosa, *Ach.;* sepulta, *Mass.;* (*sub* Thelidium) pyrenophorum, *Ach.;* (*sub* Verrucaria) margacea, *Whlnb.;* striatula *Whlnb.;* rupestris, *Schrad.*, *v.* integra, *Nyl.;* (*sub* Endococcus) gemmifera, *Tayl.*

Collema pulposum, *Bernh.;* ceranoides, *Borr.;* scotinum, *Ach.*

Leciophysma Finmarkica, *Th. Fr.*

Pyrenopsis granatina, *Smrf.*

Several, moreover, of the *genera* in the foregoing list are not represented in Greenland, *e.g.*, *Arctomia, Lecothecium, Hymenelia, Sporostatia, Polyblastia, Bacidia, Microglæna, Thelidium, Leciophysma.*

Did the necessary data exist, it would be interesting to compare the Lichen-flora of Greenland with that of the *Arctic-American Islands*—those large islands north of the American Continent, intervening between Greenland and what has hitherto been known as Russian America. But, as regards these islands, the necessary data do not exist. Almost all we know of their Lichen-flora consists of the determination by Robert Brown (of the British Museum) of the collections of Parry during his First Voyage * on Melville Island,† most of which Lichens are now in the Kew Herbarium, where I have examined them. In that herbarium I found, labelled "Melville Island (Parry)"—

Alectoria bicolor; ochroleuca; divergens.

Cetraria aculeata.

Usnea melaxantha.

Dactylina arctica.

Pertusaria glomerata.

There are a few other Lichens, foliaceous or fruticulose, which *may* have been collected on that or other of the Arctic-American Islands; but the labels are so vague in their reference to localities, that the species to which they relate cannot be safely quoted in the present category. Thus one specimen (*Umbilicaria hyperborea*) is labelled, "North-West Passage" (Parry); another, "Arctic Islets"; a third, "Parry's Voyage to the North Pole"; others, "North Pole," or "North-Polar Expedition." Such descriptions too frequently constitute *all* the information conveyed by the labels attached to Parry's Arctic Lichen-collections

* In the Appendix to the First Voyage, 1819–20, and reprinted in Brown's Botanical works, vol. i (1866), p. 250.

† There are probably other citations, *e.g.*, in the List of Dr. Sutherland's collections in the "Lady Franklin" (Captain Penny), given by Churchill Babington in "Hooker's Journal of Botany," iv. 276, to which I have not at present access. The only quotation in my note-book is *Lecanora vitellina,* on bone used as an implement by the Esquimo on Cornwallis Island.

contained in the Kew Herbarium. I have little doubt that they refer mainly, if not exclusively, to the *Spitzbergen* group of islets mentioned on page 285.

Tuckerman, in his "Synopsis," enumerates as Melville Island Lichens, following Brown's determination of Parry's collections—

Usnea melaxantha (*sub nom.* sphacelata.)
Cladonia pyxidata.
Cetraria odontella.
Stereocaulon paschale.
Parmelia stygia.
Placodium elegans.

Even to this very meagre list difficulties relating to synonymy—similar to those already pointed out under the head of the Spitzbergen collections of the same distinguished navigator, and the determinations of the same celebrated "Botanicorum Princeps"—attach themselves. Thus *Stereocaulon paschale* may be really, as in the Spitzbergen plant, *tomentosum.* I have elsewhere seen it recorded that *S. corallinum*, Fr., occurs both in Melville Island and Ross's Islet, as well as in the "barren lands" of Arctic America. This Lichen may be *S. coralloides*, Fr.; but in the "*L. Arctoi*" (p. 142), it is given as not distributed more to the north than Lapland and Nordland. Equally probably it is *tomentosum;* or, according to my own view, it and all the Arctic forms of *Stereocaulon* are referable, as mere conditions, to the type *S. paschale.*

The whole list of Melville Island Lichens contains the insignificant number of twelve species; while there is no reason to doubt the Lichen-flora of the Arctic-American Islands must be as rich, at least, as that of Spitzbergen.

Though by no means to equal extent, the Lichen-flora of Iceland is, as I have elsewhere * shown, very defective, amounting only to 147 species and varieties. It does not, therefore, any more than the Lichen-flora of Arctic America or the Arctic-American Islands, afford data for comparative generalisations in regard to the Lichens of Greenland, and the countries or islands in similar latitudes east and west of it in Europe or America.

Most singularly, there is no record of any *Economical Applications* of Lichens in Greenland,† while there is no scarcity of evidence regarding the uses to which they are applied in Arctic America, Iceland, and Scandinavia:—(1.) As food for wild or domesticated animals; (2.) As food for man; (3.) As medicines; or (4.) As dye-stuffs. Dr. Brown assures me, in more than one of his letters, that Lichens are absolutely unapplied to any useful purpose by the Greenlanders or by the native animals of Greenland. "I really believe," he says, "there are no economic uses "for Lichens in Greenland. . . . I made inquiries of all the "Danish officials, and . . . I have re-examined Rink's 'Grön-"'land geographisk og statistik' for any reference, and can

* "Northern Lichen-Flora," pp. 393–4; *vide* also p. 289 of present Memoir.

† Parallel instances of non-use of Lichens in countries in which useful species abound are given in my "Northern Lichen-Flora," p. 415.

"find none; now, Rink is Governor of South Greenland, and "has passed some 15 years in the country. Lichens may be used "in Iceland and yet not in Greenland, where the Eskimo element "is predominant. The Icelanders learned their use from their "Norse ancestors."—*Letter of July* 1868.

The author of the "Edinburgh Cabinet Library" volume on Greenland writes (p. 228), "The curious fact of finding Reindeer "in this desolate region would seem to imply that it was not all "so barren or devoid of vegetation as the portion just described." But though Crantz, in his "History of Greenland," published in 1820 (p. 61), apparently describes the Reindeer-food as consisting, in winter at least, of *Cladonia rangiferina*,* the so-called abundant "Reindeer-moss" of all arctic and sub-arctic countries, Brown says that it does not now at all constitute their food.† On the other hand, in other parts of the North Polar regions,‡ if not in Greenland, Captain Hall describes *Cl. rangiferina* as the most important terrestrial vegetation, "for on it feed the herds "of Reindeer, which are the prey of the Wolves, Bears, and "Esquimaux."§ It is only possible to reconcile such discrepancies in evidence by supposing that in certain parts of Greenland, or the Arctic regions generally, "Reindeer-moss" is abundant, and constitutes the winter or other food of the Reindeer; while in others the "moss" in question is absent or scarce, or it does not form the food, or part of the food, of that animal.

The abundance in Greenland of at least some of the Lichens that in other Arctic countries are, or have been, serviceable in supplying the wants of man or animals, and the possibility of developing the economical applications of Lichens in a country whose resources are otherwise so limited, render it desirable here to make some reference to certain of the uses elsewhere of the more abundant Arctic species.

In Arctic America the various *Umbilicariæ*, as "Tripe de roche," have played a very prominent part in the history of geographical exploration, more especially so in the record of the now well-known land-journey of Franklin, Richardson, Hood, and Back. When, in October 1821, Franklin and his companions were pushing on for the Coppermine River in a condition approaching starvation, under the joint influence of cold and hunger, that intrepid explorer writes:—"Our suffering from cold in a "comfortless canvas tent, in such weather, with the temperature "at 20 degrees, and without fire, will easily be imagined: it

---

* He also describes Hares as living partly on it (p. 66).

† *See* my "Observations on Greenland Lichens," sub *Cl. rangiferina*. A short paragraph on the food of the Greenland Reindeer will be found in Dr. Brown's paper "On the Mammalian Fauna of Greenland," in the Proceedings of the Zoological Society (of London) for 1868, No. xxiii, p. 355. (*See* above, p. 26.)

‡ Dr. Brown writes me, "Hall's 'Esquimaux Land' refers, I suspect, "wholly to the *western* shores of Davis' Strait. In Danish Greenland there "is scarcely a pure native living."

§ "Life with the Esquimaux: the Narrative of Captain Hall of the Whaling "Barque, 'George Henry,'" 2 vols., London, 1864.

" was, however, less than that which we felt from hunger. . . . .
" Weak from fasting, and their garments stiffened with the frost, " after packing their frozen tents and bed-clothes, the poor tra-" vellers again set out on the 7th. . . . . After feeding almost " exclusively on several species of *Gyrophora* (= *Umbilicaria*), a " Lichen known as 'Tripe de roche,' which scarcely allayed the " pangs of hunger, on the 20th they got a good meal by killing a " Musk-ox. . . . . On the 17th they managed to allay the pangs " of hunger by eating pieces of singed hide and a little 'Tripe " de roche.' This and some Mosses, with an occasional solitary " Partridge, formed their invariable food; on very many days " even this scanty supply could not be obtained, and their appe-" tites became ravenous. . . . . Mr. Hood was also reduced to a " perfect shadow from the severe bowel-complaint which the " 'Tripe de roche' never failed to give him. . . . . Not being " able to find any 'Tripe de roche,' they drank an infusion of the " 'Labrador tea-plant' (*Ledum palustre*, v. *decumbens*), and ate " a few morsels of burnt leather for supper. This continued to " be a frequent occurrence." *

Horace Marryat, in his "One Year in Sweden," † writes, " We now tread under foot what has served as food for men, " baked into bread, and that right often in Sweden," referring, probably, to *Cladonia rangiferina*, for he mentions *Cetraria Islandica* separately. "The peasants are ready prepared for famine. " . . . . In an old book are printed as many as 13 receipts for " what is termed 'Weed-bread,' commencing with Bark" Iceland-moss-bread, or Lichen-bread, is only one kind. Among others there are bone, grass, straw, sorrel, bran, and furze breads! The substances which give the bread its name are, in these cases, probably only the *chief* ingredients as to bulk. Consul Campbell, in his report to the British Government on the trade of Finland for the year 1867, alluding to the famine of that year in that country, says, "The bread given to support life is composed of pease-" straw, combined with Iceland-moss and a small proportion of " flour."

The "Old Bushman," in his "Spring and Summer in Lapland," ‡ writes, "That the Reindeer thrive [on *Cl. rangiferina*], " is proved by the fact that no park-fed Deer in England can look " fatter and sleeker than the Reindeer when they come down " from the fells at the end of summer; in fact, 'fat as a Rein-" deer,' is a common saying here." It would even appear to be occasionally too rich a fodder. The hair of the animal becomes frequently very brittle; it snaps across as if rotten, and falls readily from the skin. This condition is ascribed to its feeding too much on "dry moss" (p. 220). That *Cl. rangiferina* contains a considerable percentage of starchy matter is shown by the fact that quite recently a Swedish chemist has obtained *alcohol* from

* Extracts from the narrative of Franklin's First Land Expedition (1819–21), in Simmonds' "Sir John Franklin and the Arctic Regions."

† London, 1862, vol. i, p. 231, describing the Falls of Trollhättan.

‡ London, 1864, p. 173.

it in large quantity (as well as from other *amyliferous Lichens*), by converting the *starch* into sugar by heat and acids, and the sugar into alcohol by fermentation.†

Parry points out the scarcity of *Cl. rangiferina* in certain parts of the North-Polar regions, in relation to the requirements of tame Reindeer. "It would be next to impossible to procure there "a supply of provender sufficient even to keep them alive, much "less in tolerable condition, a whole winter" (p. 206). He shipped Reindeer, for sledge-drawing during his expedition, at Hammerfest, taking with him a supply of *Cl. rangiferina* as their only provender (p. 7).‡

Franklin states that his party "used the Reindeer-moss for "*fuel*, which afforded us more warmth than we expected" (p. 128).§

II. *Enumeration of the Lichens of Greenland.*

The prefixed asterisks indicate the species or varieties collected by Dr. Brown. They have been already separately enumerated in his "Florula Discoana," Trans. Bot. Soc. Edin., vol. ix, p. 454. *See above*, p. , where the localities ‖ of almost all are given.

The generic names given in parentheses are those of Th. Fries' classification of Arctic Lichens (in his "L. Arctoi," 1860, and "L. Spitsbergenses," 1867).

Gen. 1. *Ephebe.*

Sp. 1. * pubescens, *L.* Th. Fries mentions it only on authority of Nylander (Syn., p. 90).

Gen. 2. *Pyrenopsis.*

Sp. 2. hæmatopis, *Smrf.*

Gen. 3. *Collema.*

Sp. 3. * melænum, *Ach.*

4. * flaccidum, *Ach.* (Synechoblastus.)

5. saturninum, *Dicks.* (Leptogium.)

6. * lacerum, *Sw.*

var. pulvinata, *Ach.*

Gen. 4. *Calicium.*

Sp. 7. *furfuraceum, *L.* (Coniocybe.)

Gen. 5. *Sphærophoron.*

Sp. 8. *coralloides, *Ach.;* various forms (*e.g.*, *var.* isidioidea, *Linds.*); rarely fertile.

9. fragile, *L.;* frequently fertile.

Gen. 6. *Cladonia.*

Sp. 10. *pyxidata, *L.;* several forms.

† "Illustrated London News," November 7, 1868.

‡ "Narrative of an Attempt to reach the North Pole" (in 1827), by Captain Parry. London, 1829.

§ Narrative of his First Land Expedition, 1819–22, vol. ii, 2d. ed., London, 1824.

‖ On the south and east coasts of Disco Bay, the sides of the eastern end of the Waigat Strait, and at Godhavn on the south point of Disco Island.—Editor.

Sp. 11. carneola, *Fr.*; not mentioned by Th. Fries, but cited by Tuckerman (Syn. Lich. New Engl., p. 52), on authority of Fries *père*.
12. verticillata, *Hffm.*, and
**var.* cervicornis, *Ach.*; both rare.
13. *gracilis, *L.* (ecmocyna, *Ach.*); various forms.
14. *furcata, *Schreb.*, and
**var.* crispata, *Ach.*
— subulata, *L.*
— pungens, *Ach.*, and other forms.
15. *rangiferina,† *L.*, and
var. sylvatica, *Hffm.*
16. *uncialis, *L.*; various forms.
17. *amaurocrœa, *Flk.*; fruit common.
18. *cornucopioides, *L.*
19. *bellidiflora, *Ach.*
20. *deformis, *L.*; various forms.
21. digitata, *L.*
22. *degenerans, *Flk.*, and
**var.* coralloidea, *Ach.*
23. *alcicornis, *Flk.*
24. *squamosa, *Hffm.*; various forms
25. *fimbriata, *Hffm.*
26. *cyanipes, *Smrf.*

Gen. 7. *Stereocaulon.*
Sp. 27. *paschale, *L.*
28. *tomentosum, *Fr.*, and vars. or forms.
**var.* alpina, *Laur.*
— coralloidea, *Linds.*
Sp. 29. *denudatum, *Flk.*, and
var. pulvinata, *Sch.*

Gen. 8. *Thamnolia* (Cladonia).
Sp. 30. *vermicularis, *Sw.*

Gen. 9. *Usnea.*
Sp. 31. melaxantha, *Ach.*; always sterile.

Gen. 10. *Alectoria.*
Sp. 32. jubata, *L.* (Bryopogon), and
**var.* bicolor, *Ehrh.*
* — chalybeiformis, *L.*
— nitidula, *Th. Fr.*; all sterile.
33. *Thulensis, *Th. Fr.*‡
34. *divergens, *Ach.* (Cornicularia.)
35. *ochroleuca, *Ehrh.*, and
var. cincinnata, *Fr.*
— sarmentosa, *Ach.*; very rare, on branches of *Salix glauca.*
* — rigida, *Vill.*
* — nigricans, *Ach.*§

---

† Th. Fries spells this *Rhangiferina*; but the other, though probably etymologically incorrect, is the spelling generally adopted.
‡ Assigned to *A. nigricans* in Fries' "L. Spitsberg."
§ Promoted to the rank of a separate species in Fries' "L. Spitsberg."

Gen. 11. *Dactylina.*

Sp. 36. *arctica, *Br.* (Cladonia; Dufourea.) Always sterile, says Th. Fries (Arct., p. 160).

Gen. 12. *Cetraria.*

Sp. 37. *aculeata, *Ehrh.* (Cornicularia), and
**var.* alpina, *Hepp*, 360.
— muricata, *Ach.*
* — acanthella, *Ach.;* with other forms.

38. *odontella, *Ach.*

39. Islandica, *L.;* and vars. or forms
**var.* leucomeloides, *Linds.*
— platyna, *Ach.*
* — crispa, *Ach.*†
* — Delisei, *Bory.*

40. *nivalis, *L.*

41. *cucullata, *Bell.*

42. juniperina, *L.*
var. pinastri, *Scop.*

43. sæpincola, *Ehrh.* (on bark of *Betula*).

Gen. 13. *Nephroma.*

Sp. 44. *arcticum, *L.;* always sterile.

45. papyraceum, *Hffm.* (Nephromium, *Nyl.*)
var. sorediata, *Sch.*

Gen. 14. *Peltidea.*

Sp. 46. *aphthosa, *Ach.*

47. *canina, *Hffm.*; various forms. Not mentioned by Th. Fries, but cited by Tuckerman (Syn. Lich. N. Engl., p. 29), on authority of Giesecké.

48. *rufescens, *Fr.*

49. scabrosa, *Th. Fr.;* fruit.

50. *venosa, *L.*

Gen. 15. *Solorina.*

Sp. 51. *crocea, *L.*

52. saccata, *L.*, and
var. limbata, *Smrf.*

Gen. 16. *Sticta.*

Sp. 53. scrobiculata, *Scop.*

Gen. 17. *Parmelia.*

Sp. 54. *saxatilis, *L.*, and
**var.* omphalodes, *L.*
* — panniformis, *Ach.*
* — leucochroa, *Wallr.*
* — sphærophoroidea, *Linds.*

55. physodes, *L.*, and
**var.* obscurata, *Ach.*

56. *encausta, *Sm.*, and
var. intestiniformis, *Vill.*

---

† In the Edinburgh University Herbarium there are specimens labelled "*C. Boryi*, Dél., Greenland," from Fries; and "Newfoundland," from Bory, 1831.

Sp. 57. hyperopta, *Ach.* (= aleurites, *Ach.*)
58. *olivacea, *L.*; various forms.
59. *Fahlunensis, *L.*, and
var. sciastra, *Fr.*
— polyschiza, *Nyl.*
60. *stygia, *L.*; various forms.
61. alpicola, *Th. Fr.* (Arct., p. 57.)
62. *lanata, *L.*; fruit abundant.
63. conspersa, *Ehrh.*
64. centrifuga, *L.*
65. incurva, *Pers.*
66. diffusa, *Web.* (= ambigua, *Ach.*)

Gen. 18. *Physcia.*

Sp. 67. *pulverulenta, *Schreb.*
var. muscigena, *Ach.*
68. *stellaris, *L.*; various forms.
69. *cæsia, *Hffm.*, and
var. albinea, *Ach.*
70. obscura, *Ehrh.*
var. orbicularis, *Neck.*
71. lychnea, *Ach.* (Xanthoria controversa, *Mass.*, *var.* pygmæa, *Bory*, Th. Fr., p. 68.)

Gen. 19. *Umbilicaria.*

Sp. 72. Pennsylvanica, *Hffm.*
73. *vellea, *L.*
74. *spodochroa, *Hffm.*
75. anthracina, *Wulf.*
76. *arctica, *Ach.*
77. *hyperborea, *Ach.*
78. proboscidea, *L.*
79. flocculosa, *Hffm.* Not cited by Th. Fries, but mentioned as occurring in Greenland by Tuckerman (Syn. Lich. New England, 1848, p. 71).
80. erosa, *Web.*
81. polyphylla, *L.*; only in one locality.
82. *cylindrica, *L.*; various forms.
var. Delisei, *Despr.*, is frequent in Greenland according to Dr. Nylander (Scand., p. 117).
83. hirsuta, *Ach.* Walker and Mitten† cite this from Greenland (Lievely). It is not, however, mentioned in Th. Fries' "L. Arctoi" as a Greenland Lichen, and doubtfully as an Arctic-Scandinavian species, or very rare.

Gen. 20. *Pannaria.*

Sp. 84. *brunnea, *Sw.*, and
**var.* coronata, *Hffm.*
85. lepidiota, *Th. Fr.* (Arct., p. 74), and
**var.* tristis, *Th. Fr.*
86. Hookeri, *Sm.*
var. macrior, *Th. Fr.*

† Journ. Linn. Soc., Botany, vol. v, p. 87.

Sp. 87. hypnorum, *Vahl.* (Psoroma, *Fr.*, *Nyl.*)
88. muscorum, *Ach.* (Massalongia carnosa, *Dicks.*)

Gen. 21. *Squamaria.*

Sp. 89. *saxicola, *Poll.* (Placodium.)
90. *chrysoleuca, *Sm.*, and
var. opaca, *Ach.* (marked *)
* — feracissima, *Th. Fr.* (L. Spitsb., p. 18.)
91. straminea, *Whlnb.*
92. gelida, *L.*
93. *elegans, *Link.* (Xanthoria.)
94. murorum, *Hffm.*
95. fulgens, *Sw.* (Placodium.)
var. alpina, *Th. Fr.*
96. melanaspis, *Ach.* (Lecanora, *Nyl.*)
var. alphoplaca, *Whlnb.*
97. geophila, *Th. Fr.*, p. 85.

Gen. 22. *Lecanora.*

Sp. 98. *tartarea, *L.*, and
var. frigida, *Sw.* (marked *)
* — gonatodes, *Ach.*
— vermicularia, *Linds.*
* — grandinosa, *Ach.*
— thelephoroides, *Th. Fr.* (L. Spitsb., p. 21.)
99. *parella, *L.* (pallescens, *L.*), and
var. Upsaliensis, *L.* (marked *)
100. *oculata, *Dicks.* (Aspicilia); various forms, common.
101. atra, *Huds.*
102. *subfusca, *L.*, and
var. epibrya, *Ach.* (marked *)
— Hageni, *Ach.*†
103. *frustulosa, *Dicks.*
104. epanora, *Ach.*; doubtfully.
105. *badia, *Ehrh.*
106. varia, *Ehrh.*
var. symmicta, *Ach.*
* — polytropa, *Ehrh.*
* — intricata, *Schrad.*
— leucococca, *Smrf.*
107. atro-sulphurea, *Whlnb.*
108. cenisea, *Ach.*
109. glaucoma, *Ach.* (sordida, *Pers.*).
110. *bryontha, *Ach.*
111. peliscypha, *Whlnb.* (Acarospora.)
112. molybdina, *Whlnb.*
113. *smaragdula, *Whlnb.*, and
var. sinopica, *Whlnb.*
114. chlorophana, *Whlnb.*
115. *ventosa, *L.* (Hæmatomma.)

---

† Stizenberger, in "Botanische Zeitung" (1868, p. 895), mentions var. *atrynea*, Ach., as occurring in Greenland.

Sp. 116. nimbosa, *Fr.* (Dimelæna.)
117. oreina, *Ach.*
118. *turfacea, *Whlnb.* (Rinodina), and
var. depauperata, *Th. Fr.*
— roscida, *Smrf.*
119. *sophodes, *Ach.*; various forms.
120. mniaræa, *Ach.*
121. exigua, *Ach.*
122. verrucosa, *Ach.* (Aspicilia), and
var. paryrga, *Ach.*
123. *calcarea, *L.*
var. contorta, *Hffm.*
124. *cinerea, *L.*, and
**var.* Myrini, *Fr.*
— aquatica, *Fr.*, and other forms.
125. lacustris, *With.*
126. ferruginea, *Huds.* (Caloplaca.)
var. cinnamomea, *Th. Fr.*
— hypnophila, *Th. Fr.*
127. jungermanniæ, *Vahl.*, and
var. convexa, *Sch.*
128. fusco-lutea, *Dicks.*
129. *cerina, *Hedw.*
var. stillicidiorum, *Œd.*
130. aurantiaca, *Lightf.*; doubtfully.
131. crenulata, *Th. Fr.*, p. 70 (Xanthoria).
132. vitellina, *Ehrh.*
133. leucoræa, *Ach.* (Blastenia).

Gen. 23. *Urceolaria.*
Sp. 134. scruposa, *L.*

Gen. 24. *Pertusaria.*
Sp. 135. *P. paradoxa, *Linds.*

Gen. 25. *Lecidea.*
Sp. 136. contigua, *Hoffm.*, and
var. flavicunda, *Ach.*
137. *fusco-atra, *L.*
138. panæola, *Ach.*
var. elegans, *Th. Fr.*
139. spilota, *Fr.* (tessellata, *Flk.*), and
var. polaris, *Th. Fr.*
140. *lapicida, *Ach.*; rare, and only the ferruginous conditions of the ordinary form.
141. auriculata, *Th. Fr.* (Arct., p. 213).
142. alpestris, *Smrf.* ("fere *L. aggeratæ*, Mudd," says Th. Fries, L. Spitsb., p. 39). Nylander's *L. stenotera,* recorded as a Norwegian species in my "North. Lichen-Flora," p. 385, Th. Fries refers to *alpestris* as a variety (L. Arct., p. 214).
143. arctica, *Smrf.*
144. aglæa, *Smrf.*
145. *sabuletorum, *Schreb.*, and
var. muscorum, *Wulf.*

Sp. 146. *parasema, *Sch.*, and
var. enteroleuca, *Ach.*
— euphorea, *Flk.*, and other forms.
147. turgidula, *Fr.*
var. denudata, *Schrad.*
148. atro-brunnea, *Ram.*
149. armeniaca, *DC.*
var. melaleuca, *Smrf.*
150. elata, *Sch.* (= amylacea, *Ach.*)
151. pallida, *Th. Fr.* (Arct., p. 221).
152. vittellinaria, *Nyl.*
153. *disciformis, *Fr.* (Buellia.)
154. insignis, *Næg.*
**var.* muscorum, *Hepp.*
— geophila, *Smrf.*
155. *myriocarpa, *DC.*
156. *atro-alba, *Ach.*; by no means rare, according to Th. Fries (Arct., p. 231).
157. scabrosa, *Ach.*
var. cinerascens, *Th. Fr.*
158. urceolata, *Th. Fr.* (Arct., p. 233), and
var. deminuta, *Th. Fr.*; parasitic on thallus of various Lichens.
159. coronata, *Th. Fr.*, p. 205 (Rhexophiale.)
160. geminata, *Fw.* (Rhizocarpon); common.
161. *petræa, *Wulf.*, and
var. Œderi, *Ach.*, with other forms.
162. *Grœnlandica, *Linds.*
163. *geographica, *L.*, and
**var.* alpicola, *Sch.*
164. globifera, *Ach.* (Psora), and
var. rubiformis, *Whlnb.*
165. atro-rufa, *Dicks.*
166. decipiens, *Ehrh.*
167. squalida, *Ach.* (Toninia.)
168. candida, *Web.* (Thalloidima.)
169. *obscurata, *Smrf.* (Bilimbia.)
170. cumulata, *Smrf.* (Biatorina.)
171. cinnabarina, *Smrf.* (Biatora.)
172. *vernalis, *L.*
173. cuprea, *Smrf.*
174. *castanea, *Hepp.*
175. Tornœensis, *Nyl.*
176. fuscescens, *Smrf.*
177. uliginosa, *Schrad.*
178. leucoræa, *Ach.* (Blastenia.)
179. pezizoidea, *Ach.* (Lopadium); rare.
180. flavo-virescans, *Dicks.* (Arthroraphis.)
181. icmadophila, *Ach.* (Icmadophila æruginosa, *Scop.*)
182. *sanguineo-atra, *Ach.*; various forms.
183. *fusco-rubens, *Nyl.*

Sp. 184. *Discoensis, *Linds.*
185. *Campsteriana, *Linds.*
186. Friesiana, *Linds.*
187. *Egedeana, *Linds.*
188. subfuscula, *Nyl.* Recorded as a Greenland species in Th. Fries' "L. Spitsb.," p. 35, though not in his "L. Arct."

Gen. 26. *Arthonia.*
Sp. 189. trabinella, *Th. Fr.* (Arct., 240); on worked wood.

Gen. 27. *Endocarpon.*
Sp. 190. miniatum, *L.* (Dermatocarpon), and
var. complicata, *Sw.*
191. hepaticum, *Ach.*
192. Dædaleum, *Kremp.;* rare.
193. *viride, *Ach.* (Normandina, *Nyl.*)

Gen. 28. *Verrucaria.*
Sp. 194. clopima, *Whlnb.* (Staurothele.)
195. maura, *Whlnb.*, and
var. aractina, *Whlnb.*
196. ceuthocarpa, *Whlnb.*, and
var. mucosa, *Whlnb.*
197. epidermidis, *Ach.* (Arthopyrenia).
var. analepta, *Ach.;* rare.
198. pygmæa, *Körb.* (Endococcus, *Nyl.*); parasitic.
199. tartaricola, *Linds.*

Pseudo-genus. *Pyrenothea.*
Sp. 200. *P. Grœnlandica, *Linds.*

Summary.

Genus.	Number of Species and Varieties.	Genus.	Number of Species and Varieties.
1. Ephebe	1	Brought forward	76
2. Pyrenopsis	1	16. Sticta	1
3. Collema	4	17. Parmelia	21
4. Calicium	1	18. Physcia	6
5. Sphærophoron	2	19. Umbilicaria	13
6. Cladonia	23	20. Pannaria	7
7. Stereocaulon	6	21. Squamaria	11
8. Thamnolia	1	22. Lecanora	54
9. Usnea	1	23. Urceolaria	1
10. Alectoria	11	24. Pertusaria	1
11. Dactylina	1	25. Lecidea	63
12. Cetraria	14	26. Arthonia	1
13. Nephroma	2	27. Endocarpon	5
14. Peltidea	5	28. Verrucaria	8
15. Solorina	3		
Carry forward	76	Total	268

The Genera richest in *Species* are, therefore, in the order of their richness—

1. Lecidea.	5. Cetraria.
2. Lecanora.	6. Umbilicaria.
3. Cladonia.	7. Squamaria. }
4. Parmelia.	8. Alectoria. }

These are not necessarily, however, the genera richest in *individuals*—the genera, therefore, which give a character to or constitute predominant vegetation. There is insufficient evidence to show what the latter genera are. All that can be asserted, on the evidence of travellers, is, that in some localities the predominant Lichens and the prevailing vegetation are species of *Umbilicaria* or *Placodium*. There is no evidence that the *Cladoniæ* occupy the same important position, as coverers of the soil, that they do in Northern Scandinavia and Russia;* or that the *Alectoriæ*, *Cetrariæ*, and *Parmeliæ* occur in the same gregarious assemblages that I have seen them do in Norway or Iceland.†

XXXVI.—On the NATURE of the DISCOLORATION of the ARCTIC SEAS. By DR. ROBERT BROWN, F.L.S., F.R.G.S.

[Reprinted, by Permission, from the "Transactions of the Botanical "Society of Edinburgh," vol. ix., pp. 244–252. Read December 12, 1867. Revised by the Author, March 1875.]

The peculiar discoloration of some portions of the Frozen Ocean, differing in a remarkable degree from the ordinary blue or light green usual in other portions of the same sea, and quite independent of any optical delusion occasioned by light or shade, clouds, depth or shallowness, or the nature of the bottom, has, from a remote period, excited the curiosity or remark of the early navigators and whalemen, and to this day is equally a subject of interest to the visitor of these little-frequented parts of the world. The eminent seaman, divine, and *savant*, William Scoresby, was the first who pointedly drew attention to the subject, but long before his day the quaint old searchers after a North-West Passage "to Cathay and Cipango" seem to have observed the same phenomenon, and have recorded their observations, brief enough it must be acknowledged, in the pages of "Purchas—His Pilgrimes," or the ponderous tomes of Master Hakluyt. Thus Henry Hudson, in 1607, notices the change in the colour of the sea, but has fallen into error when he attributes it to the presence or absence of ice whether the sea was blue or green—mere accidental coincidences. John Davis, when, at even an earlier date, he made that famous voyage of his with

---

* *See* paper on the "Arctic Cladoniæ," p. 179.

† "Flora of Iceland," p. 24 (Trans. Bot. Soc. of Edin., or Edin. New Philosophical Journal, 1861); "Northern Lichen-Flora," p. 403, *et seq.*

the "Sunshine" and the "Moonshine," notes that, in the strait which now bears his name, "the water was very blacke and "thicke, like unto a filthy standing pool."* More modern voyagers have more equally noted the phenomenon, but without giving any explanation, and it is the object of this paper to endeavour to fill up that blank in the physical geography of the sea. In the year 1861 I made a voyage to the seas in the vicinity of Spitzbergen and the dreary island of Jan Mayen, and subsequently a much more extended one through Davis Strait to the head of Baffin's Bay, and along the shores of the Arctic Regions lying on the western side of the former gulf, during which I had abundant opportunities of observing the nature of this discoloration. At that period I arrived at the conclusions which I am now about to promulgate. In the course of the past summer [1867] I again made an expedition to Danish Greenland, passing several weeks on the outward and homeward passages in portions of the seas mentioned, during which time I had an opportunity of confirming the observations I had made six years previously, so that I consider I am justified in bringing my researches, so far as they have gone, before the Botanical Society.

1. *Appearance and Geographical Description of the discoloured Portions of the Arctic Sea.*

The colour of the Greenland Sea varies from ultramarine blue to olive-green, and from the most pure transparency to striking opacity, and these changes are not transitory but permanent.† Scoresby, who sailed during his whaling voyages very extensively over the Arctic Sea, considered that in the "Greenland Sea" of the Dutch—the "Old Greenland" of the English—this discoloured water formed perhaps one-fourth part of the surface between the parrallels of 74° and 80° North latitude. It is liable, he remarked, to alterations in its position from the action of the current, but still it is always renewed near certain localities year after year. Often it constitutes long bands or streams lying north and south, or N.E. and S.W., but of very variable dimensions. "Sometimes "I have seen it extend two or three degrees of latitude in length, "and from a few miles to ten or fifteen leagues in breadth. It "occurs very commonly about the meridian of London in high "latitudes. In the year 1817 the sea was found to be of a blue "colour and transparent all the way from 12° East, in the parallel "of 74° or 75° N.E., to the longitude of 0° 13′ East in the same "parallel. It then became green and less transparent; the colour "was nearly *grass-green*, with a shade of black. Sometimes the "transition between the green and blue waters is progressive, "passing through the intermediate in the space of three or four "leagues; in others it is so sudden that the line of separation is "seen like the rippling of a current; and the two qualities of the

* The First Voyage of M. Iohn Dauis vndertaken in June 1585. (Hakluyt's Collection.)

† Scoresby, "Arctic Regions," i., 175.

" water keep apparently as distinct as the waters of a large muddy " river on first entering the sea."* In Davis Strait and Baffin's Bay, wherever the whalers have gone, the same description may hold true—of course making allowances for the differences of geographical position, and the discoloured patches varying in size and locality. I have often observed the vessel in the space of a few hours, or even in shorter periods of time, sail through alternate patches of deep black, green, and cærulean blue; and at other times, especially in the upper reaches of Davis Strait and Baffin's Bay, it has ploughed its way for 50 or even 100 miles through an almost uninterrupted space of the former colour. The opacity of the water is in some places so great that "tongues" of ice and other objects cannot be seen a few feet beneath the surface.

2. *Causes of the Discoloration.*

These patches of discoloured water are frequented by vast swarms of the minute animals upon which the great "Right Whale" of commerce (*Balæna mysticetus*, Linn.) alone subsists, the other species of *Cetacea* feeding on Fishes proper, and other highly organised tissues. This fact is well known to the whalers, and, accordingly, the "black water" is eagerly sought for by them, knowing that in it is found the food of their chase, and therefore more likely the animal itself. From this knowledge, and from observations made with the usual lucidity of that distinguished observer, Captain Scoresby attributed the nature of the discoloration to the presence of immense numbers of *Medusæ* in the sea, and his explanation has hitherto met the acceptance of all marine-physical geographers; and for more than forty years his curious estimate of the numbers of individual *Medusæ* contained in a square mile of the Greenland Sea has become a standard feature in all popular works on zoology, and a stock illustration with popular lecturers. In 1861, and subsequently, whilst examining microscopically the waters of the Greenland Sea, I found, in common with previous observers, that not only were immense swarms of animal life found in these discoloured patches, but that it was almost solely confined to these spaces. In addition, however, I observed that the discoloration was not due to the *Medusoid* life, but to the presence of immense numbers of a much more minute object—a beautiful siliceous moniliform Diatom, and it is this Diatom which brings this paper within the ken of botanists. On several cold days, or from no apparent cause, the *Medusæ*, great and small, would sink, but still the water retained its usual colour, and on examining it I invariably found it to be swarming with *Diatomaceous* life, the vast preponderance of which consisted of the Diatom in question.

It had the appearance of a minute beaded necklace about 1-400th part of an inch in diameter, of which the articulations are

* "Arct. Reg.," i., 176. *See also* Scoresby's account of the "brown-coloured" water at 68° 26′ lat., 11° 55′ W. long., and the "yellowish-green" at lat. 70° 34′, in his "Journal of a Voyage to the Northern Whale-fishery," &c., 8vo., Edinburgh, 1823, pp. 353–356.

about $1\frac{1}{2}$ or $1\frac{1}{4}$ time as long as broad. These articulations contain a brownish-green granular matter, giving the colour to the whole plant, and again through it to the sea in which it is found so abundantly. The whole Diatom varies in length, from a mere point to 1-10th of an inch, but appears to be capable of enlarging itself indefinitely longitudinally by giving off further bead-like articulations. Wherever, in those portions of the sea, I threw over the towing net, the muslin in a few minutes was quite brown with the presence of this Alga in its meshes. Again, this summer, I have had occasion to notice the same appearance in similar latitudes on the opposite shores of Davis' Strait where I had principally observed it in 1861. This observation holds true of every portion of discoloured water which I have examined in Davis Strait, Baffin's Bay, and in the Spitzbergen or the Greenland Seas—viz., that wherever the green water occurred the sea abounded in *Diatomaceous* life, the contrary holding true regarding the ordinary blue water. These swarms of Diatoms do not appear to reach in quantity any very great depth, for in water brought up from 200 fathoms there were few or no Diatoms. They seem also to be affected by physical circumstances, for sometimes in places where a few hours previously the water on the surface was swarming with them few or none were to be found, and in a few hours they again rose. But the Diatom I found plays another part in the economy of the Arctic Seas. In June 1861, whilst the iron-shod bows of the steamer I was on board of crashed its way through the breaking-up floes of Baffin's Bay, among the Women's Islands, I observed that the ice thrown up on either side was streaked and discoloured brown; and on examining this discolouring matter I found that it was almost entirely composed of the siliceous moniliform Diatom I have described as forming the discolouring matter of the iceless parts of the icy sea. I subsequently made the same observation in Melville Bay, and in all other portions of Davis Strait and Baffin's Bay where circumstances admitted of it. During the long winter the *Diatomaceæ* had accumulated under the ice in such abundance that when disturbed by the pioneer prow of the early whalers they appeared like brown slimy bands in the sea, causing them to be mistaken more than once for the waving fronds of *Laminaria longicruris* (De la Pyl.), which, and not *L. saccharina*, as usually stated, is the common Tangle of the Arctic Sea. On examining the under surface of the upturned masses of ice, I found the surface honeycombed, and in the base of these cavities vast accumulations of *Diatomaceæ;* leading to the almost inevitable conclusion that a certain amount of heat must be generated by the vast accumulations of these minute organisms, which thus mine the giant floes into cavernous sheets. These are so decayed in many instances as to be easily dashed on either side by the "ice chisels" of the steamers which now form the majority of the Arctic-going vessels, and they get from the seamen, who too frequently mistake cause for effect, the familiar name of "rotten ice." I have since found that in noticing the *Diatomaceous* character of these slimy masses, I was forestalled by Dr. Sutherland ("Journal of a Voyage," &c., 1852, vol. i., pp. 91–96, and vol. ii., Appendix, by Dr. Dickie, pp. cxcviii., &c.). Though one Diatom, as I have remarked, predominates, yet there are, besides *Protozoa*, vast multitudes of many different species of Diatoms, as shown (*loc. cit.*)* by Dr. Dickie (now of Aberdeen).

Is it carrying the doctrine of final causes too far to say these Diatoms play their part in rendering the frozen north accessible to the bold whalemen, as I shall presently show they do in furnishing subsistence for the giant *quarry* which leads him thither?

I have spoken of the discoloured portions of the Arctic Sea as abounding in animal life, and that this life was nowhere so abundant as in those dark spaces which, as I have already demonstrated, owe this hue to the *Diatomaceæ* in question.

These animals are principally various species of *Beroidæ*, and other *Steganophthalmous Medusæ*; *Entomostraca*, consisting chiefly of *Arpacticus Kronii*, *A. chelifer* and *Cetochilus arcticus* and *septentrionalis*, and Pteropodous Mollusca—the chief of which is the well-known *Clio borealis*, though I think it proper to remark that this species does not contribute to the Whale's food nearly so much as we have been taught to suppose. The discoloured sea is sometimes perfectly thick with the swarms of these animals, and then it is that the whaler's heart gets glad as visions of "size Whales" and "oil money" rise up before him, for it is on these minute animals that the most gigantic of all known beings solely subsists. What, however, was my admiration (it was scarcely surprise) to find, on examining microscopically the alimentary canals of these animals, that the contents consisted entirely of the *Diatomaceæ* which give the sable hue to portions of the Northern Sea in which these animals are principally found! It thus appears that in the strange cycle of nature the "Whales' food" is dependent on the Diatom, so that in reality the great things of the sea depend for their existence upon the small things thereof! I subsequently found (though the observation is not new) that the alimentary canals of most of the smaller *Mollusca*, *Echinodermata*, &c. were also full of these *Diatomaceæ*. I also made an observation which is confirmatory of what I have advanced regarding the probability of these minute organisms giving off *en masse* a certain degree of heat, though in the individuals inappreciable to the most delicate of our instruments. On the evening of the 4th of June, this present year (1867), in latitude 67° 26′ N., the sea was so full of animal (and *Diatomaceous*) life, that in a few minutes upwards of a pint measure of *Entomostraca*, *Medusæ*, and *Pteropoda* would fill the towing net. The temperature of the sea was then by the most delicate instruments found to be 32·5 Fahr., and next morning (June 5th), though the air had exactly the same temperature, no ice at hand, and the ship maintained almost the same position as on the night previous, yet the surface temperature of the sea had sunk to 27·5 Fahr., and was clear of life, so much so that in the space of half an hour the

* *See* pages 319, 320; and further on.

towing net did not capture a single *Entomostracon*, *Medusa*, or *Pteropod*. I also found that this swarm of life ebbed and flowed with the tide, and that the whalers used to remark that Whales along shore were most frequently caught at the flow of the tide, coming in with the banks of Whales' food. This mass of minute life also ascends to the surface more in the calm Arctic nights when the sun gets near the horizon during the long summer day. In 1861 I was personally acquainted with the death of thirty individuals of the "Right Whalebone Whale" (*Balæna mysticetus*, L.), and of this number fully three-fourths were killed between ten o'clock p.m. and six o'clock a.m., having come on the "whaling grounds" at that period (from amongst the ice where they had been taking their *siesta*), to feed upon the animals which were then swarming on the surface, and these again feeding on the *Diatomaceæ* found most abundantly at that time in the same situations. I would however, have you to guard against the supposition, enunciated freely enough in some compilations, that the "Whales' food" migrates, and that the curious wanderings of the Whale north, and again west and south, is due to its "pursuing "its living"; such is not the case. The "Whales' food" is found all over the wandering ground of the *Mysticete*, and in all probability the animal goes north in the summer in pursuance of an instinct implanted in it to keep in the vicinity of the floating ice-fields (now melted away in southern latitudes); and again it goes west for the same purpose, and finally goes south at the approach of winter—but where, no man knows.

There are some other streaks of discoloured water in the Arctic Sea known to the whalers by various not very euphonious names, but these are merely local or accidental, and are also wholly due to *Diatomaceæ*, and with this notice may be passed over as of little importance. I cannot, however, close this paper without remarking how curiously the observations I have recorded afford illustrations of representative species in different and widely separated regions. In the Arctic Ocean the *Balæna mysticetus* is the great subject of chase, and in the Antarctic and Southern Seas the hardy whalemen pursue a closely allied species, *Balæna australis*. The Northern Whale feeds upon *Clio borealis* and *Cetochilus septentrionalis*; the Southern Whale feeds upon their representative species, *Clio australis* and *Cetochilus australis*, which streak with crimson the Southern Ocean for many a league. The Northern Sea is dyed dark with a Diatom on which the *Clio* and *Cetochilus* live, and the warm waters of the Red Sea are stained crimson with another Alga; and I doubt not that, if the Southern Seas were examined as carefully as the Northern have been, it would be found that the Southern "Whales' food" lives also on the Diatoms staining the waters of that Austral Ocean.

I do not claim any very high credit for the facts narrated in the foregoing paper, either general or specific, for really it is to the exertions of the sailor-savant, William Scoresby, that the first light which has led to the solution of the question is due, though the state of science in his day would not admit of his seeing more clearly into the dark waters of that frozen sea he knew and loved so well.

At the same time I believe that I am justified in concluding that we have now arrived at the following conclusions from perfectly sound data, viz.:—

1. That the discoloration of the Arctic Sea is due not to animal life, but to *Diatomaceæ*.

2. That these *Diatomaceæ* form the brown staining matter of the "rotten ice" of Northern navigators.

3. That these *Diatomaceæ* form the food of the *Pteropoda*, *Medusæ*, and *Entomostraca*, on which the *Balæna mysticetus* subsists.

I have brought home abundant specimens of the *Diatomaceous* masses which I have so frequently referred to in this paper, and I am now engaged in distributing them to competent students of this order, so that the exact species may be determined; but as these take a long time to be examined (more especially as Diatoms do not seem so popular a study as they were a few years ago), I have thought it proper to bring the more important general results of my investigations before you at this time, and to allow the less interesting subject of the determination of species* to lie over to another time. I have to apologise to you for introducing so much of another science, foreign to the objects of the Society, into this paper; but when the lower orders of plants are concerned, we are so near to the boundaries of the animal world that to cross now and then over the shadowy march is allowable, if not impossible to be avoided.

Finally, you will allow me to remark that, in all the annals of biology, I know nothing stranger than the curious tale I have unfolded:† the Diatom staining the broad frozen sea, again supporting myriads of living beings which crowd there to feed on it, and these again supporting the huge Whale,—so completing the wonderful cycle of life. Thus it is no stretch of the imagination to say that the greatest animal in creation,‡ whose pursuit gives employment to many thousand tons of shipping and thousands of seamen, and the importance of which is commercially so great that its failure for one season was estimated for one Scottish port alone at a loss of 100,000*l.* sterling,§ depends for its existence on a being so minute that it takes thousands to be massed together before they are visible to the naked eye, and, though thousands of ships for hundreds of years sailed the Arctic, unknown to the men who were

---

* The species principally causing the discoloration is apparently *Melosira arctica*.

† The foregoing observations on the Discoloration of the Arctic Sea have been confirmed by the German Expedition in the Spitzbergen Sea, under Koldeway; by that to Greenland, conducted by Professor Nordenskiöld.

‡ Nilsson, in his "Skandinavisk Fauna," vol i., p. 643, estimates the full-grown *B. mysticetus* at 100 tons or 220,000 lbs., or equal to 88 elephants or 440 white bears.

§ In 1867 the twelve screw steamers of Dundee only took two Whales, and the loss to each steamer was estimated at 5,000*l.*, and to the town in all at the sum I have given.

most interested in its existence, illustrating in a remarkable degree how Nature is in all her kingdoms dependent on all, and how great are little things!

NOTE.—Prof. NORDENSKIÖLD offers the following remarks on the Colour of the Arctic Seas, in his "Expedition to Greenland," "Geological Magazine," vol. ix., pp. 9, 10:—

Hudson, and other veteran mariners of the Arctic seas, mention the variety of colours that distinguish the water in certain parts, which are frequently so sharply distinguished that a ship may sail with the one side in blue and the other in greyish-green water. It was at first supposed that these colours were indications of different currents—the green of the Arctic, the blue of the Gulf-stream. Later, Scoresby affirmed that the phenomenon arose from the presence of innumerable organisms, which he seems to consider as Crustacea, in the water. This observation has since been continued, partly by the former Swedish Arctic Expedition, and partly by Dr. Brown,* during the voyages made by him in the Arctic seas as surgeon in a whaler, and as a member of Whymper's Expedition. We also endeavoured to divert the tedious monotony of the voyage by observations on this phenomenon.

The sea-water in the neighbourhood of Spitzbergen is marked by two sharply distinguished colours—greyish-green and fine indigo-blue. In the Greenland seas we also find water with a very decided shade of brown. These colours are seen most pure if one looks vertically down from the ship to the surface of the water through a somewhat long pipe. The green, or rather grey-green, water is generally met with in the neighbourhood of ice (whence it was supposed to arise from the Arctic Current); the blue where the water is free from ice; the brown, as far as I am aware, chiefly in that part of Davis Strait which is situated in front of Fiskernaes. When specimens of the water are taken up in an uncoloured glass, it appears perfectly clear and colourless, nor can one with the naked eye discover any organisms to account for the colour. But if, when the velocity of the ship allows of it (i.e., when the ship makes from one to three knots an hour), a fine insect-net be towed behind the ship, in the green and brown water, it will soon be found covered with a film of—in the former case green, in the latter case brown slime, of organic origin, and evidently the real cause of the abnormal colour of the sea-water. Just in these parts may be found swarms of small Crustacea, which live upon this slime, and in their turn, directly or indirectly, become the food of larger marine animals. The blue water, on the contrary, at least in these seas, deposits no slime upon the insect-net, and is far less frequented by Crustacea, Annelides, &c.

---

* A very interesting essay on the subject has been published by Dr. Brown. *The Farmer*, Jan. 1, 1865, p. 16. *See* above, p. 311.

than the green. Thus, as Brown, in the article above referred to, remarks, the presence of this slime, inconsiderable as it is, but spread over hundreds of thousands of square miles, is a condition necessary for the subsistence, not only of the swarms of Birds that frequent the northern seas, but also of that giant of the animal creation, the Whale, and all branches of industry dependent on Whale-fisheries.

Of these remarkable organisms Dr. Öberg collected specimens when possible, during the voyage, which it is intended hereafter to submit to a careful scientific examination, in conjunction with similar specimens from preceding expeditions. Here we need only mention that the slime itself in each particular place is formed only of a few species of Diatomaceæ, often so large that after drying the mass the siliceous frustules may be discerned with the naked eye; but, on the other hand, different parts of the ocean exhibit entirely different forms, so that, for example, the green slime in one place has sometimes not a single species identical with that in another. A long continued collection will therefore be required to explain this scanty, but nevertheless remarkable, and we may safely say important, Flora of the ocean's surface.

XXXVII.—Notes of Diatomaceæ from Danish Greenland, collected by Robert Brown. By Professor Dickie, Aberdeen.

[Reprinted, by Permission, from the Trans. Bot. Soc. Edinburgh, vol. x., pp. 65–67. Read January 14, 1869.]

Mr. Brown sent for examination small packets of material collected at various places along the coast of Danish Greenland, with a request that any Diatomaceæ they contained might be recorded.

The larger marine Algæ, from high northern latitudes, I have invariably found to yield abundance of these organisms. The species collected by Mr. Brown, as well as his special Diatomaceous gatherings, have not yet been seen by me, and therefore the present communication gives but an imperfect idea of the marine Diatomaceæ in the localities visited.

No. 1. A small mass, chiefly of *Hypnum fluitans*, in the sea at Jakobshavn, contained the following:—

Cocconeis scutellum.
Coscinodiscus eccentricus.
Stauroneis pulchella.
Hyalodiscus subtilis.
Rhabdonema arcuatum.
Cocconema cistula.
C. parvum.

The two latter are freshwater species, and were, doubtless, attached to the Hypnum before it was conveyed to the sea; the marine species seem not merely entangled among the mass, but attached to it.

No. 2. A mass of Ectocarpus, from Jakobshavn, contained the following :—

Cocconeis scutellum.
Coscinodiscus eccentricus.
C. radiatus.
Grammonema Jurgensii.
Podosphenia gracilis.

And very fine examples of two freshwater species, viz. :—

Himantidium undulatum.
Pinnularia stauroneiformis.

No. 3. A mass of Schizonema and Cladophora, floating in Jakobshavn, contained—

Cocconeis scutellum.
Synedra salina.
Podosphenia gracilis.
Podosira hormoides.
Grammonema Jurgensii.
Biddulphia ——; a few fragments, too imperfect for recognition.

The Schizonema is doubtfully referred to *S. Dilwynii*, being in a very imperfect condition.

No. 4. At a depth of three fathoms, Jakobshavn :—

Cocconeis scutellum, *var β*.
Coscinodiscus radiatus ?, fragments only.
C. eccentricus.
Campylodiscus angularis, a solitary example.
Synedra salina.
Amphiprora alata.
Surirella gemma.
Nitzschia sigma.
Navicula didyma.
N. elliptica.
N. liber.
Stauroneis pulchella.
Podosira hormoides.
Hyalodiscus subtilis.
Rhabdonema arcuatum.
Biddulphia aurita.
Schizonema ?

No. 5. A mass, chiefly of Dictyosiphon, from sea-shore, Ritenbenk, contained—

Cocconeis scutellum.
Coscinodiscus eccentricus.
C. radiatus.
Eupodiscus fulvus ?
Grammonema Jurgensii.
Podosira hormoides.

And a single example of a freshwater species, viz., *Gomphonema geminatum*.

No. 6. A mass, consisting mainly of Conferva, Melagonium, Dictyosiphon, and Lyngbya, from Godhavn, contained—

Cocconeis Grevillei.
C. scutellum.
Coscinodiscus radiatus.
C. eccentricus.
Amphora membranacea, one only.
Synedra gracilis ?
Navicula Jenneri.
N. eliptica.
Nitzschia angularis ?
Podosira hormoides.
Grammatophora serpentina.

All the species recorded here are British, with the single exception of *Hyalodiscus subtilis*, originally described by the late Professor Bailey from Halifax; found also on shores of North-west America, and now on shores of Greenland.

XXXVIII.—NOTE on some PLANTS from SMITH SOUND, collected by DR. BESSELS. By DR. J. D. HOOKER, C.B., F.R.S. 1873.

[From "A Whaling Cruise to Baffin's Bay and the Gulf of "Boothia; with an Account of the Rescue of the Crew of the "'Polaris.'" By A. H. MARKHAM, Commander R.N., F.R.G.S. 1874. Page 296.]

Captain Markham's collection contains 20 species of Flowering Plants, including four collected by Dr. Bessels in the highest latitude from which Flowering Plants have hitherto been obtained, namely 82° N. The locality appears to have been on the east side of Smith's Sound.* They are *Draba alpina*, *Cerastium alpinum*, *Taraxacum Dens-leonis*, and *Poa flexuosa*. All of them are common Arctic plants, being found on both coasts of Greenland, as well as throughout the Parry Islands. Of the other species collected by Captain Markham himself, the Arctic distribution is well known. None of them belong to the remarkable assemblage of Scandinavian plants which inhabit Greenland, and of which no other member has been found on the eastern shores of Baffin's Bay. On the other hand, one of them is a member of that smaller number which has never been found on the Greenland coast. This is the peculiar and beautiful little *Pleuropogon Sabini*, the only genus which is absolutely confined to the Arctic regions, and of which the solitary species is restricted in its distribution to the Arctic-American islands. It was discovered by Captain, now General Sir, Edward Sabine, in Melville Island, during Parry's First Voyage in 1819–1820, and is probably found in all the islands. Capt. Markham's specimen was gathered on Fury Beach.

The other species call for no special remark. They are interesting as, in several cases, coming from places where the same plant had not previously been gathered. These localities are valuable, as completing a knowledge of the area inhabited by such species, though they do not materially enlarge it.

[*See above*, Dr. Hooker's memoir on Arctic Plants, pages 205, &c.]

XXXIX.—NOTES on the ZOOLOGY, BOTANY, and GEOLOGY of the VOYAGE of the "POLARIS" to KENNEDY and ROBESON CHANNELS.

---

1. "*Nature*," July 17, 1873, vol. viii., p. 218.—"During the summer the entire extent of both lowlands and elevations are bare of snow and ice excepting patches here and there in the shape of the

* "Nature," vol. viii., No. 206, p. 487. Oct. 9, 1873.

rocks. The soil during this period was covered with a more or less dense vegetation of Moss, with which several Arctic Plants were interspersed, some of them of considerable beauty, but entirely without scent, and many small Willows, scarcely reaching the dignity of shrubs.

"The rocks noticed were of a schistose or slaty nature, and in some instances contained fossil plants, specimens of which were collected.

"Distinct evidence of former glaciers were seen in localities now bare of ice, these indications consisting of the occurrence of terminal and lateral moraines.

"Animal life was found to abound, Musk-oxen being shot at intervals throughout the winter. Wolves, Bears, Foxes, Lemmings, and other Mammals were repeatedly observed.

"Geese, Ducks, and other Water Fowls, with Plover and other Wading Birds, abounded during the summer, although the species of Land Birds were comparatively few, including, however, as might have been expected, large numbers of Ptarmigan or Snow-partridge.

"No Fish were seen, although the net and line were frequently called into play in the attempt to obtain them. The water, however, was found filled to an extraordinary degree with marine Invertebrata, including Jelly-fish and Shrimps; Seals are very abundant. Numerous Insects were observed, also especially several species of Butterflies, specimens of which were collected; also Flies and Bees, and Insects of like character."

2. "*Nature*," March 26, 1874, vol. ix., p. 405.—"Zoology and Botany.—The collections of Natural History are almost entirely lost. With the exception of two small cases containing animals, minerals, and one package of plants, nothing could be rescued. The character of the Fauna is North-American, as indicated by the occurrence of the Lemming and the Musk-ox. Nine species of Mammals were found, four of which are Seals. The Birds are represented by 21 species. The number of species of Insects is about 15, namely, one Beetle, four Butterflies, six Diptera, one Humble-bee, and several Ichneumons, parasites in caterpillars. Further, two species of Spiders and several Mites were found. The animals of lower grade are not ready yet for examination.

"The Flora is richer than could be expected, as not less than 17 Phanerogamic Plants were collected, besides three Mosses, three Lichens, and five freshwater Algæ.

"Geology.—Although the formation of the Upper Silurian Limestone, which seems to constitute the whole west coast north of the Humboldt Glacier, is very uniform, some highly interesting and important observations have been made. It is found that the land is rising, as indicated, for instance, by the occurrence of marine animals in a freshwater lake more than 30 feet above the sea-level and far out of reach of the spring-tides.

"Wherever the locality was favourable, the land is covered with Drift, sometimes containing very characteristic lithological specimens, the identification of which with rocks in South

Greenland was a very easily accomplished task. For instance, garnets of unusually large size were found in lat. 81° 30′, having marked mineralogical characteristics, by which the identity with some garnets from Fiskernaes was established. Drawing a conclusion from such observations, it became evident that the main line of Drift, indicating the direction of its motion, runs from south to north." *

3. *A. H. Markham's "Whaling Cruise,"* &c., 1874, p. 205.—"In the latitude of their winter quarters (81° 38′) Musk-oxen were met with and 26 were shot. Foxes and Lemmings were also seen; but other animals were comparatively scarce, and only one Bear was seen during the whole year. Narwhal and Walrus were not seen to the north of 79°, but Seals were obtained up to the extreme point in 82° 16′. They were of three kinds, namely, the common Greenland Seal, the Ground Seal, and the Fetid Seal. The Bladder or Hooded Seal was not met with. On the western side, it was stated by the Etah Esquimaux, that Ellesmere Land abounded with Musk-oxen; and judging from the configuration of Grinnell Land, the same abundance of animal life is to be found there also. The Birds all disappeared during the winter, though Ptarmigan and a species of Snipe made their appearance early in the spring; and in summer all the genera found in other parts of the Arctic regions were abundant.

"With the exception of a Salmon seen in a freshwater lake not far from the beach, no Fish were met with. The contents of the stomachs of the Seals they caught were found to consist of Shrimps and other small shell-fish. Dr. Bessels used to dredge on several occasions, but owing to the ice, he could seldom do so to a greater depth than 18 or 20 fathoms, the results being generally unimportant, and with the exception of a few Shrimps and other *Crustacea*, nothing of interest was procured.

"No less than 15 species of Plants, five of which were Grasses, were collected by the Doctor at their highest latitude, on which the Musk-oxen must subsist. He gave me four specimens of the *flora* of 82° N., the names of which will be found in the List of Plants in the Appendix, to which Dr. Hooker has kindly added an explanatory note. [*See above*, p. 321.]

"Mr. Chester presented me with a fossil from the Silurian Limestone of that high latitude, which is also referred to in the Appendix. [*See further on.*]

"Dr. Bessels made a fair collection of Insects, principally Flies and Beetles, two or three Butterflies and Mosquitoes; and Birds of 17 different kinds were shot in 82°, including two Sabine Gulls and an Iceland Snipe."

4. *Clements Markham's "Threshold,"* &c., 1873, p. xxii.—Wintering (1871) near their furthest point, 82° 16′ N., they found abundance of animal life, and saw much Driftwood of recent date, which must have come thence across the Polar Sea from the shores of Siberia.

* Extracted from "Nature," by permission.

XL.—GEOLOGICAL NOTES on BAFFIN'S BAY. BY J. ROSS and MACCULLOCH, 1819.

From miscellaneous specimens brought home from Davis's Strait and Baffin's Bay by Capt. JOHN ROSS, and determined by Dr. J. MACCULLOCH (Ross's "Voyage of Discovery," &c., 2nd. edit., 1819, 8vo., vol. ii., pp. 121–141), it appears that:—

(1.) At Waygatt Island (70° 36′ N. lat., 54° 40′ to 55° W. long.) and at Four Island Point (70° 46′ N. lat., 53° 3′ W. long.) there are granite, gneiss, schists, and igneous rocks, with lignite at the former place.

(2.) At the Three Islands of Baffin (74° 1′ N. lat., 57° 25′ W. long.), gneiss abounding in garnets and containing molybdena.

(3.) Cape Melville, granite and porphyry.

(4.) Bushnan Isle (36° 04′ N. lat., 65° 66′ W. long.), granite, gneiss, and mica-schist; claystone and amygdaloidal claystone.

(5.) Cape York, *Esquim.*, Inmallick (76° N. lat., 66° 46′ W. long.), porphyritic greenstone, used by the natives in cutting off their iron from the masses.

(6.) Between Cape York and Cape Dudley-Digges, including the Crimson Cliffs (from 75° 45′ to 76° 10′ N. lat., and from 67° to 68° 40′ W. long.), granite, gneiss, and mica-schist, and greenstone.

(7.) From Possession Bay and Cape Byam-Martin, on the west side of Baffin's Bay (73° 33′ N. lat., 77° 28′ W. long.), granite, gneiss (with pyrites, garnet, and green felspar), sandstones, shale, limestone, jasper, siliceous schist, and igneous rocks and agates. "The gneiss with green compact felspar appears to be of common "occurrence on this coast . . . . ; it is exactly similar to that "which occurs abundantly in the Western Isles, and more par-"ticularly on the western coast of Ross-shire, prevailing particu-"larly about Loch Ew and Loch Greinord [Laurentian]." "The "jasper, siliceous schist, and chert resemble exactly those spe-"cimens which are found in the Island of Sky, among the beds "of shale, sandstone, and limestone [Jurassic], when these are "immediately in contact with the larger masses of trap; and "probably they here also owe their origin to the same cause."

(8.) From Agnes Monument (70° 37′ N. lat., and 67° 30′ W. long.), granite, gneiss, and graywacke-schist.

(9.) The Iron mentioned above, under "Cape York," is stated by Capt. Ross to have been used by the "Arctic Highlanders" of Prince-Regent's Bay (lat. 75° 54′, long. 65° 53′), for the edges of their knives, and to have been obtained by them from the mountains near the coast, behind Bushnan Island. It was said by the natives, as interpreted by Sacheuse, to occur in several large masses or pieces, of which one in particular, harder than the rest, was part of the mountain. The Iron was cut off with a hard stone, and then beaten into small flat oval pieces. The place where the metal was found was called "Sowallick," and about 25 miles inland (lat. 76° 12′ N., long. 53° W.). This Iron Dr. WOLLASTON estimated to contain between 3 and 4 *per cent.* of *Nickel*, and Mr. FYFE found in it 2·56 *per cent.* Hence they regarded it as of *meteoric origin*. *Op. cit.*, vol. i., p. 132, p. 140, and vol. ii., pp. 181–6.

XLI.—NOTES on METEORIC IRON used by the ESQUIMAUX of the ARCTIC HIGHLANDS. By CAPTAIN (now GENERAL SIR) EDWARD SABINE, R.A., F.R.S., &c. &c. 1819.

---

1. "Quart. Journ. Lit. Sc., &c.," 1819, vol. vi., p. 369; and "Geolog. Magazine," vol. ix., p. 74, 1872.

The northern Esquimaux lately visited by Captain Ross [in August 1818] were observed to employ a variety of implements of iron; and upon inquiry being made concerning its source by Captain Sabine, he ascertained that it was procured from the mountains about 30 miles from the coast. The natives described the existence of two large masses containing it. The one was represented as being nearly pure iron, and they had been unable to do more than detach small fragments of it. The other, they say, was a stone, of which they could break fragments, which contain small globules of iron, and which they hammered out between two stones, and thus formed them into flat pieces about the size of half a sixpence, and which let into a bone handle, side by side, form the edges of their knives. It immediately occurred to Captain Sabine that this might be meteoric iron; but the subject was not further attended to till specimens of the knives reached Sir Joseph Banks, by whose desire Mr. Brande examined the iron, and he found in it more than three per cent. of nickel. This, with uncommon appearance of the metal, which was perfectly free from rust, and had the peculiar silvery whiteness of meteoric iron, puts the source of the specimens alluded to out of all doubt. The one mass is probably entirely iron, and too hard and intractable for further management; the other appears to be a meteoric stone containing pieces of iron, which they had succeeded in removing and extending upon a stone anvil."

---

2. Extract from "An Account of the Esquimaux who inhabit the "West Coast of Greenland above the Lat. 76°." By Captain EDWARD SABINE, R.A., F.R.S., F.L.S. "Quart. Journ. of Literature, Science, &c.," vol. vii., 1819, pp. 72–94. *See also* the "Geological Magazine," vol. ix., 1872, pp. 73–74.

"Each of the Esquimaux who visited us on the 10th of August [1818], and I believe each of the others whom we after saw, had a rude instrument answering the purpose of a knife. The handle is of bone, from 10 to 12 inches long, shaped like the handle of a clasped knife; in a groove which is run along the edge are inserted several bits of flattened iron, in number from three to seven in different knives, and occupying generally half the length. No contrivance was applied to fasten any of these pieces to the handle, except the one at the point, which was generally two-edged and was rudely riveted. In answer to our inquiries from whence they obtained the iron, it was at first understood that they had found it on the shore; and it was supposed to be the hooping of casks, which might have been accidentally drifted on the land. We were surprised, however, in observing the facility with which they were induced to part with their knives; it is true, indeed, that they received far better instruments in exchange, but they did not appear to attach that value which we should have expected to iron so accidentally procured. This produced some discussion in the gun-room, when it appeared that some of the officers who had been present in the cabin when the Esquimaux were questioned were not satisfied that Zaccheus' [" Sacheuse" of Captain Ross's Narrative, 1819] interpretation had been rightly understood; he was accordingly sent for afresh, and told that it was desired to know what had been said about the iron of the knives (one of which was on the table), and he was left to tell his story without interruption or help. He said it was not English or Danish, but Esquimaux iron; that it was got from two large stones on a hill near a part of the coast which we had lately passed, and which was now in sight; the stones were very hard; that small pieces were knocked off from them, and beaten flat between other stones. He repeated this account two or three times, so that no doubt remained of his meaning. In reply to other questions, we gathered from him that he had never heard of such stones in South Greenland; that the Esquimaux had said they knew of no others but these two; that the iron breaks off from the stone just in the state we saw it, and was beaten flat without being heated. Our subsequent visitors confirmed the above account, and added one curious circumstance—that the stones are not alike, one being altogether iron, and so hard and difficult to break that their supply is obtained entirely from the other, which is composed principally of a hard and dark rock; and by breaking it they get small pieces of iron out, which they beat as we see them. One of the men, being asked to describe the size of each of the stones, made a motion with his hands conveying the impression of a cube of two feet, and added that it would go through the skylight of the cabin, which was rather larger. The hill is in about 76° 10′ lat., and 64° $\frac{3}{4}$′ long.; it is called by the natives 'Sowilic,' derived from 'sowic,' the name for iron amongst these people, as well as amongst the South-Greenlander. Zaccheus told me this word originally signified a hard black stone, of which the Esquimaux made knives before the Danes introduced iron amongst them; and that iron received the same name for being used for the same purpose. I suppose that the Northern Esquimaux have applied it in a similar manner to the iron which they have thus accidentally found.

"We are informed in the account of Captain Cook's Third

Voyage, that the inhabitants of Norton Sound, which is in the immediate neighbourhood of Behring's Straits, call the iron which they procure from Russians 'shawic,' which is evidently the same word. The peculiar colour of these pieces of iron, their softness and freedom from rust, strengthened the probability that that they were of meteoric origin, which has since been proved by analysis."

XLII.—The MINERALOGICAL GEOLOGY of GREENLAND from CAPE FAREWELL to DISKO. By the CHEVALIER CHARLES L. GIESECKÉ, Professor of Mineralogy to the Dublin Society.

[Extracted from his Article "GREENLAND," in Brewster's "Edinburgh Encyclopædia," 1816.]

[In the nomenclature of his day Giesecké here refers to the highly altered or metamorphic palæozoic (Laurentian and other) strata, with their igneous rocks, as "Primitive"; and to the more distinctly bedded igneous rocks, of later data, as belonging to the "Flœtz-trap formation": some old slates and trap-rock he refers to Werner's intermediate "Transition formation."—EDITOR.]

The accumulation of the ice having rendered the interior of Greenland totally inaccessible, it can only be examined on different parts of the coast; and the promontory Cape Farewell, which is its most southern point, presents to the eye immense groups of precipitous mountain-masses, insulated, barren, and naked, sharp-pointed at the top, greatly decomposed at the surface, and cleft by the action of the snows and the ice. These rocks are intersected by narrow valleys, where immense broken and scattered masses are borne along by irresistible currents, and carried immediately to the shores where there is no low land to intercept their course.

1. The GRANITE of this island (Cape Farewell) is fine-granular, consisting of pearl-white felspar, greyish-black mica, and very little quartz of an ash-grey colour. The whole rock is very much ironshot and disintegrated. At the foot of the granite rocks occur beds of common quartz of a milk-white colour (not milk-quartz), and flesh-red felspar, with small crystals of moroxite (foliated or common apatite). In another place are found flesh-red felspar, with little quartz, common hornblende, magnetic iron-stone, and gadolinite, crystallised in longish four-sided pyramids. A bed on the east side of the promontory contains garnets in a fine-granular greyish-white rock, very much resembling the rock of Namiest, in Moravia, called by Werner "Weiss-stein" (white stone); but the crystals of garnet here are larger and perfect dodecahedrons. The granite extends from Cape Farewell to the east and south-east of the coast, *viz.*, over the islands of Staaten-

huck and Kakasoeitsiak, Alluck, and Cape Discord, to a distance of more than 400 miles. Gneiss and mica-slate lie upon it at Kippingajak, both rocks containing garnets. Talc-slate forms a large bed in it at Akajarosanik, along with actynolite, which occurs in large masses. Near the coast of Akajarosanik is the small island of Kakasoeitsiak. It consists of one hill formed of a granite rock, mixed with some hornblende, slender crystals of zirkon, and the new mineral called *allanite* (*see* Trans. Edin. Soc., vol. vi., p. 371). The rock here assumes the character of the Norwegian zirkon-syenite; but its constituent parts are of a finer grain. All the granitic mountains of the islands of Staaten-huck and Cape Farewell are surrounded by numerous very small islands, presenting round-backed or flat conical hills of primitive syenite. To the west of Cape Farewell, at a place called Niakornak, is a very extensive bed of yellowish-white felspar, crystallised in large flat six-sided prisms, the crystals being only separated by black mica, which gives to the rock a porphyritic appearance. The place is very difficultly accessible, it being harassed perpetually by the most boisterous sea, and washed by the tide at high water. Not far from this, at an elevation of about 1,000 feet, the granite is divided into immense columnar or quadrangular pieces, which, seen from a distance, present an appearance similar to the ruins of a town. The Greenlanders state that the masses were carried thither by some giants, who inhabited the country in the oldest times, and, having been sorcerers, disappeared from the earth.

As granite is the principal rock which constitutes the mountains of this vast coast, to enumerate all the places where it is found would exceed the limits of such an article as the present. Its most common colour is greyish-white, flesh-red, and tile-red; the latter colours are characteristic of the coarse-granular felspar. Magnetic iron-ore is generally found either disseminated or imbedded in the red variety. In some places, molybdena occurs, and in others graphite, imbedded in the rock. At Baal's River and at Disko Island, iron-pyrites is found; but, excepting there, the rock is not very metalliferous. Precious garnet occurs very frequently; also common schorl, tourmaline, common hornblende, jade, rock-crystal, moroxite, calcareous spar, fluor-spar, and the above-mentioned substances. Rock-crystal is only found in veins traversing the red coarse-granular variety, and appears to be contemporaneous; the vein being intimately mingled with the rock, and presenting no walls. Beds of hornblende-slate, mica-slate, felspar, and quartz rest upon it, and on the red coarse-granular granite at Kogneckpamiedluœk there is an extensive bed of red ironstone mingled with massive iron-flint ("Eisenkiesel" of Werner). At the end of the north-eastern arm of Baal's River, in the vicinity of the great continental ice, the traveller, ascending from a narrow cliff, suddenly beholds a dreadful chaos of immense columnar granitic blocks detached from each other, and heaped together in the most fantastic groups, the planes of fracture being so fresh that the points from which they are broken are distinctly observable. Places of desolation and devastation of this kind are very frequently met with in the mountains of Greenland. Most of the granitic rocks affect the needle.

2. The next rock which forms numerous mountains in this country is GNEISS. It occurs very often alternating with granite, sometimes with mica-slate. Its character or texture may be ascertained partly in the cliffs and on the shores, partly by the forms of the mountains. The granitic mountains are always more decomposed and therefore more precipitous, presenting very sharp-edged summits; the summits of the gneiss are more flat and round-backed. The texture of the gneiss is thick- and thin- slaty; its felspar generally pearl-grey and pearl-white, seldom flesh-red, fine-granular; its mica-grey, pinchbeck-brown, and blackish-brown; it contains but little ash-grey quartz. The valleys and clefts round the mountains are filled with rhomboidal fragments, many of them of immense size. The smaller fragments were used by the old Norwegians, with mica-slate, hornblende-slate, and slaty claystone to build their houses; the walls of which, although not cemented, after a lapse of several centuries still brave the power of this destructive climate. Gneiss constitutes one of the most elevated points of this extensive coast, *viz.*, the mountain *Kingiktorsoak*, situated in the 62nd degree of latitude. It is covered with mica-slate from the shore to a height of about 1,000 feet above the level of the sea, where the gneiss again becomes visible, and continues to a height of nearly 3,000 feet. The top of this mountain is similar in shape to the roof of a house, where the ridge is not much elevated. It is entirely free from snow in summer, except a few small spots, where it rests in the hollows of its summit.

The mica-slate resting upon the gneiss presents a variety of beds of hornblende-slate, whitestone (Weiss-stein) with small garnets, talc-slate with common and indurated talc, potstone, actynolite, and precious splintery serpentine. The gneiss is transversed with numerous veins of greenstone, various in thickness from one inch to six feet. The greenstone which occurs in the veins resembles basalt; but it is more crystalline in its texture, lighter in its colour, and not quite so hard. Common schorl, tourmaline, and precious garnet occur imbedded in gneiss. It contains veins of tinstone, accompanied by arsenical pyrites, wolfram, fluor, and quartz, in a firth called Arksut, situated about 30 leagues from the colony of Juliana-Hope, towards north-east. The same place is remarkable for two thin layers of cryolite, resting upon gneiss; and it is the only place where this mineral has hitherto been found. One of these layers contains the snow-white and greyish-white variety, unmixed with any other mineral. Its thickness varies from one foot to two feet and a half; and it is divided from the underlying gneiss by a thin layer of mica, always in a state of disintegration. The other variety is of a yellowish-brown colour, passing into tile-red. It occurs along with iron-pyrites, liver-brown sparry-iron-ore crystallised in rhombs, earthy cryolite, quartz, compact and foliated fluor, earthy fluor, and galena. It is remarkable that the galena is sometimes coated with a greyish-white sulphureous crust, which burns in the flame of a candle with a bluish colour, emitting a sulphureous smell.

These layers of cryolite are situated very near each other, only separated by a small ridge of gneiss, of a thickness of 27 feet; both are washed at high water by the tide, and for the most part exposed, the superincumbent gneiss having been removed. The white cryolite, seen at a distance, presents the appearance of a small layer of ice; small detached fragments have acquired, from decomposition, the shape of cubes. This mineral is called by the Greenlanders *Orsuksiksæt*, from the word *orksuk*, blubber, to which it bears some resemblance. The same name is also given by the natives to white calcareous spar.

3. MICA-SLATE is likewise one of the most common rocks in Greenland, and an inseparable companion of gneiss; there are very few instances where they are not found in the vicinity of each other, and frequently in contact. Mica-slate forms in this country a very extensive series of mountains, which never rise to a considerable height, and appear generally to rest upon gneiss. Mica-slate is frequently visible on the shores; and the gneiss itself forms also very extensive beds in it at Disko Bay, where the whitestone also occurs in beds. The Greenlandish mica-slate abounds in mica; it is generally thin-slaty, and only thick-slaty when the quartz prevails. Sometimes it has an undulating aspect; but, when this is the case, it passes into primitive clay-slate. The mica of this mica-slate is mostly greyish-black and pinchbeck-brown, passing into brownish-black, seldom silver-white. Its quartz is pearl-grey. It is sometimes mingled with nodules of pearl-grey felspar, from the size of a pea to that of an orange, and this gives it the appearance of gneiss; but they may be easily and accurately distinguished, as the mica-slate presents a surface perfectly continuous, and easily separable in the direction of the plates of the mica. The strata dip towards north-west.

Mica-slate also occurs in beds in various parts of this country. One of the most remarkable, most interesting, and most extensive, is that in the firth of Kangerdluarsuk in the 61st degree of latitude, in the district of Juliana-Hope. It extends about five miles in length and four miles in breadth; its thickness varies from 6 to 12 feet; and it contains, besides felspar, which is its principal constituent part, hornblende, augite, actinolite, sahlite, garnet, and that new mineral which has been analysed by Dr. Thomson and Professor Eckeberg, called sodalite. It is of pale apple-green, leek-green, greenish-white, and pearl-grey colour; partly massive, partly crystallised. Another mineral, which has not been analysed, occurs also with the sodalite: it is of a peach-blossom red and purple-red colour. On the shore the underlying gneiss is visible in several places. In the superincumbent mica-slate, graphite ["granite" in original] is found, of very fine texture, partly in veins, partly imbedded. Calcareous spar and fluor occur in veins, both of which are sometimes coated with a thin crust of chalcedony; also galena in small veins. Blue phosphate of iron, in detached pieces, is found on the shores. The mica-slate is generally decomposed and iron-shot, where the graphite is imbedded. In the Firth of Arksut a bed of very fine granular limestone is found in mica-slate, which resembles the Carrara marble. The beds which occur in this rock on the mountain Kingiktorsoak have been already mentioned. Hornblende-slate, forming beds in mica-slate, is found in many places.

In the 64th degree of latitude, in a firth called Ameraglik, in the south of the Danish colony Godthaab (Goodhope), a variety of mica-slate is found, which passes into talc-slate, forming a very small layer in coarse-granular granite. It is very remarkable, on account of the large groups of tourmaline which occur, imbedded or rather involved in talcose mica, and which are the largest crystals of this fine mineral that have been met with. At the end of the same firth, at Auaitsirksarbick, in the neighbourhood of the great continental glacier, the finest garnets are found. They are of a lamellar texture, and surpass the oriental specimens in colour, lustre, and hardness. At the same place, dichroite and hypersten e of a beautiful blue colour occur, along with precious garnet, in decomposed mica-slate. All the lower mountains from the 66th to the 71st degree of north latitude; and particularly all the mountains of the continent forming Disko Bay, with the greatest part of the adjacent islands, are composed of mica-slate. There is scarcely a square mile where the rock is entirely free from garnets. A large mountain in Omenaks Firth, called Sedliarusæt presents on its surface only the powder of mica-slate, and fragments of precious garnet. From the appearance of this powder, it is probable that the rock formerly contained great masses of imbedded iron-pyrites. No snow rests on the surface of this mountain in the coldest winter. The fragments of precious garnet which are found here, when clear, are the most highly prized of any on the coast. Other minerals which are found in mica-slate in Greenland are emery, on the island Kikertarsoeitsiak in South Greenland; granatite, on the island Manetsok; moroxite, in very large six-sided prisms, at Sungangarsok, in North Greenland; and dichroite in six-sided prisms on the island Ujordlersoak, in the 76th degree N. lat. Except iron-pyrites, copper-pyrites, and galena, no metal occurs in this rock.

White Stone (Weiss-stein), which has lately been determined by Werner, appears to belong to this rock. It presents a white and greyish-white granular appearance, which was formerly supposed to be compact or granular felspar. It is in this country characterised by very small and minute crystals of garnet disseminated through the whole mass. Here it is found in layers of inconsiderable extent, resting on mica-slate, very seldom on gneiss. It is also found in detached pieces.

4. Clay-slate is very seldom met with on this coast, and consequently the different beds which are characteristic of this rock, viz., flint-slate, lydian-stone, alum-slate, but rarely occur. Nevertheless, at the mouth of the Firth Arksut it forms two islands of some importance, called Arksut and Ujorbik. The colour of the slate is ash-grey and bluish-grey; its fragments present a double cleavage; and it is traversed in all directions by numerous veins of massive and crystallised quartz, massive hornstone, and sparry-iron-ore of an isabella-yellow colour. An extensive bed of flinty slate and lydian-stone rests upon it on the east side of

the Island Ujorbik. In Ameraglikfiord, in 65° 4′, there is a small island, where the clay-slate forms small layers in fine-grained granite; fine cubes of iron-pyrites, with various truncations, occur in this slate, which is greatly decomposed. Some small islands in the south-east of Disko Bay consist of clay-slate, with a variety of small beds and layers, viz., very ironshot hornblende, with small garnets, whet-slate, granular hornblende, and greenstone. This clay-slate may perhaps belong to the class denominated "Transition rocks."

5. Porphyry is very common in the south of Greenland, from Cape Farewell to the 64th degree of latitude; but it is generally found towards the interior of the continent, forming insulated rocks. In the interior of the Firth Igalikko, at Akulliaraseksoak, hornstone-porphyry is found, very distinctly stratified, and resting upon fine-grained granite, containing large crystals of reddish-white, flesh-red, and tile-red felspar, and another mineral of a talcose appearance, crystallised in six-sided prisms, and hitherto unknown. The mass of the porphyry is brownish-red, and passes in some places into claystone forming claystone-porphyry, the crystals then becoming less distinct. Hornstone-porphyry, with a few very small crystals of felspar, occurs also in an adjacent firth called Tunugliarbik. This rock rests upon Old Red Sandstone.* The porphyry is very much decomposed. It is of a brown-red colour, and is called by the natives *aukpadlrtok*, that is, blood-red rock. It contains small layers of a brown-red iron-ochre, which the Greenlanders use as a dyeing material, to embellish their utensils, and the interior of their houses, a species of luxury they have learned from the Europeans.

6. Syenite, and all the porphyritic rocks belonging to the Primitive and Transition Trap-formation, are found in great abundance in this country. Hornblende is a mineral which occurs almost everywhere. A kind of coarse-granular syenite, composed of coarse-granular Labrador-felspar and crystallised common hornblende, rests upon fine-grained granite at the mountain Illejutit, or Redekammen, 61° lat., in the neighbourhood of that extensive bed of sodalite, sahlite, and hornblende, which has been already mentioned. This Labrador-syenite occurs also at the mountain Kognek, 62° lat., upon granite of a coarser grain. In the vicinity of the mountain Kognek is a group of more than 50 islands, lying in a western direction, in Davis' Strait, and called by the natives Kittiksut, from *kitta*, west. These islands form round-backed low hills, and consist of common felspar, of yellowish-brown and leek-green colours, and common hornblende of raven-black and sometimes velvet-black colour, accompanied by small four-sided prismatic crystals of zirkon of red-brown and purple-red colour, with fine grained common magnetic ironstone interspersed, and very little black mica. In some parts of the rock allanite occurs, of a pitch-black colour. The rocks are somewhat ironshot, and disintegrated on their surface.

* The relative age of this sandstone has not been determined.—Editor.

Titanium-iron-ore is found in small layers, and fine-granular chromate of iron. The rock itself has a striking resemblance to the zirkon-syenite, found at Friedrichswærn and other places in Norway, and described by Von Buch, Esmark, and Hausmann. The neighbouring mountains have no trace of that rock. At Narsak, in the vicinity of Baal's River, brown titanite, or brunon, is found disseminated in syenite.

Granular porphyritic syenite is found at Nunarsoit (Cape of Desolation). Its stratification is not very distinct. It contains very extensive beds of coarse-grained, tile-red felspar, and common magnetic ironstone.

7. Primitive Trap (*Greenstone*).—The islands which lie between the 62° and 63° of latitude present a very complete series of the rocks that belong to the Primitive Trap-formation. The greenstone first appears at Sakkak and Ujorbik in the mouth of Arksutsfiord, where clay-slate predominates; and extends from those islands towards the east, that is, to the continent of Greenland, alternating with greenstone of a porphyritic structure (*porphyrartiger Grünstein*, of Werner), and green porphyry or *verde antico*. Another rock of slaty texture, consisting of compact felspar and hornblende, appears to be intermediate between hornblende-slate and greenstone-slate; it is here the only rock which presents very distinct stratification. The greenstone-slate covers uninterruptedly both the greenstone and the green porphyry, and appears to belong to the Transition Greenstone-formation; and perhaps the whole formation should be referred to it. It probably extends farther to the interior of the continent, as the fragments which are thrown out from the continental ice have an appearance exactly similar. Variolite is found there in small, roundish, rolled pieces. The greenstone, alternating with syenite, is found upon gneiss and mica-slate on the large island Nunarsoit.

8. Primitive Limestone, of fine-granular texture, is found only in beds and rolled pieces, and occurs very seldom in Greenland. Its beds are confined to gneiss and mica-slate, and it is mingled with minute leaves of silver-white mica, seldom with grains of quartz. It is generally accompanied by tremolite, asbestos, actynolite, sahlite, and seldom with rock-cork. Thus situated, it occurs at the Island Akudlek, at the Island Manetsok, at Kakarsoit and Kangerluluk, mountains in the vicinity of Jakobs-havn and Christians-haab, in Disko Bay.

It is very surprising, that no vestige of Flœtz limestone is found on this vast coast; nor does any petrifaction occur there. Very distinct impressions of the *Salmo arcticus* [*Mallotus Grœnlandicus*, Cuvier], with its bones very little altered, occur in detached pieces on the alluvial land, which are forming daily.* In the uppermost sandstone, which belongs to the brown coal of the Flœtz Trap-formation, fragments of *Pecten Islandicus* are found, which have undergone but little alteration.

* See Mr. Watson's translation of Dr. M. Sars's "Papers on the Fossiliferous "Nodules in the Post-Tertiary Clays of Norway, &c." Geol. Mag., vol. i., p. 158, &c.—Editor.

9. The Flœtz Trap-formation of Greenland is perhaps the most extensive that has been discovered. It begins at 69° 14′ N. lat., occupies the large island Disko, and the eastern coast of the Waygat, from Niakornak, on the northern cape of Arve-prinz Island, round the Cape Noursoak, as far as the end of the southern coast of Cornelius Bay, where it reaches the continental glacier. Hare Island in the north of Disko Island, Unknown Island in the mouth of Cornelius Bay, the islands Kakiliseit in the north of the latter, and many other northern islands consist entirely of flœtz-trap. From thence it extends over a part of the continental coast of Greenland, viz., London Coast, Svartenhuk, Ekalluit, Kangersoeitsiak, Karsorsoak, and disappears in the 76th degree under the most northern continental ice or glacier, which precludes all further investigation.

The whole Flœtz Trap-formation of Greenland, as far as it has been examined, rests on gneiss or on mica-slate, these rocks alternating continually. The underlying Primitive rocks, as well as the superincumbent Flœtz-trap, are always somewhat decomposed, where they come in contact. Trap-tuff generally rests immediately upon the Primitive rock; it consists of balls and nodules of basalt and wacké, joined together by a cement of the same substance; the centre of the balls and nodules is very often filled with mesotype, blended with massive or crystallised apophyllite, the crystals of which are sometimes penetrated by acicular mesotype. This trap-tuff scarcely presents another mineral; and the apophyllite, or ichthyophthalmite, does not occur there in any other rock. The underlying Primitive rock is very variable in its elevations, sometimes it does not surpass the level of the sea, sometimes (for instance, at Godhavn) it reaches a height of from 500 to 600 feet, which can be observed very exactly in the cliffs there. Columnar basalt lies upon trap-tuff. It presents four, five, and seven-sided columnar distinct concretions; the columns very seldom exceed a foot in diameter. This basalt does not include any mineral, except sometimes very minute spots of greyish-white glassy felspar. Wacké generally rests upon it, forming an amygdaloid with different minerals, viz., chabasite, stilbite, analcime, chalcedony, opal, heliotrope, quartz, zeolite, miemite, and basillar arragonite. At Hare Island the chalcedony is found crystallised in cubes. At Kannioak in Omenaksfiord, miemite occurs in kidneys, along with chalcedony, opal, wavellite, arragonite, and some quartz, in grey decomposed wacké. The wacké of the Flœtz Trap-formation of this country is generally intersected by small veins of iron-clay and bole. Lithomarge and green-earth occur in nodules. Olivine and augite are but seldom met with in the Flœtz-trap of Greenland. Laumonite, in a friable state, is found in very small veins, traversing wacké at Sergvarsoit, on the northern coast of Disko Island. Most of the Greenlandish basalt affects the needle very powerfully. There are generally two, and sometimes three strata of columnar basalt, and one of them forms the summit, except at Hare Island, where the summit consists of porphyry-slate resting upon wacké. The shape of the mountains is very various; some of them present pyramidal, some conical forms, and some are entirely flat. Their stratification is very nearly horizontal; and the valleys between the mountains are generally narrow. There is no doubt that some of the mountains have been separated by very recent eruptions of rapid torrents.

On some parts of Disko Island beds of brown-coal occur in Flœtz-trap; they rest upon yellowish-white coarse-grained sandstone, which is very friable. Large balls of iron-pyrites are imbedded in it. The beds of coal are generally divided from each other by strata of fine-grained sandstone, and are of very unequal thickness. In some places of the east coast of Disko Island, in the Waygat, the sandstone becomes harder, and carbonised impressions of leaves are found in it, which are similar to those of *Sorbus* and *Angelica.*

The coal of Disko Island is common brown-coal, of a slaty texture; it burns very easily, but it leaves a great residuum in the form of white ashes, which have a slaty texture, and somewhat resemble the polishing slate from Bilin in Bohemia. A very remarkable variety of brown-coal, passing into bituminous wood, occurs in a small bed at Hare Island. It is of slaty texture; and honey-yellow amber, in numerous grains of various sizes, is disseminated parallel to the cleavage of the coal. It rests upon ash-grey coarse-grained sandstone, is covered with grey common clay, and belongs undoubtedly to the newest brown-coal formation. At Koome in Omenaksfiord, native capillary and fibrous sulphate of iron, of a beautiful green colour, is found in the cliffs of the brown-coal. All the Greenland coal is subordinate to Flœtz-trap.

Alluvial land has been formed at the end of every bay and firth of the coast; and, in addition to grey and greyish-white sandy clay, it contains fragments of the neighbouring mountains. This formation is daily increasing, and contains no metallic substance, except magnetic iron-sand, with which it generally abounds.

XLIII.—On the MINERALOGY of DISKO ISLAND. By SIR CHARLES GIESECKÉ, F.R.S. Edin.

[From the Transact. Roy. Soc. Edinburgh, vol. ix., 1821, p. 263, &c. Read April 4, 1814.]

[The Wernerian terms "Primitive rocks," "Flœtz Trap," and "Flœtz "formation" are used as in the preceding paper.]

Disko Island is situated in front of a bay in the continent of Greenland, within Davis' Strait, known by the name of Disko Bay, which is sometimes called, particularly in the old Dutch charts, Sydost (South-east) Bay. This name is derived from an immense curvature, screened by innumerable islands, made in the continent by the sea. Disko Island is situated in 69° 14′ of N. latitude. It is distant from the continent towards the south 12 German miles; on the west and north it is surrounded by the sea of Davis' Strait; and on the east, it is separated by a narrow sound, distinguished by the name of Waygat by the Dutch, and by the Greenlanders *Ikareseksoak*. It stretches northward from 69° 14′ to 70° 24′; and its greatest breadth, which is from Fortune Bay on the west, to Flakkerhuk, so named by the Dutch, on the east, is 10 German miles.

The whole of Disko Island belongs to the Flœtz-trap-formation, which extends over part of the continent beyond the Waygat, and shows itself on the other side at 69° 20′ of N. lat., continuing towards and occupying the peninsula of Noousoak, which separates Disko Bay from the Bay of St. James, called by the Dutch Stikkendejakob's Bay. On the east end of this, the Flœtz-trap disappears under the stupendous glacier or ice-blink of this immense arm of the sea; and on the opposite side of it, not the smallest vestage of Flœtz-trap is to be discovered. On quitting the shore, however, towards the north, the same formation occurs, at the island of Upernavik, or Spring Island, which is formed of basalt, with immense beds of sandstone, containing veins of brown and bituminous wood-coal. Two considerable islands situated beyond the Frith, one named Ubekjendte (or Unknown) Island, and the other Hasen (or Hare) Island, belong also to the Flœtz-trap.

These islands, although now detached, all appear to have originally belonged to the same mass, and to have been torn asunder by the impetuosity of the sea, which, impelled by the winds from every quarter, runs with a force almost beyond belief. During such a tempest, I have myself seen the jaws of the great Greenland Whale, *Balæna mysticetus*, thrown to a distance of 200 feet inland upon the beach.

Beyond the Bay of St. James, towards the great Northern Cape called Svartenhuk, the Flœtz-trap is interrupted, either by the Primitive rocks, or by an immense plain covered with alluvial soil. Svartenhuk is composed of a granitic rock, with large beds of micaceous schistus, mixed with small garnets. In the adjacent bay, called Hytten, the Flœtz-trap shows itself in small hills, resting on a bed of sandstone, in which bituminous wood occurs. From this point, the continent of Greenland, which consists of granite, stretches away to the east of north, and is covered with an incredible number of small islands, called the Vrowen or Women's Islands. The base of these islands is uniformly granite or gneiss; the last sometimes, though rarely, mixed with garnets. Some of the islands are covered with beds of the Flœtz Formation, particularly Kakarsoak, the largest of the group.

To the north of Kakarsoak, in the colony of Upernavik, in lat. 72° 32′, the Flœtz-trap again disappears, and granite, alternating with gneiss, present themselves, and continue to lat. 73° 32′, at the islands of Udjordlersoak and Tessiursak. Near Cape Nullok, in Sanderson's Hope, the Flœtz-trap again appears in large masses of columnar basalt, resting on gneiss; but beyond this place, there is no farther approach, the country being covered by the Great Boreal Glacier—the Northern Iceblink.

The direction of the trap-rocks, which are here spread over such an extent of country, is almost entirely similar, being nearly horizontal, stretching from south-west to north-east. The beds of which they are composed are of a very unequal thickness; those of basalt are most prevalent. The hills composed of gneiss and granite are never highly elevated; and the Flœtz rocks are placed immediately on the gneiss, which is always slightly decomposed upon the surface, where in contact with the trap. The prismatic basalt of this district, as of that species distinguished in Germany by the name of Basaltic Greenstone (grünsteinartiger Basalt). It is almost pure, but sometimes contains a few detached specks, perhaps crystals, of felspar. I found only in one place some small grains of augite and of hornblende. The massive basalt, on the contrary, often becomes amygdaloidal by the small globules of mesotype, stilbite, and quartz which it contains.. It occurs very generally undermost, and touching the Primitive rocks, which is very rarely the case with the columnar basalt.

The trap-tuff, which is very common among the Flœtz rocks of Disko, rests also always immediately on the Primitive rocks; indeed, I never found it in any other situation in that island. It appears to me here necessary to mark two varieties of this rock, namely, that which consists almost entirely of fragments of wacké contained in a paste of the same substance in a state of decomposition; it is of a very fine grain, very soft, and almost friable. The other is composed of fragments of wacké, but more compact, and of globular pieces of basalt. When these globules are broken, the interior is occupied by geodes of crystallised apophyllite, accompanied with capillary mesotype, sometimes decomposed and reduced to powder, in which state it is known by the name of earthy zeolite. These are the only minerals I found in this globular basalt. The apophyllite I never observed in the other variety of trap-tuff, in which I discovered no simple mineral whatever, except some very small geodes of radiated zeolite. I shall distinguish the one by the name of Trap-tuff, and the other by that of Basalt-tuff. The last appears to me to be the oldest of the two, and occurs, wherever I saw it, under the other. If the tuff be entirely absent, then the amorphous basalt occupies its place; and on it rests the amygdaloid, the paste of which is of a reddish-brown colour. It is the amygdaloid of this colour in which the greatest number of minerals occur, such as stilbite, mesotype, quartz, calcedony, and igloite. When exposed to the action of the weather, this rock becomes extremely fragile, and falls in conchoidal fragments, almost like bole. It occasions, particularly in the spring season, by reason of its feeble cohesion, immense devastation. Rent by the effects of the severe frosts of winter, it falls in huge blocks into the valleys, when the basalt, deprived of its support, is precipitated in enormous masses, and to such an extent, that rivers are often impeded in their course, and the whole neighbourhood laid under water. Over this amygdaloid, a mass of ferruginous clay occurs, similar to the "Eisenthon" of the Germans, which approaches to the jaspery oxide of iron. This is again covered by amorphous basalt, separated from columnar basalt, which usually forms the summits of these hills, by another seam of the same ferruginous substance, of a brownish colour.

The mountain called Ounartorsak, near Godhavn, presents the following proportions in one of its precipices:—

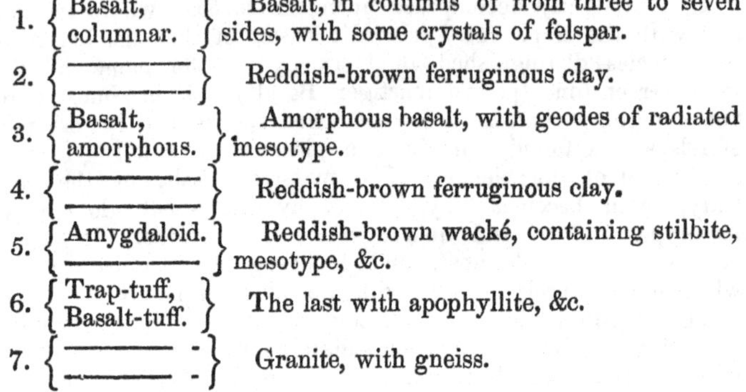

1.	Basalt, columnar.	Basalt, in columns of from three to seven sides, with some crystals of felspar.
2.	———	Reddish-brown ferruginous clay.
3.	Basalt, amorphous.	Amorphous basalt, with geodes of radiated mesotype.
4.	———	Reddish-brown ferruginous clay.
5.	Amygdaloid.	Reddish-brown wacké, containing stilbite, mesotype, &c.
6.	Trap-tuff, Basalt-tuff.	The last with apophyllite, &c.
7.	———	Granite, with gneiss.

All the basalt of Disko is magnetic. That found in the most elevated situations is most so; the fallen masses dispersed around the base of the mountains having more power over the needle than the others.

The mountains of Disko are almost all flat at the top, and at a distance present the appearance of large houses. It was only in the Waygat, and in the Bay of St. James (Omenaks Fiord), where I observed pyramidal and conical summits. Mannik, a mountain in Waygat, is terminated by an immense basaltic pyramid of four sides.

On the summits of all the mountains which I ascended, I found numerous rolled masses of Primitive rocks, often of considerable size, and of a weight beyond my power to move. These masses consisted either of granite, gneiss, mica-slate, siliceous schist, quartz, or hornstone.

Porphyry-slate is the rarest rock among those of the Trap-formation in Greenland. I ascended several of the mountains, but I found it only in two, Unknown Island and Hare Island, to the north of Disko; and there it occupied only the summit, in tables split into a thickness varying from six inches to two, affording a clear ringing sound when struck by a hammer. The Greenlanders informed me, that during tempestuous weather, even at the foot of the mountain, they often heard tones resembling those of music, and that Tornarsuk, their good and evil deity, when enraged, was the cause of them. He never, however, happened to be out of humour within my hearing.

At the foot of this immense Trap-formation of Disko, considerable beds of sandstone occur. It makes its appearance at Aukpadlartok, Akkiarut, and Imnarsoit; but the mass of greatest magnitude is at Aumarurtiksæt, where it is accompanied with beds of coal. From this spot the beds extend along the edge of the sea, by Waygat, and become very considerable at Kudlisæt, where the stratification is deposed in the following arrangement :—

——— Sandstone, sometimes with globules of pyrites.
——— Brown coal.
——— Schistose [thin-bedded] sandstone.
——— Pitch-coal.
——— Shale [Argillaceous schistus, in orig.].
——— Brown-coal.
——— Sandstone, with vestiges of Plants.

The sandstone is very light, and sometimes friable, which is also the case with the shale [clay-slate, orig.]. The vegetable impressions that occur in the lowest bed, seem to be those of the leaf of *Angelica archangelica.* The most considerable bed of coal is about 9 feet thick ; while some of the seams are not above 7 or 8 inches.

It is nearly impossible to render this coal available, as scarcely any shelter is to be found all along the Waygat for vessels of any description, while a tempest almost continually prevails in the Strait. It is the same case with the coal of Hare Island, generally known on account of the grains of amber which it contains. There it occurs under an argillaceous wacké, in the following order :—

——— Coarse conglomerate.
——— Argillaceous wacké.
——— Brown-coal, with amber.
——— Fine-grained conglomerate.
——— Sand.

I have now only to mention the simple minerals which accompany the Flœtz-Trap-formation of this country, of which the different members of the family of zeolite, its usual companion in all quarters of the globe, are the most remarkable.

1. *Mesotype.*—The most common sub-species of this mineral is the fibrous and radiated. The last is found crystallised in rectangular prisms, truncated, with pyramids of four planes.*

b. Capillary. Near Sergvarsoit in Disko there is a small cave covered with capillary mesotype, which the Greenlanders consider to be the hair of one of their magicians called Angekok. When this variety is decomposed, it forms the earthy or mealy zeolite.

* Dr. Brewster has examined the Greenland mesotype, and has found it to be an entirely different mineral from the Auvergne mesotype. In its crystalline form it resembles the Auvergne specimens, while, in its optical properties, it resembles the Iceland mesotypes. It is very remarkable that capillary crystals from Sergvarsoit have been found by Dr. Brewster to be different from the large crystals, and to be the same as those from Auvergne.

2. *Stilbite*,—in thin hexagonal tables.
   b. In quadrangular prisms, acuminated by truncated pyramids.
3. *Chabasie*,—crystallised in the primitive rhomb.
   b. In truncated rhombs.
   c. In macles.
4. *Analcime*,—crystallised in the form of leucite.
5. *Compact Zeolite*, white and red.—This mineral occurs in cavities and veins in all the rocks of the Flœtz-Trap-formation, except the basalt-tuff.
6. *Apophyllite* or Ichthyophthalme, occurs—
   a. In prisms perfectly rectangular.
   b. Also with the solid angles replaced. This variety was mistaken for mesotype and described as *Mesotype epointé.*
   c. By a curious arrangement of the particles, the crystals of apophyllite at the extremities present the shape of a barrel.* They also occur acuminated and diverging sometimes in the form of a rose. In perfect cubes, the apophyllite occurs in Greenland only in the basalt-tuff, accompanied with delicate capillary mesotype. Notwithstanding, in Faroe and Iceland it is found in wacké. This substance forms an opaque jelly in nitric acid, frothing up and exfoliating. The apophyllite also occurs in a radiated form similar to stilbite, but with a more brilliant lustre, presenting on the surface a crystallisation similar to the cock's-comb barytes.

8. *Carbonate of Lime* occurs in all rock of this formation, in cavities and veins, of a greyish-white colour, sometimes massive, sometimes crystallised in rhombs, also in pyramids of three and six planes, and in prisms of six planes. I have found it also crystallised in nearly perfect cubes.

9. *Igloite*, the arragonite of Haüy, and hard calcareous spar of Bournon, occurs fibrous, radiated, and crystallised in pyramids of three planes; also in prisms of six planes, terminating by degrees in pyramids.

10. Radiated and concentric globular mineral, of a yellowish-green colour, which I take to be *Wavellite.*

11. *Compact quartz, bacillaire,* and crystallised in prisms, in geodes.

12. *Calcedony*, massive, and very rarely in cubes. Quartz and calcedony occur in all the rocks.

13. *Opal, common,* in veins and cavities, white and yellow, particularly in basalt.

14. *Cereolite*, a mineral of a yellowish, brownish, and greenish colour, very similar to compact lithomarge.

---

* The cylindrical apophyllite, according to the experiments of Dr. Brewster who examined some specimens which I transmitted to him, differs in a remarkable manner from the apophyllite of Iceland, Faroe, Uto, and Fassa. Its optical properties he has found to be of a very curious kind.

15. *Green Earth*, lining cavities, and sometimes filling geodes.
16. *Heliotrope*, in geodes and veins in basalt.
17. *Agate*, in geodes in basalt.
18. *Felspar* in small crystals, constituting the basaltic-porphyry and porphyry-slate.
19. *Ferruginous Clay*, of a reddish-brown colour, the "Eisenthon" of Werner.
20. *Bolus*, in small veins.
21. *Bituminous Wood*, very rarely in minute beds in wacké and basalt.
22. *Brown-coal.*
23. *Pitch-coal*, above described.

The Primitive Rocks, which constitute some small islands on the south side of Disko, are very rarely accompanied with any of the simple minerals. The felspar of the granite sometimes becomes opalescent; the granite contains occasionally compact and prismatic epidote, also diallage and tourmaline; at Kangek it sometimes, but very rarely, contains some cubes of pyrites; and in one place I observed magnetic iron [ore ?], in nodules, mixed with it. In the islet of Fortune Bay, I noticed some specks of the green oxide of copper in the micaceous schistus.

XLIV. — On the CRYOLITE* of WEST GREENLAND; a Fragment of a Journal by SIR CHARLES GIESECKÉ. (Reprinted from the Edinburgh Philosophical Journal, vol. vi., 1821–22, pp. 141–4.)

Towards the end of September 1806, on returning from my mineralogical excursions around Cape Farewell and part of the eastern coast of Greenland, I was informed by one of the Greenlanders who accompanied me, that they sometimes found loose pieces of "lead" (*Akertlok* of the natives) in a frith to the northward of Cape Desolation (Nunarsoit of the Greenlanders), but he could not tell me the exact spot. Though the unfavourable season was already advanced so far, and the equinoctial gales had begun blowing so violently as to make it unadvisable to venture such a doubtful excursion, yet I resolved to go in search of the place, as we were near to the mouth of the frith in question. The name of the frith is *Arksut* (Engl., the Leeward): it was divided into two arms; that on the right of the entrance had a south-easterly, and that on the left an easterly direction. I steered up the eastern arm about 16 miles, and put on shore at different places. I already began to despair of finding lead, when I observed, at some distance, but near the shore, a snow-white spot. At first

---

* I know no name in the system of mineralogy more expressive of the external character and the fusibility of this substance than that adopted by my deceased friend Dr. Abilgaard, late Professor in the University of Copenhagen, who was the first who noticed and analysed this substance.

sight I suspected it might be a small glacier; but considering that no such thing could exist, at this time of the year, so near the sea, I landed, and I found, to my great astonishment, a bed of Cryolite, the geological situation of which had been hitherto so doubtful.

The islands which lie across or shut up, as it were, the mouth of this frith, consist of coarse-granular granite. The lofty mountain *Kognekpamiedluœt* (Engl., the clifted rock with the long tail), which rises on the left side of the entrance of the frith, and which is a landmark to the navigator, is composed of the same granite, but with overlying syenite, the felspar of which is beautifully labradoric. This granite continues uninterrupted for eight miles on both sides of the frith of Arksut, when it disappears and alternates with gneiss. This gneiss forms the shores on both sides of the frith for from seven to eight miles, to the spot called *Ivikœt* by the natives, where the cryolite is found.

The name *Ivikœt* (from *ivik*, grass) was given to this place by the Greenlanders on account of its peculiar fertility. The place was formerly visited by them during the summer season, on account of its being a good place for fishing and drying *Angmaksœt* (*Salmo arcticus*, L., the *Lodde* of the Norwegians [the Capelin]); but it was deserted 20 years ago on account of the increasing floating ice. Hence it arises that we owe the first discovery of cryolite to the Greenlanders, who, in finding it to be a soft substance, employed the water-worn rounded fragments as weights on their angling lines. In this shape, the first specimens of cryolite were sent by the Missionaries as an ethnographical curiosity to Copenhagen. It was of course incorrectly stated in some periodical papers that the cryolite was discovered by me; I only found its geological situation, and I dare say by a mere accident.

The cryolite is found, as I mentioned before, near to the shore, resting immediately upon gneiss. This rock, which here forms the shore of the frith, is under water during the tide, as well as the superincumbent cryolite, and both are very much decomposed where they are in contact with each other. The gneiss is metalliferous, and intersected by small horizontal and vertical veins of quartz, from the thickness of 1 inch to that of 3 or 4 inches, containing tinstone, accompanied by arsenical pyrites, common iron-pyrites, small particles of wolfram, and lithomarge; the whole bearing a striking resemblance to the tinstone veins in Saxony and Bohemia. The tinstone occurs massive and crystallised in imperfect octahedrons; the arsenical pyrites is partly massive, partly crystallised in oblique four-sided prisms; the iron-pyrites occurs only disseminated.

At a distance of about 120 fathoms from this spot, there is an extensive bed of large quartz crystals, similar to those found near Zinnwald in Bohemia; but they are throughout in a perpendicular position, some of them measuring a foot in length, and from 4 to 5 inches in thickness, containing small imbedded crystals of tinstone of the above-mentioned forms. This bed is intersected by a nearly vertical vein of compact fluor, of the thickness of from 6 to 7 inches. The whole is equally exposed to the tide. The fluor contains no metallic substance, but it is of a singular nature. Its colour is reddish-blue, verging towards lavender-blue; the substance is dull, soft, and presents rather blunt-edged indeterminably angular fragments. Its powder is reddish-white. It emits a strongly hepatic smell when rubbed. The common kind of compact fluor occurs along with it.

The cryolite rests upon the gneiss, which contains the substances just enumerated, and forms two distinctly different beds, which are nearly of the same dimensions, namely, 10 fathoms in length, and from 5 to 6 in breadth. The purest cryolite is that of a snow-white colour, without any intermixed foreign substance, if I except a few nearly minute spots of galena. Its colour passes gradually into greyish-white, when it approaches to the other bed. The greyish-white variety on the surface very much resembles ice which has been corroded and grooved by the power of the sun's rays. In these fissures we sometimes observe the threefold cleavage of this substance beautifully displayed. Fragments of quartz and sparry-iron-ore in rhombs sometimes occur in the greyish-white variety.

The other bed is separated from the former by an elevation of the underlying gneiss, and has a very different appearance. The snow-white and greyish-white colour is changed gradually into reddish-white, and passes, in proportion to the quantity of imbedded metallic substances, into orange-yellow and brownish-red. We find in the reddish-white variety quartz crystals and particles of flesh-red felspar; in the orange-yellow and brownish-red varieties sparry-iron-ore, iron-pyrites, copper-pyrites, and galena occur in great abundance. Sparry-iron-ore occurs massive and in rhomboidal crystals, accumulated in groups of considerable size. Its colour is always dark blackish-brown, and the surface of the crystals partly tarnished, partly decomposed. I found some of the crystals hollow, and some filled with particles of common iron-pyrites. Iron-pyrites occurs generally massive, rarely crystallised in cubes and dodecahedrons. Copper-pyrites occurs only disseminated in galena. The galena of this place has the peculiar property of melting calmly before the blowpipe into a globule, without the least decrepitation. Some fragments are covered with a yellowish-white and greenish-white coating, which, when held to a candle, burns with a blue flame and a sulphurous smell. This kind of galena presents some properties of native lead, as the sulphur appears to be elicited, and the ore reduced, by the action of the sea-water or the atmospheric air. Galena occurs here disseminated, massive, but rarely crystallised in perfect cubes, and in cubes truncated on the angles and edges.

This variety of cryolite (I may perhaps call it, in a geological view, *metalliferous cryolite*) was not known in Europe before I visited the coast of Greenland; because, owing to its decomposed state, it was not used for any domestic or economical purpose by the Greenlanders. They preferred the white variety, which, from its colour and greasy appearance, was called by them *Orksoksiksæt* (from the word *orksok*, blubber), a substance that has resemblance to blubber. I could have remained with pleasure during the whole winter on this spot, so alluring to a mineralogist; but I had to provide for twelve human beings who followed me, and who looked more for Seals than for minerals. The floating ice pressed upon us in all directions, and it was advisable to get rid of the frith and gain the open sea, as we had to clear 250 miles in a very boisterous season before we could reach our winter residence.

XLV.—On the CRYOLITE * of EVIGTOK, GREENLAND. By J. W. TAYLER, ESQ. (Reprinted, by Permission, from the Quarterly Journal of the Geological Society of London, vol. xii., 1856, pp. 140–144.)

[*See above*, Giesecké's paper on the Cryolite of Greenland, p. 341.]

. . . . Evigtok (which signifies in the Esquimaux language "a place where there is plenty") is distant about 12 miles from the Danish Settlement of Arksut, and forms a small bay in the Fjord of Arksut. It is a semi-circular space of rather low, irregular ground, surrounded by a ridge of mountains rising abruptly to the height of about 2,000 feet, making the enclosed space appear the half of a deep basin about two miles in diameter. Evigtok is noted in Greenland for its abundance of fish in the summer season; shoals of Capelins blacken the small bays, whilst thousands of Codfish swim close to the shore in pursuit of them, both of which are taken by the natives in large quantities. At the foot of the mountains and on their sides are to be found many Grouse, Hares, and Arctic Foxes. In the winter season immense flocks of Eider-ducks and other water-fowl resort to this part of the fjord. Vegetation, such as it is in Greenland, also prospers here; a miniature forest of *Salix arctica*, about 4 feet high, covers about a square mile, and *Angelica*, *Rumex*, *Taraxacum*, *Potentilla*, and other plants are met with more abundantly than is general in Greenland; the spot appearing like a garden amidst the general barrenness of a land buried deep in snow nine months out of the twelve. But Evigtok is more remarkable as being the only place in the world in which the mineral cryolite has hitherto been found.

---

* *See* Thomson's "Outlines of Mineralogy," vol. i., p. 251; and Giesecké's article "Greenland," in the "Edinburgh Encyclopædia," 1816. [*See also* Giesecké, "Edinb. Phil. Journ.," 1821–2, p. 141, &c.; Allan, "Transact. R. Soc. Edinb.," vi., p. 351; a note on the trade in cryolite, in Irminger's "Notice sur les Pêches du Danemark," &c., 1863, p. 15; Ellis on Cryolite and its Products, "Chem. News," 1868, vol. xvii., p. 173 (from "Proceed. "Americ. Pharmac. Assoc.); and Notes by Dr. R. Brown *further on*.—EDITOR.]

By reference to the horizontal section (fig. 1), two trap-veins will be seen bounding a space containing the Cryolite and the minerals accompying it. To this space I shall confine my remarks. The section is not drawn accurately to a scale, but it is about $\frac{1}{24}$ inch to the fathom.

FIG. 1.—HORIZONTAL SECTION OR GROUND-PLAN of the CRYOLITE at EVIGTOK.

(The exposure of the Cryolite is about 300 feet in length.)

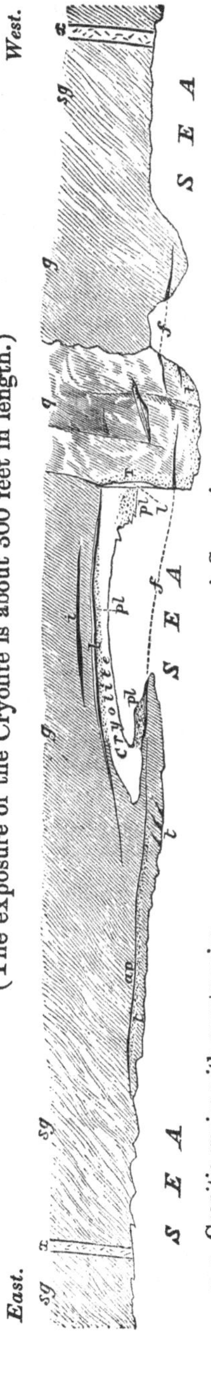

g, g. Granitic gneiss with quartz veins.
sg, sg. Schistose gneiss and hornblende-schist.
x, x. Trap-dykes.
q. Quartz-rock, with felspar, cryolite, and ores of iron, tin, lead, zinc, &c.
f, f. Fluor-spar.
i. Sparry-iron-ore.
ap. Arsenical pyrites and tin.
pl, pl. Iron- and copper-pyrites, galena, and blende.
l, l. Galena.
t, t. Tinstone.
T. Tantalite.

Starting from the western trap-vein, which is situated in schistose gneiss and hornblende schist, we find the gneiss gradually losing its slaty structure, until in the neighbourhood of the cryolite it becomes granitic, and now contains numerous metallic traces. Before arriving at the cryolite we find a wide vein of white quartz and felspar running about S.W.; the quartz and felspar are in very large masses and crystals, some crystals of quartz measuring a foot in thickness. This rock is traversed in several directions by small veins and masses of cryolite, isolated from the larger body of that mineral, in which, as well as in the rock, are to be found numerous crystals of a variety of tantalite, oxide of tin, blende, molybdenum, much galena, copper pyrites, arsenical and iron pyrites, and sparry iron ore. In this rock are many small caverns, arising from the decomposition of the felspar, and probably also from the decomposition of the cryolite, which is here porphyritic, containing crystals of felspar and quartz. The floors of these caverns are covered with loose crystals and fragments of felspar, and in some places kaolin, crystals of tinstone, and carbonate of iron. In one of these cavities is a large vein of arsenical pyrites and purple fluor-spar; also a large vein of black cryolite, containing copper- and iron-pyrites, and red felspar. Smaller cavities are found when blasting, the sides of which are completely covered with crystals of the tantalite, resembling on a large scale the crystalline cavities in amygdaloidal traps. In this quartz and felspar rock there is a remarkable vein, containing soft ferruginous clay and rolled pebbles, sparry-iron-ore, and copper-pyrites. The copper lies over the sparry-iron, and runs in fine threads between the folia of the partly decomposed iron-ore, appearing as if it had run into it in a state of solution. To this quartz and felspar rock succeeds more granitic gneiss, in which the cryolite occurs; this gneiss gradually loses its granitic character as it approaches the eastern trap-vein, where it again takes on the same slaty appearance as at the western trap-vein.

Fig. 2.—Transverse Section of the Cryolite at Evigtok.
(The width of the Cryolite is about 80 feet.)

South. *North.*

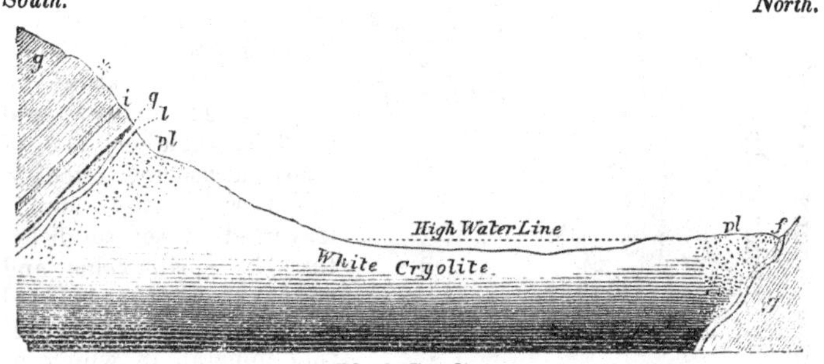

Black Cryolite.

g, g. Gneiss.
i. Sparry-iron-ore.
q. Quartz vein.
l. Argentiferous galena.
f. Purple fluor-spar.

pl, pl. Galena, copper-pyrites, blende, iron-pyrites, and carbonate of iron, scattered in cryolite.
* A fragment of cryolite was found imbedded at this spot.

We will now refer to the transverse section of the cryolite (fig. 2). The cryolite forms a bed or vein parallel to the strata, and is about 80 feet thick and 300 feet long; it dips to the south at an angle of nearly 45°, and runs nearly E. and W. In the upper wall of gneiss, about 2 feet above its junction with the cryolite, runs a vein of sparry-iron, with the same dip as the cryolite; and a layer of opaque quartz crystals lines the under side of the gneiss, between the iron-ore and the cryolite. Sometimes sinking several feet into the cryolite, but never rising into the gneiss, is a vein of argentiferous galena, containing 33½ per cent. of lead and 45 ounces of silver in the ton of ore; this was worked during the year 1854–5, and some good ore was extracted. The cryolite below this vein is impregnated for a few feet with galena, copper-pyrites, and sparry-iron-ore; but beyond, until within a few feet from the under wall of gneiss, it is quite pure and white. Within 10 feet, however, of this under gneiss, it again contains the same minerals disseminated; but is here separated from the gneiss by a vein of dark purple fluor-spar. The gneiss on both sides of the cryolite contains much fluor-spar disseminated.

The upper part of the cryolite at its junction with the gneiss is much decomposed, leaving many cavities, which contain loose crystals of sparry-iron. At a depth of about 10 feet from the surface the cryolite, although free from foreign matter, assumes a darker colour, and at 15 feet it is nearly black, and more translucent and compact; and, as the deeper we sank we found the cryolite became darker, there is reason to believe that below this depth the mineral will be found to be wholly black. As the white cryolite is only found at the surface, and bears evidence of partial disintegration by having lost some of its compactness and translucency, it is reasonable to suppose that the cryolite was originally wholly dark-coloured or black.

When the black croyolite is heated to redness, it loses about 1 per cent. moisture and acid, the whole of its colour, and part of its translucency, becoming perfectly white, like the cryolite at the surface; and from this fact we may conclude that the white colour of the cryolite at the surface has been produced by a similar cause. I consider it probable that the trap now found at each end of the cryolite has formerly overlain it, heating it superficially, and rendering it white. There are at present no remains of overlying trap between these two veins, but in this country the trap and allied rocks disintegrate most rapidly from the effects of frost. The cryolite itself has considerably decreased, from this and other causes, for I found a piece of it imbedded in the upper gneiss, more than 8 feet above the highest part of the cryolite, proving that it formerly stood at that height.

In working the lead vein we sank about 30 feet on the dip of the cryolite; it probably extends to a great depth, and exists in great quantity.

The fact of its solitary occurrence in this spot induces speculation in regard to its origin. The number of minerals, mostly crystallised, which accompany it, indicate some powerful and long-continued agency to have operated in a limited space. The few facts I have stated may suggest some opinions which may elucidate the as yet ill-understood subject of mineral veins.

The cryolite has been hitherto applied to few purposes. The Greenlanders were the first to turn it to account, which they did in a curious manner, viz., the manufacture of snuff. They grind the tobacco leaf between two pieces of cryolite, and the snuff so prepared contains about half its weight of cryolite powder. This snuff they prefer to any other. In Europe cryolite has been employed to a limited extent, but the recent discovery of the mode of preparing aluminium will probably render it a valuable ore of that metal.

XLVI.—On the VEINS of TIN-ORE at EVIGTOK, near ARKSUT, GREENLAND. By J. W. TAYLER, ESQ., F.G.S., Mining Engineer to the Greenland Mining Association.* (Reprinted, by Permission, from the Quarterly Journal of the Geological Society of London, vol. xv., 1859, pp. 606–7.)

The area over which the veins of tin extend is about 1,500 feet in length by 80 in breadth; their number is 18 or 20; and they run in various directions, some E. and W., others N.E. and S.W., or N. and S. The tin occurs also disseminated in crystals through the rocks, and accompanying the finer-grained galena and tantalite. The appearance of the veins at the surface is not very promising, the tin being in small detached crystals, scattered through the gangue (which is mostly quartz). The widest of those veins is 10 inches, the tin being 1 or 1¼ inch, occupying one side of the vein. The gangue here is felspar, quartz, sparry-iron (carbonate of iron), and fluor-spar. This vein runs E. and W. into the white cryolite. Another vein, about 200 feet west from the cryolite, is visible for about 30 paces; at the surface it is not more than ⅛ of an inch thick, but at a depth of 6 feet it is 3 inches thick. Other veins are at the surface mere strings, varying from ½ to ¼ of an inch thick. Nearly all these veins occur in a large vein or bed of felspar and quartz, some of the crystals of the latter having a diameter of 18 inches. This mass contains, in a limited space, various other minerals, such as galena, blende, copper-, iron-, and arsenical pyrites, fluor-spar, black cryolite, tantalite, molybdena, sparry-iron, zircon, &c.

There are some peculiarities connected with these tin-veins which deserve consideration, and would lead one to expect that at a greater depth they might afford a large produce. The only surface vein which can be seen sectionally (no sinkings having yet been made) widens rapidly as it descends; and in other spots sectional views of two small veins are to be seen which do not reach the surface. The most remarkable of these is a vein which I found in the floor of a small cavern. It consisted at first of sparry-iron, arsenical pyrites, fluor, &c.; a few feet deeper it changed its character, and contained good traces of copper; but water prevented us following it deeper: it is here about 15 inches wide. The minerals accompanying the copper are those usually met with in good and productive veins, besides the black cryolite, which is here peculiar to it.

These veins, in my opinion, evidently belong to the underlying granite, which appears at the surface in veins, and which is probably at no great depth below, since the tin-veins penetrate into the overlying gneiss, which dips to the south, and under the

* For Mr. Tayler's remarks on the cryolite of Evigtok, *see* Quart. Journ. Geol. Soc., vol. xii., p. 140; and *above*, p. 344.

outcrop of it the granite loses itself. It is highly probable that the cryolite forms a bed between the gneiss and granite, but partly enclosed in the gneiss. We have not yet, however, seen the bottom of the cryolite in any of our workings. This tin district differs from all others known, being associated with the cryolite, whilst tantalite seems to have taken the place of wolfram, which generally accompanies tin-ores.

XLVII.—Catalogue of a Geological and Geographical Collection of Minerals from the Arctic Regions, from Cape Farewell to Baffin's Bay, Lat. 59° 14′ N. to 76° 32′ N. By the late Sir Charles Giesecké, Professor of Mineralogy to the Royal Dublin Society.

[Abstract: from the Journal of the Royal Dublin Society, vol. iii., 1861, pp. 198–215. *See also above*, pp. 327, &c.]

The "356" Specimens, from "219" Localities, may be grouped as follows:—

I. The East Coast, for a few miles N. of Cape Farewell. Localities, Nos. 1–7.

Rocks.—Granites (red and white), syenite, mica-schist, talc-schist, hornblende-schist, basalt.

Minerals.—Tourmaline, garnet, hemispherical mica, asbestos, talc, allanite,* avanturine, and common quartz.

II. The South Coast, the Islands of Staatenhuk, and the large island off Cape Farewell. Localities, Nos. 8–37.

Rocks.—Granites (chiefly fine-grained, grey, some red and reddish, and at one place green), syenite and syenitic granite, mica-slate, hornblende-slate, schistose weiss-stein, chlorite-slate, basalt, greenstone, red jasper, hornstone, clay (along-shore).

Minerals.—Green mica, garnet, quartz, greenish quartz, dark blue quartz, with imbedded green garnets, in veins in granite; schorl and tourmaline, talcose mineral like triclasite, actinolite in talc, hornblende, pyrites, arsenical pyrites, glassy-grey felspar, indigolite, tinstone and zirkon (in coarse-grained syenite up the Pysursoak-fjord), graphite (in granite), labrador-felspar, moroxite (in reddish, coarse granite), fergusonite (in syenitic granite), calc-spar (in granite).

III. The West Coast.

1. From Christian Sound and Islands (*see* above) to the Igalikko-fjord. Localities, Nos. 38–47.

Rocks.—Granites (red and grey), mica-slate, chlorite-slate, hornblende-rock, basalt.

---

* *See also* "Experiments on Allanite, a new Mineral from Greenland," by T. Thomson, M.D., &c., Trans. Roy. Soc. Edinb., vol. vi., pp. 371–386.

Minerals.—Quartz, garnet in granite, felspar, massive epidote, siliceous sinter at hot-spring on Ounartok (Island).

2. From the Igalikko-fjord to the Kangerdluarsuk-fjord (both inclusive), and to the Tunagliarbik-fjord (not included). Localities, Nos. 48–74.

Rocks.—Granites (white, grey, and red), syenite, gneiss, chlorite-slate, hornblende-slate, red sandstone, basalt, greenstone, claystone, claystone-porphyry, ironclay, weiss-stein, iron-jasper, quartz-rock, hornstone-porphyry.

Minerals.—Talc, quartz, quartz with chlorite, hornblende, phosphate of iron, labrador-felspar, green and greenish felspar,* compact manganese in red sandstone, gieseckite and felspar in claystone porphyry (Akulliarasiarsuk, eastern branch of the Igalikko-fjord), calcite, fluor in hornblende-slate at Julianeshaab, green jasper-veins in granite; with the coarse red granite of Nunarsoout, north coast of the Igalikko-fjord, are graphite in mica-slate, yellow sparry-iron-ore, calc-spar, massive fluor, green massive felspar,* and beds of weiss-stein with sodalite † (mylopsite) and eudiolite.

3. From Tungaliarbik-fjord to Arksut-fjord (the latter not included). Localities, Nos. 75–107.

Rocks.—Granites (chiefly grey, both south and north of Cape Desolation), syenite at the fjord and zirkon-syenite at Cape Desolation, greenstone, red sandstone, blue schistose jasper.

Minerals.—Prehnite and red felspar in greenstone, magnetite, skorodite?, hornblende, apophyllite, phosphate of iron, zirkon and chromate of iron in syenite, labrador and adular felspar in syenitic granite, corundum in granite, blue vein-quartz (at Ilundeoe).

4. From Arksut-fjord to Fredrikshaab (Pamiut) and to Ikertok Bay. Localities, Nos. 108–189.

Rocks.—Granite, gneiss, and mica-slate, predominating; syenite at Arksuts Storoë and Pamiut.

Minerals.—Green jasper, quartz, allanite (in granite at Kingiktorsoak), talc, precious serpentine and amianthus, axestone, magnetite in serpentine, moroxite (in mica-slate, Pamiut), garnet (in mica-slate); at Ivitæk and neighbourhood (Arksut-fjord), tinstone in vein-quartz, arsenical pyrites, and white, yellow, and earthy cryolite, with sparry-iron-ore, galena, copper-pyrites, quartz, and compact fluor.

5. From the neighbourhood of Ikertok Bay and Lichtenfels to Buxe-fjord. Localities, Nos. 133–151.

Rocks.—Granite,‡ gneiss, mica-slate, siliceous schist (quartz-slate), hornblende-slate, greenstone, siliceous conglomerate.

Minerals.—At Fiskernaes: garnets in gneiss, blue quartz, titanite with quartz and calcite, allanite in felspar-veins, sapphirine in mica-slate and with tremolite and quartz.

---

* *See* Dr. MacCulloch's note, p. 324.

† *See* "A Chemical Analysis of Sodalite, a new Mineral from Greenland," by T. Thomson, M.D., &c., Trans. Roy. Soc. Edinburgh.

‡ In the northern? part of this district, for several successive localities, "granite, alternating with quartz and mica-slate" (gneiss?), seems to predominate.

6. From the Buxe-fjord to the neighbourhood of Godthaab. Localities, Nos. 152–193.

Rocks.—Granite (grey and red), syenitic granite, gneiss, mica-schist and talcose mica-slate, talcose rock with grains of felspar, syenite, basalt.

Minerals.—Variously coloured quartz, tourmaline, tremolite, actinolite with yellow mica, hornblende, asbestos, amianthus with rhomb-spar, glassy actinolite (diallage), sahlite, augite, anthophyllite and varieties, allanite in granite, garnet in mica-slate and weiss-stein, mica with amianthus, olivine with mica, pargasite in calc-spar, felspar, green felspar, scapolite, epidote, emerald, moroxite in granite and mica-slate, indigolite with quartz, mundic in clay, pyrites in talc, titanite in quartz, felspar, and sahlite, molybdenite, siliceous sinter.

7. From Godthaab to Narkseitsiak (both inclusive), N. of Baal's River. Localites, Nos. 194–219.

Rocks.—Red and grey granites (chiefly grey at Baal's River, &c.), mica-slate, talc-rock, axestone, sandy clay, and clay with nodules of *Mallotus Grœnlandicus.*

Minerals.—Quartz, chalcedony, hornstone, adularia, tourmaline, actinolite, tremolite, sahlite, mica, talc, potstone, garnet in granite and mica-schist, allanite in red granite.

Mr. J. W. Tayler * adds (p. 215), the following notes :—

Loc. 46. *Sardlock.* Fergusonite, titanic iron, and beryl in coarse-grained granite.

„ 59. The gieseckite has been searched for here repeatedly without success.

„ 62. Julianshaab. A pebble of phosphate of lead in gravel.

„ 96. *Nanaitiak,* an island close to *Nunarsoit.* Copper-ore in chloritic schist, and amazon-stone (green felspar).

„ 111. *Ivikaet, Ivigtout,* or *Evigtok.* Tantalite in fine crystals, a new hydrous silicate of zirconia and yttria, fluor, blende, auriferous arsenical pyrites, and galena containing 58 ozs. of silver to the ton.

„ 116, 117, 118. Allanite, potstone.

„ 120. Greenlanders say that large bones are to be found high up on this mountain (Kingiktorsoak or Tindingen).

„ 126 (Pamiut). Copper-pyrites, hypersthene.

Avigait, near *Loc.* 187. Much allanite.

Near *Godthaab.* Selenide of lead and copper.

---

* *See above,* pp. 344 and 348.

XLVIII.—On the GEOLOGICAL and GLACIAL PHÆNOMENA of the COASTS of DAVIS' STRAIT and BAFFIN'S BAY. By P. C. SUTHERLAND, M.D., late Surgeon in the Arctic Expeditions.*

[Reprinted, with Permission, from the Quart. Journ. Geol. Soc., London, vol. ix., 1853, pp. 296–312. Read June 1, 1853.]

From Cape Farewell to Cape Atholl.—The Danish settlers in Greenland have pretty accurately laid down the geological character of the eastern coast of Davis' Straits from Cape Farewell, about lat 60°, to Cape Shackleton, about lat 74°.† Beyond this latitude and down the west side of Davis' Strait, the coast is almost unknown, from the difficulty experienced in approaching the land by the Whaling and Discovery Ships, the only ships that ever attempt to reach it.

Commencing at Cape Farewell‡ we find the crystalline rocks§ (granite, gneiss, &c.) forming a rugged and pinnacled coast, intersected by fiords of great length, in which the tide is generally very rapid, and the water is of considerable depth. The coast indeed appears as if composed of a cluster of islands varying much in size and lying in front of the great glacial plateau constituting the continent of Greenland.

Disco Island, Black Hook, &c.—Proceeding northward we find Disco Island, on the 70th parallel of latitude, to be chiefly composed of trap-rock. Viewing this island from a distance of ten miles, it presents a succession of steps, and appears to be made up of a number of truncated cones, planted so closely together that the bases of all meet; some of them, at the level of the sea, bounding long and winding valleys, and others at every intermediate elevation, until the top itself is reached at a height of from 2,000 to 5,000 feet. At its southern extremity hypogene rocks (granite, &c.) occur, from the sea-level to an elevation of about 100 feet, and passing beneath the trappean formation. In South-east or Disco Bay several clusters of islands are observed, all of which appear to be composed of the same crystalline rocks. On the S.E. and N.E. shores of Disco Island, the N. shore of the Waigat Strait, Hare Island, the S. shore of Omenak Fiord, Upernivik Naes (North-east Bay), and in the neighbourhood of the Black Hook, on the 72nd parallel of latitude, coal (lignite) has been found to a

---

* *See also* Dr. Sutherland's "Journal of Capt. Penny's Voyage," &c. With Appendix, 2 vols., 1852. And Capt. Inglefield's "Summer Search for Sir J. "Franklin," &c. With Appendix, 1853.—EDITOR.

† See Rink's Geology of West Greenland, 1852, Trans. Roy. Soc. Denmark.

‡ The author refers in this paper to numerous sketches presented by him to the Geological Society; these references are omitted here.—EDITOR.

§ Copper, tin, lead, and silver ores have been discovered in the vicinity of Julianes-Haab, about a degree north-west of Cape Farewell; and at Upernivik, about lat. 71°, graphite of tolerable purity occurs in abundance.

considerable extent and of rather tolerable quality. The specific gravity of the coal is 1·3848, and the following analysis of its proximate ingredients, made by Dr. Fyfe, Professor of Chemistry, King's College, Aberdeen, upon a specimen obtained from the same source as that now in the Museum of this Society, enables us to judge of its value and purity:—

Volatile matter		50·6	100·00
Ash	9·84		
Fixed carbon................	39·56		
		49·40	

I have not myself visited the beds of this mineral, but from the recent elaborate researches of Dr. H. Rink, the enterprising Danish traveller, it appears that sandstone is associated with this coal.

At Cape Cranstoune, situate on the north side of North-east Bay (Omenak Fiord), and immediately adjacent to the above two localities, the trap-rocks again occur, and thence extend northward, apparently in one unbroken series, as far as Proven, in lat. 72° 20′. Northward of this to Cape York, lat. 76°, with one or two slight exceptions, in lat. 73° 20′ and lat. 74°, the numerous islands and every part of the coast that protrudes from beneath the glacier are composed of gneiss and granite.

Capes York and Atholl.—At Cape York, lat. 76°, and on to Cape Atholl, thirty to forty miles further north, although differing in outline, owing to the glacial accumulation, from Disco Island and other well-known parts of the coast to the southward, the rocks can be referred with certainty to the same trappean formation. Specimens of greenstone-porphyry were taken from the cliffs at Petowak, near Cape Atholl.

Wolstenholme Sound to Cape Hatherton.—Northward of Cape Atholl we find, in the entrance of Wolstenholme Sound, a flat island (Saunders Island), which from its distinctly ***stratified*** appearance suggests the commencement of a different series of rocks. And eastward of the same cape, on the south shore of this Sound, the strata are seen cropping out with a dip to the south-west. This is at variance with what we observe in Saunders Island, about twelve miles N.N.W., for there the strata are perfectly horizontal. At North Omenak a sandstone, or slaty quartzose grit, with a dip of about 15° to W.S.W., occurs interstratified with greenstone-porphyry; and it is very probable that Mount Dundas, a tabular hill with a talus, is also composed of igneous rock. At the top of Wolstenholme Sound, in the same bluff, the strata, dipping about south-west, vary in their inclination from 10° to 25° or 30°.

In Granville Bay, about twenty miles farther north, the strata are at one place but little out of the horizontal, and at another the dip is about 45° to the north-west, and at another we have strata somewhat curved. In the entrance of Granville Bay several small islands occur which are probably formed of trap-rock. In Booth Sound, lat 77°, near Cape Parry, there is a very remarkable bell-shaped rock (Fitzclarence Rock), of a dark colour and rising in an isolated form to a height of probably 500 or 600 feet, as if from out of a comparatively level spit of ground this also appears to consist of similar rock. From Cape Parry (lat 77° 5′) north-eastwardly to Bardin Bay (lat. 77° 20′), in the south shore of Whale Sound, the strata incline a little to the S.W., and in many places they are somewhat curved. Still farther to the north-east they have a general dip of 30° to S.W., and they are intersected by irregular dark-coloured dikes of igneous rocks. One of these dikes rises in the form of a rough peak above the outline of the strata. In the entrance of Bardin Bay, the ship, drawing ten to twelve feet of water, struck upon a rock, which from the depth of the water (fifty to sixty fathoms) within a couple of hundred yards, may be a second protrusion of the same dike above the stratified rocks. Specimens of quartzose grit were obtained from the low point on the north-east side of Bardin Bay, and were taken from strata inclining W.S.W. at a general angle of 15°, but a little curved. In them we recognise the same sandstone as that of North Omenak, about sixty miles to the southward. A specimen of syenitic porphyry was taken from the shoulder of the hill in the vicinity of the dike above-mentioned. In other parts of Whale Sound (in Northumberland, Herbert, and Milne Islands) the strata are perfectly horizontal; and at Cape Saumarez, on the same coast, but thirty miles further north, the same strata can be traced from one cliff to another in conformable and horizontal lines over many miles. At Cape Alexander, the eastern boundary of the entrance of Sir-Thomas-Smith's Sound, in lat. 78° 15′, we again find the strata somewhat curved; but about seven miles farther north (a few miles south of Cape Hatherton), they are so regularly and horizontally piled one on another, that from their peculiar appearance they have received the name of the Crystal-Palace Cliffs. A small island, lying in front of a glacier two miles southward of Cape Alexander, appears to be composed of a dark rough-grained sandstone, similar to that found in Whale Sound. The strata are somewhat indistinct from the large disintegrated fragments that occupy the surface; they appear, however, to incline to the westward at an angle of ten or fifteen degrees.

West Coast of Baffin's Bay. Smith's Sound.—The west shore of Smith's Sound, from Victoria Head, beyond the 79th degree of latitude, to Cape Isabella near the 78th, as well as the coast leading southwardly to Jones' Sound, is so inapproachable from the drifting pack-ice in the season for navigation, that I fear we shall not soon have specimens of the rocks by which the character of so large a portion of the coast can be determined; and it is, moreover, everywhere so covered by the glacier, that the outlines of mere protrusions of the land, taken at a distance of ten to twenty miles, scarcely afford the materials for correct results. From its greater height in many parts than the adjacent, opposite shore, and also from its rugged, in some cases even pinnacled, contour, thus resembling the coast at Cape Farewell, it probably consists for the most part of crystalline rocks.

Jones' Sound, and North Devon.—Similar appearances obtain (with some local exceptions) along the north and south shore of Jones' Sound, the Cobourg and neighbouring islands, and the eastern coast of North Devon.

Lancaster Sound to Cumberland Sound [from 74° to 64° N. lat.]—On the opposite shore of Lancaster Sound, at Cape Walter Bathurst, the crystalline rocks are again recognised, and from this point they occupy the whole coast southward to Cumberland Strait,* and probably considerably beyond it. To this, however, I believe there is one exception at Cape Durban, on the 67th parallel, where *coal* has been found by the whalers; and also at Kingaite, two degrees to the south-west of Durban, where, from the appearance of the land as viewed from a distance, *trap* may be said to occur on both sides of that inlet. Graphite is found abundant and pure in several islands situate on the 65th parallel of latitude in Cumberland Straits, on the west side of Davis' Straits.

Silurian District of the Georgian Islands, &c.—The above-mentioned extensive development of crystalline rocks is flanked to the westward by an equally, if not much more, extensive tract of Silurian rocks, the limits of which as yet we have been unable to ascertain. The chief, indeed, it may be said, the only navigable channel through which this Silurian district has yet been reached is Lancaster Sound; it is probable, however, we may find it continuous to the westward with the American series of the same rocks. Through the labours of Prof. Jameson and Mr. König, thirty years ago, and of Mr. Salter only very recently, some of the numerous Silurian fossils peculiar to North-Somerset, North-Devon, and the North-Georgian Islands, have been described from the fragmentary specimens brought home by the ships engaged in the discovery of these places during the last thirty years.†

Drift Deposits.—On Cornwallis and Beechey Islands in Barrow Straits, west of Lancaster Sound, deposits containing existing arctic sea-shells occur at every elevation up to nearly 1,000 feet,—the greatest height attained by any part of that district. On the undulating slopes and along the raised beaches of this Silurian district of the North-Georgian Islands, &c. occur travelled materials, such as fragments of anthracite, greenstone, quartz, serpentine, gneiss, and granite, but all of such small size that their mode of conveyance to their present position is clearly referable to the action of *coast-ice* (previous to the elevation of the land), such as at the present day occupies the comparatively shallow seas in the inlets and channels of that district.

On the Greenland side of Davis' Strait, on the contrary, we find immense travelled boulders of gneiss and granite resting on the islands and the coast, which have been brought there at former periods by floating icebergs, previous to the elevation of the coast above the sea-line. The probable causes of these differences of

---

* "Cumberland Straits" of Baffin, its original discoverer at the end of the sixteenth century; "Hogarth Sound" of Capt. Penny, who rediscovered it in 1839; and "Northumberland Inlet" of Capt. Wareham in 1841.

† See Appendix to Sutherland's Journal of Capt Penny's Voyage, 1852, 2 vols. 8vo. See also Mr. Salter's Paper, *infra*, p. 312. The Rev. Mr. Longmuir, of Aberdeen, found numerous specimens of the genus *Rhynchonella* in the ballast of the "Prince Albert," a ship recently returned from Batty Bay, Prince-Regent's Inlet, on the eastern shore of North Somerset.

ice-action on these two opposite coasts are explained in the subsequent observations.

It may be noticed also, that, from the observations of Dr. Pingel and Capt. Graah, the west coast of Greenland presents evidence of its now undergoing the process of gradual submersion.

Glacial Conditions.—At Cape Farewell the fiords run so far into the interior, that none of the icebergs escaping into them from the great inland glacier ever reach Davis Strait, and if the navigator meets with icebergs in the neighbourhood of this promontory, they must have drifted to it from other sources. As we advance northward along the coast of West Greenland, and thus diminish the annual mean temperature both of the sea and of the atmosphere, we find the glacier approaches nearer and nearer the coast-line, until in Melville Bay, lat. 75°, it presents to the sea one continuous wall of ice, unbroken by land, for a space of probably seventy or eighty miles. To the southward of Melville Bay, there are numerous outlets for the ice in the coast, and they vary in breadth from two or three up to fifteen or twenty miles. To have a correct idea of the glacier accumulation in Greenland, we must imagine a continent of ice flanked on its seaward side by a number of islands, and in every other direction lost to vision in one continuous and boundless plain. Through the spaces between these apparent islands, the enormous glacial accumulations slowly seek their passage to the sea and send off an annual tribute of icebergs to encumber, to cool, and to dilute the waters of the adjoining ocean.

The average height or depth of the ice at its free edge in these intervals, or valleys, between the projecting points of coast is 1,200 or 1,500 feet, of which about one-eighth, or 150 feet, will be above water. In some of the valleys, however, the depth is upwards of 2,400 feet. This may be considered to be satisfactorily ascertained, for the Esquimaux around South-east Bay, lat. 68°, while pursuing halibut-fishing during the winter months, require lines of three hundred fathoms to reach the bottom at the foot of the glacier near Claushaven. In South-east (Disco) Bay, and also in North-east Bay (Omenak Fiord), we meet with the icebergs that draw the greatest depth of water, but those of the greatest cubic contents occur in Melville Bay and in several smaller bays to the southward of it. At Cape York, lat. 76°, although the glacier there is the northward continuation of the glacier in Melville Bay, its protrusions into the sea never exceed 50 to 60 feet above the sea-level; and in some places it does not enter the sea in a continuous mass, but, having descended over the brow of the cliff, it breaks off and slips down into the sea over the rocks, scratching and scoring them in a very marked manner. This is very well seen at Cape York, 76° N. lat. where the free edge of the ice is upwards of 20 feet thick, and at least 100 feet above the sea-level; the inclination of the abraded part of the coast being about 43°. But it is much better seen on the west side of Baffin's Bay, at Cape Fitzroy, on the south side of Jones' Sound, and at Cape Bowen, Pond's Bay, where the free edge of the ice is at least 50 feet thick, and about 200 feet above the sea-level. Although many hundred miles of coast intersected by glaciers were examined in the late voyage of the

"Isabel," under the command of Capt. Inglefield, R.N., these cited localities were the only places, with one or two very trifling exceptions, where this interesting phænomenon of powerful abrading action was observed. I believe it can be so far accounted for by the steepness of the inclination, but chiefly by the greater friability (diminished plasticity) of the ice from the diminished temperature.

One cannot easily determine why the icebergs that come from the glaciers at, and to the northward of, Cape York and on the west side of Davis' Straits, are of less dimensions generally than elsewhere. At Cape York, where we have a new formation of rocks (trappean) commencing, and further northward in the same coast, it is probably owing to the comparative shallowness of the valleys and to a diminished supply of snow from the greater intensity of the cold. On the west coast, from Victoria Head to Jones' Sound, although the land has almost a perfect icy casing, the icebergs that are sent off are by no means large, and this, as in the other case, may arise from the decrease of evaporation with the decrease of temperature. Again, from Jones' Sound southward, there cannot be such extensive accumulations of ice as on the opposite and more northern shore of Greenland, although the rocks in both cases are of the some character generally, for the reason, I believe, that the vapour-bearing stratum of air coming from the southward, over an extensive tract of land, contributes but scantily to the growth of the glacier on the former as compared with the latter, which is liberally supplied by the vapour-charged currents going northerly from the North Atlantic and Davis' Strait. But it is still more difficult to account for the entire absence of glaciers on the *Silurian rocks* westward of Lancaster Sound. Why the snow and rain falling on the land around Barrow Strait and its tributary inlets and bays should all escape into the sea in running streams of water every year during the two short months and a half of June, July, and August, while that falling on the coasts of Davis' Strait makes its escape as hard, but yielding ice, after a lapse of many ages, is a question worthy the attention of the student of physical phænomena.

The annual mean temperature in the creeks and inlets of Barrow Strait is several degrees lower than that in corresponding latitudes on the shores of Davis Strait; and even at Wolstenholme Sound, nearly two degrees higher latitude, the annual mean temperature is nearly three degrees higher than at Melville Island. This, however, will not throw light upon our difficulty. The ranges of temperature will probably prove more useful. A few degrees above the freezing point of water would settle the question. We know that the sea exerts a wonderful influence in rendering the climate temperate, as well as in reducing the ranges of temperature. Upon this theory, so clearly illustrated in Sir Charles Lyell's "Principles of Geology," (7 Edit. ch. vi.) the summer in the neighbourhood of Barrow Strait ought to be hotter than in Davis' Strait. And such we find it, as far as our limited observations can be made available. The month of July 1851, at Cornwallis Island, was found to be three degrees warmer than the same month of the preceding year in a corresponding latitude on the east side of

Davis' Strait. This difference is certainly small, but still it is on the favourable side; and when we associate with it the different structure of the rocks and also the diminished supply of vapour during the winter months, we have a faint approximation to the true cause why the glacier preponderates so largely in one direction, while it is entirely absent in another. The fact too that large sections of the coast-ice, before it was generally detached from the land, became dissolved by the streams discharging the melting snows of the North-Georgian Islands into the sea, may be taken as an additional proof that the summer heat was positively higher than was necessary for the conversion of snow into water.

Glaciers.—The travels of Prof. J. Forbes and of Agassiz in the Alps have so fully established the true theory of the descent of glaciers, which is applicable also to Greenland, as to render any remarks on this head almost unnecessary. The introduction of extraneous matter into the substance of the ice to be borne along must be the same in every country. And so also must be the deposition of moraines at the angles where the glacier begins to protrude beyond the land, whether they occur at the sea-level, or at rapid turnings at higher elevations. This deposition arises from the dissolution of a portion of the ice rich in earthy matter consequent upon increased freedom of exposure to the action of the sun, and also from mechanical displacement of the rocky matter by the advancing mass of the glacier. This was remarkably well seen at the north side of the Petowak glacier, near Cape Atholl, both at the sea-level and at an angle two miles further up the side the glacier.

The concentric and wavy appearance of the glacier-surface so often noticed in the Alps, is remarkably well seen in the vicinity of Cape Saumarez and of Cape Alexander, and also in Bardin Bay.

Both Prof. Forbes and Agassiz agree in attributing the roughness and irregularity of the surface of the glacier to the inequalities of the bottom over which it has to pass, more especially in cases where the action of the sun has not been distributed irregularly by means of accumulations of extraneous matter. This is frequently exemplified in the Arctic regions; and, as in the Alps, large crevasses are the result when a protruding mass is slipping imperceptibly over a convex or ledged surface.*

Although there certainly is a relation between the upper and lower surfaces of a plastic glacier, even when it may be upwards of 2,000 feet in thickness, still I must confess that in my opinion we can scarcely attribute the regularly pinnacled appearance of many a large iceberg and magnificent glacier to this cause. Some glaciers and icebergs, again, are so flat and smooth on the upper surface, that one can hardly conceive a rocky bottom beneath a

* This is very well seen in the glacier of Petowak, and also in a glacier at Cape Fitzroy, on the south shore of Jones' Sound. These crevasses are not unfrequently filled up with mud, &c. brought down by debacles and other means from the land on either side, and then they become frozen, thus cemening the whole mass firmly together, and perhaps forming part of the future iceberg so long as a few cubic feet of it remain undissolved.

glacier to be equally smooth. On the north side of Cape Clarence in the north shore of Jones' Sound, during the late voyage of the "Isabel," I observed that one portion of the surface of a flat but extensive glacier, that protruded several miles into Glacier Strait, was exceedingly smooth, while another portion of it was so rough and pinnacled that to walk over it would have been impossible. This roughness must be attributed to some peculiar atmospheric cause, or to the difference of temperature between the surface and the interior of the glacier.

Eight feet is the depth to which a minimum temperature of $-45°$, a monthly mean of $-30°$, or an annual mean temperature of $+2·5°$, extended the freezing-point of water through freshwater ice on a lake of two fathoms' depth (Kate-Austin's Lake) in lat. 74° 40′ and long. 94° 16′. If we can presume the heat-conducting power of ice formed on the surface of water, and of glacier-ice, to be the same, then the temperature of the interior of the glacier below the above depth, with the same minimum or mean annual degree of cold, would be about $+32°$. The surface exposed to any alternation of heat and cold, from the freezing-point to — 45° or many degrees lower, would necessarily become contorted and disturbed by contraction and expansion, even supposing its base or supporting part were standing still. Of this we had unexceptionable proofs in the condition of the surface of the ice on the lake already noticed. But when we take into account that the whole bulk of the glacier, except a few feet of its upper surface, retains its plasticity and continues its downward motion, it need not be wondered that the latter, hard and friable, assumes a broken-up appearance. This view, however, does not fully satisfy us, not being universally applicable.

Following the example of Mr. Christie, one of the Secretaries of the Royal Society,* during a winter in Barrow Strait, I performed a number of experiments by submitting water in a strong iron bottle to various temperatures, from $+32°$ to — 45°. While the temperature to which the bottle containing the water was exposed did not descend more than eight or ten degrees below the freezing point, the column of ice, ascending through the orifice or "fuze hole," and always amounting to about one tenth of the whole mass of water used, retained its cohesive property so perfectly that without being broken, and although only half an inch in diameter, the whole apparatus weighing four to five pounds could be raised by its means, and sometimes even inverted. But at lower temperatures the ascending column escaped with a slight crepitating sound, and frequently with explosive reports, accompanied each by a sudden propulsion of a portion of it to a distance of several feet; it was so friable too that it separated into discs of half or a quarter of an inch in thickness, and sometimes crumbled to fragments between the fingers. The important points, relative to the plasticity of ice, contested some years ago by Prof. J. Forbes and Mr. Hopkins come within this field of research; they are well known, and need not be recounted here.

---

* Lyell's Principles of Geology, Seventh Edition, p. 226.

Icebergs.—From what has been observed in the Alps, it may be considered a settled question that the downward motion of the glaciers is constant and comparatively unaffected by low temperatures applied to the surface, especially when the depth of the solid ice amounts to several hundred feet. In the Alps, and even within the tropics, they travel great distances from the snow-clad heights, until frequently they gradually descend into the regions habitable by man, where they undergo dissolution by the increase of temperature. In Greenland, after descending into the sea through the valleys, they retain their hold of the land * until the buoyant property of water upon ice comes into operation, and then they give birth to icebergs, sometimes of enormous dimensions.† The constant rise and fall of the tide exerts great power in detaching these floating ice-islands. By it, a hinge-like action is set up as soon as the edge of the glacier comes within its influence, and is carried on, although the surface of the sea for many leagues around is covered with one continuous sheet of ice. After summer has set in and somewhat advanced, the surface-ice either drifts away or dissolves, and then we have winds prevailing in a direction contrary to what they had been during the cold season of the year; and the result of this is a great influx of water into Davis' Strait, which causes tides unusually high for other seasons of the year, and which in their turn set at liberty whole fields of icebergs, then to commence their slow southward course. In August 1850 the number set free in a deep fiord near Omenak, North-east Bay, so occupied the navigable passage out of the harbour at that settlement, that the Danish ship which had but a few weeks previously entered the harbour was in great danger of being detained for the winter. In the same month in 1852, the whole of the coast southward from Melville Bay, extending over a space of 180 miles in length and probably 12 to 15 miles in breadth, was rendered perfectly unnavigable by any means whatever. When we sailed along that portion of the coast about the middle of August in the season of 1852, we were astounded by the constant booming sounds that issued from whole fields of floating icebergs, often bursting and turning over. To me the change appeared to be remarkable, for I spent the months of June and July of 1850 in company with a whole fleet of whalers there, sailing safely in the very place which now we could no more enter with our ship than navigate her through the city of London, half submerged in the sea, and all the houses tumbling about and butting each other as in an earthquake. At Cape York one could count nearly two hundred icebergs in a semicircle of twelve miles, all of which appeared to have been quite recently detached from the glacier; and in the upper part of Wolstenholme Sound, the icebergs, that had come off from the three protruding points of the glacier entering it, were so closely planted together, that it was not

* Some of these glaciers of Northern Greenland push forward into the sea to the extent of from one to three miles.

† For the description of an immense iceberg, 200 feet high above the sea and two miles in length, see Sutherland's Journal, vol. i. pp. 61, 62.

without some difficulty and even danger we advanced among them, although aided by steam.

Action of glaciers on the sea-bottom.—The effect of bodies of such dimensions on the rocks and mud at the bottom must be as extensive as it is important. While passing up the Strait early in the season, one rarely sees sea-weed floating in the water, but at a period somewhat later, after these natural reapers have sallied out to mow down their crops, we meet with whole rafts of the produce of the sub-marine forests of these regions floating down the straits. The stems of *Laminaria* are often found abraded, and their roots contain shells and other animals, some of which appear to have participated in the violent action that liberated the plants they sought as a protection. In every part of Davis' Strait, from Cape Farewell to Smith Sound, on either side or mid-channel, from two to two hundred fathoms, wherever the dredge has reached the bottom, these animals have been found to exist, in spite of iceberg action in its most intense form upon their rocky or muddy habitats. Ascidians and Cirrhipeds, and many other animals which attach themselves to the rocks at considerable depths, are often found. The Echinoderms, which we know are too slow in their motions to escape danger, swarm in those seas. A species of Sea-Urchin (*Echinus neglectus*) and Brittle Stars have been taken up from depths varying from ninety to two hundred fathoms in Melville Bay, and from various other depths in all parts of the Strait. Shells also occur, but they are sometimes found broken, as some of the species of *Mya, Saxicava, Cardium, Pecten*, and many others, taken from depths of seven to one hundred fathoms, will show.* Except from the evidence afforded by plants and animals at the bottom, we have no means whatever to ascertain the effect produced by icebergs upon the rocks. Doubtless when they contain earthy and stony matter they must scratch and groove the rocks "as the diamond cuts the glass," and when they are impelled along a muddy bottom, they cannot fail to raise moraines and leave deep depressions in its otherwise smooth surface. But it will be well to bear in mind that when an iceberg touches the ground, if that ground be hard and resisting, it must come to a stand; and the propelling power continuing, a slight leaning over in the water, or yielding motion of the whole mass, may compensate readily for being so suddenly arrested. If, however, the ground be soft, so as not to arrest the motion of the iceberg at once, a moraine will be the result; but the moraine thus raised will tend to bring it to a stand. We can more readily conceive this from the fact that the power which impels icebergs is applied to about the upper third or fourth part of their whole bulk.

Another mode of action is sometimes exhibited by the iceberg, by which its triturating and ploughing force is locally brought into play with immense effect. Icebergs resting on the bottom, and situated at the edge of the fixed surface-ice (that which is attached to the land), when pressed upon by loose and drifting

* For an account of the Fauna of these seas, see Appendix, vol. ii. of 'Sutherland's Journal." Also Catalogues, &c. reprinted *above*.

floes of large size, are frequently subjected to a rotatory motion, extending sometimes to three-fourths of a circle, or even a complete revolution.

The conveying power of icebergs is so well known to geologists, that I need make but few remarks on this subject. As a general rule, the source of all the foreign matter they contain is the land on both sides of the glacier. It may, however, be received from other sources. I have often thought that the fragments of a huge iceberg, acquiring a state of quiescence after separating into several masses in one of its fearfully grand revolutions, had turned up mud and other earthy matter from the bottom. This, however, is doubtful; for we can hardly conceive it possible that anything extraneous can adhere to hard and brittle ice passing rapidly through the water during the iceberg's revolution. Icebergs are sometimes floated so close along a bold and overhanging rocky coast, as to touch the perpendicular cliffs and to remove disintegrating fragments of the rock. Another, probably the most common, of these *unusual* modes of receiving débris, is from coast-ice, which, impelled by the winds and tides, is often piled up with its load of pebbles, sand, and mud against icebergs. The foreign substances thus cast upon the surface of an iceberg must necessarily be precipitated to the bottom at the first revolution it undergoes.

The quantity of rocky matter which ice is capable of floating away can be estimated from the specific gravity of both substances. Taking 2·5 as the density of granite and ·92 as that of ice, an iceberg half a mile in breadth, a mile in length, and 200 feet high above the water (dimensions, we may observe, by no means out of the average) will convey a load of one hundred and forty millions tons weight. Some of the icebergs seen in Davis' Strait are so charged and impregnated with earthy matter, that by inexperienced persons at a distance they may be mistaken for masses of solid earth. And we often observe large boulders, of perhaps one hundred tons each, lying on the surface of icebergs, or sometimes imbedded deeply in the ice.

By far the greatest number of these floating masses dissolve in Davis' Strait, and deposit their earthy contents throughout its extent. Some of them, however, find their way into the Atlantic, and appear disposed to push far to the southward into the Temperate zone. As Sir Charles Lyell and others have remarked, where the greatest number of these undergo dissolution there the deposition of rocky matter is most active, consisting of angular and rounded fragments, together with sand and mud, a great part of which materials are probably from sources of very opposite character.*

Coast-ice.—Ice forming on the surface of sea-water is also well known as an agent of importance in conveying away to considerable distances the materials of the sea-coast. With strong gales the ice in the Arctic Seas is driven in upon the coasts with great force, and, if the bottom about the low-water-mark is composed of loose

* See also Col. Sabine's Observations, Brit. Assoc. Rep., Trans. Sect., 1843.

gravel or mud, moraines are raised to a height of several feet. The wind ceasing and high tides proving favourable, the ice again withdraws from the coast, carrying with it large accumulations of the loose shingle of the beach, which it deposits in the surrounding seas, after travelling several hundred miles. The moraines it had raised are not wholly obliterated, and as winter proceeds to hem the coast with a fringe of ice, they cause an irregularity in the surface of the latter by the rise and fall of the tides, which results in a large portion of their contents, mud, sand, shingle, and perhaps also traces of animal and vegetable matter, being included in the new ice formation. This process ceases altogether only with the return of summer, and then the coast-ice, varying in thickness from two or three up to twenty or more feet, according to the degree of cold, the stillness of the water, and the extent of the rise and fall of the tides, is subject, in some localities at least, to the power exerted by debacles in loading it with foreign matter. Thus freighted, it withdraws from the shore when the straits and inlets open out, and drifts many hundred miles before it is dissolved by the action of the sun and the water, and yields itself and its carefully bound cargo to the sea. We find this occurring every season on the south shore of the North-Georgian Islands; but from the testimony of numerous travellers,* it occurs on a magnificent scale at the entrances of the great American and Siberian rivers which discharge their waters into the Arctic Seas.

Polar Currents.—The necessity there is for currents into the Polar Seas to keep up their mean salinity will become obvious when we reconsider the vast amount of fresh water which enters them in the form of icebergs from the glaciers. That there are currents out of the Polar regions is sufficiently clear; were there no such currents, evaporation alone from the surface of the sea, the greatest part of which is generally covered with ice, would fail to remove the excess carried by the annual crop of icebergs; and then we should have an icy pile ever growing and gradually extending into the Temperate zone. The difference of temperature observed by the navigator in the waters of the eastern and western shores of the North Atlantic, amounting, as it does, to nearly 30° of Fahrenheit's thermometer in lat. 59° during the warmest months of the year, affords the best possible proof of the existence of currents in the two directions we have indicated. In Davis' Strait, although on a much smaller scale, there is also a difference in the temperature of the sea on its two shores. On several occasions during the late expeditions in search of Sir John Franklin, while the ships were crossing that strait from east to west a fall of a few degrees was observed. This accounts pretty accurately for the fact that the east shore during a great part of the year keeps clear of ice, while the opposite is for the most part encumbered; and the greater mildness of the climate on the east side arises from the same cause. Allusion need not be made here to the late President's Paper on the temperature of the North of

* Principles of Geology, Seventh Edition, page 86.

Europe,* as the analogy between the North Atlantic and Davis' Strait with respect to currents will easily occur to us.

The specific gravity of the water also assists in determining the direction of the currents in the Polar Seas. During Captain Inglefield's late voyage, it was found to decrease as we approached Cape Farewell and advanced northward and westward in Davis' Strait.†

In the Atlantic, long. 30°, lat. 56° 30′, 24th October 1852, the natural temperature being 48° Fahr., the density at 60° Fahr. was 1·02808.

At Cape Farewell on the 31st July, natural temperature 33°, density 1·0245.

Davis' Strait, lat. 68°, sixty miles off the coast of West Greenland, August 11th, natural temperature 40°, depth fifty fathoms, density 1·0265.

Close to Cape York, lat. 76°, August 21st, natural temperature 30°, sea-water-ice and icebergs abundant, depth fifty-four fathoms, density 1·0215.

About two miles off Cape Alexander, Smith's Sound, lat. 78° 20′, long 71°, August 27th, natural temperature 32° no ice but in the vicinity of the glaciers in the coast, depth 154 fathoms, density 1·02516.

In Jones' Sound, lat. 76° 11′, long. 83° September 1st, natural temperature 30°, sea-water-ice and icebergs present but not abundant, density 1·02451.

About two miles off Cape Fitzroy, Lady-Anne's Strait, Jones' Sound, lat 75° 35′, September 2nd, natural temperature 30°, sea-water-ice thirty to forty feet thick, most abundant, no bottom, 150 fathoms, density 1·0235.

And off Cape Walsingham, lat. 66° 34′, long. 60° 50′, October 12th, natural temperature 30°, density 1·0245.

Slight as the differences in these densities may appear to be, in my own estimation they are assignable to no other cause than the increased saltness of the water on the east shore, consequent upon a tendency of the water to advance *from* the southward, and the diminished salinity resulting from the dilution of the water moving *to* the southward. The above cases taken by chance, I have cited from the observations made every day at noon; and, although the results at Cape York and Cape Farewell do not bear out the impression conveyed by the whole, we may still presume that impression to be safe in a general point of view. The exception at Cape Farewell arises in all probability from a diversion of the great Arctic Current which flows round that promontory, and carries into that part of the Strait ice and drift-wood which may have come southward from great distances in the Greenland Seas; and in this respect it may be taken as a proof of the dilution of the water of that current consequent upon its burden of comparatively freshwater ice. Again the exception at Cape York may arise from some local cause, such as the presence of an unusually large number of icebergs. On our own coasts the mean density of the sea is often disturbed by the discharges of rivers and small streams. An evening of rainy weather in November of the past year reduced the density of the sea in Stromness harbour from 1·0285 to 1·0235, and eighteen hours of heavy rain on the 17th of the same month reduced that of the water of St. Margaret's Hope, Frith of Forth,

* On the causes which may have produced changes in the earth's superficial temperature. By W. Hopkins, Esq., M.A., F.R.S., &c., Quart. Journ. Geol. Soc. vol. viii., pp. 56 *et seq.*

† Compare Forchhammer's Observations on the Currents and Salinity of the Polar Seas in the Reports and Transactions of the British Association, 1846.

from 1·0245 to 1·0185. Before, however, this theory of a northerly seeking current in Davis' Strait along the eastern shore can be accepted, we must get over the difficulty arising from the position of the great Arctic Current in the North Atlantic. This current sweeps southward across the entrance of Davis' Strait and prevents the ingress at the surface of any water from the Atlantic, except such as the current itself would supply. The Rev. Dr. Scoresby suggests the idea that two currents may arise from the existence of two strata of water varying in temperature.* The question then arises as to the order of superposition. If sea-water independent of its saline ingredients follows the law of expansion peculiar to water from 40° to 32°, one current at a temperature of 36° may pass over another at 44°, and if we separate the extremes eight degrees more, the coldest is still the most buoyant, for, even although it is sea-water, if in a state of tolerable quiescence a portion of it will have become congealed. It is a well-known fact that the process of congelation separates the saline from the watery particles. I have often observed sea-water freezing when the immersed thermometer stood at 32°, and the ice produced at the time was found to contain little more than a trace of saline matter. But there seems to be no reason why this separation should be confined solely to the act of congelation, since it is owing to the universal law of contraction observed in obedience to cold by, I believe, everything in nature except water itself, and that only between the temperatures of 40° and 32°. This may appear somewhat at variance with the experiments of Erman, as quoted by Sir Charles Lyell† ; the latter, however, acknowledges the possibility of the colder and more diluted water of the Arctic Current passing over the warmer and more saline waters of the Gulf Stream. Until our knowledge of the physical changes peculiar to these high latitudes extends, such phænomena as the above must remain more or less obscure ; at present we may rest assured that a meeting and commingling of waters differing in point of saltness and temperature takes place in the entrance of Davis' Strait, and to this causing sudden and decided meteorological changes may be attributed, in great measure, the extreme violence of the storms experienced by navigators when they approach Cape Farewell.

Sea-bottoms and Soundings.—Presuming then upon the existence of currents into the Arctic Seas which may assist the action of the sun in dissolving icebergs and sea-water-ice, we are in a position to consider the extent and character of deposits and accumulations of drift material or "till" now forming in the track of these conveying agents. At the confluence of two opposite currents the largest amount of foreign matter will be deposited, for there icebergs and coast-ice are brought to a stand in the eddies. and are liable to be detained until they are dissolved. In such cases submarine ridges and mounds begin to grow above the general level of the sea-bottom, and they may continue to increase until the surface of the water is reached.

* Principles of Geology, 7th edit., page 97. † *Ibid.*

A bank in latitude 67° and 68° off the coast of West Greenland, well known to the whaling and cod-fishing vessels by the name "Reefkoll or Riscoll Bank," seems to answer this description.* The depth of water on the highest part of it does not exceed fifteen fathoms. It appears to be composed of angular fragments of rock and other materials brought down by icebergs and coast-ice. This, however, can only be inferred from the sounding line, and from the rough usage to which the lines of the whalers are submitted when they attack and get fast to their prey in its neighbourhood. Its limits can be defined almost at all times by the clusters and groups of small icebergs that take the ground upon it; and like other banks of a similar character but less extensive on the same coast, it is exceedingly fertile in shoals of Cod-fish and Halibut which frequent it in the months of May, June, July, and August. These and other fishes, including myriads of Sharks, may pass the whole year upon it; but this we have not as yet had the means of putting to the test.

In other parts of Baffin's Bay and Davis' Straits the bottom is composed of fine mud, sand, rounded and angular fragments of rock, shells, and marly deposits resulting from minute subdivision of calcareous, phosphatic, and siliceous animal and vegetable matter all of which have been brought up in the dredge. In the neighbourhood of islands composed of crystalline rocks the bottom was often found to be rocky; but, as might be expected, numerous depressions were filled with sand and shells. From a depth of twenty-five to thirty fathoms at the Hunde Islands,† South-east Bay, lat. 68°, the dredge passed over a loose and softish deposit, and brought up a quantity of dark-coloured rather finely divided matter resembling peat, which appeared to have been the result of the decomposition of *Fuci* at the bottom. In some cases the roots, being the hardest and most enduring parts, could be detected.

Organic Remains deposited in the Arctic Seas.—Diatomaceæ are exceedingly abundant within the Arctic Circle. Mud from almost every locality has not failed to yield considerable varieties; but the most productive source is the surface-ice when undergoing decay. It often occurred to me that these microscopic forms may be accumulating in a state of great purity, and to a considerable extent, in some of the highly favourable localities so common in

---

* Mr. E. Whymper states—"On the voyage up Davis Strait (in 1867) we were becalmed off Rifkol, a noted landmark, and anchored on some banks in 18 fathoms. These banks have certainly been greatly increased, if not originated, by the deposition of matter from the icebergs of the Jakobshavn ice-stream. At the time we were anchored a large number of small bergs were aground upon them, breaking up and revolving all around. We took the opportunity to put down the dredge; and although we only worked from the ship's side and consequently over a very limited amount of bottom, we brought up in two or three hauls fragments of granite, gneiss (some with garnets), syenite, quartz, hornblende, greenstone, and mica-slate. The sounding-lead showed a fine sand bottom, and the anchor-flukes brought up fetid mud" (Brit. Assoc. Rep. for 1869, p. 2).—Editor.

† For the Entomostraca, Foraminifera, and Algæ of these dredgings, *see above*, pages 166, 192, and *further on*.—Editor.

Davis' Strait. In many of the sheltered bays, where the water is still and the ice dissolves without drifting much about, a brownish slime, consisting of nothing but these forms, occupies the whole surface of the water among the ice, which, after the latter has all disappeared, becomes rolled into rounded pellets by the rippling of the water, and ultimately sinks to the bottom.* This process of deposition extending over thousands of years would produce accumulations scarcely second to those of the "berg-mehl" of Sweden, or of the "tripoli" of the Isle of France, &c.

In addition to such varied materials as we have indicated, the accumulation of "Till" will contain abundant remains of animals high in the order of creation. Of all parts of the ocean this is the most frequented by the large Cetacea and the Seals. The numbers of the former are very great, and that of the latter almost beyond comprehension. Their bones must be strewed on the bottom, and thus they will become constituents of the growing deposit. It may also contain the enduring remains of other animals. Every Arctic traveller is aware of the fact that Polar Bears are seen on the ice at great distances from the land; and my own experience bears testimony to the fact that not unfrequently they are found swimming when neither ice nor land is in sight. The Arctic Fox and, I believe, also the Wolf, and certainly the Esquimaux Dog, animals not generally known to take the water, are set adrift upon the ice and blown out to sea, where they perish when the ice dissolves. And cases are known, although perhaps not recorded, in which human beings have been blown away from the land upon the drifting floes, and never heard of. Two persons to my own knowledge have thus disappeared from the coast of West Greenland. One of them, however, reached the opposite side of Davis' Strait, where he spent the remainder of his life among his less civilised brethren. And the ships engaged in the whaling on the west side of this Strait sometimes have a deed of humanity to discharge by taking from the drifting pack-ice a group of Natives. I have not alluded to the remains of Reindeer and other ruminants of these regions, for the reason that I believe they frequent the ice much less than those that have been mentioned, and consequently are much less liable to be drifted away. It is highly probable, however, that their bones, as well as human remains and works of art, sometimes reach the bottom of the Arctic Seas, the ice of rivers and deep inland bays being the conveying agents.

---

* For an account of the Diatomaceæ of these seas, see Prof. Dickie's "Notes on the Algæ," in the Appendix to Dr. Sutherland's "Journal," vol. ii. p. cxcv. *et seq.*; also *above*, page 319, and *further on*.—Editor.

XLIX.—FOSSILS from the WEST COAST of KENNEDY CHANNEL. By Prof. F. B. MEEK.

[From Silliman's American Journal of Science and Arts, ser. 2, vol. xl., 1865, pp. 31–34. *See* also Hayes's "Open Polar Sea," &c., 8vo., London, 1867, p. 341.]

These were collected by Dr. I. I. HAYES at Cape Leidy, Cape Frazer, and other points of the west coast of Kennedy Channel.

1. Zaphrentis Hayesii, Meek.
2. Syringopora, sp.
3. Favosites, sp.
4. Strophomena rhomboidalis, Wahl.
5. Strophodonta Headleyana, Hall?
6. Strophodonta Beckii, Hall.
7. Rhynchonella, sp.
8. Cœlospira concava, Hall.
9. Spirifera, sp.
10. Loxonema? Kanei, Meek.
11. Orthoceras, sp.
12. Illænus, sp.

Dr. Meek regards these as closely allied to species found in the "Catskill Shaly Limestone"* of the "Lower Helderberg" group of the State of New York.

L.—On the MIOCENE FLORA of NORTH GREENLAND. By Prof. OSWALD HEER. Translated by ROBERT H. SCOTT.

[From the Report of the Thirty-sixth Meeting of the British Association for the Advancement of Science, held at Nottingham in August 1866; Transactions of the Sections, pp. 53–55; 1867.†]

The Royal Dublin Society is in possession of a rich collection of fossil plants, which have been brought from the Arctic regions by Capt. Sir F. Leopold M'Clintock and Capt. Philip H. Colomb at various times, and have been presented by those gentlemen to the Museum of the Society. I am indebted to the kindness of Mr. Robert H. Scott, Hon. Sec. of the Royal Geological Society of Ireland, for a sight of these specimens, as the Royal Dublin Society has been induced to entrust the whole collection to me for examination. Before I received these, Dr. J. D. Hooker had entrusted to me specimens which had been presented to the museum at Kew by Dr. Lyall and Dr. Walker. In this latter collection I discovered seven determinable species, which are also to be found among the specimens of the Dublin collection. In this I find 63

---

* Upper Silurian; the "Delthyris Shaly Limestone," Hall, Geol. Report, p. 144; Mather, Geol. Report, p. 345; Bigsby, Quart. Journ. Geol. Soc., vol. xiv., p. 370, &c.—EDITOR.

† Also in the "Journal of the Royal Dublin Society," vol. v. No. 36, pp. 69–85, with rather more details and some notes by Captains Inglefield, Colomb, and M'Clintock, and an extract from Giesecké's Journal.

recognisable species. If we add to this the additional species mentioned by Brongniart and Vaupel, we obtain a total of 66 species.

All the specimens of the Dublin and Kew collections come from Atanekerdluk, as do also the specimens which Capt. E. A. Inglefield brought home, of which he deposited a portion in the Museum of the Geological Survey, and retained a portion in his own hands. The former have been kindly sent to me by Sir Roderick Murchison, while I have obtained the latter through the goodness of their owner.

Fossil Plants from Atanekerdluk.—Atanekerdluk lies on the Waigat, opposite Disco, in lat. 70°. A steep hill rises on the coast to a height of 1,080 feet, and at this level the fossil plants are found.* Large quantities of wood in a fossilised or carbonised condition lie about. Captain Inglefield observed one trunk thicker than a man's body standing upright. The leaves, however, are the most important portion of the deposit. The rock in which they are found is a sparry-iron-ore, which turns reddish-brown on exposure to the weather. In this rock the leaves are found, in places packed closely together, and many of them are in a very perfect condition. They give us a most valuable insight into the nature of the vegetation which formed this primæval forest.

The catalogue which I append to this paper † will give a general idea of the flora of this forest of Atanekerdluk; but before we proceed to discuss it I must make a few remarks.

(1.) The fossilised plants of Atanekerdluk cannot have been drifted from any great distance. They must have grown upon the spot where they are found. This is proved—

(*a.*) By the fact that Capt. Inglefield and Dr. Rink observed trunks of trees standing upright.

(*b.*) By the great abundance of the Leaves, and the perfect state of preservation in which they are found. Timber, hard fruits, and seeds may often be carried to a great distance by ocean currents; but leaves always fall to pieces on such a long journey, and they are the more liable to suffer from wear and tear the larger they are. We find in Greenland very large leaves, many of which are perfect up to the very edge. It is, however, difficult to work them out from a stone which splits very irregularly, and consequently we can hardly exhibit the entire leaves in a perfect condition.

(*c.*) By the fact that we find in the stone both fruits and seeds of the trees whose leaves are also found there. Thus of *Sequoia Langsdorffii* we see not only the twigs covered with leaves, but also cones and seeds, and even a male catkin. Of *Populus, Corylus, Ostrya, Paliurus,* and *Prunus* there are leaves and some remains of fruit, which could not be the case if the specimens had drifted from a great distance.

---

* For detailed sections by Prof. Nordenskiöld and Dr. Brown, *see further on.*—EDITOR.

† For a later and more perfect catalogue, *see further on.*—EDITOR.

(*d.*) By our finding remains of Insects with the leaves. There is the elytron of a small Beetle, and the wing of a good-sized Wood-bug (probably belonging to the family of the *Pentatomidæ.*)

(2.) *The Flora of Atanekerdluk* is Miocene.* Of the 66 species of North Greenland, 18 occur in the Miocene deposits of Central Europe. Nine of these are very widely distributed both as to time and space, viz., *Sequoia Langsdorffii*, *Taxodium dubium*, *Phragmites Oeningensis*, *Quercus Drymeia*, *Planera Ungeri*, *Diospyros brachysepala*, *Andromeda protogæa*, *Rhamnus Eridani*, and *Juglans acuminata.* These are found both in the Upper and Lower Molasse of Switzerland, while some species, viz., *Sequoia Couttsiæ*, *Osmunda Heerii*, *Corylus Macquarrii*, and *Populus Zaddachi*, have not as yet been noticed in the Upper Molasse. From these facts it seems probable that the fossil forest of Atanekerdluk flourished in that high northern latitude at the earlier Miocene epoch.

(3.) *The Flora of North Greenland is very rich in species.* This is evident from the great variety of plants which the specimens exhibit. Although the amount of material obtained from Atanekerdluk is of small extent compared with that which has come from the Swiss localities, yet many of the slabs contain four or five species, and in one instance even eleven. Atanekerdluk has only been twice visited,† so that we have only got a glimpse of the treasures buried there, and which await a more careful search. At Disco and Hare Island there are extensive beds of brown coal, in whose neighbourhood we may fairly expect to find fossil plants. Professor Göppert mentions three species from Kook (Kome) in lat. 70° N., (nearer 70° 30′) *Pecopteris borealis*, *Sequoia Langsdorffii*, and *Zamites arcticus*,‡ which last he has described in the Neues Jahrbuch für Mineralogie, &c., 1866, pp. 130 and 134.

(4.) *The Flora of Atanekerdluk proves, without a doubt, that North Greenland, in the Miocene epoch, had a climate much warmer than at present. The difference must be at least* 30° F.

Professor Heer discusses at considerable length this proposition. He says that the evidence from Greenland gives a final answer to those who objected to the conclusions as to the Miocene climate of

---

* The fossil Plants here referred to were obtained high up on the hill-side; lower down, at the foot of the hill, Nordenskiöld subsequently discovered remains of a Cretaceous Flora (*see further on*); and Prof. Heer suggests that *Eocene* Plants also are to be found in that section.—EDITOR.

† This was written in 1866. In 1869 Prof. Heer communicated to the Royal Society of London (Phil. Trans. for 1869,) a memoir descriptive of additional specimens from Atanekerdluk and Kudlisaet (Noursoak and Disco) collected by Messrs. E. Whymper and R. Brown, in the summer of 1867. A further collection, made by the Swedish Expedition in 1870, is described in the K. Sv. Vet. Akad. xiii. No. 2, 1874. A general resumé is given in the 3rd. vol. of his "Flora fossilis arctica," 1875.

‡ Prof. O. Heer subsequently proved that both *Pecopteris borealis*, Brong., and *Zamites arcticus*, Göppert, are Cretaceous species, from Kome, on the Noursoak Peninsula; whilst *Sequoia Langsdorfii* (Brongniart) is of Miocene age and common in Disco Island.—EDITOR.

Europe drawn by him on a former occasion. It is quite impossible that the trees found at Atanekerdluk could ever have flourished there if the temperature were not far higher than it is at present. This is clear, first, from many of the species, of which we find the nearest living representatives 10° or even 20° of latitude to the south of the locality in question. Some of the species are quite peculiar, and their relationship to other forms is as yet in doubt. Of these the most important are a *Daphnogene* (*D. Kanii*) the genus *Macclintockia*, and a *Zamites*.* The *Daphnogene* had large thick leathery leaves, and was probably evergreen. *Macclintockia*, a new genus, comprises certain specimens belonging perhaps to the family of the *Proteaceæ*. The *Zamites* is also new. Inasmuch as we know no existing analogues for these plants we cannot draw accurate conclusions as to the climatal conditions in which they flourished. It is, however, quite certain that they never could have borne a low temperature.

If, now, we look at those species which we may consider as possessing living representatives, we shall find that on an average the highest limit attainable by them, even under artificial culture, lies at least 12° to the southward. This, however, does not give a fair view of the circumstances of the case. The trees at Atanekerdluk were not all at the extreme northern limit of their growth. This may have been the case with some of the species; others, however, extended much further north, for in the Miocene Flora of Spitzbergen, lat. 78° N., we find the Beech, Plane, Hazel, a Poplar (*Populus Richardsoni*), a Fir, and the *Taxodium* of Greenland; and in lat. 79° N., a Lime and *Populus arctica.* For the opportunity of examining these specimens I am indebted to Professor Nordenskiöld. At the present time the Firs and Poplars reach to a latitude 15° above the artificial limit of the Plane, and 10° above that of the Beech. Accordingly we may conclude that the Firs and Poplars which we meet at Atanekerdluk and at Bell Sound, Spitzbergen, must have reached up towards the North Pole so far as there was land there in the Tertiary Period. The hills of fossilised wood found by M'Clure and his companions in Banks' Land (lat. 74° 27′ N.) are therefore discoveries which should not astonish us; they only confirm the evidence as to the original vegetation of the Polar regions, which we have derived from other sources. The Professor then proceeds to say that the course of reasoning which led him to the conclusion that the Miocene temperature of Greenland was 30° F. higher than its present temperature would be fully developed in his work "On the Fossil Flora of the Polar Regions," which will contain descriptions and plates of the plants discovered in North Greenland, Melville Island, Banks' Land, Mackenzie River, Iceland, and Spitzbergen, and which he hopes to publish at an early date.

He then selects *Sequoia Langsdorffii*, the most abundant of the trees at Atanekerdluk, and proceeds to investigate the conclusions as to climate deducible from the fact of its existence in Greenland.

* *See* the more perfect list further on.—EDITOR.

Sequoia sempervirens, Lamb (Red-wood), is its present representative, and resembles it so closely that we may consider *S. sempervirens* to be the direct descendant of *S. Langsdorffii*. This tree is cultivated in most of the botanical gardens of Europe, and its extreme northern limit may be placed at lat. 53° N. For its existence it requires a summer temperature of 60° F. Its fruit requires a temperature of 65° F. for ripening. The winter temperature must not fall below 31° F., and that of the whole year must be at least 50° F. Accordingly we may consider the isothermal of 50° as its northern limit. This we may then take as the northern temperature of the *Sequoia Langsdorffii*, and 50° F. as the absolute minimum of temperature under which the vegetation of Atanekerdluk could have existed there.

The present annual temperature of the locality is about 20° F. Dove gives the normal temperature of the latitude (70° N.) at 16° F. Thus Greenland has too high a temperature ; but if we come further to the eastward we meet with a temperature of 33° F. at Altenfiord. Even this extreme variation from the normal conditions of climate is 17° F. lower than that which we are obliged to assume as having prevailed during the Miocene period.

The author states that the results obtained confirm his conclusions as to the climate of Central Europe at the same epoch (conf. Heer, Recherches sur le Climat et la Végétation du Pays Tertiaire, p. 193), and shows at some length how entirely insufficient the views of Sartorius von Waltershausen are to explain the facts of the case.

Herr Sartorius would account for the former high temperature of certain localities by supposing the existence of an insular climate in each case. Such suppositions would be quite inadequate to account for such extreme differences of climate as the evidence now under consideration proves to have existed.

Professor Heer concludes his paper as follows:—

I think these facts are convincing, and the more so as they are not insulated, but confirmed by the evidence derivable from the Miocene Flora of Iceland, Spitzbergen, and Northern America. These conclusions, too, are only links in the grand chain of evidence obtained from the examination of the Miocene Flora of the whole of Europe. They prove to us that we could not, by any re-arrangement of the relative positions of land and water, produce for the northern hemisphere a climate which would explain the phenomena in a satisfactory manner. We must only admit that we are face to face with a problem whose solution in all probability must be attempted and, we doubt not, completed by the astronomer.

Appendix by Editor.

Memoirs and Reports on the Fossil Plants of Greenland, &c.

1867.—On the Miocene Flora of North Greenland. By Professor Oswald Heer. Translated by R. H. Scott. Brit. Assoc. Report for 1866, Trans. Sections, p. 53. 1867.

1867.—On the Miocene Flora of North Greenland. By Professor Oswald Heer. Translated by Robert H. Scott, M.A. With Notes by E. A. Inglefield, P. H. Colomb, Sutherland, and M'Clintock, and extract from Giesecké's Journal. Read November 19, 1866. The Journal of the Royal Dublin Society, vol. v., for 1867, pp. 69–85.

1868.—Flora fossilis arctica. I. Die fossile Flora der Polarländer, enthaltend die in Nordgrönland, auf der Melville-Insel, im Banksland, am Mackenzie, in Island und in Spitzbergen entdeckten fossilen Pflanzen, von Dr. Oswald Heer, Professor am Polytechnikum und an der Universität in Zurich. Mit einem Anhang über versteinerte Hölzer der arctischen Zone, von Dr. Carl Cramer, Professor am Polytechnikum in Zurich. (Mit 1 Karte und 50 Tafeln.) 4to. Zurich, 1868.

1870.—Report of Proceedings to obtain a Collection of Fossil Plants in North Greenland for the Committee of the British Association. By Edward Whymper. Brit. Asoc. Report for 1869, pp. 2–8. 1870.

1870.—Report on the Fossil Plants collected by Mr. Whymper (and Mr. Brown) in North Greenland in 1867. By Professor O. Heer. Brit. Assoc. Report for 1869, pp. 8–10. 1870.

1871.—Flora fossilis arctica. II. Die foss., &c., Zweiter Band, enthaltend: 1. Fossile Flora der Bären-Insel (Kongl. Svensk. Vetensk. Akad. Handl., vol. ix., No. 5, 1871). 2. Flora fossilis Alaskana (K. Sv. Vet. Ak. Handl., vol. viii., No. 4, 1869). 3. Die miocene [miocäne] Flora und Fauna Spitzbergens (K. Sv. Vet.-Ak. Handl., vol. viii., No. 7, 1870). 4. Contributions to the Fossil Flora of North Greenland (Phil. Trans. Roy. Soc., London, 1869). Mit 59 Tafeln. 4to. Winterthur, 1871.

1872.—Notice of Heer's Flora fossilis arctica (vol. ii.). Communicated by R. H. Scott, F.R.S. Geol. Mag., vol. ix., pp. 69–72. 1872.

1871.—Prof. E. A. Nordenskiöld. Redogörelse för en Expedition till Grönland År 1870; Öfversigt af K. Vet.-Akad. Förhandl., 1870, No. 10. Separately published, 1871.

1872.—Prof. E. A. Nordenskiöld. Expedition to Greenland in 1870. Geol. Mag., vol. ix., pp. 289, 354, 409, 449, & 516. 1872.

1875.—Flora fossilis arctica. III. Die foss., &c., Dritter Band, enthaltend: 1. Beiträge zur Steinkohlenflora der arctischen Zone (K. Sv. Vet. Ak. Handl., vol. xii., No. 3, 1874). 2. Die Kreideflora der arctischen Zone (K. Sv. Vet. Ak. Handl., vol. xii., No. 6, 1874). 3. Nachträge zur miocenen [miocänen] Flora Grönlands (K. Sv. Vet. Ak. Handl., vol. xiii., No. 7, 1874). 4. Uebersicht der miocenen [miocänen] Flora der arctischen Zone, 1874. Mit 49 Tafeln. 4to. Zurich, 1875.

1875.—R. Brown. Geological Notes on the Noursoak Peninsula, Disco Island, &c. Transact. Geol. Soc., Glasgow, vol. v. 1875.

LI.—NOTICE of HEER'S "FLORA FOSSILIS ARCTICA" (CARBONIFEROUS FOSSILS of BEAR ISLAND and SPITZBERGEN, and CRETACEOUS and MIOCENE PLANTS of SPITZBERGEN and GREENLAND). Communicated by ROBERT H. SCOTT, F.R.S., &c.

[Reprinted, with Permission, from the Geological Magazine, Vol. IX., No. 2, February 1872, pp. 69–72.]

Carboniferous: Bear Island and Spitzbergen.—In vol. ii. of his "Flora fossilis arctica," Professor Oswald Heer has treated of the Fossil Flora of Bear Island, and shown that it belongs to the Lower Carboniferous Formation, of which it forms the lowest beds (named by him the "Ursa Stage"), close to the junction with the Devonian.* The Yellow Sandstone of Kiltorcan in Ireland, some of the Grauwacke of the Vosges and the southern part of the Black Forest, and some strata near St. John in New Brunswick, belong to the same group. In the summer of 1870 two young Swedish naturalists (Wilander and Nathorst) discovered this same formation in the Klaas Billen Bay of the Eisfiord in Spitzbergen, and brought home fine specimens of *Lepidodendron Veltheimianum* and *Stigmaria ficoides.* It has also been found in West Greenland; for Prof. Nordenskiöld tells us that the Swedish Expedition, which went to Disco, in the course of last summer, to fetch the meteorite, weighing about 20 tons, which he discovered at Ovifak in that island, has brought home fossil plants of true Carboniferous age.

The Carboniferous formation, therefore, has been extensively developed in the Arctic regions, for it occurs also in the Parry Islands and in Siberia; on the Lena it approaches the Arctic Circle. These facts show us that at the Carboniferous epoch there was a great extent of land near the North Pole, covered with a vegetation closely resembling that of our own latitudes at the same period. Of 18 species of fossil plants at Bear Island, only 3 are peculiar to it, the others are common to the European localities (such as *Lepidodendron Veltheimianum, Knorria imbricata,* &c.); and, from the fact that they are as fine and as well developed in the northern as in the southern deposits, it is evident that no great difference of climate could have prevailed between the two localities.†

Tertiary: Spitzbergen.—In Spitzbergen we have, besides the Miocene Flora and Fauna, an important Diluvial formation. 132 species of Miocene plants have been found, mostly in Eisfiord (lat. 78° N.), but some in King's Bay (lat. 78° 56′ N.). The chief form here is an *Equisetum* (*E. arcticum*); but it is sur-

* *See also* Quart. Journ. Geol. Soc. vol. xxvii., p. 1, and xxvii., pp. 161–173

† Prof. Heer has worked out this idea very fully in his paper on Bear Island, and traced the alternations of rise and fall of the land, which probably occurred during the later part of the Palæozoic period.

prising to find a Lime (*Tilia Malmgreni*), an Arborvitæ (*Thuites Ehrenswaerdi*), a Juniper, and two Poplars nearly on the 79th parallel of latitude. The Flora of the Eisfiord is much richer, especially that of the black shales of Cape Staratschin, where we find 26 Conifers belonging to the *Abietineæ*, the *Cupressineæ*, and the *Taxodieæ*. Several of these species are represented not only by leaves, but by their flowers and fruit. The chief forest-trees were a *Sequoia* (*S. Nordenskioldi*), of which we have leaves, twigs, and seeds, *Libocedrus Sabiniana*, and *Taxodium distichum*. Of the last-named the collection contains, not only the twigs clothed with leaves, but the male and female flowers, the scales, and seeds; so that not even the delicate catkins are wanting to identify this tree with that which is now growing in the Southern States of North America. No one can possibly doubt that the tree grew where its remains are now found. *Libocedrus Sabiniana* is also well represented by its peculiar seeds; it was the most graceful tree in Spitzbergen, and its nearest congeners are now found in Chili. Of other trees, Poplars are the most common, with the Birch, Hazel, and Snowball (*Viburnum*); but we are not so much surprised at finding them as two large-leaved Oaks, the Ivy, and a Walnut.

This Flora has the greatest resemblance to that of North Greenland and the other Arctic localities; but several species extend southwards into Europe. On the whole, this Miocene Flora bears evidence of a far greater contrast of climate between Europe and the Arctic regions at that epoch, than the Lower-Carboniferus plants show for their period. All the tropical and even sub-tropical forms are wanting. These facts show us that great changes of climate must have occurred, and it will be interesting to trace when these first began to show themselves.

Cretaceous: Greenland.—The Cretaceous Flora of the Arctic regions throws important light on this point, and our knowledge of it has been largely enriched by the discoveries of the Swedish Expedition of 1870. When the first volume of the "Flora arctica" appeared, Prof. Heer could only speak of a few specimens belonging to this epoch, which had been found at Kome, on the north side of the Noursoak Peninsula. Prof. Nordenskiöld has, however, paid great attention to these fossils, and has discovered several new localities for them on the same coast. They are found in black shales, apparently, from the character of the fossils, belonging to the Lower Cretaceous—the Urgonian, for they resemble the Flora of Wernsdorf, in the Carpathians. Among forty-three species already determined, Prof. Heer finds twenty-four Ferns, five Cycads, eight Conifers, three Monocotyledons. Only one fragment is dicotyledonous, a Poplar leaf, and it is the *oldest dicotyledonous plant that has hitherto been discovered.* Among the numerous Ferns the *Gleichenia* is the most common type, but *Marattiaceæ* and *Sphenopteris* are not rare. Of Cycads we have *Zamites*, with very fine leaves, and *Podozamites Hoheneggeri* (known from Wernsdorf in the Carpathians). It is striking that Sequoias and Pines approaching closely to Tertiary types appear among the Conifers.

The plants of the black shales of the south side of Noursoak Peninsula have a different character. Nordenskiöld has found them at two points (Atane, and on the shore below Atanekerdluk, the well-known Miocene locality). The number of species is about equal to that found in the Lower Cretaceous just referred to; but their type is almost totally different, and it indicates that they belong to the Upper Cretaceous. *Sequoia* again predominates among the Conifers, and fortunately cones were found as well as twigs. With them were found a *Thuites* and a *Salisburea* (?). Cycads are much less common than in the Lower Cretaceous beds, only one (*Cycadites Dicksoni*) having been discovered. Among the Ferns, though these are common (eleven species) only two *Gleicheniæ* were found instead of six; other forms, such as *Marattiaceæ*, *Adiantum*, and *Dictyophyllum*, have disappeared. The predominant forms are Dicotyledons, of which there are twenty-four, of various genera and species; many of them have not yet been absolutely determined. But there are three species of Poplar, one Fig (leaves and *fruits*), one *Myrica*, one *Sassafras*, one *Credneria*, with two *Magnoliæ*.

These facts show us that here, as in Central Europe, the Lower Cretaceous Flora consists principally of Ferns, Conifers, and Cycads; while in the Upper Cretaceous Dicotyledons appear. The climatological changes which produced so important modifications in the types of vegetable life must have been as extensive in high as in lower latitudes. If we examine into the climatic character of the Lower Cretaceous Flora, we find it to be almost tropical, as will be seen from the predominant forms of vegetation. The same is true of the Flora of Wernsdorf in the Northern Carpathians, so that in this respect the Lower Cretaceous Flora resembles the Carboniferous Flora. The comparative rarity of Gleichenias and Cycads, and the disappearance of *Marattiaceæ*, might point to a change of climate for the Upper Cretaceous; but the presence of *Ficus* renders this doubtful, so that we cannot decide whether the change of climate occurred during the Cretaceous or the Tertiary period in Greenland; at all events, the Flora of the former epoch has a more southern character than that of the latter.

Miocene: Greenland.—Besides these fossils, Nordenskiöld has brought over a large series of Miocene Plants from various localities. The most interesting of these are from a deposit, which is separated by beds of basalt, some 2,000 feet thick, from the Lower Miocene plant-bearing strata, and which, though still Miocene, are much later in age.

[According to the succession of strata, Nordenskiöld divides the Miocene plant-bearing formation into three groups :—

I. The *lowest* ("Upper Atanekerdluk") consists of sand, sandstone with shale, coal-seams, and clayironstone. To this belongs the upper portion of the Atanekerdluk section, with its rich fossil flora, Lower Miocene in character. Also found at Iglosungoak and Isungoak on Disco Island.

II. The *middle* or "Ifsorisok group" of plant-bearing sand, shale, coal, and brown clayironstone, lying between basalts, tuffs, and lavas, several thousand feet in thickness, and approximately in the middle of the trap-formation. This group is found at :—

1. Netluarsuk, N.W. of Atane, between Noursoak and Noursak, at the mouth of the Waigat, and near the N.W. end of the Noursoak Peninsula, about 1,000 feet above the sea. Sand, shale, coal, and brown clayironstone, with plant-remains, between basalts.

2. Ifsorisok, N.E. of Netluarsuk (about 70° 40′ N. lat.), 12 miles from the coast, and about 2,250 feet above the sea. A soft sandy clay, alternating with thin coal-seams, and containing the plants, rests on basalt, which further inland forms high mountains. The Kinnitak, between Niakornak and Ekkorfat, is the nearest, and reaches the height of 6,000 feet, and apparently consists wholly of eruptive rock.

3. Asakak. Not far from Kome, on the north side of the Noursoak Peninsula, is the Asakak Glacier, and among the stones on its surface carbonised and silicified wood abounds, also some fragments of coarse sandstone containing Miocene Plants. The place of origin could not be discovered.

III. The *uppermost* group consists of some sand and clays, on the south coast of Disco, lying on and in the basalt, which there overlies gneiss; and it was probably contemporary with the last of the great post-cretaceous volcanic eruptions of the district. The fossil plants from Puilasok, having an Upper-Miocene character, belong to this series, and occur in black or dark-grey sand, or soft sandy shaly clay; but Prof. Heer thinks that the plants from the clayironstone of Sinifik, on the same coast, must be somewhat older. The soft sandstones and sandy shales of Puilasok, with their thin irregular coal-seams, are represented by Nordenskiöld's section, at p. 4 of O. Heer's "Nachträge," &c., as lying in nearly horizontal layers to the height of 200 feet, on and against the eroded slope of horizontally stratified basalt and basalt-tuff.—O. Heer, "Nachträge zur mioc. Flora Grönland's," 1874, pp. 3, 4.]

East Coast of Greenland.—The German Expedition has brought from the East Coast of Greenland some vegetable fossils, many of which are, however, only undistinguishable carbonaceous traces. Lieutenant Payer, however, brought some specimens from Sabine Island which could be identified. They belong to *Taxodium distichum* and *Populus arctica*, with a fragment which probably belongs to *Diospyros brachysepala*. These trees have been discovered in West Greenland, and the two first-named in Spitzbergen also, so that they probably flourished over the entire district from the west coast to Spitzbergen. In his paper on Spitzbergen, Prof. Heer had remarked that we might expect to find the plants which were common to the West Coast of Greenland and to Spitzbergen on the East Coast of Greenland also. This anticipation has now been confirmed by the discovery of these two species, and it may fairly be expected that the fossiliferous sandstones and marls of Germania Mountain in Sabine Island contain many of the missing forms.

LII.—The MIOCENE FLORA and FAUNA of the ARCTIC REGIONS. By the Rev. Dr. OSWALD HEER, F.M.G.S., Professor of Botany, University of Zurich, &c.

[From "Flora fossilis arctica," vol. iii., 1875.]

Miocene Plants have been found in Spitzbergen from 77½° to 78⅔° N.L., in West Greenland from 70° to 71° N.L., in East Greenland at 74½° N.L., in Iceland between 65° and 66°, on the Mackenzie (Bear-Lake River) at 65°, and in Banks' Land at 74° 27′ N.L.

1. In *Spitzbergen* Miocene Plants have been collected at six places; namely, at the Scott Glacier in Recherche Bay* (77½° N.L.); at Cape Lyell at the entrance of Bell Sound; at Cape Staratschin at the entrance of the Ice-fjord (78° 5′ N.L.); at Cape Heer near Greenhaven; and in King's Bay, 78° 56′ N.L.

We now know 179 species of plants from these localities; 34 from the Scott Glacier, 51 from Cape Lyell, 9 from Bell Sound, 115 from Cape Staratschin (from black shale and sandstone), 15 from Cape Heer, and 16 from King's Bay.

Equisetum arcticum, Taxodium distichum, Populus Richardsoni, P. arctica, Corylus Macquarrii, Platanus aceroides, and *Hedera Macclurii* occur at nearly every one of the places, and must have spread over all the old land. With other species common to several of the localities, they show that the strata containing them belong to one formation. The new discoveries show also that these beds are Lower Miocene. We find 40 of the Spitzbergen species in the Miocene of Europe; 23 in Switzerland, of which 19 occur in our Lower Freshwater Molasse.

2. In *West Greenland* Miocene Plants have been collected at 10 localities. Excepting Asakak, they are on the shores of the Waigat, some in Disco, some in the Noursoak Peninsula. The most important is the upper part of the hill at Atanekerdluk, where 124 species have been got. In the "Supplemental Remarks on "the Miocene Flora of Greenland," in this vol. of the "Flora foss. arct.," it is shown that during the long period of the great Basalt-formation of Greenland no special change in the vegetation took place. The plants of Ifsorisok, Netluarsuk, and Sinifik agree generally with those of Upper Atanekerdluk, although during their deposition 2,000 feet thickness of basalt was formed. The deposit at Puilasok only, which lies on the basalt, shows some difference in its plant-remains, some of which resemble those of our Upper Molasse (Œningen beds).

Of the 169 Miocene species from Greenland, 69 occur in Europe;

* The new localities for these fossils at the Scott Glacier, Cape Lyell, and Cape Heer were discovered by Prof. Nordenskiöld in the summer of 1873. The plants then collected will be published by-and-by; they increase the Miocene Flora of Spitzbergen by 47 species.

42 in Switzerland. Of these there are 35 in the Lower and 24 in the Upper Molasse. Thus, excepting those from Puilasok, the Greenland fossil plants, like those of Spitzbergen, accord chiefly with those of our Lower Miocene.

3. In *East Greenland* MM. Payer and Copeland have obtained some fossil plants at Sabine Island (about 74½° N.L.): remains of *Taxodium distichum*, *Populus arctica*, and *Diospyros brachysepala*, all known from West Greenland, and the first two common in Spitzbergen. The 35 species which are common to Spitzbergen and West Greenland probably existed all over the intermediate region, and their remains may yet be found at Sabine Island.

4. The Lignite- or Browncoal-formation of *Iceland* is widespread, and plants have reached me from five places:—Briamslock (about 65½° N.L.), Hredavatn (64° 40′ N.L.), Langavatsdalr, Sandafell (about 65⅓° N.L.), and Husawick (65° 40′ N.L.), yielding 42 species, of which 18 belong to the Miocene of Europe.

5. From the *Mackenzie River district* we have still only the 17 species of Miocene Plants described in vol. i. of "Flora foss. arct."

6. Still further off is *Banks' Land* 74° 27′ N.L.), with its remarkable wood-hills, yielding five Conifers and a Birch. One of its Pines (*Pinus Macclurii*) has been found also in Greenland by Nordenskiöld, confirming my conjecture of the Miocene age of these accumulations of wood ("Flora arct.," i., p. 20.)

7. Altogether there are 353 species from these Arctic regions. [A table of distribution in families is given, p. 5.] There are 31 Cryptogams, 53 Gymnosperms, 55 Monocotyledons, 65 Apetalæ, 16 Gamopetalæ, 88 Polypetalæ, and 45 of doubtful alliance.

Several species of these Miocene Plants can now be traced from Spitzbergen, by Greenland, to the Mackenzie in N.W. America, namely:—

Taxodium distichum.	Salix Raeana.
Glyptostrobus Ungeri.	Corylus Macquarrii.
Sequoia Langsdorfii.	Platanus aceroides.
Populus arctica.	Hedera Macclurii.

The Swamp Cypress, Poplar, Hazel, and Plane are among the most abundant of those both in Greenland and Spitzbergen; *Sequoia* and *Glyptostrobus* abound in Greenland, and have lately been found in Spitzbergen, but appear to be more limited there. Either wide-spread or abundant in both of these countries are—

Lastræa Stiriaca.	Quercus Grœnlandica.
Taxites Olriki.	Q. platania.
Phragmites Œningensis.	Andromeda protogæa.
Populus Richardsoni.	Nordenskiœldia borealis.
Salix varians.	Paliurus Colombi.
Carpinus grandis.	Rhamnus Eridani.
Fagus Deucalionis.	

Of the Arctic species, 97 occur in the Miocene Flora of Europe. There is 27½ *per cent.* (more than ⅓) of these Arctic plants common to the Miocene of Europe. The per-centage for Spitzbergen is about 22, for Greenland about 41, for Iceland and the Mackenzie 40. In the most northern regions, in Spitzbergen, it is therefore the least; in Greenland it is almost twice as great, and it increases southward.

Of all the European fossil floras, that of the Baltic comes nearest to the Arctic. Of the 71 species known from the Samland and the Browncoal of Rixhöft (near Dantzic, about 55° N.L.), there are 38 (54 *per cent.*) in the Arctic fossil flora. [Other comparisons and illustrative remarks follow; and at pages 13–24 a table is given, of which the following is an abstract.]

List of the Miocene Plants of the Arctic Regions.

[From Prof. O. Heer's "Uebersicht der miocene Flora der arctische Zone," 1874; pp. 13–24.]

Spitzbergen = Sp.; Greenland = G.; Iceland – I.; MacKenzie River = M.

I.—Cryptogamæ.

Fungi.

Sphæria arctica, H. Sp.
S. annulifera, H. Sp., G.
S. pinicola, H. Sp.
S. hyperborea, H. Sp.
Dothidea borealis, H. I.
Sclerotium Dryadum, H. I.
S. Cinnamomi, H. G.
S. populicola, H. G.
Rhytisma induratum, H.? I.
R. boreale, H. G.
Polyporites Sequoiæ, H. G.

Algæ.

Muensteria deplanata, H. Sp.

Musci.

Muscites Berggreni, H. Sp.
M. subtilis, H. G.

Filices.

Polypodiaceæ.

Adiantum Dicksoni, H. Sp.
Woodwardites arcticus, H. G.
Lastræa Stiriaca, (Ung.) Sp., G.
Sphenopteris Miertschingi, H. G.
S. Blomstrandi, H. Sp., G.
Aspidium Meyeri, H. G.
A. Heerii, Ett. G.
A. ursinum, H. G.
Pteris Œningensis, Ung. G.
P. Rinkiana, H. G.
P. Sitkensis, H. G.
Pecopteris gracillima, H. G.

Osmundaceæ.

Osmunda Heerii, Gaud. G.
O. Torelli, H. G.

Equisetaceæ.

Equisetum boreale, H. G.
E. Winkleri, H. I.
E. arcticum, H. Sp.

II.—Phænogamia.

Coniferæ.

Taxineæ.

Taxites Olriki, H. Sp., G.
T. validus, H. G.
Ginkgo adiantioides (Ung). G.
Torellia rigida, H. Sp.
T. bifida, H. Sp.

Cupressineæ.

Juniperus rigida, H. Sp.
Biota borealis, H. G.
Libocedrus Sabiniana, H. Sp.
L. gracilis, H. Sp.
Thuites Ehrenswardi, H. Sp.
Cupressinoxylon Breverni, Merkl. G.

Cupressinoxylon ucranicum, Gp.? G.
C. pulchrum, Cr. Banksland.
C. polyommatum, Cr. Banksland.
C. dubium, Cr. Banksland.

Taxodieæ.

Widdringtonia Helvetica, H. G.
Taxodium distichum-miocænum, H. Sp., G., M.
T. Tinajorum, H. Sp.
Glyptostrobus europæus, Brgn. G.
G. Ungeri, H. Sp., G., M.
Sequoia Langsdorfii, (Brongn.) Sp., G., M.
var. *b.* striata. Sp.
„ *c.* acuta. Sp.
„ *d.* obtusiuscula. Sp.
„ *e.* abrupta. Sp.
„ *f.* angustifolia. Sp.
S. disticha, H. Sp.
S. brevifolia, H. Sp., G.
S. Nordenskiœldi, H. Sp., G.
S. Couttsiæ, H. G.
S. Sternbergi, Gp., sp. G. I.,

Abietineæ.

Pinus montana, Mill. Sp.
P. polaris, H. Sp., G.
P. Thulensis, Steenstr. I.
P. Martinsii, H. I.
P. cyloptera, Sap. Sp.
P. stenoptera, H. Sp.
P. macrosperma, H. Sp.
P. abies, L. Sp.
P. Ungeri, Endl. Sp.
P. Loveni, H. Sp.
P. Macclurii, H. G., Banksland.
P. (Tsuga) microsperma, H. I.
P. æmula, H. I.
P. Dicksoniana, H. Sp.
P. Malmgreni, H. Sp.
P. brachyptera, H. I.
P. (Picea, Don.) impressa, H. Sp.
P. hyperborea, H. Sp., G.
P. Ingolfiana, Steenstr. I.
P. Steenstrupiana, H. I.
P. Armstrongi, H. Banksland.
Pinites latiporosus, Cram. Sp.
Pinites cavernosus, Cram. Sp.
P. pauciporosus, Cram. Sp.
P. Middendorfianus, Gp. Boganida.
P. Baerianus, Gp. Taimyrland.

Gnetaceæ.

Ephedoites Sotzkianus, Ung. Sp.

GRAMINEÆ.

Phragmites Œningensis, H. Sp., G.
P. multinervis, H. G.
Poacites Mengeanus, H. G.
P. Nielseni, H. G.
P. avenaceus, H. Sp.
P. hordeiformis, H. Sp.
P. Friesianus, H. Sp.
P. lævis, A. Br. Sp.
P. æqualis, H. (læviusculus, H. *olim*). Sp.
P. effosus, H. Sp.
P. sulcatus, H. Sp.
P. parvulus, H. Sp.
P. Torelli, H. Sp.
P. argutus, H. Sp.
P. trilineatus, H. Sp.
P. bilineatus, H. Sp.
P. lepididus, H. Sp.

CYPERACEÆ.

Cyperus arcticus, H. Sp.
C. Sinifikianus, H. G.
Carex rediviva, H. I.
C. Andersoni, H. Sp.
C. Berggreni, H. Sp.
C. hyperborea, H. Sp.
C. misella, H. Sp.
C. ultima, H. Sp.
C. antiqua, H. Sp.
C. Noursoakensis, H. Sp., G.
Cyperacites borealis, H. Sp., G.
C. Zollikoferi, H.? G.
C. Islandicus, H. I.
C. nodulosus, H. I.
C. microcarpus, H. G., I.
C. strictus, H. Sp.
C. argutulus, H. Sp.
C. trimerus, H. Sp.

JUNCACEÆ.

Juncus antiquus, H. Sp.

SMILACEÆ.

Majanthemophyllum boreale, H. Sp.
Smilax grandifolia, Ung. G.
S. Franklini, H. M.
S. lingulata, H. G.

TYPHACEÆ.

Sparganium Valdense, H. I.
S. stygium, H. G.
S. crassum, H. Sp.

AROIDEÆ.

Acorus brachystachys, H. Sp.

NAJADEÆ.

Potamogeton Nordenskiœldi, H. Sp.
P. Rinkii, H. G.
P. dubius, H. G.
Najas striata, H. Sp.
Caulinites borealis, H. I.
C. costatus, H. G.

ALISMACEÆ.

Alisma macrophyllum, H. Sp.
Sagittaria ? difficilis, H. Sp.
S. ? hyperborea, H. Sp.

IRIDEÆ.

Iris latifolia, H. Sp.
Iridium Grœnlandicum, H. Sp., G.

SALICINEÆ.

Populus balsamoides, Gp Sp.
P. Richardsoni, H. G., Sp.
P. Zaddachi, H. Sp., G.
P. curvidens, H. Sp.
P. arctica, H. Sp., G., M.
P. Hookeri, H. Sp., M.
P. mutabilis, H. G.
P. Gaudini, Fisch ? G.
P. retusa, H. Sp.
P. sclerophylla, Sap. G.
Salix macrophylla, H. Sp., I.
S. varians, Gp. Sp., G.
S. Ræana, H. Sp., G., M.
S. Grœnlandica, H. G.
S. elongata, O. Webb. G.
Salix longa, A. Braun. G.
S. tenera, A. Braun. G.
Liquidambar europæum, Al. Braun. G.

MYRICAEGÆ.

Myrica acuminata, Ung. G.
M. borealis, H. G.
M. lingulata, H. G.
M. grosseserrata, H. G.
M. acutiloba, Brgn. (Dryandra, *olim.*) G.

BETULACEÆ.

Alnus nostratum, Ung. G.
A. Kefersteinii, Gp. Sp., I.
Betula macrophylla, Gp., sp. Sp., I.
B. prisca, Ettingsh. Sp., I.
B. Forchhammeri, H. I.
B. Miertschingi, H. G.
B. Macclintocki, Cr. Bnksld.

CUPULIFERÆ.

Ostrya Walkeri, H. G.
Carpinus grandis, Ung. Sp., G.
Corylus Macquarrii (Forb.). Sp., G., I., M.
C. Scotii, H. Sp.
C. insignis, H. G.
Fagus Deucalionis, Ung. Sp., G., I.
F. dentata, Ung. ? G.
F. macrophylla, Ung. G.
Castanea Ungeri, H. G.
Quercus Drymeia, Ung. G.
Q. furcinervis (Rossm.). G.
Q. Lyelli, H. Sp., G.
Q. elæna, Ung. Sp.
Q. Grœnlandica, H. Sp., G.
Q. Olafensi, H. G., I., M.
Q. platania, H. Sp., G.
Q. Steenstrupiana, H. G.
Q. Laharpii, Gaud. G.
Q. spinulifera (venosa, *olim*), H. Sp.
Q. atava, H. G.

ULMACEÆ.

Ulmus Braunii, H. Sp.
U. diptera, Steenstr. I.
Planera Ungeri, Kov. G., I.

MOREÆ.

Ficus ? Grœnlandica, H. G.

PLATANEÆ.

Platanus aceroides, Göpp. Sp., G., I., M.
P. Guillelmæ, Göpp. G.

POLYGONEÆ.

Polygonum Ottersianum, H. Sp.

CHENOPODIACEÆ.

Salsola arctica, H. Sp.

ELÆAGNEÆ.

Elæagnus arcticus, H. G.
Elæagnites campanulatus, H. Sp.

THYMELEÆ.

Daphne persooniæformis, O. Webb. G.

LAURINEÆ.

Sassafras Ferretianum, Mass. G.

PROTEACEÆ.

Hakea (?) arctica, H. G.

ARISTOLOCHIEÆ.

Aristolochia borealis, H. G.

SYNANTHEREÆ.

Cypselites sulcatus, H. Sp.
C. incurvatus, H. Sp.

ERICACEÆ.

Andromeda protogæa, Ung. Sp., G.
A. Narbonensis, Sap. G.
A. Saportana, H. G.
A. denticulata, H. G.

EBENACEÆ.

Diospyros brachysepala, Al. Braun. G.
D. Loveni, H. G.

GENTIANEÆ.

Menyanthes arctica, H. G.

ASCLEPIADEÆ.

Acerates veterana, H. G.

ALEACEÆ.

Fraxinus denticulata, H. G.
F. (?) microptera, H. Sp.

RUBIACEÆ.

Galium antiquum, H. G.

CAPRIFOLIACEÆ.

Viburnum Whymperi, H. Sp., G.
V. Nordenskiœldi, H. Sp.
V. macrospermum, H. Sp.

ARALIACEÆ.

Aralia Browniana, H. G.
Hedera Macclurii, H. Sp., G., M.
Cornus rhamnifolia, O. Web. Sp.
C. macrophylla, H. Sp.
C. orbifera, H. Sp.
C. hyperborea, H. Sp., G.
C. ramosa, H. Sp.
C. ferox, Ung. G.
Nyssa arctica, H. Sp., G.
N. reticulata, H. Sp.
N. europæa, Ung. Sp.
Nyssidium Ekmani, H. Sp.
N. crassum, H. Sp.
N. oblongum, H. Sp.
N. Grœnlandicum, H. G.
N. fusiforme, H. Sp.
N. lanceolatum, H. Sp.

AMPELIDEÆ.

Vitis Islandica, H. I.
V. Olriki, H. G.
V. arctica, H. G.

HAMAMELIDEÆ.

Parrotia pristina, Ett. Sp.

RANUNCULACEÆ.

Helleborites marginatus, H. Sp.
H. inæqualis, H. Sp.

MAGNOLIACEÆ.

Magnolia regalis, H. Sp.
M. Nordenskiœldii, H. Sp.
M. Inglefieldi, H. Sp.
Liriodendron Procacinii, Ung. I.

MENISPERMACEÆ.

Coculites Kanii, H. G.
Macclintockia Lyallii, H. G.
M. dentata, H. G.
M. trinervis, H. G.
M. ? tenera, H. Sp.

NYMPHÆACEÆ.

Nymphæa arctica, H. Sp.
N. Thulensis, H. Sp.

MYRTACEÆ.

Callistemophyllum Moorii, H. G.

TILIACEÆ.

Tilia Malmgreni, H. Sp.
Nordenskiœldia borealis, H. Sp., G.
Grewia crenata, H. Sp.
G. crenulata, H. Sp.
G. obcvata, H. Sp.
Apeibopsis Nordenskiœldii, H. G.

STERCULIACEÆ.

Pterospermites spectabilis, H. G.
P. alternans, H. G.
P. integrifolius, H. G.
P. dentatus, H. M.
Dombeyopsis Islandica, H. I.

ACERINEÆ.

Acer otopteryx, Gp. G. ?, I.
A. arcticum, H. Sp.
A. Thulense, H. Sp.
A. angustilobum, H. G.
A. inæquale, H. Sp.

SAPINDACEÆ.

Kœlreuteria borealis, H. Sp.

ILICINEÆ.

Ilex macrophylla, H. G.
I. longifolia, H. G.
I. reticulata, H. G.

CELASTRINEÆ.

Evonymus amissus, H. G.
Celastrus cassinefolius, Ung. Sp.
C. Greithianus, H. Sp.
C. firmus, H. G.

RHAMNEÆ.

Zizyphus borealis, H. G.
Paliurus Colombi, H. Sp., G.
P. borealis, H. G.
Rhamnus Eridani, Ung. Sp., I., G.
R. brevifolius, A. Br. G.
R. Gaudini, H. G.

ANACARDIACEÆ.

Rhus Brunneri, H. I.
R. bella, H. G.
R. arctica, H. G.

JUGLANDEÆ.

Juglans acuminata, A. Braun. G.
J. bilinica, Ung. I.
J. paucinervis, H. G.
J. Strozziana, Gaud. G.
J. denticulata, H. G.
J. albula, H. Sp.

POMACEÆ.

Sorbus grandifolia, H. Sp.
Cratægus antiqua, H. Sp., G.
C. Warthana, H. G.
C. Carneggiana, H. Sp.
C. oxyacanthoides, Gp. Sp.
C. glacialis, H. Sp.

ROSACEÆ.

Rubus ? scabriusculus, H. Sp.
Fragaria antiqua, H. Sp.

AMYGDALEÆ.

Prunus Scotii, H. G.
P. Staratschini, H. Sp.

LEGUMINOSÆ.

Colutea Salteri, H. G.
Leguminosites arcticus, H. G.
L. Thulensis, H. Sp.
L. vicioides, H. Sp.
L. longipes, H. G.

Incertæ sedis.

Phyllites liriodendroides, H. G.
P. membranaceus, H. G.
P. rubiformis, H. G.
P. celtoides, H. G.
P. evanescons, H. G.
P. acutilobus, H. I.

Phyllites tenellus, H. I.
P. vaccinioides, H. I.
P. aceroides, H. M.
P. hyperboreus, H. Sp.
Antholites amissus, H. M.
Carpolithes cocculoides, H. G.
C. potentilloides, H. G.
C. follicularis, H. G.
C. sulculatus, H. G.
C. pusillimus, H. G.
C. Najadum, H. I.
C. geminus, H. I.
C. borealis, H. Sp., I.
C. symplocoides, H. G.
C. sphærula, H. G.
C. lithospermoides, H. G.
C. bicarpellaris, H. G.
C. seminulum, H. M.
C. caudatus, H. Sp.
C. singularis, H. Sp.
Carpolithes funkioides, H. Sp.
C. pulchellus, H. Sp.
C. rosaceus, H. Sp.
C. oblongo-ovatus, H. Sp.
C. clavatus, H. Sp.
C. ovalis, H. Sp.
C. nuculoides, H. Sp.
C. circularis, H. Sp.
C. deplanatus, H. Sp.
C. planuisculus, H. Sp.
C. læviusculus, H. Sp.
C. annulifer, H. Sp.
C. impressus, H. Sp
C. lateralis, H. Sp.
C. apiculatus, H. Sp.
C. oblongulus, H Sp.
C. minimus, H. Sp.
C. poæformis, H. Sp.
C. tenue-striatus, H. Sp.

List of Fossil Animals from the Arctic Miocene Formation.

[Flora foss. arct., i., p. 129, 130; and p. 484–5 (Phil. Trans., 1869); and Kong. Sv. Vet. Akad. Handl., xiii., No. 2 (Flora foss. arct., iii.), p. 25.)]

Insecta: Coleoptera.

Trogosita insignis, H. Atanekerdluk.
Chrysomelites Fabricii, H. Atanekerdluk.
Cistelites punctulatus, H. Atanekerdluk and Puilasok.
C. minor, H. Puilasok.
Cercopidium rugulosum, H. Atanekerdluk.

Orthoptera.

Blattidium fragile, H. Atanekerdluk.

Rhynchota.

Pentatoma boreale, H. Atanekerdluk.

Mollusca.

Cyclas, sp. Atanekerdluk.

LIII.—The CRETACEOUS FLORA and FAUNA of GREENLAND.

[From Professor O. HEER's Memoir on the Cretaceous Flora, &c., Kongl. Sv. Vet. Akad. Handl., vol. xii., part vi., pp. 5–7 and 16–18. 1874. *See above*, p. 375; and Prof. Nordenskiöld's Memoir, *further on*.]

1.—FOSSIL PLANTS from the "KOME-FORMATION" (LOWER CRETACEOUS) on the NORTH COAST of the NOURSOAK PENINSULA, including the Localities KOME, PATTORFIK, KARSOK, AVKRUSAK, ANGIARSUIT, and EKKORFAT.

FERNS.

Asplenium Dicksonianum, H.
A. Johnstrupi, H.
A. Nordenskiœldi, H.
A. Boyeanum, H.
Sphenopteris fragilis, H.
S. Johnstrupi, H.
S. grevillioides, H.
Scleropteris bellidula, H.
Adiantum formosum, H.
Aneimidium Schimperi, H.
Baiera arctica, H.
B. grandis, H.
Oleandra arctica, H.
Acrostichites Egedeanus, H.
Pecopteris Andersoniana, H.
P. borealis, Brong.
P. arctica, H.
P. hyperborea, H.
P. Bolbrœana, H.
Gleichenia longipennis, H.
G. Gieseckiana, H.
G. Zippei, H.
G. Thulensis, H.
G. rotula, H.
G. rigida, H.
G. comptoniæfolia, Deb.
G. Nordenskiœldi, H.
G. gracilis, H.
G. acutipennis, H.
G. nervosa, H.
G. delicatula, H.
G. micromera, H.
Dictyophyllum Dicksoni, H.
Danæites firmus, H.
Osmunda petiolata, H.
Jeanpaulia borealis, H.
J. lepida, H.
Sclerophyllina cretosa (Shk.).
S. dichotoma, H.

Gleichenia Nauckhoffii, H., was found, together with *Gl. Zippei* and *Gl. rigida*, in large rolled blocks of quartzose sandstone (brown within and yellowish without), at Ujarasusuk on the north coast of Disco, south of, and nearly opposite to, Atanekerdluk.

[Here also *Caulopteris punctata* (Sternb.) was found under similar conditions; *see* page 388.—EDITOR].

SELAGINES.

Lycopodium redivivum, H.

CALAMARIÆ.

Equisetum amissum, H.
Equisetites Grœnlandicus, H.
E. annularioides, H.

CYCADACEÆ.

Zamites speciosus, H.
Z. arcticus, H.
Z. borealis, H.
Z. acutipennis, H.
Z. brevipennis, H.
Pterophyllum concinnum, H.
P. lepidum, H.
Glossozamites Schenkii, H.
Anomozamites cretaceus, H.

CONIFERÆ.

Torreya Dicksoniana, H.
T. parvifolia, H.
Inopelis imbricata, H.
Thuites Meriani, H.
Frenelopsis Hoheneggeri (Ett.).
Cyparissidium gracile, H.
Glyptostrobus Grœnlandicus, H.
Sequoia ambigua, H.
S. Reichenbachi (Gein.).
Sequoia rigida, H.
S. gracilis, H.
S. Smittiana, H.
Pinus lingulata, H.
P. Peterseni, H.
P. Crameri, H.
P. Eirikiana, H.
P. Olafiana, H.

GLUMACEÆ.

Poacites borealis, H.
Cyperacites hyperboreus, H.
C. arcticus, H.

CORONARIÆ.

Eolirion primigenium, Sch.

SALICINEÆ.

Populus primæva, H.

INCERTÆ SEDIS.

Fasciculites Grœnlandicus, H. (? Eolirion).
Carpolithes Thulensis, H. (Monocotyledon.)

2.—FOSSIL PLANTS from the "ATANE FORMATION" (UPPER CRETACEOUS) on the SOUTH COAST of the NOURSOAK PENINSULA.

Those marked with an asterisk (*) are also found in the Kome Formation.

Asplenium Foersteri, Det.
A. Nordstrœmi, H.
Pecopteris striata, Stb.
* P. arctica, H.
P. Pfaffiana, H.
P. denticulata, H.
P. argutula, H.
P. Bohemica, Cord.
P. Kudlisetiana, H.
* Gleichenia Zippei (Cord.)
G. acutiloba, H.
* G. gracilis, H.
Osmunda Œbergiana, H.
Cycadites Dicksoni, H.
Otozamites Grœnlandicus, H.
Salisburea primordialis, H.
Thuites Pfaffii, H.
Widdringtonites subtilis, H.
* Sequoia Reichenbachi (Gein.).
* S. rigida, H.
S. fastigiata, (Stb.)
S. subulata, H.
Pinus vaginalis, H.
P. Quenstedti, H.
P. Staratschini, H.
Arundo Grœnlandica, H.

Sparganium cretaceum, H.
Zingiberites pulchellus, H.
Populus Berggreni, H.
P. hyperborea, H.
P. stygia, H.
Myrica Thulensis, H.
M. Zenkeri (Ett.).
Ficus protogæa, H.
Sassafras arctica, H.
Proteoides longus, H.
P. crassipes, H.
P. vexans, H.
P. granulatus, H.
Credneria, sp.
Andromeda Parlatorii, H.
Dermatophyllites borealis, H.
Diospyros prodromus, H.
Myrsine borealis, H.
Panax cretacea, H.
Chondrophyllum Nordenskiold, H.
C. arbiculatum, H.
Magnolia Capellinii, H.
M. alternans, H.
Myrtophyllum Geinitzii, H.
Metrosideros peregrinus, H.
Sapindus prodromus, H.
Rhus microphylla, H.
Leguminosites prodromus, H.
L. phaseolites, H.
L. cassiæformis, H.
L. Atanensis, H.
L. coronilloides, H.
L. amissus, H.
Phyllites linguæformis, H.
P. lævigatus, H.
Carpolithes scrobiculatus, H.

3.—NOTE.—CAULOPTERIS PUNCTATA (Sternberg) a CRETACEOUS FOSSIL, according to MR. CARRUTHERS, F.R.S.

At page 7 of his "Beiträge zur Steinkohlen-Flora der "arktischen Zone" (K. Sv. Vet. Akad. Handl., vol. xii., No. iii., 1874), Prof. O. HEER describes several fragments of a fossil Fern-stem, found in rolled sandstone at Ujarasusuk, on the north coast of Disco, where also some rolled blocks of sandstone containing real Carboniferous plants (*Sigillaria*, &c.) have been found. The Fern-stem above mentioned Prof. Heer refers to the same species to which Sternberg's "*Lepidodendron punctatum*" from Kaunitz in Bohemia belongs. This was entered in systematic works as having come from the Coal-measures of that country, but Dr. FRITSCH, of Prague, has assured Mr. CARRUTHERS, F.R.S., of the British Museum, that it was derived from the Cretaceous (Upper Greensand) beds of Bohemia, in which also two allied forms* have been discovered by M. Dormitzen (*see* "Geological Magazine," vol. ii., 1865, p. 485.) Another specimen of this Tree-fern, which Mr. Carruthers refers to Goeppert's *Caulopteris* (as being an older and better generic term than Presl's "Protopteris"), was discovered of late years in the Upper Greensand of Shaftesbury, Wiltshire, and has been described by Mr. Carruthers in the "Geol. Mag.," *loc. cit.*, as *Caulopteris punctata*, with evidence showing that it belongs to the same species as Sternberg's specimen.

With these facts before us, and observing that the Disco specimens of *Caulopteris* (*Protopteris*) *punctata* (Sternb.) figured by Prof. O. Heer (*op. cit.*, pl. 5, figs. 1, 2, & pl. 6), are in their *yellowish-brown* colour decidedly different from the *grey* Carboniferous fossils, Mr. CARRUTHERS has suggested to the EDITOR that

---

* *Alsophilina Kauniciana*, Krejči, and *Oncopteris Nettwalli*, Kr.

these rolled specimens of Tree-ferns from the Disco shore should be registered as "Cretaceous," having been derived from the *Cretaceous* Formation, whether in the immediate neighbourhood or at a distance. Atanekerdluk, with its Cretaceous, underlying Tertiary, strata, is due north on the other side of the Waigat.—EDITOR.

4.—FOSSIL INSECTS from the CRETACEOUS STRATA of GREENLAND. From Prof. O. HEER'S Memoir on the Cretaceous Flora of Greenland, K. Sv. Vet. Ak. Handl., vol. xii., No. 6, p. 91, 92, 120. 1874.

Rhynchophora.

Archiorhynchus angusticollis, Heer. Kome.
Curculionites cretaceus, Heer. Kome, with leaves of *Pinus Crameri.*

Myriopod.

Iulopsis cretaceus, Heer. Lower Atanekerdluk.

LIV.—ACCOUNT of an EXPEDITION to GREENLAND in th Year 1870. By Prof. A. E. NORDENSKIÖLD, Foreign Correspondent Geol. Soc. Lond., &c. &c. &c.

[Reprinted, with Permission, from the "Geological Magazine," vol. ix., 1872.]

PART I. ("Geol. Mag.," vol. ix., p. 289, &c.)

After explaining the reasons for the Expedition, and enumerating the several voyages of discovery to East Greenland, from 1579 to 1869, with some observations on the present state of the Greenlanders, Prof. Nordenskiöld proceeds to state:—

We took up our night-quarter, the 12th of July, at Manermiut, the 13th at Kangaitsiak, the 14th, 15th, and 16th on islands in Auleitsivikfjord. On the 17th we at length arrived at the northern side of the glacier which shoots out from the inland ice, and occupies the bottom of the northern arm of Auleitsivikfjord, that is to say, the spot selected as the starting-point for our journey over the ice.

The tract through which we passed, like the whole west coast of Greenland south of the basalt region, bears a strong resemblance to the Scandinavian peninsula, and that resemblance is not the result of any accident, but of a similar geological formation, and a similar geological history. The surface of Greenland, like that of Scandinavia, is for the most part occupied by stratified crystalline rock (gneiss, hornblende-schist, hornblende-gneiss, mica-schist, etc.), crossed by dykes and veins of granite, which even bear the same peculiar minerals which distinguish the Scandinavian granite-veins; and, as in the case of our mountains, the mountains of these regions have once been covered with glaciers, which have left unmistakeable marks of their presence in the boulders, which are met with scattered high up on the sides of the mountains, in the rounding off, in the polishing and grooving of the surface, and in the deep fjords, evidently scooped out by glaciers, which distinguish the western coasts of both Scandinavia and Greenland. There is, however, this difference, that whereas the glacial period of Scandinavia belongs to an age long past, that of Greenland, though it is receding,* still continues. While, in fact, numberless indications show that the inland-ice has in ancient times covered even the skerries round the coast, these are now so free from ice that a traveller in most places has to advance several miles into the country before reaching the border of the present inland-ice. It is at least certain that wherever any one hitherto has penetrated into the land he has met with its border,† and in all instances has seen it from some neighbouring mountain-top, rising inwards with a gradual and regular ascent, till it levels undistinguishably hill and dale beneath its frozen covering, like the waves of a vast ocean.

Of this inland-ice the natives entertain a superstitious fear, an awe or prejudice, which has, in some degree, communicated itself to such Europeans as have long resided in Greenland. It is thus only that we can explain the circumstance, that in the whole thousand years during which Greenland has been known, so few efforts have been made to pass over the ice farther into the country. There are many reasons for believing that the inland-ice merely forms a continuous ice-frame, running parallel with the coast, and surrounding a land free from ice, perhaps even in its southern parts woody, which might perhaps be of no small economical importance to the rest of Greenland. The only serious attempt that has hitherto been made, in the parts of Greenland colonized by Danes,‡ to advance in that direction was made by—

A Danish expedition, fitted out for the purpose in 1728.—A Danish governor, Major Paars, with an armed company, artillery,

---

* Certainly receding, although the inland ice sometimes makes its way to the sea, and thus tracts which have been free from ice are again covered. We have an example of this in the ice-fjord of Jacobshavn, of which more hereafter.

† I have, however, met with persons in Greenland who do not consider it as fully proved, that the inland-ice really does form an inner border to the whole of the external coast. Many Danes have resided several years in Greenland without ever having seen the inland-ice.

‡ Dr. Hayes's remarkable journey, in October 1860, over the fields of ice that cover the peninsula between Whale Sound and Kennedy Channel (78° N.L.), was performed, not upon the real inland-ice, but upon a smaller ice-field connected with the inland-ice, like the ice-fields at Noursoak peninsula. The character of the ice here seems to have differed considerably from that of the real inland-ice. Hayes ascended the glacier at Port Foulke, on the 23rd of October, and advanced on foot, the first day 5, the second 30, the third 25 miles, in all 60 English miles. He was here forced to return, in consequence of a storm. The height of the spot where he turned back above the level of the sea was 5,000 feet (*The Open Polar Sea*, by Dr. I. I. Hayes, pp. 130–136).

etc., was that year sent from Denmark, to Greenland, and took with him, among other things, also horses, with which it was intended to ride over the mountains, in order to rediscover, by an overland course, the lost (East) Greenland. The horses, however, died, either during the voyage out or shortly after their arrival in the country ; and thus this expedition, really magnificent, but prepared in entire ignorance of the real nature of the country, was abandoned.

Dalager's attempt, 1751.—This year the Danish merchant Dalager made an attempt, in about 62° 31′ latitude, to advance in the beginning of September over the inland ice to the east coast. In the first volume of Kranz's "History of Greenland"* there is a short description of this journey, interesting, among other reasons, as recording an instance of a glacier, which since Greenland has been an inhabited land has forced its way forward and closed the entrance of a previously open fjord. We find further from that account, that Dalager, partly on foot and partly in a canoe, in company with five natives, reached the border of the inland ice near the bottom of a deep fjord situated north of Fredrikshaab. For two days they continued their journey over the ice, but succeeded during this time in advancing only eight English miles to some mountain summits rising above the ice-field, where a reindeer hunt was undertaken. Dalager would willingly have continued the journey a day or two longer, but was unable to do so, partly because the two pairs of boots taken with them for each person were so cut to pieces by the ice that they walked "as good " as barefoot," partly because the cold at night was so severe that their limbs became stiff after a few hours of rest. On the other hand, the route chosen by Dalager seems not to have been interrupted by very many or deep chasms—in the beginning of the journey the surface of the ice was even "as smooth as a street in " Copenhagen." Further on, however, it was extremely rough.

E. Whymper's expedition, 1867.—All that I know about this expedition is, that Mr. Whymper, in company with Dr. R. Brown, three Danes, and a Greenlander, endeavoured to make their way upon the inland ice with dogs immediately to the north of the ice-fjord at Jacobshavn, but that they turned back again on the second day, after having proceeded only some few miles. The reason of this was probably the unfitness of dogs for such a purpose.

It was originally my intention to renew these attempts, but on conversing in Copenhagen with Messrs. Rink and Olrik, who had formerly been Inspectors in North Greenland, as also with several other persons who had visited Greenland, I found all so unanimous in considering further advance over the inland ice as impossible, that I determined not to risk the whole profit of the summer on an

* I have not had access to Dalager's original account. "Grönlandske Relationer, indehaaldende Grönländernes Liv og Levnet, deres Skicke og Vedtägter, samt Temperament og Superstitioner, tillige nogle korrte Reflexioner over Missionen, sammenskrevet ved Fredrickshaabs Colonia i Grönland," by Lars Dalager, Merchant.

undertaking of the kind beforehand disapproved of by everybody. Nevertheless, I was unwilling entirely to abandon my plan, and determined therefore to make a little attempt at a journey on the inland ice only of a few days' extent.

If the inland ice were not in motion, it is clear that its surface would be as even and unbroken as that of a land field. But this, as is known, is not the case. The inland ice is in constant motion, advancing slowly, but with different velocity in different places, towards the sea, into which it passes on the west coast of Greenland by eight or ten large and a great many small ice-streams. This movement of the ice gives rise in its turn to huge chasms and clefts, the almost bottomless depths of which stop the traveller's way. It is natural that these clefts should occur chiefly where the movement of the ice is most rapid, that is to say, in the neighbourhood of the great ice-streams, and that, on the other hand, at a greater distance from these the ground should be found more free from cracks. On this account I determined to begin our wanderings on the ice at a point as far distant as possible from the real ice-fjords. I should have preferred one of the deep "Strömfjords" (stream-fjords) for this purpose, but as other business intended to be carried out during the short summer did not permit a journey by boat so far southward, I selected instead for my object the northern arm of the above-mentioned Auleitsivikfjord, which is situated 60 miles south of the ice-fjord at Jakobshavn, and 240 miles north of that of Godthaab. The inland ice, it is true, even in Auleitsivikfjord reaches to the bottom of the fjord, but it only forms there a perpendicular glacier, very similar to the glaciers at King's Bay in Spitzbergen, but not any real ice-stream. There was accordingly reason to expect that such fissures and chasms as might here occur would be on a smaller scale.

On the 17th July, in the afternoon, our tent was pitched on the shore north of the steep precipitous edge of the inland ice at Auleitsivikfjord. After having employed the 18th in preparations and a few slight reconnoitrings, we entered on our journey inwards on the 19th. We set out early in the morning, and first rowed to a little bay situated in the neighbourhood of the spot occupied by our tent, into which several muddy rivers had their embouchures. Here the land assumed a character varied by hill and dale; and further inward it was bounded by an ice-wall sometimes perpendicular and sometimes rounded, covered with a thin layer of earth and stones; near the edge, only a couple of hundred feet high, but then rising at first rapidly, afterwards more slowly, to a height of several hundred feet. In most places this wall could not possibly be scaled; we however soon succeeded in finding a place where it was cut through by a small cleft, sufficiently deep to afford a possibility of climbing up with the means at our disposal, a sledge, which at need might be used as a ladder, and a line originally 100 fathoms long, but which, proving too heavy a burden, had before our arrival at the first resting-place been reduced one-half. All of us, with the exception of our old and lame boatman, assisted in the by no means easy work of bringing over mountain, hill, and dale, the apparatus of the ice expedition to this spot, and, after our dinner's rest, a little further up the ice-wall. Here our followers left us. Only Dr. Berggren, myself, and two Greenlanders (Isak and Sisarniak) were to proceed farther. We immediately commenced our march, but did not get very far that day.

The inland ice differs from ordinary glaciers by, among other things, the almost total absence of moraine-formations. The collections of earth, gravel, and stone, with which the ice on the landward edge is covered, are in fact so inconsiderable in comparison with the moraines of even very small glaciers, that they scarcely deserve mention, and no larger, newly formed ridges of gravel running parallel with the edge of the glacier are to be met with, at least in the tract visited by us. The landward border of the inward ice is, however, darkened, we can scarcely say covered, with earth, and sprinkled with small sharp stones.

Here the ice is tolerably smooth, though furrowed by deep clefts at right angles to the border—such as that made use of by us to climb up. But in order not immediately to terrify the Greenlanders by choosing the way over the frightful and dangerous clefts, we determined to abandon this comparatively smooth ground, and at first take a southerly direction parallel with the chasms and afterwards turn to the East. We gained our object by avoiding the chasm, but fell in instead with extremely rough ice. We now understood what the Greenlanders meant, when they endeavoured to dissuade us from the journey on the ice, by sometimes lifting their hands up over their heads, sometimes sinking them down to the ground, accompanied by to us an unintelligible talk. They meant by this to describe the collection of closely heaped pyramids and ridges of ice over which we had now to walk. The inequalities of the ice were, it is true, seldom more than 40 feet high, with an inclination of 25 to 30 degrees; but one does not get on very fast, when he has continually to drag a heavily laden sledge up so irregular an acclivity, and immediately after to endeavour to get down uninjured, at the risk of getting broken legs, when occasionally losing one's footing on the here often very slippery ice in attempting to moderate the speed of the downward rushing sledge. Had we used an ordinary sledge, it would immediately have been broken to pieces; but, as the component parts of our sledge were not nailed but tied together, it held together at least for some hours.

Already the next day we perceived the impossibility under such circumstances of dragging with us the 30 days' provision with which we had furnished ourselves, especially as it was evident that, if we wished to proceed further, we must transform ourselves from draught- to pack-horses. We therefore determined to leave the sledge and part of the provisions, take the rest on our shoulders, and proceed on foot. We now got on quicker, though for a sufficiently long time over ground as bad as before. The ice became gradually smoother, but was broken by large bottomless chasms, which one must either jump over with a heavy load on the back (in which case woe to him who made a false step), or else make a long circuit to avoid. After two hours' wandering, the region of clefts was passed. In the course of our journey, however, we very frequently met with portions of similar ground, though none of any very great extent. We were now at a height of more than 800 feet above the level of the sea. Farther inward the surface of the ice, except at the occasionally securring regions of clefts, resembled that of a stormy sea suddenly bound in fetters by the cold. The rise inwards was still quite perceptible, though frequently interrupted by shallow valleys, the centres of which were occupied by several lakes or ponds with no apparent outlet, although they received water from innumerable rivers running along the sides of the hollow. These rivers presented in many places not so dangerous, though quite as time-wasting, a hindrance to our progress as the clefts; they did not occur so often, but the circuits to avoid them were much longer.

During the whole of our journey on the ice we enjoyed fine weather, frequently there was not a single cloud visible in the sky. The warmth was to us, clad as we were, quite sensible; in the shade, near the ice of course, but little over zero; higher up, in the shade, as much as 7° or 8°; but in the sun 25° to 30° Centig. After sunset the water-pools froze, and the nights were very cold. We had no tent with us, and, although our party consisted of four men, only two ordinary sleeping sacks. These were open at both ends, so that two persons could, though with great difficulty, with their feet opposite to each other, squeeze themselves into one sack. With rough ice for a substratum, the bed was so uncomfortable that, after a few hours' sleep, we were awakened by cramp; and as there was only a thin tarpaulin between the ice and the sleeping sack, the bed was extremely cold to the side resting on the ice, which the Greenlanders, who turned back before us, described to Dr. Nordström by shivering and shaking throughout their whole bodies. Our nights' rests were, therefore, seldom long; but our midday rests, during which we could bask in a glorious warm sun-bath, were taken on a proportionately larger scale, whereby I was enabled to take observations both for altitude and longitude.

On the surface of the inland ice we do not meet with any stones at a distance of more than a cable's length from the border; but we find everywhere, instead, vertical cylindrical holes, of a foot or two deep, and from a couple of lines to a couple of feet in section, so close one to another that one might in vain seek between them room for one's foot, much less for a sleeping-sack. We had always a system of ice-pipes of this kind as substratum when we rested for the night, and it often happened, in the morning, that the warmth of our bodies had melted so much of the ice, that the sleeping sack touched the water, wherewith the holes were always nearly full. But, as a compensation, wherever we rested, we had only to stretch out our hands to obtain the very finest water to drink.

PART II. ("Geol. Mag.," vol. ix., p. 355, &c.)

These holes in the ice, filled with water, are in no way connected with each other, and at the bottom of them we found everywhere, not only near the border, but in the most distant parts of the inland ice visited by us, a layer, some few millimètres thick, of grey powder, often conglomerated into small round balls of loose consistency. Under the microscope, the principal substance of this remarkable powder appeared to consist of white angular transparent grains. We could also observe remains of vegetable fragments; yellow, imperfectly translucent particles, with, as it appeared, evident surfaces of cleavage (felspar ?); green crystals (augite); and black opaque grains, which were attracted by the magnet. The quantity of these foreign components is, however, so inconsiderable, that the whole mass may be looked upon as one homogenous substance. An analysis by Mr. G. Lindström of this fine glacial sand gave—

Silicic acid - - - -	62·25
Alumina - - - -	14·93
Sesquioxyd of Iron - - -	0·74
Protoxyd of Iron - - -	4·64
Protoxyd of Manganese - -	0·07
Lime - - - -	5·09
Magnesia - - - -	3·00
Potassa - - - -	2·02
Soda - - - - -	4·01
Phosphoric acid - - -	0·11
Chlorine - - - -	0·06
Water, organic substance (100° to red heat) - - - - -	2·86
Hygroscopic water (15° to 100°) -	0·34
	100·12

After long digestion with sulphuric acid only 7·73, and with muriatic acid 16·46 per cent. was dissolved. The remainder was entirely white after heating to redness. The analysis gives the atomic relation—

$$2\,\dot{R} + \dddot{Al} + 7\,\ddot{Si} + \dot{H}$$

or the formula—

$$2\,\dot{R}\,Si^2 + \dddot{Al}\,Si^3 + \dot{H}$$

Specific gravity = 2·63 (21°). Hardness inconsiderable, crystallization probably monoclinic.

The substance is not a clay, but a sandy trachytic mineral, of a composition (especially as regards soda) which indicates that it does not originate in the granite-region of Greenland. Its origin appears therefore to me very enigmatical. Does it come from the basalt-region? or from the supposed volcanic tracts in the interior of Greenland? or is it of meteoric origin? The octahedrally crystallised magnetic particles do not contain any traces of nickel. As the principal ingredient corresponds to a determinate chemical formula, it would perhaps be desirable to enter it under a separate class in the register of science; and for that purpose I propose for this substance the name Kryokonite (from κρύος and κόνις).

When I persuaded our botanist, Mr. Berggren, to accompany me in the journey over the ice, we joked with him on the singularity of a botanist making an excursion into a tract, perhaps the only one in the world, that was a perfect desert as regards botany. This expectation was, however, not confirmed. Dr. Berggren's quick eye soon discovered, partly on the surface of the ice, partly in the above-mentioned powder, a brown polycellular Alga, which, little as it is, together with the powder and certain other microscopic organisms by which it is accompanied, is the most dangerous enemy to the mass of ice, so many thousand feet in height and hundred miles in extent. The dark mass absorbs a far greater amount of the sun's rays of heat than the white ice, and thus produces over its whole surface deep holes which greatly promote the process of melting. The same Plant has no doubt played the same part in our country; and we have to thank it, perhaps, that the deserts of ice which formerly covered the whole of northern Europe and America have now given place to shady woods and undulating corn-fields. Of course, a great deal of the grey powder is carried down in the rivers, and the blue ice at the bottom of them is not unfrequently concealed by a dark dust. How rich this mass is in organic matter is proved by the circumstance, amongst others, that the quantity of organic matter in it was sufficient to bring a large collection of the grey powder, which had been carried away to a distant part of the ice by sundry now dried-up glacier-streams, into so strong a process of fermentation or putrefaction, that the mass, even at a great distance, emitted a most disagreeable smell, like that of butyric acid.

Dr. BERGGREN has communicated the following notice* of the Microscopic Organisms met with on the Inland Ice.

"One of the species of Algæ met with on the inland ice occurred in such vast quantities, that the surface of the ice throughout larger or smaller tracts was tinted with a peculiar colour. Two others seemed exclusively to belong to the fine sand, which is found either in the form of a thin covering on the surface of the ice, or as a more or less thick layer at the bottom of the pipe-like holes that appear in the surface. The first-mentioned species, occurring copiously, does not require any such substratum, but is found principally on the sides of ice-hills, where the water from the melting ice filtered itself out between the little inequalities of the surface.

"The most copiously represented species has the form of a short thread, not spreading out in branches, but consisting of a single row of cells; the number of cells in each thread is 2, 4, 8, or at most 16. Threads of 4 and 8 cells are most common. The species very frequently appears only as a single cell. The threads are usually a little bent, sometimes, when the number of cells is 16, forming a complete semicircle. The number 2 or its multiples

* A more detailed account, accompanied by drawings, of these remarkable Algæ will hereafter be published in the "K. Vet. Akademiens Öfversigt."

taken as the standard for the number of cells in the separate threads is accounted for by the regular continuous bisection of the cells, whereby their propagation proceeds. The connexion between the cells is the looser the older the partitions become, as the older membranes assume a looser consistence. In a thread of 16 cells, the connexion between the eighth and ninth cells is soon broken, and in the two threads thus resulting the connexion between the fourth and fifth cells is weaker than that between the second and third or the sixth and seventh. The threads therefore often lie bent at an angle. The diameter of the cells is 0·008 — 0·012 mm., and their length 0·016 — 0·040 mm. Individual cells may sometimes attain a length 0·055 mm. and a breadth of 0·015 mm., whereas a great number of other single cells are met with of very small dimensions, from spherical forms of only 0·006 mm. diameter to those of ordinary form and size. As the ends of the cells, where they are joined together, are rounded, there is, of course, a contraction between them, which becomes more and more conspicuous as the connexion between them is loosened by time. The membrane is thin and hyaline, and its outermost layer (the remnants of the membranes of the mother-cells altered after division) is of an almost slimy consistence, whereby the cells are for some time kept together. The contents of the cells are in part concealed by a dark purple-brown colouring-matter, which in dried cells is immediately drawn out on wetting them. The centre of the cells is occupied by an oblong or cylindrical mass of chlorophyll, of somewhat irregular contour, in the extremity of which two nuclear rounded bodies are imbedded, which in general cannot be perceived by the eye till the colouring-matter has been removed by means of reagents. We sometimes meet with four such bodies in a cell, sometimes only one: the former a result of accidentally checked division of the cells; the latter of such division having lately taken place. In the liquid of the cells a number of small grains are found, which are for the most part collected round the periphery of the cell or at its ends.

"Judging from the construction of the cells, and the manner of their multiplication, the Alga before us appears to belong to the *Conjugatæ;* but as I have not succeeded in discovering fructification in it, it would be rash to decide to which genus it is to be referred. The thread-like rows of connected cells agree with the *Zygnemaceæ;* whereas, on the other hand, an unmistakable likeness to the *Desmidiaceæ*, especially *Cylindrocystis*, and the nearly related genera, is indicated by the strongly marked divisions into multiples of two, and by the tendency of the rows of cells to fall asunder, as far as the destructibility of the uniting cell-membranes permits, into parts consisting of cells united in pairs, which however is seldom possible, in consequence of the greater energy possessed by the power of multiplying the cells. As the above-mentioned small single cells, which occur in great numbers, are much less in diameter than those cells which arise from the bi-section of the threads, they have perhaps a different origin from these latter, although the researches which I have hitherto been enabled to devote to this subject have not furnished any illustration of it.

If these daughter-cells originate in the division of the spores, the above-mentioned supposition with respect to the systematic place of the species being correct, the stage of conjugation and spores, in some period of its development, ought to be found. Two rare forms of peculiarly constructed cells ought not to be passed unnoticed. I have sometimes found the extreme cell in a thread considerably more swelled than the others, more elliptic in form, also provided with a thicker membrane, and with the contents of the cell more coarse-grained. I once found one of the middle cells in a thread thus transformed, and on two occasions I have met with single cells of the same kind. I also once met with a cell of very peculiar construction. It had the usual form, but was unusually large, with a long mass of chlorophyll, as usual, in the midst, and the granular matter grouped rather towards the ends of the cell; and in it there were about twenty larger or smaller spherical bodies. Four of these lay arranged at each end of the cell, and were almost entirely opaque, of a dark-brown colour, and in appearance much resembled the smaller cells of *Protococcus nivalis*. The others were translucent, with sharply defined contours. As our knowledge of the nature of these bodies is confined to what is here stated, the fuller explanation of their significance must be reserved for future investigation.

"In places similar to those in which this species occurs, and often in company with it, *Protococcus nivalis* was met with.

"Amidst the fine gravel upon the ice, but to a trifling amount, there are small green cells, sometimes united in little groups, sometimes isolated, which appeared to belong to *Protococcus vulgaris*.

"*Scytomena gracile*, on the other hand, is everywhere met with in great profusion, wherever the gravel either lies in thinly scattered grains on the surface of the ice, or forms more or less thick layers. The threads lie either alone, or united in small bunches, as they join together at the lower part, and bend backwards higher up. They are neither stiff, S-shaped, or forming a curve of several undulations, and yellowish-brown in colour. Their length is very various; their breadth generally about 0·009 mm."

At our mid-day rest on the 21st we had reached latitude 68° 21′ and 36′ longitude east of the place where our tent was pitched, and a height of 1,400 feet above the level of the sea.

Later in the day, at our afternoon rest, the Greenlanders began to take off their shoes and examine their little thin feet—a serious indication, as we soon perceived. Isak presently informed us, in broken Danish, that he and his companion now considered it time to return. All attempts to persuade them to accompany us a little farther failed; and we had, therefore, no other alternative than to let them return, and continue our excursion without them.

We took up our night quarters here. The provisions were divided. The Greenlanders, considering they might, perhaps, not be able to find our first depôt, were allowed to take as much as was necessary to enable them to reach the tent. We took cold provisions for five days. The remainder, together with the excellent photogen portable kitchen, which we had hitherto carried with us, were laid up in a depôt in the neighbourhood, on which a piece of tarpaulin was stretched upon sticks, that we might be able to find the place on our return; which, however, we did not succeed in doing, though we must have passed in its immediate vicinity.

Dr. Berggren and I then proceeded farther inward. The Greenlanders turned back.

At first we passed one of the before-mentioned extensive bowl-shaped excavations in the ice-plain, which is here furrowed by innumerable rivers, often obliging us to make long circuits; and, when to avoid this, we endeavoured to make our way along the margin of the valleys, we came, instead, upon a tract where the ice-plain was cloven by long, deep, parallel clefts running true N.N.E.—S.S.W., quite as difficult to get over as the rivers, and far more dangerous. Our progress was accordingly but slow. At twelve o'clock on the 22nd we halted, in glorious, warm, sunny weather, to make a geographical determination. We were now at a height of 2,000 feet, in latitude 68° 22′, and in a longitude of 57′ *of arc* east of the position of our tent at the fjord.

During the whole of our excursion on the ice we had seen no animals except a couple of Ravens, which on the morning of the the 22nd, at the moment of our separation, flew over our heads. At first, however, there appeared in many places on the ice traces of Ptarmigans, which seemed to indicate that these birds visit these desert tracts in by no means inconsiderable flocks. Everything else around us was lifeless. Nevertheless, silence by no means reigned here. On bending down the ear to the ice, we could hear on every side a peculiar subterranean hum, proceeding from rivers flowing within the ice; and occasionally a loud single report like that of a cannon gave notice of the formation of a new glacier-cleft.

After taking the observations, we proceeded over comparatively better ground. Later in the afternoon we saw, at some distance from us, a well-defined pillar of mist, which, when we approached it, appeared to rise from a bottomless abyss, into which a mighty glacier-river fell. The vast roaring water-mass had bored for itself a vertical hole, probably down to the rock, certainly more than two thousand feet beneath, on which the glacier rested.

The following day (the 23rd) we rested in latitude 68° 22′ and 76′ of arc longitude east from the position of our starting point at Auleitsivik.

The provisions we had taken with us were, however, now so far exhausted, that we were obliged to think of returning. We determined, nevertheless, first to endeavour to reach an ice-hill visible on the plain to the east, from which we hoped to obtain an extensive view; and, in order to arrive there as quickly as possible, we left the scanty remains of our provisions and our sleeping sack at the spot where we had passed the night, taking careful notice of the ice-rocks around; and thus we proceeded by forced march, without encumbrances.

The ice-hill was considerably further off than we had supposed

The walk to it was richly rewarded by an uncommonly extensive view, which showed us that the inland ice continued to rise towards the interior, so that the horizon towards the east, north, and south was terminated by an ice-border almost as smooth as that of the ocean. A journey further (even if one were in a condition to employ weeks for the purpose—which want of time and provisions rendered impossible to us) could therefore evidently furnish no other information concerning the nature of the ice than that which we had already obtained; and even if want of provisions had not obliged us to return, we should hardly have considered it worth while to add a few days' marches to our journey. Our turning-point was at a height of 2,200 feet above the level of the sea, and about 83′ of longitude, or 30 miles west of the extremity of the northern arm of Auleitsivikfjord.

On departing from the spot where we had left our provisions and sleeping sack, we had, as we supposed, taken careful notice of its situation; nevertheless, we were nearly obliged to abandon our search as vain—an example which shows how extremely difficult it is, without lofty signals, to find objects again on a slightly undulating surface everywhere similar, like that formed by the inland ice. When, after anxiously searching in every direction, we at length found our resting place, we ate our dinner with an excellent appetite, made some further reductions in our load, and then set off with all haste back to the boat, which we reached late in the evening of the 25th.

At a short distance from our turning-point, we came to a copious, deep, and broad river, flowing rapidly between its blue banks of ice, which were here not discoloured by any gravel, and which could not be crossed without a bridge. As it cut off our return, we were at first somewhat disconcerted; but we soon concluded that—as in our journey out we had not passed any stream of such large dimensions—it must at no great distance disappear under the ice. We therefore proceeded along its bank in the direction of the current, and before long a distant roar indicated that our conjecture was right. The whole immense mass of water here rushed down a perpendicular cleft into the depths below. We observed another smaller, but nevertheless very remarkable, waterfall the next day, while examining, after our mid-day rest, the neighbourhood around us with the telescope. We saw in fact a pillar of steam rising from the ice at some distance from our resting-place, and, as the spot was not far out of our way, we steered our course by it, in the hope of meeting—judging from the height of the misty pillar—a waterfall still greater than that just described. We were mistaken: only a smaller, though nevertheless tolerably copious, river rushed down from the azure-blue cliffs to a depth from which no splashes rebounded to the mouth of the fall; but there arose instead, from another smaller hole in the ice, in the immediate vicinity, an intermittent jet of water, mixed with air, which carried hither and thither by the wind, wetted the surrounding ice-cliffs with its spray. We had then here, in the midst of the desert of inland ice, a fountain, as far as we could judge from the descriptions, very like the geysers which in Iceland are produced by volcanic heat.

In order, if possible, to avoid the district of ice-rocks, which on our journey out had required so much patience and exertion, we had in returning chosen a more northerly route, intending to endeavour to descend from the ice-ridge higher up on the slip of ice-free land, which lies between the inland ice and Disko Bay. The ice was here, with the exception of a few ice-hillocks of a few few feet high, in most places as even as a floor, but often crossed by very large and dangerous clefts, and we were so fortunate as immediately to hit upon a place where the inclination towards the land was so inconsiderable that one might have driven up it four-in-hand.

The remainder of the way along the land was harder, partly on account of the very uneven nature of the ground, and partly on account of the numerous glacier-streams which we had to wade through, with the water far above our boots. At last, at a little distance from the tent, we came to a muddy glacier-stream, so large that, after several failures, we were obliged to abandon the hope of finding a fordable place. We were, therefore, obliged to climb high up again upon the shining ice, so as to be able to find our way down again further on, after passing the river; but the descent on this occasion was far more difficult than before.

Laborious as this journey along the land was, it was, nevertheless, extremely interesting to me in a geological point of view. We passed in fact over ground that had but lately been abandoned by the inland ice, and the whole bore such a resemblance to the woodless gneiss-districts in Sweden and Finland, that even the most sceptical persons would be obliged to admit that the same formative power had impressed its stamp on both localities. Everywhere rounded, but seldom scratched, hills of gneiss* with erratic blocks in the most unstable positions of equilibrium, occur, separated by valleys with small mountain-lakes and scratched rock-surfaces. On the other hand, no real moraines were discoverable. These, indeed, seem to be in general absent in Scandinavia, and are, generally speaking, more characteristic of small glaciers than of real inland ice.

The border of the ice is, as indicated in Figs. 1 and 2, p. 402, everywhere sprinkled with smaller boulders, partly rounded, partly angular; but the number of these is so inconsiderable, that, when the ice retires, they give rise only to a slope covered with boulders; not to a moraine, similar, for example, to that which the little Assakak glacier in Omenakfjord drives before it. The little earth-bank, which at most places collects at the foot of the glacier,

* For the preservation of a scratched rock-surface it is necessary that it should be protected by a layer of water, clay, or sand, from the destructive effects of frost, and more especially from those of Lichens. The finest scratches disappear in a few years from a mountain slab, the position of which is favourable to the growth of Lichens, but are, on the contrary, preserved where Lichen-vegetation cannot develop itself—as, for example, when the rock is, for a time in the spring, covered with water.

is frequently washed away again by the glacier-streams and rain. We often find at the foot of the glacier, as in Fig. 2, ponds or lakes in which a freshwater glacial clay, containing angular blocks of stone, scattered around by small icebergs, is deposited.

Figs. 1, 2, and 3. Inland Ice abutting on Land.
A. Inland Ice; B. Solid Rock; C. Small collection of earth at the foot of the Glacier; D. Lake; E. Separate blocks of Ice.

(The Woodcuts illustrating this Memoir have been kindly lent by the Publishers of the *Geol. Mag.*)

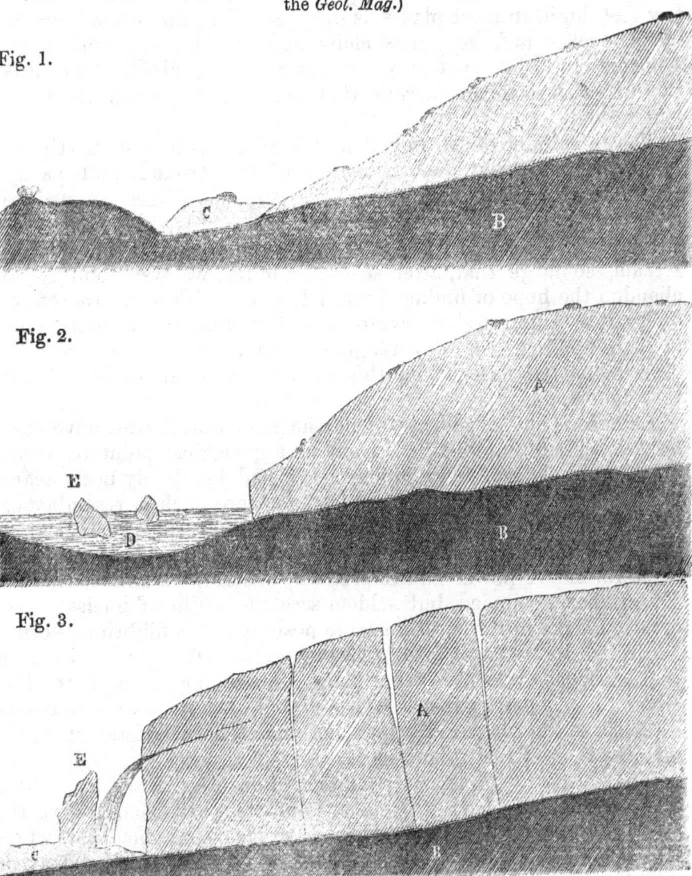

It is a common error among geologists to consider the Swiss glaciers as representing on a small scale the inland ice of Greenland, or the inland ice which once covered Scandinavia.* The real glacier bears the same relation to inland ice which a rapid river or brook does to an extensive and calm lake. While the glacier is in perpetual motion, the inland ice, like the water of a

* Switzerland was probably never quite covered with real inland ice; its glaciers have, however, been considerably more extensive than they now are.

lake, is comparatively at rest, excepting in those places where it streams out into the sea by vast but short glaciers. If one of these glaciers, through which the ice-lake falls out into the sea, pass over smooth ground where the ocean's bottom gradually changes into land without any steep breaks, steep precipitous

Fig. 4.

Fig. 4. Inland Ice (A) extending into the Sea (D) and terminating in a steep edge, 100 to 200 feet high.

glaciers are produced, from which indeed large ice-masses fall down, but do not give rise to any real iceberg. But if the mouth of the fjord be narrow, the depth of the outlying sea great, and the inclination of the shore considerable, the result will be one of those magnificent ice-fjords which Rink so admirably describes, and which we, later in the course of our journey, had an opportunity of visiting. The following diagram will illustrate this more clearly.

1,500 ft.

Fig. 5.

Fig. 5. Inland Ice abutting on the bottom of an ice-fjord, *i.e.*, a fjord in which real icebergs are formed.

True icebergs are formed only in those glaciers which terminate in the manner indicated in Fig. 5; though pieces of ice of considerable dimensions may fall from a steep precipice (Fig. 4). These various kinds of glaciers occur not only in Greenland, but also in other ice-covered polar lands, *e.g.* in Spitzbergen, though on so much smaller a scale than in Greenland, that one never meets in the surrounding waters with icebergs at all comparable in magnitude with those of Davis Strait.

In Spitzbergen, and probably also in some parts of Greenland, the ice passes into the sea in the following manner.

Fig. 6. Inland Ice abutting on a mud-bank.

As I have already remarked in the account of the geological relations of Spitzbergen,* this last-mentioned kind of termination of inland ice towards the sea is met with only either in those places where the limits of the inland ice rapidly recede, or where the ice breaks for itself a new channel or way to the sea. This is, for example, the case with Axels Glacier in Bell Sound, which, when I first visited the spot in 1858, had an edge like that indicated in Fig. 6, but which a couple of years latter filled the whole of the harbour lying before it, and is now terminated in the manner shown in Fig. 5.

The great denuding effect of the glaciers has been, as is known, proved by numerous and accurate investigations. Greenland also offers examples of this in the long and deep fjords that indent its coasts, and which, if they run parallel to ante-glacial depressions of the earth's crust, yet, as the smoothed, scratched, and grooved rocks, and the erratic blocks strewn high up upon the slopes show, have been widened, formed, and cleansed from earth, gravel-beds, and looser sedimentary mountain-detritus by the operation of the glaciers. The mere effect of the immovable inland ice cannot be anything like so great. Nevertheless, here also the earth and the layers of gravel are completely washed away by the rapid glacier-streams running under the ice. The subjacent original rock is thus exposed, and perhaps to some extent worn away, especially in places where the ice passes over layers of limestone, sandstone, or slate. Its original depressions, filled during the older geological periods, therefore re-appear, and often form—when the ice-covering has again retired—the basins of those beautiful lakes which characterize all glacial lands. To assume that the whole lake-basin

* (?)

has been scooped out during the Glacial Period is, however, evidently a mistake; and equally erroneous is the form in which it is customary to clothe the theory of the origin of Alpine lakes. But when we take into consideration how rapidly (even within historical periods) a lake is filled and converted, first into a morass, and then to a level and dry plain, we easily see the reasonableness of the following proposition:

We meet with lakes only in those places where, from some cause or other, during the latest geological periods, depressions or excavations have taken place in the crust of the earth; and since, among more generally operating causes than this, we know only of the volcanic and glacial powers, it is natural to conclude that modern (not filled up) lake-basins only occur where the strata, in consequence of volcanic activity, have fallen in, or where the ice has ground to powder, and the glacier-streams have swept away, the looser earth and rocks situated nearest to the surface of the earth.

On observing Tessiursarsoak from the heights nearest to the spot where we had first descended from the glacier, we had perceived that its appearance had changed in a remarkable manner; its surface was bright as a looking-glass, and so thickly covered with ice that our first impression was that we had an arm of the inland ice before us. On arriving at the tent we discovered the cause of this. During our absence the inland ice had launched or deposited ice in such quantities that the whole bay was almost choked with it, and the Greenlanders were very uneasy, for fear partly of our being inclosed, and partly of the violent waves caused by the deposition. They were therefore very glad when, immediately on our arrival at the boat, we declared our readiness to start on the following day.

In order to be in time to meet the Inspector—who just at this time was expected to visit the colonies around Disko Bay in a commodious yacht, whence he was to sail through the Waigat up to Upernivik, and who had offered us a place on board as far as our routes were the same—we had agreed with several kayak men from Ikamiut and the surrounding districts, that on an appointed day they were to meet us at the Tessiursarsoak. Our intention was to have the whale-boat dragged over the low neck of land which at Sarpiursak separates the innermost part of the north arm of Auleitsivikfjord from Disko Bay, and thus entirely to avoid the long circuit round Kangeitsiak. At the appointed time we saw a whole flotilla of these small, elegant, and light kayaks approaching our tent. We immediately started, and, as soon as the necessary dram of welcome had been distributed to the canoe-men, rowed over to the other side, where Dr. Öberg, with the crew of the zoological boat and a number of other men, awaited us. We were now a large body of men, but Greenlanders are neither strong nor inclined to unusual exertions, and we were obliged to let our people row the whale-boat all the way round, while with our effects we passed directly over to Sarpiursak, where two other whale-boats lay at our disposal.

According to Dr. Rink, the interior of the fjord we had just left had never before been visited by Europeans, and even natives only visit it in summer to hunt and fish, usually in an "umiak," which is carried over the neck of land. It is seldom that they row from the mouth to the end of the fjord. They are afraid of the violent currents which the tide produces in the long narrow estuary, and which, as the Greenlanders several times, with horror in their countenances, informed us, when we wished to take advantage of the favourable but violent current to get on faster, had once swallowed up two "umiaks," with all the men, women, and children on board. There must now, however, be but very little to be got by hunting there; at, least, during the whole of our journey we saw no Reindeer. But there are persons still living who remember the time when thousands of Reindeer were killed in these parts for the sake of the skins only. This abundance of game enticed a few families to settle there also during winter, and one meets in several places traces of old houses. The shores of the fjord are occupied by gneiss hills separated from each other by valleys, in which Grass and Lichen grow plentifully, thus affording copious pasture for such Reindeer as may occasionally stray thither. This is an event which has now become rare; but many maintain that the good times may return, for, according to their account, the Reindeer make periodical migrations, sometimes appearing at a particular place in vast numbers, and then suddenly disappearing, and there are many persons who connect this account with that of an inland tract free from ice, or even with the story of wild inhabitants with European features in the interior of the country. To us the visit to this fjord was of interest, partly because we hoped thus to become acquainted with the true, unmixed Greenlander scarcely in contact with civilization, and partly for botanical reasons. We hoped in fact here, far from the moist fogs of the ocean, to find a far richer vegetation than on the outer coast. A very small tree was said to have been transplanted hence to the clergyman's garden at Egedesminde. This anticipation of the botanist was, however, not confirmed, at least not to the amount expected. The flora was indeed richer and the Willow-bush larger than at Egedesminde, but not so rich nor so large as in the more northerly but fertile basalt-region of Disko, which is traversed by subterranean streams of warm water. The Insect fauna, on the other hand, appears to be somewhat richer here than on the coast; at least we collected the best harvest of Insects that we had during the whole summer on the 17th of July, on a little island in Tessiursarsoak, and the time we spent at the foot of the inland ice was, although in other respects extremely pleasant, embittered to a degree—of which those who have not experienced it cannot form an idea—by countless swarms of Gnats. The Greenland Gnat is like ours, but its bite is far more venomous, though at first not particularly painful. One is therefore usually too incautious at first, and exposes oneself to twenty or thirty gnat-bites in the face at once. A few hours later one's face becomes unrecognizable with the lumps and swellings caused by the bites, and this is followed by pain and fever, especially at night, which hinder sleep, and are almost enough to drive one mad.

The inland ice, in former times, evidently covered the whole of

Auleitsivikfjord, together with the surrounding valleys, mountains, and hills. The ice has accordingly, during the last thousand or hundred thousand years, considerably retired. Now, on the contrary, its limit in these parts is advancing, and that by no means slowly. Of late years the rowing of an "umiak" in Tessiursarsoak has been rendered difficult by ice-blocks fallen from the glaciers, which is said not to have been the case formerly; and one of our rowers, Henry Sissarniak, even affirms that he rowed without obstruction seven years ago round an island, which now forms a peninsula jutting out from the margin of the inland ice. Many similar examples in North Greenland are adduced: thus, for example, the glacier that issues into Bläsedal, near Godhavn, has, since the time when Dr. Rink mapped that place, according to the statement of Inspector Smith, advanced much farther into the valley,—in the fjords around Omenak the ice has advanced considerably within the memory of man,—a path formerly often frequented between Sarfarfik and Sakkak * is now closed by inland ice, etc., etc. I shall have occasion hereafter to mention a similar case in the ice-fjord at Jakobshavn. In a word, there can be no doubt that in many parts of North Greenland the inland ice is certainly gaining ground; but I nevertheless think that the conclusion drawn by many persons, that the whole coast of North Greenland will, at no very distant period, be again covered with ice, is somewhat too hastily made. These persons, in observing the phenomena relative to this subject, not only seem to have forgotten to register the examples occasionally adduced by the Greenlanders of a retiring of the ice —a less striking and therefore less observed phenomenon, but they have also attributed far too great weight to an experience extending only over a few years, which may perhaps have been peculiarly unfavourable. On the contrary, the extensive, rounded, polished, and grooved border of land, which almost everywhere separates the inland ice from the extreme coast, shows plainly that the inland ice has in many places during the last geological period retired several miles. That this border-land has been uncovered later even than that at Spitzbergen is proved by this fact among others, viz., that not one of the numberless small sea-basins in North Greenland, in spite of the suitableness of the locality for moss-vegetation, has yet become filled with turf, even to the depth of a few feet; and this indicates that the slip of ice-free land is but a child of yesterday. It is true that "turf" is the Greenlander's principal winter fuel, but what he means by that name is, in almost all instances, merely an earth consisting of rotten moss, grass-roots, guano, and refuse, which to the

* At Sakkak ("Sunside," Giesecké) the great valley which runs into the heart of the Noursoak Peninsula is drained by a small stream that appears to divide the gneiss of the mainland from the trap-formation of Noursoak. The glaciers on both Disco and Noursoak appear, from this place, to be steadily advancing, "so much so that their progress can be noted year by year," as noted long ago also by Giesecké. The glaciers to the south, however, are, in Mr. Whymper's opinion, decidedly shrinking. (Mr. E. WHYMPER, Brit. Assoc. Rep. for 1869, pp. 2 and 3.)—EDITOR.

depth of a few inches is soon formed on the skerries and islands in the sea, and serves the sea-fowls as places of incubation. The greatest part of the Greenlander's turf-beds are situated on gulls-hillocks ("maagetuer"), and have, therefore, geologically speaking, nothing in common with what we mean by turf-deposits. It was accordingly impossible for me to collect, as I had desired, by an examination of the older turf-beds, materials for determining the latest Post-tertiary changes of climate that have taken place in Greenland. But instead, we find here many other deposits, which serve at least to give an indication of the changes that the animal world has undergone during the Glacial Period.

PART III. (Geol. Mag., vol. ix., p. 409, &c.)

Before proceeding to give an account of these changes in the Fauna of Greenland, I wish to draw attention to the possibility which exists in these parts of obtaining a comparison between the units of geological and historical chronology, that is, if, by collecting observations and reports from many different localities, it be possible to determine certain limits for the velocity with which the border of the inland ice moves. One may arrive at the lower limit by the following considerations. The breadth of the slip of border-land at Auleitsivikfjord is about 60 miles, or 350,000 ft. The annual retreat can, of course, never exceed the thickness* of the covering that yearly melts, divided by the sine of the inclination of the icy surface, which in the places passed by us was nowhere less than 30°. It is hardly probable that during a summer in Greenland an ice-layer of more than 10 ft. can melt away, so that a yearly retreat exceeding $\frac{10}{\sin 30^{\circ}} = 20$ ft. is not to be thought of. This would give for the time that has been required for the uncovering of the outer strip of land at Auleitsivikfjord a period of at least 17,000 to 18,000 years. But this number is evidently too low, for neither the yearly falls of snow nor the advance of the ice-mass has been taken into account, as they of course ought to be; and yet we have here to do with a geological period which undoubtedly forms but a small fraction of the interval that has elapsed since the first appearance of Man.

The point at Sarpiursak forms a very level and extensive plain, elevated about 60 to 150 feet above the sea, covered with a vegetation of "Lyng," Moss, and Sedge, too scanty to conceal the clay which forms the bottom of the plain. Similar formations in many other places along the shores of Disko Bay and Auleitsivikfjord

* Estimated at right angles to the surface of the ice. The annexed cut shows this more clearly. If G is the surface of the ice in *e.g.* 1870, and G′ the same surface in 1871, then AG′ is the thickness of the layer that has melted; and the distance the ice has receded is = AG′: sin V. The angle V is, of course, determined by the relation between the velocity of melting and the velocity with which the ice flows out of the higher parts of the glacier.

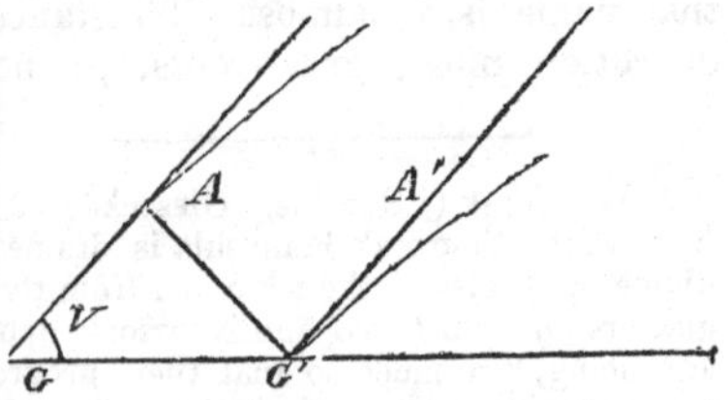

have given rise to vast clay-beds, which attracted attention long ago in these parts so ill-supplied with clay. Our Greenlanders even mentioned that they contained petrified shells and "Angmaksäter" (the Capelin *). These fossils are also mentioned by Dr. Rink in his work on North Greenland; and he adds, that a collection which he had sent home had been examined by Dr. O. A. L. Mörch, who found the shells partly to belong to species still existing on the coasts of North Greenland, partly to more southern forms. As the collection of materials for forming a judgment relative to the changes in the climate of the polar regions was one of the principal objects of the purely scientific part of our expedition, it was natural that we should pay especial attention to these circumstances.

Older glacial† fossils occur in N. Greenland in two different formations, namely, either imbedded in clay (south of the Waigat), or else at Pattorfik in a somewhat hardened basaltic sand, becoming a basalt-tuff. The material of the clay-beds has evidently been deposited by the glacier-rivers whose muddy water everywhere bursts out from under the inland ice; but in general the deposits are marine, which proves that these regions, in the course of the present glacial period, have been elevated at least 100 feet.

The Danes, on the other hand, who have long resided in Greenland, declare most decidedly that a depression is now taking place in most parts of the country. Herr Einar Hansen, who has for 19 years lived in the colony of Omenak, says that even in that short period he has clearly seen this; and it is still more evident when we refer to the statements left by Herr Hansen's predecessor relative to its height 60 years ago. The situation of the blubber-house at Fredrickshaab, as well as many other facts in South Greenland, shows the same. At Godhavn, in Disko, on the contrary, a rise is said to be taking place. It would be an important service if these circumstances, to which attention has been called by Pingel, Brown, and others, were fully investigated, with an accurate collection of all data relating to the subject; and proper bench-marks fixed in appropriate spots among the skerries along the coast of Greenland.

As at the present time the glacial clay, covered with muddy water, is poorly supplied with animal life, so the similar clays deposited in ancient times present but a scanty variety of fossils. In the clay-beds at Auleitsivikfjord, for example, we could only find a few shells of *Saxicava arctica*, and in the thick clay-beds of Sarpiursak we at first sought in vain for any remains of animal life; these, however, were very numerous on the shore. Bivalves still united, inclosing, and often inclosed in, a hardened

* "Angmaksæt," Giesecké, &c.; "Angmaksak," Reinhardt, &c.

† Of course one finds in many places, at about the level of the sea, modern deposits, with sub-fossil shells identical with forms now living. From these formations those of which we are now speaking differ by the great age of the latter, and by the very different type of shell-remains found therein. This is especially the case with the shell-deposits at Pattorfik, which appear to me to belong to the earliest part of the glacial period of Greenland. A very considerable, but lately formed, bank of shell-earth, with bones of Whales and Walruses, alternating with beds of Sea-weed, occurs at Saitok, at the mouth of Disko-fjord. Unfortunately we had only time to examine it cursorily.

mixture of sand and clay, and flat or ring-shaped claystones, containing remains of Fish, Ophiuræ, Crustacea, etc., were found there in great numbers, for the sea is constantly washing away a clay bank, 60 feet high, and fossils and claystones are left on the shore. The fossils in the clay itself are but few; but the claystones form a separate layer, in which they lie close together. Similar fossils, together with a few Gasteropods were collected by Dr. Öberg at the foot of a clay-bank, South Leerbugt, near Claushavn.

The fossils at Pattorfik were large and with thicker shells. They are found at a height of from 10 to 100 feet above the sea-level, imbedded in greyish-green basaltic sand, in part hardened into a kind of basalt-tuff. This is especially the case in the neighbourhood of Shells, and accordingly they were most easily discovered by breaking up the hard round nodules that are imbedded in the rest of the mass. These nodules are often so hard and tough, that they cannot be broken with an ordinary hammer. Besides these this "basalt-tuff" contains large rolled blocks of stone, indicating that at the time of the formation of these layers a glacial period prevailed in the region.

The fossils* brought home by us from these parts have been examined by Professor S. Lovèn, who gives the following list of them.

SUBFOSSIL SPECIES of ANIMALS collected in GREENLAND during the Expedition of 1870.

	Pattorfik.	Sarpiursak.	Leerbugt.	Tessiur-sarsoak.
Mya truncata, L.	+	+	+	
Mya arenaria, L.	+	+	+	
† *Cyrtodaria siliqua*, Spgl.	+	—	+	
Saxicava arctica, L.	+	—	+	+
† *S. Norvegica*, Spgl.	+			
Lyonsia arenosa, Möll.	—	+		
Tellina sabulosa, Spgl.	+	+	+	
T. tenera, Leach	+	+		
Astarte corrugata, Br.	+	+		
A. elliptica, Br.	+	—	+	
A. striata, Leach	—	+		
Cardium Islandicum, Chem.	+	+	+	
C. Grœnlandicum, Chem.	+	+	+	
Leda pernula, M.	+	+		
Yoldia truncata, Br.	—	+	+	
Y. hyperborea, Lovèn.	—	+	+	
Mytilus edulis, L.	+			
Pecten Islandicus, L.	+			
Tritonium undulatum, Möll.	—	—	+	
T. Grœnlandicum, Chem.	—	—	+	
T. hydrophanum, Hancock	+	+		
Natica clausa, Sow.	—	+		
Idothea Sabinei, Kröyer	—	+		

All are species still living in the Arctic Seas. Those marked with † are called "fossil" by Dr. Rink; perhaps not found living in the Greenland waters.

* Krantz in his work speaks of fossil shells at Godthaab, *which are nowhere else found in these parts.*

After passing some time at Sarpiursak in collecting fossils, we removed to Christianshaab, and thence onward to Leerbugten, south of Claushavn. By means of certain arrangements made by the Inspector, we were enabled to make a particularly interesting tour inland, to the extremity of one of the largest ice-fjords in Greenland—the ice-fjord of Jakobshavn.

This fjord is found inserted on very early maps of Greenland, though generally as a sound uniting the North Atlantic with Baffin's Bay. It is now known that the supposed sound is only a deep fjord, filled throughout its whole length with huge icebergs, which completely close the fjord, not only to ships, but also to whale-boats and umiaks, nay, even to kajaks (canoes). The shores of the fjord are therefore uninhabited, and seldom visited. A tradition exists among the Greenlanders, that the fjord was in former times less obstructed by ice, and was consequently a good hunting and fishing place; and this is confirmed by the older maps of the fjord, but especially by the numerous remains of old dwellings, which are still met with along the shores, not only of the principal fjord, but of its southern arm, Tessiursak, now completely barricaded by icebergs and inaccessible from the sea (not to be confounded with the fjord Tessiursarsoak which we had just left). Tessiursak itself is still tolerably free from ice, and is easily reached by dragging an umiak over the point which separates the western shore of Tessiursak from the ocean. For such a purpose, however, a traveller must take his umiak with him, partly because he cannot obtain any boat at the now deserted Tessiursak, partly because about half-way over the point he meets with a lake, to go round which would be a considerable circuit.

On our arrival at Leerbugten, we found, in consequence of the Inspector's excellent arrangements, a Greenland family there to meet us, and the women's boat, or "umiak," lay drawn up upon the shore. The journey over the point was immediately commenced. Six men took the roomy umiak upon their shoulders, others took our instruments, and provisions for us and our people for two days. The way was first over a high ridge, which separates the sea from the lake, on the shore of which the Greenlanders had pitched their summer tent. Here we rested awhile, and tried the temperature of the water (12° Centigr.), by a bathe in the lake, to the great astonishment of the Greenlanders. We then rowed over the lake in the umiak, took it up and carried it on our shoulders over another point, steeper but shorter than the former, and clothed just at this time in all the colours that the Flora of the extreme north can offer. On the other side of this point was water again, not however fresh, but salt: it was the above-mentioned southern arm of the Jakobshavn ice-fjord. The umiak was again launched, and, after a row of a few hours, interrupted by hunting after young Seagulls, we reached the spot where Tessiursak falls into the main ice-fjord very near its inner extremity. Here the water that was free, or nearly free, from ice terminated, and we had to make our way along the southern shore of the ice-fjord for a distance, not indeed long, but dangerous, on account of the masses of ice driven hither and thither by the violent currents near the shore.

Further out the fjord was completely covered with lofty sharp-pointed icebergs, some of which stood so firmly on the ground that the stream could only move them at flood-tide. Others, which did not draw so much water, were carried hither and thither by the currents, and it is difficult to describe in words the deep booming and scraping which took place when these were driven against each other or against the still mightier masses aground. A loud report sometimes gave notice of the splitting of an iceberg, which was usually followed by a violent undulation reaching to the shore. It is not surprising that the Greenlanders do not like to make long voyages in such waters. Neither did we long continue our row. Just on the other side of a headland formed by a high steep gull-hill, bordering the mouth of Tessiursak, were the remains of an old house, which formed the terminus of our journey. Here we rested for the night, and returned next day by the same route by which we had come. We employed our time partly in an examination from the tops of the neighbouring hills of the vast iceberg-factory that lay at our feet, and partly in a careful investigation of the remains of the dwellings left desolate for a century, perhaps many centuries, where we now rested.

I have already given a profile of this glacier (Fig. 5), from which it may be seen that it is impossible to draw any definite line of boundary between the inland ice and the sea. The glacier is, in fact, as its profile indicates, for a considerable distance, probably several miles, from its end broken up into icebergs, the original situation of which has, by the continual advance of the ice, been entirely disturbed, so that they are thrown in confusion one over the other. Even at the mouth of the fjord these icebergs are as closely packed as when they formed a part of the glacier, and most of them perhaps always aground. It is at a considerable distance further on that they are separated from each other, so far at least as to allow the surface of the water to be seen between them.

Even if there had been time to take topographical measurements, it would not have been possible for me to state how many hundred yards the house we now visited is from where the fjord and inland ice meet. It is certain that at present the distance is not very great, and the appearance of the environs must have been very different when Kaja—such is said to have been the name of the locality—was an inhabited place. That it was so for a long period is shown by the magnitude of the kitchenmiddens, and by the numerous remains of houses and graves. Either the level of the water in the fjord has risen or the land has sunk considerably since that time. It is not probable that the situation of a house would be chosen so close to the shore that not even a canoe could find room in front of the dwelling.

As a Greenlander now seldom resides at any distance from the Danish-trading stations, one finds in numberless places along the coast old deserted dwelling-places. They are recognizable at a distance by the lively verdure, arising from the rich vegetation, which the remnants of fishing and hunting prey scattered round the cottages or tents has produced. On taking a few spadefuls of earth, or on examining the walls of the new houses,—generally built with turf taken from these spots,—one everywhere finds the earth and grass-roots mixed with the bones of the animals which the Greenlanders hunt. The animals killed by the men are in fact cleansed by the women beside or in the cottage itself, and the refuse after the cleansing or the meal is thrown away—seldom far from the cottage-door. Even now, in the course of years, a heap is frequently collected as truly circular as if it had been drawn with a pair of compasses round the door as a centre. On examining its contents, it is found to consist of a black, fat earth, formed of decayed refuse—frequently bits of bone gnawed asunder and broken, shells, especially those of *Mytilus*, lost or broken household goods, etc. This bone-mixed earth most likely contains, like guano, not only considerable quantities of phosphoric acid, but also ammoniac salts, and it may happen that the trade of Greenland may find in this a valuable article of export.

As the kitchenmidden dates from the Stone Age in Greenland,—which undoubtedly extended beyond the epoch at which the whalers first began to visit these coasts,—we find in it arrow-heads, skin-scrapers, and other instruments of various kinds in stone, and especially a quantity of stone-flakes knocked off in forming the instruments, easily recognizable, not only by their form, but by their consisting of stones—chalcedony, agate, and especially green jasper (called by the Greenlanders "angmak")—not met with in the gneiss formation, but only at certain spots in the basalt region of Disko or the peninsula of Noursoak. One sometimes finds smaller instruments of clear quartz, also half-wrought crystals of the same mineral. Everything shows that the material was carefully chosen among such minerals as united the necessary hardness *with absence of cleavage and a flat conchoidal fracture.* Among minerals in general, the different varieties of quartz (rock-crystal, agate, chalcedony, flint, and jasper) are the only ones which fully satisfy these conditions; and it is therefore almost exclusively these minerals that the various races of men have chosen for making their *chipped* (not ground) stone instruments.

The two largest of the old house-sites, among which we were now resting, lay so near the sea that their bases were washed by the water. A small stream had found its way through one of them, and had thus not only exposed a section of the kitchen-midden, but also subjected a part of it to a washing process, in consequence of which bits of bone and other heavier objects lay clean-washed at the bottom of the channel and in the hollows of the gneiss slabs of the shore. These were carefully examined, and a number of stone instruments and stone chips were collected. There were no traces of iron; but we found a small oval perforated piece of copper, which had evidently once served as an ornament. At the largest site a tolerably thick circular stone wall, 8 or 10 feet high, and 26 in section, was still distinguishable, divided into two unequal portions by a party-wall. The entrance seems to have led into the larger of these areas, judging from the extensive kitchenmidden just outside it. In one of the other heaps of bones a flat stone was found, so large as to require the united efforts of several Greenlanders to turn it. They declared that the workshop for the fabrication of stone instruments must have been situated on that spot, and expected accordingly to find a great quantity of chips in its vicinity, which, however, the result of their searches did not confirm.

The kitchenmidden outside the large cot rested on a low slab of gneiss, separated from it by a thin layer of turf, in which were no trace of any pieces of bone, and which had therefore been formed before the place was inhabited. In other respects this turf, of which specimens were taken away, was perfectly like the earth which was mixed with bones and stone-chips. Here, there were no *Mytilus* shells, though these are everywhere else found around Greenland dwellings: an indication that formerly the inhabitants were not obliged to have recourse to this species of famine-food.

To discover the various animal forms that had here been the prey of the hunter, Dr. Öberg collected a quantity of bones, in which work the Greenlanders took a lively interest, usually determining with great certainty the species to which the pieces of bone had belonged.

The following species could be ascertained:

Cervus tarandus.
Ursus maritimus.
Trichechus rosmarus.
Cystophora cristata.
Phoca barbata.
Phoca Grœnlandica.
„ *hispida.*
„ *vitulina.*
Delphinapterus leucas.

Even if we suppose that this spot was first inhabited shortly after the Esquimaux entered Greenland by Smith's Sound, its age will still be scarcely more than 500 years, a period generally too short to show marks of the slow but continuous changes to which the organic world is subjected. Neither do the kitchenmiddens of Kaja contain any other forms of animals than those still living on the coast of Greenland. Nevertheless we obtain here an interesting confirmation of the changes that the ice-fjord has undergone. The Walrus, *Phoca barbata,* and *Cystophora cristata* no longer venture into this long ice-blockaded fjord; and even the Bear has now become so scarce in the colonies of North Greenland south of the Waigat that most of the Danes resident in those parts have never seen it. The remnants of bones in the kitchenmiddens on the other hand prove that these animals were abundant there formerly, and are consequently an evidence that the fjord at Jakobshavn was less filled with ice than now. The uniform agreement of the older maps in placing here a strait, extending completely across Greenland,* indicates that it is only within the last few centuries that this fjord has been converted into an ice-fjord, and that accordingly the same phenomenon, though on a larger scale, has taken place here as in the northern harbour of Bell Sound, Spitzbergen. Krantz mentions a similar case with reference to the ice-fjord north of Fredrickshaab in South Greenland.

At all the old house-sites in Greenland one meets with graves, and such is the case here. The grave usually consists of a cairn, built of moderate-sized stones, in the middle of which an oblong excavation, about the length of a man, and covered with a large flat stone, forms the chamber. In these we usually find the skeletons of several persons, so that the grave has been a sort of family tomb. Peculiar small chambers close beside the real grave-chamber form store-rooms for the deceased's outfit for the next world. We find here arrow-heads, scraps of leather, bone, stone or iron knives, water-ladles, bits of stone pans, lamps, pieces of flint, bows, models of canoes, oblong smoked pieces of pebble-stones, small wooden staves (according to the statement of the Greenlanders, dipped in oil, and to be used as torches), &c., &c. In a similar grave-chamber at Fortune Bay I found a number of glass beads, evidently of European origin, beads of bone, flint-points, and some rusty nails (these last probably the most costly among the valuables, which the male or female potentate resting in the grave was to take with him or her to the other world). A Greenlander gave to Dr. Öberg a pair of blinkers, or, more intelligibly speaking, snow-spectacles, made of wood found in a grave. The proprietor would seem to have suffered from weak eyes, and to have been afraid of the reflection of the light from the snow-fields in the abode of the blessed.

It seems to be usually assumed, that whatever iron is met with among the Greenlanders is either of meteoric origin, or has come from the original Northmen colonists, or from the merchants and whalers of modern times. This assumption appears to me erroneous. First, as regards meteoric iron, it is certainly met with in Greenland, as in all other lands which have been but a short time inhabited by man; in other countries it has been used up during the period when iron was more valuable than gold. The meteoric iron that has hitherto been found in Greenland is, however, generally too hot-short, cold-short, and brittle, to be otherwise than exceptionally used; and even if a piece of better quality should be

* The views we got of the land inwards from a high mountain near Kaja showed clearly, however, that the often repeated story of a strait passing completely across Greenland has arisen from a misunderstanding of the Greenlanders' accounts of the long narrow fjord. We received from the Greenlanders at Auleitsivikfjord a similar account of the southern arm of that fjord; but on questioning them more closely, it appeared that they only meant that the distance to the extremity of the fjord was, according to their notions, immensely great. Krantz (in the middle of the last century) speaks of the fjord as quite full of ice. It was then long before Giesecké's time, when, according to Brown, "this inlet was quite open for boats" (Quart. Journ. Geol. Soc., xxvi. p. 684).

met with, I cannot see how the Greenlanders, with the tools they at present possess, could possibly forge an arrow-point out of a piece of iron weighing a couple of pounds. But, on the other hand, since the time when ships first began to cross the Atlantic, a wreck may now and then have been carried by the current on to the coast of Greenland, sometimes far up Baffin's Bay. We were able to verify an example of this. During our stay in North Greenland, a fragment of a small schooner or brig drove on shore at Disko, between Diskofjord and Mellanfjord. As soon as notice of the matter was given, the Greenlanders in the neighbourhood made an accurate inventory of everything on board that could be turned to any useful purpose. They found bread and sundry other provisions, also potatoes, but no paper or any indication of the name the ship had once borne, or the nation to which it had belonged, further than that the brass bolts by which the timbers were fastened together bore the stamp "Skultuna;" they were therefore from the Swedish brass-foundry of that name, and it is perhaps probable that the vessel itself was either Swedish or Norwegian. It was a two-masted vessel of 100–150 tons burden, according to the estimate of the Danes, and, according to the Greenlanders, could take a cargo equal to about half that of a three-master. The timbers were of oak, the outer covering of pine, the sides were not strengthened to resist ice, the stern was round "as a Dutchman's." The Greenlanders asserted that undoubtedly the ship was neither a whaler nor intended to sail amongst ice; and there is not the slightest reason to doubt the accuracy of their judgment, which is most sagacious in such matters. We have then here an example of a wreck drifting hither from the southern seas. Similar events must of course have often happened before, and what an abundance of iron the wreck of a ship supplies to a Greenland colony with its limited wants, is evident from the quantity of iron lying, at our visit, scattered around the houses in Godhavn, and obtained from whalers that had been stranded there in the preceding year. Here again was evidence of the Greenlander's improvident character. It never entered the mind of any one of them, out of all that quantity of iron—sufficient perhaps to supply the wants of the Greenlanders for a century—to preserve more than what he for the moment required; and if the regular exportations from Europe were to cease, the colony would again in a few years have to go back to the bone-knife, the bow, and the stone implements.

For bone-knives, such as are sometimes found in old graves, the edge of which is formed by an iron plate let into a groove in the bone, a piece of an iron hoop of a barrel, that may have washed ashore, may easily enough have been used; an old worn-out iron knife would have been less fit for the purpose. These iron-edged bone-knives are therefore by no means always remnants from the time when the iron brought into the country by the Northmen in the beginning of the present millennium had begun to be scarce; but merely examples of the Greenlanders' way of turning to use for their simple wants, in the most appropriate manner, any objects that may come in their way.

At Kaja persons have been buried, not only in ordinary graves, but in low caves formed at the foot of neighbouring steep cliffs of gneiss by huge blocks of rock fallen from the mountain one over another. Most graves in the vicinity of the colonies have been long ago plundered by searchers after antiquities. This was not the case in this distant locality; nevertheless, all that we found in the graves was a pair of water-ladles and arrow-heads. On the other hand, as has been already said, a rich harvest was gathered at the sites of the old houses.* Some skulls were also taken, the Greenlanders not appearing to object to this, and it being a matter of the greatest scientific interest to obtain perfectly authentic skulls of the original inhabitants of Greenland before any mixture of race had taken place.

On the 31st July we returned to Leerbugten, where we were obliged to divide our little expedition into two parties. It was of interest to the geologists to visit as many places along the coast as possible, even if it were only for a few hours, whereas the botanist and the zoologist for their researches, and especially for the preservation of their collections, were obliged to remain at least some days at each place. Dr. Berggren and Dr. Öberg therefore now went together, to collect from the bottom of Disko Bay, and from its mountainous shores, the fauna and flora of the place. Dr. Nordström and I, on the other hand, hastened to the Basalt region, to seek new materials for the climatological history of the extreme north, in the coal-, sand-, and clay-beds to be met with there. The harvest we gathered was rich beyond our expectations.

In the first volume of his work on Greenland, Krantz has introduced some notices of the mineralogy of the country, whence we find that the coal-beds of Disko were then (1765) already known. A statement of the Greenlanders is moreover adduced, that in certain distant parts all sorts of fishes were to be found turned into stone. Some years later the surgeon Brasen, who in 1767 made a voyage to these parts for his health, collected a quantity of minerals, of which a catalogue is given in the third volume of Krantz's work. This catalogue contains 25 items, including different varieties of quartz, granite, graphite, pot-stone (steatite), pumice (of which it is justly remarked, that it has been brought hither by the currents from Iceland), and so forth. In the beginning of the next century (1806–1813) C. Giesecké—who was first an actor, afterwards a mineralogist with the title of "Bergsraad," and lastly professor in Dublin, and Knight, made extensive mineralogical excursions on the coasts of Greenland. Giesecké himself has published but little of his observations,† though carefully kept journals of his travels are preserved in manuscript at Copenhagen. Numerous and important new dis-

---

* Stone implements of various kinds were collected and purchased by us at several other places, so that the collection we brought home consisted of above 1,000 specimens. Dr. Öberg made the richest harvest at Kikertak.

† In Brewster's Edinburgh Encyclopædia, vol. x. pp. 481–502, under the word "Greenland," is an article written by Giesecké, containing, among other things, some short notices of the mineralogy of that country. There is also a work by him on Cryolite in the Edin. Philos. Journal, vi., 1822. *See above.*

coveries prove that his researches were carried out in a true scientific spirit, and with a completeness and accuracy the like of which but few of the old civilized lands of Europe could at that time produce. Even North Greenland was visited by Giesecké. Here he discovered, among other things, fossil Plants at Kome * and on the east coast of Disko,† and furnished several instructive sections. Subsequently (1838) the coal-beds of North Greenland were, by order of the Danish Government, examined by J. C. Schythe, though, as it appears, chiefly for technical purposes. A more important event for geological science was Dr. Rink's four years' residence (1848–1851) in North Greenland, during which time he visited many parts of the Basalt region, whence rich collections were taken home, among which may be mentioned fossil trunks of trees from several places, as also fossils from Kome, described in Heer's "Flora fossilis arctica." Some years later a Dane, Jens Nielsen, residing at Atanekerdluk, discovered magnificent Miocene fossils there, a large number of which were collected, when Captain Inglefield, in company with Captain Colomb, and Mr. Olrik, the Inspector of North Greenland, visited the place in July 1854.

These strong proofs of a formerly warm climate up in the neighbourhood of the Pole aroused astonishment in all who saw them. More collections were made, partly by Inspector Olrik,‡ partly by other officials of the Danish Trade. Also Prof. Torell, Dr. Walker, Dr. Lyall, and others brought home not inconsiderable collections from their travels in Greenland.

The importance of this discovery to the history of our globe was, however, first taught by means of Heer's "Flora fossilis arctica," in which these fossils are described, together with similar fossils collected during the English Franklin Expeditions from the most northerly archipelago of America, by Prof. Steenstrup from Iceland, and by the Swedish Polar Expeditions from Spitzbergen. The British Association had already (1867), at the instance of Mr. Robert H. Scott, F.R.S., sent out an expedition to make new researches in this geologically interesting quarter. These were entrusted to Messrs. Whymper and Brown ;§ but in

* Giesecke's Journal. Heer's Flora fossilis arctica, p. 7.

† The above-mentioned article in Brewster's Edinburgh Encyclopædia, p. 493.

‡ Mr. Olrik's collections were given partly to the University Museum at Copenhagen, partly to Capt. M'Clintock, who, on his return in 1859, passed Disko, and, on returning home, presented them to the Royal Society in Dublin, the same institution to which Capt. Colomb had presented his collections. Capt. Inglefield's collections were given partly to the Geological Survey in London; Dr. Walker's and Dr. Lyall's (from the eastern side of Disko, near the sea-level) to the Botanical Museum at Kew; Prof. Torell's to the National Museum at Stockholm; Mr. Whymper's and Mr. Brown's to the British Museum. The collections from Spitzbergen and of the expedition of 1870 will be divided between the Museums of Stockholm and Gottenburg.

§ See Oswald Heer, "Contributions to the Fossil Flora of North Greenland, being a Description of the Plants collected by Mr. Edward Whymper (and Dr. Brown) during the summer of 1867."—Phil. Transactions of Roy. Soc., vol. 159, part ii., p. 445. 1870.

consequence of a combination of unfavourable circumstances, the new researches were confined to the already well-examined locality of Atanekerdluk and the opposite shore of the Waigat. The new collections thus indeed completed the knowledge we already possessed of the Flora of the Miocene Period in the extreme north, but they opened no new views of the periods which immediately preceded and followed it.

As I had in 1858, and especially in the Spitzbergen Expedition of 1868, the opportunity of contributing in some measure to the climatic history of the extreme North, this question interested me in the highest degree. It was especially desirable to collect materials from the Cretaceous beds at Kome, and to obtain, if possible, fossil Plants belonging to the long periods between the Fern-forests of the Cretaceous and the Beech- and Plane-woods of the Miocene Epoch ; as well of the ages intervening between the last-mentioned era and the present time. This was the object of the tours made by Dr. Nordström and myself during the remainder of the summer.

Aug. 1. We departed in the Inspector's yacht, with our own whale-boat in tow, from Sandbugten to Flakkerhook, where the Inspector took leave of us, promising to meet us again at Atanekerdluk. We rowed, touching at a number of intermediate places to collect plant-fossils, past Mudderbugten, round Isungoak, to Ujarasusuk, whence I passed, in a boat obtained from the Danish officer, to Ritenbenk's coal-mine, north of Kudliset, and then crossed the Waigat to Atanekerdluk. Dr. Nordström stopped a little longer to collect more fossils at Ujarasusuk, and thence sailed in somewhat rough weather direct to our appointed place of meeting. On this now uninhabited spot we all met on the 5th of August. On the 9th we rowed farther, to Mannik, Atane, Noursak, and Noursoak, where we remained a couple of days (August 12 and 13).

The time there was employed partly by a visit to the coal-beds of Netluarsak, situated high up in the basalt beds between the two last-mentioned places. From Noursoak the Inspector continued his journey to Upernivik, while we rowed along the shore of Omenakfjord, touching at Niakornet,* Ekkorfat, Karsok, and other places, to Pattorfik. From Niakornet and Karsok two trips were made into the interior ; to coal-beds at Ifsorisok and to the famous graphite-bed at Karsok. From Pattorfik we rowed over the fjord, though densely packed with icebergs, to Omenak, where we arrived on the 20th of August. Here we were detained by the ice a couple of days, during which we were lodged in the most hospitable manner by the local Colonial Governor, Mr. Boye.

On the 22nd, in the afternoon, we rowed over to the Assakak glacier, and the following day onward to Kome, whence we went on board a ship lying there belonging to the Greenland Trade, in which, in the evening of the 24th, we set sail for Godhavn, where we arrived on the 30th, and whence some excursions were made to the spot where the Meteoric Iron was discovered a Ovifak ;

* "Niakornak," on the map accompanying Prof. Nordenskiöld's memoir.

to Saitok, at the mouth of the Disko fjord; and to Puilasok and Sinnifik. Shortly after our arrival at the last-mentioned place (Sept. 3), we received a Kayak express from Godhavn with the news that war had broken out, which induced us to hasten back to the colony in order to avail ourselves of the first opportunity to return to Europe. As no vessel was just then lying there, nor was any expected to arrive at Godhavn for the next few days, I immediately passed over to Egedesminde. Dr. Nordström remained at Godhavn, awaiting Drs. Öberg and Berggren, to return home with them. At Egedesminde I went on board the brig Thialfe, commanded by Captain Brockdorff. Contrary winds prevented our departure till the 23rd of September, and the passage was slow in consequence of storm and unfavourable winds, so that it was not till the 2nd of November that I could land at Elsinore.

During the whole period of our boat-excursions in Greenland we had, with the exception of one rainy night, a constantly clear sky and a favourable sailing breeze, circumstances, which greatly facilitated our movements, and rendered it possible in so short a time to investigate at least the principal geological features of that remarkable tract, and to collect extensive series of plant-fossils from above twenty separate localities, and belonging to five widely separated geological horizons.

Like previous similar collections from the Arctic regions, these have been transmitted for examination to Prof. Oswald Heer, of Zurich, and I venture to hope that, when duly interpreted, they will give us an idea of the changes of climate these regions have undergone since the epoch when serious variations of climate first took place upon the globe. I will offer a few remarks on the geognosy of these interesting beds.

The basalt or, as it is also called, trap-formation, probably extends completely across Greenland north of the 69th degree of latitude; at least Scoresby found, in his remarkable visit to the eastern coast of Greenland, trap with the impression of Plants* at many places along the extent of coast visited by him. It is possible that the same formation may continue under the sea to Ireland, and thence, partly in a more northerly direction by Jan Mayen to Spitzbergen, partly in a southern direction from Jan Mayen, by the Faroe islands, to the Hebrides and Ireland.† The same eruptive formation extends also westward over a vast part of Franklin's Archipelago, perhaps even to the volcanic tracts at Behring's Strait. These basalt beds probably originated from a volcanic chain, active during the Tertiary Period, which perhaps indicates the limits of the ancient polar continent, in the same

* Scoresby's collections from these parts seem to have been lost. On the other hand the last German Expedition to East Greenland brought back collections of plant-impressions, which have also been placed for investigation in the hands of Prof. Osw. Heer.

† The agreement between the basalt formations of Greenland and the British Islands, both as regards the character of the rocks and the age of the beds, seems to be perfect.

manner as is now the case with the eastern coast of Asia and the western of America, thus confirming the division of land and water in the Tertiary Period, which upon totally different grounds has been supposed to have existed.

This formation appears most developed in North Greenland to the large island of Disko and the peninsulas of Noursoak and Sortenhook [Svartehuk ?], where it occupies an area of about 7,000 square miles, with a vertical section of 3,000 to 6,000 feet.

Here these eruptive rocks are divided into beds, which, between Godhavn and Fortune Bay, rest immediately upon the gneiss ; but on the coast of Omenakfjord, between Ekkorfat and Kome, upon sand- and clay-beds belonging to the Cretaceous age. To the east of Godhavn, at Puilasok and Sinnifik, we meet with sand- and clay-beds lying between, not under, the basaltic rocks, and accordingly newer than some of the latter. The fossils in these beds belong to the Tertiary Period. It follows, then, *that the eruptions, which have given rise to these vast beds of basalt, have taken place subsequently to the commencement of the Cretaceous, and have ceased before the termination of the Tertiary Period.*

In the preceding pages I have intentionally spoken of basaltic strata or beds. In almost every place where I have had the opportunity of examining it, the Greenland basalt is so stratified that one is forced to admit that it is only exceptionally that we have to do with masses of lava, but for the most part with sedimentary beds of volcanic ashes and volcanic sand, which in the course of thousands of years have become hard and assumed a crystalline structure.

Decided lava-streams I have scarcely observed ; even large or small dykes are not so common as one might expect ; and, where they are found, the mass of lava has produced scarcely any effect upon the loose beds of sand or clay, or the basalt that it has pierced.

No volcanoes, either extinct or active, are met with in these parts, although circular depressions in the basalt plateau, caused by glaciers or brooks, may, when carelessly observed, easily be mistaken for true craters. It is, of course, quite natural that great cavities in the interior of the earth must arise in the places whence the great eruptions have issued, which have produced the basalt region of Greenland ; and that these in their turn must, within a short period, be followed by the destruction of the superjacent volcanic cone. The place or places where these old volcanoes once rose high over the surrounding plains will therefore now most probably correspond with the greatest depths in the neighbouring sea.

At Godhavn the lowest strata resting immediately upon the gneiss formation (*e.g.* outside Bläsedalen) consist of a basalt-tuff or breccia, containing various species of zeolites (according to Giesecké only apophyllite), next comes columnar basalt, free from zeolites, then again basalt-tuff with zeolites, alternating with true basalt. A coarse crystalline dolerite, very similar to the Spitzbergen hyperite, forms at Atanekerdluk, near the shore, a hill several thousand feet high.

The basalt beds are 50 to 100 feet thick, and may be traced for miles along the shores, often separated from each other by thin layers of red basaltic clay. Sometimes the layers are crossed by dykes of a hard, fine-grained basalt.

Not only dykes, but also basalt beds, on the cooling of the melted mass, or during the drying and crystallizing process which the volcanic ashes have undergone in their transformation to basalt, have been broken into regular columns, mostly hexagonal. Brännvinshamn, Skarffjäll, Kudliset, and other places on Disko and the peninsula of Noursoak, afford examples of this kind of basaltic structure, comparable in magnificence with Staffa and other geologically famous European localities.

Volcanic eruptions, as has been above remarked, no longer occur in this region. Yet, in consequence of the rapidity with which basalt is destroyed, layers of basaltic sand constantly collect on the shores—beds which, in the course of thousands of years, may, under favourable circumstances, harden into a rock not distinguishable from real basalt, unless perhaps it be that, as these beds are deposited in the sea, they may contain marine fossils, which tuffs of the real basalt formations do not. Such a hardened fossiliferous basaltic sand occurs at Pattorfik, in Omenakfjord, and between that place and Sarfarfik. This stratum, which has already been described, is, however, evidently far more recent than the newest beds of the real basalt. *See* p. 409.

Young as are the colonies in these parts, tradition can nevertheless adduce sundry examples of the rapidity with which basalt rocks are destroyed. It is difficult to induce a Greenlander to penetrate by boat into the inner parts of the three fjords which cut into the west coast of Disko Island. The reason of this is said to be, that on one occasion a whole house with all its inhabitants was crushed by a sudden fall of a basalt rock. At Godhavn, on the brow of a basalt mountain, there were formerly twelve huge projecting elevations, called "the twelve apostles." Of these there is now but one remaining.

In the immediate neighbourhood of Godhavn the basalt either extends quite down to the sea, or lies immediately upon the gneiss formation, which there forms the shore-cliffs. On rowing from this point further to the east, as soon as Skarffjället is passed,* beds of sand or sandstone are found nearest the shore, increasing in thickness as we approach the Waigat, so that at Flakkerhook and Isungoak they form mountains of 1,500 to 2,000 feet high, frequently crowned with a perpendicular basalt diadem. The same formation is met with on the other side of the Waigat at Atanekerdluk. Further north-west in the strait, however, the conformable sandstone and basalt sink again, so that before

---

* Some of these beds (at Puilasok and Sinnifik) nearest Godhavn are however more recent than the great basalt formation, *i.e.*, stratified *between*, not *under*, some the rocks of this formation.

we arrive at Noursak the basalt reaches the sea-level. Beyond that point the peninsula is entirely occupied by basalt-beds, terminating in terraces, between which no sand-layers can be discovered from the shore. But at a height of from 1,000 to 2,000 feet above the sea we find here, also, sedimentary formations of sand, clay, coal, &c., but very thin, and therefore, for the most part, concealed by basaltic detritus.

Further inward, the shore of Omenakfjord is occupied exclusively by basalt, extending beyond Niakornet; but afterwards we again meet with a formation similar to that of Atanekerdluk, though of a widely different age, and resting, not upon basalt, but upon gneiss. These strata belong to the Lower Cretaceous series. Here the basalt strata no longer extend down to the water; and the beach pebbles farther inward are again of gneiss. But the glaciers that extend downwards from the interior continually carry with them blocks and columns of basalt, indicating that the lofty inland mountains are composed of that rock; and that there also it is interstratified with Tertiary beds, is shown by the plant-remains which lie, mixed with pieces of basalt, on the surface of the Assakak glacier.

Here also was found a specimen of wood inclosed in basalt; but, with this exception, all the fossil plants have been found in the coal-bearing sandstones and clay-beds which are associated with the basalt in Greenland. I have no doubt that organic remains will be found in the red basalt clay that lies between the real basalt beds, though we had not time to look for them.

The fossils in the sedimentary strata of the Trap-formation * in Greenland consist exclusively of Plant-remains, and fragments of two or three Insects and Freshwater Mollusca; there are no traces of Marine Mollusca nor vertebrate animals.

An extensive continent occupied this portion of the globe at the time when these strata were deposited; and the abundance of sandy strata, furthermore, seems to indicate that, during the Cretaceous and Tertiary periods, this was a vast sandy desert [?], varied only by oases of inconsiderable extent. *At that time there were no glaciers in these parts.* For the sand-beds contain no traces of any such erratic blocks or large boulders as always accompany and characterise the glacial formations, and such as are met with even in the loose clay-beds, of glacial origin, which, where a subsequent denudation has taken place, cover the beds of basalt and Tertiary sand. I ought however to mention that in places where both the modern glacial formation and a part of the subjacent Tertiary sand have been washed away, sections often occur, which, on a cursory examination, seem to indicate that the Tertiary sand contains a vast quantity of erratic granite and gneiss blocks. But wherever time permitted us to make a careful investigation, or

* I have preserved this name as used in Greenland as a common denomination for the Cretaceous formation, dolerite, diabase, basalt, the Tertiary strata included in basalt, and the strata at Sinnifik and Puilasok, probably deposited shortly after the cessation of the eruption of the basalt.

Fig. 7.

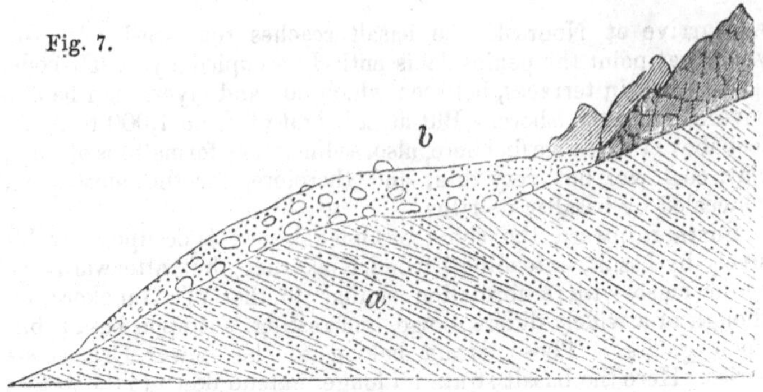

Fig. 7.—*Section before any modern denudation had taken place.*
a. Tertiary strata without erratic blocks.
b. Glacial strata with erratic blocks.

where, as is the case in most of the places where plant-remains are found, fresh sections are exposed, it is evident that these blocks have been washed down from superjacent glacial strata of more recent date (*b*), and in no wise belonged originally to the Tertiary strata (*a*), on which they now lie. The accompanying Figs. 7 and 8 show this clearly:—

Fig. 8.

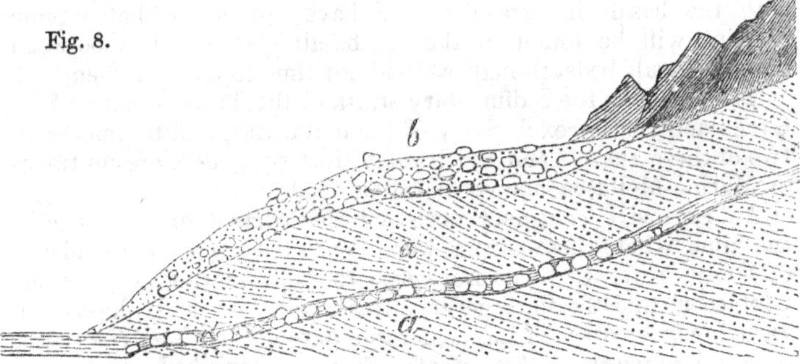

Fig. 8.—Section along a modern mountain-stream (*c*– *c*′), with blocks derived from the glacial deposits at the surface.
a, *b*, as in fig. 7.

These Tertiary beds therefore do not afford any evidence that the favourable climatic circumstances of the Tertiary era have been interrupted by a separate glacial period, which has subsequently disappeared. The Cretaceous, Miocene, and recent sand-beds are in outward appearance perfectly alike; and, if a new elevation should expose the sand-beds now in process of formation in many places at the bottom of the Waigat, it would be very difficult, wherever they are destitute of organic remains, to distinguish these from the Cretaceous sand-beds at Kome, or the Miocene beds at Atanekurdluk, Isungoak, &c.

It was formerly supposed that all the coal-beds of Greenland belong to the same geological period. Heer's important discovery that the strata at Kome and Atanekerdluk belong to two widely different periods, showed that this is not the case. Subsequently a rolled block was found on the shore of Disko containing an impression of a *Sigillaria*. This stone, however, appears either to have been brought hither as ballast or by ice. At least, we could not discover anywhere in these parts beds belonging to the old Carboniferous Period.* Heer's discovery was not only confirmed by our researches last summer, but we also discovered Plant-remains from one or two geological horizons quite new for Greenland.

In the description of these I follow the chronological order, beginning with the oldest.

I.—*The Kome Strata* (older division of the Cretaceous formation, according to Heer).

By this name I designate a sedimentary, coal-bearing formation, occurring here and there between Kome and Ekkorfat, on the coast of the Noursoak Peninsula, S.W. of Omenak. The name is taken from the place where the chief coal-bed is found, and which, in all probability, yielded the plant-impressions brought away by Giesecké and Rink. These strata, however, occur not only at Kome, but all along the above-mentioned coast, with the exception of a few interruptions by gneiss hills. The Kome strata rest immediately upon undulating gneiss-beds, probably filling up old valleys and depressions between them. Higher up the gneiss is covered by eruptive rock. The strata generally lie tolerably horizontal; but sometimes with a dip inwards of as much as 20°. They are most developed in the neighbourhood of the two extremities, Ekkorfat and Kome, where the thickness exceeds 1,000 feet.

As the fossil Plants occur almost exclusively in the lowest strata, we cannot, without a careful examination of the few fossils we have brought home from the upper strata, decide whether the whole of this vast series of strata belong to the same geological formation or not. It is, however, probable that the upper portion, distinguished by its thick coal-bed, belongs to the next division.

Most of the Kome strata consist of sand or soft sandstone, often interstratified with shale † and coal-bands. The shale is generally mixed with sand, and, as it were, thoroughly corroded by acids; and in these cases it is so loose that the fossil Plants it may contain can scarcely be preserved. Fortunately there is also found, especially in the neighbourhood of the lowest coal-beds, a harder, sometimes argillaceous, sometimes talcose [?], shale, with numerous impressions, chiefly of Ferns and Coniferæ (not

* Fossils really belonging to the Coal-period have been since found by Dr. Nauckhoff, at Kudliset (Expedition of 1871).

† Among the somewhat foreign words and expressions revised in this reprint, the word "slate" (Skifer, *Swed.*, Schiefer, *Germ.*, &c.) is replaced by "shale" (Skiferlager, *Swed.?*), used in the nomenclature of English geologists for the laminated clays, of various consistence, which constitute so large a portion of these Tertiary and Cretaceous formations. The word "slate" is applied by geologists in England only to roofing-slate and some metamorphic schists.—EDITOR.

only twigs, but cones, seeds, and leaves). The leaves especially occur in abundance, generally transformed into a dark-brown, semi-transparent, parchment-like substance, resembling the vegetable parchment produced by the action of sulphuric acid on lignite. Some beds occur in which these leaves are so numerous that they form a flexible felt (which can almost be unravelled), woven of leaves and other similarly altered remnants of plants. It is possible that this fossilization depends upon the action of the acid gases which have come forth during the volcanic eruptions and condensed themselves in the waters of the locality, and that the condition of the fossil leaves is thus connected with the extremely corroded appearance of the shale and sandstone.

The most important of the coal-beds * occur in the upper part of the strata at Kome; but bands of coal are interstratified with the shale in many other places. They are not very extensive, however, though sufficient to provide some Greenland households with the few tons of coal they want in the year. At present, according to the statement of the Governor of the Colony, coal is thus collected, not only at Kome, but also at Sarfarfik, Pattorfik, Avkrusak, and, less frequently, at Ekkorfat.

To this, or rather to a still more recent formation, belongs also the remarkable layer of graphite at Karsok, and probably the layer of graphite at Niakornet. One has to pass over a somewhat extensive tract of the subjacent gneiss before arriving at the sedimentary strata, which appear, with a steep inclination, on the bank of the Karsok River at a height of 840 feet. Afterwards, slopes of basalt, boulders, gravel washed down from the mountains, &c., continue, until, at a height of 1,150 feet, one arrives at a terrace covered with gravel, in which a few angular pieces of graphite may be discovered, and fragments of a hard sandstone impregnated with coal. In consequence of the unfitness of our Greenland assistants for real labour, our attempts to dig through the strata of gravel and reach the graphite bed were unsuccessful; but we were informed by Capt. G. N. Brockdorff—master of the ship which, in 1850, was to have taken a cargo of graphite to Europe, and which actually carried over about five tons of that mineral—that the graphite here forms a horizontal bed eight to ten inches thick, covered with clay, sand, and angular fragments of sandstone. This interesting graphite bed does not contain any organic remains; but, as both the underlying Cretaceous strata and the graphite lie horizontally and near each other, and the latter is situated about 300 feet higher up [?], it is evident that *the Graphite at Karsok belongs either to the Cretaceous or to a still later period.*

---

* As stated above, the coal-beds probably do not belong to the *Lower*, but to the *Upper* Cretaceous (the Atane) beds.

PART IV. ("Geol. Mag.," vol. ix., p. 449, &c.)

Somewhat farther to the west of Karsok, and about 50 feet higher up, occurs another similar stratum, containing a mass of graphite, so soft that it may be cut with a knife. This spot was not, however, accurately examined. A similar stratum, of graphite imbedded in sand and clay, occurs also at a very great height above the sea at Niakornet; but time did not admit of our visiting it.

The graphite from Karsok is perfectly compact, without any signs of cleavage. On being heated, some pieces decrepitate violently and yield water. An analysis by Dr. Nordström gave:

	I.	II.	III.
Carbon -	93·70	95·68	95·42
Hydrogen -	0·69	0·22	0·27
Ash - -	4·92	3·60	3·60
	99·31	99·50	99·29

Part of the loss was probably oxygen. The ash contained peroxide of iron, alumina, and 50 per cent. of silica; so that even these analyses indicate that this mineral is much nearer pure graphite, with which it fully agrees in appearance, than the coal that is usually found in these formations.

In the strata belonging to this division we found plant-remains at the following places:—

1. *Ekkorfat.*—The strata here rest upon a red gneiss, which has a tendency to break off in scaly flakes, thus forming rounded hills on the coast. Nearest to the gneiss, at an inconsiderable distance from the shore, a little above the level of the water:

(1.) (Lowest.) Hard sandstone, unfossiliferous (60 feet).

(2.) Carbonaceous shale, with sandstone and coal-bands, interstratified with thin layers of leaves of Coniferæ (30 feet).

(3.) Hard red and white sandstone (300 feet).

(4.) Red sandstone, with bands of shale, and ripple-marked (30 ft.).

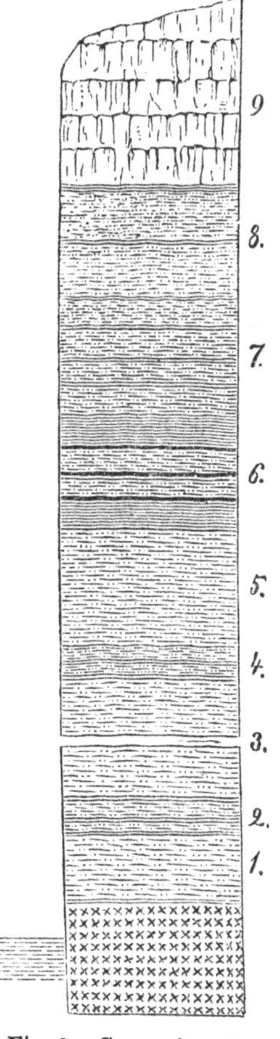

Fig. 9. Succession of strata at Ekkorfat.

(5.) Hard grey sandstone, almost like porphyry, inclosing round nodules of small stones and fragments of coal (100 feet).

(6.) Alternating layers of sandstone and carbonaceous shale, with seams of coal, layers of harder shale, impressions of leaves, etc. (100 feet).

(7.) Black shale and grey sandy shale with sandstone seams, no fossils (300 feet).

(8.) Sandstone of uniform yellow colour, the upper part, for a depth of 200 feet, interstratified with grey shale, sandstone, and coal seams (300 feet).

(9.) Basalt.

2. *Angiarsuit.* — Yellow sandstone, interstratified with grey shale, with seams of coal and impressions of plants; the same stratum as No. 8 at Ekkorfat (Fig. 9). At Ekkorfat the strata, with the exception of occasional irregularities, dip S.W., so that nearer Karsok the yellow sandstone (8) reaches to the level of the sea. We thus had an opportunity of collecting fossils from this stratum, at a place called by the natives Angiarsuit; and these decidedly belong to the same formation as the fossils from the lower strata at Ekkorfat.

3. *Avkrusak.*—Fine impressions of plants are found here, near the shore, immediately under the sandstone, in horizontally stratified shale.

4. *Karsok.*—The coast-land here, as has been mentioned above, is occupied by gneiss rocks, which, at a height of eight or nine hundred feet, are covered by a layer of shale containing fine impressions of Ferns. The shale, however, at a short distance is covered by gravel, so that the formation is exposed here only for a very limited distance, close to the Karsok river.

5. *Pattorfik.*—For a distance of six English miles from Karsok the coast towards the fjord is occupied by gneiss; but on the other side of the river, at Pattorfik, first shales and then sandstones reappear close to the shore; the former with particularly beautiful fossils, found principally in the beds nearest the gneiss. No extensive sections are however to be met with here, for the perpendicular exposed cliff, some yards above the sea-level, is covered with detritus of basalt, often hardened to a tuff-like mass, and inclosing the large subfossil shells mentioned above (p. 409).

6. *Kome*, or more properly *Kook.*—The former name, though grammatically wrong, ought however to be retained, as having been already introduced into science. The lowest portion of these strata forms on the shore an abrupt terrace, from 80 to 150 feet high. Higher up the strata terminate in a gravel-covered slope, scored by a number of deep ravines, which offer very clear sections of the various strata of the formation, for the most part nearly horizontal, or slightly dipping inwards. The series is as follows (beginning at the top):—

On the brow of the hill	*Basalt.*
About 1,500 to 1,200 feet above the level of the sea.	*Thick banks of gravel*, concealing the strata.

1,200 to 1,000 feet above sea-level.	*Shale.* *Sandstone.*
1,000 to 750 feet above sea-level.	*Shale.* *Sandstone.* *Shale, with seams of coal,* and a few plant-impressions. *Sandstone.*
750 feet above sea-level - -	*A thick stratum of coal.* *Shale, with layers of sand.* *Sand.* *Shale.* *Sand.* *Sandstone, very loose.*
150 feet above sea-level - -	*Carbonaceous shale, with bands of sand and coal.* *A coal seam.* *Shale, with abundance of impressions of plants.* *Strata not exposed.* *Gneiss.*

This section was taken in a ravine opening into the centre of Kome Bay. The finest impressions of Plants, however, occur in the neighbourhood of the house-sites, not far from the limit of the gneiss, which here forms a high mountain, immediately east of the river (Kook), which on that side seems to mark the limit of the Lower Cretaceous beds of Greenland.

Thick as the Lower Cretaceous strata are, they are now visible only over a small area, as they merely fill the valleys between the gneiss hills near the coast. The strata at Kome are separated by gneiss hills from the strata at Pattorvik, and these again in the same manner from those of Karsok, Angiarsuit, Avkrusak, and Ekkorfat. The main mass of the formation, which evidently once extended over Omenak Fjord, has been washed away. Whether or not it extended inward, into Noursoak peninsula, under the basalt, it is impossible to say with certainty, as several of the deeper valleys are filled with ice. I think, however, that this is extremely probable, although the real Kome strata seem to be wanting at Atanekerdluk. They may possibly reappear between the last-mentioned place, and the gneiss formation at Takkak. Calcareous strata are entirely absent in the Greenland Cretaceous, and it is useless to look for marine fossils there: everything shows that what we here have before us is a fresh-water deposit.

The fossils are most numerous and best preserved in the lowest strata, and consist principally of Ferns and Coniferæ. Leaves of Coniferæ and other Plant-remains are also met with, although rarely, in the upper strata; but these, in consequence of their friability, can hardly be preserved. As regards these fossils, Prof. Oswald Heer has made the following communication:—

"All the places where these remains have been discovered (Kome, Avkrusak, Angiarsuit, Karsok, Ekkorfat, Pattorfik) have the same Flora, the character of which is marked * by numerous Ferns, among which the *Gleicheniæ* (*Gleichenia Rinkiana, Zippei, Gieseckiana*) play the chief part; by a remarkable Cycad (*Zamites arctica*), magnificent leaves of which are found; by a large number of Conifers (*Pinus Crameri, Sequoia Reichenbachii, Widdringtonia gracilis*, etc.); and, in addition to this, by the almost total absence of Dicotyledons. The fine new discoveries tend to confirm my opinion, already expressed,† that this Flora belongs to the Lower Cretaceous, in all probability to the Urgonian stage. This is particularly shown by the beautiful Cycad, *Glossozamites Hoheneggeri* [?], discovered at Kome. The Greenland collections contain many specimens which resemble the plants from Wernsdorff, belonging to the Urgonian, and have exactly the same character as those from Kome. Among the most remarkable new species from the Greenland Lower Cretaceous, a fine *Tæniopteris*, n. sp., an *Adiantum* (both from Avkrusak), and an elegant new *Sequoia* from Pattorfik, deserve special mention." *See* the Lists, *above*, pp. 386, &c.

II.—*The Atane strata* (Upper Cretaceous, according to Heer).

These strata occur on the south side of the Noursoak Peninsula, between Atanekerdluk and Atane, and probably also farther on towards the north on the eastern side of the Waigat. Some few, and not clearly determinable, vegetable remains from Kome (750–1,100 feet above the sea), and from the strata situated nearest the sea-level at Kudliset‡ (Ritenbenk's coal-mine), probably belong to this formation, which contains more shale than either the subjacent Cretaceous strata or the superimposed Miocene beds, besides sand and soft sandstone, but no limestone. The thickest coal-beds in Greenland—as well those at Atane (the richest I have seen in Greenland) as those near the sea-level at Ipiit, and probably also those 750 feet above the sea at Kome—belong to this period. This is also probably the case with the strata inclosing retinite (not *amber*) at Hare Island. Small nodules of resin, however, occur in the Greenland Miocene.

Fig. 10.

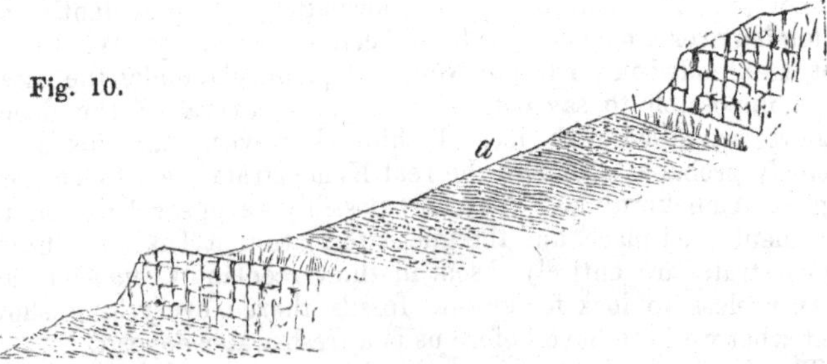

Fig. 10.—Lower series of strata at Atanekerdluk ("Lower Atanekerdluk," Heer).

---

* For the later determinations of these characteristic Plants, *see* Prof. Heer's Lists in his "Kreideflora," &c., reprinted *above*, pages 386, &c.—EDITOR.

† Heer's "Flora fossilis arctica."

‡ The upper strata in the neighbourhood of Kudliset are Miocene.

Fig. 10 shows the succession in the lower portion of the section seen at Atanekerdluk. The mass of the formation in this lower slope consists of very fine black shale (*a*), resembling the shale at Cape Starastschin, in Spitzbergen, containing a quantity of plant-remains, which, however, it is very difficult to preserve, in consequence of the brittleness of the shale. There are no marine fossils whatever here, so that it is evidently a freshwater formation.

At Atane the adjoining cliffs nearest to the water's edge are concealed by stone and gravel, consisting partly of sandstone and partly of basalt and basalt-breccia containing zeolite. Over these we have :

At 450 feet, horizontal strata of hard sandstone.

At 600 feet, shale, which soon alternates with sandstone.

At 650 feet, a thick coal-bed resting upon fine shale, with impressions of Plants (Upper Cretaceous) and particles of resin. Then again shale, often interstratified with coal-beds of considerable thickness.

At 900 feet, a coal-bed two feet thick, from which, on the side left bare by the ravine a white salt has fretted out (sulphate of alumina). On this is a sandstone 50 feet thick, then shale, and over that sandstone again, and lastly basalt.

On the fossils from these places Professor Heer remarks: "The fossils from the lower strata at Atanekerdluk belong probably to the Upper Cretaceous. This appears from :—

"1. The presence of a remarkable Cycad (*Cycadites Dicksoni*). It is true that this is not altogether consistent with the supposition that these impressions belong to the Eocene formation [?]; but at any rate no Cycad, and especially no *Cycadites*, has hitherto been found in strata belonging to the Eocene epoch.

"2. The frequent occurrence of Ferns.

"3. The occurrence of a *Sequoia* scarçely distinguishable from *Sequoia Reichenbachii*;

"4. And of a *Credneria*, of which, however, only fragments are before us.

"On the other hand, this Flora differs entirely from that at Kome, especially by the presence of pretty numerous dicotyledonous leaves, which are, moreover, quite unlike the Greenland Miocene plants. The investigation of these fossils presents serious difficulties, as the greater part of them are those of full-bordered leaves with a complicated nervation offering but few fixed points of discrimination. One leaf seems to agree with *Magnolia alternans*, Heer, from the Upper Cretaceous of Nebraska.

"These dicotyledonous leaves indicate the Upper Cretaceous formation, but to which of its sub-divisions the lower strata at Atanekerdluk are to be assigned can only be determined by a closer investigation. This new flora is, at any rate, one of the greatest discoveries of the Expedition of 1870, opening, as it does, for North Greenland an entirely new geological horizon, which shows that in the Arctic regions, as in Europe, *Dicotyledonous*

Plants do not occur in the Cretaceous beneath the Gault, whereas immediately above it they appear in a great variety of forms. In North Greenland then, as well as in Europe and America, the vegetable world underwent great changes during the course of the Cretaceous age."

III.—*The Miocene Formation.*

During the Miocene Period masses of basalt, sand, and clay, to a depth of many thousand feet, were piled together in the district of Greenland we are now considering; and by far the greater part of the rocks on Disko Island and Noursoak Peninsula belong to that epoch. The Greenland Miocene strata (of sedimentary and eruptive origin) may be arranged under three divisions, namely:—

(*a*) *Lowest.* Sand or soft sandstone, with shale, coal-bands of slight thickness, and ferruginous clay-beds, very rich in impressions of plants.

(*b*) Basalt, Tuff, and Lava, several thousand feet in thickness, usually as regularly stratified as sand-beds, often alternating with basalt beds. At about the middle of this basalt formation layers of fossiliferous clay, sand, and ferruginous clay, of limited thickness, are met with.

(*c*) Loose layers of sand, and one or two bands of clay, deposited on the southern coast of the Isle of Disko, *between* the basalt rocks, and therefore of more recent date.

From all these localities, separated from each other by basalt strata, 2,000 feet thick, numerous fossils have been collected, indicating, according to Heer, the Miocene Period. As the strata are, nevertheless, in geological respects widely different from each other, I give an account of each separately.

III. *a.*—*Upper Atanekerdluk strata.*—At Atanekerdluk we meet with fossils from two different stages, namely: (1) between 300 and 400 feet above the sea, shales with thin sand-beds and coal-seams (*e*, fig. 12, further on), with fossils imbedded in black shale and belonging to the Upper Cretaceous (the Atane strata described above, p. 430); and (2) thick sand-beds, with occasional shales (*c*, *d*), containing but few fossils. At 1,000–1,200 feet these layers of sand begin to be interstratified with a ferruginous clay, which, as well as the sandstone close to it, is remarkably rich in impressions of plants. The greatest part of the fossils that have been brought home from Greenland belong to this locality, of the discovery and scientific examination of which I have already given a succinct account. Here I will only add a few words on the hitherto imperfectly, and in part, inaccurately, described geognostic relations of the place.

By the name "Atanekerdluk," the Greenlanders designate a little peninsula, 400 feet high, connected with the mainland by a small isthmus, in the southern part of the Waigat, and forming a projection from the cliffs of Noursoak, which are bold everywhere else, and rise to 3,000 feet even close to the coast. This place was formerly the seat of a Greenland colony, round a Danish "outpost" (Utliggare), but is now uninhabited. Deserted house-sites and paths, which in Greenland remain unobliterated for a great length of time,* and a number of graves, still serve to remind us of the now dead or scattered little colony. The peninsula itself is formed of a rusty-brown, rather coarse-grained dolerite, composed of two species of felspar (labradorite and sanidin ?), titaniferous iron, in thin hexagonal laminæ, and augite. In this it differs from the genuine Greenland basalt and basalt-tuff, although it evidently only forms the oldest link of the vast volcanic and plutonic chain of rocks of North-west Greenland. At the steep cliffs on the western side of the peninsula one can see even that dolerite is lying on sandstone of the same loose character as the superjacent sand and sandstone beds.

Immediately on the other side of the low isthmus, which rises only a few feet above the water, uniting the peninsula with the mainland, we first meet with the above-described Atane strata (*e*); then follows sand, after which a basalt bed again, covered by layers of sand alternating with shale, and crossed by vast plutonic veins (a, a', a'', a'''), which seem not to have had the smallest influence on the sand through which they have passed. Only here and there a grain of sand is found melted, or rather rusted, into the surface of the dyke.† The upper part of the dyke generally forms a ridge standing up from the surrounding loose layers of earth. Between the layers of shale we find one or two small seams of coal, and in the sand here and there a carbonised stem of a tree, but no real impressions of leaves, until we come to a height of 1,200 feet above the sea.‡ Here commences sand or sandstone, with clay, covered by shale, and interstratified with thin beds of ferruginous clay-rock (*b*), often divided into large or small lenticular masses, and extremely rich in Miocene fossils. These occur not only in the ferruginous clay, but also in the surrounding somewhat hardened sandstone, and may perhaps be obtained from this sandstone in greater perfection than from the extremely hard and unmanageable ferruginous clay. We often found in the sandstone nodules and flat ellipsoids of ferruginous clay so full of remains of plants, especially on the surface, that it looks as if these nodules, before they had been imbedded in the sand and hardened, had been rolled in a heap of leaves. The ferruginous clay has, when newly broken, a dark-grey fracture, which, by exposure to the air and the polishing effect of the sand, acquires a polish and a brick-brown colour. Pieces of it are plentifully scattered about in the confined locality where these vegetable remains occur. In the same sandstone, a little south of the spot where the impressions of leaves are met with, may be found at the edge of the glen, very deep at

* Rink mentions paths still remaining in districts uninhabited since the time of the old Northmen colonists, and we ourselves could clearly distinguish at Kaja the paths round the long-deserted house-sites there.

† The remarkably slight effect which the eruptive rock has produced on the surrounding layers of sand astonished Mr. Brown also.

‡ 1084 Inglefield; 1175 mean of six measurements with the aneroid by Whymper; 1203 by the aneroid used by the Expedition of 1870.

this spot, trunks of trees, the tops of which rise above the sand, or form black spots in the white sand. An excavation was made in our presence, and we saw, as the annexed woodcut indicates, the roots branch out in an underlying clay-bed. There can, therefore, be no doubt that these trunks once grew in the place where they are now found. Above these strata is sand, then a thick stratum of basalt, over which sand again, and lastly a basalt bed, perhaps 2,000 feet thick, and, as far as one can judge from a distance, not interstratified with layers of sand or shale.

FIG. 11.

Fig. 11.—Bituminised tree trunk at Atanekerdluk.

At Atanekerdluk itself the strata follow the direction of the strait (or, more correctly speaking, strike true N.N.W. S.S.E.*), and the slope, as indicated in the following sections, taken from a ravine the direction of which was at right angles to the shore, is 8°–32° E.N.E. Further up in the strait the strata gradually sink, so that the capping of basalt reaches down to the surface of the sea a little north of Atane. The perturbations at Atanekerdluk, therefore, seem to have been only local; and, on the whole, the strata may be said to lie nearly horizontal, with a slight dip to N.W.

This Miocene formation has evidently in former times extended completely over the Waigat to Disko Isle, at the south-east angle of which it attains its greatest thickness. One may here see from the sea sandhills of 2,000 or 3,000 feet high, often, but not always, containing basalt-beds. The chief substance of the mountain consists of vast horizontal sand-beds, interstratified with thinnish beds of clay, and occasional horizontal coal-bands, with carbonised stems of trees, sometimes in their original position and of considerable size. A stem of this kind, two feet in diameter, was, for example, seen in a rock in the district about Mudderbugten. The quantity of carbonised stems is often so great that the Greenlanders collect and use them as fuel. Silicified tree-stems are also met with, though more rarely. The greatest number of impressions of leaves occur, both on the western shore of the Waigat, and at Atanekerdluk, almost invariably in a hard, grey, ferruginous clay-rock turning red by exposure to the atmosphere ("Atanekerdlukstone"), which forms either peculiar beds, one or two inches thick and a few fathoms in extent, or lenticular masses in sand or clay, or small balls in huge, almost spherical sandstone

* Mean of several observations made in the ravine along the side of which I ascended the slope. Brown gives E. and W. as the direction. The difference probably arises from the circumstance that the magnetic perturbations at Atanekerdluk are of a local nature, and thus different in different ravines.

nodules, segregated in the sand by the infiltration of some conglomerating medium, and often of remarkably regular form and some yards in section. Atanckerdlukstone, like the nearest sands and clays always contains remains of leaves, which may then either form small separate layers or an isolated nodule in

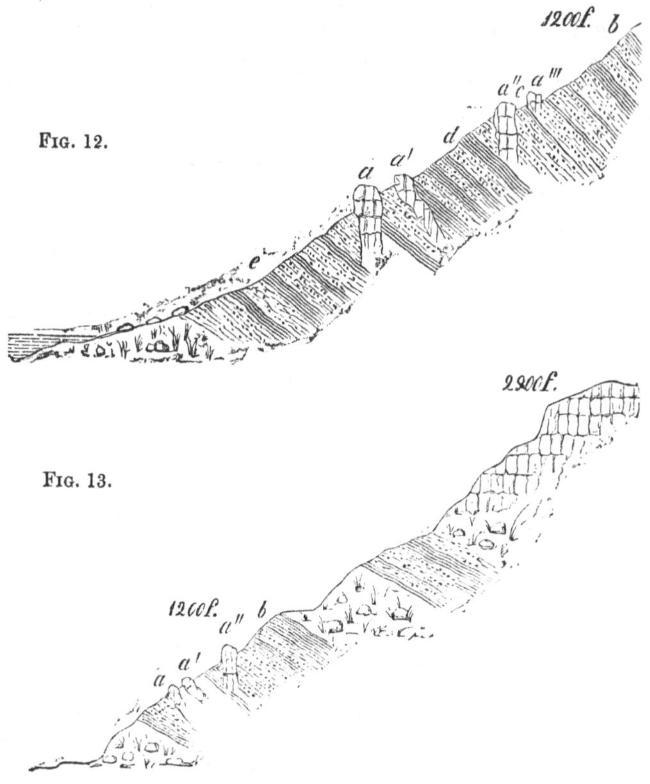

Figs. 12 and 13.—Section of the strata at Atanekerdluk. (See also fig. 10.) The scale of fig. 13 is about half that of fig. 12.

the sand, some few inches in diameter; whereas it would be vain to look for impressions of leaves in the more distant sand-beds. Coal-beds worth working probably do not occur in this horizon of the Miocene; at least the layers at Atane, the largest coal-beds at Kome, and at Ipiit near Kudliset, seem to belong to *the Upper Cretaceous*, while the strata at Netluarsak, Isorisok, the coal in the high fells at Skandsen and Assakak, belong *to the middle, not the lower, horizon of the Miocene of Greenland.* Probably also the coal-beds at Hare Island belong to the Upper Cretaceous formation, as I have already observed (*above*, p. 430).*

* Dr. Nauckhoff's and Dr. Pfaff's discovery of *Sigillaria* makes it possible that the Coal of the Coal-formation occurs at Ujarasusuk.

From the Lower Miocene strata at Disco we collected fossils at Flakkerkuk, and near Mudderbugten, Isungoak, Ujarasusuk, and Iglosungoak. These localities are not to be compared with Atanekerdluk for richness in fossils.

III. *b.—Ipsorisok strata.*—By this name I designate the thinnish layers containing fossils that occur imbedded in the basalt of the high hills. Such strata have been met with at—

Netluarsuk, between Noursoak and Noursak. A little north of Atane the basalt sinks down to the surface of the sea, and from a distance it is impossible to discover in the very regularly stratified basalt-beds, ending at the shore with a vertical section of several thousand feet, any sand or shale beds. Neither do the Greenlanders know of any other coal-beds in that neighbourhood than one which is met with at Netluarsuk, at an elevation of about 1000 feet. The strata are here, for a distance of a few dozen feet, exposed at a steep gorge between the basalt hills. They seem to be of trifling thickness, and consist of alternating beds of from 0·2 to 2 inches thick of sand, coal, shale, and a ferruginous clay, different in appearance from the ferruginous clay at Atanekerdluk, though, like it, full of fossils, chiefly of Fir leaves and twigs, mixed with clay or coal. Among these fossils occur not only leaves and cones, but also seeds. The coal consists almost exclusively of flattened and carbonized stems.

Ifsorisok, about twelve miles from the coast, and 2250 feet above the sea. We visited the spot from Holländarbugten or Itiblit, situated a little to the north of Niakornet. Some distance from the coast we first find thick layers of a rock, which appears to be a much changed siliceous slate. Afterwards the path proceeds up steep slopes of basalt detritus and basalt rocks, or (at 2300 feet) extensive plains, covered with the same material, and, at the period of our visit, free from snow, though hardly clothed with any vegetation. Here one has to pass long distances over weathered and crumbling slabs of basalt, which show that the underlying rocks are everywhere composed of eruptive masses. From these plains considerable basalt hills rise further inward, among which Kinnitok—a lofty mountain-ridge between Niakornet and Ekkorfat—is the largest. This mountain is probably 5000 or 6000 feet high, and, seen from a distance appears also to be composed entirely of the eruptive rock common in these parts.

Somewhat beyond the spot where one passes the highest point of the plains are some shallow valleys. In the slope of one of them is the spot which formed the object of our visit. The place betrays itself by large and small pieces of coal lying mixed with the basalt detritus ; and, on digging here, sedimentary strata, consisting of coal-seams some inches thick, sandy clay, and fine, grey, hardened clay are discovered. The clay contains impressions of plants, and among the coal flattened and imperfectly carbonized tree-stems are met with. Silicified wood is also found in the gravel. The schists are evidently of no great thickness, but regularly stratified with a dip of about 10° towards the north.

Assakak.—Immediately south of Kome River, Noursoakland, the coast consists of lofty gneiss rocks, between which a number of glaciers project. One of these, Assakak glacier, has long been celebrated for the charred tree-stems lying scattered on the surface of the ice. The glacier itself does not reach down to the sea, but is separated from the shore by a low foreland, covered with boulders of gneiss, and passing without any discoverable line of demarcation into the glacier, which is there also itself covered with gravel. The gravel, however, here principally consists of angular fragments of basalt, among which pieces of charred wood may be here and there remarked. Higher up the mass of charred or silicified wood increased considerably, and was often piled together, as if by human hand. It was, however, easy to satisfy oneself that this was not the case, but that the coal came from some stratum in the neighbourhood of the glacier, on the surface of which it now lay scattered, chiefly at a height *estimated* by me at about 300 feet. The nearest high mountains surrounding the glacier seemed to consist of gneiss, hornblende-slate, etc. A thick fog prevented us from seeing far inward, and induced us to defer an excursion we had intended in that direction, which probably, as far as the object of finding the stratum from which the pieces of wood had come is concerned, would not have been crowned with success. In fact, it is probable that the fragments of wood belong to a Tertiary stratum *beneath* the glacier. After a careful search, pieces of clay and sandstone were found, containing remnants of plants exactly similar to the fossils at Ifsorisok, whence I draw the conclusion, that the strata, whence the coal has originated, were about contemporaneous with those of Ifsorisok and Netluarsuk.

The strata of this horizon are separated from the Lower Miocene strata at Atanekerdluk by basalt-beds several thousand feet thick, for the formation of which an immense lapse of time must have been required; and one would accordingly expect to find here remains of a vegetation very different from the Miocene vegetation of Atanekerdluk; but this is not the case. According to Professor Heer, the fossils in both these places have a purely Miocene stamp. As evidence of this, Professor Heer adduces the presence of *Sequoia Langsdorfii*, at Ifsorisok, and that of *Taxodium distichum*, *Glyptostrobus Europæus*, and *Chamæcyparis Massiliensis* at Netluarsak.

IV.—*The Sinnifik strata.*

At Godhavn the basalt rests immediately upon gneiss, but only a little way to the east the eruptive rock reaches the sea-level; and in rowing hither along the southern shore of Disko, we pass cliffs of basalt-tuff and basalt, often (as, for example, at the Brännvinshamnen) broken up in the most splendid manner into hexagonal basalt columns, basalt grottoes, and basalt arches. On the other side of BDededalen the basalt first begins to be interstratified with sand and shale beds, which probably are the beginnings of those vast sandy strata that meet us on both sides of the entrance to the Waigat.

Further on, at Puilasok and Sinnifik, the shore itself consists of sandstone, with very thin shales, here and there interrupted by basaltic cliffs, with a worn and smooth surface. The sandstone around the cliffs is not in appearance distinguishable from the sand still heaped by the action of wind and wave around the basalt rocks on the shore. Everything seems to show, that in many places hereabouts,* we have before us sand-beds deposited between basalt rocks. In this case these layers are more recent than the whole basalt formation; and the fossils they contain, imbedded partly in an extremely brittle shale, partly (at Sinnifik) in hard marl-nodules resembling those at Atanekerdluk, but not containing very much iron, are of interest as indicating the limit of the period during which this tract was the scene of the vast volcanic eruptions which have given rise to the basaltic masses of North-west Greenland. These fossils consist, at Puilasok, of fragments of leaves of broad-leaved trees (*Salix*, *Myrica*, *Plantanus aceroides*, *Cratægus antiqua*, etc.); at Sinnifik, of leaves of both broad-leaved trees and Coniferæ (*Sequoia Langsdorfii*, *Taxites Olrikii*, *Populus arctica*), and, according to Heer, bear constant witness to a Miocene, perhaps an Upper-Miocene epoch. If this be so, the volcanic agency in these parts commenced during the Cretaceous and terminated previously to the close of the Miocene period. The basalt-beds in the Cretaceous and Lower (Greenland) Miocene are, however, quite trifling in comparison with those which cover the Miocene deposits at ¡Atanekerdluk, Ujarasusuk, Isungoak, etc. Accordingly in these (Greenland) districts the volcanic action attained its greatest intensity in the Middle Miocene.

During our involuntary stay at Godhavn, I made an excursion, in company with some comrades, in a boat manned by Greenlanders, to the spot whence the Rudolph meteoric iron was supposed to have been taken, namely, the old whaling-station of Fortune Bay, in the neighbourhood of Godhavn. On arriving there, I ordered the Greenlanders to look after *heavy, round, rusty-brown stones, which I knew would certainly be found somewhere thereabout.* It was in vain. No meteoric stones, or rather pieces of meteoric iron, were on this occasion found; but before leaving the spot I again repeated to the Greenlanders, that pieces of iron of the nature described *were most unquestionably to be met with somewhere in that neighbourhood,* and I promised them a reward, if they could discover them, against my return in the autumn.

When, at the end of August, we returned from Omenak to Godhavn, one of the Greenlanders communicated to me, with many lively gestures to express their size, shape, etc., that they had

* In this neighbourhood we even meet with sand seams beneath basalt.

decidedly hit upon the stones I had described. A small specimen was shown, which confirmed the statement.

The place where the iron masses were found was not, however, at Fortune Bay, but one of the shores most difficult of access in the whole coast of Danish Greenland, namely Ovifak, or the Blue Hill, which lies quite open to the south wind, and is inaccessible in even a very moderate sea, between Laxe-bugt and Disko-fjord. I scarcely need mention that this discovery completely altered the plan for our further excursions. Our intention had been to employ the rest of our sojourn in Greenland in an examination of the basalt formations between Skandsen and Godhavn, and we had therefore, immediately on our arrival at Godhavn, hired two whale-boats manned with Greenlanders, with a view to rowing in short day-journeys with them along the coast of Disko to the eastward of Godhavn. These boats, on the morning when the discovery of the meteorites was made, lay ready and provisioned on the strand. We immediately set sail, and, favoured by a tolerably good wind, we sailed westward to Ovifak, where we arrived the same evening before sunset. The sea was calm, so that it was possible to land, and the very stone at which we lay to was itself a piece of meteoric iron, probably the largest piece yet known. On searching more carefully we further discovered two large and a great number of smaller pieces of meteoric iron scattered over an area of a few square fathoms in the vicinity of the large stone.

The meteorites lay as on the accompanying map and section,* between high and low water, among rounded blocks of gneiss and granite, at the foot of a vast basalt slope, from which, higher up, the horizontal basalt-beds of Mount Ovifak project. Sixteen metres from the largest iron block a basalt ridge, a foot high, rises from the detritus on the shore, and could be followed for a distance of four metres, and is probably part of the rock. Parallel with this and nearer to the sea is another similar ridge, also about four metres long. *The former contained lenticular and discoidal blocks* of nickel-iron, like meteoric iron, in external appearance, chemical nature, and relation to the atmosphere (weathering). On being polished and etched this iron exhibited fine Widmanstädtian figures. The native iron lay imbedded in the basalt, separated from it at the most by a thin coating of rust. Moreover, in that basalt, in the neighbourhood of the blocks of native iron, nodules of hisingerite were found, evidently formed by the oxidation of the iron, as also small imbedded particles of nickel-iron.

The meteorites themselves were of various colours, from that of tombac (pinch-beck) to rusty-brown; and in some places at least they had a metallic lustre on the surface. Here and there one could discover on their surface, and in the iron nearest the sur-

* This Map (Plate VIII., fig. 1 and 2), inserted at p. 355, "Geol. Mag." for August, with Part II. of Prof. Nordenskiöld's paper, is not reproduced here.—EDITOR.

face, pieces of basalt, or fragments of a crust of basalt, perfectly similar to the basalt in the above-described ridge. The inner part of the iron mass contained no basalt, and as far as analysis has yet been able to discover, scarcely any traces of silica. In the neighbourhood of the smaller stones the sand and gravel were rusty with the effects of the weathering of the meteorites, yet their upper surface was usually pretty pure, but the under surface generally rusty. The larger stones were strongly polar-magnetic, so that the upper part of the stones attracted the north, the lower part the south pole of the magnetic needle.

Within the area represented on the map above-mentioned not exceeding 50 square metres, the following blocks of meteoric iron were found by the Expedition of 1870:

Fig. 14.—The three largest Meteoric Stones.* From a sketch made on the spot by Dr. Th. Nordström in 1870.

		kilog.
1.	Ovally rounded. Greatest diam. above ground 2 metres, smallest 1·7 m. Probable weight -	21,000
2.	Nearly spherical. Greatest and least diam. above ground, 1·3 and 1·27 met. Probable weight -	8,000
3.	Somewhat conical. Greatest and least diam. above ground, 1·15 and 0·85 met. Probable weight -	7,000
4.	Oval stone, weighing - -	142
5.	A drop-shaped stone, weighing -	96
6.	Another: now belonging to the British Museum;	about 87
7.	A stone, weighing - - - - -	„ 54
8.	A stone „ - - - -	81
9.	A stone „ - - - - -	about 42
10.	A stone - - - - -	18
11.	A stone - - - - - -	24
12.	A stone, which immediately after our arrival home fell to dust; originally weighing -	about 54
13.	A smaller stone, weighing - -	6·4
14.	Another „ - - -	3·4
15.	Another „ - - -	2·5
	Several lenticular pieces of iron from the basalt vein, 3–4 inches thick, weighing altogether - -	about 100

* Nos. 1, 2, and 3 of the list: brought to Europe by the Swedish Greenland Expedition of 1872, under command of Capt. Baron von Ober.

The Ovifak iron is extremely crystalline and brittle, so that smaller pieces may be broken with a hammer; and, with the exception of the little bits of basalt on or near the surface, it is not mixed with any silicates visible to the naked eye. The iron from the basalt ridge differs from the other by a rougher fracture and greater toughness. With the naked eye one can seldom discover any nodules of troilite or iron-sulphide. In the weathered detritus, on the other hand, a few black magnetic grains were found, with strongly reflecting facettes and octahedral surfaces, which on examination we found to be magnetite. When cut and polished, the different specimens varied very greatly; on some of them parts, yellow as brass, of troilite were discernible, and the polished surface of the metal itself appeared, when the light fell on it in a certain direction, divided into rounded parts, of different brilliancy and shades of colour. Other pieces seemed to form a perfectly homogeneous aggregate of crystal needles of carburetted nickel-iron. The Widmanstädtian figures were visible after etching on some, but not all, of the specimens. *These were particularly distinct on the iron from the above-mentioned basalt ridge.* In general the iron was so hard that they would not undertake at the ironworks to saw through any of the larger balls, in consequence of which I know no more of the internal character of the meteoric iron than what I could ascertain from the specimens which fell to pieces.

PART V. ("Geological Mag.," for November 1872, p. 516.)

Notwithstanding the very inconsiderable amount of sulphur it contains, this Greenland iron has a remarkable tendency to fall to pieces by the action of the air. The weathering depends on an oxidation, probably produced by a quantity of chlorine contained in the iron, and its great porosity; nevertheless, some of the phenomena connected with the weathering still appear to me inexplicable. I shall therefore somewhat more fully detail the observations and experiments made towards explaining this very disagreeable circumstance.

The Ovifak meteoric iron does not fall to pieces at the place where it was found, though sometimes washed by the sea, sometimes left bare; but on the shore it was preserved at the temperature of the sea, which varies but little during the whole year.

Even during the passage, when the masses lay packed in wooden chests in the hold, and were exposed to a very moist atmosphere and at a temperature but little above freezing-point, the unbroken stones did not suffer perceptibly; whereas almost all the fragments packed in the same manner split into pieces, more particularly those which I had preserved in the heated cabin.

From some of the pieces of iron sea-green drops oozed out, which afterwards became reddish-brown by the action of the atmosphere. They contained protochloride of iron with traces of sulphate.

One of the larger pieces, which, after our return home, was placed in a room of ordinary temperature, soon began to crack on its surface, and ultimately, when unpacked two months later in Stockholm, crumbled to a reddish-brown powder, consisting partly of a fine rust-powder, partly of angular bits of iron, rusty on the surface, and varying in magnitude from the size of a pea to that of a hemp-seed. An entirely unchanged, and therefore, on fresh fracture still metallic, portion of Stone 4, began at one corner to rust, swell and crumble, while the remainder of the iron remained unaltered. The rust spread itself like a fungous growth over the rest of the piece, and extended itself to the interior, which thereupon swelled and crumbled like an efflorescent salt. During this time the weight of the piece of iron increased.

Weight of a fragment of iron when packed	-	29·935 gr.
" " after 129 days	-	30·143 gr.
Weight of the unchanged iron	- -	24·529 gr.

so that 5·406 gr. had weathered away to a rusty-brown powder and during this time had increased in weight 0·208 gr. or 3·8 per cent.

In a hermetically sealed glass tube the iron was completely unchanged.

In a glass tube, that had been hermetically sealed, but in which a fine crack had taken place in cooling, the iron continues to crumble.

In a eudiometer over mercury, the iron in a few days absorbed a considerable amount of oxygen, in consequence of which the mercury rises in the tube.

In alcohol, the iron does not crumble. In water, it rusts, but does not appear to fall to pieces.

In air dried by sulphuric acid the crumbling process takes place slowly.

Varnishing does not fully protect these pieces of iron from weathering, not even if immersed in warm copal-varnish. I thought at first that the cracking was the result of the contracting and shrinking of the mass, but this is not the case. On the contrary, the cracking is caused by dilatation. With what force this operates may be judged from the fact, that a piece of iron, on which chisel and saw are used without effect, is broken or bent by the decomposition of the mass. In general, cracks first appear at right angles to the surface of the stone; these diverge as from a centre, and, at a depth of a few lines below the surface of the stone, meet a crack that runs parallel with the surface, which, by the swelling of the overlying crust, is soon formed into a little dome, sometimes an inch in height. In the meantime the overlying crust is raised, doubled up and broken in a manner which bears a striking likeness to the doubling of the stratified rocks by the so-called eruptive forces,—that is, if one supposes that the cracks, instead of being empty, are filled with detritus, which gradually hardens to an "eruptive" rock.

When fragments of the largest stone, weighing 134 gr., were heated to redness, they parted with nearly two litres of gas, or about 100 times the volume of the iron, as also a considerable amount of water, which, like the gas, had a bituminous smell.

The gas was clearly no primary constituent, but was formed partly by the decomposition of organic matter in the meteorite, partly by the reducing operation of compounds containing carbon on the oxide of iron in the meteorite, which was found to be completely reduced at the termination of the experiment. On the iron being dissolved in chloride of mercury, only a trifling quantity of gas was emitted, probably coming from the pores in the iron. In hydrochloric and nitric acid the meteoric iron is dissolved, leaving in some cases a residue containing much carbon, in others very little residue at all. The gas that escapes during solution in hydrochloric acid has a most penetrating smell, probably due to some hydrocarbon. On dissolving in acid Ovifak iron which has been heated to redness in air or oxygen, there often remains a flocky, voluminous, brown material soluble in warm, but hardly soluble in cold, water, and very easily dissolved in ammonia, forming a dark-brown, almost opaque fluid. The same material is obtained from the carbon that remains after the solution of the iron in acids. It can again be precipitated by means of acids from the ammoniacal solution, though not quite completely, so that the acid solution is also brown, but of a very light tint. This material is a humus-like compound, which probably did not originally exist in the meteorite, but arises from the solution of the carboniferous iron in acids.* This humus-like body can be decomposed with difficulty by long boiling in strong nitric acid or chlorate of potash and hydrochloric acid.

The following analyses have been made of this iron from Ovifak:—

I. Analysis of a fragment from one of the large stones, by A. E. Nordenskiöld. II. Analysis of a specimen of more compact iron, by Th. Nordström. III. Analysis of iron with conspicuous Widmanstättian figures from the basalt ridge, by G. Lindström:—

	I.	II.	III.
Iron - -	84·49	86·34	93·24
Nickel - -	2·48	1·64	1·24
Cobalt - -	0·07	0·35	0·56
Copper - - -	0·27	0·19	0·19
Alumina - - -	hardly perceptible	0·24	——
Lime - - -		0·48	——
Magnesia - - -	0·04	0·29	some traces.
Potash - -	scarcely enough to weigh.	0·07	0·08
Soda - - -		0·14	0·12
Phosphorus - - -	0·20	0·07	0·03
Sulphur -	1·52	0·22	1·21
Chlorine - - -	0·72	1·16	0·16
Silicic acid - - -	scarcely perceptible	0·66	0·59
Insoluble portion - -	0·05	4·37	
Carbon, Organic matter, Oxygen, and Water (loss)	10·16	3·71	C. 2·30 H. 0·07
	100·00	100·00	99·79

* A similar substance, obtained by dissolving iron containing carbon, has been mentioned by Berzelius, in *Afhandl. i Fysik, Kemi och Mineralogi.* When iron containing carbon is dissolved in hydrochloric acid of proper

I. Contained scarcely any traces of silicic acid, alumina, or lime. The iron was, therefore, entirely free from silicates, although large lumps of basalt were firmly rusted on to the surface of the meteorite, and one or two fragments of basalt surrounded with iron could be observed within the iron near the surface. Even before heating to redness, I. emitted a good deal of water and gas, as much apparently as amounted to about 100 times the volume of the iron—that is to say, considerably more than the iron examined in Analyses I. and II.; this explains the large loss in I. The specific gravity of I. was ascertained, from two (porous) fragments of some grammes weight, to be 6·36 and 5·86. The smaller specific gravity here arises evidently from the large quantity of carbonaceous matter that is contained in this iron. Nordström obtained the specific weight of II. from two experiments on small pieces=7·05 and 7·06. Lindström found the specific gravity of III. at 17° C. to be equal to 6·24. The iron employed in Analysis II. was less crystalline and more compact than that used in Analysis I. It was hard to break, and small grains could be hammered flat without disintegration. In Analyses II. and III. the materials examined were in external appearance precisely alike, and I therefore consider it as probable that the material of II. also was from the basalt ridge, although it had afterwards crumbled apart.

IV. Analysis of the silicate that remained undissolved in Analysis II., by Dr. Th. Nordström. V. Analysis of a piece of basalt firmly rusted on to the surface of the largest meteorite, by Dr. Th. Nordström:—

	IV.	V.
Silicic Acid	61·79	44·01
Alumina	23·31	14·27
Sesquioxide of Iron	1·45	3·89
Protoxide of Iron	—	14·75
Magnesia	2·83	8·11
Lime	8·33	10·91
Potash } (loss included)	2·29	0·97
Soda } (loss included)		2·61
	100·00	99·52

VI. and VII. Analyses of the carbonaceous matter in the iron of II. by Nordström. 33·0479 gr. gave, after first treating with

strength and temperature, not only is this humus-like matter generated, but hydro-carbons also, and (according to a statement made to me by Prof. Eggertz) even fluid hydro-carbons, the atomic composition of which is very complicated. We have here, then, a method for attempting the synthesis of organic substances from their inorganic components unemployed hitherto, as far as I am aware, in synthetic organic chemistry. *Iron containing carbon was pointed out by Berzelius in* 1818 (Aph. i. Fysik, Kemi, etc., vol. v. p. 534) *as an inorganic material which might serve as a means for the synthetical formation of organic compounds.*

chloride of copper, and afterwards with chloride of iron, 4·79 per cent. of a carbonaceous matter containing 42·58 per cent. ash. An elementary analysis of this carbonaceous matter, deducting the ash, gave—

	VI.	VII.
Carbon - - -	63·59	63·64
Hydrogen - -	3·26	3·55
Oxygen (loss included) -	33·15	32·81
	100·00	100·00

This substance is not soluble in either alcohol, ammonia, or potash, and evidently consists of a mixture of organic matter, water, and carbon.

The discovery at Ovifak is remarkable, not only as the largest discovery of meteoric iron hitherto known to have been made, but also as of that which is richest in carbon, excepting the carbon powder that fell at Hessle. Add to this, the remarkable circumstances, that lenticular and discoidal pieces of native iron occur *at the same place* in the underlying basalt, and that basalt in pieces of considerable size, at numerous spots, forms a crust on the larger meteorites, and even sometimes has been driven through the surface into the iron. Nevertheless, in spite of this, it appears to me that there cannot be a doubt of the really meteoric origin of the large masses. Their form, their composition, their appearance, sufficiently indicate this. To explain the occurrence of meteoric iron together with basalt we must then assume:

That (1) the ridges FG and GH (*see* map*) on the shore are not really in connexion with the rock, but are only fragments of one large meteorite of 20 to 40 feet in diameter, formed principally of a mass of basalt-like matter, with balls of iron disseminated through it, that has fallen at this spot. This assumption would, however, be too hazardous, and is rendered improbable by the circumstance that the basalt that surrounds the meteoric iron is perfectly similar to the variety of Greenland basalt forming the rocks of the locality.† The greatest part of the mass in which the iron particles are scattered is, however, very unlike genuine basalt, and in external appearance rather resembles the meteoric stone from Tanacera Pass, in Chili. Time has not yet permitted a more accurate investigation.

Or that (2) the whole fall of meteoric iron took place during the period when the piling up of these Greenland basalt rocks was in progress, *i.e.* during the latter portion of the Cretaceous and the beginning of the Tertiary periods. Some of the pieces of meteoric iron have fallen to iron-dust, and filled cracks in the basalt, where they have again hardened into the iron above de-

* Published with Part II. in "Geol. Mag.," August, p. 355.

† Only the basalt in some parts of the ridge FG and GH, but not the basalt from other districts of Disko and Noursoak, contains native iron.

scribed as found in the ridge FG. Of similar origin are also the particles of native iron in the basalt lying nearest the iron, which occasionally has a conglomerate-like structure.

As considerable masses of iron, of a composition probably very similar to that of meteoric iron, without a doubt occur in the interior of the earth, it may be suggested that the Ovifak iron may be of telluric origin, and that it has been, together with the plutonic rocks, thrown up during the eruptions that have given rise to the vast strata of basalt in this neighbourhood. But not only does the fully marked meteoritic form of the many iron pieces militate against this supposition, but also the circumstance that the iron in question—as the facts of its containing organic matter, its porosity etc., show—*has evidently never been heated even to a temperature of a few hundred degrees.*

Neither is it possible that these masses of iron can have arisen from the reduction by gases developed in connexion with basalt eruptions of a ferruginous mineral. Iron-pyrites cannot be reduced by these means ; and no oxide-of-iron-mineral containing nickel, and at the same time almost free from lime and silica, is known. The formation of the iron from chloride of iron, erupted from the interior of the earth and since reduced, can hardly be supposed. The explanation I have given above, that the iron is the result of an unusually rich Miocene fall of meteoric iron, seems, therefore, to me most plausible.

Öberg was fortunate enough to meet with a piece of meteoric iron from the neighbourhood of Jakobshavn. He received the piece, which weighed 7½ Skålpund (7lb Avoird.), from Dr. Pfaff, of Jakobshavn. This piece, which is now preserved in the Riks Museum at Stockholm, is an oval lump, with a somewhat rough surface, consisting principally of very hard, tough iron, not crumbling. On being sawn through, it presented the appearance of a mass of iron grains welded together, here and there impregnated with a basalt-like black silicate. On etching, fine Widmanstättian figures are obtained. We have not had time to analyse it, and I need not therefore dwell longer on the description of it, especially if, as is greatly to be wished,* the three larger iron blocks at Ovifak should be brought home, in which case I shall be enabled to give a complete account of all the Greenland discoveries of iron, together with more analyses. I will here simply enumerate the discoveries of iron hitherto made on the western coast of Greenland.

(1.) *Ross' and Kane's discovery of Iron in Davis Strait.*—According to these famous polar navigators, the Esquimaux in North Greenland make knives and instruments of iron from some large blocks situated probably somewhere to the north of Upernivik. *See* above, p. 324.

---

* As I have above mentioned, the Swedish Government sent for this purpose an expedition to Greenland in 1872, which succeeded in bringing home not only the three meteorites of 21, 8, and 4 tons, but also several smaller ones of from 4 to 200 kilogr.

(2.) *Rink's discovery of Iron at Niakornak, Jakobshavn District.*—In 1847 Rink found in the possession of some Greenlanders an iron ball, which they said they had found in a plain covered with boulders near the mouth of the Anorritok River. It weighed 21lb, with a specific gravity of 7·02. Analysed by Forchammer. Crumbling scarcely perceptible.

(3.) *Rudolph's discovery of Iron at Fortune Bay.*—A piece of iron weighing 11,844 gr. was found by Colonial Governor Rudolph among ballast that had been taken in at Fortune Bay. The iron crumbles much, and belongs probably to the same fall as the iron found at Ovifak.

(4.) *Fiskernäs.*—A small piece of metallic iron was found by Rink at Fiskernäs in South Greenland. The iron was declared by Forchammer to be of meteoric origin.

(5.) *The Pfaff-Oberg Iron from Jakobshavn.*

(6.) *The Iron discovered at Ovifak.*

Lastly it should be mentioned, that the old northern chronicles state, that during the time the old colonies existed in Greenland, so violent a shower of stones once happened that several churches and other buildings were destroyed.

It is remarkable that Giesecké, in his many years of travel in Greenland, should not have met with any meteoric iron, whereas he mentions that huge balls of iron-pyrites were found in the sand-beds of the basalt formation. We also met with some such nodules at an elevation of two hnndred feet above the sea, between Ujarasusuk and Kudliset. They were as much as from 3 to 4 feet in diameter, spherical, and lay loose in the sand close to a basalt dyke. Nevertheless, they did not contain pyrites, but a mineral (not yet analysed) like magnetic pyrites, of a very unusual appearance. (*See* above, p. 335.)

LV.—On Meteoric Irons found in Greenland. By Walter Flight, D.Sc., F.G.S., of the Department of Mineralogy, British Museum; Assistant Examiner in Chemistry, University of London.

[Reprinted, by Permission, from the "Geological Magazine," new series, vol. ii., Nos. 3 and 4, March and April, 1875, pp. 115 and 152.]

Meteoric Irons found August, 1870.—Ovifak (or Uigfak) near Godhavn, Kekertarssuak or Island of Disko, Greenland; Lat. 69° 19′ 30″ N.; Long. 54° 1′ 22″ W.*

The interesting story of the discovery of these enormous masses, by Prof. Nordenskjöld is already known to the readers of the

* A. E. Nordenskjöld, Redogörelse för en Expedition till Grönland År 1870; *K. Vet.-Akad. Förh.*, 1870, 873. (See translation in *Geol. Mag.*, IX. 289, *et seq.*)—D. Forbes, *Abstract Proc. Geol. Soc.*, No. 238, November 8th, 1871;

Geological Magazine through a translation of his original memoir. While exploring in Danish Greenland in 1870, his attention was directed to the possibility that meteorites might be met with in Disko Island, by the accidental discovery of a block of meteoric iron in some ballast which had been taken in at the old whaling station at Fortuna Bay, near Godhavn. and he urged the Greenlanders to search the district for masses of that metal. He proceeded to explore Omenak and other islands north of Disko, and, on his return to Godhavn at the end of August in the same year, not only learned from the Greenlanders that masses such as he sought for had been found, but he was shown a specimen of meteoric iron in confirmation of their statement. They were discovered, not at Fortuna Bay, but further eastward along the shore at Ovifak, between Laxe-bugt * and Disko Fjord, a spot than which there is none more difficult to reach along the whole of the coast of Danish Greenland, as it lies open to the south wind, and is inaccessible in even a very moderately rough sea. Nordenskjöld at once chartered two whale boats, manned by Greenlanders, and set sail for Ovifak, where, the sea being calm, they were able to land, and the stone at which they lay to proved afterwards to be the largest block of meteoric iron that they were to discover.

As the readers of this Magazine are already familiar with the description which Nordenskjöld gives (*see above*) of the condition under which these masses are found, we may break off here to consider the more recently published report of Nauckhoff, the Geologist of the Expedition of 1871, of the peculiar geological characters of the rocks at Ovifak (Blåfjell, or Blue Cliffs) with which they are associated.

Chem. News, November 17th, 1871. — A. E. Nordenskjöld, Remarks on Greenland Meteorites ; *Abstract Proc. Geol. Soc.*, December 20th, 1871.— T. Nordström, *Öfv. Vet.-Akad. Förh.*, 1871, 453. See also *Geol. Mag.*, VIII. 570, and IX. 88.—A. E. Nordenskjöld, Les Météorites ; *Revue Scientifique*, 1872, ii. [2], 128.—G. A. Daubrée, *Compt. Rend.*, lxiii. 1268 ; *Compt. Rend.*, lxxiv. 1542 ; *Compt. Rend.*, lxxv. 240.—E. Ludwig, *Min. Mitt.*, 1871, i. 109. —E. Hébert, Séance Soc. Géol. de France, February 5th, 1872 ; *Revue Scientifique*, i. [2], 858.—É. de Chancourtois et M. Jennatez, Séance Soc. Géol. de France, February 19th, 1872 ; *Revue Scientifique*, i. [2], 905.—G. A. Daubrée, Séance Soc. Géol. de France, May 20th, 1872 ; *Revue Scientifique*, i. [2], 1169 ; *Amer. Jour. Sc.*, iii. 71 and 388.—F. Wöhler, *Nachricht. K. Gesell. Wiss. zu Göttingen*, 1872, No. 11, 197 ; *Pogg. Ann.*, cxlvi. 297 ; *Ann. der Chem.*, clxiii., 247 ; *Nachricht. K. Gesell. Wiss. zu Göttingen*, 1872, No. 26 ; *Ann. der Chem.*, clxv. 313.—G. Rose, *Zeit. Deutsch. Geol. Gesell.*, xxiv. 174.—G. von Helmerssen, *Zeit. Deutsch. Geol. Gesell.*, xxv. 347.— C. Rammelsberg, Ueber die Meteoriten (*Samm. Wiss. Vorträge*), pages 14 and 18.—C. W. Blomstrand, *Ber. Deutsch. Chem. Gesell.*, iv. 987.— G. Nauckhoff, *Svenska Vet. Akad. Handl.*, 1872, i. No. 6 ; *Ber. Deutsch. Chem. Gesell.*, vi. 1463 ; *Mineralogische Mittheilungen*, 1874, 109.—G. Tschermak, *Mineralogische Mittheilungen*, 1874, 165 ; *Der Naturforscher*, 1874, Nos. 49–52.—J. Lawrence Smith, *Compt. Rend.*, lxxx. 301.—For a map of Disko see *also Geographical Mag.*, February, 1875.

* See *Geol. Mag.*, 1872, vol. ix. Pl. VII. In this map two bays called "Laxebugt" are given ; the one mentioned above is situated to the south of Disko Fjord.

and is in every respect like that of the large loose blocks. Moreover, like them, it unfortunately possesses the property of exuding a yellow liquid (ferrous chloride), and of weathering away. It was noticed that these inclosed masses had their major axes parallel to the direction of the ridge, and that they were, in a way, connected with each other by little veins of weathered iron.

Nordenskjöld states that the large free blocks of metal had a tombac to rusty-brown colour, and, when found, exhibited metallic lustre on parts of their surface. Here and there, fragments of basalt, similar to that of the ridge, were found adhering to them. The inner parts contained none of the rock, and his analyses detected the presence of little silicic acid. They were strongly polar, the upper surface attracting the north, the lower side the south pole of the magnetic needle.

The iron of the larger masses is crystalline and brittle, so that pieces can readily he removed with a hammer ; the metal of the ridge is tougher, and has a rougher fracture. The presence of troilite was rarely detected in the detritus ; a few black magnetic grains were met with, which, by their octahedral faces, were recognized to be magnetite.

The characters of the polished sections of the different masses differ greatly ; in some the surface shows rounded areas of varying brightness and shades of colour, with parts of a brassy yellow (troilite) ; others are more homogeneous, or appear to be made up of fine prisms of "carburetted nickel-iron." Some, not all, exhibit figures when etched.

Though containing little sulphur, the Greenland irons, since they have been brought to Europe, have shown a marked tendency to crumble to pieces. On the shore at Ovifak, sometimes exposed to the wash of the waves, sometimes left high and dry, but preserved at the constant temperature of the sea, which varies little throughout the year, the masses apparently underwent little change. Already during the passage, however, many fragments crumbled away, and when unpacked at Stockholm two months later, and placed in a room of ordinary temperature, others broke up into a reddish-brown powder. A freshly fractured lustrous surface of one of the masses commenced in one corner to rust, expand, and crumble away ; while the remainder experienced no change, till at length the oxidation extended into the interior and the whole fell to pieces. In a hermetically sealed glass tube the iron is preserved unchanged ; but in another tube with a fine crack oxidation continued. In alcohol no change takes place ; in air, dried by sulphuric acid, the change is greatly impeded. Attempts to preserve them by coating them with varnish were of slight avail. The cracking is caused by dilatation, and takes place with such force that masses of metal, on which chisel and saw were without effect, are broken and bent out of shape during oxidation.

Nordenskjöld found that a fragment of the largest iron, when heated to redness, gave off more than 100 times its volume of a gas which had a bituminous smell. It was evidently gas not simply occluded by the metal, but was produced by the decomposition of "the organic matter in the meteorite," through the reducing action of those compounds on the oxide of iron associated with them. When such iron is treated with mercury-chloride but little gas is evolved; in aqua-regia it dissolves, leaving in some cases a carbonaceous residue, in others very little residue of any kind; by the action of hydrochloric acid a gas is given off which has a penetrating odour resembling that of some hydrocarbon. By treatment with acid a humus-like compound appears to be generated, which is soluble in ammonia, insoluble in acid, and can be oxidized only with difficulty by long boiling with very strong acids.

In Nordenskjöld's paper are given the earliest analyses of these irons:

I. Fragment of one of the large iron masses: this specimen evolved more gas than II. and III. Specific gravity=5·86—6·36. Analysed by Nordenskjöld. II. Fragment of iron, more compact and less crystalline than I., probably from the basalt ridge. Small grains were observed to be malleable. The specimen from which this was taken subsequently crumbled away. Specific gravity=7·05—7·06. Analysed by T. Nordström. III. Fragment of iron from the basalt ridge, which exhibited well-marked Widmanstättian figures. In external appearance this iron exactly resembled II. Specific gravity=6·24. Analysed by G. Lindström. *See* above, p. 443.

	I.	II.		III.
Iron	84·49	86·34		93·24
Nickel	2·48	1·64		1·24
Cobalt	0·07	0·35		0·56
Copper	0·27	0·19		0·19
Phosphorus	0·20	0·07		0·03
Sulphur	1·52	0·22		1·21
Chlorine	0·72	1·16		0·16
Alumina	trace	0·24		—
Lime	trace	0·48		—
Magnesia	0·04	0·29		trace
Potash	trace	0·07		0·08
Soda	trace	0·14		0·12
Silicic acid	trace	0·66		} 0·59
Insoluble portion	0·05	4·37		}
Carbon, Organic Matter, Oxygen, and Water	10·16	3·71	{ Carbon	2·30
			{ Hydrogen	0·07
	100·00	99·93		99·79

Nordström analysed the carbonaceous residue of the compact iron II., after digestion with double chloride of copper and sodium, and iron chloride, and found, when a quantity of ash is deducted, that it is composed of:

Carbon	63·59	63·64
Hydrogen	3·26	3·55
Oxygen (by difference)	33·15	32·81
	100·00	100·00

These numbers yield no satisfactory atomic ratios, and it is not improbable that the carbon is present in two allotropic modifications, as well as a constituent of a complex organic compound.

In 1872 two interesting papers were published by Wöhler on the results of his examination of this iron, especially that from the ridge. The specimen he chose for examination came from a vein of metal, several inches wide and some feet in length, which was inclosed in a rock "that presents a marked difference in "composition from the basalt-breccia whence it protrudes." He describes this iron as bearing a close resemblance to grey cast iron; it has a bright lustre, is very hard, is quite unalterable in air, and has a specific gravity=5·82. Nordenskjöld, as we have seen, extracted gas from the metal of the larger masses by heating it. Wöhler finds that the iron of the vein evolves more than one hundred times its volume of a gas that burns with a pale blue flame, and is carbonic oxide, mixed with a little carbonic acid. The "iron," in fact, contains a considerable amount of carbon, as well as a compound of oxygen; and, according to Wöhler can at no time have been exposed to a high temperature. After it has been heated, the iron becomes brighter, and, though more soluble in acid, it still leaves a carbonaceous residue. A fragment heated in dry hydrogen, with a view to determine the amount of oxygen present, formed a quantity of water, and lost 11·09 per cent. of its weight. "It contained, in other words, 11·09 per "cent. of oxygen." It is not stated whether the water corresponded in weight to that amount of oxygen. Hydrochloric acid acts but slowly and imperfectly on this metal, evolving first sulphuretted hydrogen, and then hydrogen possessing the odour of a hydrocarbon; and leaves a black granular magnetic powder, which, though insoluble in cold acid, generates, on the application of heat, a gas with a strong odour of a hydrocarbon, leaving a residue of amorphous sooty carbon and slightly lustrous graphitic particles. In iron-chloride the "iron" dissolves without evolution of gas; about 30 per cent. of a black residue remaining, which, after having been dried at 200° C., lost by ignition in hydrogen 19 per cent. of its weight, water being produced. It is now very readily attacked by acid, evolves sulphuretted hydrogen, and gives a residue of nearly pure carbon in powder or in graphitic scales. Iron-chloride and acid appear, therefore, in the main, to remove the free metal only, and to be without action on the compounds with sulphur and oxygen. The ultimate composition of the specimen he analysed is as follows:—

Iron	80·64	Sulphur	2·82
Nickel	1·19	Carbon	3·69
Cobalt	0·47	Oxygen	11·09
Phosphorus	0·15		
			100·05

Wöhler was disposed to regard the oxygen, constituting so considerable a portion of an apparently metallic mass, as present in the form of a diferrous oxide, Fe_2O, were it not that, according to this view, there would be no iron provided for combination with the sulphur and carbon. As, however, Nordenskjöld found magnetite in or near other Ovifak irons, Wöhler, regards the substance constituting the veins as an intimate mixture of magnetite, of which there would be 40·20 per cent., with metallic iron, of which there would then be 46·60 per cent., the sulphide, carbide, and phosphide, as well as the alloys with nickel and cobalt, and some carbon in isolated particles. The latter probably undergo no change when the magnetite and carbide, by the action of heat, generate carbonic oxide.

A specimen of the iron from the basalt has also been investigated by Daubrée; he describes it as having a metallic lustre and being nearly black. He found its composition to be :—

	I.
Iron in the free state - - -	40·940
Iron in combination - - -	30·150
Carbon in the free state - - -	1·640
Carbon in combination - - -	3·000
Nickel - - - - -	2·650
Cobalt - - - - -	0·910
Phosphorus - - - -	0·210
Arsenic - - - - -	0·410
Sulphur - - - - -	2·700
Silicium - - - - -	0·075
Nitrogen - - - - -	0·004
Oxygen - - - - -	12·100
Water (hygrometric) - - -	0·910
Water in combination - - -	1·950
Chromium, copper, &c. - - -	1·010
Calcium sulphate, chloride, &c. - -	1·354
	100·013

In his second paper he gives analyses of two more specimens :

II. Light-grey iron, possessing metallic lustre. It is not homogeneous, as it might be assumed to be from its lustre and colour. When crushed in a mortar, it is divided into two parts: the one crumbles to fine powder, the other is flattened into plates, requiring much trituration to break them up. III. Metallic grains mechanically separated from the rocky portion in which they were distributed. These spherules exhibit figures, when etched, and contain silicate distributed in very fine particles throughout their mass; in one rounded fragment the silicic acid of this silicate amounted to 11·9 per cent. of the total constituents.

	II.	III.
Iron in the free state - -	80·800	61·990
Iron in combination - -	1·600	8·110
Carbon in the free state - -	0·300	1·100
Carbon in combination - -	2·600	3·600
Silicium - - - -	0·291	—
Water - - - -	0·700	—
Calcium chloride - -	0·233	0·146
Iron chloride - - -	0·089	0·114
Calcium sulphate - -	0·053	0·047
Copper - - - -	trace.	trace.

It will be seen that specimen III. is not less rich in carbon than I., and that specimen II. also contains a considerable quantity. Specimen I. is distinguished from II. by a large proportion of combined iron. By treatment with alcohol, calcium-chloride was extracted and determined in I.; with cold distilled water, the soluble salts were removed from II. and III. I. contains more lime-sulphate and less chloride than II. and III.

These meteoric masses are distinguished by the amount of carbon, free and combined, which they contain; by the presence of a large proportion of iron in combination with oxygen, but in what state of oxidation is not clearly ascertained; and by the occurrence of soluble chlorides and sulphates, especially calcium-sulphate, throughout their structure. No salt of potassium has been detected in them, nor, which is very remarkable, has sodium-chloride been found, although carefully sought for. The intimate distribution of these salts through the Ovifak iron is certainly an indication that they must be numbered among the original constituents of these meteorites.

Daubrée noticed that specimen II. showed a marked tendency to absorb water and to rust away; a few days sufficed to make this apparent. The local nature of the oxidation he attributes to the irregular distribution of the deliquescent salts. Among these compounds, instead of iron-chloride, to the action of which the decay of meteoric iron has usually been ascribed, calcium-chloride appears to play the most prominent part. In support of this view it may be remarked that No. II. iron, the one most liable to change, is that containing the greatest proportion of this salt, the amount being six times that met with in No. I. iron.

Calcium- and magnesium-sulphates were noticed by Daubrée to form constituents of the Orgueil Stone, and the latter salt is also present in the aerolites of Kaba and Alais. All these are carbonaceous meteorites. May the calcium-sulphate of these irons, as well as that of the above-mentioned aerolite, be a product of the oxidation of a calcium (magnesium) sulphide such as occurs in the meteorite of Busti, which stone also contains, among other constituents, augite and metallic iron?

The greater stability which these masses exhibited so long as they were in polar latitudes is no doubt due to the reduced tension of aqueous vapour; had they fallen in regions further south and been exposed to a milder climate, they would without doubt have long since fallen to powder.

In his second paper Wöhler points out the probability of the No. II. iron, which Daubrée examined, being of the same kind as that which he himself analysed. He remarks that, although Daubrée found this variety of the metal to show a tendency to oxidise even in a few days, his specimen had remained bright and unchanged after it had been a year in his collection.

Nauckhoff, whose exhaustive examination of the rocks associated with the Ovifak irons we shall immediately turn to consider, analysed the spangles and spherules which can be removed by a magnet from the rock that occurs in rounded masses in the basalt ridge, and of which the composition is given in the table of his analyses under III. Some of these spangles could be pulverized only with difficulty, and were readily flattened out; the spherules, though so hard that a sharp steel file would scarcely touch them, were easily crushed. They had the following composition:—

Iron	58·25	Alumina	1·45
Nickel	2·16	Nickel and cobalt oxides	0·44
Cobalt	0·30	Magnesia	0·33
Copper	0·13	Lime	0·50
Hydrogen	0·28	Soda	0·09
Carbon	1·64	Potash	trace.
Sulphur	0·16	Residue	6·07
Chlorine	0·16		
Magnetite	30·42		
Silicic acid	0·26		102·64
Phosphoric acid	trace.		

In the basalt of the ridge, of which an analysis is given under II. in the same table, a compact, very brittle, yellow, or slightly brown mineral occurs in thin flakes, sometimes in nodules of the size of a pea; it is invariably penetrated and usually surrounded by a mineral resembling hisingerite, to which attention will presently be directed. The mineral has a hardness of 5 to 5·5, and easily fuses before the blowpipe, with evolution of sulphurous acid, to a magnetic regulus. It has the composition:

			Equivalent Ratios.	
Iron	52·94	57·91	2·068	2·258
Nickel	5·06	5·53	0·190	
Copper	trace	trace	—	
Sulphur	33·41	36·56	2·285	
Silicate	8·59	—	—	
	100·00	100·00		

These numbers give the formula (Fe,Ni)S, or that of the iron (nickel) monosulphide or troilite, which has hitherto only been met with in meteorites.

Intimately associated with the troilite, and evidently a product of its oxidation and further alteration, is the mineral already mentioned, the fresh fracture of which is of a light olive-green colour, that by exposure to the air soon becomes brown, and after some days turns quite black.

Its specific gravity is 2·919; and its composition:

		Oxygen.
Silicic acid	31·70	16·90
Iron sesquioxide	51·49	15·44
Iron protoxide	3·81	0·85
Water	15·56	12·05
	100·56	

These numbers indicate the formula:

$$FeO,SiO_2+3(2Fe_2O_3,3SiO_2)+14H_2O$$

as that of the mineral. Nauckhoff, however, draws attention to the rapidity with which the oxidation of the pulverised mineral takes place: five days after the analysis was made the per-centage of iron-protoxide in another portion had fallen to 3·47, and after three weeks to 1·55. The original unchanged mineral was probably a hydrated ferrous silicate.

The following rocks from Disko island have been examined by Nauckhoff:

I. Section of a six-sided basalt column from Brededal, east side of Skarfvefjell and about 10′ E. of Godhavn; showing compact dark greyish-green ground-mass with crypto-crystalline texture; under the microscope, crystals of a felspar, augite, and magnetite are recognised. Fusible before the blow-pipe.— II. Basalt from the east side of the ridge at Ovifak, where the iron and breccia were found. Fusible before the blowpipe.— III. Rock occurring in rounded masses, with green foliated crust, in the basalt ridge, and inclosing spangles and spherules of iron, some 6—7 mm. in diameter; these exhibit Widmannstättian figures. Appears to be a very finely granular mixture of a felspar with a small amount of a green mineral, probably augite, and imperfectly crystallised magnetite, which latter usually surrounds the spangles of iron; olivine is only occasionally met with, in grains the size of a pea. Melts with difficulty before the blowpipe.— IV. Very hard brown-coloured mass inclosing rock in which iron spangles are found; it closely resembles III. The ground-mass consists of a felspar, probably anorthite, the crystals of which are occasionally large, and show marks of twinning, and a great number of reddish octahedra closely resembling spinel. Small particles of a greenish mineral, having the appearance of augite, are also to be distinguished. Spangles of iron are very rarely found in the felspar; and magnetite is apparently absent. Melts very slowly before the blowpipe. —V. Rounded lump of grey rock from the basalt ridge; it was covered with a dark-green vesicular crust, from 15 to 20 mm. thick. Through the ground-mass, which appears to consist of a felspar, were disseminated numerous brilliant greyish scales, besides some very black magnetite or graphite. Augite sparsely distributed; abundance of red spinel in some parts, none in others. Melts with great difficulty before the blowpipe.—VI. The dark greenish-brown crust of V., closely resembling that of the rounded masses III. It consists of a felspar inclosing a brown and a green augite-like mineral, and, in places, clusters of granules of spinel. Melts with great difficulty before the blowpipe. —VII. Light-grey foliated rock from Ovifak, the exact circumstances of the occurrence of which are not known. The ground-mass consists of a mixture of a felspar with a grey, finely foliated mineral with graphitic lustre. Red spinel is met with abundantly in both constituent minerals. This variety of rock, like those from the ridge, is covered with a rust-like crust. It breaks easily, and always parallel to the scales. Before the blowpipe it melts with difficulty on the edges.— VIII. Compact, slightly weathered breccia, filling a fissure two to three inches wide in the basalt ridge parallel to which it runs. It is a black granular mass, devoid of metallic lustre, and incloses fragments, some with edges sharp and angular others with the corners rounded, of a rock exactly like that forming the ridge.— IX. Loose, much weathered breccia, from the top of the ridge, in irregularly shaped fragments. It can be broken in pieces with the hand, is much rusted, and closely resembles the oxidation of the metal blocks. Like the preceding specimen, it incloses rounded fragments of the rock forming the ridge. The specific gravity is about midway between that of iron and of magnetite.— X. The broken-up basalt, resembling that of the ridge, inclosed in the weathered breccia IX.

	I.	II.	III.	IV.	V.	VI.	VII.	VIII.	IX.	X.
Silicic acid - -	49·18	48·04	42·72	34·72	36·59	44·94	37·92	1·04	0·81	41·25
Titanic acid - -	0·52	0·39	trace	—	—	—	—	—	—	0·34
Phosphoric acid -	0·13	0·07	trace	—	—	—	—	0·12	0·12	—
Iron sesquioxide -	5·52	6·89	1·64	4·88	—	—	—	—	—	16·18
Alumina - -	13·52	13·13	16·01	31·83	19·18	22·20	32·36	2·31	2·92	13·06
Chromium oxide -	—	—	—	—	—	—	0·08	—	—	—
Magnetite - -	—	—	—	—	—	—	—	52·51	77·39	—
Iron protoxide -	10·31	11·14	14.27	5·53	14·85	9·45	4·02	—	—	10·78
Manganese protoxide.	0·28	0·11	trace	—	0·29	—	0·19	—	—	0·25
Nickel and Cobalt oxides.	—	—	—	—	—	—	—	1·17	0·82	—
Magnesia - -	6·83	5·17	7·93	9·35	7·24	4·98	2·86	0·02	trace	6·41
Lime - - -	11·51	10·87	10·10	10·19	8·73	11·01	11·57	0·30	0·20	7·97
Soda - - -	1·84	2·83	1·65	1·00	0·79	1·86	1·48	0·08	0·11	1·54
Potash - - -	0·06	0·06	0·13	0·27	trace	0·06	trace	trace	trace	0·03
Iron - - -	—	—	4·57	0·09	5·01	1·11	—	28·36	7·73	—
Nickel - - -	—	—	0·44	—	0·25	—	trace	1·22	1·81	—
Cobalt - - -	—	—	trace	—	trace	—	trace	0·30	0·33	trace
Copper - - -	trace	—	trace	—	trace	trace	trace	0·08	0·30?	trace
Hydrogen - -	—	0·25?	0·30?	0·29?	0·31?	0·31?	0·24?	0·38	0·51	0·49
Carbon - - -	—	0·79	0·30	0·53	2·55	3·35	6·90	3·52	2·33	0·86
Sulphur - -	—	0·98	0·32	—	trace	trace	0·77	0·34	trace	trace
Chlorine - -	trace	trace	0·08	0·12	0·23	0·20	trace	trace	0·14	0·25
Water - - -	0·34	—	—	—	—	—	—	—	—	—
Residue - -	—	—	—	—	—	—	—	9·64	3·71	—
	100·04	100·72	100·46	98·80	96·02	99·47	98·39	100·39	99·23	99·41
Specific Gravity -	3·016	3·024	3·169	2·942	3·141	2·927	2·761	4·560	6·570	3·353

Tschermak examined two microscopic sections of the Ovifak rocks, and compared them with sections of the meteorites of Jonsac, Juvinas, Petersburg, and Stannern, which consist chiefly of augite and anorthite, with little or no nickel-iron; they form a class which G. Rose termed "eucritic." Both sections exhibit a crust, as meteorites possess; it is, however, so altered by oxidation, that it is not possible to determine whether it is the fused crust usually noticed on a meteorite. The crystals of felspar, which, according to Nauckhoff's analyses, must be regarded as anorthite, are fully developed; they penetrate the augite, iron, and magnetite, and must evidently have been formed before them. They are completely transparent, and have but few and large cavities, which are filled, partly with black granules, partly with a brown substance of irregular form; some traversing the length of the crystals are filled with a transparent glassy substance. The augite is of a light greenish-brown hue, traversed here and there by flaws; it fills gaps between the other constituents, as has been often observed in dolerites and diabases, and encloses individual black grains. In the section containing iron the colourless felspar encloses a black or brown substance running the length of the crystals, or dust-like fine black granules, or larger round transparent bodies of a violet colour, which may be the mineral Nauckhoff regards as spinel. Side by side with the felspar, brown grains, less numerous than in the former section, are seen, and these are probably augite. Black particles, moreover, occur, which by reflected light appear to be semi-metallic, and are probably magnetite, as well as others that are likewise black, but devoid of lustre, which seem to be graphite. A few small grains of troilite were also recognised. In the second section, which bore a general resemblance to the first, the felspar crystals were larger, the matrix being made up of finer crystals. In some of the felspar crystals cloudy pale-brown patches were observed, which, when viewed with a higher power, were found to be due to numberless, minute, elongated, inclosed granules lying in parallel position, or to others that were shorter and more rounded. These appearances recall those noticed in eucritic meteorites, like that of Jonsac, except for the fact that the inclosed particles are of smaller size. The larger cavities in the felspar are filled in the same manner as in the other rock-section from Ovifak. The structure of eucritic meteorites is tufaceous; that of the Ovifak rock very compact. This distinction, however, has often been observed in meteorites. Many chondritic meteorites are tufaceous; while others, having similar chemical composition, like the aerolites of Lodran and Manbhoom, are compact and crystalline. The augite of the Ovifak rocks has not the characteristically filled cavities observed in that of certain eucritic meteorites; but in the augite of some meteorites, as those of Shergotty and Busti, for example, they are equally wanting.

The meteorites of Ovifak in some respects resemble the carbonaceous meteorites, though they differ greatly from them in other characters: especially in the appearance of both metallic and rocky portions. They form a new type in the series of meteoric rocks, and fill the gap that has hitherto separated the carbonaceous from other meteorites.

If some differences are to be traced between the remarkable rocks and irons of Ovifak and known meteorites, others still greater present themselves, when we compare the Greenland masses with terrestrial rocks, even with the basalts and diorites, near which it might be proposed to class them, on account of the occurrence in them of magnetite, and of the crystalline arrangement of their silicates. Iron has not hitherto been found as metal inclosed in basalt, except on very rare occasions (as by Andrews in the basalt of Antrim,* and then only in fine particles, and apparently not alloyed with nickel and cobalt), while troilite is a meteoric mineral, and has never been met with in a terrestrial rock.

But if the weight of evidence favours the assumption that these masses are of meteoric origin, there remain the following considerations to which attention has been drawn by Rammels-

* A. E. Reuss detected the presence of iron in some Bohemian basalts by Andrews' method. (*Kenngott's Uebersicht Result. Min. Forschungen*, 1859, 105.)

berg, supporting the view that they may possibly have been erupted.

Of the rocks composing the globe, the greater portion accessible to us have been modified by the action of water. There is one class of which this cannot be said: the molten masses brought to the surface by volcanos, the various rocks we term "lava." How ever they may differ as regards constituent minerals, they have amongst them a family resemblance, and it is with them that the meteoric rocks may be compared. The old lavas of Iceland and Java consist of augite and anorthite, as do the meteorites of Juvinas, Jonsac, and Stannern. The "bombs" of the prehistoric volcanos of the Eifel are composed of olivine, augite, bronzite, and chromite, minerals that are commonly met with in meteorites. Hence arises the question: Are these masses, so similar in their lithological characters to the meteorites, samples perhaps of the the inner unchanged nucleus of our planet? Does the original mass of the earth differ in point of magnitude only from the fragments which yield to its attraction?

The mean density of the earth is greater than that of the minerals composing the rocks of the outer crust. The volcanic rocks and the meteorites, which in point of chemical constitution are basic, are alike denser than this crust. The presence of metallic iron, a characteristic feature of meteorites, points to the absence of water and free oxygen as one of the essential conditions for their formation. Terrestrial rocks rarely contain iron, but it is replaced by an oxidised form of iron—magnetite. Only in combination with platinum is it found in the metallic state. May the rocks of the interior of our globe contain this, the most important of all the metals, in an uncombined condition?

It has been pointed out by Daubrée that a region like Greenland, where doleritic rocks cover so wide an area, appears in a marked degree to present the conditions necessary and favourable for the upheaval of masses from very considerable depths.

Another phase of the question to which he directs attention should also be mentioned. It appears not improbable that the basalt of Greenland, which contains more than 20 per cent. of iron-oxide, may during eruption have undergone reduction such as he imitated in his laboratory some years since. This theory is the more admissible from the fact that in the region under consideration, between Lat. 69° and 72°, numerous large beds of lignite, as well as graphite, occur, especially in the Island of Disko, in which Ovifak is situated.

In a paper on the anomalous magnetic characters of iron-sesquioxide prepared from meteoric iron, communicated in February last to the French Academy, Dr. Lawrence Smith announces that the investigation of this iron, on which he is at present occupied, has convinced him that the Ovifak metallic masses are of terrestrial origin.

The fact, observed by Nordenskjöld and Wöhler, of the evolution of a large amount of gas by Ovifak iron when heated, led these observers to the conclusion that it could never have been exposed to a high temperature. Tschermak, however, points out that this phenomenon has only been observed in experiments conducted at ordinary pressure, and it must not be forgotten, he maintains, that these masses, though surrounded by a heated medium, were at the same time subjected to the superincumbent pressure of a vast layer of fluid basalt. They may, moreover, have originally had a different composition, and the oxygen which plays so essential a part in the gaseous evolution, may have been taken up subsequently during exposure to the atmosphere.

Daubrée draws attention to a reaction, mentioned by Stammer, and thoroughly investigated by E. L. Gruner (*Compt. Rend.* xxiii. 28; xxiv. 226), that, in the presence of iron-oxide, or even of iron under certain circumstances, carbonic oxide breaks up, depositing carbon, partly in combination with iron, partly in intimate mixture with iron-oxide; and that this reaction, which has been found to occur at 400°, does not take place at very high temperatures.

Nordenskjöld's paper is illustrated with a plan of the shore at Ovifak, where the irons were found, and with a sketch made on the spot by Nordström of the three largest masses (*above*, p. 440), showing them partly immersed; and in a plate are representations of seven of the blocks—one showing very distinctly the manner in which the metal is rent during oxidation. Nauckhoff has appended to his paper in the *Mittheilungen* a drawing of the gangue, indicating the position of the smaller pieces of iron and the breccia. Four excellent photographs of the larger masses have been published by the Hofphotograph Jaeger, in Stockholm.

One of the largest blocks, weighing 10,000 lbs., was offered for sale in New York for 12,500 dollars in gold, and smaller specimens at eight dollars per lb.

As is well known, implements of meteoric iron have from time to time been found in the possession of the Esquimaux (*above*, p. 324, &c.). Some recent specimens, inserted in bone handles, from Esquimaux kitchen middens, were described by Steenstrup at the *Congrès international d'Anthropologie et d'Archéologie prehistoriques à Bruxelles* (Session de 1872). For figures of these implements see also *Matériaux pour l'histoire primitive et naturelle de l'Homme*, 9 Année, 2e Série, Tome IV. 2e Livraison, 1873, p. 65, pl. 7.

CRYOCONITE found 1870, July 19th–25th, on inland ice, east of Auleitsivik Fjord, Disko Bay, Greenland.—Meteoric metallic particles found in snow, which fell (1) 1871, December, Stockholm; (2) 1872, March 13th, Evoia, Finnland; (3) 1872, August 8th, Lat. 80° N., Long. 13° E.; (4) 1872, September 2nd, Lat. 80° N., Long. 15° E.*

Early in December 1871, there was a heavier fall of snow in the neighbourhood of Stockholm than any that had occurred there within the memory of living persons; and it presented to Norden-

---

* A. E. Nordenskjöld. Redogörelse för en Expedition till Grönland År 1870, p. 28. See also *Geol. Mag.* ix. 356, (*above*, p. 395); *Compt. Rend.*, lxxvii. 463; *Jour. Prakt. Chem.*, ix. 356; *Pogg. Ann.*, cli. 154.

skjöld an opportunity of determining whether the snow brought cosmical matter to the earth's surface. A cubic metre of apparently pure snow, collected towards the end of the fall, left on melting a small black residue. From some of this substance, when heated, a liquid product distilled over; a portion when burnt left a red ash; while a magnet extracted particles which, when rubbed in an agate mortar, exhibited metallic characters, and on being treated with acid proved to be iron. Although the possibility must be admitted that this material may have been derived from the chimneys and iron roofs of the city, already covered with a thick layer of snow, the result was sufficiently interesting to make it desirable that a similar experiment should be tried with snow falling remote from towns. For this purpose snow was collected on the 13th March 1872, by Dr. Karl Nordenskjöld at Evoia, in Finnland, to the north of Helsingfors, and in the centre of a large forest. It was taken from off the ice of the Rautajerwi, at a spot which is separated by a dense wood from the houses of that northern station. When melted, this snow yielded a soot-like residue, which under the miscroscope was found to consist not only of a black carbonaceous substance, but white or yellowish-white granules, and from it the magnet removed black grains, which when rubbed in a mortar were seen to be iron. Here again the material was too small in amount to allow of a determination of the presence of nickel and cobalt; in other words, to establish the meteoric origin of the metal. The Arctic Expedition of 1872 presented an opportunity for the collection of snow in a region as far removed as possible from human habitation. On the 8th August, the snow covering the drift-ice at Lat. 80° N. and Long. 13° E. was observed to be thickly covered with small black particles, while in places these penetrated, to a depth of some inches, the granular mass of ice into which the underlying snow had been converted. Magnetic particles were abundant, and their power to reduce copper-sulphate was established. Again, on the 2nd September, at Lat. 80° N. and Long. 15° E., the ice-field was found covered with a bed of freshly fallen snow, 50 mm. thick, then a more compact bed 8 mm. in thickness, and below this a layer 30 mm. thick of snow converted into a crystalline granular mass. The latter was full of black granules, which became grey when dried, and exhibited the magnetic and chemical characters already mentioned; they amounted to 0·1 to 1·0 millegramme in a cubic metre of snow. Analysis of some millegrammes enabled Nordenskjöld to establish the presence of iron, phosphorus, cobalt, and probably nickel. The filtrate from the iron-oxide gave a small brown precipitate, which gave a blue bead with borax. The portion insoluble in acid consisted of fine angular colourless matter, containing fragments of Diatoms. This dust from the polar ice North of Spitzbergen bears a great resemblance to the remarkable substance, cryoconite,* which was found in Greenland in 1870, very evenly

* A. E. Nordenskjöld. An Account of an Expedition to Greenland; *Geol. Mag.* vol. ix. p. 353. *See also* above, p. 395.

distributed in not inconsiderable quantity on shore-ice, as well on ice thirty miles from the coast and at a height of 700 metres above the sea. The dust of both localities has probably a common origin.

The cryoconite is chiefly met with in the holes of the ice, forming a layer of grey powder at the bottom of the water filling the holes. Considerable quantities of this substance are often carried down by the streams which traverse the glacier in all directions. The ice-hills which feed these streams lie towards the east, on a slowly rising undulating plateau, on the surface of which not the slightest trace of stone or larger rock-masses was observed. The actual position of this material, to which Nordenskjöld has given the name of cryoconite (κρύος ice, and κόνις dust), in open hollows on the surface of the glacier, precluded the possibility of its having been derived from the ground beneath.

The grey powder contained a not inconsiderable amount of organic matter, which, even at the low temperature of the ice, undergoes putrefactive decomposition. A quantity, amounting to from two to three cubic metres, which was lying in the dried-up bed of a glacier-stream, emitted a very offensive odour, bearing some resemblance to that of butyric acid.

When examined with the microscope, the chief constituent of this powder appears to consist of colourless, crystalline, angular, transparent grains, among which are a few yellow and less transparent. Some had distict cleavage-surfaces, and were possibly a felspar; other crystal fragments, having a green colour, were probably augite; while other black, opaque particles could be removed with a magnet. These foreign constitutents, however, are present in so small a quantity that, if all the white grains consist of one and the same mineral, it may be regarded as homogeneous. The specific gravity of this mineral is 2·63; the hardness apparently inconsiderable, and the form probably monoclinic. It resists the action of acids; by long digestion with sulphuric acid 7·73 per cent., with hydrochloric acid 16·46 per cent. were dissolved. Lime-carbonate was not present. According to Lindström's analysis, it consists of—

Silicic acid - -	62·65	Potash - -	2·02
Phosphoric acid -	0·11	Soda - -	4·01
Alumina - -	14·93	Chlorine - -	0·06
Iron oxide - -	0·74	Water (hygroscopic)	0·34
Iron protoxide -	4·64	Organic matter and combined water* -	2·86
Manganese protoxide	0·07		
Lime - -	5·09		
Magnesia - -	3·00		100·12

This composition corresponds with the formula:

$$2RO,SiO_2+Al_2O_3,3SiO_2+H_2O.$$

The origin of this cryoconite is highly enigmatical. That it is not a product of the weathering of the gneiss of the coast is shown

* This passed off when the mineral was heated to temperatures ranging from 100° to a red heat.

by its inferior hardness, indicating the absence of quartz, the large proportion of soda, and the fact of mica not being present. That it is not dust derived from the basalt area of Greenland is indicated by the subordinate position iron-oxide occupies among the constituents, as well as by the large proportion of silicic acid. We have then to fall back on the assumption that it is either of volcanic or cosmical origin.

That dust may be carried immense distances has been well established. Darwin* refers to instances of its having fallen on ships when more than a thousand miles from the coast of Africa, and at points sixteen hundred miles distant in a north and south direction. If the Greenland dust were volcanic, it would probably have been wafted from Iceland or Jan Mayen, or some as yet unknown volcanic region in the interior of Greenland. Nordenskjöld found it to bear the closest resemblance, under the microscope, to the ash of Vesuvius (1822), and to a specimen of that which fell at Barbadoes and probably came from St. Vincent. Looked at in the mass, however, it is at once seen that the volcanic ash is of a brownish red; the cryoconite is grey. The magnet when placed in contact with the Vesuvian ash failed to extract anything; out of that from Barbadoes it drew magnetic particles, which, however, were not metallic, nor did they contain nickel or cobalt.

The cryoconite, nevertheless, whencesoever it comes, contains one constituent of cosmical origin. Nordenskjöld extracted, by means of the magnet, from a large quantity of material, sufficient particles to determine their metallic nature and composition. These grains separate copper from a solution of the sulphate, and exhibit conclusive indications of the presence of cobalt (not only before the blowpipe, but with solution of potassium-nitrite), of copper, and of nickel, though in the latter case with a smaller degree of certainty, through the reactions of this metal being of a less delicate character. Moreover, ammonia removes from cryoconite a humus-like substance which, among other characteristics, in its powers of resisting powerful oxidising reagents, closely resembles the organic compound found in the residue of Ovifak iron after treatment with acid.

Hail, which fell at Stockholm in the autumn of 1873, was found by Nordenskjöld to contain grey metallic particles that reduced copper from its sulphate. Although the roofs of the buildings surrounding the Academy, in the courtyard of which these hailstones fell, are of iron, the grains were rounded, and of light colour, instead of a reddish-brown. The observation is of sufficient interest to allow of its being placed on record.

It has been shown that small quantities of a cosmical dust, containing iron, cobalt, nickel, phosphorus, and carbonaceous substances, fall with other atmospheric precipitates on the earth's surface. Nordenskjöld, in his paper, alludes to the theory, already advanced, we believe, by Haidinger, that this deposit may play an

* C. Darwin. *Journal of Researches of a Naturalist*; *Voyage of H.M.S. "Beagle,"* new ed., 1870, p. 5.

important part in the economy of nature in supplying phosphorus to soils already exhausted by the growth of crops. His observations, moreover, are of value through the light they throw on the theories of star-showers, auroræ, &c. The small but continuous increase of the mass of our planet, which appears to take place, may lead students of geology to modify the view at present held, that from the time of the first appearance of vegetable and animal life upon our planet it has undergone no change, in a quantitative sense; in other words, that the geological changes which have occurred have been confined to a difference in the distribution of material, and not to the introduction of new material from without.

When the instances of the fall of soot-like particles, blood-rain, sulphur-showers, &c., which have from time to time been described, are considered, the view pronounced by Chladni, that these phenomena are due to the precipitation of large quantities of cosmical dust, appears of great import. The black carbonaceous substances which fell with the Hessle meteorites, and coated some of them, may be quoted as an illustration. Some meteorites, moreover, are so loose and friable in texture that they are very readily reduced to powder, as the Ornans meteorite (1868, July 11th), while that which fell at Orgeuil (1864, May 14th) breaks up when placed in water. If this stone had not fallen on a day when the atmosphere was dry, portions, if not the whole of it, would probably have reached the earth's surface in the form of powder. These atmospheric deposits may have a very varied composition. The dust which fell in Calabria, in 1817,* contained chromium. The red rain that fell at Blankenberg, in Flanders,† in 1819, owed its colour to the presence of cobalt-chloride.

In 1872 three papers were published in the *Comptes Rendus*,‡ on the origin of polar auroræ, which called forth one from Baumhauer,§ where he refers to a theory as to their origin propounded in his thesis *De ortu lapidum meteoricorum* (Utrecht, 1844). After having shown the connexion which apparently exists between the planets, their satellites, the comets, the shooting-stars, the meteorites ("*qui, pour moi, sont de petites planètes*"), and the zodiacal light, a disc of asteroids or cosmical matter massed together near the sun, he gives expression to the following views respecting the polar auroræ: Not only solid masses, large and small, but clouds of "uncondensed" matter probably enter our atmosphere (probabile etiam est nebulas materiei primigeniæ sine nucleo condensato in atmosphæram venire). If from our knowledge of the chemical composition of the stones and irons which

* L. Sementini. *Atti della Reale Acad. delle Scienze*, 1819, i. 285; *Gilbert's Ann.*, lxiv. 327.

† Meyer and Van Stoop, *Gilbert's Ann.*, lxiv. 335.

‡ Le Maréchal Vaillant, *Compt. Rend.*, lxxiv. 510 and 701.—J. Silbermann, *Compt. Rend.*, lxxiv. 553, 638, 959, and 1182.—H. Tarry, *Compt. Rend.*, lxxiv. 549.

§ E. H. Von Baumhauer, *Compt. Rend.*, lxxiv. 678.

fall to the earth's surface, we may draw any conclusion respecting the chemical constitution of these clouds of matter, it appears possible that, as many of these stones consist partly, and the irons almost entirely, of iron and nickel, the attenuated cloud-like matter may also contain a considerable proportion of these magnetic metals.

Let such a cloud, the greater part of the constituents of which have magnetic characters, approach our earth, which we have been taught to regard as a great magnet: it will evidently be attracted towards the poles of this magnet, and, penetrating our atmosphere, the particles which have not been oxidised and are in a state of extremely fine division will, by their oxidation, generate light and heat, the result being the phenomenon which we term a polar aurora. Observations have shown that the seat of these phenomena is about, not the geographical, but the magnetic poles. Not a few facts, even at that time, could be advanced in support of the theory, which assumes the occasional presence of metallic particles in the higher regions of our atmosphere. More than once such particles had been discovered in a fall of hail. Eversmann * found in the hailstones which fell on the 11th June, 1825, at Sterlitamak, 200 wersts from Orenburg, Siberia, crystals of a compound of iron and sulphur, in which Hermann found 90 per cent. of that metal.† In hail which fell in the province of Majo in Spain on the 21st June 1821, Pictet ‡ found metallic nuclei which were proved to be iron; and the hail which fell in Padua on the 26th August 1834, was observed to contain nuclei of an ashy grey colour. The larger ones were shown by Cozari § to be attracted by the magnet, and to contain iron and nickel. "It would," wrote Baumhauer, "be very interesting, in verification of this theory of the origin of polar auroræ, to detect in the soil of polar areas the presence of nickel." This theory, which at the time it was promulgated appeared so rash that it met with severe criticism by the great Berzelius,‖ has gained support from recent researches; among others, the discovery by Heis of the simultaneity of boreal and austral auroræ, the relation between the auroræ and the meteor-showers, the perturbations of the telegraph-lines, which not only accompany, but forecast, an auroral display; and the identity of the light, principally that of

* E. Von Eversmann, *Archiv für die gesammte Naturlehre,* iv. 196.—A. Neljubin, *Archiv für die gesammte Naturlehre*, x. 378.—R. Hermann, *Gilbert's Ann.*, lxxvi. 340.

† Though Von Baumhauer cites this instance, it does not appear that the metallic character of the "crystals" was fully established in this case. Neljubin found them to consist of 70 per cent. iron-oxide, and 17·5 per cent. of other metallic oxides. In fact, this substance appears to have been an impure limonite, like that which fell at Iwan, in Hungary, on the 10th of August 1841, and was probably not meteoric.

‡ Pictet, *Gilbert's Ann.*, lxxii. 436.

§ D. L. Cozari, *Ann. Sc. Regn. Lomb.*, 1834, Nov. e Dec.; *New Ed. Phil. Jour.*, xxxvii. 83.

‖ *Jahresbericht*, xxvi. (1847), 386.

the green portion of the spectrum, in zodiacal and auroral light, as established by Respighi.*

In connexion with this subject, reference should be made to the discovery by Reichenbach some years since of the presence of nickel in soils. From the Lahisberg in Austria, a conical hill some 300 to 400 metres in height, and covered to the summit with beech-trees, he took samples of soil from the thick underwood, and found therein traces of nickel and cobalt. Other specimens from the Haindelberg, Kallenberg, and Dreymarcksteinberg, adjacent hills, yielded the same results, and that from the Marchfeld plain also revealed traces of nickel. These hills consist of beds of sandstone and limestone, and are quite free from metallic veins. It has already been suggested that impoverished soils may have their fertilising powers renewed by the precipitation of cosmical matter containing phosphorus.

LVI.—Abstract of "Geological Notes on the Noursoak Peninsula, Disco Island, and the Country in the Vicinity of Disco Bay, Greenland, by Dr. Robert Brown, F.L.S., F.R.G.S., etc." "Transact. Geological Soc. of Glasgow," vol. v., 1875. By Permission.

I.—Introduction.

The author prefaces these notes on the geology of Mid-Greenland,† with a succinct account of the history of geological discovery in Greenland; namely, (1.), by Sir C. L, Giesecké, early in this century, who collected largely, and left many published and MS. observations; (2.), Dr. Pingel,‡ 1828; (3.), Dr. H. Rink,§

* Respighi, *Compt. Rend.*, lxxiv. 514.—The green ray is that known as 1241 in Kirchhoff's scale; and near it is another of less brilliancy, 1826 in the same scale.

† Greenland is divided by the Danish Government, who maintain a monopoly of the trade up to 73° N. lat., into the North and South Inspectorates; the former embracing all the coast-line north of lat. 66°, the latter all south of that parallel. The region Dr. Brown is describing may be termed, Middle Greenland, he says, as it lies between the parallels of 68° 30′ and 71° N. lat.

‡ With Capt. Graah. *See* "Undersögelses-Reise til Ostkysten af Grönland efter Kongl. Befaling udfört i Aarene, 1828-31, 1832." Translated into English by G. G. Macdougall, 1837. *See also* "Proceed. Geol. Soc. London," vol. ii. p. 208.

§ "Om den geographiske Beskaffenhed af de danske Handels-distrikter i Nordgrönland": Udsigt over Nordgrönlands Geognosie (Det Kongl. danske Vidensk. Selskab. Sk., 3 Bind, 1853, p. 71). In all, Dr. Rink resided 22 summers and 16 winters in Greenland. Also "Journal of the Royal Geographical Society," vol. xxiii. and vol. xxviii.; "Naturhistorisk Tidsskrift" (3rd series), 1852, &c.; "Grönland geographisk og statistisk beskrevet," 2 vols.; and "Eskimoiske Eventyr og Sagn oversatte efter de Infödte fortælleres Opskrifter og Meddelelser."

1852, &c.; (4.), Later explorers in Expeditions from 1852 to 1867, especially Sutherland, Inglefield, Kane, and Hayes; C. J. M. Olrik, Whymper, and Brown, 1867; Nordenskjöld, Berggren, Nordström, and Oberg,* 1870; J. G. Rohde and K. J. V. Steenstrup,† 1872; (5.), Scoresby and the German Expeditions, for East Greenland.

II.—General Geology of Greenland (p. 8).

1. *Primitive and Metamorphic Rocks.*—These are chiefly gneiss, mica-schist, hornblende-schist, syenite, &c., pierced by granitic veins, and are the most widely distributed of all the Greenland formations, extending from the extreme north to the south of Greenland, with the exception of a few patches occupied by traps and Miocene beds and the Igalliko Sandstone. They reach to 4,000 feet in height, and form the rock almost everywhere seen protruding from beneath the "inland" or continental ice which covers the interior of the country. The aspects of the country occupied by this rock are, rounded bosses of hills, ice-planed, with mossy valleys or glaciers protruding to the sea. Rarely does it form the jagged peaks and more varied scenery of the trap-rocks, though sometimes rising into high hills.

In this formation are found all the economic minerals of Greenland; such as Steatite, used to make the lamps and other domestic ware of the natives, though this manufacture is greatly on the decrease, and Cryolite, a mineral of great and increasing value. The mineral was originally brought to the notice of mineralogists by Giesecké;‡ but it was not until a comparatively recent period, chiefly owing to the energy of Mr. Tayler of London, now of Prince Edward's Island, and the Messrs. Thomsen of Copenhagen, that its value has been properly developed. The mineral derives its name from χρύος, ice, from its property of melting in the flame of a candle. It is a crystalline mineral, varying in colour from snow-white when pure to almost black when mixed with extraneous substances, probably, according to Eriglok, graphite, as the colouring matter disappears when the mineral is heated. In chemical composition it is a double fluoride of sodium and aluminium: Aluminium, 13·0; Sodium, 32·8; Fluorine, 54·2. It is valuable as an ore of aluminium, and is now exported in great quantities to America and Europe, chiefly to Copenhagen, where it is applied to a variety of purposes. In 1861 30 ships were sent from Greenland with cryolite. From a recent report by Mr. Paul Quale we learn that the chief veins at present worked are situated at Iviktout or Ivigtot on the south side of Arsuk Fjord, in South

* "Redogörelse för en Expedition till Grönland År 1870," af A. E. Nordenskjöld (Öfversigt af K. Vet.-Akad., Förh. 1870, No. 10; and as a separate publication, Stockholm, 1871). It has been translated, apparently by Professor Nordenskjöld himself, in "The Geological Magazine," July, August, September, October, and November, 1872.

† Petermann's "Geograph. Mittheil," 1874, p. 142; and "Om de Kuhlförende Dannelser paa Öen Disko, Hareöen og Syd-Siden af Nügssuak's Halvöen i Nord-Grönland," Af K. J. V. Steenstrup, &c. "Vidensk. Medd. fra den Naturh. Forening i Kjöbenhavn, 1874, No. 3-7.

‡ *See above*, p. 341.—Editor.

Greenland, in 61° 13′ N. lat. and 48° 9′ W. long. (Green.) The surface of the vein of cryolite was originally covered with a layer of earth, clay, and gravel, which being removed, the vein is exposed. Its greatest length is about 600 feet, breadth about 150 feet, and the extent about 53,000 cubic feet. It is composed of two parts, one close to the fjord, the other to the east, and separated by a rock from 5 to 15 feet high and about 100 feet broad; it is contained in gneiss crossed in several parts by veins or layers running from north-west to south. The principal mineral in the vein is, of course, cryolite, but as constant accompaniments are found quartz, ironstone, lead-ore, sulphide of iron, sulphur, arsenic, and tin-stone, though the last two are very rare. These substances are not evenly distributed through the mass; but the cryolite is partly found quite clean and unmixed, while larger and smaller quantities are found containing more or less of the above-mentioned minerals. A strip of the vein only a few feet broad along the south and south-west boundary of the same is conspicuous for its richness in iron-stone and quartz, but especially for lead-ore and sulphide of iron. The surrounding mountain is also in certain places, viz., along the south-western boundary, strongly impregnated with ores of lead, arsenic, tantalite, molybdena, and tin. In the cryolite, according to Mr. Quale, are found pieces of the surrounding rock, both of granite and trap, and it therefore appears that the cryolite is newer than the gneiss and the trap. The mine during the winter is filled with ice and snow, and all blasting must therefore cease from November to April. In the summer and autumn, from May to October, the mine can be worked, except under unusually severe climatic conditions, and in this time, with a gang of about 50 men, about 5,000 tons of cryolite can be had. There is considerable difficulty in keeping the workmen in health; but from what I can learn there need be no immediate fear of a failure of the mineral, though its distribution is exceedingly local, being almost entirely confined, except in small quantities, to the fjord mentioned. Excepting a small amount found at Miask in Siberia, the mineral is confined to Greenland.*

* In addition to the report quoted (Smithsonian Report, 1866) and a report issued "On the Resources of Iceland and Greenland" by the United States Government (1868), for further particulars regarding the geology and economic value of cryolite, we may refer to the work of Rink already mentioned, and the three following works in Danish :—L. Jacobsen's "Et Aar i Grönland," with sections and plans (1862); E. Bluhme's "Fra et Ophold i Grönland," 1863–1864 (1865), and V. Vallö's "Grönland" (No. xxxiv. of the "Danske Folkeskrifter," 1861), as well as certain papers by Lieut. Bluhme (of which the above work by him is merely a reprint), in the Danish Magazine "Fra alle Lande," vol. i. Many details of the chemical and other characteristics of cryolite and its derivative aluminium may be found in numerous reports and communications scattered through various scientific periodicals, by Wohler, H. St. Clair Deville, H. Rose, Bunsen, Charles N. A. Tissier, J. W. Tayler, C. Brunner, Salvétat, Salm-Horstmar, Erdmann, Heeren and Karmarsch, N. Debreeq, Mallet, Baff, H. Masson, Kobell, Bondet, Schötter, Hirzel, Degousse, Fabian, Christofle, Sauerwein, J. Thomsen, Scherer, A. Strange, R. Wagner, Dallo, C. Hegemann, and others. Deville has published a separate work, "De l'Aluminium" (Paris, 1859), and Tissier, a "Guide Pratique de la Fabrication de l'Aluminium."

Graphite is found in considerable quantities at Karsok and Niakornak, on the Noursoak Peninsula, and on some of the islands lying off the Greenland coast, but hitherto it has not been found profitable to export it to Europe.

The basalt or trap extends right across the country north of 69° N. lat., and appears to date from Tertiary periods. It is particularly well developed in Disco Island and the Noursoak peninsula, covering an area of some 7,000 square miles, and rising in some places to a height of 6,000 feet. Near Disco these traps overlie the gneiss. Properly speaking, the Disco Bay traps are not masses of lava, but "consolidated beds of ashes and volcanic sand," which in time by pressure have assumed a crystalline character, a fact pointed out also by Nordenskjöld. In the peculiarly disintegrating climate of Greenland the trap speedily decomposes. With these trap-rocks are almost invariably associated the Miocene and Cretaceous beds.*

2. *The Red Sandstone.*—This was originally discovered and described by Pingel,† and is only found in the fjords of Igalliko and Tunnudleorbik in about 61° N. lat. It is probably Devonian, but owing to the absence of fossils the age of this interesting formation cannot be accurately determined. The rock is, for the most part, firm and hard in structure, and is composed of fused quartz particles. It appears to be very local, but perhaps is partly concealed under the continental ice.

3. *Carboniferous.*—The coal long known in Greenland belongs to a period much more recent than the true "Carboniferous." In 1870 and 1871, however, the Swedish Expedition, as well as my friend Dr. Georg Pfaff, the inspecting surgeon, for many years resident in Greenland, discovered *Sigillaria,*‡ a *Pecopteris* (?), &c., on the Disco shore at Ujarasasuk. These fossils, probably belonging to the true Carboniferous period, have, however, been as yet only found in travelled blocks, not in rocks *in situ,* so that the existence of the Carboniferous formation in Greenland is still problematical. They may have drifted in ice from the Carboniferous strata of Melville Island. One of these plants, sent me by Dr. Pfaff, I submitted to the inspection of Mr. Robert Etheridge, jun., F.G.S., Palæontologist to the Geological Survey of Scotland, who says: "The fossil is unfortunately only a blackened cast of "the Fern itself, all the vegetable matter having apparently "vanished, except along the rachis, where a little carbonized "material still remains. Its characters are rendered yet more "indistinct from the nature of the matrix, a fine-grained grit. The "Fern appears to be allied to those figured by Heer under the "name of *Gleichenia* (*G. Gieseckiana,* Hr., *G. Zippei,* Corda, "*G. Rinkiana,* Hr., &c.), but I think not identical with any of "them; more probably a new species of either *Gleichenia* or

---

* For an excellent description of the basalt of Greenland see Nordenskjöld, "Geol. Mag." vol. ix., pp. 421, &c.

† "Afhand. Videnskab. Selskab. Skrifter," 1828.

‡ Noticed also by Prof. O. Heer in his "Beiträge zur Steinkohlen-Flora der arktische Zone," p. 10, *note.*—Editor.

" *Pecopteris.* The whole of the nervation has disappeared, so " that I think it would be somewhat hazardous to describe it as a " new species simply on the two specimens in your possession."

4. *Cretaceous.*—This formation in Greenland has only been recently separated from the Miocene formation, with which it is associated. It is, as far as we yet know, only found in the vicinity of Kome or Koke, on the Omenak Fjord, in about 70° N. lat., though traces have been found on Disco and elsewhere.

5. *Miocene.*—This formation (included by Rink* in his " Trap-formation med kulförende Lag "), in West Greenland, is confined to the vicinity of the Waigat Strait and the west coast of Disco Bay, and is the chief subject of this paper.

6. *Quaternary Deposits.*—These consist of peats at present in process of slow formation, and of beds of clay, with scattered boulders, shells, and other marine remains, the exact counterpart of the laminated fossiliferous, brick-clays in Scotland. These are described by the author in the " Quarterly Journal of the Geological Society of London " (February 1871), and in " Das Innere der Grönland," Petermann's " Geog. Mittheil.," 1871.

III.—THE MIOCENE BEDS OF GREENLAND (p. 13).

The existence of fossil plants of relatively recent date in strata on the shores of the Waigat Strait was known to Giesecké; but he was unaware of their exact age and significance—supposing them to be identical with, or nearly allied to, the ordinary *Sorbus aucuparia* and *Angelica officinalis*, now growing on Disco Island.† It was not until MM. Olrik and Rink had sent some of the specimens to Europe that their nature was apprehended. Subsequently through the exertions of Captain (now Admiral) Inglefield, R.N., Dr. Lyall, R.N., and Dr. David Walker of the " Fox " Expedition —aided by M. Olrik, several specimens reached England, and these with those already in the Copenhagen Museum were, with the other Arctic plants, described in Professor Heer's very important work, " Flora fossilis arctica " (1867). A general resumé of his results was given by him in the " Journal of the Royal Dublin Society," vol. v. pp. 69, &c., for 1867.

After explaining the circumstances of his visit to Noursoak, in company with Mr. E. Whymper, Mr. A. P. Tegner, two other Danes, and several natives, in 1867, Dr. R. Brown proceeds to give notes from his journals.

1. *Atanakerdluk*, Lat. 70° 2′ 30″ N., Long. 52° 15′ 3″ W.—The rock seen on the shores of the mainland when coasting from Jakobshavn (Ilulissat) to this locality is the gneiss, sometimes rising into hills 1,400 feet in height, but generally in low rounded hills, with mossy valleys; the hills covered with boulders, and smoothed by ice-action. The rocks are bare, except where coated with a mossy covering of Lichens, while here and there, on

---

* " Grönland," vol. ii., App., p. 139. See further on, p. 497.
† See " Florula Discoana," above, p. 256.

places where a peaty soil has accumulated, the Arctic Birch (*Betula nana*), Willow (*Salix*, various species), *Andromeda*, *Vaccinium*, or Crowberry (*Empetrum*) creep in such unwonted profusion that in the height of summer some more favoured spots seem almost to support the florid title which Erik Ranthri gave to the country he was the first European to land on. Everywhere the dreary marks of present or former ice-action are apparent.

The first exposure of trap-rock noticed was at Akpaet (Ritenbenk, Lat. 69° 45′ 34 N., Long. 51° 7′ W.), appearing also on Arve-Prince Island. Such Tertiary igneous rocks continue as far as 76° N. Lat. To the south older traps are known to exist.

At Atanekerdluk,* on the north coast of the Waigat,† the shore consists of a strip of sandy beach, backed by lofty trap cliffs, sloped and ravined, from 3,000 to 4,000 feet high, and traversed with white streaks of Miocene beds. The chief locality for fossils here is on the side of a hill,‡ about 1,154 feet above the sea, mostly in fragments of hard ferruginous shale, resisting disintegration better than the associated shales and sandstones.§

Dr. Brown, ascending the ravine of the Ekadluk stream, made the following section (p. 21) :—

Section along the Course of the Stream at Atanekerdluk: General Strike of the Strata E.N.E., Dip as noted.

1. 6 feet (?): sandstone, dip 32° (concealed by débris, and probably resting on trap).
2. Shaly coal (?), 10 inches.
3. Seam of lignitic coal, 14 inches.
4. Shale with faint vegetable impressions, 16 inches.
5. Hard clayey shale, broken into splintery angular pieces, 10 inches.
6. 20 feet: black shale, faint vegetable impressions.

---

* Esquimaux : "Rocks in the Sea"; basalt rocks forming a skerry off the shore.

† Waygatz of the old Dutch; Waigattet of the Danes; Ikareseksoak of the natives.

‡ The most northern of three peaks overlooking the bay.—WHYMPER, "Brit. Assoc. Rep. 1869," p. 3.

§ A trench cut here, apparently in the talus of the hillside, gave the following section to Mr. E. Whymper, "Brit. Assoc. Rep. 1869," p. 4:—

1. Light grey fine sand	- 1 foot 7 inches.
2. Grey fine sand -	- 8 inches.
3. White fine sand -	- 8 inches.
4. Grey fine sand -	- 9 inches.
5. White fine sand -	- 6 inches.
6. Yellow sand (not penetrated).	

The plant-impressions were found in pieces of hard red clay in the uppermost bed or on the surface; and also from beds 2, 3, and 4, in softer and more brittle shales, and in lumps of hard iron-grey clay.

Mr. E. Whymper's Report on the Geology of this locality is given in the "Brit. Assoc. Report" for 1869, p. 1, &c., and, with Notes by Dr. R. Brown, in Prof. Heer's Memoir on the fossil plants from Greenland in the "Phil. Trans." for 1869, p. 446, &c. *See also* Nordenskjöld's Memoir, above, p. 432. —EDITOR.

7. 16 feet: fine-grained sandstone, with brownish specks dip 26°.

8. Shaly seam, interstratified with sandstone in thin layers, 6 inches.

9. 6 inches: sandstone in thin layers.

10. 1 inch: shale.

11. 18 inches: fine white sandstone.

12. 8 inches: soft shale.

13. 11 inches: shaly coal, dip 28°.

14. Soft black shale, 3 feet.

15. Soft white sandstone, 3 feet.

16. 4 inches: soft black splintery shales.

17. 2½ feet: hard sandstone.

18. 3 feet: white sandstone.

19. 2½ feet: splintery shale.

20. 10 inches: soft sandstone shale.

21. 4 feet: splintery shale; the section is here much hidden by débris.

22. White soft sandstone, 3 feet.

23. Splintery shales, 6 feet.

24. Fine-grained brownish sandstone, 5 feet. Dip 27½°.

25. Shaly beds, intermixed with some thin sandstones and hard shale, covered with débris. I estimated the depth of these beds to be about 150 feet.

26. 4 feet: white sandstone.

27. 60 feet: splintery shale.

28. 5 feet: white sandstone, dip 43°.

29. 8 feet: shale.

30. 16 feet: brown sandstone.

31. 3 feet: shales.

32. 5 feet: white sandstone.

33. 28 feet: black splintery shales.

34. 4 feet: shales.

35. Brown sandstone, exposed here and there, but the section for about 150 feet is concealed by débris.

36. White sandstone was here exposed in either cliff, about 30 feet as far as could be seen. Dip 40°.

Here a dyke of trap (A) intersected the beds right across the ravine, in a course N. 43° E. Mag., standing out in wall-like masses of an average breadth of 12 feet, the angle about 5°.

37. Brownish sandstone. Dip as above. Partly concealed by débris. Where in contact with the trap, this sandstone was but little altered. In some cases it enclosed small pieces of the trap. This sandstone has altogether, on either side of the dyke, an average thickness of 30 feet.

38. Here, on the opposite (south) side of the ravine, were several feet of shales interbedded with seams of coal and sandstone. Not traced on the north side on account of débris.

39. Hard fine sandstone, 3 feet.

40. Soft crumbling shales, 16 feet.

41. Thin seam of sandstone; scaly sandstone, with black

shales; thin sandstone; laminated sandstone; layers of sandstone; and irregular very thin black shale.

42. The section here becomes somewhat indistinct, but shows about 40 feet of gritty white sandstone, with brownish stains at the lower portion, and intermixed with an irregular seam of conglomerate. Dip 27°.

43. Reddish sandstone, with an alkaline efflorescence.

44. 40 feet: splintery shales.

45. 3 feet: greyish sandstone, with specks of oxide of iron.

46. Black splintery shale, thin.

47. 5 feet: sandstone.

48. The section is continued on the south side of the ravine, the north side being obscure. Here 8 feet of shales and sandstones, alternating in beds of from 1 to 2 inches in thickness, were exposed. The topmost layer is black and soft, like shaly coal, about 4½ feet. Dip 28°.

49. 7 feet: brown sandstone, irregular.

50. Slaty shale, 4 feet.

51. 7 feet: hard black sandstone shales, with indistinct vegetable impressions.

52. 20 feet: shaly seams, alternating with thin layers of sandstone, one 16 inches thick.

53. 6 feet: white sandstone, with (near the top) slight layer of black matter, looking like coal, and shaly matter, mixed as in bed 52.

54. 30 feet: irregular seams of brownish sandstone and shales, and 1 inch of soft shaly coal.

55. 6 feet: brownish sandstone, dip about 39°.

56. Shales, &c., 35 feet.

57. White gritty sandstone, dip 37°.

Here another trap dyke (B) cuts across at an angle of 8°; the stream tumbles over it in a waterfall. Course of this dyke is S.W. and N.E.; average breadth 6 feet. The rock contiguous to it (on the further side) is 57 bed, but the dyke also cuts across 58, 59, and a portion of 60 bed.

58. Black shale, with irregular seams of sandstone.

59. Six feet: white sandstone. At 59 the section is again continued on the north side of the ravine, where it is more distinct than on the south side.

60. 3 feet: black splintery scaly shales.

61. 3 feet: white sandstone.

62. 10 feet: irregularly bedded shales and sandstone mixed—faint fucoid-like impressions.

63. 25 feet: white sandstone; 31° dip.

64. An irregular series of thin beds, as follows:—

4 feet: splintery shale; 4 feet: white sandstone; 8 feet: irregular shales and sandstone; 4 feet: white sandstone; black shale; 4 feet: white sandstone; and 6 feet: irregular shales.

65. 25 feet: white sandstone.

66. 10 feet: black splintery shale.

67. 18 feet: shales and brownish sandstone with 1 inch seams of shale.

68. 20 feet: soft shales.

69. 12 feet: hard grey sandstone.

70. 5 feet: laminated shales.

71. 10 feet: hard greyish sandstone.

72. 3 feet: shales and sandstones; and 6 feet: sandstone.

73. 8 feet: laminated shales and sandstones.

74. 8 feet: hard grey sandstone.

75. 80 feet: irregular shales, sandstones, &c., so high up as to be indistinct; and below covered by débris.

76. Coal, 1 foot.

77. White sandstone, 2 feet.

78. Splintery shales, 4 feet.

79. Hard greyish sandstone, 20 feet. A narrow spur from trap dyke (C) runs perpendicularly through this.

80. Irregular seams of sandstone, shale, &c., about 25 feet.

81. Hard grey sandstone, mostly washed by the torrent.

Between this and 82 a trap dyke (C) runs. Its course is N. Mag., rising in the direction of the Atanekerdluk basaltic headland, intersecting the beds at an acute angle, and about at right angles to "A" and "B" dykes.

82. Irregular beds of sandstone shales, to about 45 feet.

83. Grey sandstone, 8 feet.

84. Brownish, hard sandstone, 4 feet.

85. Black shaly coal (?), 6 feet.

86. White gritty sandstone, with perhaps a layer of shale intervening, 60 feet.

Leaving the ravine, Dr. Brown crossed to a ravine, which unites with the main valley at dyke "C," and thence he continued the section:—

87. Here were exposed, from the bed of the stream to near the pinnacled summit of the hill, about 80 feet of brown sandstone shales, clayey iron-stone, or siderite (in which the fossil impressions are chiefly found), coal, &c. In ascending order they were as follows:—

a. Splintery shales, 12 feet; *b.* Bed of coal, 2 feet; *c.* Hard shales intermixed with gritty sandstone, much obscured by débris; *d.* White gritty sandstone; *e.* 2 feet of the sideritic shale—the same as the fossil leaves are found in; out of reach; *f.* Irregular shales, 10 feet; *g.* About 20 feet of irregular shales with irregular layers of coal; and *h.* Shales, &c., with small bits of lignitic coal, irregularly scattered throughout, about 10 feet.

88. Many thin beds of sandstones and shales, thus:—

1. A few inches of sandstone; 2. Thin layer of shale; 3. Thin sandstone, same as 1; 4. 1 inch of sideritic shale, same as the fossil impressions are in; 5. Sandstone, as before, 1 inch; 6. Shale, 1 inch; 7. Sandstone, 2 inches; 8. Shale, with pieces of coal; 9. Sandstone, thin; 10. Thin clay ironstone, same as fossils are contained in; 11. Thin shale; 12. Sandstone, thin; 13. Thin shale; 14. Thin seam of coal; 15. Sandstone, 1 foot thick; 16. Shales, 6 inches, with 17. A broken thin layer of coal in the

middle; 18. Layer of clayey siderite with fossils, 2 inches; 19. Sandstone, 1 foot; 20. Shale, with fragments of coal, same as 6; 21. Excessively splintery shales with very faint fossil impressions; 22. Thin seam, sandstone, with a broken seam of coal, 4 inches; 23. Splintery sideritic ahales with faint fossil impressions; 24. 1 foot: coarse-grained white sandstone, as before; 25. Splintery shale, as in 23; 26. Gritty sandstone, 14 inches; 27. Splintery shales, 18 inches (as in 23), with faint vegetable impressions; 28. Gritty sandstone, with little specks of coal, and stained with iron in patches; 29. 3 feet: splintery shale, bluish-black and reddish, with imperfect vegetable impressions; 30. A seam of coal and shale, 1 inch; 31. Gritty sandstone, stained, 4 feet; 32. Brittle shales, with faint impressions of leaves, &c.; 33. 4 feet: gritty sandstone, with several irregular thin layers (not exceeding 2 inches) of shale as before; 34. 2 feet: splintery shale with faint vegetable impressions; 35. White gritty sandstone, 8 inches; 36. 1 foot: gritty sandstone; 37. Thin shale; 38. 2 feet: gritty sandstone; and 39. Shales, about 3 inches.

89. Various shales and sandstones; possibly 100 feet; dipping 33°, with E. strike; obscured with débris, containing much of the thin, hard, shaly clay iron-stone, with leaf marks.

90. Irregular beds of sandstone and shale again, consisting of—

1. Fine sandstone, 10 inches; 2. Splintery shale, as before, 5 feet; 3. Shaly sandstone, 10 inches; 4. Shale, 6 feet; 5. Shaly sandstone (as 3), 8 inches; 6. Splintery shale, 8 feet; 7. Shaly sandstone (as 3), 10 inches; 8. Splintery shales, 12 feet; 9. Shaly sandstone, with cherty iron-clay, shale, siderite ("Atanakerdluk-stone" of the Greenland Danes), 12 feet; 10. Splintery shale, 8 feet; 11. Broken layer of sandstone shale (as 3), on the summit of the hill.

Beyond this, trap rocks predominate over the country; dykes of variously weathered configuration, such as Rink's Obelisk, forming prominent objects. Travelled blocks of syenite and greenstone lie about.

In Prof. Heer's paper "On the Miocene Flora" in the "Journal of the Royal Dublin Society" (*l. c.*), a sketch is given by Commander Colomb, R.N., of certain fossil stems which he and Captain (Admiral) Inglefield, R.N., saw in this locality at the time of their visit; but these could nowhere be seen. There had apparently been a landslip, and they are doubtless buried under the debris.*

Dr. Brown further examined another section of the same strata, but rather lower down in the series, about 1,007 yards S.E. by S. (Mag.), from House at Atanekerdluk, in the cliff facing the sea (p. 32.)

1. Brown sandstone, 2 feet, dip 33° (on the top); 2. Shale and shaly coal, irregular, average 8 inches; 3. Seam of coal, 2 inches; 4. Shale, 6 inches; and 5. Whitish sandstone, stained with brown, 2 feet exposed.

---

* Nordenskjöld describes bituminised stems found in this locality by him with the roots so permeating the soil as to leave no doubt but that they grew *in situ*. *See above*, p. 434.

In the shale, and in the sandstone where it joins the shale, faint and imperfect impressions of plants appeared. Probably bed No. 5 is No. 1 of the former section; but it is obscured by débris and pieces of trap fallen from the dyke "A," which here extends right across the strata. It becomes fully exposed further on, to a thickness of 6 feet, and is interstratified with sandstone and shale, and succeeded by 16 inches shaly lignite, with faint impressions. This is topped by 8 inches of white sandstone, succeeded by 3½ feet of grey and black shales; 4 feet of white soft sandstone, interstratified at the top with 2 inches of sandstone and shale; and finally by black shales, here dwindling away to 6 feet, though doubtless, from their position and dip, one of the thick beds exposed up the ravine. The bottom of this is a kind of coal, united with the sandstone beneath by 2 inches of shale. A mile and a half (English) further along the beach the trap dyke comes down perpendicularly through the strata, and sends a transverse vein at 61° between the sandstone and the shaly sandstone. North of this, another perpendicular vein was previously met with, coming down from dyke "A" in the same direction (S.E), but not so well marked. The strata are slightly tilted at the line where the dyke spur cuts through them; on the northen side the dip being 47° S., and on the other side 61°, dipping slightly northward. The strike of all the strata is, however, the same (N.E. Mag.). The white sandstone resumes its former dip on the other side, and there does not seem to be deflected by the dyke. This gritty sandstone does not seem to have been much metamorphosed, though doubly in contact with the trap. In some places the trap seems to take the character of the sandstone, being soft, crystalline, and easily broken with the hand. On the other hand, the grit in contact with the trap does not always seem to differ from that in the normal state, but in most cases to have been melted by the heat, a thin glaze being formed on the surface.

The section extends to about 1,180 feet (by aneroid), and is continued nearly uniformly N. and S.; though on the surface facing the sea the strata were denuded in many places. In some other places, northward in the cliffs with greater height, a similar section, with a similar uniformity and alternation, could be traced, probably even to 1,200 feet.*

2. *Ouiarasuksumitok.* Lat. 69° 51′ 2″ N., Long. 52° 19′ 6″ W. (Nordenskjöld.)—In the cliff at the water-edge, the section exhibited sandstone, a small seam of shale, and 2 feet of poor lignite, the whole resting on trap. Ouiarasuksumitok means the place where the coal is mined out, and was chiefly built for the convenience of the coal-mines at Ounartok. It is also called Ujarar-

---

* From the examination of some fossils from the lower strata at Atanekerdluk, Dr. Heer is of opinion that probably they belong to the Upper Cretaceous. The fossils on which he founds this opinion are *Cycadites Dicksoni*, the frequent Ferns, a *Sequoia* allied to *S. Reichenbachii*, and a *Credneria*. For further descriptions of Atanekerdluk, see also Nordenskjöld's Memoir (above, p. 432), Heer's "Les Régiones polaires du Nord" ("Bibl. Univ.," Jan. 1867, p. 51), and his great work already mentioned.

susuk.* M. Olrik found fossil leaves in the sandstone here. It was in a boulder in this locality that Dr. Pfaff found a *Sigillaria* and *Pecopteris* (?) of Carboniferous age † (see above, p. 470).

3. *Ounartok (Ritenbenk Coal-mine).*—Here, nearly opposite to Atanekerdluk, shales rest on trap (?) and are covered by white sandstone; this is apparently covered by coal (for along the beach there was a face of 3 feet of lignitic coal exposed, though the floor was of coal, and therefore the seam appeared to extend to a still greater thickness), covered by 1½ foot of shale, this by 10 feet of coarse brownish grit, then 2 feet of hard brownish sandstone, finally 4 feet hard grey sandstone, coloured in some places by iron. The dip of these strata was about 33°, the strike easterly (across the Waigat). Both the shaly sandstone and grit contained innumerable impressions of stems, interlaced in every direction, more particularly in the shales; but there were no *leaves*, and most of the stems in the shales appeared only like lines of charred wood. Ounartok was probably once a native "house-place," and much débris has been brought down by the mountain-stream close by. In this débris, Gudeman and Pavia, two of the native boatmen, found fair impressions of leaves, though not hitherto found in the stratum itself here; and Gudeman discovered the fruit of the *Magnolia* (*M. Inglefieldi*, Hr.). Dr. Brown found that these fossils had come from the little stratum of hard brown sandstone, which, as well as the sandstone and shales already noted, had yielded some fossil impressions. These comprehended dicotyledonous leaves (*Populus*, probably *P. Richardsoni*, Hr.) and fronds of *Aspidium Meyeri*, Hr. and *A. Heerii*, Ett.; but it was was very difficult to obtain pieces of the rock on account of the superincumbent mass of strata.

4. *Kudlisaet.* Lat. 70° 5′ 35″ N.‡—Here were green mossy slopes, but the vegetation on the whole was on this side much more backward than on the Noursoak shore (August). An *Arabis* and *Stellaria Edwardsii*, R. Br., were frequent on the dry places, but little of *Hierochloe borealis*, so abundant elsewhere. The glaucous Willow (*Salix glauca*) was common, and *Elymus arenarius* gave a link-like character to the sandy shore, while a little marsh at the mouth of a little stream was thick with *Equisetum arvense* and *E. variegatum.*

---

* In Professor Heer's paper in the "Phil. Trans.," l.c., Ounartok is erroneously called by the two names of Ujararsuk and Ouiarasuksumitok, while that of Ounartok applied to the locality where the coal is at present mined is never mentioned.

† Here also were found the remains of *Caulopteris punctata*; Cretaceous according to Mr. Carruthers, F.R.S. *See above*, p. 388.—EDITOR.

‡ In the map attached to Professor Nordenskjöld's memoir, already quoted, Kudlisaet (Kudliset) is apparently applied to a locality in lat. 70°, while the locality of Kudlisaet is marked as Ritenbenk Coal-mine ("Ritenbenks Kolbrott"). A locality still farther to the south, on a stream, is marked on Nordenskjöld's map as "Igdlokungoak." It is in about the latitude of Ounartok; but as I had no opportunity, adds Dr. Brown, of taking observations to determine the latitude (as apparently Professor Nordenskjöld had not either), I cannot be certain; as we were very careful, however, in getting the exact names from our native boatmen, all inhabitants of the immediate vicinity, I cannot but consider that they are correctly given by us.

Section southward along the Shore at Kudlisaet.

1. Alluvium and débris of rock from cliffs behind; 20 feet.
2. 5 feet: brown gritty sandstone.
3. 14 inches: hard grey sandstone.
4. 1 foot: hard sandy shale, with vegetable impressions, &c.
5. 14 inches of poor coal, exposed by men attempting to work it at a place where a little stream breaks through the strata.
6. Shales, 2 to 3 feet.
7. Hard sandstone, 3 feet.
8. Shales, sandstones, &c., irregular; 2 feet.
9. Hard sandstone, 1 foot.
10. Shales, 2 feet.
11. Hard grey sandstone, *with pieces of coal in it.* About 14 feet of this is exposed. The dip is here 45° to the N., strike E. across the Waigat. Scattered along the beach and on the slopes are great blocks of breccia-like material, apparently the peculiar basalt referred to already (see p. 470), which have rolled from the mountains.

Geology of Heer's Creek.—In the stream, called by Dr. Brown "Heer's Creek," a little north of the last locality, Mr. Andersen, of Ritenbenk, had found fossil stems. Gudeman, who had accompanied him, soon pointed them out. They were lying in fragments in the stream, mixed up with pieces of sandstone containing impressions of leaves. The stems were much broken, but their dicotyledonous character was quite apparent. The section where these stems had rolled down (as it afterwards appeared) out of the coal, showed, from the level of the stream upward, the following beds:—

1. (Bottom) 4 feet: splintery shales; 2. 1 foot: hard gritty sandstone; 3. 4 feet: mixed shales and sandstone; 4. 1 foot: coal; 5. 1½ foot: shale, with faint impressions of leaves, stems, &c.; 6. 2½ feet: coal enveloping siliceous stems lying apparently horizontally E. and W. and N. and S. (Mag.), in fragments; brownish outside; mostly without the bark; 7. Shales, 18 inches; 8. Sandstone, with leaves, &c., 3 feet; 9. Thin soft shale; 10. 1 foot: coal; 11. 4 feet: shale, soft and splintery; 12. 1 foot: coal; 13. 3 feet: soft, splintery, brownish shales; 14. Whitish gritty sandstone (?); 15. Soil.

(P. 42.) The Miocene strata seem to continue down the coast to very near Godhavn, and coal appears at various places. At one place, called Skandsen (The Battery) by the Danes, appear regular basaltic columns. The north-east of Godhavn is built on a low syenitic island. Syenite also appears on the opposite side on the main island, backed by great fells of trap rising to the height of between two and three thousand feet.

No calcareous beds have as yet been met with in either the Miocene or the Cretaceous strata of Greenland; and they appear to have been of freshwater origin.

IV.—Greenland Coal.

(P. 42.) The coal of Noursoak and Disco Island was used by some of the English Arctic Expeditions (Inglefield and M'Clintock's, at Kudlisaet and Atanakerdluk). Many years ago it was mined for use in the Disco-Bay Settlements, but for 30 years the use of it was discontinued. Since Mr. Andersen has had charge of Ritenbenk, the workings have been resumed for a few days every summer, and he takes out yearly about 300 Danish töndes (barrels about the size of a sugar hogshead), which costs at the rate of 48 skillings, or about 1*s.* 1½*d.* per tönde. The coal is like all Tertiary coal—rather poor, little coherent, and breaking with a cubical fracture, though bright and glistening in colour. It abounds in a small species of retinite, varying from a mere speck to pieces the size of a marble. It gives off less smoke than English coal. It gives as little ash as wood, and the heat-giving power Mr. Anderssen estimates as being only one half of English coal. He generally mixes it with English coal, and finds the arrangement work well in every respect; for in the "General Taxt-List" for Greenland in 1867 English coal is charged at the rate of 3 rd. 48 sk. (or 7*s.* 10½*d.*) per cask of 4·677 bushels.

Two analyses of this coal, one made by the late Prof. Fyfe, of Aberdeen, and in the Appendix to "Inglefield's Summer Search after Sir J. Franklin" (p. 151), and another by Mr. T. W. Keates, of London, in the "Philosophical Transactions" for 1869, p. 449,* are here placed side by side for comparison :—

——	Keates.			Fyfe.	
Specific gravity	1·369	—	—	1·3848	
Gaseous and vol. matter	—	45·45	—	—	50·60
Moisture	—	·75	46·20	—	—
Sulphur	—	—	·55	—	—
Coke { Fixed, carbon	—	47·75	—	—	39·86
Coke { Ash	—	5·50	53·25	—	9·54
			100·00		100·00
This lignite contains a trace of bitumen; the coke is non-caking, and of little use.					

In addition to the localities mentioned, coal has been found by the natives at various places on the coast. For instance, Dr. Rink † mentions the following places, and Giesecké several others, where coal could be conveniently dug :—1. Atanakerdluk; 2. Patoot;

* A third analysis, by Dr. Wartha, is given in Heer's "Flora fossilis arctica," p. 5.

† "Grönland," vol. i., pp. 172–178.

3. Atane; 4. Kordlutok; 5. Nulluk Cape; 6. Ekkorgvœt; 7. Mibersteen Fjeld; 8. Pattorfik; 9. Sarfarfik; 10. Kome; 11. Upernavik Cape; 12. Innerit Fjord; 13. Hare Island; 14. Ritenbenk Coal-mine (various places on the Disco shore of the Waigat); 15. Skandsen; 16. Makkak; 17. Igligtsiak.

The first five localities named are on the Noursoak shore of the Waigat; the next five are on the Noursoak shore of Omenak Fjord, and most likely belong to the Cretaceous formation, certainly not altogether to the Miocene; the twelfth locality is on Svarte Huk Peninsula, in Upernavik district; while the three last-named places are on the southern shore of Disco Island—Igligtsiak being only 16 English miles from the Settlement of Godhavn, to which the natives sometimes bring the coal on their sledges in winter.

NOTE.—The maps accompanying Prof. Nordenskiöld and Dr. Brown's memoirs comprise that portion of the West Coast of Greenland lying between the 69° and 71° parallels of North latitude.

The very indented coast-line of the country, intersected by numerous fjords bearing many islands, bends outwards in a north-westerly direction to form the Peninsula of Noursoak at about 70° N. lat. Between it and the adjacent mainland to the south lies the Bay of Disco, so called from the large Island of Disco, which lies due south of Noursoak, from which it is separated by the Waigat Strait.

The mainland, as well as a small portion of the eastern end of the Noursoak Peninsula, and the southern extremity of Disco (at Godhavn) is mainly composed of gneiss or granitic rocks; but the peninsula itself, as well as the Disco Island, consists chiefly of Miocene or of Cretaceous formations underlying Traprock.

In the Island of Disco the Miocene beds are represented as extending from the gneissic promontory of Godhavn on the south, round the eastern side to nearly the northern point; and on this coast, proceeding in the above direction, Puilasok, Sinnifik, Skandsen, Isungoak, Ujarasusuk, Igdlokungoak, and Kudlisæt are known as localities whence fossil plants can be obtained.

In the Noursoak Peninsula the Miocene beds lie on its southern coast from Sakkak, past Atanekerdluk, to beyond Mannik, where plant-fossils have been found; and at Atanekerdluk, the position of which is marked by a definite promontory, is the Ekadluk ravine from which Dr. Brown's principal section is taken. Netluarsuk and Ifsorisok, near the northern point of Noursoak, are also fossiliferous.

The Cretaceous strata appear below the trap along the northern coast of Noursoak, extending from Ekkorfat past Angiarsuit Avkrusak, Pattorfik, Kome, and Assakak, to nearly the neck of the peninsula, where the overlying trap appears to unite with the gneiss of the mainland. All the places above mentioned are rich in fossils.

Cretaceous fossils have also been obtained from below the Tertiary strata at Atane, on the Waigat, and in the lower part of the section at Atanekerdluk.

Hare Island, at the northern entrance of the Waigat, and Omenak, an island in the Omenak Fjord, the boundary of the peninsula on the north, consist apparently of trap-rock.

LVII.—Dr. PINGEL on the GRADUAL SINKING of PART of the SOUTH-WEST COAST of GREENLAND.

[Reprinted, with Permission, from the Proceedings of the Geological Society of London, vol. ii., 1833-1838, p. 208; read Nov. 18, 1835.]

. . . . . . The first observations which led to the supposition that the west coast of Greenland had subsided were made by Arctander between 1777 and 1779. He noticed, in the firth called Igalliko (lat. 60° 43′ N.), that a small, low, rocky island, about a gunshot from the shore, was almost submerged at spring tides, yet there were on it the walls of a house 52 feet in length, 30 feet in breadth, 5 feet thick, 6 feet high. Half a century later, when Dr. Pingel visited the island, the whole of it was so far submerged that the ruins alone rose above the water.

The colony of Julianeshaab was founded at the mouth of the same firth in 1776; and near a rock, called the Castle by the Danish colonists, are the foundations of their storehouse, which are now dry only at very low water.

The neighbourhood of the colony of Frederickshaab (lat. 62° N.) was once inhabited by Greenlanders, but the only vestige of their dwelling is a heap of stones, over which the firth flows at high water.

Near the well-known glacier which separates the district of Frederickshaab from that of Fiskenæs, is a group of islands called Fulluartalik, now deserted; but on the shore are the ruins of winter buildings, which are often overflowed.

Half a mile to the west of the village of Fiskenæs (lat. 63° 4′ N.), the Moravians founded, in 1758, the establishment called Lichtenfels. In 30 or 40 years they were obliged once, perhaps twice, to move the poles upon which they set their large boats, called "umiak" or women's boats. The old poles still remain as silent witnesses, but beneath the water.

To the north-east of the mother colony, Godthaab (lat. 64° 10′ N.), is a point called Vildmansnæs by St. Egede, the venerable apostle of the Greenlanders. In his time, 1721-1736, it was inhabited by several Greenland families, whose winter dwelling remains desolate and in ruins, the firth flowing into the house at high tide. Dr. Pingel says that no aboriginal Greenlander builds his house so near the water's edge.

The points mentioned above Dr. Pingel had visited, but he adds, on the authority of a countryman of his own highly deserving of credit, that at Napparsok, 10 Danish miles (45 miles English) to the north of Nye-Sukkertop (lat. 65° 20′ N.), the ruins of ancient Greenland winter houses are to be seen at low water.

Dr. Pingel is not aware of any instance of subsidence in the more northern districts, but he suspects that the phenomenon reaches at least as far as Disco Bay, or nearly to 69° north lat.

NOTE.—Some facts relating to the Rise and Fall of the Greenland Coast have been collected by DR. R. BROWN in the "Quart. Journ. Geol. Soc.," xxvi., pp. 690–692 (June 1870). Reprinted in the *Royal Geograph. Soc. Manual*, published for the Arctic Expedition.

LVIII.—RECENT ELEVATIONS of the EARTH'S SURFACE in the NORTHERN CIRCUMPOLAR REGION. By HENRY H. HOWORTH.

[Reprinted in part, with Permission, from the Journal of the Royal Geographical Society, London, vol. xliii., 1873, pp. 240–263.]

. . . . . . It is well known that Greenland is subject to a movement of oscillation, the northern portion of it being in process of elevation, and the southern of depression, the axis of the movement being variously placed between the parallels of 74 and 77. I will quote a passage from Dr. Kane's travels: "The "opportunity I had to-day of comparing the terrace and boulder-"lines of Mary River and Charlotte-Wood Fiord enables me "to assert positively the interesting fact of a secular elevation "of the crust commencing at some as yet undetermined point "north of 76°, and continuing to the great glacier and the high "northern latitudes of Grinnell Land. This elevation is con-"nected with the equally well-sustained depression of the Green-"land coast south of Kingutak." * Again: "The depression of "the Greenland coast which I had detected as far north as "Upernavik is also going on here (*i.e.*, the Crimson Cliffs). "Some of the Esquimaux huts were washed by the sea or torn "away by the ice that had descended with the tides. The turf "too, a representation of very ancient growth, was cut off even "with the water's edge, giving sections 2 feet thick. I had "noticed before such unmistakable evidence of the depression "of this coast. Its converse elevation I had observed to the "north of Wolstenholme Sound. The axis of oscillation must "be somewhere in the neighbourhood of latitude 77°." † M'Clintock says: "It has been abundantly proved by the exis-"tence of raised beaches and fossils that the shores of Smith's "Sound have been elevated within a comparatively recent period." He then goes on to show that this elevation has probably ceased in the very latest times, and concludes that at Upernavik the land

---

* Vol. ii. p. 80. † Op. cit. p. 277. See also "The Open Polar Sea," p. 402.

has sunk, as is plainly shown by similar ruins, over which the tides now flow.*

Crossing Baffin's Bay to the American coast, we have little difficulty in proving that the axial line previously spoken of extends into that continent. Thus, in regard to Labrador: "From all the indications noticed casually by us, such as the " portion of beaches apparently very recently raised above the " sea-level, so as to be just beyond the reach of the waves, the " land is slowly gaining on the sea. The Rev. C. C. Campbell, " minister at Caribou Islands, in the Straits of Belle Isle, also " informs me that this is his impression, gained both from his " observations and from information given by the settlers. To this " last source Mr. J. F. Campbell is indebted for the statement in " his 'Frost and Fire' that the Coast of Labrador is slowly " rising." † In Chimmo's account of his visit to the north-east coast of Labrador, he mentions many reefs, &c., not marked on the maps.‡ These were probably, therefore, recent elevations.

In regard to Newfoundland Mr. Moreton says "that there is " much bare protruding rock in all parts of the island presenting " everywhere a rounded, worn, and water-washed appearance, " such as can only be produced by their having once been part " of the ocean-bed. Large boulders of stone of different character " from all the rock around are lodged in all parts. Some of the " most remarkable are on the highest lands. A recent, and I " suppose still proceeding, uprising of the whole island from the " sea is very observable, and many proofs of it have been brought " to my notice. For instance, a narrow tickle at the head of " Greenspond Harbour, in which the water now is scarcely deep " enough for a punt passing, was in the memory of aged people " sufficient for the passage of large fishing-boats called shallops. " At Purchard's or Pilchard's Island and in Twilling-gate Harbour, " rocks now above water are remembered as formerly sunken " rocks, over which it was possible and usual to row small boats. " In many places, from the same causes, the fishermen cannot " now let their boats ride in the same water where their fathers " were wont to moor them. I have been told of similar changes " in Trinity Harbour." § Going somewhat further south, Mr. Hopkins says: "Two hundred and fifty years ago Sir Francis " Drake sailed into Albemarle Sound through Roanoke Inlet, " which is now a sandbank above the reach of the highest tide; " only seventy years ago it was navigable by vessels drawing 12 " feet of water." ‖

It is clear, however, from Lyell's observations,¶ that we are here on the borders of an area of subsidence which extends along

* "M'Clintock's Journal," pp. 76, 77.

† Packard's "Glacial Phenomena of Labrador," &c., "Memoirs of Boston Nat. Hist. Society," vol. i. part 2, p. 229.

‡ "Journal of the Royal Geographical Society," vol. xxxviii. p. 258.

§ Ibid., vol. xxxiv. pp. 264–5.

‖ "World before the Deluge," p. 22.

¶ "Principles of Geology," 11th edition, vol. i. p. 563.

the United States coast as far south as Florida, an area which I shall describe more particularly in a future paper. I will now adduce the facts which make it clear that the elevatory movement is shared by the whole Arctic border-land of America.

In Franklin's voyage in 1819, 1820, and 1821, he mentions having found much Drift-wood in the estuary of the Copper-Mine River. He also picked up "some decayed wood far out of reach "of the water." He adds that the Copper-Mine River itself brings down no drift-wood.* In his second voyage along the Arctic Sea he describes the coast from the Mackenzie River to the Rocky Mountains as very shallow and full of shoals and reefs. Inside some of the latter was brackish water, as was also the water in pools at some distance inland; piles of wood were also thrown up far from the coast.† While Franklin surveyed the coast westward, Dr. Richardson did the same to the east. The latter says: "On the coast from Cape Lion to Point Keats there is a line of "large drift-timber, evidently thrown up by the waves, about 12 "feet in perpendicular height above the ordinary tides." He shortly afterwards mentions that in the Polar Sea when cumbered with ice such waves are impossible, and as his journey was in the hot season, and the sea was then crowded with hummocks, the inference that the drift-wood was thrown up by the waves is inadmissible, and the line of drift-wood 12 feet above the sea-level is only a parallel to the numerous other cases. The vast sheet of shallow and brackish water, 140 miles long and 150 broad, which is separated from the Polar Sea by low banks and spits of sand, and is called by Dr. Richardson "Esquimaux Lake," formed, there can be little doubt, very recently, as that traveller suggested, a bay of the Polar Sea, and is an example of the creation of huge brackish lakes by a sea which is constantly contracting, such as are common in the eastern borders of the Caspian. "M'Clure "found shells of *Cyprina Islandica* at the summit of the Cox-"comb Range, in Baring Island, at an elevation of 800 feet above "the sea level. Captain Parry has also recorded occurrences of "*Venus* (probably *Cyprina Islandica*) in Byam-Martin Island, "and in the recent voyage of the 'Fox' the surgeon found the "following sub-fossil shells at Port Kennedy, at elevations of "100 feet to 500 feet:—*Saxicava rugosa*, *Tellina proxima*, "*Astarte arctica* (*borealis*), *Mya Uddevallensis*, *Mya truncata*, "*Cardium* sp., *Buccinum undatum*, *Acmea testudinalis*, and "*Balanus Uddevallensis*."‡

Speaking of the eastern part of Melville Island, Parry says: "One of the 'Hecla's' men brought to the boat a narwhal's horn, "which he found on a hill more than a mile from the sea, and "which must have been carried there by the Esquimaux or by "bears (!) . . . . Sergeant Martin and Captain Sabine's "servant brought down to the beach several pieces of fir-tree "which they found nearly buried in the sand at the distance of

* "Narrative," p. 357. † Ibid., p. 134.

‡ "Appendix to M'Clintock's Narrative."

"300 or 400 yards from the present high-water-mark, and not "less than 30 feet above the sea-level. We found no indication "of this part of the island having been inhabited, unless the "narwhal's horn above alluded to be considered as such." * Again, speaking of the northern part of Melville Island, near Point Nias, two pieces of drift-wood were also found on the beach 10 or 20 feet above the present level of the sea, both of pine, one 7½ feet long and 3 inches in diameter, and the other much smaller. Both were partly buried in sand and their fibres so decayed as to fall to pieces on being laid hold of.† Again, speaking of the west of Melville Island: "The land gains upon the sea, "as it is called, in process of time, as it has certainly done here, "from the situation in which we found the drift-wood and the "skeletons of Whales." ‡

King William Island is rather low, the western shore extremely so, and bears evidence of a gradual and tolerably recent upheaval from beneath the sea.§ These extracts from the Arctic voyages might be extended, but they will suffice to show what is generally recognised, that the archipelago north of the American continent shows, wherever examined, signs of current elevation. We may now continue our survey along its western coast.

So long ago as 1778, Captain Cook makes the following remark about the coast of Behring's Straits, near Cape Denbigh:—"After "breakfast a party of men were sent to the peninsula for brooms "and spruce. . . . . It appeared to me that this peninsula "must have been an island in remote times, for there were marks "of the sea having flowed over the isthmus. And even now, it "appeared to be kept out by a bank of sand, stones, and wood "thrown up by the waves. By this bank it was evident that the "land was here encroaching upon the sea, and it was easy to "trace its gradual formation." ‖

In describing the journey of Captain Krenitzin and Lieutenant Levashef in 1768–69, Coxe says: "The 'St. Catherine' wintered "in the Strait of Alasca, and was drawn into shoal water. The "instructions set forth that a private ship had in 1762 found "there a commodious haven, but the captain looked for it in vain. ". . . . On surveying this strait and the coast of Alasca "many craters were observed in the low grounds close to the "shore, and the soil produced few plants. May not this allow "the conjecture that the coast had undergone considerable "changes, even since the year 1762?" ¶

In Whymper's account of his journey to Alaska, I find the following passage:—"The island of St. Michael's is covered with "moss and berries, resting sometimes on a bed of clay, but more "commonly on a porous lava rock. The formation apparently

---

* "Parry's Voyage in 1819–20," p. 68. † Ibid., p. 193.

‡ Ibid., p. 235.

§ M'Clintock in "Journal of the Royal Geographical Society," vol. xxx. p. 10.

‖ "Cook's Voyages," edition of 1842, vol. ii. p. 344.

¶ "Coxe's Russian Discoveries," p. 251.

" extends to the Youkon. The Indians have a tradition that the " island was upheaved from the sea, an occurrence at least pos- " sible. A large rock in the chain of the Aleutian Islands, known " to the Russians as the Bajaslov Volcano, rose from the sea in " 1796." Zagoskin says: " That the spot where the fort (*i. e.*, " Fort Youkon) now stands has been covered by the sea within " the memory of the Indians living at the date of his visit in 1842 " and 1843." Again: " The entire country is sprinkled over " with remains of Pliocene Mammals, *Elephas* (?), *Ovibos mos-* " *chatus*, &c. Beds of marl near Fort Youkon contain fresh- " water shells still living in the vicinity." * Mr. Grant tells us that in Vancouver's Island a raised sea-beach with scanty sandy soil is mentioned as extending with a breadth of from 300 to 500 yards all along the north-east end of the harbour of Port St. Juan.†

In a paper on the Beaches of British Columbia, by Mr. Begbie, I find the following paragraph: " Changes of level are now going " on in a gradual way in some parts of the colony. At a point " near Frazer River, 13 miles south of Quesnelle, and again on " that creek an affluent of Bonaparte River, I have noticed beaver- " dams on a slant,—abandoned dams, of course. A beaver-dam is " never known to give way, never built on a stream that runs dry " in summer, and is, of course, as level as the surface of the water " it is meant to retain. There had been no violent commotion, " for the dams were all quite perfect. No water was now " running there. The old watercourse still visible and many " cotton-trees still growing, perhaps 30 years old, but no signs of " living Beavers."‡

To prove that this movement of the northern coasts of America is shared by the interior of the country, we must examine the great series of lakes that form such a notable feature in the physical geography of that continent.

Captain Back says that the country from the Great Slave Lake to the Polar Sea is strewn with boulders, &c., and has evidently not been long reclaimed from the sea.§

The country forming the Hudson's Bay Territory is covered with erratic boulders, and many patches of Pleistocene deposits, containing marine shells of the present Arctic species (*Mya truncata, Saxicava rugosa, &c.*). The whole country is too flat for these boulders to have been the débris of glaciers. They were most probably left by floating ice and icebergs when the land was submerged. The cliffs of Lake Winnipeg contain fresh-water shells still living in the lower waters, such as *Unio*, *Helix*, *Pupa*, &c., often raised more than 100 feet above the present levels of the streams, and appear to be ancient lake- or river-terraces, leading to the belief that the existing series of lakes from the St. Lawrence northward were once united in one or more vast

* " Journal of the Royal Geographical Society," vol. xxxviii.

† Ibid., vol. xxvii. p. 285.

‡ Ibid., vol. v. p. 132.

§ Ibid., vol. vi. p. 1.

fresh-water seas. A subsidence of 400 feet would make Lake Ontario discharge its waters by the Mohawk and Hudson into the Atlantic, convert Lake Champlain into a maritime strait, and form islands of the States of New York, New England, and Maine, New Brunswick and Nova Scotia; a subsidence of one-fourth of this would carry the waters of the Missouri and the Upper-Churchill and Mackenzie Rivers into Lake Winnipeg, and convert the plain country bordering the Rocky Mountains into an inland sea. The raised beaches of Lake Superior are 100 feet above the present level.*

" On Lake Superior, in Canada, deposits face the lake in the " shape of bare earth-banks and terraces. They are all the pro- " duce of the lake when standing at a higher level. . . . . On " Lake Huron are successive belts of water-worn erratics of large " size, one above another, with a few yards interval between each. " On the summit of a cliff 100 feet high, Colonel Delafield informs " me there is a range of water-worn stones, regularly strewn as " on a beach, for 200 feet in length. These instances of remains " of ancient deposits might be greatly multiplied, as they are " very usual in this lake when the vegetation permits them to " be seen." I have extracted this passage from a very interesting paper by Dr. Bigsby on Canadian Erratics,† which describes similar traces as existing in nearly all the lakes of North America. His *résumé* of the evidence states that " the Canadas, in common " with all the western and northern parts of the United States, " are mapped out by irregular concentric rings of terraces and " ridges, sometimes hundreds of miles in circuit, which enclose " the beds (with or without water) of lakes and ponds more or less " closely. The mouths of rivers here and there break through " these rings, and the rivers themselves are also bordered with " terraces . . . . . . The terraces are the margins of former " bodies of water much loftier and larger than those now existing. " These ancient lakes have been more or less emptied by the " elevation of their beds, an elevation taking place perhaps very " extensively, slowly, and variously" (p. 236).

Having shown by the evidence of the lakes and rivers (those gauges of level by which alone we can test the change of level that is progressing in a country) that the interior of the northern part of the American continent is rising as well as the coast, we will now pass on to an examination of the remaining half of the northern circumpolar regions comprised in Europe and Asia.

The remarkable changes that have taken place in Scandinavia, in illustration of our subject, are among the elementary facts of geology. They have given rise to an extensive literature, somewhat fierce in its controversial bitterness. The question has been complicated by a difficulty which arises in many

---

* Isbister's " Geology of the Hudson's Bay Territory, &c.", " Quart. Journal, Geological Society," vol. xi. p. 497.

† " Quart. Journal, Geological Society," vol. vii. pp. 215, &c.

other districts, namely, that we seem to have arrived at a critical turning-point in the world's history, where areas which have long been rising have become quiescent or even begun to sink again. I cannot enter into the details of the Scandinavian controversy which have been collected by Sir Charles Lyell, in his 'Bakerian Lecture' on this subject and elsewhere, but will content myself with quoting the more striking authorities, with whom I agree. Early in the last century Celsius expressed his opinion that the waters both of the Baltic and Northern Ocean were gradually subsiding, and from numerous observations inferred that the rate of depression was about 40 Swedish inches in a century. In support of this position he alleged that there were many rocks both on the shores of the Baltic and of the Ocean known to have been once sunken reefs and dangerous to navigators, but which were in his time above water; that the Gulf of Bothnia had been gradually converted into land, several ancient ports having been changed into inland cities, small islands joined to the continent, and old fishing-grounds deserted as being too shallow or entirely dried up. He also maintained that in the time of the ancients Scandinavia was what they described it to be, namely, an island, and that it became a peninsula some time between the days of Pliny and the ninth century.* This view was opposed by several writers. Playfair, in 1802, accepted the views of Celsius, and argued that the change was due to the rise of the land. In 1807 Von Buch, after returning from a tour in Scandinavia, announced his conviction that the whole country from Frederickshall in Norway to Abö in Finland, and perhaps as far as St. Petersburg, was slowly and insensibly rising. He was led to these conclusions principally by information obtained from the inhabitants and pilots, and in part by the occurrence of marine shells of recent species which he had found at several points on the coasts of Norway above the level of the sea. He also mentions the marks set on the rocks.† These discoveries induced several Swedish philosophers to have certain rocks grooved at the level of the water in calm weather, with the date of each added. In 1820 and 1821 the marks were examined by the officers of the pilotage-service, who reported to the Royal Academy of Stockholm. From this Report it appeared that along the whole coast of the northern part of the Gulf of Bothnia the water was lower than formerly. New marks were at the same time made. In 1834 Sir Charles Lyell made an elaborate survey of the district, and published the result, as the "Bakerian Lecture," in the "Philosophical Transactions" for 1835. He reports that in the interval between 1821 and 1834, the land appeared to have risen in certain places north of Stockholm 4 or 5 inches, and he convinced himself during his visit to Sweden, after conversing with many civil-engineers, pilots, and fishermen, and

* Lyell's "Principles of Geology," 9th edition, p. 520.

† Trans. of "Von Buch's Travels," 387, quoted in Lyell, op. cit.

after examining some of the ancient marks, that the evidence formerly adduced in favour of the change of level, both on the coasts of Sweden and Finland, was full and satisfactory. *Inter alia* he mentions rocks and boulders strewn over the shoals, which have been observed to increase in height and dimensions within the previous half-century. Some formerly known as dangerous sunken rocks are now only hidden when the water is the highest. Similar points have grown to long reefs, while others have been changed from a reef annually submerged to a small islet on which a few lichens, a fir-seedling, and a few blades of grass attest that the shoal has at length been fairly changed into dry land. Long fiords and narrow channels, once separating wooded islands, have been deserted by the sea within the memory of living witnesses on several parts of the coast. It is well known that the southern extremity of Scania is sinking; the proofs will be collected in another paper. On the eastern or Baltic side of Sweden Sir Charles Lyell found the first unmistakable evidence of rising at Calmar, in 56° 41′ N. lat. The foundations of the castle there, which had originally been subaqueous, were found to have risen 4 feet in four centuries. At Stockholm there were found striking proofs of change since the Baltic acquired its present tenants—Testacea, found there 70 feet above the sea-level, being identical with those now found in the adjacent sea at Sodertelji. A little further south, strata of sand, clay, and marl, more than 100 feet high, and containing shells of species now inhabiting the Bothnian Gulf, were found. The three lakes of Husar, Ladu, and Uggel, which formerly (temp. Charles IX.) constituted the Gulf of Fiskartorp, had grown much shallower, and in part become dry land. At Upsala, 40 miles N.N.W. of Stockholm, brackish-water plants were found in meadows where there are no salt-springs; proof that the sea has recently retired. The Marsh at Oregrud, 40 miles north of Upsala, had risen 5½ inches in the interval since 1820. At Gefle, 40 miles to the N.W., are low pastures, where the inhabitants' fathers remembered boats and even ships floating. At Pitea, in the Bothnian Gulf, the land had gained a mile in 45 years; at Lulea, a mile in 28 years; and at Tornea it was advancing rapidly, according to M. Reclus,* at the rate of 5 feet 3 inches in a century.

These facts, which might be multiplied, suffice to show that the Baltic coast of Sweden, north of about the 56th parallel, has been recently rising from the sea.

M. Reclus argues that the Baltic communicated but recently with the North Sea by a wide channel, the deepest depressions of which are now occupied by the Lakes Mälar, Hjelmar, and Wener, considerable heaps of oyster-shells being found in several places on the heights commanding these lakes. Similar beds are

* "The Earth," vol. ii. p. 622.

found round the Gulf of Bothnia.* From von Baer's researches it would seem that oysters cannot live and grow in water holding more than 37 parts in a thousand of salt, or less than 16 or 17 in a thousand. The waters of the Baltic now do not contain more than 5 parts in a thousand, and yet the beds of oysters prove that both the Baltic and the inland lakes were once as salt as the North Sea. M. Reclus argues this saltness could only come from some former strait which occupied the depressions in which the Swedish engineers have dug out the Trolhatta Canal. Besides, he says, when the sluices were being constructed, there were found not far from the cataracts, and at a height of 40 feet above the Cattegat, various marine remains mingled with relics of human industry, boats, anchors, and piles.† Sir Charles Lyell says similar oyster-beds have been found further inland on the borders of Lake Wener, 50 miles from the sea, at an elevation of 200 feet, near Lake Rog-varpen. Similar beds have also been discovered on the southern shores of Lake Mälar, at a place 70 miles from the sea.‡ So that we may take it as proved that the great Swedish lakes are the remains of a very recent marine strait, separating Scania from the mainland. The shores of the Cattegat afford ample evidence of upheaval.

The greater part of Denmark is either stationary or sinking; but according to Forchhammer, the terminal point of Jutland, bounded by an ideal line tending obliquely from Fredericshavn towards the north-west, rises 11·70 inches in a century. The amount is here probably exaggerated.

We will now turn to the coast of Norway. Here we approach evidently a boundary-line between rising and sinking land. "Professor Keilhau, of Christiania," says Sir Charles Lyell, "after collecting the observations of his predecessors respecting former changes of level in Norway and combining them with his own, has made the fact of a general change of level at some unknown, but, geologically speaking, modern period (that is, within the period of the actual testaceous fauna) very evident. He infers that the whole country from Cape Lindernas to the North Cape, and beyond that as far as the fortress of Vardhuus, has been gradually upraised, and on the south-east coast the elevation has amounted to more than 600 feet." The same author tells us that marine fossil shells of recent species have been collected from inland places near Drontheim. On the other side Mr. Everest has shown that the island of Munkholm, an insulated rock in the harbour of Drontheim, has remained nearly stationary for eight centuries. Brongniart and Sir Charles Lyell both found beds of recent shells raised 200 feet above the sea at Capellbacken, all the species being identical with those now inhabiting the contiguous ocean. The former also found *Balani* adhering to the rocks above the shelly deposit, showing that the sea had remained

* These beds of shells have since been traced by Erdmann to Sinde, at the head of a lake of that name, 130 miles west of Stockholm, at the height of 230 feet above the sea.

† Reclus, op. cit. vol. ii.

‡ Lyell, op. cit. 527–9.

there a long time. This was verified by Lyell in 1834, at Kured, about 2 miles north of Uddevalla, at a height of more than 100 feet above the sea. He says these Barnacles adhered so firmly to the gneiss, that he broke off portions of the rocks, with the shells attached. Similar deposits of shells are found at the island of Orust, opposite Uddevalla. Between Gothenburg and Uddevalla, and on the islands of Marstrand and Gulholmen, similar proofs may be studied,* proving that we are here on the borders of a doubtful line.

In 1844 M. Bravais showed that in the Gulf of Alten, in Finmark, the most northern part of Norway lying to the north of Lapland, there are two distinct lines of upraised ancient sea-coast, one above the other.†

From Finmark we may naturally step across to Spitzbergen, an island which is notoriously rising from the sea at a rapid rate. I find the following passage as early as 1646:—These mountains (twenty-two mountains of Spitzbergen) increase in bulk every year, so as to be plainly discoverable by those that pass that way. Leonin was not a little surprised to discover upon one of these hills, about a league from the seaside, a small mast of a ship, with one of its pulleys still fastened to it. This made him ask the seamen how that mast came there, who told him they were not able to tell, but were sure they had seen it as long as they had used that coast. Perhaps formerly the sea might either cover or come near their mountain, where some ship or other being stranded, this mast is some remnant of that wreck.‡ Parry, in an account of his journey towards the Pole, page 126, refers to the vast quantities of Drift-wood stranded on the Spitzbergen coast above high-water-mark.

In the 16th vol. of the "Quart. Journal of the Geological Society," Mr. Lamont tells us that he found great quantities of Drift-wood on all the Thousand Islands, as well as on the south coast of the Spitzbergen main—some of it much worm-eaten, much of it lying at least 30 feet above high-water-mark. He nowhere found any wood *in situ*. On all parts of Spitzbergen and its islands, visited by him, he found numerous bones of Whales far inland and high above high-water-mark. One large piece of a jawbone, found by himself in October 1859, was discovered 40 feet above the sea. It was part of an entire skeleton, which lay half buried in moss, about half a mile from the sea, in Walter-Thymen's Strait. There was also a terrace of trap-rocks higher than the moss intervening between the latter and the sea. On one of the Thousand Islands he counted eleven very large jaw-bones, along with many bones forming other parts of the Whale's skeleton, all lying close together in a slight depression, about 10 feet above the sea level. On the same island he saw what he took to be a further proof of the recent upheaval of the land. This was a sort of furrow or trench, 100 yards long by 3 or 4 feet deep, and 3 or 4

* Lyell, op. cit. passim. † Lyell, "Principles," 11th edition, 194.

‡ "Account of Greenland," by La Peyrère, in "Churchill's Voyages," vol. ii.

feet broad, ploughed up among the boulders, and presumed to be done by icebergs. It was on a gentle slope, about 20 feet above the sea, and extended from north-east to south-west, exactly the run of the current-ice at the present day.* The seal-fishers told Mr. Lamont that the land was rising, and that the Right Whale had forsaken the Spitzbergen seas, which had become too shallow for it.

The German Expedition of 1869 also found heaps of Drift-wood 20 feet high above high-water-mark on the south-east shore of Spitzbergen.

East of Nova Zembla Captain Mask, who made a journey there in 1871, found the barren and sandy islands known as the Gulf-Stream Islands. In the spot where these now are, the Dutch, in 1594, found and measured a sandbank in soundings of 18 fathoms, showing an upheaval here of 100 feet in 300 years. In the same year Captain Nils Johnson landed in the country called Wiche Land in the map of 1617, situated about 30° east longitude and 78° north latitude. He says that the shores there to a distance of 100 miles inland, and to a height of about 20 feet above high-water-mark, are covered with drift-wood.†

We may now return to the mainland of Europe, and continue our survey eastwards.

Pennant long ago observed that the White Sea and the Baltic were but recently joined together by a strait. He says the Lakes Sig, Onda, and Wigo form successive links from the Lake Onega to the White Sea. The Lake Siama almost cuts Finland through from North to South. Its northern end is not remote from Lake Onda, and the southern extends very near to the Gulf of Finland, a space of nearly 40 Swedish, or 260 English miles. These were probably part of the bed of the ancient Streights (*sic*) which joined the White and Baltic Seas.‡ Great portions of Finland, which is known to the natives as Suomenia, or the land of swamp, has all the character of a recently emerged land. It is sprinkled over with lakes separated by flats of sand covered with moss. The level of some of these lakes is rapidly falling, which means that the land is rising. We are told this especially of the River Vosca and the Lake Samia, of which it is the only feeder. In the spring of 1818, Lake Souvando, on the west of Lake Ladoga, broke down the isthmus that separated it; its waters were lowered 5·026 fathoms, and much land was left dry.

Sir Charles Lyell tells us, that on the coast of Finland, as on that of Sweden, the fishermen have traditions that what is now dry land was in their fathers' days water. The surface of Finland generally is covered with traces of a prodigious diluvial revolution in recent times.

MM. de Keyserling, Murchison, and de Verneuil have found at points 250 miles to the south of the White Sea, on the banks of the Dwina and the Vaga, beds of sand and mud containing several

---

* "Quart. Journ. Geol. Soc.," vol. xvi. p. 428.
† "Ocean Highways," pp. 247 and 292.
‡ "Appendix to Arctic Zoology," p. 23.

kinds of shells similar to those which inhabit the neighbouring seas, and so well preserved that they had not lost their colours.*

In some notes on the ice between Greenland and Nova Zembla, by Captain Jansen,† he says, quoting an experienced navigator called Thenius Ys: "Drift-wood, though there is plenty on the "beach, is found far above this mark, and so remarkably high, "that I do not understand how it is brought there." Again, speaking of Captain Wm. de Vlaningh, who sailed along the north and north-east coasts of Nova Zembla in 1664: At a considerable height he found on a rock, on the smallest of the three Islands of Orange, a very large tree that three or four men could not lift. This tree was rotten. The tree lay much too high to have been brought there by water, perhaps by a waterspout he says.

The two islands of Nova Zembla are each divided from north to south by a prolongation of the Ural Mountains, but they consist chiefly of a marshy moss-clad plain. It has lately been found that there are saline lakes in these islands.‡

M. de Middendorf states that the ground of the Siberian tundras is in a great part covered with a thin coating of sand and fine clay, exactly similar to that which is now deposited on the shores of the Frozen Ocean. In this clay, too, which contains in such large quantities the buried remains of Mammoths, there are also found heaps of shells perfectly identical with those of the adjacent ocean. Far inland, besides, traces of drift-wood are seen, the trees which once grew in the forests of Southern Siberia: these trees, having been first carried into the sea by the current of the rivers, have been thrown up by the waves on the former coast, which are now deserted by the sea.§ Our chief authority for the shores of the Arctic Sea is Von Wrangel, and from his travels I shall quote freely:—"In 1810, Hedenstrom went across the "Tundra direct to Utsjansk. He says, on the Tundra, equally "remote from the present line of trees, among the steep sandy "banks of the lakes and rivers are found large Birch-trees, com-"plete with bark, branches, and roots. At first sight they appear "well preserved, but on digging them up they are found to be in "a thorough state of decay. On being lighted they glow, but "never burst into flame; the inhabitants use them for fuel; they "call them Adamoushina, or, 'of Adam's time.' The first living "Birch-tree is not now found nearer than three degrees to the "south, and then only as shrubs." Again, in 1811, Samukof reports that he found the skulls and bones of various animals in the interior of Kotelnoi Island, and that both there and in New Siberia he found large trees partially fossilized. These islands have apparently all been recently submerged, for it is reported that the greatest stores of Mammoth ivory are now got from the sand-banks which are constantly appearing near the Bear Islands;

* "Reclus, op. cit.," vol. ii. p. 627.
† "Proceedings of the Royal Geographical Society," vol. ix. p. 163.
‡ Maltebrun: Geog., vol. ii. p. 394.
§ Reclus, "The Earth," vol. ii. pp. 627, 628.

the barren surface of the latter, a conglomerate of bones, stones, and ice, has all the character of a recently recovered sea-bottom. Wrangel tells us that ribs of Whales are often found on the west coast, and that Whales are now very seldom seen on the Siberian coast, while in the 18th century their appearance there was much more frequent. The only cause for this desertion that I can suggest is that assigned by the Spitzbergen fishermen, namely, that the sea is becoming too shallow for the Whale. "The shores of "the Polar Sea, from the Lena to Behring's Straits, are for the "most part low and flat. In winter it is hard to say where land "ends and sea begins. A few versts inland, however, a line of "high ground runs parallel with the present coast, and formerly "no doubt constituted the boundary of the ocean. This belief is "strengthened by the quantity of drift-wood found on the upper "level, and also by the shoals that run far out to sea, and will no "doubt become dry land." * Again: "At several places along "the coast we found old weathered drift-wood at the height of "two fathoms above the present level of the sea, while the fresh "drift-wood lay on a lower level. This indicates change of level." Again: "Captain Sarytschew says the winter-dwellings erected "by Laptef on the bank where his vessel was driven on shore "lead to the belief that the channel must formerly have been on "that side. At present there is no water there for a vessel of "any size, and even a boat can only approach at high water. At "low water the shoal runs three versts out to sea." † Diomed Island, described by Chalavrof in 1760, and by Laptef at a later date, no longer exists: it now forms a part of the main. The same voyagers describe the east coast of the Swatoi Moss as very sinuous: it is now very straight, the sinuosities having meanwhile disappeared. These facts will suffice to prove that so far as we have any evidence, the whole Siberian coast, as far as Behring's Straits, is rising from the sea.

In Mr. Grieves' translation of the "History of Kamtchatka," I find it stated, in the description of Behring's Island and the adjacent island, that, 30 fathoms higher than the sea-mark, lie wood and whole skeletons of sea-animals which have been left by the sea.‡ He speaks of one of the rivers at Ochotsk as being now dry; this is probably caused by upheaval. And in describing the Penschinska Sea, he says he had seen "trees which are not to be "found in the country hanging out of the earth, and more than "seven feet below the surface; whence (he says), it may be concluded that all these barren, boggy places, where at present "there are no woods but shrubs and stunted Sallows and Birches, "were once covered with water, which has decreased by degrees "here, as it has on the north-eastern coast." §

Quite recently Russian travellers have discovered on the coast of the great island of Saghalien heaps of modern Shells, lying not far from the shore on beds of marine clay, and also former bays,

* Von Wrangel, Sabine's translation.

† Ibid., cvii.

‡ Page 54

§ Von Wrangel, Sabine's translation, pp. 59–61.

which are now converted into lakes or salt-marshes. In like manner it has been proved that the regions of the Amur are gradually being upheaved, for, in order to maintain its level, the river has constantly to hollow out its bed between the cliffs; and on the plateau by the river-side semicircular sheets of water may still be seen, which are evidently former windings of the Amur.* . . . . . . [Alterations of level and evidences of desiccation in the Japanese and Chinese regions, in Central Asia, and the Persian and Caspian areas are next noticed.] . . . . .

I have now completed a rough and slight survey of the great mass of land that surrounds the North Pole, and have shown that, so far as we have any evidence, that great mass is undergoing a general' movement of upheaval; or, to be perfectly correct, we find on it traces in all directions that there has been a movement of upheaval since there was any subsidence; and in those areas, which are accessible enough to enable us to experiment, as in Sandinavia, &c., we find that the movement is going on now at a greater or less rate. This general movement of Circumpolar land having its focus apparently near the Pole, has no doubt been coincident with a corresponding revolution in other physical phenomena, such as climate, the distribution of magnetism, &c. . . . . There is one fact which is very obvious . . . . . that in the vast area over which we have shown that there are traces of upheaval, there is not, so far as I know, a single volcano. If the ancient theory that volcanoes are due to the eruptive forces of the earth be true, this fact requires explanation. . . . . .

LIX.—Rocks and Minerals of Greenland. By Dr. H. Rink. (From "Tillæg No. 7," of Dr. H. Rink's "Beskrivelse af Grönland," 1857 [This ought to have followed "Sutherland," at p. 368, in order of date.]

In the "Primitive" rocks, which constitute a great portion of Greenland, granitic and gneissic varieties, with or without hornblende, and with garnets and magnetite, are common. Trap-rocks also occur with them. As bedded masses in these granitoid rocks, hornblende-schist, with garnet, is most common (shown in Rink's sketch, at p. 140, as traversing the Omenak Hills). Actinolite, asbestos, and pyrites occur in this schist. Dolomite, with tremolite and other minerals, occurs in the Omenak, Christianshaab, and Egedesminde districts. Anthophyllite, with actinolite, is noticeable at Upernavik; where also occurs a felspathic stratum with quartz, dichroite, and garnet. As veins in the granitic rocks, red felspar is found, with magnetite, apatite, allanite, zircon, and pyrites; also white felspar and tourmaline. Epidote also, with iron-glance and calcite, is disseminated in the rock, and barytes

---

* Reclus, vol. ii. p. 660.

occurs in layers. In the gneiss are pyrites and magnetite; and graphite at Upernaviks Lange and Omenaks Storoe. Greenstone, granite, or norite, also diabase, graphic granite, granulite, and slates are met with in different parts of the coast and mainland.

The red quartzose sandstone of Igalliko Fjord and Tunnudliorbik Fjord is local; it is pierced by porphyry veins.*

(The great Trap-formation,† with Brown-coal, has been more fully described by NORDENSKJÖLD and others since the date of RINK's Memoir, and the alluvial formations will also be found more fully noticed in the same papers. *See* above.)

RINK notices the following minerals and their places of occurrence: Quartz, siliceous sinter, jasper, olivine, felspar, adularia, opalescent adularia, labradorite, amazon-stone, scapolite, pumice (from Jan Mayen), gieseckite, nephrite, sodalite, eudialite, zeolites, mica, chlorite, talc, serpentine, hornblende, actinolite, smaragdite, tremolite, augite, asbestos, crocidolite (loc. incog.), clay, clay-slate, barytes, garnet (various), dichroite, epidote, zircon, emery, beryl, tourmaline, saphirine, allanite, gadolinite, fergusonite, calcite, dolomite, fluor, cryolite (with a plan of the locality where it is worked‡), tungstate (tungspath), sparry-iron-ore, malachite, apatite, magnetite, specular-iron-ore, brown iron-stone, yellow ochre, titaniferous iron-ore, tin-stone, wolfram, native sulphur, native iron, pyrites, arsenical pyrites, smaltine, copper-pyrites, galena, copper-glance, molybdenite, blende, graphite. (*See* also Giesecké's list of rocks and minerals above, p. 349.)

* It is sometimes called "Old Red Sandstone;" but it may be of Cambrian age.—EDITOR.

† Rink gives sketches,—1. Of Innerit (south side) with its long cliffs of horizontal sandstone, and beds of lignite, capped with great beds of trap; 2. Of a part of the gneiss county, near Proven, overlain by massive trap-rocks on one part, and showing a great, isolated, *crater-like* hollow in the trap in another.

‡ *See* above, Giesecké, Tayler, &c.

§ II.

PARRY ISLANDS AND EAST-ARCTIC AMERICA.

LX.—NOTE on ANIMAL LIFE in the NORTH-WEST PARRY ISLANDS (Melville Island, Prince-Patrick Land, and the Polynia Islands). From "CAPT. M'CLINTOCK'S Reminiscences of ARCTIC ICE-TRAVEL in search of SIR JOHN FRANKLIN" (Third Expedition). With Permission: Journal Royal Dublin Society, vol. x. pp. 236–7.

"A comparison of 300 thermometric observations made upon this journey and those simultaneously registered at Bridport Inlet, in latitude 74° 56′ N., shows a difference of 3·5° of lower temperature for any more northern position, for which, as a mean, we may assign the parallel of 76½° N., being about 100 miles north of the ship (H.M. "Intrepid"). The means of temperature thus compared were 18·5° and 22°.

"These observations having been made between 12th April and 15th July (1852), sufficiently account for the diminished vegetable growth, and consequent decrease of animal life upon the land; whilst the absence of the Polar Bear is significant of a similar scarcity of the frozen deep."

"As bearing upon the distribution of animal life I subjoin a record of all that were shot or seen." Abstract:—

—	April 4 to May 13. — Melville Island.	May 14 to June 26. — PrincePatrick I.	June 26 and 30. — Emerald Island.	July 1-19. — Melville Island.
Musk-oxen - -	59	5	0	30
Rein-deer - -	29	8	13	74
Hares - - -	1	1	0	2
Seals - -	0	2	1	15
Gulls - - -	0	12	7	34
Brent Geese -	0	20	0	107
Ducks - - -	0	5	0	18
Ptarmigan - -	16	37	0	12

"No traces of Bears were found. A few Wolf tracks were seen, but only on Melville Island. No traces of Oxen, Deer, Foxes, or Ptarmigan beyond the 77th parallel, except in one instance, when a decayed bone of a Deer and traces of a Fox were found. Up to 77° N. Fox tracks were frequently seen, although we never saw the animal. Lemmings were tolerably numerous whenever there was vegetation.

"Three kinds of Gulls. The Ivory Gulls (*Larus eburneus*) were the earliest to arrive, and were found furthest north; they began to lay eggs before the thaw commenced; eight only were seen, and all of them upon Prince-Patrick's Land. Seventeen

Glaucous and twenty-eight Skua Gulls (*Lestris parasiticus*), the latter chiefly on Melville Island. Of the Ducks, three were long-tailed (*Anas glacialis*), and the other twenty-two were King Ducks (*A. spectabilis*). Several Snow-buntings, sparingly, but universally distributed; four or five Red Phalaropes; two Sea-Snipes; a Raven; and a bird supposed to be a Snowy Owl, complete the list."

[H.M.S. "Resolute," stationed at Dealy Island, Bridport Inlet, Melville Island, obtained from 3 Sept. 1852, to 9 Sept. 1853:—

Musk-oxen, 114.	Bears, 6.	Geese, 128.
Rein-deer, 95.	Wolves, 3.	Ducks, 229.
Hares, 146.	Ptarmigan, 711.	Plover, 16.

BELCHER'S "Last of the Arctic Voyages," 1855, vol. ii. p. 155.]

LXI.—MAMMALIA of the PARRY ISLANDS and NEIGHBOURING COASTS.

From:—

I. A Supplement to the Appendix of Captain Parry's Voyage for the Discovery of a North-west Passage in 1819–20. 4to. London, 1824. Mammalia, by Captain Sabine, pp. clxxxiii–cxcii.

II. Appendix to Captain Parry's Journal of a Second Voyage, &c., in 1821–23. 4to. London, 1825. Quadrupeds and Birds, by John Richardson, M.D., pp. 287–341.

III. Journal of a Third Voyage of Discovery, &c., in 1824–5. 4to. London, 1826. Zoology; Mammals, by J. C. Ross, R.N., &c., pp. 92–95.

R. Narrative of a Second Voyage in search of North-west Passage, &c., in 1829–33. By Sir John Ross, C.B., &c. 4to. London, 1835. Zoology, by Captain J. C. Ross, R. Owen, &c., pp. vii–xxiv.

The species and genera have been critically revised in later works. *See* other Catalogues in this "Manual."

I. II. III. R. Ursus maritimus, L. *Esquimaux*, Nonnook or Nennook. *Greenl.*, Nennok.

I. II. R. Gulo luscus, L. *Esq.*, Kablee-aree-oo (Parry's 2nd Voyage). *Esq. of Melville Peninsula*, Kab-le-a-rioo (Ross). *Esq. of Boothia Felix*, Kā-ĕ-wēēk. Remains throughout the winter as far North as 70°.

I. II. R. Mustela erminea, L. *Esq.*, Terree-ya.

I. II. Canis lupus, L. *Esq.*, Amărŏk.

R. C. lupus occidentalis, Rich.

I. II. III. R. C. lagopus, L. *Esq. of Melville Peninsula*, Tērree-ānee-ărĭoŏ. *Esq. of Delcome and Coppermine River*, Terreganncœuck. *Greenl.* Terienniak, Kakkortak.

II. R. C. lagopus, var. β. fuliginosus, (Shaw). *Greenl.* Kernektak.

II. Arvicola Grœnlandica, (Traill). *Esquimaux* ?, Ow-in-yuk.

I. II. III. A. Hudsonia (Forster).

II. R. A. trimucronata, Rich.

II. R. Arctomys Parryii, Sab.

I. II. III. R. Lepus glacialis, Leach.

I. II. III. R. Cervus tarandus, L. *Greenl.*, Tukta (*male*), Pangnek (*female*), Kollowak ; *young*, Norak. *Esq.*, Tooktoo.

I. II. III. Ovibos moschatus, L. *Esq.*, Oo-ming-mak. *Greenl.*, Omimak or Umimak, *and of the natives of Wolstenholm Sound.*

II. III. R. Phoca fœtida, Müll. *Greenl.*, Neitsek. *Esq.*, Neith-keek or Neitiek (*middle-sized*, Kairolik ; *young*, Ibbeen).

II. R. P. barbata, Fabr. *Esq.* (Ross), Oo-ge-ook. *Esq. of the Welcome River*, Ogg-œook. *Esq. of Melville Peninsula*, Ogūke. *Grenl.*, Urksuk.

II. R. P. Grœnlandica, Fabr.

I. P. vitulina, L.

I. II. III. Balæna mysticetus, Lacépède. *Esq.*, Aggă-uĕek (Yookai, *whalebone*).

I. B. physalus, L.

I. II. III. R. Monodon monoceros, L. *Esq.*, Keina-lov-a.

I. II. III. R. Delphinapterus beluga, Lacép. *Esq.* (Ross), I-we-ak.

I. II. III. R. Trichecus rosmarus, L. *Esq. of Melville Peninsula*, Ei-ŭ-ĕk. *Esq. of the Welcome River*, Ej-ee-werk.

LXII.—Fishes of the Parry Islands and Neighbouring Regions.

---

1. From Parry's 1st and 3rd Voyages (**I.**, pp. ccxi.–ccxiv., and **III.**, pp. 109–112), and Ross's 2nd Voyage (**R.**, pp. xlvi.–xlix.). Described by Sabine, Ross, and Richardson, 1824, 26, 35.

[For Dr. Lütken's revised Catalogue of the Arctic Fishes, *see* above, pp. 115–122.]

R. Cyclopterus minutus.

I.. R. Liparis communis.

III. R. Ophidium Parrii, Ross.

III. R. O. viride, Fabr.

R. Gadus morrhua. *Esquimaux of Boothia*, O-wuk.

R. G. callarias. *Esquimaux of Boothia*, Il-lit-toke.

I. III. R. Merlangus polaris, Leach.

I. M. carbonarius.

I. R. Blennius polaris, Sab.

I. R. Cottus quadricornis, Bloch. *Esquimaux of Boothia,* Kan-ny-yoke.
I. III. R. C. polaris, Sab.
R. Pleuronectes hippoglossus.
R. Salmo Rossii, Richardson.
R. S. alipes, Rich.
R. S. nitidus, Rich. *Esquimaux,* Augmalook.
R. S. Hoodii, Rich. *Esquimaux,* Masamacush.

2. From WELLINGTON SOUND (CHANNEL), collected by Sir E. BELCHER, and described by SIR JOHN RICHARDSON, in Belcher's Last of the Arctic Voyages, ii., Appendix, pp. 347–376; pl. 23–30, 1855.

Cottus glacialis, Richardson.
Phobetor tricuspis, (Reinhardt).
Gasterosteus insculptus, Rich.
Gunellus fasciatus, (Bloch).
Lumpenus nubilus, Rich.
Lycodes mucosus, Rich.
Gymnelis viridis, (O. Fab.).
G. viridis, *var.* unimaculatus, Rich.
Merlangus polaris, Leach and Sabine.

LXIII.—INSECTS and ARACHNIDS from GREENLAND, the PARRY ISLANDS, and Neighbouring Lands, collected during PARRY'S First and Third Voyages, and ROSS'S Second Voyage, and determined by the REV. WM. KIRBY, LIEUT. J. C. ROSS, and MR. JOHN CURTIS. 1825, 26, 35.

I. PARRY'S FIRST VOYAGE, Appendix, &c., pp. ccxiv.–ccxix.

INSECTS. By the REV. W. KIRBY.

Bombyx Sabini, K.
Bombus arcticus, K. = *Apis alpina,* Fab.
Ctenophora Parrii, K.
Chironomus polaris, K.

SPIDER.

Salticus ? Melvillensis, K.

II. PARRY'S THIRD VOYAGE, Appendix, &c., pp. 112–115.

INSECTS. By J. C. ROSS.

Simulium reptans (Fab.).
Ctenophora Parrii, K.
Pedicia rivosa (Fab.).
Culex caspius, Pallas.
Melitæa Tullia (Fab.).
Bombyx Sabini ?, K.
Bombus arcticus, K.
Formica rubra, L.

SPIDERS.

Salticus scenicus, (Fab.).
Dysdera erythrina, Latr.
Oxyopes variegatus, Latr.
Lycosa saccata (Fab.).

III. Ross's Second Voyage, Appendix, &c., pp. lx., *et. seq.*

Insects. By John Curtis, Esq., F.L.S.

Order, Coleoptera.

Fam., *Dytiscidæ.*

Colymbetes mœstus.

Order, Dermaptera.

Forficula. Taken 23 June, 1831 ; scarce ; under stones.

Order, Hymenoptera.

Fam., *Ichneumonidæ.*

Ichneumon Lariæ. Infesting the larvæ of *Laria Rossii*, early in July.
Ephialtes, sp.
Campoplex ? arcticus.
Microgaster unicolor.

Fam., *Formicidæ.*

Myrmica rubra, Linn. Numerous, under stones.

Fam., *Apidæ.*

Bombus Kirbiellus.
B. polaris.

Order, Trichoptera.

Fam., *Phryganeidæ.*

Tinodes ? hirtipes.

Order, Lepidoptera.

Fam., *Papilionidea.*

Colias Boothii.
C. chione.
Hipparchia Rossii.
H. subhyalina.
Melitæa Tarquinius.
Polyommatus Franklinii.

Fam., *Bombycidæ* vel *Arctiidæ.*

Laria Rossii.
Eyprepia hyperborea.

Fam., *Noctuidæ.*

Hadena Richardsoni.

Fam., *Phalænidæ.*

Psychophora Sabini.
Oporabia punctipes.

Fam., *Tortricidæ.*

Orthotænia Bentleyana.
O. septentrionana.
Argyrotoza Parryana.

Order, Hemiptera.

Fam., *Acanthidæ.*

Acanthia stellata.
Pedeticus variegatus.

Order, Diptera.

Fam., *Culicidæ.*

Culex caspius, Pall.

Fam., *Tipulidæ.*

Chironomus polaris, Kirby.
C. borealis.
Tipula arctica.

Fam., *Syrphidæ.*

Helophilus bilineatus.

Fam., *Muscidæ.*

Tachina hirta.
Anthomyia dubia.
Scatophaga apicalis.
S. fucorum, Fall.

LXIV. — Some MARINE INVERTEBRATA enumerated in PARRY's 1st, 2nd, and 3rd Voyages (**I.** Sabine, Kirby; **II.** J. C. Ross, Dr. Fleming; **III.** J. C. Ross); Ross's 1st* (**L.** Leach, 1819) and 2nd(**R.** J. C. Ross and R. Owen) Voyages, 1824, 25, 26; 1819; 1835.

[For the Vertebrata, *see above*, pages 1-93, and 499: the Birds are included in Prof. Newton's catalogue, pages 94, &c. For the Fishes, *see* pp. 115–122, and 500. The Mollusca are included in Dr. Mörch's catalogue, pp. 124–135; and the Tunicata in Dr. Lütken's list, pp. 138–139. The other groups are revised in Dr. Lütken's catalogues, *see above*, pp. 146–191. The native names and the localities in these older lists are points of interest.]

TUNICATA.

I. Ascidia globifera, Sab. West coast of Davis Strait, lat. 70°; trawl.

R. Boltenia reniformis, MacLeay. 70 fathoms, Elizabeth Harbour.

R. Cystingia Griffithsii, MacLeay.† Fox Channel and Felix Harbour.

PYCNOGONIDA.

I. III. Nymphum glossipes, Fab.
I. III. N. hirsutum, Sab.
I. Phoxichilus proboscideus, Sab.
} North-Georgian Isles, at ebb tide (**I.**).

CRUSTACEA.

(Those from Ross's 2nd voyage (**R.**) were taken at Port Bowen, Prince-Regent Inlet.)

I. III. Idotea entomon (Pall.). Melville Island, at ebb tide (**I.**); wide-spread (**III.**).

I. III. I. Baffini, Sab. W. coast of Baffin's Bay, 20 fathoms; trawl (**I.**).

I. R. Gammarus nugax (Phipps). Polar Sea, at ebb tide (**I.**).

I. III. R. Talitrus ampulla (Phipps). Polar Sea and Davis Strait; trawl (**I.**).

---

* The importance of Sir JOHN ROSS's Deep-sea Soundings in Baffin's Bay, whereby several of the Invertebrates were obtained, is strongly insisted on in Dr. WALLICH's "The North-Atlantic Sea-bed" (Part I., p. 79); 4to. Van Voorst, London, 1862. The deep-sea forms specially referred to are—*Hippolyte*, sp., *Gammarus Sabini*, *Nereis*, *Phyllophora*, *Lepidonotus Rossii*, and *Gorgonocephalus articus*.

† See also Trans. Lin. Soc., xiv. 1825.

I. III. R. Gammarus boreus, Sab. } Polar Sea, at ebb tide
I. III. R. G. loricatus Sab. } (**I.**).
L. I. III. R. G. Sabini, Leach, 1819. Baffin's Bay (**L.** and **I.**).
I. III. Amphithoe Edvardsii (Sab.). W. coast of Davis Strait; trawl (**I.**).
I. Talitrus Cyaneæ, Sab. Parasitic on *Cyanea arctica.*
I. III. R. Crangon Boreas (Phipps). W. coast of Davis Strait (trawl) and Melville I. (dredge) (**I.**).
I. R. Sabinea septemcarinata (Sab.). W. coast of Davis Strait; trawl (**I.**).
I. III. Alpheus aculeatus (Fab.). Melville Is. (**I.**).
I. R. Hippolyte polaris (Sab.). Melville I., 50 fathoms (**I.**).
R. H. aculeata (Fab.).
R. H. Sowerbei, Leach.
R. H. borealis, Ross.
III. R. Mysis flexuosus, Müll. *Esq.*, Il-le-ak-kak. (**R.**)
R. Themisto Gaudichaudii, R. & O.
R. Acanthonotus cristatus, R. & O.
R. Acanthosoma hystrix, R. & O.
III. Caprella scolopendroides, Lam.
III. Cyamus Ceti (Fab.).
III. Nebalia glabra ?, Lam.
L. Balanus arcticus, Leach, 1819. Baffin's Bay.
I. B. glacialis, Gray, 1824.

Annelida.

I. Nais ciliata, Müll. North-Georgian Islands.
I. Polynoe cirrata et P. scabra (Fab.). Melville Island (September).
L. Nereis phyllophora, Leach, 1819. Baffin's Bay.
L. Lepidonotus Rossii, Leach, 1819. Baffin's Bay.
I. Spirorbis nautiloides (Lin.).
I. S. spirillum (Lin.).

Echinodermata.

L. Gorgonocephalus (Euryale) arcticus, Leach, 1819. Baffin's Bay.
I. Ophiura texturata, Lam. }
I. III. O. fragilis, Lam. }
III. O. filiformis (Müll.). }
I. Asterias papposa, Fab. } Davis Strait; trawl (**I.**).
I. A. rubens, Fab. }
I. A. violacea, Müll. 12-18 fathoms. }
I. A. polaris, Sab. Melville Island.
I. Echinus saxatilis, Lin.

CŒLENTERATA.

I. Beroe ovum, Fab. Baffin's Bay.
I. B. cucumis, Fab. Baffin's Bay and New Georgia.
I. III. B. pileus, Fab. Davis Strait and Baffin's Bay.
I. III. Dianæa glacialis, Sab. Baffin's Bay and adjacent seas.
I. III. Cyanea arctica, Sab. Barrow Strait and Polar Sea.
I. Flustra angustiloba, Lam.
II. F. pilosa, Solander.
II. Cellaria loriculata, Sol.
II. C. ciliata, Sol.
II. Tubularia?
II. Sertularia argentea, Sol.
II. S. cupressina, Sol.
II. S. repens, Sol.
II. Plumularia bullata, Fleming.
II. Millepora pumicosa, Sol.
II. M. tubulosa, Sol.

PROTOZOA.

II. Spongia infundibuliformis, Lin.
II. S. parasita, Montagu.

LXV.—FISHES from Port Kennedy, lat. 72° N., long. 94 W. (From DR. WALKER'S Memoir "On ARCTIC ZOOLOGY," &c. Journ. Royal Dublin Soc., vol. iii. 1860, p. 67.)

Cottus polaris, Sab. A few specimens dredged up in 5–15 fathoms.

Lumpenus nubilus, Richardson. A single mutilated specimen.

Gymnelis viridis, Fabr. Two small specimens, from the beak of a Glaucous Gull.

Salmo Hoodii, Rich. A few specimens, 3 to 11 inches in length, from a freshwater lake at the head of Port Kennedy.

LXVI.—ARCTIC MOLLUSCA, obtained during the Voyage of the "FOX." From DR. WALKER'S Memoir on Arctic Zoology: Journ. Royal Dublin Soc., vol. iii. 1860, pp. 70–72.

Rossia palpebrosa? Several *beaks*, probably of this species, were found in the crop of a Gull.

Fusus tortuosus, Reeve. A few dead specimens at 100 fathoms in Melville Bay.

Buccinum Donovani, Gray. Godhavn, 20 fathoms.

B. cyaneum, Chemn. Melville Bay, 140 fathoms; Port Kennedy (lat. 72° N., long. 94° W), 15 fathoms.
B. undulatum. Port Kennedy, 15 fathoms.
B. hydrophanum, Hancock. Port Kennedy, 15 fathoms.
B. Grœnlandicum, Hanc. Port Kennedy, 15 fathoms.
B. tenebrosum, Hanc. Godhavn, 10–20 fathoms.
B. ciliatum, Hanc. Port Kennedy, 10 fathoms.
B. plicosum, Hanc. Melville Bay, 100 fathoms.
Trophon clathratus, L. Godhavn, 20 fathoms.
T. clathratus, var. scalariformis. Godhavn, 20 fathoms.
Mangelia turricula, Mont. Godhavn, 20 fathoms; Melville Bay, 80–140 fathoms.
M. Trevelliana, Turton. Port Kennedy, 15 fathoms.
M. rufa, Mont. Godhavn, 15 fathoms; Port Kennedy, 10 fathoms.
Natica pusilla, Say? Gould. Godhavn, Fiskernaes, and Port Kennedy, from 5 to 20 fathoms.
Turritella lactea, Möll. Melville Bay, 80–100 fathoms.
Littorina tenebrosa, Mont. Godhavn, 20 fathoms.
Scalaria Grœnlandica, Sow. Godhavn, 20 fathoms.
Margarita cinerea, Couth. Godhavn, 15 fathoms.
M. arctica, Leach. Port Kennedy, 10 fathoms, very abundant.
M. umbilicalis, Sow. Cape York, 10 fathoms; Port Kennedy.
M. undulata, Sow. Godhavn, 15 fathoms.
Lottia testudinalis, Gray. Godhavn; Fiskernaes; Melville Bay.
Chiton marmoreus, Fab. Port Kennedy and Godhavn, 10–20 fathoms.
Bulla sculpta, Reeve. Port Kennedy, 10 fathoms.
Cylichna alba, Lovèn. Near Cape York, 15 fathoms.
Saxicava arctica, L. Godhavn, Fiskernaes, Melville Bay, and Cape York, 5 to 140 fathoms.
S. rugosa, L. Reefkol and Godhavn, 10–25 fathoms.
Mya arenaria, L. Godhavn and Melville Bay, 10–120 fathoms.
M. truncata, L. Reefkol; Fiskernaes; Godhavn.
M. truncata, var. Uddevallensis. Reefkol; Fiskernaes; Godhavn.
Tellina proxima, Brown. Godhavn; Melville Bay.
Astarte compressa, Mont. Reefkol; Godhavn; Melville Bay; Port Kennedy; 10–140 fathoms.
A. arctica, Gray. Melville Bay; Port Kennedy.
A. elliptica, Br. Melville Bay; Port Kennedy.
A, sulcata, DC. Godhavn.
A. Warhami, Hancock. Godhavn.
A. fabula, Reeve. Port Kennedy, 15 fathoms.
Cardium Grœnlandicum, Chemn. (Young.) Godhavn; Melville Bay.
C. Islandicum, Chemn. Godhavn; Melville Bay.
Montacuta bidentata, Gould.? Cape York. Small and unique.
Mytilus edulis, L. Fiskernaes; Godhavn; Melville Bay.
M. edulis, var. elegans. Melville Bay.
Velutina lævigata. Melville Bay, 100 fathoms.
Nucula tenuis, Mont. Godhavn and Melville Bay, 10–20 fathoms.
N. Portlandica. Melville Bay, 80 fathoms.

N. truncata, Br. Port Kennedy, 15 fathoms.
N. nitida, Sow. Godhavn.
Leda caudata, Don. Godhavn; Melville Bay.
L. minuta. Godhavn.
L. pygmæa, Münster. Godhavn.
Yoldia lucida. Melville Bay, 140 fathoms.
Modiola nigra, Gray. Cape York; Port Kennedy.
Crenella discors, L. Crimson Cliffs.
C. decussata, Mont. Melville Bay.
C. glandula. Melville Bay, 75 fathoms.
Pecten Islandicus, Müll. Godhavn; Reefkol; Melville Bay.
P. Grœnlandicus, Sow. Cape York; Melville Bay.
Hypothyris psittacea, Chemn. Reefkol; Godhavn; Melville Bay, 140 fathoms. Numerous.

PTEROPODA.

Clio borealis. Baffin's Bay; Melville Bay.
Limacina arctica. Baffin's Bay; Melville Bay.
Hyalæa tridentata and Deadora pyramidata in the North Atlantic.

LXVII.—INSECTS and ARACHNIDS from PORT KENNEDY, lat. 72° N., long. 94° W., June and July 1859, and from POND'S BAY; taken during the Voyage of the "Fox." From Dr. D. WALKER'S "Notes on the Zoology of the last Arctic Expedition under Captain Sir F. MCCLINTOCK, R.N., &c." Journal Royal Dublin Soc., vol. iii., pp. 72–74.

I. INSECTA.

Order COLEOPTERA.

Fam. Feroniidæ, Lat.

Platyderus nitidus, Kirby.

Order HYMENOPTERA.

Fam. Tenthredinidæ, Leach.

Nemetus intercrus, St. Farey.

Fam. Apidæ, Leach.

Bombus frigidus, Smith.
B. terricola, Kirby.

Order LEPIDOPTERA.

Fam. Argynnidæ, Dup.

Melitæa Tarquinius, Curtis.

Fam. Arctiidæ, Leach.

Arctia Americana, Harris. A larva.

Fam. Hadenidæ, Granby.

Hadena Richardsoni, Curtis.

Fam. Larenitidæ, Green.

Psychophora Sabini, Kirby.

Fam. Tortricidæ, Steph.

Cheimatophila, n.s.?

Order DIPTERA.

Fam. Chironomidæ, Hal.

Chironomus polaris, Kirby.
C. aterrimus (?), Meig.
C. sp.?

Fam. Tipulidæ, Lat., Leach.

Tipula artica, Curtis.
Limnobia, n.s.?

Fam. Muscidæ.

Eurigaster hirtus (Curtis).

Anthomyia dubia, Curtis.
A. 2 spp. ?

Fam. Empidæ, Leach.

Microphorus drapetoides, Barnston.
Scatophaga pubescens, Barnston.

Order THYSANURA.

Fam. Poduridæ, Burns.

Podura sp.
Pond's Bay lat. 72° *N., long.* 76° *W., August* 1858.

Order LEPIDOPTERA.*

Hadena Richardsoni, Curtis.

Order DIPTERA.

Tipula arctica, Curtis.

II. ARACHNIDA.†

PHALANGIDÆ.

Fam. Opilionidæ, Herbst.

Opilio scabripes, sp. nov.

ARANEIDÆ.

Fam. Lycosidæ, Sundis.

Lycosa andrenivora, Walck.
L. blanda, Koch. Immature.

Fam. Lingphidæ, Black.

Nereine longipalpis, Sundervoll.
Walckenæra sp. ?

LXVIII.—ARCTIC CRUSTACEA and PYCNOGONIDÆ, collected by English Expeditions. From Dr. D. WALKER'S Memoir on Arctic Zoology, Journ. Royal Dublin Soc., vol. iii., 1860, pp. 68, 69.

[*See* Dr. LUTKEN'S Catalogues, *above*, at pp. 146 and 163, for the revised nomenclature of these and many other Arctic species.]

CRUSTACEA.

Decapoda.

Hyas coarctata, Leach. Godhavn, 69° N., 20 fathoms, M'Clintock.

Pagurus pubescens, Kröyer. Godhavn, M'Clintock.

Hippolyte borealis, Owen. Ross' 2nd Voy. Belcher.‡ Port Kennedy (lat. 72° N., long. 94° W.), 10 to 15 fathoms, M'Clintock.

H. aculeata, Fabr. Parry's 1st and 3rd. Voy. Ross's 2nd Voy. In a Cod's maw, lat. 66° 34′, long. 55° 8′, Penny. Belcher. Port Kennedy; Melville Bay, 68 to 140 fathoms; M'Clintock.

---

* The *Hadena*, *Tipula*, &c., came, apparently, from Port Kennedy; but it is not clear, from the insertion of "Pond's Bay," &c., after "Podura," whether the succeeding groups came from Pond's Bay or from Port Kennedy. —EDITOR.

† For the PYCNOGONIDÆ, *see* further on, p. 510.

‡ The Crustacea obtained in Capt. Sir E. Belcher's voyage, 1852–4, were dredged between Beechey Island and Northumberland Sound, generally in depths not exceeding 30 fathoms. (Belcher's "Last of the Arctic Voyages," vol. ii., p. 400.)

H. polaris, Sab. Parry's 1st Voy. Ross's 2nd Voy. Assistance Bay, 7 to 15 fathoms, Penny. Belcher. Port Kennedy, 10 to 15 fathoms, M'Clintock.

H. Sowerbyi, Leach. Ross's 2nd Voy. Near Cape York, 10 to 25 fathoms, M'Clintock.

H. Belcheri, Bell. Belcher.

Crangon borëus, Sab. Parry's 1st and 3rd Voy. Ross's 2nd Voy. Assistance Bay, 12 to 15 fathoms, Penny. Belcher. Port Kennedy, from the stomach of a Bearded Seal, M'Clintock.

C. septem-carinatus, Sab. Parry's 1st. Voy. Ross's 2nd Voy. Assistance Bay, 7 to 15 fathoms, Penny. Melville Bay, 110 fathoms, M'Clintock.

Cumadæ.

Alauna Goodsiri, Bell. Belcher.

A. uncinata, Baird. Assistance Bay, 7 to 15 fathoms, Penny.

Stomapoda.

Mysis, flexuosus, Müll. Parry's 3rd Voy. Ross's 2nd Voy. Union Bay, Beechey Island, 15 fathoms, Penny. Port Kennedy, M'Clintock.

M. ? Fabricii ? Kr. Belcher.

Amphipoda.

Gammarus Sabinii, Leach. Parry's 1st and 3rd Voy. Ross's 2nd Voy. W. Greenland, 4 to 20 fathoms, Penny. Belcher.

G. loricatus, Sab. Parry's 1st and 3rd Voy. Ross's 2nd Voy. Belcher. Port Kennedy, M'Clintock.

G. locusta, Mont. Port Kennedy, M'Clintock.

G. Kroeyeri, Bell. Belcher.

G. boreus, Sab. Port Kennedy, M'Clintock.

G. (Lysianassa) nugax, Fabr. Parry's 1st and 3rd Voy. Ross's 2nd Voy. W. coast of Greenland, 4 to 20 fathoms, Penny.

Amphitoë Jurinii, Kr. ? Belcher.

A. læviuscula, Kr. Belcher.

A. Edwardsii, Sab. Parry's 1st and 3rd Voy. Ross's 2nd Voy. Assistance Bay, 7 fathoms, Penny. Near Cape York, 15 fathoms, M'Clintock.

Acanthostoma hystrix, Owen. Ross's 2nd Voy. Belcher. Near Cape York, 15 fathoms, M'Clintock.

Lysianassa lagena, Kr. Belcher.

L. Vahlii, Kr. Port Kennedy, M'Clintock.

L. appendiculata, Kr. Port Kennedy, F. M'Clintock.

Stegocephalus ampulla, Phipps, Kr. Ross's 2nd Voy. Belcher. Port Kennedy, 10 fathoms, M'Clintock.

S. inflatus, Kr. Assistance Bay, 7 fathoms, Penny.

Anonyx, sp. Penny.

Metoecus Cyaneæ, Sab. Parry's 1st Voy.

Themisto arctica, Kr. Ross's 2nd Voy. Port Kennedy, in a Seal's stomach, M'Clintock.

Isopoda.

Arcturus Baffini, Sab. Parry's 1st and 3rd Voy. Union Bay, Beechey Island, 15 fathoms, Penny. Belcher. Near Cape York, 15 fathoms, M'Clintock.

Idotæa entomon, Latreille. Parry's 1st and 3rd Voy. Assistance Bay, 7 fathoms, Penny. Belcher. Port Kennedy, 10 fathoms, M'Clintock.

Læmodipoda.

Caprella cercopoides, Adam White. W. coast of Greenland, lat. 73° 16′ N., long. 57° 16′ W., 4 to 20 fathoms, Penny.

C. spinifera, Bell. Belcher.

Entomostraca.

Nebalia Herbstii, Leach. Parry's 3rd Voy.

Arpacticus Kronii, Kr. Davis Strait, lat. 66° 34′, long. 55° 8′, Penny.

A. chelifer, Baird. Davis Strait, lat. 73° 20′, long. 57° 16′, Penny.

Bradycinetus Brenda, Baird. Assistance Bay, Barrow Strait, Penny.

Cetochilus septentrionalis, Baird. North Sea, Penny. (Not Arctic.)

C. arcticus, Baird. Off Cape Desolation, lat. 71° 21′, in lat. 64° 19′, and Melville Bay, Penny. Melville Bay, M'Clintock.

Cyclopsina sp. ?, Baird } Davis Strait, lat. 73° 20′, long. 57° 16′,
Cyclops sp. ?, Baird } Penny.

Pycnogonidæ.

Nymphon grossipes, Fabr. Parry's 3rd Voy. Near Cape York, 20 fathoms, and Melville Bay, 80 to 100 fathoms, M'Clintock.

N. hirtum, Fabr. Parry's 3rd Voy. Near Cape York, 20 fathoms, and Melville Bay, 80 to 100 fathoms, M'Clintock.

N. crassipes, White. Union Bay (dredged), Penny.

N. hirtipes, Bell. Belcher.

N. robustum, Bell. Belcher.

N. sp. ?, White. Union Bay, Beechey Island, Penny.

LXIX.—Echinodermata, Cirripedia, and Actiniæ, collected during the voyage of the "Fox." (From Dr. Walker's Memoir "On Arctic Zoology," &c. Journ. Roy. Dublin Soc., vol. iii. 1860, p. 70.)

I. **Echinodermata.**

Alecto glacialis. Dredged, 80–95 and 140 fathoms, in Melville Bay.

Ophiura texturata. Godhavn, 15 fathoms; Crimson Cliffs, 140 fathoms; Melville Bay.

O. alvida. Port Kennedy, 10 fathoms, lat. 72° N., long. 94° W.
O. echinulata. Port Kennedy, 10 fathoms.
O. fasiculata. Port Kennedy; Crimson Cliffs.
Ophiocoma, sp. n. Port Kennedy, 10 fathoms.
O. nigra. Crimson Cliffs.
Uraster violaceus (?), Young. Port Kennedy, 10 fathoms; Melville Bay, 80 fathoms ?.
Echinus neglectus. Godhavn, 15 fathoms; Melville Bay, 100 fathoms; Port Kennedy, 8 fathoms.
Solaster endeca. Port Kennedy, 10 fathoms.
S. papposus. Reefkol, 20 fathoms.

II. **Cirripedia.**

Balanus porcatus, DC. Godhavn; Fiskernaes; Reefkol; Port Kennedy.
B. scoticus. Godhavn.

III. **Anthozoa.**

Actinia candida. }
A. digitata. } Fiskernaes; Godhavn; Melville Bay.

LXX.—MOLLUSCA from WEST GREENLAND and the PARRY ISLANDS. From the Appendix to Dr. P. C. SUTHERLAND'S Journal of Captain Penny's Voyage, &c., 1852, pp. cci. ccii.

The Mollusca of the Arctic Regions are enumerated in the Revised Catalogue by Dr. MÖRCH, *above*, pp. 124–135; and most, if not all, are described in Mr. J. GWYN JEFFREYS' "British Con- " chology, or an account of the Mollusca which now inhabit the " British Isles and the surrounding seas," vol. i—v.; 1862–69, 8vo.

Conchifera.

Modiola (Lanistina) discors.
Astarte arctica.
A. Spitzbergensis.
Nucula (Yoldia) arctica.
N. (Leda) fluviatilis.
N. radiata.
N. cordata.
Hiatella arctica.
H. minuta.
Tellina (Psammobia) fusca.
T. calcarea.
Cardium (Aphrodite) Grœnlandicum.
Saxicava rugosa.
Mya truncata.
Pandora glacialis.
Montacuta substriata.

} Abundant; Assistance Bay and the shores of Barrow Strait 7–20 fathoms.

Gasteropoda.

Buccinum glaciale. Assistance Bay, 7–10 fathoms.
B. cyaneum. West Greenland, 15–20 fathoms.
Turbo corneus, Keiner. Assistance Bay, 7–15 fathoms.
Trichotropis costellata. Assistance Bay, 15 fathoms.
Margarita undulata.
M. glauca.
M. artica.
M. Vahlii.
M. umbilicalis.
} Abundant; Assistance Bay and West Greenland, 12–20 fathoms.
M. helicina. West Greenland, 15–20 fathoms.
Bulla corticata. Assistance Bay, 7–10 fathoms.
Chiton lævigatus. Barrow Straits, 12–15 fathoms.
Patella rubella.
P. cerea.
Lottia testudinalis.
} Assistance Bay, 12–15 fathoms.

Cephalopoda.

Philonexis sp.? On ice, Melville Bay, July 1850.
Sepia, sp.? In stomach of Narwhal, Melville Bay, July 1850.

Annelid.

Sagitta bipunctata. Large in Davis Strait, and abundant everywhere.

LXXI.--The Results of some DREDGINGS made at GOODHAAB, WEST GREENLAND, by Dr. G. C. WALLICH in 1860. "The North-Atlantic Sea-bed," &c., by G. C. WALLICH, M.D., F.L.S., F.G.S., 1862 (4to, London), p. 102.

[This ought to have been inserted in § 1, under "West Greenland."]

I. From 50–100 fathoms.

CRUSTACEA.

Hyas arenarius.

MOLLUSCA.

Hiatella arctica.
Cardium Grœnlandicum.
Pecten Islandicum.
P. pectinatum.
Mytilus edulis.
Trophon Fabricii.
T. Gunneri.
Margarita striata.
Volutomitra Grœnlandica.
Astarte elliptica.

ANNELIDA.

Pectinaria.
Spio.
Cirratulus.
Syllis.
Terebella.

COTTIDÆ.

Cottus glacialis, Rich.

II. From 100–250 fathoms.

CRUSTACEA.

Hippolyte polaris.
Gammarus arcticus.
Caprella linearis (on Algæ).

MOLLUSCA.

Acmea testudinalis.
Chiton mermoreus.
Natica Grœnlandica.
Lima subauriculata (East Coast, in sounding).

TUNICATA.

Boltenia picta, n.s.

ECHINODERMATA.

Echinus sphæra.
E. neglectus.
Ophiocoma bellis.
O. granulata.
O. Goodsiri.
Ophiura texturata.
Solaster papposus.

ANNELIDA.

Seppula contortuplicata.
Terebella (several species).
Spirorbis communis.

LXXII.—Some ARCTIC ASCIDIANS and ECHINODERMS. (From Dr. Sutherland's Journal, &c., Appendix. 1852.)

ASCIDIA, pp. ccxii.–ccxiii.

Pelonaia corrugata, Forbes. Assistance Bay.
Pelonaia sp. ? Assistance Bay.
Dendrodon sp. ? Assistance Bay.
Phallusia Sutherlandi, Huxley. Assistance Bay.

ECHINODERMATA, pp. ccxi–ccxii.

Echinus neglectus. Assistance Bay, 7–15 fathoms.
Ctenodiscus polaris. Assistance Bay.
Uraster violaceus ?, Young. Assistance Bay.
Solaster papposus, Young. Assistance Bay.
Ophiura fasciculata, Forbes. Assistance Bay.
O. glacialis, Forbes. Assistance Bay.
O. sericea, Forbes. Assistance Bay.
Ophiocoma echinulata, Forbes. Assistance Bay.
Cucumaria Hyndmani. Assistance Bay, 7–10 fathoms.
C. fucicola. Assistance Bay.
Chirodota brevis, Huxley. Assistance Bay.

LXXIII.—SOME ARCTIC INSECTS, &c. From the Appendix to Dr. Sutherland's Journal of Captain Penny's Voyage in Baffin's Bay, &c., pp. ccvii–ccxi. 1852.

Podurellæ.

Desoria arctica, Adam White. Near *D. saltans*, Ag. (*D. glacialis*, Nicolet), abundant on some glaciers in Switzerland. *D. arctica* was found by Dr. Sutherland abundantly on Nostoc in the neighbourhood of Assistance Bay, Barrow Strait, July 1851 ("Journal of a Voyage " in Baffin's Bay and Barrow Strait," &c. pp. 201–207).

Acaridæ.

Ixodes Uriæ, Ad. White. Parasitic on the Loon (*Uria troile*).

SPIDER.

Micraphantes arcticus, Ad. White. Assistance Bay, among the vegetation.

LXXIV.—PLANTS from Barrow Strait and Davis Strait, &c. Named by SIR W. J. HOOKER, K.H., D.C.L., F.R.S., &c. From DR. SUTHERLAND'S "Journal of a Voyage," &c., 1852, vol. ii., Appendix, pp. clxxxix–cxc.

DICOTYLEDONES.

Ranunculus frigidus, Willd. Assistance Bay, south end of Cornwallis Land.

Papaver nudicaule, L. Assistance Bay.

Cochlearia fenestralis, Br. Assistance Bay.

Parrya arctica, Br. Assistance Bay.

Cardamine bellidifolia, DC. Assistance Bay.

Braya glabella, Richardson. Assistance Bay.

Draba rupestris, Br. Assistance Bay.

D. glacialis, Adams, var. Assistance Bay.

D. alpina, L. Assistance Bay.

Arenaria Rossii (?), Br. Assistance Bay.

A. rubella, Hook. Assistance Bay.

Cerastium alpinum, L., and var. glabatum. Assistance Bay and Bushnan Island.

Stellaria longipes, Goldie. Northumberland Inlet or Hogarth Sound, and Assistance Bay.

Lychnis apetala, L. Assistance Bay.

Potentilla nana, Lehm. Berry Island (73° 14′ N. Lat., 56° 50′ W. Long.), and other islands in Davis Strait.

Dryas integrifolia, L. Assistance Bay; Berry Island and adjacent islands.

Cruciferæ?

Epilobium latifolium, L. Northumberland Inlet.

Saxifraga pauciflora (?), Stev. Bushnan Island.

S. oppositifolia, L. Assistance Bay and Berry Island.
S. nivalis, L. Assistance Bay.
S. cernua, L. Northumberland Inlet and Assistance Bay.
S. cæspitosa, L. Assistance Bay.
S. flagillaris, Willd. Assistance Bay.
S. tricuspidata, Dc. Northumberland Inlet.
S. hirculus, Dc. Northumberland Inlet.
Pyrola rotundifolia, L. Northumberland Inlet.
Cassiopeia tetragona, Don. Bushnan Island.
Vaccinium vitis-idæa, L. Bushnan Island.
V. uliginosum, L. Northumberland Inlet.
Arctostaphylos alpinus, Spr. Northumberland Inlet.
Polygonum viviparum, L. Assistance Bay.
Oxyria reniformis, L. Assistance Bay and Northumberland Inlet.
Empetrum nigrum, L. Northumberland Inlet.
Salix cordifolia, Parsh. Assistance Bay.
S. arctica, Pall. Assistance Bay and Bushnan Island.
Juncus biglumis, L. Assistance Bay.
Carex Hepburnii, Boott. Berry Island.
Luzula hyperborea, Br. Berry Island, Davis Straits, and Bushnan Island.
Eriophorum polystachyum, L. Assistance Bay.
Phippsia monandra, Trin. Assistance Bay.
Alopecurus alpinus, Sm. Bushnan Island, Assistance Bay.
Poa cenisia, Al. Bushnan Island, Assistance Bay.
Hierochloë alpina, Wahl. Bushnan Island.
Woodsia glabella, Br. Berry Island and other islands in Davis Straits.

LXXV.—ARCTIC ALGÆ collected in Davis Strait, Baffin's Bay, Barrow Strait, and Wellington Channel, by DR. SUTHERLAND during CAPTAIN PENNY'S Expedition, 1850-51. By Dr. DICKIE. From DR. P. C. SUTHERLAND'S "Journal of a Voyage," &c., 1852, vol. ii., Appendix, pp. cxci–cc.

MELANOSPERMEÆ.

Sporochnaceæ.

Desmarestia aculeata, Lamour. 3 fathoms, N. lat. 73° 20′, W. long. 57° 20′.

Laminarieæ.

Laminaria saccharina, De la Pyl. Assistance Bay, 15 fathoms N. lat. 74° 40′.
L. fascia, Ag. Union Bay, dredged.
Agarum Turneri, Post. and Rupr. Assistance Bay, 15 fathoms. Union Bay (between Beechey Island and Cape Spencer), dredged.

Ectocarpeæ.

Chætopteris plumosa, Kütz. 3 fathoms, N. lat. 73° 20′, W. long. 57° 20′.

RHODOSPERMEÆ.

Halymenieæ.

Dumontia sobolifera, Lamr. 3 fathoms, N. lat. 73° 20′, W. long. 57° 12′.

Polysiphonieæ.

Polysiphonia urceolata, Grev. On *Ch. plumosa.*

Corallineæ.

Melobesia polymorpha, Harv. On stones, 15 fathoms. Union Bay.

Kallymenia Pennyi, n.s. Harvey MS. Assistance Bay, 15–20 fathoms.

CHLOROSPERMEÆ.

Confervaceæ.

Cladophora lanosa, Kütz. 2–6 fathoms, N. lat. 73° 20′, W. long. 57°.

C. sp. nov, ? On driftwood, lat. 66° 53′.

Conferva melagonium, Web. & Mohr. Assistance Bay. Attaining 5 feet in length.

C. glacialis, Kg. ? Matted crust on stones in a stream, Prospect Hill, winter quarters, south end of Cornwallis Land.

C. ærea, Dillw. ? With others, N. lat. 73° 20′, W. long. 57° 16′.

Ulothricheæ.

Ulothrix zonata, Kg.
U. æqualis, Kg.

Scytonemeæ.

Sirosiphon ocellatus, Kg.

} Wet rock on island. N. lat. 73° 20′, W. long. 57° 16′ and 57° 20′.

Rivularieæ.

Rivularia microscopica, n.s. Very minute; on *Enteromorpha compressa.*

Oscillatorieæ.

Oscillatoria sp. ? Thin green crust on limestone, Seal Island, Wellington Channel.

Ulvaceæ.

Prasiola arctica, n.s. with *Nostoc microscopicum*; winter quarters.

Enteromorpha compressa, Hook. Pools on shore above sea-level, Beechey Island, Assistance Bay, and Baring Bay.

Nostochineæ.

Nostoc microscopicum, Carm. On stones in stream, winter quarters.

N. Sutherlandi, n.s. Winter quarters. South-Cornwallis Land.

N. arcticum, Berk. On wet and boggy slopes around Assistance Bay. When the ground becomes frozen, the plant is loosened off the surface, and often blown away; thus it is found far out at sea on the ice, and sometimes full of *Poduræ* which may have hatched from ova laid in it on land. Edible. *Nostoc edule*, Berk. & Mont., is used as food in China.

Palmelleæ.

Hæmatococcus minutissimus, Hass. Wet stones on island, north side of Baring Bay, N. lat. 75° 49′.

Protococcus nivalis, Ag. Red crust on stones in stream, winter quarters, July 1851. Also on snow and ice as usual.

DESMIDIEÆ.

Cosmarium crenatum, Ralfs, and C. pyramidatum, Bréb. On *Nostoc Sutherlandi*. Found also by Dr. Dickie at a considerable elevation in N.E. Scotland.

Arthrodesmus minutus, Kütz. With *Diatoma flocculosum*, in fresh water (melting snow), 100 feet above the sea, on island, N. lat. 73° 20′, W. long. 57°, 22 June, 1850.

DIATOMACEÆ.

Dr. Dickie states, "At my request, made previous to the departure of the Expedition, Dr. Sutherland paid special attention to the colouring matters of ice and sea-water; samples of such from different localities were carefully collected and forwarded for my inspection. They were found to consist almost solely of *Diatomaceæ*; and in some instances freshwater forms were detected, though rather sparingly, intermixed with others exclusively marine. This is not surprising when we consider the copious discharges of fresh water from the land, occasioned by the melting of snow and ice during the brief summer.

"The contents of the alimentary canal of examples of *Leda*, *Nucula*, and *Crenella* dredged in Assistance Bay, consisted of mud in a fine state of division, including also numerous *Diatomaceæ* identical with those colouring the ice and the water. Though not a new fact, it is one of some interest in relation to the existence of animal life in those high latitudes. Where *Diatomaceæ* abound certain *Mollusca* obtain sure supplies of food; these in turn are the prey of Fishes; these last contribute to the support of marine Mammals and Birds." . . . "Many of the species enumerated have also been found in other parts of the world; and this confirms the ideas entertained respecting their wide distribution, and the very general diffusion of these minute organisms."

From water, due to melting snow and ice, on island, N. lat. 73° 20′, W. long. 57°, 100 feet above the sea.

1. Achnanthes minutissima, Ag.
2. Diatoma flocculosum, Ag.
3. Eunotia monodon, Ehr.
4. E. diodon, Ehr.
5. Navicula affinis, Ehr.
6. N. lanceolata, Ehr.
7. N. n. sp. ?

Nos. 1, 2, and 5, occur at high altitudes in Scotland.

Schizonema Grevillii, Ag. ? On *Desm. aculeata.*

Micromega Stewartii, n. s. Dredged; N. lat. 73° 20′, W. long. 57° 16′.

Grammonema Jurgensii, Ag. On *Desm. aculeata*, 2 fathoms, with myriads of minute Crustacea.

Melosira arctica, n. s. Mixed with last two, and giving brown tinge to the water in Melville Bay, off the Devil's Thumb, in threads of mucilaginous consistency, infested with Microzoa, July 11, 1850.

Triceratium striolatum, Ehr. 15 fathoms. Union Bay, Sept. 3.

Fragilaria, n. s.? Round brown pellicles in the sea. N. lat. 73° 20′, W. long. 57° 16′.

From water in which the *Kallymenia* above-mentioned (from Assistance Bay) had been macerated; some freshwater, but the majority marine, forms:

Amphora hyalina, Kg.
Cocconeis borealis, Ehr.
Coscinodiscus striatus, Kg.
C. minor, Ehr.
C. subtilis, Ehr.
C. sp. n. ?
* † Cyclotella, sp. n.
* † Cymbella helvetica, Kg.
Epithemia zebra, Kg.
E. Westermanni, Kg.
Gomphonema acuminatum, Ehr.
G. curvatum, Gr., var. marinum.
Grammatophora stricta, Ehr.
*† Grammatophora anguina, Ehr.
Navicula quadrifasciata, Ehr. ?
* N. didyma, Ehr.
* † Odontella obtusa, Kg.
O. aurita, Kg.
* Rhabdonema minutum, Kg. ?
* † Stauroneis aspera, Kg.
† Synedra curvula, Kg. ?
* † S. pulchella, Sm.
* † Triceratium striolatum, Ehr.
Pleurosigma prolongatum, Sm.
P. elongatum, Sm.
P. fasciola, Sm.

Those marked * were also obtained from the washings of *Desmarestia* and *Chætopteris* dredged in N. lat. 73° 20′, W. lang. 57° 20′.

Those marked † were also met with in the washings of *Agarum Turneri* from Union Bay.

From a slimy substance on the surface of the water and under the ice, N. lat. 72° 15′.

Grammonema Jurgensii, Ag.
Pleurosigma Thuringica, Kg.
P. fasciola, Sm.
Navicula, n. sp. ?
Surirella, n. sp, ?
Triceratium striolatum, Ehr.

Similar, from N. lat. 73° 17′.

Achnanthidium delicatulum, Kg.
Cocconeis rhombus, Ehr.
Coscinodiscus marginatus, Ehr.
Melosira arctica, n. sp.
Navicula oxyphyllum, Kg.
N. Thuringica, Kg.
Nitzchia n. sp. ?
Synedra pulchella, Sm.

From N. lat. 73° 40′, W. long. 57°, July 1851, among rotten ice.*

Denticula obtusa, Kg. ?	Pleurosigma fasciola, Sm.
Melosira arctica, n. s.	P. angulatum, Sm.
Navicula oxyphyllum, Kg.	

In Hinkson's Bay, N. lat. 73° 50′, W. long. 57°, July 2, 1850.

Amphora hyalina.	Melosira arctica.
Amphipsora alata, Kg.	Navicula oxyphyllum.
Ceratoneis closterium, Ehr.	Nitzchia, n. sp. ?
Cocconeis rhombus.	Pleurosigma fasciola.
Denticula obtusa.	P. Thuringicum.
Dictyocha gracilis, Kg.	Schizonema, n. sp.
Grammonema Jurgensii.	Triceratium striolatum.

In lat. 75° 42′, May 1850.

Cocconeis rhombus ?	Odontella aurita.
Navicula didyma.	Rhabdonema minutum ?
N. semen, Ehr.	Stauroneis aspera.
N. n. sp. ?	Synedra curvula.
Nitzchia, n. sp.	S. pulchella.
Odontella obtusa.	Triceratium striolatum.

The colouring matter of Arctic ice is due sometimes to *Algæ* decomposed and triturated by ice. The littoral species are few, kept down by the abrasion of ice. The olive and red *Algæ* are rare. Of the five olive *Algæ*, four are British, and one (*Agarum*) is exclusively American. Of the three red *Algæ*, one (*Polysiphonia*) is common in Britain, the *Dumontia* is American, and the third is new. Of the green *Algæ* there are six marine, and fourteen from fresh water or moist places ; and about a third are British. Of the three Desmids in this collection two are British, and the *Arthodesmus* has been found in France and Germany. The Diatoms are relatively numerous.

LXXVI.—On Algæ collected in Cumberland Sound, by Mr. James Taylor, with remarks on the Arctic Species in general. By G. Dickie, M.D., F.L.S.

[With Permission, from the Proceed. Lin. Soc. Botany, vol. ix., pp. 235–243. 1866. Read 15 June, 1865. (Somewhat abridged.)]

[This article should have been inserted in § I. under "West Greenland," &c.; but its close connexion with the preceding article, by the same Author, and its general conclusions, allow of its insertion in this place.]

Cumberland Sound, an arm of Davis Strait, on the west side, commences about lat. 65° 10′ N., long. 64° 40′ W., and is about

* The so called "rotten" condition of the ice is described by Brown as being due to the intermixture of Diatomaceous slime. *See above*, pp. 314–315.—Editor.

90 miles long, and 30 broad. It has many small islands along its shores, and is indented by numerous bights. It is frequented by whalers, and the Kikerton Islands, near the head of the Sound are their head-quarters. The tide rises, it is said, 30 feet, forming a "bore"; but in the open strait it rises only 6 or 7 feet. Animal life abounds in the upper part of the Sound; and some of the *Algæ* are abundant and large. The mean temperature of the air during 40 days, through August to September 9, 1861, was 35·5° Fahr. that of the sea 32·7°; observations made while the ship was in the pack-ice.

I.—Melanospermeæ.

Fucaceæ.

Fucus vesiculosus, L.
F. nodosus, L.

Sporochnaceæ.

Desmarestia aculeata, L.

Laminariaceæ.

Agarum Turneri, P. & R.
Laminaria longicruris, De la P.
L. saccharina, Lam.
L. fascia, Ag.

Dictystaceæ.

Dictyosiphon fœniculaceus, Grev.
Chorda lomentaria, Grev.
Punctaria plantaginea, Grev.

Chordariaceæ.

Chordaria flagelliformis, Ag.
Ralfsia deusta, C. Ag.
Elachista fucicola, Fries.

Ectocarpaceæ.

Chætopteris plumosa, Lyngb.
Sphacelaria arctica, Harv.
S. cirrhosa, Ag.
Ectocarpus littoralis, Harv.

II.—Rhodospermeæ.

Rhodomelaceæ.

Odonthalia dentata, Lyngb.
Rhodomela subfusca, Ag.
Polysiphonia arctica, J. Ag.

Rhodymeniaceæ.

Euthora cristata, J. Ag.
Rhodymenia palmata, Grev.

Cryptonemiaceæ.

Ahnfeldtia plicata, J. Ag.
Kallymenia Pennyi, Harv.
Halosaccion ramentaceum, J. Ag.
H. dumontioides, Harv.

Ceramiaceæ.

Ptilota serrata, Kütz.
Callithamnion Americanum, Harv.
C. sparsum, Harv.
C. Rothii, Lyngb.

III.—Chlorospermeæ.

Ulvaceæ.

Ulva latissima, L.
Enteromorpha intestinalis, L.
E. clathrata, Grev.
E. compressa, Grev.

Confervaceæ.

Cladophora arctica, Kütz.
C. lanosa, Kütz.
Rhizoclonium riparium, Kütz.
Chætomorpha melagonium, Kütz.
C. tortuosa, Kütz.

Oscillatoriaceæ.

Calothrix pilosa, Harv.
Oscillatoria, sp.? Freshwater.

Nostochineæ.

Nostoc arcticum, Berk.

[For other enumerations of Arctic Algæ (including *Diatomaceæ*), in this Manual, *see* pp. 239, 255, 276, 280, 319, 515; and further on.]

Other Arctic Algæ, from beyond lat. 60°, noticed by other observers, Harvey ("Nereis-Boreali-Americana); Agardh ("Species Algarum"), &c.:—

Fucus distichus, L.
F. serratus, L.
Alaria Pylaii, Grev.
Laminaria digitata, Lam.
Asperococcus echinatus, Grev.
Myrionema strangulans, Grev.
Elachista flaccida, Aresch.
Ectocarpus Durkeei, Harv.?
Rhodomela lycopodioides, Ag.
R. gracilis, Kütz.
Polysiphonia urceolata, Grev.
Corallina officinalis, Lam.
Melobesia polymorpha, L.
Desseria sinuosa, Ag. Found in lat. 75° N.
Rhodophyllis veprecula, J. Ag.
Phyllophora interrupta, Grev.
Ceramium rubrum, Ag., var. δ, virgatum.
Ulva bullosa, L.
Ulva crispa, Lightf.
Porphyra vulgaris, Ag.
Cladophora rupestris, Kütz.
Chætomorpha Piquotiana, Mont.
Harmotrichum Carmichaelii, Harv.
H. boreale, Harv.
H. Wormskioldii, Kütz.
Mougeotia, sp.?
Oscillatoria corium, Ag.
Nostoc muscorum, Vaucher.
N. microscopicum, Carm.
N. verrucosum, Vaucher.
Scytonema myochroum, Ag.
Sorospora montana, Harv.
Hæmatococcus frustulosus, Harv.
Tyndaridea anomala, Harv.

Summary.

Of *marine Algæ* there have been found beyond 60° N. lat., along the shores of Davis Strait and Baffin's Bay, and their branches—

Melanospermeæ	25	63 species.
Rhodospermeæ	22	
Clorospermeæ	16	

The Families represented are the *Fucaceæ*, *Sporochnaceæ*, *Laminariaceæ*, *Dictyotoceæ*, *Chordariaceæ*, *Ectocarpaceæ*, *Corallinaceæ*, *Sphærococcoidæ*, *Rhodymeniaceæ*, *Cryptomeniaceæ*, *Ceramiaceæ*, *Ulvaceæ*, and *Confervaceæ*.

Collections have been few, and many localities have yet to be explored.

Some of the *Algæ* occur in the greatest profusion. Masses consisting chiefly of *Fucus vesiculosus*, *Desmarestia aculeata*, *Laminaria longicruris*, and *Alaria*, float about in the summer, set free by the action of icebergs. The multitudes of *Diatomaceæ*, abounding everywhere in the Arctic Seas and on ice-floes, supply abundant material for the support of animal life. In the alimentary canal of various Molluscs brought home in spirits, I have invariably found abundance of Diatoms; and masses comparable with "sodden biscuit," can be gathered from the ice-floes; and these I find to be pure Diatomaceæ.

Not a few of the species of *Algæ* occur also on the west coast of America, as shown by Prof. Harvey's remarks on Dr. Lyall's collection, in a recent volume of the Linnean Society's Proceedings.

Of the species enumerated above, the following are truly Arctic:

Sphacelaria arctica.	Polysiphonia arctica.
Kallymenia Pennyi.	Phyllophora interrupta.
Halosaccion dumontioides.	

General conclusions as to species peculiar to east or west side of the Strait cannot yet be arrived at; but *Callithamnion Americanum* and *Kallymenia Pennyi*, are as yet known only on the west side.

Judged by the number of species enumerated in the "Flora Antarctica," the Arctic Seas, may be said to be rich in *Algæ*.

Three products, usually plentiful in certain marine *Algæ* of low latitudes, are also abundant in some of the Arctic species, namely, the iodides, mannite, and carbonate of lime. *Melobesia* and *Corallina officinalis*, of course contain the last. *Laminaria saccharina*, as its name implies, yields mannite, which appears as a white efflorescence when the plant is dried, without having been washed; and *L. longicruris* contains a still larger quantity. The presence of iodides I have also tested in Arctic specimens. A filtered and concentrated solution from a few grains of ash, with nitric acid and starch, is generally sufficient to give, in a rough way, proof of the amount of iodides, as indicated by tints varying from dark-blue, through purple, to pale rose; and where iodides are in small proportion, the last colour may not appear until after a few minutes·

Fucus nodosus - - -	Pale purple.
F. vesiculosus - - -	Pale purple.
Dictyosiphon fœniculacens -	Very dark blue.
Laminaria longicruris - -	Pale rose.
Ptilota serrata - - -	Pale purple.
Ahnfeldtia plicata - -	Pale rose.
Polysiphonia arctica - -	Pale rose.
Ulva latifolia - - -	Pale rose, slowly.
Conferva melagonium - -	Pale purple.

LXXVII.—Lichens from Barrow and Davis Straits, collected by Dr. P. Sutherland, during Capt. Penny's Arctic Voyage in the "Lady Franklin." By the Rev. Churchill Babington, M.A., &c.

From "Hooker's Journal of Botany," vol. iv., pp. 276-8. 1852.

Lichenes.

Cladonia rangiferina, Hoffm. Near Cape York.
C. pyxidata, Fries. Near Cape York.
Everina ochroleuca, Fries. Near Cape York.
E. divergens, Fries. Near Cape York.
Dufourea ramulosa, Hook. (?)

Cetraria nivalis, Fries. Near Cape York.
Parmelia triptophylla, var. Schraderi, Fries. Assistance Bay.
P. fulgens, Sw. (?) Assistance Bay.
P. coarctata, Sw. (?) Assistance Bay.
P. stygia, L., var. lunata, Fries. (*Cornicularia lanata*, Auctt.)
P. elegans, Ach., and var. α, miniata, Schær. The latter near Assistance Bay.
P. aurantiaca, var. γ, calva, Fries. Assistance Bay.
P. aquila (?), Ach. } Cornwallis Island, on a bone implement.
P. vitellina }
P. pulverulenta, Ach., var. Fries. (*P. pityrea*, Ach.)
Umbilicaria hyperborea, Hoffm.
U. proboscidea, L.. var. α, Fries (two forms).
U. vellea, Fries, var. lecidina, Bab. Near Cape York.
Urceolaria scruposa, Ach. (?). Assistance Bay.
Lecidea geographica, var. contigua, Fries. Assistance Bay.
L. atroalba, Ach. Assistance Bay.
L. lapicida (?), Fries. Assistance Bay.
L. confluens, Fries et Auct. pr. p. Assistance Bay.
L. contigua, var. calcarea, Fries. Assistance Bay.
L. vesicularis, var. globosa, Fries. Assistance Bay.
L. sp. Assistance Bay.

Alga.

Protococcus nivalis. Assistance Bay.

LXXVIII.—An ACCOUNT of the PLANTS collected by Dr. WALKER in GREENLAND and ARCTIC AMERICA, during the EXPEDITION of Sir FRANCIS M'CLINTOCK, R.N., in the Yacht "Fox." By J. D. HOOKER, Esq., M.D., F.R.S., F.L.S., &c. (Read June 21, 1860.) Reprinted, by Permission, from the Proceedings of the Linnean Society, Botany, vol. v., 1861, pp. 79–89.

On the termination of Capt. M'Clintock's memorable voyage, the Plants collected by Dr. Walker, surgeon and naturalist to the Expedition, were placed in my hands by that officer for determination, together with some accurate notes of the localities, and of the temperature of the soil and air to which they are exposed in their native habitats. Though containing no absolute novelties amongst Flowering Plants,* Dr. Walker's herbarium is a particularly interesting one, both from the thorough manner in which that officer explored the localities he visited, and from the proximity of one of his stations (Port Kennedy, in the Boothian peninsula) to the magnetic pole. The florula of that province is further im-

* Amongst the Cryptogamic plants are two Algæ of great rarity, and three new Fungi.

portant, as affording a means of determining the western and eastern limits respectively of several Arctic, Western American, and Greenland plants.* A glance at the northern circumpolar chart shows that the peninsula of Boothia is placed in a very central position amongst the Arctic-American islands, the Botany of the eastern, western, and northern of which has been investigated by many indefatigable and intrepid officers, whilst of the central districts, and especially of Boothia itself, nothing has hitherto been known.

The total number of species brought by Dr. Walker is about 170, of which nearly 100 are Flowering Plants. Of these, only 46 Flowering Plants, and 58 Cryptogamic, were collected at Port Kennedy; most of the remainder were gathered either on the coasts of Greenland—at Frederickshaab and Godthaab, south of the Arctic circle, and at Disco (and Godhavn), Fiskemær, and Upernavik, north of that circle—or in Pond's Bay and Lancaster Sound, to the west of Baffin's Bay. As these are all botanically well-known localities, I shall make no further remarks on them here, observing only that Dr. Walker's plants from these quarters have been of great use to me in drawing up a general account of the whole Arctic flora which I shall have the honour of laying before this Society,† and I shall confine my attention at present to the Port Kennedy flora.

Port Kennedy is situated in latitude 72° N., and is 250 miles north of that part of the Arctic-American coast which was traversed in 1839 by Dease and Simpson (who made careful collections), and about as far south of the Parry Islands, which have been thoroughly explored by General Sabine, Admiral Sir James Ross, Dr. Lyall, and many other officers.

The country about Port Kennedy would at first sight appear to be favourable to Arctic vegetation in many ways. It is uncovered by snow from July 1st to October 1st. The soil is not unfavourable, and there are ravines, lakes, marshes, and sea-beach, offering both shelter and varied conditions for plants; but yet the flora seems to be considerably poorer than that of any of the surrounding islands—Melville Island containing no less than 67 Flowering Plants. Dr. Lyall's Wellington-Channel herbarium contained 50, all collected north of latitude 76° N.; Dr. Anderson and Herr Miertsching obtained 108 species on Banks Land and the adjacent islands, in latitude 70°–74°; whilst Dr. Rae got 78 species on Prince-Albert, Victoria, and Wollaston Lands, in latitude 66°–69°. On the west coast of Baffin's Bay, between the Arctic Circle and Lancaster Sound, 80 have been collected.

Comparing Dr. Walker's herbarium with those to the north, east, and west, I find the following contrasts:—

* *See* Dr. Hooker's Memoir on Arctic Plants, Trans, Linn. Soc., xxiii., 1861, reprinted (in part), *above*, pp. 197 *et seq.*—Editor.

† The Memoir referred to in the foregoing note.

In Melville Island the following species occur which were not found by Dr. Walker at Port Kennedy:—

Ranunculus hyperboreus, Rottb., var. (Sabini, Br.).
R. auricomus, L., var. (affinis, Br.).
Caltha palustris, L., var. (arctica, Br.).
Draba Lapponica, DC.
Parrya arctica, Br.
Cardamine bellidifolia, L.
Stellaria longipes, *Goldie* (Edwardsii, Br.).
Phaca astragalina, DC.
Oxytropis Uralensis, DC. var., (artica, Br.).
Chrysosplenium alternifolium, L.
Sieversia Rossii, Br.
Taraxacum officinale, DC. (var. palustre).
Arnica montana, L.
Senecio palustris, L., var. (Cineraria congesta, Br.).
Nardosmia corymbosa, Hk.
Antennaria alpina, Br.
Deschampsia cæspitosa, P.B., var. (brevifolia, Br.).
Trisetum subspicatum, P.B.
Hierochloe alpina, R. & S.
H. pauciflora, Br.

The Port-Kennedy plants not found in Melville Island are the following:—

Epilobium latifolium, L.
Chrysanthemum integrifolium, Rich.
Cassiopeia tetragona, Don.
Pedicularis capitata, Ad.
P. hirsuta, Willd.
Salix reticulata, L.

On the western shores of Baffin's Bay, between Pond's Bay and Herne Bay, the following Port-Kennedy plants appear to be absent:—

Stellaria humifusa, Rottb.
Saxifraga flagellaris, L.
Pedicularis sudetica, L.
Pedicularis capitata, Ad. (Western limit).
Dupontia Fischeri, Br.

Lastly, comparing Dr. Walker's Port-Kennedy collection with Dr. Anderson's and Herr Miertsching's from the Western Polar Islands and Banks Land (lat. 71°–75°), I find the following in Dr. Walker's which are absent in the Western Islands:—

Stellaria humifusa, Rottb.
Arenaria verna, L., var. (rubella, Br.).
Saxifraga rivularis, L.
Juncus biglumis, L.
Luzula arcuata, Wahl.
Pleuropogon Sabini, Br.
Phippsia algida, Br.

Of these the *Pleuropogon* and *Phippsia* are the most peculiarly Arctic plants. The others, except the *Stellaria*, were more probably overlooked in Banks Land, though the collections from there appear to be so complete that this is hardly likely.

[The Catalogue of Flowering Plants is omitted, the species being incorporated in Dr. Hooker's list above, at pages 225 *et seq.*]

Musci.

(Determined by W. Mitten, Esq., A.L.S.)

Aulacomnion turgidum, Schw. Cape Osborne.
Bryum nutans, Schreb. Cape Osborne.
Pogonatum alpinum, Brid. Cape Osborne.

Hepaticæ.

Jungermannia Starkii, Funk. Cape Osborne.

Algæ.

(By Dr. Dickie, Professor of Botany, Queen's University, Belfast.)

Agarum Turneri, Post. & Rupr. Port Kennedy.
Laminaria saccharina, Lamour. Port Kennedy. Attains a length of 20 feet.
Rhodymenia interrupta, Grev. Port Kennedy.
Dredged up. Only a single specimen of this plant was known previously. It was brought from the Arctic regions by Lieut. Griffiths and preserved in Mrs. Griffiths's herbarium.
Kallymenia Pennyi, Harv. Port Kennedy.
The specimens, though wanting the point of attachment, exhibit the general outline of *K. Dubi.*, but with the margin more or less laciniate. Previously known only from fragments brought home by Dr. Sutherland.*
Gymnogongrus plicatus, Kg. Port Kennedy.
Callithamnium Americanum, Harv. Port Kennedy.
Conferva melagonium, Web. & Mohr. Port Kennedy.
Enteromorpha compressa, Grev. ? Port Kennedy.
Nostoc verrucosum, Vauch. Port Kennedy. Fresh water.
N. arcticum, Berk. Port Kennedy. Fresh water.
N. muscorum, Ag. ? Port Kennedy. Fresh water.
Rivularia Pisum, Ag. Port Kennedy. Fresh water.
Scytonema myochroum, Ag. Port Kennedy. Fresh water.
Soraspora montana, Harv. Port Kennedy. Fresh water.
Hæmatococcus frustulosus, Harv. Port Kennedy. Fresh water.
Tyndaridea anomala, Ralfs. Port Kennedy. At the beach-line.

Fungi.

(By the Rev. M. J. Berkeley.)

Marasimus arcticus, n. sp., Berk. Frederikshaab.
Agaricus furfuraceus, P. Godhavn and Port Kennedy.
A. vaginatus, Bull. Godhavn and Port Kennedy.
A. cyathiformis, Bull. Port Kennedy.
A. umbelliferus, L. Port Kennedy.

---

* *See above*, page 516.

A. allosporus, Berk. Port Kennedy.
Hygrophorus coccineus, Fr. Godhavn.
Illosporium carneum, Fr. Port Kennedy. Very scarce.

LICHENS.*

(Determined by W. MITTEN, Esq., A.L.S.)

Collema furvum, Ach. Port Kennedy. Rare. In wet places and bed of lake.
Sphærophoron coralloides, Ach. Port Kennedy. Rare. Lieveley.
Cladonia deformis, Hoffm. Port Kennedy.
C. pyxidata, Fries. Port Kennedy. Scarce. Lievely and Cape Osborne.
C. gracilis, Hoffm. Greenland.
C. rangiferina, Hoffm. Greenland and Frederikshaab.
C. bellidiflora, Schœr. Greenland.
Stereocaulon botryosum, Ach. Port Kennedy. Very common. Cape Osborne, Leively, and Greenland.
Alectoria orchroleuca, Nyl. Port Kennedy, Lievely, and Cape Osborne.
A. jubata, Ach. (chalybeiformis). Lievely.
Dufourea madreporiformis, Ach. Port Kennedy.
D. arctica, Hk. Pond's Bay.
Cetraria Islandica, Ach. Lievely.
Solorina crocea, Ach. Lievely.
S. saccata, Ach. Port Kennedy. Very scarce.
Platysma nivalis, Nyl. Port Kennedy and Frederikshaab.
P. juniperinum, Nyl. Port Kennedy.
Parmelia saxatilis, Ach. Port Kennedy and Lievely.
P. incurva, Fries. Port Kennedy and Lievely.
P. conspersa, Ach. Port Kennedy.
P. stygia, Ach. Port Kennedy.
P. lanata, Nyl. Lievely.
Physcia pulverulenta, Fries. Port Kennedy.
P. candelaria, Nyl. Cape Osborne.
Umbilicaria hyperborea, Hoffm. Port Kennedy. Not common.
U. hirsuta, DC. Port Kennedy. Rather rare. Lievely.
U. proboscidea, DC. Port Kennedy. Most common of *Umbilicariæ*.
U. cylindrica, Ach. Lievely.
Squamaria gelida, Nyl. Port Kennedy. Rare.
Placodium murorum, DC. Port Kennedy.
P. elegans, DC. Port Kennedy. Abundant. Cape Osborne.
Lecanora tartarea, Ach. Port Kennedy. Abundant. Lievely and Cape Osborne.
L. subfusca, Ach. Port Kennedy.
L. chlorophana, Ach. Port Kennedy.

---

* The Arctic Lichens are treated of *above* in Dr. W. LAUDER LINDSAY'S Memoir, pages 284 *et seq.* *See also* BABINGTON on SUTHERLAND'S Collection, *above*, p. 522; and further on.—EDITOR.

L. frustulosa, Ach. Port Kennedy. Scarce.
L. cerina, Ach. Port Kennedy. Scarce.
L. ventosa, Ach. Port Kennedy, Lievely, and Cape Osborne.
L. vitellina, Ach. Port Kennedy. Scarce.
Lecidea vesicularis, Ach. Port Kennedy. Scarce.
L. lapicida, Fries. Port Kennedy.
L. rupestris, Ach. Port Kennedy.
L. globifera, Ach. Port Kennedy.
L. petræa, Ach. Port Kennedy.
L. geographica, Ach. Port Kennedy and Cape Osborne.
L. alpicola, Wahl. Port Kennedy. Universal.
Urceolaria scruposa, Ach. Port Kennedy.

I append a *résumé* of the important observations made by Dr. Walker on the temperature of the air and earth, and the average covering of snow; and, to render them more complete, I have extracted and meaned the monthly temperatures of Boothia, of from three to four years' observations, published in Sir John Ross's "Voyage." The observations of these officers correspond to a remarkable degree, the approximate mean annual temperature, according to Ross, being + 2°·5, and by Walker (interpolating August as 28°) + 1°·0. The high mean temperature of the soil at 2 feet 2 inches depth is very remarkable, and that of the surface of the earth below the snow, which depends much on the temperature of the subsoil, and is of great influence upon the vegetation, is still more remarkable.

Appendix.—Observations at Port Kennedy on the Temperature of the Soil, &c. By Dr. Walker.

On the 14th September 1858, so soon as it appeared probable that we should winter at Port Kennedy, I sunk a brass tube 2 feet 2 inches vertically in the ground, and inserted a padded thermometer.

The ground at the time of sinking the tube was frozen from 6 inches below the surface, and it was with great difficulty that I could get the tube sufficiently far down. The soil (surface) was similar to that strewn over land, but from below 6 inches it was of a yellowish mud. The thermometer used was one of very small bore, with a long stem finely graduated (it had been prepared for taking the temperature of trees).

From the 18th to the 29th September no register was made, as the ship was not in port; also from the 10th to the 28th March 1859, as I was absent from the ship, travelling. The minimum temperature registered was +0·5, on March 10th, 1859. The lowest may be assumed at zero, on the 16th March.

The register was continued until June 18th, when water entered the tube, and the thermometer was frozen to the side, so that it could not be detached. Column 2 gives the register of the thermometer. Column 3 gives the depth of the overlying snow, which was always greater than the average quantity over the land. On the 17th January 1859, a tube was placed 1 foot 1 inch deep in a mixture of shingle and earth; in this a thermometer was placed. The position of the ground was such that scarcely any snow lay upon it, the constant strong winds removing it almost as soon as deposited.

Column 4 gives the register of this thermometer.

February 12th 1859.—A tube was placed horizontally on the surface of the ground beneath the snow lying over the place where thermometer No. 1 was sunk, and the temperature as shown by this thermometer (column 5) was registered until all the snow disappeared. Column 6 gives the mean temperature of the air for the day on which the registers of the different thermometers were taken.

Column 7 gives the mean temperature of the air for the number of days or hours intervening between the registering of the thermometers.

All the registers of the different thermometers are corrected so as to reduce them to that of the standard.

Date.	Thermometer buried 2 ft. 2 in.	Depth of snow.	Thermometer buried 1 ft. 1 in.	Temp. of surface below snow.	Mean of air on days of observation.	Mean of air from that of intervals.	Ross.
1858.		ft. in.					
September - -	+30·9	—	—	—	+25·0	+24·5	+25
October - -	+24·4	0 6	—	—	−9·6	−8·4	−9
November - -	+15·8	4 0	—	—	−12·0	−13·6	−9
December - -	+12·0	4 5	—	—	−34·9	−33·7	−16
1859.							
January - -	+6·2	5 8	−21·5	—	−34·0	−33·4	−26
February - -	+2·3	5 8	−24·8	−3	−32·7	−34·8	−32
March - - -	+0·7	6 0	−16·7	−3·8	−19·9	−17·4	−27
April - - -	+1·6	6 6	−8·1	+0·7	−1·6	−5·4	−3
May - - -	+3·5	6 0	+8·1	+4·9	+16·4	+14·4	+14
June - - -	+7·5	3 6	+27·3	+26·0	+34·2	+35·0	+34
July - - -	+31·7	0 0	+39·0	—	+40·7	+39·9	+36
August - -	? +40·0	? 0 0	?	—	? +30	+28·0	+25
Mean - -	+14·7	—	—	—	+1·8	+1·0	+2·5

LXXIX.—On some Arctic Diatomaceæ, collected during the voyage of the "Fox." By the Rev. Eugene O'Meara. (By Permission, from the Journal Royal Dublin Soc., vol. iii. pp. 59–60, 1860.)

In this account of Diatomaceæ found amongst mixed sediments of bottles in which larger specimens had been brought home from various Arctic localities, namely, Port Kennedy, long. 94° W., lat. 72° N., Bellot Strait, Brentford Bay, off the Crimson Cliffs,

Cape York, and elsewhere. The following list and remarks are given :—

Achnanthes longipes.
Amphipleura sigmoidea.
Amphiphora alata.
A. constricta.
Amphora cymbifera.
A. hyalina.
A. robusta.
A. proboscidea.
A. tenera.
Bacillaria paradoxa.
Biddulphia aurita.
Campylodiscus parvulus.
C. angularis.
Cocconeis scutellum.
C. scutellum, var. β.
C. splendida.
C. nitida.
C. ornata.
C. ovalis.
C. pseudomarginata.
Coscinodiscus radiatus.
C. minor.
C. punctulatus.
Cymbella scotica.
Denticula fulva.
Doryphora Boeckii.
Eupodiscus fulvus.
E. crassus.
Fragillaria virescens.
Grammatophora marina.
G. serpentina.
Gomphonema marinum.
Hyalodiscus parallelus.
Melosira nivalis.
M. nummuloides.
Navicula elliptica.
N. cryptocephala.
N. inconspicua.
N. didyma.
N. lyra.
N. cuspidata.
N. minutula.
N. cluthensis.
N. minor.
Nitzchia angularis.
N. hyalina.
N. lanceolata.
N. parvula.
N. sigma.
N. distans.
Pinnularia interrupta.
P. cyprina.
P. lata.
P. latistriata.
P. affinis.
P. Johnsonii.
P. directa.
P. panduriformis.
P. peregrina.
Pleurosigma lanceolatum.
P. angulatum.
P. prolongatum.
P. strigosum.
P. delicatulum.
P. fasciola.
Podosira hormoides.
Podosphenia Ehrenbergii.
Rhabdonema arcuatum.
R. minutum.
Rhizolenia styliformis.
R. calcar-avis.
Schizonema crucigerum.
Striatella unipunctata.
Stauroneis crucicula.
S. pulchella.
S. pulchella, var. β.
Surirella ovata.
S. fastuosa.
Synedra tubulata.
S. fulgens.
S. gracilis.
S. arcus.
Tabellaria floculosa.
Triblionella constricta.

"In addition to the forms detailed in the foregoing catalogue, there were ten species which I could not identify with any figured in the books which up to this time I have been able to consult, but without research I would not presume to describe them as new to science.

"In conclusion I would take the opportunity to remark that it appears to me a very important and interesting fact that so many

British species of Diatoms should have been found in these scanty gatherings from the Arctic regions, their generic and specific characters having been in no perceptible degree affected by conditions so different as exist between the climate of the Arctic Ocean and that of our own waters."

LXXX.—Some PLANTS of the ARCTIC AMERICAN ARCHIPELAGO. From "A Whaling Cruise to Baffin's Bay and "the Gulf of Boothia, and an Account of the Rescue "of the Crew of the 'Polaris.'" By ALBERT HASTINGS MARKHAM, Commander R.N., F.R.G.S., 1874. 8vo. London.

Appendix B.—Arctic Plants collected by Capt. A. H. Markham, R.N., F.R.G.S. 1873.

Name.	
Ranunculus glacialis (L.)	Fury Beach; Elwyn Inlet.
Papaver alpinum (L.)	Fury Beach; Elwyn Inlet; Navy-Board Inlet.
Lychnis apetala (L.)	Fury Beach.
Stellaria Edwardsii (R. Br.)	Elwyn Inlet; Fury Beach.
Dryas octopetala (L.)	Navy-Board Inlet.
Saxifraga cæspitosa (L.)	Fury Beach.
S. nivalis (L.)	" "
S. flagellaris (Willd.)	" "
S. oppositifolia (L.)	Port Leopold; Elwyn Inlet.
Pedicularis hirsuta (L.)	Navy-Board Inlet; Elwyn Inlet.
Juncus biglumis (L.)	Fury Beach.
Salix arctica (R. Br.)	14′ S. of Cape Garry.
Alopecurus alpinus (L.)	Fury Beach.
Festuca ovina (L.) var.	6′ S. of Cape Garry.
Pleuropogon Sabini (R. Br.)	Fury Beach.

Lichens.

Platysma junipernus (L.)	6′ S. of Cape Garry; Fury Beach.
Alectoria ochrolenæ (Ehrh.)	Fury Beach.

LXXXI.—ON ARCTIC SILURIAN FOSSILS. By J. W. SALTER, Esq., F.G.S. (Reprinted, by permission, from the Quart. Journ. Geol. Soc., vol. ix., 1853, pp. 312–317.)

A considerable number of limestone fossils were brought home by the officers and gentlemen engaged in the late Arctic Expedition (1850–51), which have added very materially to our knowledge of the geology of those Polar regions.

The rocks along the coasts of Barrow Strait and the shores of Prince-Regent's Inlet were already partly explored by Parry, and described by Prof. Jameson and Mr. König, and a few fossils, brought away in ballast from Prince-Leopold's Island by Sir James Ross, had been detected at Woolwich by Capt. James, R.E., and mentioned in Prof. Ansted's "Geology." But the present collections, from the entrance of Wellington Channel, are far more extensive than those before named, and the localities have been exactly marked.

Her Majesty's squadron, under command of Capt. Austin, collected specimens of fossiliferous limestone in Assistance Bay, Cape Riley, and Beechey Island, and in Griffiths and Somerville Islands; and some travelling parties brought home a specimen of the same limestone with fossils from Cape Walker, still farther to the southwest. Our thanks are especially due to Capt. Ommanney and Mr. Donnet, who have presented their specimens to the Museum of Practical Geology, and to Sir John Richardson, who has permitted us to examine those brought home by Mr. Pickthorne of the "Pioneer."

In addition to these, Capt. Penny and his companions discovered that the same rock with fossils extends up both sides of the strait, and covers the islands at the mouth of the newly discovered Queen's Channel. These researches were prosecuted more particularly by Dr. P. C. Sutherland, assisted very willingly by the seamen; and to Sir John Richardson we are indebted also for the examination of these fossils.

Of the following list of organic remains (described in the Appendix to Dr. Sutherland's "Journal"*) some are known European fossils. Among these the common Chain-coral (*Halysites catenulatus*), the *Favosites Gotlandica* and *F. polymorpha*, and the *Atrypa reticularis* are well-known cosmopolitan species. There are some others more doubtfully European forms; and three, among which is the *Pentamerus conchidium*, appear to be identical with Swedish species in the Wenlock Limestone of Goth-

---

* *See* the "Journal of a Voyage in Baffin's Bay and Barrow Strait in the "years 1850–1, performed by H.M. ships 'Lady Franklin' and 'Sophia,' "under the command of Mr. William Penny, in search of the missing crews "of H.M. Ships 'Erebus' and 'Terror,' &c. By Peter C. Sutherland, M.D., "&c., Surgeon to the Expedition." London, 2 vols. 8vo., 1852.

APPENDIX:

Drift-wood, Dr. J. Richardson, pp. cxxii.–cxxx. Meteorology, pp. cxxxi.–clxxxvi.

Vascular Plants, pp. clxxxix.–cxc., Sir W. Hooker, K.H., F.R.S., &c. Included in Prof. Hooker's Table, pages 225, *et seq.*

Algæ, Dr. Dickie, pp. cxci.–cc. *See* page 515.

Zoology: Mollusca, pp. cc.–cciv., Dr. P. C. Sutherland. *See* page 511.

—— Crustacea and Insects, pp. ccv.–ccxi., Arthur Adams and Adam White. *See* pages 508 and 514.

—— Ascidians, Prof. Huxley, pp. ccxi.–ccxiii. } *See* page 513.

—— Echinoderms, Prof. E. Forbes, pp. ccxiii.–ccxvi. } *See* page 513.

Geology, J. W. Salter, pp. ccxvii.–ccxxxiii.—EDITOR.

land. The rest are new to me either as occurring in Europe or America.

The general resemblance with the fossils of our own Upper-Silurian rocks is very considerable, and, in the absence of many characteristic Lower-Silurian genera, the identity of several species with Upper-Silurian forms, and the great prevalence of Corals, I think we are quite warranted in placing these strata in the upper division of the Silurian system.*

The shores at the entrance of Wellington Strait, at Cape Riley, Beechey Island, and Cornwallis and Griffiths' Islands, contain the following fossils, most of which are figured in the work above quoted :—

Crustacea.

1. Encrinurus lævis, *Angelin?* (Sutherland's "Journal," Appendix, pl. 5, f. 14).
2. Proetus, sp. (*l. c.*, pl. 5, f. 15).
3. Leperditia (baltica, *Hisinger*, sp., *var.*) arctica, *Jones*, *l. c.*, pl. 5, f. 13.

Mollusca.

4. Lituites, n. sp. Allied to *L. articulatus*, Sow., but with more numerous whorls.
5. Orthoceras Ommanneyi, n. sp. (*l. c.*, pl. 5, f. 16, 17).
6. —— species, with distant septa and central siphuncle.
7. —— species, imperfect, but distinct from both the preceding.
8. Murchisonia, sp. (*l. c.*, pl. 5, f. 18). Like *M. gracilis*, Hall.
9. —— larger species, with numerous whorls (*l.c.*, pl. 5, f. 19).
10. Euomphalus, a small species.
11. Modolia (or Modiolopsis). An oval and flattish form.
12. Strophomena Donneti, n. sp. (*l. c.*, pl. 5, f. 11, 12).
13. ——, sp. The same as at Leopold's Island.
14. Orthis, large flat species. Griffiths' Island.
15. Spirifer crispus, *Linn?* (*l. c.*, pl. 5, f. 8).
16. ——, sp., very like *S. elevatus*, Dalm.
17. Chonetes lata, *Von Buch?*
18. Pentamerus conchidium, *Dalm.* (*l. c.*, pl. 5, f. 9, 10). Coarse-ribbed and fine-ribbed varieties.
19. Rhynchonella phoca, n. sp. (*l. c.*, pl. 5, f. 1–3).
20. ——, sp. (*l. c.*, pl. 5, f. 5).

Radiata.

Encrinites are so abundant at Cape Riley that in several places the rock is composed entirely of their detritus. A species of *Actinocrinus* is the only fragment at all recognisable.

Corals are very abundant; twenty or more species having been observed. Among them there are :—

21. Ptychophyllum.
22. Cystiphyllum.

---

* These researches by Mr. Salter are referred to and verified in Murchison's "Siluria," last edit., 1867, pp. 440–441.—Editor.

23. Cyathophyllum.
24. Strephodes Pickthornii, n. sp. (*l. c.*, pl. 6, f. 5).
25. —— ? (Clisiophyllum ?) Austini, n. sp. (*l. c.*, pl. 6, f. 6).
26. Clisiophyllum, sp. (*l. c.*, pl. 6, f. 7).
27. Aulopora, sp.
28. Favistella reticulata, n. sp. (*l. c.*, pl. 6, f. 2).
29. —— Franklini, n. sp. (*l. c.*, pl. 6, f. 3).
30. Favosites polymorpha, Goldfuss (*l. c.*, pl. 6, f. 9). Branched and amorphous varieties.
31. —— Gotlandica, *Linn.*, var.
32. ——, sp.
33. ——, sp.
34. Cœnites (Limaria), sp. Abundant at Beechey Island.
35. Halysites catenulatus, *Linn.* (*l. c.*, pl. 6, f. 11).
36. Syringopora, sp.
37. Heliolites (Porites), sp. Rare.
38. Columnaria Sutherlandi, n. sp. (*l. c.*, pl. 6, f. 8). Beechey and Seal Islands.

Proceeding up Wellington Channel, at Point Eden, on the south side of Baring Bay, Dr. Sutherland found a new Coral,—

39. Arachnophyllum Richardsoni, n. sp. (*l. c.*, pl. 6, f. 10); like some Carboniferous forms.

And at the south-west end of Seal Island, a rock in Baring Bay, in white crystalline limestone,—

Encrinurus lævis, and
Leperditia (balthica, *His.*, sp., *var.*) arctica, *Jones*, above noticed.

40. Atrypa reticularis (*Lin.*, *l. c.*, pl. 5, f. 7). Small, abundant.
41. Rhynchonella Mansonii, n. sp. (*l. c.*, pl. 5, f. 5).
42. —— sublepida, *De Vern.* ? (*l. c.*, pl. 5, f. 6).
43. Fenestella (*l. c.*, pl. 6, f. 1). Small species, same as at Leopold Island ?
44. Crotalocrinius; stem only; like *C. rugosus*, Miller.
45. Calophyllum phragmoceras, n. sp. (*l. c.*, pl. 6, f. 4); a Cup-coral with large flat diaphragms.

Dr. Sutherland followed the margin of the strait to its north-eastern angle in lat. 76° 20′, and Capt. Stewart continued along the shore until he reached the new Queen's Channel, long. 97°. Along the whole coast the same limestone rocks were visible; and, from its peculiar uniform castellated appearance, it could be traced by the eye to extend still farther up the sides of that inlet. In the meantime Captain Penny and his crew were exploring the islands which separate this channel from the Wellington Strait, and both in Hamilton Island and Dean-Dundas Island abundance of fossils were seen. The necessity of abandoning their boat prevented their bringing them away; but one of the seamen, James Knox, contrived to roll up a Trilobite and a Bellerophon in the corner of his shirt, and bring them back to Dr. Sutherland. They were from Dundas Island, in lat. 76° 15′, the most northerly point of the new continent from whence fossils have been brought. One of them is the—

Encrinurus lævis, before mentioned; the other—

46. Bellerophon nautarum, n. sp. (*l. c.*, pl. 5, f. 20); so named in honour of the crews of the "Lady Franklin" and "Sophia."

Returning to Barrow Strait, the ballast from Leopold's Island, before quoted, yielded:—

Favosites polymorpha. Abundant.
—— Gotlandica.
Fenestella, sp. Probably the same as above, No. 43.
Strophomena, sp. Same as No. 13.
Rhynchonella phoca. As above, No. 19.
—— sublepida, *De Vern.* var.
47. ——, sp. With simple plaits; distinct from No. 20.

The existence of this great formation of Upper-Silurian limestone along the shores of Prince-Regent's Inlet is rendered all but certain from the notes furnished by Prof. Jameson and Mr. König in the Appendices to Capt. Parry's Voyages. From their accounts the coasts are occupied by a "transition limestone" of an ash-grey or yellowish and grey colour, often fœtid, and sometimes crystalline or compact. It is described as filled with Zoophytes and Shells, and in certain parts, as noticed by Mr. König, quite made up of the detritus of Encrinites, the fragments of which are so comminuted that it might readily be mistaken for a granular limestone. He also found in it the Chain-coral.

Prof. Jameson gives a list of organic remains from Port Bowen, which, by modernising the nomenclature of the fossils, would agree well with those from the north shore of Barrow Strait, and indeed he has himself identified them. And he mentions that this same rock extends eastward to Cape York, Admiralty Inlet, and even occurs at Possession Bay; while in a southerly direction it was found as far as the Regent's Inlet was explored. We have seen that a similar limestone occurs at Cape Walker, Russell Island, and from the general low character of the shores that stretch to the west (explored lately by Capt. Ommanney), it is probably continued along them.

The north-eastern shores of Lancaster Sound are composed, at least in part, of igneous and crystalline rocks; but from the commencement of the table-land at Powell's Inlet all along the coast to the Wellington Channel, a uniform appearance of the shore (the cliffs appearing like fortifications) indicates the presence of the same tabular strata of limestone.

Dr. Sutherland, who is well acquainted with the appearance of the limestone cliffs, and who had the advantage of communicating with the different exploring parties, is without doubt of its continuity along this coast.

Again, the limestone of Melville Island, according to Mr. König, contains *Favosites* and *Terebratulæ*.

Dr. Conybeare adds to them *Catenipora* and *Caryophylliæ*; and this is just the aspect which the fossils of the limestone we have described would present on a cursory examination. In Melville

Island, however, it is connected with a sandstone and coal-formation, with a Carboniferous flora; and, as this sandstone contains *Trilobites*, *Encrinites*, and *Aviculæ*, we may hesitate, in the absence of authentic specimens, to extend the Silurian limestone so far.*

We may now, then, definitely colour the shores of Wellington Channel and Barrow Strait, except the eastern entrance of the latter (which is occupied by igneous or crystalline rocks), as Upper Silurian; and, on the return of the expedition under Capt. Belcher, the limits of this formation will be no doubt greatly extended. I may mention that coal or lignite was picked up at Byam-Martin Island, and that a fragment of it occurred in the detritus 350 feet above the sea, at Kate-Austin's Lake, Cornwallis Island. Also at Griffiths' Island and Browne Island fragments of *iron* [native?] were found.

In conclusion, it is worth while to observe the occurrence of Pleistocene deposits with marine Shells of existing Arctic species (*Mya truncata*, *Saxicava rugosa*, &c.), which were found on every elevation up to 500 feet on Beechey and Cornwallis Islands.

LXXXII.—On the OCCURRENCE of numerous FRAGMENTS of FIR-WOOD in the ISLANDS of the ARCTIC ARCHIPELAGO; with Remarks on the ROCK-SPECIMENS brought from that Region. By SIR RODERICK IMPEY MURCHISON, D.C.L., F.R.S., V.P.G.S., Director-General of the Geological Survey. Reprinted, with Permission, from the Quart. Journ. Geol. Soc., vol. xi., 1855, pages 536–541. (Read June 13, 1855.)

On the present occasion I cannot attempt to offer any general, still less any detailed description of the rocks and fossils of the north-western portion of that great Arctic Archipelago whose shores were first explored by Parry and Sabine. The specimens they brought home from Melville Island, and which were described by Mr. König, first conveyed to us the general knowledge of the existence there of fossiliferous limestones and other rocks analogous to known European types in Scandinavia. Since those early days the voyages of Franklin, and of the various Officers who have been in search of our lamented friend, have amplified those views, and have shown us that over nearly the whole of the Arctic Archipelago these vast islands possess a structure similar to that of North America. We shall soon, I

* In the collection of rocks from Melville Island, in the Society's Museum, a specimen of the compact limestone and a Coral (*Favistella Franklini*) occur; the latter was "collected by Lieut. Liddon, second in command in the "Expedition of 1819–20," and presented by Dr. Granville, F.G.S.

believe, be made acquainted with the characters of the specimens collected by the Expedition under Sir Edward Belcher, who is preparing a description of the natural-history products of his survey. My chief object now is to call attention to the remarkable fact of the occurrence of considerable quantities of wood, capable of being used for fuel or other purposes, which exist in the interior, and on the high grounds of large islands in latitudes where the Dwarf Willow is now the only living shrub.

Before I allude to this phenomenon, as brought to my notice by Capt. M'Clure and Lieut. Pim, I would, however, briefly advert to a few rock-specimens collected by the latter officer in Beechey Island, Bathurst Land, Eglinton Island, Melville Island, Prince-Patrick's Island, and Banks Land, where he joined Capt. M'Clure; specimens which we ought to value highly, seeing that they were saved from loss under very trying circumstances.

From this collection, as well as from other sources to which I have had access, as derived from Voyages of Parry, Franklin, Back, Penny, Inglefield, and the recent work of Dr. Sutherland,* I am led to believe that the oldest fossiliferous rock of the Arctic region is the Upper Silurian, viz., a limestone identical in composition and organic contents with the well-known rocks of Wenlock, Dudley, and Gothland.

No clear evidence has been afforded as to the existence of Devonian rocks, though we have heard of red and brownish sandstone, as observed in very many localities by various explorers, and which possibly may belong to that formation. Thus, in North Somerset, to the south of Barrow Strait, red sandstone is associated with the older limestones. Byam-Martin Island was described by Parry as essentially composed of sandstone, with some granite and felspathic rocks; and, whilst the north-eastern face of Banks Land is sandstone, its north-western cliffs consist (as made known by Capt. M'Clure) of limestone. But, whilst in the fossils we have keys to the age of the Silurian rocks, we have as yet no adequate grounds whereon to form a rational conjecture as to the presence of the Old Red Sandstone, or Devonian group.

True Carboniferous *Producti* and *Spiriferi* have been brought home by Sir E. Belcher from Albert Land, north of Wellington Channel; and hence we may affirm positively that the true Carboniferous rocks are also present. Here and there bituminous shale and coal are met with; the existence of the latter being marked at several points on the general chart published by the Admiralty. With the Palæozoic rocks are associated others of igneous origin and of crystalline and metamorphosed character. Thus, from Eglinton Island to the south of Prince-Patrick's Island, first defined by the survey of Capt. Kellett and his officers, we see concretions of greenstone associated with siliceous or quartzose rocks and coarse ferruginous grits; and in Princess-Royal Island, besides the characteristic Silurian limestones, there are black basalts and red jaspers, as well as red rocks less altered by heat, but showing a passage into jasper. Highly crystalline

* *See above*, pp. 532, &c.

gypsum was also procured by Lieut. Pim from the north-western shores of Melville Island. In the collection before us we see silicified stems of Plants, which Lieut. Pim gathered on various points between Wellington Channel on the east, and Banks Land on the west. Similar silicified Plants were also brought home by Capt. M'Clure from Banks Land, and through the kindness of Mr. Barrow, to whom they were presented, they are now exhibited, together with a collection made by Capt. Kellett, which he sent to Dr. J. E. Gray, of the British Museum, who has obligingly lent them for comparison.

I had requested Dr. Hooker to examine all those specimens which passed through my hands, and I learn from him that he will prepare a description of them, as well as of a great number from the same region which had been sent to his father, Sir W. Hooker, associated, like those now under consideration, with fragments of recent wood.

Of Secondary Formations no other evidence has been met with except some fossil bones of Saurians, brought home by Sir E. Belcher, from the smaller islands north of Wellington Channel, and of these fossils Sir Edward will give a description. Of the old Tertiary rocks, as characterised by their organic remains, no distinct traces have, as far as I am aware, been discovered, and hence we may infer that the ancient submarine sediments, having been elevated, remained during a very long period beyond the influence of depository action.

Let us now see how the other facts, brought to our notice by the gallant Arctic explorers who have recently returned to our country, bear upon the relations of land and water in this Arctic region during the quasi-modern period when the present species of Trees were in existence. Capt. M'Clure states that in Banks Land,* in latitude 74° 48′, and thence extending along a range of hills varying from 350 to 500 feet above the sea, and from half a mile to upwards inland, he found great quantities of wood, some of which was rotten and decomposed, but much of it sufficiently fresh to be cut up and used as fuel.

Whenever this wood was in a well-preserved state, it was either detected in gullies or ravines, or had probably been recently exhumed from the frozen soil or ice. In such cases, and particularly on the northern faces of the slopes where the sun never acts, wood might be preserved any length of time, inasmuch as Capt. M'Clure tells me he has eaten beef which, though hung up in his cold larder for two years, was perfectly untainted.

The most remarkable of these specimens of well-preserved recent wood is the segment of a tree which, by Capt. M'Clure's orders, was sawn from a trunk sticking out of a ravine, and which is now exhibited. It measures 3 feet 6 inches in circumference. Still more interesting is the cone of one of these Fir-trees which he brought home, and which apparently belongs to an *Abies* resembling *A. alba*, a plant still living within the Arctic circle. One of Lieut. Pim's specimens of wood from Prince-Patrick's

* *See* Prof. O'Heer's Memoir, *above*, p. 379.

Island is of the same character as that just mentioned, and in microscopical characters much resembles *Pinus strobus*, the American Pine, according to Prof. Quekett, who refers another specimen, brought from Hecla-and-Griper Bay, to the Larch. In like manner Lieut. Pim detected similar fragments of wood, two degrees further to the north, in Prince-Patrick's Land, and also in ravines of the interior of that island, where, as he informed me, a fragment was found like the tree described by Capt. M'Clure, sticking out of the soil in the side of a gully.

We learn, indeed, from Parry's Voyage that portions of a large Fir-tree were found at some distance from the south shore of Melville Island, at about 30 feet above high-water mark, in latitude 74° 59′ and longitude 106°.

According to the testimony of Capt. M'Clure and Lieut. Pim, all the timber they saw resembled the present driftwood so well known to Arctic explorers, being irregularly distributed, and in a fragmentary condition, as if it had been broken up and floated to its present positions by water.* If such were the method by which the timber was distributed, geologists can readily account for its present position in the interior of the Arctic Islands. They infer that at the period of such distribution large portions of these tracts were beneath the waters, and that the trees and cones were drifted from the nearest lands on which they grew. A subsequent elevation, by which these islands assumed their present configuration, would really be in perfect harmony with those great changes of relative level which we know to have occurred in the British Isles, Germany, Scandinavia, and Russia since the great Glacial Period.† The transportation of immense quantities of timber towards the North Pole, and its deposition on submarine rocks, is by no means so remarkable a phenomenon as the wide distribution of erratic blocks during the Glacial Epoch over Northern Germany, Central Russia, and large portions of our island, when under water, followed by the rise of these vast masses into land. If we adopt this explanation, and look to the extreme cold of the Arctic region in the comparatively modern period during which this wood has been drifted or preserved, we can have no difficulty of accounting for the different states in which the timber is found. Those portions of it which happen to have been exposed to the alternations of frost and thaw, and the influence of the sun, have necessarily become rotten; whilst all those fragments which remained enclosed in frozen mud or ice would, when brought to light by the opening of the ravines or other accidental causes, present just as fresh an appearance as the specimens now exhibited.

The only circumstance within my knowledge which militates against this view is one communicated to me by Capt. Sir Edward Belcher, who, in lat. 75° 30′, long. 92° 15′, observed on the east side of Wellington Channel the trunk of a Fir-tree standing vertically, and which, being cleared of the surrounding

---

* *See* Scoresby on Drift-wood, "Northern Whale-fishery," p. 19, 1823.

† *See* Mr. H. H. Howorth's Memoir, reprinted *above*, p. 483.

earth, &c., was found to extend its roots into what he supposed to be soil.*

If from this observation we should be led to imagine that all the innumerable fragments of timber found in these polar latitudes belonged to trees that grew upon the spot, and on the ground over which they are now distributed, we should be driven to adopt the anomalous hypothesis that, notwithstanding physical relations of land and water similar to those which now prevail (*i.e.*, of great masses of land high above the sea), trees of large size grew on such *terra firma* within a few degrees of the North Pole! A supposition which I consider to be wholly incompatible with the data in our possession, and at variance with the laws of isothermal lines. If, however, we adopt the theory of a former submarine drift, followed by a subsequent elevation of the sea-bottom, as easily accounting for all the phenomena, we may explain the curious case brought to our notice by Sir Edward Belcher, by supposing that the tree he uncovered had been floated away with its roots downwards, accompanied by attached and entangled mud and stones, and lodged in a bay, like certain "snags" of the great American rivers. Under this view, the case referred to must be considered as a mere exception, whilst the general influence we naturally draw is, that the vast quantities of broken recent timber, as observed by numerous Arctic explorers, were drifted to their present position when the islands of the Arctic Archipelago were submerged. This inference is indeed supported by the unanswerable evidence of the submarine associates of the timber, for from the summit of Coxcomb Range in Banks Land, and at a height of 500 feet above the sea, Capt. M'Clure brought home a fine large specimen of *Cyprina Islandica*, which is undistinguishable from the species so common in the Glacial Drift of the Clyde; whilst Capt. Sir E. Belcher found the remains of Whales on lands of considerable altitude in lat. 78° north. Reasoning from such facts, all geologists are agreed in considering the shingle, mud, gravel, and beaches, in which animals of the Arctic region are imbedded in many parts of Northern Europe, as decisive proofs of a period when a glacial sea covered large portions of such lands; and the only distinction between such deposits in Britain and those which were formed in the Arctic circle is, that the wood which was transported to the latter has been preserved in its ligneous state for thousands of years through the excessive cold of the region.

P.S. Since the above was written Capt. Collinson transmitted to me an instructive collection of rock-specimens collected during his survey. Most of them show the great prevalence of crystalline rocks along the north coast of America.

---

* This trunk of a White Spruce, standing dead in North Devon, is noticed in Belcher's "Last Arct. Voy.," i. p. 380, with a note by Dr. Hooker at p. 381.—EDITOR.

LXXXIII.—NOTES on some ROCK-SPECIMENS from the ARCTIC-AMERICAN ARCHIPELAGO, by R. ETHERIDGE, ESQ., F.R.S. In the "Whaling Cruise to Baffin's Bay and the Gulf of Boothia, and an account of the rescue of the crew of the 'Polaris';" by ALBERT HASTINGS MARKHAM, Commander, R.N. F.R.G.S., 1874. 8vo., London.

Appendix C. List of the Geological specimens collected by Captain H. MARKHAM, R.N., F.R.G.S., and examined by R. ETHERIDGE, Esq., Museum of Practical Geology.

UPERNAVIK.—1. Syenite, much resembling the Laurentian* series of Cape Wrath (Sutherlandshire). 2. Crystals of felspar, also like those in the Sutherlandshire rocks. 3, 4. Quartz-rocks.

ELWYN INLET.—Piece of quartz-rock and quartzite.

CAPE HAY.—Two pieces of limestone, extremely like that of Durness in N.-W. Sutherlandshire, of Llandeilo age (Lower Silurian). Two specimens of *Saxicava rugosa*, from 150 feet above the sea-level.

NAVY-BOARD INLET.—Specimens of "fundamental gneiss," like that of Cape Wrath; hornblende-rock, mica-schist, quartzite, and magnesian limestone.

PORT LEOPOLD.—Syenite, felspar, and quartz, like the Cape Wrath rocks. An alternation of limestone and sandstone probably Silurian. Gneissose rocks, much the same as the "fundamental gneiss" of N.W. Sutherlandshire. A specimen showing Annelide tracks in fine-grained sandstone.

FURY BEACH.—Specimens of gneiss, hornblende, quartz, and gneissose rocks, much like the "fundamental series" of Sutherlandshire. Argillaceous limestone, with the following fossils of Upper-Silurian age:—*Favosites* (two specimens); *Athyris* (two specimens); *Holopella.*

CAPE GARRY.—Hornblende; quartz-rock, stained red; crystals of calcareous spar; concretionary limestone. Limestone containing several fossils of uncertain age. *Chonetes* and *Terebratula* of the Upper-Silurian age.

Several of the specimens, having been picked up on the beach, are much waterworn.

---

* The occurrence of rocks referable to what are now known as *Laurentian* is noticeable also in Macculloch's list of Ross's Specimens, *above*, p. 324. Pingel's Igalliko quartzite and some of the quartz-rocks of the above list may be of the succeeding *Cambrian* age.—EDITOR.

LXXXIV.—On the GEOLOGY of the PARRY ISLANDS and neighbouring Lands By the Rev. Professor SAMUEL HAUGHTON, M.D., F.R.S., &c. (With Permission.)

[From Notes and Appendices accompanying M'CLINTOCK's "Reminiscences," &c., in the Journ. Royal Dublin Society, vol. i., 1857, pp. 195, 210–214, 239–250 (with six Plates and a Map); and vol. iii., 1860, pp. 53–58 (with four plates).]

(1.) Vol. 1, p. 195, in M'CLINTOCK's "Reminiscences," &c., First Expedition.

1. **Garnier Bay, North Somerset,** Lat. 74° N., Long. 92° W.
   Cromus arcticus, Hgt. J .R. D. S., i. t. 6., f. 1–5. (*Encrinurus lœvis*, according to Salter.
   Atrypa phoca, Salter. Sutherland's Journal, t. 5, f. 1–3; J.R.D.S., i. t. 5, f. 3, 4, 7.
   A. reticularis, Sow.
   Cyathophyllum helianthoides, Goldfuss. J. R. D. S., i. t. 8, f. 1, 2.
   Heliolites porosa (Goldf.). J. R. D. S., i. p. 247, t. 10, f. 5; also near *Cape Bunny*.
   H. megastoma (M'Coy).
   Columnaria Sutherlandi, Salter.

2. **Port Leopold,** Lat. 73° 50′ N., Long. 90° 15′ W.
   Limestone, containing numerous fossils of the Upper-Silurian type: *Calamopora ?*, *Gotlandica*, Goldf., J. R. D. S., i. p. 248, t. 11, f. 3.; *Rhynchonella cuneata* (?), Dalman, *Cyathophyllum*, sp.
   Dark, earthy limestone containing multitudes of casts of *Loxonema Macclintocki*, Haughton, J. R. D. S., i. t. 5, f. 2, 5; 1,100 feet above the sea-level at *North-east Cape*.
   Selenite and fibrous gypsum, from shaly beds in the cliff.

(2.) Vol. i. p. 210.—In M'Clintock's "Reminiscences," Second Expedition.

1. **Griffith's Island,** Lat. 74° 35′ N., Long. 95° 30′ W.
   Cromus arcticus, Hgt. J. R. D. S., i. t. 6, f. 1–5.
   Orthoceras Griffithi, Hgt. J. R. D. S., i. t. 5, f. 1.
   Orthoceras, sp., with lateral siphuncle.
   Loxonema Rossi, Hgt. J. R. D. S., i. t. 5, f. 6, 8–11; iii. t. 4, f. 6.
   Macrocheilus, sp.
   Strophomena Donneti, Salter. Sutherland's Journal, t. 5, f. 11, 12.
   Atrypa phoca, Salter. Sutherland's Journal, t. 5, f. 1–3. J. R. D. S., i. t. 5, f. 3, 4, 7.
   Atrypa, ribbed sp.
   Polyzoon.

Calophyllum phragmoceras, Salter. Sutherland's Journal, t. 6, f. 4.
Syringopora geniculata, Phil. J. R. D. S., i. t. 11, f. 2.
Crinoidal Limestone.

2. **Beechey Island,** Lat. 74° 40′ N., Long. 92° W.
Orthoceras, sp.
Atrypa phoca, Salter; constituting a dark limestone.
Loxonema, spp.; constituting at places a pinkish and whitish limestone.
Loxonema, sp., Salter. Sutherland's Journal, t. 5, f. 10.
Atrypa, ribbed sp.
Syringopora reticulata, Goldf.
Calophyllum phragmoceras, Salter.
Cyathophyllum cæspitosum, Goldf.
C. articulatum, Edw. & H.
Favosites Gotlandica (?), Linn.
F. alveolaris (?), Bl.
Favistella Franklini, Salter. Sutherland's Journal, t. 6, f. 3; J. R. D. S., i. t. 11, f. 1. Abundant at Cape Riley.
Clisiophyllum Salteri, Sutherland. [*Op. cit.*, t. 6, f. 7.
C. Austini? (Salter). *Strephodes*, Sutherland's Journal, t. 6, f. 6; J. R. D. S., t. 10, f. 2.
Chætetes arcticus, Hgt. J. R. D. S., i. t. 10, f. 3, 4.
Cyathophyllum, sp.
Crinoidal Limestone.

3. **Assistance Bay, Cornwallis Island,** Lat. 74° 30′ N., Long. 94° W.
Cromus arcticus, Hgt.
Orthoceras Ommaneyi, Salter. Sutherland's Journal, t. 5, f. 16, 17; J. R. D. S., i. p. 249, t. 11, f. 5.
Pentamerus conchidium, Dalman. Sutherland's Journal, t. 5, f. 9, 10. Pentamerus Limestone.
Cardiola Salteri, Hgt. J. R. D. S., i. t. 7, f. 5.
Syringopora geniculata, Phil. J. R. D. S., i. t. 11, f. 2.

4. **Cape York, Lancaster Sound,** Lat. 73° 50′ N., Long. 87° W.
Chætetes? sp. The same as "*Favosites Gotlandica,*" from Beechey Island.

5. **Possession Bay,** S. of the entrance into **Lancaster Sound,** Lat. 73° 30′ N., Long. 77° 20′ W.
Brown earthy limestone, fœtid when struck; closely resembling the limestone of Cape York, Lancaster Sound.

6. **Hillock Point, Melville Island,** Lat. 76° N., Long. 111° 45′ W.
Productus sulcatus, var. borealis, Hgt. J. R. D. S., i. t. 7, f. 1–4.
Spirifer arcticus, Hgt. J. R. D. S., i. t. 9, f. 1.

7. **Cape Lady-Franklin** (?), North Coast of **Bathurst Island,** Lat. 76° 40′ N., Long. 98° 45′ W.

Spirifer arcticus, Hgt.

Lithostrotion basaltiforme, Phil. J. R. D. S., i. p. 249, t. 11, f. 6.

8. **Ballast Beach, Baring Island,** Lat. 74° 30′ N., Long. 121° W.

Wood fossilized by brown hæmatite; structure distinct.

Cone of Spruce-Fir, fossilized by brown hæmatite.

9. **Princess-Royal Island,** in **Prince-of-Wales' Strait,** between Baring or Banks' Island and Prince-Albert Land, Lat. 72° 45′ N., Long. 117° 30′ W.

Nodules of clay-iron-stone, partly converted into brown hæmatite.

Native copper, in large masses, procured from the Esquimaux in Prince-of-Wales' Strait.

Pisolitic brown hæmatite.

Greyish-yellow sandstone, like that of Cape Hamilton and Byam-Martin Island.

Terebratula aspera, Schlotheim. J. R. D. S., i. p. 248, t. 11, f. 4.

10. **Cape Hamilton, Baring Island (Bank's Land).** Lat. 74° 15′ N., Long. 117° 30′ W.

Greyish-yellow sandstone, like that found *in situ* in Byam-Martin Island.

Coal, brownish, of lignaceous texture, consisting of very thin interstratified layers of brown coal and jetty black glossy coal; with a woody ring under the hammer. These characters hold good for all the known Arctic coal except the Tertiary coal of Disco, W. Greenland.

11. **Cape Dundas, Melville Island.** Lat. 74° 30′ N., Long. 113° 45′ W.

Coal (as above.)

12. **Cape Sir-James-Ross, Melville Island,** Lat. 74° 45′ N., Long. 114° 30′ W.

Sandstone, passing into blue quartzite.

13. **Cape Providence, Melville Island,** Lat. 74° 20′ N., Long. 112° 30′ W.

Yellowish-grey sandstone, and clay-iron-stone, passing into pisolitic hæmatite.

A specimen of Crinoidal Limestone, like that of Griffiths' Island; but the present currents and drift-ice could not bring it thence, as their set is constant from the W.

14. **Winter Harbour, Melville Island,** Lat. 74° 35′ N., Long. 110° 45′ W.

Yellow and grey fine-grained sandstone.

15. **Bridport Inlet, Melville Island,** Lat. 750° N., Long. 110° 45′ W.
   Coal with impressions of *Sphenopteris.*
   Ferruginous-spotted white sandstone.
   Clay-iron-stone, passing into brown hæmatite.

16. **Skene Bay, Melville Island,** Lat. 75° N., Long. 108° W.
   Bituminous coal, laminated; associated with brown crystalline limestone (with cherty beds), and grey-yellowish sandstone, passing into brownish-red sandstone.

17. **Hooper Island, Liddon's Gulf, Melville Island,** Lat. 75° 5′ N., Long. 112° W.
   Nodules of pure and heavy clay-iron-stone; associated with the usual ferruginous fine-grained sandstone and coal.

18. **Byam-Martin Island,** Lat. 75° 10′ N., Long. 104° 15′ W.
   Yellowish-grey sandstone, *in situ,* containing a ribbed *Atrypa,* allied to *A. primipilaris,* Von Buch, and *A. fallax* of the Carboniferous rocks of Ireland.
   Reddish limestone, with fragmentary *Atrypæ,* like the last.
   Coal, of the usual quality.
   Fine-grained red sandstone, passing into red slate [?].
   Boulders of scoriaceous hornblendic trap.

19. **Graham-Moore's Bay, Bathurst Island,** Lat. 75° 35′ N., Long. 102° W.
   Coal, of the usual quality.

(3.) Vol. i., p. 211.—In M'Clintock's "Reminiscences," &c. Second Expedition.—Coal, sandstone, clay-iron-stone, and brown hæmatite were found along a line stretching E.N.E. from Baring Island [Banks' Land], through the middle and S.-E. part of Melville Island, Byam-Martin Island, and the southern half of Bathurst Island [towards Grinnell Land and the Victoria Archipelago]. Carboniferous Limestone, with characteristic fossils, was found along the north coast of Bathurst Island, and at Hillock Point, Melville Island.

Vol. i., p. 199.—The sandstone of Byam-Martin Island is of two kinds, one red, finely stratified, passing into purple slate, and very like the red sandstone of Cape Bunny, North Somerset, and some varieties of the red sandstone and slate found between Wolstenholme Sound and Whale Sound, West Greenland, lat. 77° N. The other sandstone of Byam-Martin Island, is fine, pale-greenish, or rather greyish-yellow, and not distinguishable in Land-specimens, from the sandstone of Cape Hamilton, Baring Island [Banks' Land]. It contains numerous shells and casts of a Brachiopod, closely allied to *Terebratula primipilaris,* Von Buch, found abundantly at Gerolstein in the Eifel. On the whole, Dr. Haughton inclines to the opinion that the sandstones, limestone, and coal of Byam-Martin Island, and the corresponding rocks of Melville Island, Baring Island [Banks' Land], and Bathurst Island, are low down in the Carboniferous System;

and that there is in these northern Coal-fields no sub-division into "red sandstone," "limestone," and "coal-measures," such as prevail in the West of Europe.

If the different points where coal was found be laid down on a map, we have in order, proceeding from the S.W., Cape Hamilton, Baring or Banks' Island; Cape Dundas, Melville Island, south; Bridport Inlet and Skene Bay, Melville Island; Schomberg Point, Graham-Moore Bay, Bathurst Island; a line joining all these points is the outcrop of the coal-beds of the south of Melville Island, and runs E.N.E.

Vol. i., pp. 213, 214.—At Cape Lady-Franklin, and at many other localities along the north shore of Bathurst Island, Carboniferous fossils in limestone, clay-iron-stone balls, passing into brown hæmatite, cherty limestone, and earthy fossiliferous limestone, with the same species of *Atrypa* as at Byam-Martin Island, were found in abundance by Sherard Osborn, Esq., Commander of H.M.S. "Pioneer."

(4.) Vol. i., p. 223.—M'Clintock's "Reminiscences," &c. Third Expedition.—On landing [at Point Wilkie], I found the beach low, composed of mud with the foot-prints of animals frozen in it. A few hundred yards from the beach there are steep hills, about 150 feet in height, and upon the sides of these, in reddish limestone casts of fossil Shells abound. Inland of these the ordinary pale Carboniferous sandstone and cherty limestone reappeared. The fossils are all small, and of only a few varieties, some being Ammonites, but the greater part Bivalves. They differed from any I had met with before, and the rock was almost brick-red. I picked up what appeared to be fossil bone (*Ichthyosaurus*?), only part of it appearing out of the fragment of the rock. Point Wilkie appears to be an isolated patch of Liassic age, resting upon Carboniferous sandstones and limestones, with bands of chert, of the same age as the limestones and sandstones of Melville Island. The eastern shore of Intrepid Inlet is composed of this formation; while the western, rising into hills and terraces, is of the underlying Carboniferous epoch. At the western side of Intrepid Inlet I found upon the ice a considerable quantity of white asbestos, but did not ascertain whence it had been brought.

Vol. i., p. 238.—M'Clintock's "Reminiscences," &c. Third Expedition. (Also M'Clintock's "Fate of Franklin," &c.; Haughton's Appendix. 1859.)

Wilkie Point, Prince-Patrick's Land, Lat. 76° 20′ N., Long. 117° 20′ W.

Lias Fossils.

Ammonites Macclintocki, Haughton. J. R. D. S., i. t. 9, f. 2–4. (*See* also Dr. Haughton's *Manual Geol.*, 1865, p. 141.)
Monotis septentrionalis, Hgt. J. R. D. S., i. t. 9, f. 6, 7.
Pleurotomaria, sp. J. R. D. S., i. t. 9, f. 8.
Univalve, cast. J. R. D. S., i. t. 9, f. 5.
Nucula, sp.

(5.) PROF. HAUGHTON on the Geological Results of the Voyage of the "Fox." *Op. cit.*

Vol. iii., p. 53. During the voyage of the "Fox," and the sledge-journeys made by her officers from Bellot's Strait, in the autumn of 1858 and spring of 1859, many fossils were found, which were brought home by Captain M'Clintock, and presented to the Royal Dublin Society. These specimens complete our knowledge of the south and east shores of North Somerset, of the east and west shores of Boothia Felix, of King-William's Land, and to some extent, of the south and east shores of Prince-of-Wales Island. I have also received through Captain M'Clintock, a few specimens collected by Sir E. Belcher, in the extreme north of the Queen's and Belcher Channels. By the aid of these I have been enabled still further to complete the geological map of the Arctic Archipelago, which I published in vol. i., of this Society's Journal in February 1857.

The whole of North Somerset, Boothia Felix, King-William's Island and Prince-of-Wales Land is thus proved to be of Silurian age, although the evidence as to whether it is of Upper- or Lower-Silurian age is contradictory, as characteristic fossils of both epochs are found throughout the whole area.

From Sir E. Belcher's specimens it is evident that the Carboniferous rocks, including seams of ironstone and coal, extend through Buckingham Island, to the north-east, in the Victoria Archipelago. The following lists contain the names of the principal specimens identified :—

1. **North Cornwall.** Lat. 77° 30′ N., long. 95° W.

This land lies to the N. of Exmouth Island, on which the vertebræ of *Ichthyosaurus* were found. (*See* Belcher's "Last Arct. Voy.," ii., p. 389.) The specimens appear to be stones collected on the ground. They consist of earthy limestone, quartz-pebbles, and a kind of granite not found elsewhere in the Arctic Regions; it is composed of quartz, black mica, and grey translucent felspar. Sir E. Belcher states that the ravines consist of sandstone, and contain large masses of clay-iron-stone, septaria, nodules of iron-pyrites, and coal, not *in situ*; together with sub-fossil shells, strewed on the ground.

2. **Buckingham Island.** Lat. 77° 10′ N., long. 91° W.

Composed of clay-iron-stone and dark ferruginous shales. The specimens of coal and iron-stone brought from this island by Sir E. Belcher bear a striking resemblance to those found by M'Clintock in Liddon Gulf, Melville Island; and it is worthy of remark that a line drawn from Liddon Gulf to Buckingham Island coincides with the strike of the Carboniferous beds laid down on the map [not reproduced here]. Boulders of a binary granite, composed of grey quartz and pinkish-red felspar, were also found on this island.

3. **Village Point.** Lat. 76° 50′ N., long. 97° W.
   Coal of the usual description found in the Parry Islands was found at this point, which lies in the line joining Buckingham Island with Liddon Gulf.

4. **Depôt Point, Grinnell Land,** Lat. 77° 5′ N.
   Carboniferous Limestone.
   Lithostrotion basaltiforme, Phil. Belcher's Voyage, t. 36, f. 4; Journ. R. D. S., i. t. 11, f. 6.
   Zaphrentis ovibos, Salter. Belcher's Voy. t. 36, f. 5.

5. **Depôt Bay, Bellot's Strait.** Lat. 72° N., long. 94° W.
   Silurian Dolomite, white, saccharoid, with large rhombohedral crystals of calcspar.
   Maclurea arctica, Hgt. J. R. D. S., iii. t. 3, f. 1, 2.
   Cyathophyllum helianthoides, Goldfuss.

6. **Cape Farrand, East side of Boothia,** Lat. 71° 38′ N., long. 95° 35′ W.
   Silurian limestone; grey, earthy.
   Atrypa phoca, Salter. Sutherland's Journal, t. 5, f. 3.
   Loxonema Rossi, Hgt. J. R. D. S., i. t. 5, f. 6, 8, 9, 10, 11.
   Favosites Gotlandica, Goldf. (*F. Niagarensis*, Hall.)
   Atrypa, ribbed sp.
   Cyrtoceras, sp.

7. **West shore of Boothia.** Lat. 70° to 71° N. Containing the Magnetic Pole.
   Earthy limestone, and cream-coloured chalky dolomite.
   Atrypa phoca, Salter. Sutherland's Journ. t. 5, f. 1–3; J. R. D. S., i. t. 5, f. 3, 4, 7.
   Loxonema Rossi, Hgt.
   L. Salteri. Sutherland's Journ. i. 6, f. 18.
   Favistella Franklini, Salter. J. R. D. S., i. t. 11, f. 1; Sutherland's Journ., t. 6, f. 3.

One of the most remarkable facts brought to light by M'Clintock's geological exploration of the Arctic regions during the Voyage of the "Fox," is the occurrence of dolomite or magnesian limestone, covering large areas in almost horizontal beds. It abounds in fossils; and is an almost pure dolomite, or union of carbonates of lime and magnesia in equal atomic proportions. To my mind this fact is of as much, if not more, importance in identifying the Silurian strata of Boothia Felix, King-William's Land and Prince-of-Wales Land, as any identification of fossils could possibly be. Considered in reference to Dr. Bigsby's papers on the Silurian rocks of New York, recently published by the Geological Society of London,* it is a fact of great interest.† The following is the composition of the *Dolomite* from the western shore of Boothia Felix; those of the other dolomites are given in their proper places.

---

* Quart. Journ. Geol. Soc. London, vol. xiv. pp. 335, 427; xv. p. 251; 1858–9.

† Dr. H. Rink has noted the existence of a *Dolomite* low down on the west coast of Greenland, *above*, p. 496.—Editor.

Dolomite from Boothia Felix.

	Per cent.	Atoms.
Argil - - - -	2·22	
Peroxide of iron and alumina - -	0·29	
Carbonate of lime - - -	54·92	110
Carbonate of magnesia - -	42·57	97
	100·00	

Carbonate of lime : carbonate of magnesia :: 110 : 97=1·134 : 1.

8. **Fury Point,** Lat. 72° 50′ N., Long. 92° W.
   Silurian Limestone; grey, earthy.
   Cromus arcticus, Hgt. J. R. D. S., i. t. 6, f. 1–5.
   Maclurea arctica, Hgt. J. R. D. S., iii. t. 3, f. 1, 2.
   Stromatopora concentrica, Goldf.
   Cyathophyllum helianthoides (Goldf.).
   Petraia bina (Lonsd.).
   Favosites Gotlandica, Goldf. (*F. Niagarensis*, Hall).
   Cyathophyllum cæspitosum, Goldf.
   Favistella Franklini, Salter.
   Strephodes Austini, Salter. Sutherland's Journ., t. 6, f. 6.
   Atrypa phoca, Salter.
   Chætetes lycoperdon, Hall. J. R. D. S., iii. pl. 4, f. 4.

9. **Prince-of-Wales Land,** Lat. 72° 38′ N., Long. 97° 15′ W.
   Silurian Limestone, grey and earthy; near the junction with the red arenaceous limestone.
   Favosites Gotlandica, Goldf. (*F. Niagarensis*, Hall; *Chætetes arcticus*, J. R. D. S., i. t. 11, f. 3, 4.)
   Stromatopora concentrica, Goldf.
   Cyathophyllum, sp.

The Dolomite found by Capt. Young on the extreme west of Prince-of-Wales Land has the following composition:—

	Per cent.	Atoms.
Argil - - - - -	6·93	
Peroxide of iron and alumina -	0·65	
Carbonate of lime - - -	46·80	94
Carbonate of magnesia - -	45·62	108
	100·00	

Carbonate of lime : carbonate of magnesia : : 94 : 108=0·870 : 1.

10. **West Coast of King-William's Island.**
   Earthy limestone, grey; and cream-coloured compact dolomite.
   Loxonema Rossi. J. R. D. S., i. t. 5, f. 6, 8–11; iii. t. 4, f. 6.
   Catenipora escharoides, var. agglomerata, Hall. J. R. D. S., iii. t. 4, f. 5. Dolomite.

Ormoceras crebriseptum, Hall. J. R. D. S., iii. t. 1, f. 3.
Maclurea arctica. J. R. D. S., iii. t. 3, f. 1, 2. Dolomite.
Atrypa, sp.
Syringopora geniculata.
Clisiophyllum, sp.
Orthis elegantula.
Huronia vertebralis, Stokes. J. R. D. S., iii. t. 2, f. 1, 2. Dolomite.
Receptaculites Neptuni, Defrance. J. R. D. S. iii. t. 3, f. 3.

Composition of the Dolomite full of *Catenipora escharoides*) from the western side of King-William's Island:—

	Per cent.	Atoms.
Argil, peroxide of iron and alumina -	5·40	
Carbonate of lime - - -	52·91	106
Carbonate of magnesia - -	41·67	94
	99·98	

Carbonate of lime : carbonate of magnesia : : 106 : 94 = 1·127 : 1.

6. NOTE.—The GEOLOGY of the PARRY ISLANDS, &c. From the Map and descriptions given by the Rev. Prof. Dr. S. HAUGHTON.

The eastern portion of *North Devon* consists of **Crystalline (granitic and other) rocks,** corresponding with those round about *Wolstenholme Sound* (from Cape York to Granville Bay), and the Carey Isles, on the opposite coast of Baffin's Bay. The western part of *North Devon,* and the *Cornwallis, Griffiths,* and *Lowther* Islands further to the west, and, to the south, the northern portion of *Cockburn Land, North Somerset* (excepting some crystalline rocks on the coast of *Peel Sound,* with offstanding islands), all *Boothia, King-William's Land,* and *Prince-of-Wales Land,* are **Silurian.** *Banks'* (*Baring*) *Land,* the northern portion of *Eglinton, Melville,* and *Bathurst* Islands (approximately marked off by 76° parallel), and all *Byam-Martin I.,* are occupied by the **Carboniferous Sandstone, Limestone, Ironstone, and Coal** (" Ursa Stage," O. Heer); and these reach away to the N.E., into the *Victoria Archipelago* and *North Cornwall.* The **Carboniferous Limestone** constitutes the eastern part, at least, of *Prince-Patrick Island* (exclusive of a small Jurassic patch), the northern third of *Eglinton Island,* most part of the two northern promontories of *Melville Island,* the three northern promontories of *Bathurst Island,* and the north-eastern part of *Grinnell Land* (about 77° N., 96° W.).

Jurassic (Lias) rocks occur in two localities: (1) at *Point Wilkie,* on the Eastern coast of *Prince-Patrick Island;* and (2) between *Grinnell Land* and *North Cornwall,* at *Exmouth Island,* 77° 10′ N., 96° W. Belcher's "Last of the Arctic Voyages," vol. i. p. 106 ; ii. p. 391.

LXXXV.—GEOLOGY OF THE PARRY ISLANDS, &c.

(From Sir EDWARD BELCHER'S "Last of the Arctic Voyages," 2 vols. 1855.)

1. **Arctic Carboniferous Fossils,** collected by the Expedition under SIR E. BELCHER, C.B., 1852–4. Described by J. W. SALTER, Esq., F.G.S., "Last of the Arctic Voy.," vol. ii. pp. 377 *et seq.* 1855.

Fusulina hyperborea, n. sp., p. 380, t. 36, f. 1–3. Depôt Point, Grinnell Land (77° N., 95° W.).

Stylastræa inconferta, Lonsdale; p. 381, t. 36, f. 4. Depôt Point.

Zaphrentis ovibos, n. sp., p. 382, t. 36, f. 5. Depot Point; and similar specimens from Princess-Royal Island, and the entrance of Jones's Sound.

Clisiophyllum tumulus, n. sp., p. 383, t. 36, f. 6. Depôt Point.

Syringopora aulopora, sp., p. 385, t. 36, f. 7; on large corals. Depôt Point.

Fenestella arctica, n. sp., p. 385., t. 36, f. 8. Depôt Point.

Spirifer Keilhavii, Von Buch, p. 387, t. 36, f. 9, 10, 11. Depôt Point; and in red limestone, Exmouth Island.

Productus cora, D'Orbigny, p. 387, t. 36, f. 12. In red limestone, Exmouth Island.

Productus semireticulatus, Martin, var. frigidus, nov., p. 388, t. 36, f. 13, 14. Depôt Point.

2. FOSSIL BONES from EXMOUTH ISLAND.

"The Last of the Arctic Voyages," &c. By Sir Edward Belcher, C.B., 2 vols., 1855. Vol. i. page 106.

The following extract is descriptive of Exmouth Island, in Lat. 77° 16′ N., Long. 96° W., of which a sketch (Pl. IV., facing page 105) is given.

"The formation is red sandstone, capped about 20 feet on the " summit by fossiliferous limestone, in which some large Bivalves " (Pectens, &c.), and some bones were found, unfortunately " broken before they were brought to me. Beneath this lime- " stone, the rock is swinestone to about three-quarters from the " base, the entire height being 567 feet. In the sandy bed of one " of the large gullies a large ball of iron-pyrites was found, at " first mistaken by one of my crew for a six-pound shot, and " brought to me as belonging to one of the missing ships. Some " very slight traces of coal were noticed at the wash of the sea, " but none *in situ* on the Island."

PROFESSOR OWEN, F.R.S., F.G.S., describes, in vol. ii., page 389 (Appendix), some remains (vertebræ and pieces of ribs) of *Ichthyosaurus* (pl. 31), discovered in this Island. Sir E. BELCHER remarks at page 391, with reference to these, that they were found 570 feet above the sea-level, and above the limestone (about 114 feet thick) covering the "friable disintegrating sandstone," which forms the base of the island. The strata have a westerly dip of 7°.

[At page 379, vol. ii., J. W. SALTER, Esq., F.G.S., states that the Carboniferous *Productus Cora, Spirifer Kielhavii*, &c., were found on the top of Exmouth Island (in the limestone), and that "close upon this again" lay the Ichthyosaurian bones.]

The hasty examination given under the circumstances led to the conjecture that fossils [Lias ?] were only to be found in the area marked out by Exmouth, Table, Ekins, and Princess-Royal Islands, and that portion of the mainland lying between the latter and Cape Briggs ; but this coast, at Depôt Point and P.-Royal I., consists of Carboniferous Limestone (*above*, pp. 548, 551).

3. NORTH CORNWALL, &c.

Vol i. p. 111. The ravines are deeply channelled out of a very friable sand-stone, and in the bottom were noticed large masses of clay-iron-stone, septaria, and nodules of iron-pyrites. Coal was also found, but disseminated, and impossible to trace *in situ*.*

Bivalves, apparently of *recent origin*, and having the cartilage-hinge perfect, were abundant. . . . . . . Birds may have placed these shells; but with . . . this climate, prevailing ice, and the scarcity of animal life especially, this is scarcely credible . . . . This friable sandstone and sand, interspersed on the surface with boulders of granite, and almost *garnet masses*, constitutes the principal features of the land on West Cornwall.

While constructing a cairn at Glacier Bluff in N.-E. Grinnell Land, Northumberland Sound (vol. i. p. 125,) gold was discovered in "a heavy piece of quartz."†

Vol. i. pp. 316–320. At Buckingham Island (Victoria Archipelago), spits of gravel radiating from the embouchure of an old river-valley, now dry, are mentioned, and other evidence of former rivers and lakes in Grinnell Land and North Devon, and of submergence of mountain-ranges, when "Whales and other objects "were deposited on elevations of 500 feet and upwards." Terraces, mud-flats, gravel-ridges, pebbles of chalcedony, and carcases and bones of Whales at 80 or 500 feet elevation are also alluded to; and the invariable dip of the strata at about 5° to northward is noticed.

In some cases vertebræ and skulls of Whales have been set up as land-marks. Such are not to be mistaken for drifted remains.

(Dr. Walker of the "Fox" Expedition mentions a Whale's skeleton found at an elevation of 80 feet.)

At Mount Parker (Belcher, vol i., p. 261), the "head and pro- "bably the entire skeleton of a Whale," were discovered at a height of 500 feet above the sea level.

At Cape Disraeli (i. p. 266) an embedded Whale was found.

---

* Loose coal was found also to the south in Grinnell Land, and some bituminous shale further to the south-west (pages 40 and 97).

† Gold was also brought home by Dr. Rae from the Arctic Regions.

LXXXVI.—NOTES ON POLARIS BAY. BY DR. BESSELS.*

From the "Bulletin de la Soc. de Géographie," March 1875. Paris. pp. 291–9.

The geological formation of "Polaris Bay" and its neighbourhood consists of Silurian limestone with few fossils.

At an elevation of 1,800 feet, not only Drift-wood was found, but Shells (*Mya*, &c.), of species still existing in the neighbouring seas.

In small lakes, numerous in that region, were found *marine* Crustacea living in fresh water; and thus bearing witness to the uprise of the northern coasts of Greenland.

Erratic blocks occur in great numbers, different in character from the rocks of the surface. Among them are blocks of granite, gneiss, &c., derived from South Greenland, and certainly not transported by glaciers, but by floating ice, showing that the current in Davis' Strait was to the North formerly, and not to the South as now.

NOTE.—Mr. C. E. DE RANCE, F.G.S., has lately given a series of papers in *Nature* (Macmillan, London), vol. xi., pp. 447, 467, 492, and 508 (April 1875), on the geological, glacial, and other phenomena of the Arctic Regions. In a condensed form he here presents what is known of the Ice of Greenland, the Cryolite, the Cretaceous and Miocene formations of Mid-Greenland, the geology of East Greenland, and the Ovifak Meteorites. The geology of the Arctic-American Archipelago and of Grinnell Land (Hayes, &c.), the glacial conditions and geology of Spitzbergen and Bear Island, the extent of the Carboniferous deposits in N.-W. America, and the former continental areas, are also treated of. Lastly, some remarks are offered on the homotaxeous relationship of the Tertiary and Cretaceous Formations of Greenland with those of other countries.

---

* For some other notes on the results of the Voyage of the "Polaris," *see above*, pages 321–3.

§ III.

East Greenland, &c.

LXXXVII.—The Mammals and Fishes of East Greenland. By Dr. W. Peters. 1874. [("Die zweite deutsche Nordpolarfahrt,"* &c., vol. ii. pp. 157–174, 2 pls.)

These Lists are extracted without the synonyms and descriptions given by Dr. Peters.

A. MAMMALIA.

Feræ.

1. Ursus (Thalassarctos) maritimus, Lin. Ice Bear.
2. Mustela (Putorius) erminea, Lin., var. Putorius novæboracensis, De Kay. Ermine.
3. Canis (Vulpes) lagopus, Lin. Polar Fox. Ice Fox. Rock Fox.

Pinnipedia.

4. Odobænus rosmarus, Lin. Walrus (Pl. I, fig. 1, 2). All along the coast, especially at Sabine and Walrus Islands. In the stomachs of some were 500 or 600 of the animals of *Mya truncata* without the shells. The young Walrus remains two years with its mother, until its tusks are grown long enough to be used in grubbing up the shell-mud at the sea-bottom.
5. Phoca grœnlandica, Müller. Greenland Seal, Sadler. The stomachs of some contained remains of Crustaceans and Fish; others had only *Crustacea* (*Gammarus arcticus* and *Themisto*).
6. Phoca barbata, Müller. Bearded Seal.
7. Cystophora cristata, Exleben. Hooded Seal.

Glires.

8. Myodes torquatus, Pallas. Lemming.
9. Lepus glacialis, Leach. Polar Hare (Pl. II., fig. 1).

Pecora.

10. Cervus tarandus, Lin. Reindeer. From Kuhn Isl. on the N. down to the bottom of the Fjord (Franz-Joseph), and on Sabine Island, Reindeer seen in herds of 20 or less; most numerous in the southern

---

* Die zweite deutsche Nordpolarfahrt in den Jahren 1869 und 1870 unter Führung des Kapitän Karl Koldewey. Herausgegeben von dem Verein für die deutsche Nordpolarfahrt in Bremen: zwei Bände.— Zweiter Band. Wissenschaftliche Ergebnisse. Mit 31 Tafeln und 3 Karten. Erste Abtheilung (Botanik und Zoologie). 8vo., Leipzig. 1874. Dr. A. Pansch's "Anthropologie," p. 144, is not here noted, that subject coming within the scope of the Royal Geograph. Society's Manual. The Birds (pp. 178–239) are included in Prof. Newton's Catalogue, *above*, p. 94.

district. One killed in the middle of June on Sabine Isl. had already its summer coat nearly complete. The antlers were still covered with hair, and had not sprouted at the points. Quite young Deer were not seen.

11. Ovibos moschatus, Zimmermann. Musk-ox. Wandering all along the coast in herds of about 10 to 20. On the eastern half of Shannon, Pendulum, and Sabine Islands, they seem only in summer to occur singly. They were seen by the end of March on the south-west part of Shannon Isl. and on the mainland; therefore they probably remain there all the winter. A calf, some days old, was met with on the 26th March, and several calves were with the herds in the beginning of June, from which a rough skin could still be got. Traces of the *Ovibos* were observed both on the plains and the hills. They feed on grass, herbs, and mosses.

Cetæ.

12. Balæna mysticetus, Lin. Greenland Whale.
13. ? Balænoptera boops, Eschricht.
14. Monodon monoceros, Lin. Narwhal.
15. ? Delphinus globiceps, Cuv.

B. PISCES (pp. 169–174).

Cataphracti.

1. Cottus hexacornis, Richardson.
2. C. porosus, Cuvier, Val.
3. Icelus hamatus, Kröyer.

Discoboli.

4. Liparis gelatinosus, Pallas (Pl. I., fig. 3).

Gadini.

5. Gadus glacialis, Peters, n. sp. Sabine Island.

Salmonini.

6. ? Salmo Hoodii, Richardson. Upper freshwater lake on Sabine Island.

LXXXVIII.—Remarks on some Skulls of Eskimo Dogs. By Hermann von Nathusius. 1874.

("Die zweite deutsche Nordpolarfahrt," &c., vol. ii. pp. 175–177.)

[Extract, pp. 176, 177.]

"It is particularly interesting to notice the variability in this series of skulls, which most probably belonged to animals of one and the same race. We may well suppose that the Dogs under notice were to some extent congenerous; and, as the extinct Eskimo could not have been in active commerce with other tribes, they could scarcely have crossed their Dogs with another breed. The eight skulls, however, indicate more or less striking differences in those parts which in Dogs are especially variable, irrespective, of course, of those due to relative age, as, for example, the occipital crests.

"The nasal bones either pass higher up into the forehead than the frontal edges of the maxillary, or they do not reach a line which these edges of both maxillaries touch; the orbital borders are more or less raised, and therewith the frontal foranina are more or less large; the forehead between the orbital processes is more or less deeply concave, or nearly flat; the orbits are smaller or larger; the zygomatic arch is more or less wide and high, but only one of the skulls has this perfect. These are all characters by which to distinguish the Dogs from Wolves. The nasal bones are all broken at their distal ends.

"Thus this little collection contributes to the observation that the domestic animals, particularly the Dog, are extremely variable as to the form of skull within limits of a race-type. The skulls are sufficiently well preserved to show that none had a third molar in the upper jaw, a feature not so uncommon as usually thought in studying Dogs' heads."

LXXXIX.—The TUNICATA of EAST GREENLAND. By Prof. Dr. C. KUPFFER. 1874.

("Die zweite deut. Nordpolarfahrt," vol. ii. pp. 244, 245.)

Cynthia villosa (O. Fabricius). Germania Harbour.
C. Adolphi, n. sp. Shannon Island.
Descriptions and synonyms are given in full by the author, *loc. cit.*

XC.—The MOLLUSCS, WORMS, ECHINODERMS, and CŒLENTERATES. By Dr. KARL MÖBIUS. 1874.

("Die zweite deut. Nordpolarfahrt," vol. ii. pp. 246–961, 1 pl.)

These Lists have been extracted without the descriptive remarks given by the author.

MOLLUSCA (pp. 248–253).

Gasteropoda.

1. Chiton albus, L. Fabr. F. Gr., p. 422. Forbes & H., pl. 62, f. 12. Jeffreys, Br. Conch., V., pl. 56, f. 3. *Distrib.* Spitzbergen to the Kattegat and Britain, Massachusetts (Gould & Binney). To 550 faths.

2. Lepeta cæca, Müll. Zool. Dan., I., 12, pl. 12, f. 1–3. Jeffreys, B. C., III., 252; V., pl. 58, f. 6, 7. Walrus Island, 25 faths. *Distr.* Circumpolar, Sitka, North Japan Sea (Schrenck), Spitzbergen to the Kattegat and Britain.

3. Trochus grœnlandicus, Chemn. Conch. Cab., V., 108, f. 1671; F. & H. Brit. Moll., pl. 68, f. 1, 2. Sabine, Jackson, and North Shannon Islands, Germania Harbour, 2–30 faths. *Distr.* Labrador to Massachusetts, White Sea to the Kattegat and Britain.

4. Trochus helicinus, Fabr. F. Gr., p. 393. F. & H. Br. M., pl. 68, f. 4, 5, and pl. CC., f. 4. Jeffr. B. Conch., III., p. 295; V., pl. 61, f. 4. Sabine, Jackson, North Shannon, and Walrus Islands, 427 faths. *Distr.* Circumpolar, Massachusetts, Sea of Japan, Norway to the Kattegat and Britain.

5. Pleurotoma pyramidalis, Ström. (pl. 1, f. 1, 3). Ström. Nye Saml. Kong. Danske Vid. Selsk. Skrifter, III., 1788, p. 296, f. 22, *Buccinum pyramidale.* Mörch, Moll. Grönl., calls it *Pleurotoma pyramidale* and refers to *Defrancia Vahlii* as a synonym. Beck in Möller's Index Molluscorum Groenl., p. 86. Kröyer, Naturh. Tidsskrift, IV., 1842–3. Möller's diagnosis applies to the foregoing. Sabine, Jackson, and Shannon Islands, 4–30 fathoms. *Distr.* Greenland to Massachusetts, Spitzbergen to Norway (Bergen).

6. Fusus propinquus, Alder. F. & H. Br. M., pl. 103, f. 2. Jeffr. B. Conch., pl. 86, f. 3. Middendorf, Malac. Ross., II., p. 471, pl. 4, f. 13, *Tritonium islandicum,* var. *sulcata.* Sabine and Clavering Islands, Germania Harbour, 2–20 faths. *Distr.* Russian Arctic to the Kattegat and Ireland.

7. Buccinum undatum, L. Syst. Nat., ed. 12, p. 1204. F. & H. B. Mol., III., p. 401, pl. 109, f. 3, 5, pl. LL., f. 5. Jackson and Clavering Islands, 4 faths. *Distr.* Circumpolar, North Atlantic on the European and North American coast, Mediterranean, Okotsk Sea.

8. Scalaria grœnlandica, Chemn. Conch. Cab., XI., f. 1878–9. Kiener, Gen. Scalaria, p. 18, *Sc. platicostata,* pl. 7, f. 21. F. & H. Br. M., pl. 70, f. 5, 6. North Shannon Island, 30 faths. *Distr.* North Arctic, Norway (to Bergen).

9. Natica clausa, Brod. et Sow. Beechey's Voy., pl. 34, f. 3, and pl. 37, f. 6. Gould (Binney) Invert. Massach., 1870, p. 343, f. 612. North Shannon, Sabine, and Jackson Islands, Clavering Beach, 30 faths. *Distr.* Circumpolar, Sea of Japan, Finmark.

10. Cylichna cylindracea, Penn. (pl. 1, f. 4–9.) F. & H. Br. M., pl. 114 B., f. 6. Jeffr. B. Conch., pl. 93, f. 4. Jackson Island, 4 faths. *Distr.* Finmark, Canaries, Mediterranean, 3–160 faths.

11. Clione limacina, Phipps, Voy. North Pole, 1773. Martens, Spitzb. u. Grönl. Reis., 1671, p. 128, "See-Gots-Pferd" (Neptune's horse), pl. P., f. *f*. Eydoux et Souleyet, Voy. "Bonite," Moll. Atlas, pl. 15 *bis*, f. 1–19, *Clio borealis*, Brug. Rang, Descr. Gen. Pter. et esp. Clio, Ann. Sc. n., 1825, V., p. 285, pl. 7, f. 2, *Cl. Miquelonensis* of Newfoundland. *Distr.* West Greenland, Massachusetts.

Lamellibranchia.

1. Modiolaria discors, L. Syst. Nat., ed. 12, p. 1159. F. & H. Br. M., II., p. 195, pl. 14, f. 5, 6. Shannon, Sabine, and Clavering Islands, 4–30 faths. *Distr.* Circumpolar, North America, North Japan Sea, Mediterranean, North Sea, West Baltic.
2. Cardium grœnlandicum, Chemn. Conch. Cab., VI., pl. 19, f. 198. Sow. Conch. Man., p. 70, f. 123. Gould (Binney), Invert. Massach., p. 144, *Aphrodite grœnlandia*, f. 454. *Distr.* West Greenland, Massachusetts, Behring Straits.
3. Astarte borealis, Chemn. (*Ast. arctica*, Gould). Chemn. Conch. Cab., VIII., pl. 39, f. 412. Philippi, Abbild. u. Besch. Conch., II., pl. 1, f. 12. Gould (Binney), Invert. Massach., f. 433, *Astarte semisulcata*. Meyer u. Möbius, Fauna der Kieler Bucht, II., p. 1. Shannon, Sabine, Clavering, and Jackson Islands, 3–10 faths. *Distr.* Arctic Sea, from Behring Straits to Lapland, Norway, Baltic (to Bornholm).
4. Astarte sulcata, Da Costa, Br. Conch., p. 192. F. & H. Br. M., I., p. 452, pl. 30, f. 5, 6, *A Damnoniensis*. *Distr.* Circumpolar, Sea of Okotsk, Canaries, North-east America, Kiel and Flensburg Harbours.
5. A. compressa, Montagu, Test. Br. Suppl., p. 43, pl. 26, f. 1. F. & H. Br. M., I., p. 464, pl. 30, f. 1–3. Jackson Island. *Distr.* North Arctic, North-east America, Spitzbergen, Norway, Britain, Kiel Harbour.
6. A. crebricostata, Forbes, Ann. N. H., XIX., 1847, p. 98, pl. 9, f. 4. Gould (Binney), Invert. Massach., p. 126, f. 440. Shannon Island, 30 faths. *Distr.* West Greenland, Norway, North-east America.
7. Venus astartoides, Beck. Middend. Mal. Ross., III., p. 572. Phil. Abb. u. Beschr. Conch., III., p. 61, pl. 9, f. 4. Gould (Binney), Inv. Mas., p. 136, f. 447, *Tapes fluctuosa*. Shannon and Jackson Islands, 4–30 faths. *Distr.* West Greenland, Massachusetts, Seas of Okotsk and Japan.
8. Mya truncata, L. Syst. Nat., ed. 12, p. 1112. F. & H. Br. M., I., p. 163, pl. 10, f. 1–3. Dr. Pansch found in the stomach of a Walrus 500 bodies of this Mollusc, and only one piece of shell. Near the ice-hole where the Walrus came out lay heaps of shells. Sabine Island, 10–20 faths. *Distr.* Circumpolar, Sea of

Okotsk, North-east America, Britain, Bay of Biscay, Norway, West Baltic.

9. Saxicava rugosa, L. Syst. N., ed. 12., p. 1113 (*rugosa*), p. 1116 (*arctica*). F. & H. Br. M., I., p. 141, pl. 6, f. 7, 8. Shannon Island, 30 faths. *Distr.* Circumpolar, Japan and China Seas, North-east America, Sitka, Mediterranean, Canaries, North Sea, West Baltic.

BRACHIOPODA.

1. Terebratula psittacea, Gmel. Middend. Mal. Ross., III., pl. 11, f. 11–17. Gould (Binney), Invert. Mass., p. 210, f. 501. Jackson Island. *Distr.* West Greenland, Massachusetts, Spitzbergen, Finmark.
2. T. cranium, Müller, Zool. dan. Prodr., p. 249. Jeffr. Br. C., II., p. 11.; V., pl. 19, f. 1. Shannon Island, 30 faths. *Distr.* Norway, Shetland, Finmark to the Kattegat.

VERMES (pp. 253–258).

Annelides.

1. Polynoë cirrosa, Pallas. Sabine Island. *Distr.* Finmark to the Kattegat, Britain, W. Greenland, and Spitzbergen, 3–120 faths.
2. P. cirrata, P. Sabine Island and Clavering Strait, 4–12 faths. *Distr.* Circumpolar, Baltic to Sitka.
3. Nereis diversicolor, Müll. Shannon Island. *Distr.* Norway, North Sea, Baltic.
4. N. pelagica, L. The *Heteronereis* form has 22 segments in the anterior body (Rathke and Malmgren noted 16, Oersted 20), and 65 posteriorly. *Distr.* West Baltic to Finmark, Spitzbergen, Iceland, W. Greenland.
5. Leipoceras uviferum, nov. gen. et. sp. [Loc. ?]
6. Scolophus armiger, Müll. Sabine Island. *Distr.* Spitzbergen to the Baltic, North France.
7. Travisia Forbesii, Johnst. Sabine Island. *Distr.* West Greenland, Spitzbergen to the West Baltic, Scotland.
8. Scalibregma inflatum, Rathke. [Loc. ?] *Distr.* The Kattegat to Spitzbergen, Scotland, W. Greenland, 5–280 faths.
9. Thelepus circinatus, Fabr. Sabine Islands, 20 faths. *Distr.* Mediterranean, Britain, Kattegat to Finmark, Iceland, Spitzbergen, and W. Greenland.
10. Protula media, Stimpson. SabineIsland, 20 faths. *Distr.* Grand Manan (Bay of Fundy, 45° N. lat.).
11. Serpula spirorbis, Müll. Shannon Island. *Distr.* W. Greenland, North Sea, Baltic.

12. Chone infundibuliformis, Kröyer. Sabine Isl., 2½ faths. *Distr.* Finmark, Spitzbergen, W. Greenland, 15–40 faths.

Gephyrea.

Priapulus caudatus, Lam. (Ehlers.) [Loc. ?]

Turbellaria.

Polystemma roseum, Müll. Clavering Straits, 15 faths. *Distr.* Norway, the Sound, W. Baltic.

Nematodes.

Ascaris mystax, Zed. (*marginata*, Rud.). From the intestine of *Canis lagopus*, November 1869.

Cestodes.

1. Tetrabothrium anthocephalum, Rudolphi. From the intestine of *Cystophora cristata*, July 1869. Fabricius found it in *Phoca barbata.*
2. Tænia expansa, Rud. From intestine of *Ovibos moschatus.* Has musky odour. Found in many Ruminants of all zones.
3. T. cœnurus, Küchenmeister. In intestine of *Canis lagopus*, in September, November, and December, at many places in East Greenland.

Supplement.—In the North Sea Dr. Pansch collected *Epibdella hippoglossi*, Müll., from the skin of *Hippoglossus vulgaris* and *Rhombus maximus;* and *Ascaris clavata*, Rud., in the stomach and first intestine of *Gadus morrhua.*

ECHINODERMATA (pp. 258–260).

Holothurioidea.

1. Myriotrochus Rinkii, Steenstrup. Germania Harbour, 2 faths., October 1869. *Distr.* W. Greenland, down to 10 faths.

Echinoidea.

1. Echinus Droebrachiensis, Müll. Clavering Isl., 15 faths. *Distr.* Circumpolar, Newfoundland, Gulf of Georgia, White Sea, Kamtschatka, Sea of Okotsk, North Cape to the Sound, Britain.

Asterioidea.

1. Asteracanthion albulum, Stmps. Sabine Isl. *Distr.* W. Greenland, Grand Manan (Bay of Fundy).
2. Ophioglypha robusta, Ayres. 26 faths. [Loc. ?] *Distr.* W. Greenland, Massachusetts, Iceland, Spitzbergen, Norway to the Sound, Great Britain.
3. Ophiocten sericeum, Forb. 26 faths. *Distr.* East and West Greenland, Spitzbergen.

4. Asterophyton eucnemis, Müll. and Troschel. One young individual. *Distr.* W. Greenland, down to 1,000 faths.

CŒLENTERATA (p. 260).

1. Actinia nodosa, Fabr. One small specimen.
2. Briareum grandiflorum, Sars. On *Hornera lichenoides*, L. *Distr.* Oexfjord in Finmark, Arendal (German Baltic Exped., 1871).

XCI.—The Crustacea of East Greenland. By Dr. R. Buchholz. 1874.

(Die zweite deutsche Nordpolarfahrt, vol. ii. pp. 262–399; with 15 plates).

The List of genera and species, with localities, here compiled omits the synonyms and descriptions given in full by the author.

These Crustacea were collected by Dr. Pansch during the two German Expeditions on the coast of N.E. Greenland, and in the Arctic Ocean and North Sea; thus—

Decapoda, 13 (comprising 3 new species).
Isopoda, 3 (Bopyridæ; 1 new or little known).
Amphipoda, 27 (comprising 2 new species).
Phyllopoda, 1.
Copepoda, 8.
Cirrhipedia, 1.

Of these 55 species, there are—

(1.) Peculiar to the Arctic region (including Finmark and Nordland):—

Amphipoda, 17 }
Isopoda, 2 } 26.
Decapoda, 7 }

(2.) Belonging to the Norwegian Coast:—

Amphipoda, 12 }
Decapoda, 5 }
Isopoda, 2 } 26.
Copepoda, 6 }
Cirrhipedia, 1 }

(3.) Belonging to the English coast (Spence Bate, Westwood, Bell, and Baird):—

Amphipoda, 5 }
Isopoda, 2 }
Phyllopoda, 1 } 16.
Copepoda, 7 }
Cirrhipedia, 1 }

(4.) Belonging to the Baltic, only 5; namely,—
Gammarus locusta.
Amathilla Sabini.
Harpacticus chelifer.
Diaptomus castor.
Balanus porcatus.

DECAPODA.

Macroura.

Crangonidæ, Milne-Edw.

1. Crangon boreas (Phipps). Sabine Isl., 10–27 faths. Jackson Isl., 4 faths. (S. and W. Greenland. Spitzbergen).

Palæmonidæ.

1. Hippolyte incerta, n. sp. [Loc. ?]
2. H. turgida, Kröyer. Cape Wynn, 5 faths. Sabine Isl., 20–110 faths. North Shannon Isl., 30 faths. (S. and W. Greenland. Spitzbergen. Norway.)
3. H. Phippsii, Kr. Cape Wynn, 5 faths. Sabine Isl., 27 faths. (S. and W. Greenland, Spitzbergen, Finmark.)
4. H. polaris (Sabine). Sabine Isl., 20–100 faths. Cape Wynn, 5 faths. Shannon Isl., 2 faths. (W. and S. Greenland. Spitzbergen. Finmark. Norway.)
5. H. borealis, Owen. Sabine Isl., 20–100 faths. Cape Wynn, 5 faths. (S. and W. Greenland, Norway.)
6. H. aculeata (Fabr.). Abundant on the East Greenland coast. Sabine Isl., 10–120 faths. Cape Wynn, 5 faths. Shannon Isl. (S. and W. Greenland.)
7. H. Panschii, n. sp. North Shannon Isl., 30 faths.

Peneidæ, M.-Edw.

1. Pasiphaë glacialis, n. sp. Surface of sea, 74° N. lat.

Mysidæ.

1. Mysis oculata (Fabr.). Cape Phillip-Brooke, 3 faths. Sabine Isl., 4 and 10 faths. (S. and W. Greenland. Spitzbergen.)

Thysanopoda, M.-Edw.

1. Thysanopoda norvegica, Sars. Cape Wynn, 5 faths. (Norway.)
2. Th. Raschii, Sars. In the pack ice, 175 faths. (Norway.)

Brachyura.

1. Corystes Cassivelaunus, Penn. From stomach of Codfish, North Sea. (Norway.)

ISOPODA.

Bopyridæ.

1. Gyge hippolytes (Kröyer). In *Hippolyte polaris.* (S. and W. Greenland. Spitzbergen. England.)
1. Phryxus abdominalis (Kr.). In *Hippolyte turgida ;* also in *H. Gaimardi* elsewhere. (S. and W. Greenland. Spitzbergen. Finmark. Norway. England.)
1. Leptophryxus Mysidis, gen. et sp. nov. In *Mysis oculata.*

AMPHIPODA.

Lysianassidæ, Dana.

1. Anonyx lagena, Kr. Common along the E. Greenland coast. Sabine Isl., 5–20 faths. Germania Harbour, 3 faths. Jackson Isl., 24 faths. (S. and W. Greenland. Iceland. Spitzbergen. White Sea. Finmark. Norway. Britain.)
2. A. littoralis, Kr. Cape Wynn, 5 faths. Germania Harbour, 3 faths. (Open sea. S. and W. Greenland. Spitzbergen. Finmark.)
3. A plautus, Kr. Cape Wynn, 3 faths. Germania Harbour, 3 faths. Sabine Isl., 10 faths. (Open sea. S. and W. Greenland. Spitzbergen. Norway. Britain.)

Syrrhoinæ, A. Boeck.

1. Syrrhoe crenulata, Goës. Sabine Isl., 5–10 faths. (Spitzbergen. Norway.)

Pardaliscinæ, A. Boeck.

1. Pardalisca cuspidata, Kr. North Shannon, 30 faths. (S. and W. Greenland. Spitzbergen. Norway.)

Leucothoinæ, Dana.

1. Eusirus cuspidatus, Kr. Sabine Isl., 20–110 faths. (Open sea. S. and W. Greenland. Open sea ? Spitzbergen. Finmark. Norway.)
1. Amphithonotus aculeatus (Lepechin). North Shannon, 30 faths. Cape Wynn, 3 faths. (S. and W. Greenland. Spitzbergen. White Sea. Finmark. Norway ?)
1. Tritopsis fragilis (Goës). Sabine Isl., 10 faths. Cape Wynn, 3 faths. (S. and W. Greenland. Spitzbergen.)

Œdicerinæ, Lilljeborg.

1. Œdicerus borealis (A. Boeck). One of the most abundant species on the coast. Sabine Isl., 10 faths. Germania Harbour. (Spitzbergen. Finmark.)
2. Œ. lynceus, Sars. With the foregoing. Sabine Isl., 10 faths. Germania Harbour, 3 faths. (S. and W. Greenland. Iceland. Spitzbergen. Finmark.)

Pleustinæ.

1. Pleustes panoplus (Kr.). Sabine Isl., 20–110 faths. (S. and W. Greenland. Spitzbergen. Finmark. Norway.)

1. Parapleustes gracilis, nov. gen. et. sp. Sabine Isl., 10 faths.

Iphimedinæ, A. Boeck.

1. Vertumnus serratus (Fabr.). North Shannon, 30 faths. (S. and W. Greenland. Iceland. Finmark. Norway.)

Gammarinæ.

1. Gammarus locusta (L.). The most general and abundant species. Sabine Isl., 10-20 faths. Germania Harbour, 3 faths. (S. and W. Greenland. Open sea. Iceland. Spitzbergen. White Sea. Finmark. Norway. Britain. Baltic.)
1. Amathilla Sabini (Leach). Wide-spread. Sabine Isl., 10-110 faths. Germania Harbour. (S. and W. Greenland. Spitzbergen. Finmark. Norway. Britain. Baltic.)
2. A. pinguis (Kr.). One of the common species. Sabine Isl., 10 faths. Cape Wynn, 3 faths. North Shannon, 30 faths. (S. and W. Greenland. Spitzbergen.)

Atylinæ, Lillj.

1. Atylus carinatus (Fabr.). High Arctic; common on N.E. Greenland coast. Sabine Isl., 10–110 faths. Germania Harbour, 3 faths. (S. and W. Greenland. Spitzbergen.)
2. A Smittii (Goës). North Shannon, 30 faths. (Spitzbergen. Finmark.)
1. Acanthozone hystrix (Owen). North Shannon, 30 faths. (S. and W. Greenland. Spitzbergen. Finmark. Norway.)
1. Paramphithoë inermis (Kr.). Very general and common on N.E. Greenland coast. Sabine Isl., 10–120 faths. Cape Wynn, 3 faths. Germania Harbour, 3 faths.; and other places. (S. and W. Greenland. Open sea).
2. P. fulvocincta (Sars). Very common high up on the N.E. Greenland coast. Sabine Isl., 4–110 faths. Germania Harbour. Cape Wynn, 3 faths. Shannon Isl. (S. and W. Greenland. Open sea. Spitzbergen. Finmark.)
3. P. megalops, n. sp. Sabine Isl., 10 faths. Germania Harbour. Shannon Isl.

Ampeliscineæ, Sp. Bate.

1. Ampeliscus Eschrichtii, Kr. Germania Harbour. Sabine Isl., 10 faths. (S. and W. Greenland. Iceland. Spitzbergen. Finmark.)

Podocerineæ, A. Boeck.

1. Podocerus anguipes (Kr.). Common. Sabine Isl., 10–110 faths. Germania Harbour, and elsewhere. (S. and W. Greenland. Spitzbergen. Finmark. Norway.)

Corophinæ, Dana.

1. Glauconome leucopis, Kr. North Shannon, 30 faths. Germania Harbour. (S. and W. Greenland. Spitzbergen. Finmark.)

Hyperidæ, Dana.

1. Themisto libellula (Mandt). Common everywhere on the surface of the sea in astonishing quantities. (S. and W. Greenland. Open sea. Spitzbergen. Finmark. Britain.)

Caprellinæ, Leach.

1. Ægina spinifera (Bell). [Loc. ?] (Spitzbergen. Norway.)

PHYLLOPODA.

Nebaliadæ, Baird.

1. Nebalia bipes (O. F.). Germania Harbour, 3 faths. Sabine Isl. Jackson Isl. Shannon Bank, 150 faths. (S. and W. Greenland. Spitzbergen ? Finmark ? Norway ? Britain.)

COPEPODA.

Calanidæ.

1. Cetochilus septentrionalis, Goodsir. Wide-spread and enormously abundant. Open sea. Circumpolar ? Large individuals netted at 170 faths. (S. and W. Greenland. Spitzbergen. Finmark ? Norway ? Britain.)

1. Diaptomus castor, Jurine. Got from the tide-hole, by the side of the ship, February 1870. (Norway. Britain. Baltic.)

Harpactidæ, Claus.

1. Harpacticus chelifer (O. F. Müll.). Common. Sabine Isl., 10 faths., &c. (Finmark. Norway. Britain. Baltic.)

1. Tisbe furcata (Baird). Common. Sabine Isl., 10 faths., &c. (Finmark. Norway. Britain.)

1. Cleta minuticornis (Müll.). [Loc. ?] (Britain.)

Peltidideæ, Claus.

1. Zaus spinosus, Claus. Sabine Isl., 10 faths., &c. (Norway. Britain.)
2. Z. ovalis (Goodsir). [Loc. ?] (Norway. Britain.)

Cyclopidæ, Dana.

Thorellia brunnea, A. Boeck. Sabine Isl., 10 faths. (Norway.)

COPEPODA PARASITA.

Caligidæ, Milne-Edwards.

1. Lepeophtheirus Hippoglossi, Kr. North Sea, on the gills of *Pleuronectes rhombus* and of *Hippoglossus*.

Lernæopodidæ.

1. Brachiella rostrata, Kr. With the foregoing.

CIRRHIPEDIA.

1. Balanus porcatus, Da Costa. [Loc. ?] (S. and W. Greenland. Finmark. Norway. Britain. Baltic.)

Appendix.

PYCNOGONIDA.

1. Nymphon grossipes, O. F. North Shannon. (S. and W. Greenland. Spitzbergen. Norway.)
2. N. mixtum, Kr. [Loc. ?] (S. and W. Greenland. Spitzbergen.)
3. N. hirtum, O. F. North Shannon. (S. and W. Greenland. Spitzbergen.)

XCII.—The Arachnides of East Greenland. Extracted from the Memoir by Dr. L. Koch in "Die zweite deutsche Nordpolarfahrt," vol. ii., part i., 1874, pp. 400–403, 1 plate.

Lycosa aquilonaris, n. sp. (Pl. I.).

Lycosa aquilonaris belongs to C. Koch's subgenus *Leimonia*, and is very nearly allied to *Lyc. septentrionalis*, Westring (" Araneæ Suecicæ," p. 469), as yet known only in Norway. It is also very similar to *L. sociata*, O. Fabricius (F. Gr., p. 228), mentioned also by Thorell in his paper "On some Spiders from Greenland" (Öfvers. K. Vet.-Akad. Förh., No. 2, 1874).

[The Pycnogons are enumerated above.]

XCIII.—The Hymenoptera and Diptera of East Greenland. Extracted from the Memoir by Dr. A. Gerstäcker, with Notes by Dr. A. Pansch, in "Die zweite deutsche Nordpolarfahrt," vol. ii., pp. 404–406. 1874.

Hymenoptera.

1. Bombus pratorum, L. Common to the whole of northern and middle Europe. Often seen in E. Greenland

flying and crawling, but not numerous, nor in the swarms spoken of by Scoresby as seen farther south.

2. Cryptus sponsor, Fab. Cape Broer-Ruys, on the grass.
3. Limneria difformis, Gravenhorst. Cape Borgen, Shannon Island, on the ground.

Diptera.

1. Tipula truncorum, Meig. All along the coast on warm days.
2. Echinomyia ænea, Stæger. [Loc. ?]
3. Cynomyia alpina, Zetterstedt. Possibly this species is founded on small individuals of *C. mortuorum*, L. Sabine Island.
4. Calliphora grœnlandica, Zet. North Shannon Island, &c. Flies were plentiful in the autumn of 1869, disappearing with the first frost in September. *C. grœnl.* appeared on 26th May 1870, when the temperature first rose above freezing (max. $+0{\cdot}8^\circ$; min. $-$ $4{\cdot}3^\circ$ R. Afterwards, as the temperature rose to $-$ $4{\cdot}9^\circ$ R., Flies suddenly abounded, both the great *Cal. grœnl.* and smaller Flies, on shipboard, on the mainland, and up on the mountains. First egg found on June 7th, and in the middle of the month carrion swarmed with maggots.

XCIV.—The LEPIDOPTERA of EAST GREENLAND. Extracted from the Memoir by Captain ALEX. VON HOMEYER, with Notes by Dr. HERRICH-SCHÆFFER and Dr. WOCKE, in "Die zweite deutsche Nordpolarfahrt," vol. ii., pp. 407–410. 1874.

---

1. Argynnis polaris, Boisd. 1 ♂. Labrador. Lapland? N.E. Siberia ?

2. A. chariclea, Schnd. 1 ♂, 2 ♀. W. Greenland. Labrador. N. Lapland.

3. Colias hecla, Lef. 2 ♂, 1 ♀. W. Greenland. N. Lapland.

4. Larentia polata, Hübner (Zutr., f. 805, 806). 5 ♂. W. Greenland. N. Lapland. Labrador.

5. Geometra, sp. ? A caterpillar crawling under loose roots of herbage, Sabine Isl.

6. Dasychira grœnlandica, Wocke, n. sp., very near to *D. Rossii* of Labrador, and perhaps only a casual black variety.

XCV.—The HYDROIDA and BRYOZOA [POLYZOA] of EAST GREENLAND. Extracted from the Memoir by Dr. G. H. KIRCHENPAUER in "Die zweite deutsche Nordpolarfahrt," vol. ii., part i., pp. 411–428. 1874.

CLASS CŒLENTERATA. SUB-CLASS HYDROZOA.

Order HYDROIDA. Suborder TECAPHORA.

Fam. **Campanularidæ.**

1. Lafæa fruticosa (Hincks), Sars. North Shannon. *Distr.* Iceland. Tromsoe. North Cape. Bergen. Britain? Bass Straits?
2. Campanularia, sp. indet. North Shannon.

Fam. **Sertularidæ.**

3. Sertularella tricuspidata (Alder), Hincks. [Loc.?] *Distr.* W. Greenland. Spitzbergen. Vardoe. North Cape. Iceland. Straits of Belle Isle. Britain. New Zealand? Bass Straits?
4. Sertularia, sp. (nova?) North Shannon.

CLASS MALACAZOA. SUB-CLASS ACEPHALA (or CLASS VERMES. SUB-CLASS GEPHYREA).

Order BRYOZOA [POLYZOA]. Sub-order CHEILOSTOMATA.

Fam. **Cellularidæ.**

1. Menipea arctica, Busk. North Shannon. *Distr.* W. Greenland, Assistance Bay (Sutherland, Busk), and elsewhere. Arctic Ocean (200 faths., Smitt). Belgium?
2. M. Smittii, Norman. North Shannon. *Distr.* Spitzbergen (50 faths. and less).
3. Scrupelocellaria inermis, Norman. [Loc.?] *Distr.* Shetland. W. Greenland? Spitzbergen?

Fam. **Membraniporidæ.**

4. Membranipora Flemingii (Smitt), Busk. [Loc.?] *Distr.* Spitzbergen. Arctic, European, and Adriatic seas. Australia?
5. M. minax, Busk. [Loc.?] *Distr.* Shetland.
6. M. lineata (L.), Busk. Sabine Isl. *Distr.* W. Greenland (Lamouroux). European seas.
7. Lepralia hyalina (L.), W. Thomson. Sabine Isl. on *Fucus*. *Distr.* W. Greenland. Spitzbergen. Scandinavia. Heligoland. England. Falkland Isles. California. Probably in all seas.
8. L. Landsborowii (Smitt), Johnst. North Shannon on *Hornera* and *Cellepora lepralioides*. *Distr.* W. Greenland. Spitzbergen. Norway. England.

9. L. Smittii, nov. North Shannon on *Algæ*. *Distr.* Spitzbergen. Norway. Britain? Ægean Sea? Falkland Isles? New Zealand?

10. L. Peachii, Johnst. Sabine Is. on a stone with *L. sinuosa*. *Distr.* Spitzbergen. Norway. Britain. (Deep water.)

11. L. sinuosa, Busk. Sabine Isl. (as above), and elsewhere on *Proboscina increpata*. *Distr.* Shetland on Mussel shells. (Like *L. areolata*, Busk, from the Straits of Magellan.)

12. L. pertusa (Esper), Busk? [Loc.?] *Distr.* General?

13. Hemeschara (?) contorta, n. sp. North Shannon.

Fam. **Celliporidæ.**

14. Celleporella lepralioides, Norm. [Loc. ind.] on *Hornera lichenoides*. *Distr.* Shetland.

15. Cellepora scabra (Fabr.). Sabine Isl. and elsewhere. *Distr.* North Sea and Arctic.

16. C. incrassata, Lamarck. Common, but loc. ind. *Distr.* W. Greenland. Spitzbergen. Finmark.

17. C. Skenei (Ellis), Johnst. [Loc.?] *Distr.* Spitzbergen. Norway. Britain.

Fam. **Escharidæ.**

18. Eschara cervicornis, Lamarck. [Loc.?] *Distr.* W. Greenland. Spitzbergen. Finmark. Norway. Shetland. Britain. Adriatic.

Suborder CYCLOSTOMATA.

Fam. **Horneridæ.**

19. Hornera lichenoides (L.), Smitt. [Loc.?] Four varieties. *Distr.* General.

Fam. **Diastoporidæ.**

20. Diastopora hyalina (Flem.), Smitt. Sabine Isl. North Shannon Isl. *Distr.* Arctic, Atlantic, Mediterranean, and Adriatic Seas.

Fam. **Tubuliporidiæ.**

21. Idmonea atlantica, Smitt. North Shannon. *Distr.* Arctic and Atlantic Oceans.

22. Phalangella flabellaris (Fabr.), Smitt. North Shannon on *Hemeschara contorta*. *Distr.* W. Greenland. Spitzbergen. Norway. Britain. France. Adriatic.

Fam. **Lichenporidæ.**

23. Discoporella verrucaria (L.), Smitt. Sabine Isl. on *Alga*. *Distr.* Atlantic, North Sea, and Arctic.

24. D. hispida (Flem.), Smitt. Loc. ind. on *Hornera*. *Distr.* Arctic. North Sea. Britain. France. Mediterranean. Adriatic. Red Sea? Pacific?

Suborder CTENOSTOMATA.

Fam. **Halcyonelleæ.**

25. Alcyonidium hirsutum (forma membranacea), Smitt. North Shannon, on *Hornera*. *Distr.* Arctic Ocean. North and European Seas.
26. A. gelatinosum, Smitt? North Shannon on *Hornera*, an indeterminable little specimen.

XCVI.—The SILICEOUS SPONGES of EAST GREENLAND From Dr. OSKAR SCHMIDT'S Memoir in "Die zweite "deutsche Nordpolarfahrt," vol. ii., part i., pp. 429-433, 1 plate. 1874.

---

1. Cacospongia, sp. Interesting, as pure keratose Sponges are rare in the North.
2. Chalinula, sp.
3. Reniera, sp.
4. Isodictya infundibuliformis, Bowerbank.
5. Thecophora semisuberites, Schmidt.
6. Desmacidon anceps, n. sp. } Spicules figured in pl. 1.
7. Esperia intermedia, n. sp. } Spicules figured in pl. 1.
8. E. fabricans, n. sp. } Spicules figured in pl. 1.

XCVII.—The CALCAREOUS and SOFT SPONGES. Extracted from Dr. E. HAECKEL'S Memoir in "Die zweite deutsche "Nordpolarfahrt," vol. ii., part ii., pp. 434–436. 1874.

---

1. Ascaltis Lamarckii, Hæckel. North Shannon. North Atlantic. Mediterranean. Florida.
2. Sycaltis glacialis, H. North Shannon. Spitzbergen.

Halisarca Dujardinii, Johnst. [Loc. ind.] Norway.

XCVIII.—The MICROSCOPIC ORGANISMS of the NORTH-POLAR ZONE, both on LAND and in DEEP SEA; from the Materials brought home by the "Germania;" magnified 300 times. By Dr. Ch. G. EHRENBERG. 1874. "Die zweite "deutsche Nordpolarfahrt," vol. ii., pp. 437–467; four plates.

I. The soundings taken in the First Voyage, 1867, were partly off the coast of East Greenland (about 73° to 75° N. lat. and 12° to 17° 40′ W. long.), but mainly south and west of Spitzbergen. From these Prof. Ehrenberg determined (p. 443) 21 Polygastrica [Diatomaceæ], 16 Polythalamia [Foraminifera], 2 Polycystina, 26 Zoolitharia, Phytolitharia, and Geolithia [spicules of Sponges, &c.], and one soft plant-remain.

Of the minute terrestrial organisms of Spitzbergen are mentioned (p. 444):—

(1.) Nine Polygastrica [Infusoria and Arcella], revived at Berlin among the Mosses from Spitzbergen; also two living Nematoidea (*Anguillula*), and one Rotifer (*Callidina alpium*), and an egg.

(2.) Dead specimens: *Difflugia, Eunotia, Fragilaria, Pinnularia* (2), *Stauroneis*, and an Acarid.

Synedia ulva and *Spongolithis acicularis*, possibly terrestrial forms, got from the sea in 1841, are to be added to the above.

II. For the Second Voyage, 1869–70, 15 soundings of ooze and mud at sea off East Greenland (from 13 to 1,319 fathoms), and 2 muds at 3 fathoms, Sabine Island, are worked out (pp. 445, &c.). Also some gatherings from the land-surface, fresh water, and glaciers (pp. 450, &c.), giving chiefly *Bacillaria*, with some *Arcellina* and *Cryptomonadina* (especially from the Shannon Brook). Some greenish fine mud (possibly bird-droppings) from floating ice yielded *Coscinodiscus* and *Spongolithis.*

The two muds from Sabine Isl., south of the glacier brook, gave many *Bacillaria* and some Spongoliths. Altogether (both Expeditions) Dr. Ehrenberg enumerates (marine) 82 *Polygastrica* [Diatoms, &c.], 6 *Polycystina*, 37 *Polythalamia* [*Foraminifera*], and 43 Phytoliths, Geoliths, and Zooliths [Sponge spicules, &c.]; (terrestrial) 68 *Polygastrica* [Diatoms, Infusoria, &c.], 2 Nematoids, 2 Rotifers, an Acarid, and some Phytoliths.

Several new species are figured in the four plates. Those in Pl. I., Foraminifera, appear to be according to the nomenclature of English naturalists :—

Fig. 1. Nonionina; 2. Planorbulina; 3. Planorbulina; 4. Discorbina, *D. rosacea; 5. Pulvinulina auricula;* 6. Nonionina; 7. Cassidulina; 8, 9. Pulvinulina, near *P. Karsteni;* 10, 11. *Discorbina globularis;* 12. *Virgulina Schreibersii*, var.; 13. *Virgulina Schreibersii*, var.; 14. *Pulvinulina punctulata;* 15. Planorbulina; 16. Discorbina, with germs; 17, 18. Planorbulina; 19. Nodosaria.

These [resemble in their *facies* the North-Atlantic and Arctic Foraminiferal Fauna figured and described in "Phil. Trans.," vol. clv., 1865. (*Above*, pp. 192, 193.)

Of the 16 figured species, Dr. Ehrenberg states (p. 464) that 12 are new and known only in the Polar zone.

The following were obtained from more than 1,000 feet depth :—

Rotalia Hegemanni	=	Planorbulina.
R. ibex	=	Planorbulina.
R. microtis	=	Pulvulina auricula.
Nonionina crystallina	=	Discorbina ?
Aristerospira discus	=	Discorbina Karsteni, *var.*
A. cucullaris	=	Discorbina globularis.
Strophoconus hyperboreus	=	Virgulina Schreibersii, *var.*
Planulina profunda	=	Planorbulina.

Yellowish soft animal substance still filled the shells of *Aristospira borealis* (*Nonionina*), *A. corticosa* (*Pulvinulina punctulata*), and *Nonionina Koldeweyi;* from 168–510 feet.

XCIX.—PLANTS from EAST GREENLAND, collected by CAPT. GRAAH, and determined by PROF. HORNEMANN. 1832.

[Narrative of an Expedition to the East Coast of Greenland, &c., under command of Capt. W. A, Graah, &c., translated from the Danish by G. G. Macdougall, &c., 1837, Appendix II., pp. 177–8.]

I. PLANTS found in QUEEN-MARIA'S VALLEY, N. Lat. 63° 38′, W. Long. 41° 35′.

Hippuris vulgaris, Linn., var. tetraphylla.
Veronica saxatilis, L. High up among the cliffs.
Elymus arenarius, L.
Alchemilla alpina, L.
Polygonum viviparum, L.
Epilobium origanifolium, L. High up among the cliffs.
E. latifolium, L.
E. angustifolium, L., var. denticulata. Less than on the West Coast.
Erica cærulea, Willdenow.
Vaccinium uliginosum, L. The leaves larger than on the West Coast.
Saxifraga stellaris, L.
Cerastium alpinum, L.
Potentilla nivea, L., var. Perhaps a new species; the under surface of its leaves not white. Not yet found on the West Coast.
P. retusa, Retz. High up among the cliffs.
Rubus saxatilis, L. High up among the cliffs.
Euphrasia officinalis, L. Eight inches: one to three inches on the West Coast.
Bartschia alpina, L. High up among the cliffs.
Arabis alpina, L.
Erigeron alpinus, L.; and var. uniflora, Spreng.
Gnaphalium alpinum, L. High up among the cliffs.
G. sylvaticum, L., var. furcata, Wahlenb. High up among the cliffs.
Betula nana, L. High up among the cliffs.
Carex, sp. Without flower.
Salix glauca, L.
S. herbacea, L.
Equisetum arvense, L.
Aspidium fragile, Swarts.
Lycopodium alpinum, L.

II. PLANTS found at the ISLAND of KEMISAK; N. Lat. 63° 35′, W. Long. 41° 35′.

Veronica saxatilis, L.
Alchemilla alpina, L.
Campanula rotundifolia, L., var. uniflora.
Epilobium latifolium, L.
Polygonum viviparum, L.
Saxifraga cæspitosa, var. grœnlandica, Retz.
S. stellaris, L., var. pygmæa. Not yet found on the West Coast. The flower and fruit are, as in many Greenland plants, large in proportion to the stalk and leaves.
Stellaria humifusa, Rottbœll.
Cerastium alpinum, L.
Lychnis alpina, L.
Ranunculus hyperboreus, L. Larger than on the West Coast.
Thymus serpyllum, L., var. prostrata.
Bartschia alpina, L.
Draba muricella, Walhlenb.
Leontodon taraxacum, L., var.
Erigeron alpinus, L.; and var. uniflora.
Hieracium alpinum, L.

C.—BOTANY of the EAST COAST of ARCTIC GREENLAND. Extracted from "Die zweite deutsche Nordpolarfahrt,"* &c., Vol. ii., Part I., 1874.

1.—PLANTÆ VASCULARES (pp. 12–61). By DR. FR. BUCHENAU and DR. W. O. FOCKE.

[The names and localities only are here extracted, without the remarks and descriptions.]

I. **Ranunculacea.**

1. Ranunculus glacialis, L. At many damp places on the islands. Noticed by Scoresby and Sabine.
2. R. auricomus, L. Slopes of the Kaiser-Franz-Joseph Fjord.
3. R. nivalis, L. Common. Sabine, Clavering, Little-Pendulum, and Jackson Islands, and Cape Broer-Ruys. Collected by Sabine also.
4. R. pygmæus, Wahlenb. Little-Pendulum and Jackson Islands. In W. Greenland, 60° 43′–72° 48′, according to Lange.

II. **Papaveracea.**

5. Papaver nudicaule, L. Common, especially at Sabine Island. Noticed also by Scoresby and Sabine.

III. **Crucifera.**

6. Arabis petræa, L. Probably from the K.-Franz-Joseph Fjord. *Arabis alpina* was found by Scoresby.
7. Cardamine bellidifolia, L. Sabine and Little-Pendulum Islands. W. Greenland, 60°–72° 48′, according to Lange.
8. Vesicaria arctica, R. Br. Heidelbeerberg on the K.-Fr.-J. Fjord, at 400-800 ft. W. Greenland, 69° 40′–70° 41′, Lange.
9. Draba arctica, Vahl. (Flor. dan. t. 2294). Pendulum, Clavering, Jackson, and Sabine Islands; the K.-Fr.-J. Fjord.
10. D. Wahlenbergii, Hartm. (Flor. dan. t. 1420). Sabine and Clavering Islands; the Fjord.

---

* Botany, &c. of East Greenland, pp. 3–137 :
Preface by Dr. Fr. Buchenau, pp. 3–4.
Climate and Plant-life in East Greenland. By Dr. Adolf Pansch, pp. 5–11.
Vascular Plants. By Dr. Fr. Buchenau and Dr. W. O. Focke, pp. 12–61.
Mosses. By Dr. Karl Müller, pp. 62–74.
Lichens. By Dr. G. W. Körber, pp. 75–82.
Algæ. By Oberfinanzrath G. Zeller, pp. 83–87.
Fungi. By Dr. H. F. Bonorden and Herr L. Fuckel, pp. 88–96.
Drift Woods. By Prof. Dr. Gregor Kraus, pp. 97–137.

11. D. alpina, L. Sabine, Clavering, Shannon, and Jackson Islands; Cape Broer-Ruys; the Fjord.
12. D. rupestris, R. Br. (Flora dan. t. 2421). *D. hirta*, L., *B. alpicola*, Wahlenb. Flor. lapp., p. 175. The Fjord; Clavering and Jackson Islands.
13. D. muricella, Wahlb. (*nivalis*, Liljeb. nec DC.). Sabine Island.
14. Cochlearia fenestrata, R. B. (?). Walrus, Sabine, and Little-Pendulum Islands. Hooker (Scoresby), *C. anglica* et *C.* sp. ? (Sabine), *C. fenestrata* ?

IV. **Caryophyllea.**

15. Silene acaulis, L. Frequent in thick tufts. Sabine, Little-Pendulum, and Jackson Islands; Cape Broer-Ruys. Collected also by Scoresby and Sabine.
16. Wahlbergella (Lychnis) apetala (L.), Fries. In damp grass and among wet mosses. Sabine, Clavering, and Jackson Islands, Cape Broer-Ruys. Found by Sabine also.
17. Melandrium affine, Vahl. Sabine and Clavering Islands; the Fjord. W. Greenland, 65° 38′–72° 48′, Lange. Probably also under the name *Lychnis dioica*, var. *nana*, Hooker, in Sabine's collection.
18. M. triflorum, (R. Br.) Vahl. Sabine and Shannon Islands. W. Greenland, 60° 50′–72° 48′, Lange.
19. Arenaria ciliata, L. On fertile damp earthy bottoms, moraines, &c. Sabine, Clavering, and Jackson Islands; Cape Broer-Ruys; slopes of the K.-Fr.-J. Fjord.
20. Alsine rubella, Wahlenb. The typical form in Sabine Island. var. γ *Gieseckii*, Lange in Rink (*Alsine Gieseckii*, Hornemann, Flora danica, t. 1518) at many places; Sabine and Jackson Islands; Cape B.-Ruys; the Fjord. Found by Sabine also.
21. Al. biflora, Wahl. Sabine Isl. W. Greenland, 60°–72° 48′, up to 2,040 feet above the sea, Lange.
22. Halianthus peploides,* (L.) Fries. var. *oblongifolia*, J. Lange, MS. In thick herbage at the foot of steep cliffs; Clavering Island; near Cape Borlase-Warren. Mentioned by Sabine. W. Greenland, 60°–69° 14′, Lange.
23. Stellaria longipes, Goldie. *St. Edwardsii*, R. Br., *St. nitida*, Hooker (Scoresby). Wide-spread. On dry turf and among other plants. Sabine, Shannon, and Jackson Islands, Cape Broer-Ruys; the Fjord at about 600 ft. high. Collected by both Scoresby and Sabine.
24. St. humifusa, Rottb. In meadows. Sabine Island, noticed by Sabine and Graah. Not rare in W. Greenland.

St. cerastioides, L., also was found by Sabine.

* This is *Honckenejа peploides* of Dr. Hooker's list at page 227.

25. Cerastium alpinum, L. var. lanata. Very general. Found also by Scoresby, Sabine, and Graah.
C. latifolium mentioned in Scoresby's list, is now regarded as quite doubtful.

V. **Rosacea.**

26. Dryas octopetala, L. Very general, and forming thick clumps on the ground, as in the Alps. Var. *integrifolia,* Vahl, occurs in W. Greenland according to Lange.
27. Potentilla pulchella, R. Br. (Flor. dan. t. 2234.) Clavering Isl. The typical form occurs also in Arctic America and in W. Greenland; and a distinct variety in Spitzbergen.
28. P. nivea, L. On sunny slopes; Sabine and Jackson Islands; the shore of the Fjord. In Sabine collection, var. α and β (Hooker). W. Greenland, 64°–72° 48′, Lange.
29. P. emarginata, Pursh (Flor. dan. t. 2291). The most common plant, at most of the places visited. Probably the same as this are,—Scoresby's *Pot. verna,* L. (Hooker); also Sabine's *Pot. nivea,* var. (Hooker); and Graah's *Pot. retusa,* Retz (Hornemann.) W. Greenland, 66° 50′–72° 48′, J. Lange.

VI. **Onagrariacea.**

30. Epilobium latifolium, L. Wide-spread, especially on new bottoms, moraines, &c. One of the finest flowers of Greenland. Seen also by Scoresby, Sabrine, and Graah.

VII. **Crassulacea.**

31. Sedum Rhodiola, DC. Cape Mary, Clavering Isl. W. Greenland, 60°–67°, Lange.

VIII. **Saxifragacea**

32. Saxifraga oppositifolia, L. A wide-spread Arctic plant, met with nearly everywhere, as high up as 77° N. Lat. Found by Scoresby and Sabine also.
33. S. cæspitosa, L., var. grœnlandica (L.). Nearly everywhere, except at the Fjord. Found also by Scoresby, Sabine, and Graah.
34. S. cernua, L. Nearly everywhere. Scoresby and Sabine also.
35. S. rivularis, L. Clavering Isl. and Cape Broer-Ruys. Found by Sabine.
36. S. nivalis, L. Everywhere, especially in wet muddy places. Found by Scoresby and Sabine.
37. S. hieracifolia, W. & K. Cape B.-Ruys: Spitzbergen and Arctic lands.
38. S. Hirculus, L., var. alpina, Engler. On muddy bottoms and in wet mossy clumps. The group of Pendulum Islands, Mackenzie Bay. Found by Sabine.

39. S. flagellaris, Willd. Near glacier-brooks, &c. In the Pendulum Group.
40. S. aizoides, L. Moraine-earth and swamps of the K.-Fr.-J. Fjord, forming thick turf and clumps.

Sax. stellaris, L. (*foliosa*, R. Br.). Also was found by Sabine.

IX. Composita.

Gnaphalium alpinum, L., found by Scoresby, is wanting in this collection.

41. Arnica alpina, Murr. (*A. angustifolia*, Vahl, Flor. dan. t. 1524). On turfy slopes, very general; common at Clevering I., Jackson I., Mackenzie Bay. Slopes of the Fjord. Found also by Scoresby and Sabine. W. Greenland, Lange.
42. Erigon eriocephalus, J. Vahl (Flor. dan. t. 2299). South coast of Clavering I. Noticed by Scoresby, Sabine, and Graah.

E. compositus, Pursh, found by Sabine, was not collected in this Expedition.

43. Taraxacum phymatocarpum, J. Vahl (Flor. dan. t. 2298). At nearly every place visited. Graah mentions *Leontodon taraxacum*, var.

X. Campanulacea.

44. Campanula uniflora, L. Pendulum Islands. Noticed by Sabine. Wide spread in West Greenland.
45. C. rotundifolia, L., var. arctica, J. Lange (Flor. dan. t. 2711). On the slopes of the Fjord, abundant at 600–800 feet, with *Pyrola*. Graah met with it below 63° N. Lat. In W. Greenland, 68°–70°, J. Lange.

XI. Vacciniacea.

46. Vaccinium uliginosum, L. Wide spread, on the Islands and at the Fjord. Found also by Scoresby and Sabine. In W. Greenland, with var. also.

XII. Ericacea.

47. Andromeda tetragona, L. A very wide-spread Arctic plant. Clevering I., Shannon I., the Fjord, &c. Scoresby and Sabine also.
48. Arctostaphylos alpina, Spreng. On the alluvium of Eleonore Bay; the K.-Fr.-J. Fjord, &c. Among grasses, reeds, mosses, and *Vaccinium*. Lange has found it near Sukkertoppen, W. Greenland.
49. Rhododendron lapponicum, L. Rare. Coal Island; the K.-Fr.-J. Fjord. Found by Sabine. More common in W. Greenland.

Pedum palustre was found by Sabine, but not by this Expedition.

XIII. Pyrolacea.

50. Pyrola rotundifolia, L., var. arenaria, Koch. Very plentiful at 600–800 feet, on grassy slopes, at the

K.-Fr.-J. Fjord. Another form, *P. grandiflora*, Rad., Auct., abounds in W. Greenland.

XIV. **Empetracea.**

51. Empetrum nigrum, L. Coal I.; Mackenzie Bay. Found by Scoresby. W. Greenland, 60°–72° 48′.

XV. **Polemoniacea.**

52. Polemonium humile, Willd. (*P. acutiflorum*, Willd., *P. pulchellum*, Bung., *P. capitatum*, Eschsch., *P. Richardsoni*, Hook. et Arn., *P. pulcherrimum*, Hooker). Very characteristic for East Greenland, where it is general; but it is not known in S. or W. Greenland.

XVI. **Scrophulariacea.**

Veronica alpina was gathered by Scoresby, but not by this Expedition.

53. Euphrasia officinalis, L. Rare, at one spot under the Bürgermeister Crags, in Jackson Isl.
54. Pedicularis hirsuta, L. Common. Found by Sabine. Common also in W. Greenland.

XVII. **Plumbaginacea.**

55. Armeria maritima, Willd. (*A. Sibirica*, Turcz. in DC. Prod., XII., 678; Flor. dan. t. 2769). Rare. Sabine Island? Found by Sabine also.

XVIII. **Polygonacea.**

54. Oxyria digyna (L.), Campd. A very wide-spread Arctic plant and very gregarious; at nearly all places visited. Also Scoresby and Sabine.

Kœnigia Islandica, L., found by Sabine in E. Greenland, and more common in W. Greenland, was not collected by this Expedition.

XIX. **Betulacea.**

58. Betula nana, L., var. genuina, Regel. Forming a thick bush on the edge of the moraine, 800–1,000 ft., on the slope of the K.-Fr.-J. Fjord.

XX. **Salicacea.**

59. Salix arctica, Pallas. At all the localities visited. Also observed by Scoresby ("*Salix* aff. *glaucœ* et *limosœ*"), Sabine, and Graah.

XXI. **Juncacea.**

60. Luzula hyperborea, R. Br. A characteristic Arctic plant, found at most of the localities. Also by Scoresby ("*Luzula arcuata*, Hooker"), and Sabine.
61. Juncus biglumis, L. Sabine Isl.; Cape Broer-Ruys; the Fjord.
62. J. triglumis, L., var. Copelandi, Buchenau. Moraine at the K.-Fr.-J. Fjord.
63. J. castaneus, Sm. K.-Fr.-J. Fjord.

XXII. **Cyperacea.**

64. Carex rupestris, All. Rare. Jackson Isl.; Cape B.-Ruys. Wide-spread in Greenland.
65. C. nardina, Fries. Rare. Clavering and Jackson Islands. More common in W. Greenland.
66. C. fuliginosa, Sternb. et Hoppe. Bottoms with streams. Clavering and Sabine Islands; K.-Fr.-J. Fjord. Middle-sized specimens are from widely in the Alps.
67. C. subspathacea, Wormsk. One specimen from the Fjord. W. and S. Greenland, 60°–62° N. Lat., Lange.
68. C. rigida, Good. One specimen, Cape Broer-Ruys. Widely spread in W. Greenland.
69. Krobesia caricina, Willd. Moraines at the Fjord. W. Greenland, 64°–72° N. Lat.
70. Elyne spicata, Schrad. Cape B.-Ruys. In W. Greenland, wide-spread.
71. Eriophorum polystachum, L. Sabine Isl.; Clavering Isl.; Cape B.-Ruys; the Fjord, &c. Collected by Sabine also.
72. E. Scheuchzeri, Hoppe (*capitatum*, Host). Clavering Isl.; Cape B.-Ruys; the Fjord. Scoresby and Sabine.

XXIII. **Graminea.**

73. Alopecurus alpinus, Sm. Common everywhere, in damp low-lands. Also by Scoresby and Sabine.
64. Calamagrostis purpurascens, Br. (Flor. dan. t. 2523). With *Catabrosa latifolia* at the K.-Fr.-J. Fjord. The tallest grass plant in East Greenland. Not rare in W. Greenland.
75. Hierochloa alpina, R. et S. Shannon, Sabine, and Jackson Islands, &c.
76. Deschampsia brevifolia, R. Br. Rare. Cape Phillip Broke. By Sabine also. Limited to East Greenland.
77. Trisetum subspicatum, P. de B. A rather common Arctic grass. Jackson Isl.; Clavering Isl.; Cape Broer-Ruys. Scoresby and Sabine. S.-W. Greenland.
78. Catabrosa (Phippsia) algida, (Sol.) Fries. Sabine and Shannon Islands. Wide-spread in W. Greenland.
79. C. (Colpodium) latifolia, (R.Br.) Fries. Cape Broer-Ruys; the Fjord. W. Greenland, 70°–72° 48′. The next tallest grass after *Calam. purpurescens.*
80. Poa abbreviata, R. Br. Clavering Isl. K.-Fr.-J. Fjord.
81. P. arctica, R. Br. (and varieties). General and abundant. Probably *P. laxa* of Scoresby's and Sabine's collections (Hooker) is a variety. Spitzbergen.
82. P. cæsia, Sm. General and abundant.
83. P. annua, L. (?)

    Glyceria (*Poa*) *angustata*, (R. Br.) Fr. (near to *Gl. maritima*, M. & K.), in Sabine's collection, was not sent home by this Expedition.
84. Festuca brevifolia, R. Br. Forming turf. Jackson I.; Clavering I.; Cape Broer-Ruys (this is *Festuca ovina*, var., of Sabine's coll., Hooker). W. Greenland.

85. Festuca (?). Viviparous, forming thick turf. Sabine I. Scoresby has a *Festuca vivipara*; and Lange *F. ovina*, var. *vivipara* in W. Greenland.

XXIV. **Filices.**

86. Woodsia Ilvensis, R. **Br.** The slopes of K.-Fr.-J. Fjord, at 700 feet. Dr. Kuhn:—*W. Ilvensis*, R. Br., emended, var. *hyperborea*, R. Br.; *forma pilosella* (Ruprecht) Milde, Fil. Europ. p. 162 (*Woodsia pilosella*, Rupr. Beitr. III., p. 54, t. Spec. origin.!
87. Cystopteris fragilis, Bernh. Jackson I.; Clavering I.; the Fjord. Noticed by Sabine and Graah. Dr. Kuhn:—*C. fragilis*, Bernh., 1, *forma arctica*; 2, *forma lobulato-dentata*, Milde, Fil. Europ. p. 148. K.-Fr.-J. Fjord. Common on both mountains and plains in Germany.

XXV. **Equisetacea.**

88. Equisetum scripoïdes, Michx. In mud and under *Vaccinium ulig.*, on the east side of Sabine Island near the shore. At some places in W. Greenland.
89. Eq. arvense, L., var. boreale (Bongard). In wet places, Sabine I., Cape Broer-Ruys, &c.

2.—*List of the Plants collected by the Crew of the "Hansa," in South Greenland, at the end of their long ice-journey under Captain P. F. A. Hegemann.* Pp. 59–61.

1. Thalictrum alpinum, L.
2. Ranunculus acer, L.
3. Coptis trifolia, Salesb.
4. Cardimine pratensis, L.
5. Draba incana, L.
6. Cochlearia officinalis, L., var. arctica.
7. Viola Muehlenbergiana, Ging., β. minor, Hook. (Flor. dan. t. 2710.)
8. Viscaria alpina, (L.) Fries.
9. Stellaria cerastioides, L.
10. Cerastium alpinum, L.
11. C. triviale, Lk.
12. Lathyrus maritimus, Fr.
13. Alchemilla alpina, L.
14. A. vulgaris, L.
15. Potentilla tridentata, Pursh.
16. Sedum rhodiola, DC.
17. Saxifraga aizoon, Jacq.
18. S. oppositifolia, L.
19. S. cæpitosa, L.
20. S. nivalis, L.
21. Antennaria dioica, Gärtn.
22. Taraxacum officinale, L., var. palustre (DC.)
23. Vaccinium uliginosum, L.
24. Andromeda hypnoides, L.
25. Azalea procumbens, L.
26. Rhododendron lapponicum, L.
27. Ledum latifolium, Ait.
28. Phyllodoce cærulea, Gr. & God.
29. Pyrola rotundifolia, L.
30. Thymus serpyllum, L., var. boreale, Lange.
31. Veronica alpina, L.
32. Bartsia alpina, L.
33. Pinguicula vulgaris, L.
34. Armeria maritima, Willd.
35. Plantago borealis, Lge.
36. Rumex acetosa, L.
37. Polygonum viviparum, L.

38. Empetrum nigrum, L.
39. Betula intermedia, Thomas (*see* Babington, Revis. Flora Iceland, Journ. Lin. Soc. 1870, xi. 45; Flor. dan. t. 2852.)
40. B. nana, L., γ, intermedia, Regel.
41. Salix myrsinites, L.
42. Streptopus amplexifolius, DC.
43. Platanthera Kœnigii, Lindl.
44. Juncus trifidus, L.
45. Carex rarifolia, Sm.
46. C. nigritella, Drejer.
47. Scirpus cæspitosus, L.
48. Eriophorum capitulum, Host.
49. Phleum alpinum, L.
50. Poa pratensis, L., var. arctica, J. Lange.
51. P. alpina, L.
52. Aspidium lonchitis, Sw.
53. Polystichum spinulosum, DC.
54. Lycopodium selago, L.
55. L. annotinum, L.
56. L. alpinum, L.
57. Equisetum arvense, L.
58. Amanita (Agaric Fungus, p. 90.)

3.—REMARKS on the FLORA of EAST-ARCTIC GREENLAND (BUCHENAU).—Pp. 20–21.

The vegetation of East-Arctic Greenland agrees in all important points with the well-known flora of West Greenland. On the south coast of East Greenland, Kemisak Island, 63° 37′ N. Lat., examined by Graah, is the most northern spot whose vegetation is thoroughly known. Between Kemisak and Kaiser-Franz-Joseph Fjord lie nearly nine degrees of unsearched territory. Scoresby landed at only one intervening point; but even reckoning the Sound named after him as part of the known northerly range of coast, there remains a gap of seven degrees between the extreme points waiting for exploration. We see, therefore, why the flora of the southern part, and that of the northern part of East Greenland differ so greatly. Whether within that great space of the nine, or of the seven degrees, a sudden limit or a gradual transition between these two botanical regions is to be found, is quite unknown. As far as we know, the south part of East Greenland has no plant but what the west coast possesses. The northern flora of the east coast, on the contrary, is characterised by some peculiarities. The following plants have been found on the northern, but neither on the southern coast of East Greenland nor in West Greenland :—

1. Ranunculus glacialis, L. (very doubtful for West Greenland.)
2. R. auricomus, L.
3. Dryas octopetala, L. (forma typica!)
4. Saxifraga hieracifolia, W.K.
5. S. Hirculus, L.
6. Pyrola rotundifolia, L., var. arenaria, Koch.
7. Polemonium humile, Willd.
8. Juncus triglumis, L., var. Copelandi, Buchenau.
9. Deschampsia brevifolia, R. Br.

Of these plants some are represented in West Greenland by similar species or varieties :—

1. Ranunculus glacialis, L., by R. alpinus, L.
2. Dryas octopetala, L., by Dr. integrifolia, J.Vahl.
3. Pyrola rotundifolia, L., var. arenaria, Koch., by R. rotundifolia, L. var. grandiflora, Radde.
4. Juncus triglumis, L. var. Copelandi, Buchenau, by J. triglumis, L. (typicus.)

Of the characteristically North-eastern Greenland plants, the following live in Spitzbergen and its islands :—

1. Ranunculus glacialis, L.
2. Dryas octopetala, L.
3. Saxifraga hieracifolia, W.K.
4. Saxifraga Hirculus, L.
5. Polemonium humile, Willd.

To the west of Greenland, on Melville Island, in the Arctic-American archipelago are :—

1. Ranunculus auricomus, L.
2. Saxifraga hieracifolia, W. & K.
3. Saxifraga hirculus, L.
4. Deschampsia brevifolia, R. Br.

All these occur also in Arctic Europe. *Ranunculus auricomus, Sax. hieracifolia, S. Hirculus, Pol. humile,* and *Desch. brevifolia,* are wide-spread Arctic plants, whose absence in West Greenland is more remarkable than their presence in East Greenland. On the other hand, *Ranunc. glacialis* is a European, Alpine, and Scandinavico-Spitzbergen species stretching to North-eastern Greenland, but not over other Arctic regions. *Dryas octopetala* is represented in the Arctic-American archipelago, in West Greenland, and in many other northern stations, by *Dr. integrifolia,* wanting in Europe. Should the occurrence of this species in Spitzbergen during the Post-pliocene period (Heer, Flor. foss. arct. II., 2, p. 91, t. 16, f. 69), be substantiated, *Dr. octopetala* must be regarded as having migrated from Europe, and *Dr. integrifolia* as a supplanting race.

The variety *arenaria* of *Pyrola rotundifolia,* distributed in Middle Europe on sand-dunes, in company with littoral plants on the sea-coast and in the Valais (Switzerland), is still the same in the Arctic regions. The new varieties *Juncus triglumis,* var. *Copelandi,* B., *Draba muricella,* var. *Panschii,* B., and *Saxifraga hieracifolia,* W. & K., have been added to the Arctic-East Greenland Flora by the Second German Expedition.

CI.—The Mosses of East-Arctic Greenland. By Dr. Karl Müller. Die zweite deutsche Nordpolarfahrt, vol. ii., pp. 62–74. 1874.

[The names only of genera and species are here enumerated.]

Hypnum sarmentosum, Wahl.
H. Schreberi,* Willd.
H. nitens, Schreb.
H. stramineum, Dicks., var. laxifolium.
H. stramineum, Dicks., var. fluitans.
H. julaceum, Vill.
H. apiculatum, Thed.
H. cirrhosum, Schw.
H. plumosum, Sw.
H. salebrosum, Hoffm. ?
H. Mildeanum, Schpr.
H. polygamum, Schpr.
H. uncinatum,* Sw.
H. fluitans, var. pseudostraminea, C. Müll.
H. Wilsoni, Schpr.
H. revolvens, Sw.
H. pratense, K.
H. Lamulosum, var. julacea.
H. revolutum, Lindb. (H. Heufleri, jur.).
H. molle, Dicks., var.
H. chryseum, Hsch.
Mnium affine, Bland. Diœcious.
M. subglobosum, Br. Eur., var. pusilla. Hermaphrodite.
M. (aulacomnion) turgidum, Wahl.
M. (aulacomnion) palustre, Hdw., and var compacta.
Timmia Austriaca, Hdw.
T. Megapolitana, Hdw., and var.
Bryum pseudotriquetrum, Schw., var.
B. celophyllum, R. Br., and var. *compacta*.
B. teres, Lindb.
B. rutilans, Brid.
B. nitidulum, Lindb.
B. cernuum, Br. & Sch.
B. Archangelicum, Schpr.
B. Algovicum, Sendtn.
B. arcticum, Br. & Sch.
B. Brownei, Schpr.
B. bimum, Schreb.
B. cirrhatum, H. & H.
B. demissum, Hook.
B. Ludwigii, Spr., var. gracilis.
B. nutans, Schreb.
B. annotinum, Hdw.
B. crudum, Schreb.
Meesea tristicha, Br. & Sch.
M. longiseta, Hdw., var. luxurians.
Bartramia ithyphylla, Brid.
B. fontana, Sw.
B. cæspitosa, Wils., var. compacta.
Conostomum boreale, Sw.
Spachnum mnioides, L., var. compacta.
S. Wormskioldii, Sw.
Polytrichum commune, L.
P. polare, C. Müll.
Encalypta procera, Br. Eur.
E. rhabdocarpa, Schw.
Grimmia apocarpa, Hedw.
G. unicolor, Grev.
G. (Dryptodon) Panschii, C. Müll., n. sp.
G. (Rhacomitrium) canescens, C. M.
G. lanuginosa, C. M., and var. arctica.
Gnembelia arctica, C. M., n. sp.
Barbula ruralis, Hdw.
B. leucostoma, R. Br. ?
Trichostomum rubellum, Rabenh., var. dentata.
T. (Desmatodon) Laureri, Schultz.

* Brought also from S.W. Greenland (Julianshaab) by the crew of the "Hansa."

Distichium capillaceum, Br. Eur.
Leptotrichum flexicaule, Hpe.
Angstrœmia Wahlenbergii, C. Müll.
Dicranum strictum, Schl., var. compacta.
D. arcticum, Schpr.
Weisia curvirostris, Syn. Musc.

Clavering, Shannon, Walrus, Sabine, and the Pendulum Islands, Cape Borlase-Warren, Mackenzie Bay, and the shores of the great Kaiser-Franz-Joseph Fjord are the chief localities.

CII.—The Lichens of East-Arctic Greenland. By. Dr. G. W. Körber. Die zweite deutsche Nordpolarfahrt, vol. ii. pp. 75–82. 1874.

[Only the names of genera and species are here enumerated.]

Usnea melaxantha, Ach.
Stereocaulon paschale, var. gracilenta, Th. Fr.
S. alpinum, Laur.
Cladonia rangiferina, L., et silvatica, Hoffm. From Julianshaab, S.-W. Greenland; brought by the crew of the "Hansa."
Cetraria nivalis, L.
Peltigera rufescens, Fr.
Imbricaria stygia, β, et lanata, Ach.
I. alpicola, Th. Fr.
I. olivacea, L.
Parmelia muscigena, Ach.
Physcia parietina γ, ectanea, Ach. (*Ph. fallax*, Hepp.)
Gyrophora anthracina, Wulf. (*forma tesselata*, Ach.)
G. cylindrica, L.
G. arctica, Ach. (*G. proboscidea* β, arctica, Kbr.).
G. Tramitziana, Kbr., n. sp.
G. Koldeweyi, Kbr., n. sp.
Amphiloma elegans, Lk. Very abundant.
A. murorum β, miniatum, Hoffm.
Acarospora peliscypha, Wahl. (*rugulosa*, Kbr.).
Candelaria vitellina, Ehrh.
Calopisma Jungermanniæ (Vahl.), Th. Fr.
C. aurantiacum ζ, holocarpum, Ehrh.
C. mydaleum, Kbr., n. sp.
Rhinodina turfacea, Wahl., et β, microcarpa, Hepp.
R. mniaræa, Ach.
R. Panschiana, Kbr., n. sp.
Lecanora Hageni, Ach.
L. subfusca α, 5 bryontha (Ach.), Kbr.
L. subfusca, var. allophana.
Atrosulphurea, Wahl.
Aspicilia calcarea α*, ochracea, Kbr.
A. rosulata, Kbr., n. sp.
Psora rubiformis (Wahl.).
Blastenia fuscolutea, Dcks.
Biatora polytropa, Ehrh. Plentiful.
Bilimbia Regeliana (Hepp.), Kbr.
Buellia stigmatea, Ach.
B. Copelandi, Kbr., n. sp.
B. Payeri, Kbr., n. sp.
Lecidella sabuletorum, Schr., et forma *depauperata.*
L. subuletorum, var. æquata, Flk.
L. goniophila, Flk.
L. Hansatica, Kbr., n. sp.
Rhizocarpon geographicum, var. alpicola, Wahl.
R. inops, Kbr., n. sp.
Sporastatia Morio, Ram.

Orphniospora Grœnlandica, Kbr., n. gen. et sp.
Raphiospora flavovirescens, Dcks.
Eudopyrenium rufescens, Ach.
E. dædaleum, Kmph.
Pertusaria glomerata, Schl.
Tichothecium pygmæum, Kbr.

Localities as with the Mosses, *above*, p. 583.

CIII.—The ALGÆ of EAST-ARCTIC GREENLAND. By Oberfinanzrath G. ZELLER. (Die zweite deutsche Nordpolarfahrt, ii. pp. 83-87. 1874.)

[Only the names of families, genera, and species, are here extracted.]

LYNGBYEÆ.

Lyngbya glutinosa (?), Ag.

CONFERVACEÆ.

Chætomorpha melagonium, Kützing. (*Conferva melag.*, W. & M.)
Rhizoclonium litoreum, Kg. (*Zygnema lit.*, Lyngb.).

ECTOCARPEÆ.

Ectocarpus ochracens, Kg.
Stypocaulon scoparium, Kg.

ENTEROMORPHEÆ.

Enteromorpha ranulosa, Hook.

MESOGLÆACEÆ.

Chordaria flagelliformis, Ag.

SPOROCHNEÆ.

Desmarestia aculeata, Lamx.

LAMINARIEÆ.

Laminaria phyllitis, Lx. (According to Agardh, possibly only a young form of *L. saccharina.*)
L. digitata, Lx. (*Hafgygia dig.*, Kg.

FUCEÆ.

Fucus vesiculosus, L., var. evanescens?

CALLITHAMNIEÆ.

Ptilota serrata, Kg. (*Pt. plumosa* δ, *serrata.*) Also from Julianshaab S.-W. Greenland.

CYSTOCLONIEÆ.

Cystoclonium purpurascens, Kg. Only from Julianshaab.

TYLOCARPEÆ.

Coccotylus Brodiæi, Kg. (*Phyllophora Brod.*, Harv.)

POLYSIPHONIEÆ.

Polysiphonia stricta, Grev.

DELESSERIEÆ.

Phycodrys sinuosa, Kg. (*Delesseria*, Lx.), and var. lingulata, Ag.

DICTYOSIPHONEÆ.

Dictyosiphon fœniculaceus, Grev.

Kützing's system is followed by the author.

The localities are Walrus Island, Cape Wynn, and Sabine Island.

CIV.—Fungi, from East-Arctic Greenland. By Dr. H. F. Bonorden. From "Die zweite deutsche Nor polarfahrt, "vol. iii., pp. 88–90, 1874.

[Names only extracted.]

Lycoperdon fuscum, Bon.
Lycoperdon bovista, Fr.
Paxillus griseo-tomentosus, Fr.
Agaricus simiatus, Fr. ?
Agaricus, sp. ?
Leptonia, sp. ?
Lactarius, sp. ?
Leucosporus, sp. ?

From South Greenland.

Amanita (Agaricus vaginatus, Ag.).

CV.—Parasitic Fungi from East-Arctic Greenland. By L. Fuckel. From "Die zweite deutsche Nord-polarfahrt," vol. ii., pp. 90-96; with a Plate. 1874.

[Herr Fuckel's descriptions of these Endophytic Fungi are not extracted.]

I. **Uredinei.**

1. Melampsora salicina, Tul. On *Salix arctica*, Sabine Isl. and elsewhere.

II. **Pyrenomycetes.**

2. Pleospora hyperborea, sp. n. (Pl. I. f. 1). On dry leaves of *Andromeda tetragona*, Shannon I.
3. Pl. arctica, sp.n. (Pl. I., f. 2). On dry stems of *Epilobium latifolium*, Kaiser-Franz-Joseph's Fjord.
4. Pl. paucitricha, sp. n. (Pl. I., f. 3). On dead leaves of *Salix arctica*, K.-Fr.-J. Fjord.
5. Pl. dryadis, sp. n. (Pl. I., f. 4). On dry leaves of *Dryas octopetala*, Clavering and Sabine Island.
6. Pl. herbarum, Tal. On dry leaves and stems of *Polemonium humile*, Sabine I.
7. Sphæria nivalis, sp. n. (Pl. I., f. 5). On dry leaves and stems of *Epil. latifolium*. K.-Fr.-J. Fjord.
8. Sph. arctica, sp. n. (Pl. I., f. 6). Wide spread in the Arctic regions; abundant on *Poa cæsia* and other grasses, Sabine I. and K.-Fr.-J. Fjord.
9. Ceratostoma foliicolum, sp. n. (Pl. I., f. 7.) On dry leaves of *Salix arctica*, Sabine I. and elsewhere.
10. Cystipora capitata, sp. n. (Pl. I., f. 8.) On dry branches of *Salix arctica*.
11. Phoma drabæ, sp. n. (Pl. I., f. 9.) On dry stems of *Draba*, Clavering I.
12. Rhizomorpha arctica, sp. n. (Pl. I., f. 10.) On rotten roots of *Salix arctica*. K.-Fr.-J. Fjord.
13. Xylographa arctica, sp. n. (Pl. I., f. 11.) On hard barkless wood of *Salix arctica*. K.-Fr.-J. Fjord.

CVI.—Drift-wood of East-Arctic Greenland. From the Memoir by Dr. Gregor Kraus, in "Die zweite deutsche Nordpolarfahrt," vol. ii., pp. 97–132. 1874.

Of the 25 pieces of Drift-wood and two small pieces of Bark, submitted to the Author, one specimen was found in the K.-Franz-Joseph Fjord, and the others near the Pendulum Islands. He submitted them to an exhaustive botanical examination.

Twenty-two of the 25 Woods were Coniferous four certainly belonging to Larch (*Larix sibirica*, Lederb. ?), and probably 13 others; and 5 to *Larix Sibirica*, Led., or *Picea obovata*, Led., probably the latter. Of the two bits of Bark one belongs to Larch, and the other possibly to *Picea excelsa*. Of three remaining specimens, two are Alder, possibly *Alnus incana*, L.; and one probably *Populus tremula*, L.

That Siberia is the source of the E.-Greenland Drift-wood, as it is (according to Agardh) of that of Spitzbergen, Dr. Kraus regards as very probable. He adds his opinion, moreover, that there is strong support for Grisebach's hypothesis, that the Flora migrated from Siberia, by Novaya Zemlya and Spitzbergen, to Greenland and Iceland, in succession.

CVII.—Geology of East-Arctic Greenland.

1.—General Account of the Geology, by Fr. Toula. From the Proceedings of the Imperial Geological Institute, February 20, 1872.

[Communicated by Count A. G. von Marschall, C.M.G.S., &c. *See also* "Die zweite deutsche Nordpolarfahrt."*]

Dr. Toula reported that the geological and palæontological specimens collected in East Greenland, between 73° and 76° N. lat., by Lieut. Julius Payer and Dr. Copeland, had been examined by Dr. F. Lenz and Dr. Fr. Toula, under the superintendence of Prof. F. von Hochstetter. Sabine, Pendulum, and Kuhn Islands were best represented in this collection. The whole continent of Greenland, between 73° and 76° N. lat., constitutes

---

* Zweiter Band. Wissenschaftliche Ergebnisse. Zweite Abtheilung. Leipzig, 1874.

1. Preface, by Ferd. von Hochstetter. P. 471.
2. Geology of East Greenland between 73° and 76° N. lat., with a geological Sketch-map. P. 475.
    (1.) General sketch of the geological constitution of E. Greenland. By Dr. Franz Toula. P. 475.
    (2.) Description of the local Geology in E. Greenland. By Dr. O. Lenz. P. 481.
3. Description of Mesozoic Fossils from Kuhn Island. By Dr. Fr. Toula. (With two plates.) P. 497.
4. Analysis of Rocks from E. Greenland, by Herr John Stingl, in Prof. Bauer's Laboratory. P. 508.
5. Fossil Plants from E. Greenland. By Prof. O. Heer. (With a plate.) P. 512.

a crystalline "massif," cut into by fjords extending far inward. In the above-named islands these crystalline rocks are associated with Mesozoic and Cainozoic deposits. Some of these islands are of volcanic nature, such as Shannon, where crystalline rocks exist only on the N.E. extremity, and Sabine and Pendulum, both entirely composed of Basalts (Dolerites and Anamesites) and of Basaltic Tuffs. The Basalts extend along a line striking N.E to S.W. They begin on Shannon and continue through Sabine, the protruding peninsula between False Bay and the Tirol Fjord, the east portion of Clavering Island and Jackson Island, as far as the coast between this last-named island, Cape Broer Ruys, and Cape Franklin. According to Lieut. Payer's statements, the Basalts constitute immense cappings, spreading in form of plateaux, seldom bearing volcanic cones of any height. They occur under the form of Dolerites (crystalline granular aggregations of Labradoric felspar, pyroxene, and magnetic oxydule of iron), or of fine-grained Anamesites (genuine peridotic Basalts), occasionally of amygdaloid tuff-like or scoriaceous Basalts. The Amygdaloids generally include Zeolites (Chabasite very frequently), and double-refractive Calcite. Along the coast of Flat Bay the Anamesite takes the form of walls, 8 feet high, and is columnar in structure, the columns measuring 5 to 7 feet in length and 1½ to 2 feet in breadth.

Miocene beds exist from the S. end of Hochstetter's Foreland to S. from Cape Seebach, in a height between 300 and 500 feet, along the foot of a crystalline ridge. They become narrower as they advance northward, where they take the form of yellowish, fine-grained Sandstones with moulds and casts of a Cytherea-like Bivalve. A quartzose Sandstone with calcareous cement exists on Sabine Island. The Sandstones of the S. side of Mount Germania include shaly beds with *Taxodium distichum miocænicum*, which occurs also in a blackish-brown Shale, and in the grey shaly Sandstones W. of Mount Germania. In these beds have been found leaves of *Populus arctica* and *Diospyros brachysepala*, indicating these beds to be coeval with the Miocenes of Atanokerdluk (W. Greenland, 7° N. lat.), Iceland, and Spitzbergen.*

Mesozoic deposits.—Jurassic Marls and Sandstones were met with on the E. and S. side of Kuhn Island. Calcareous Sandstones with organic remains are found on the S. coast of False Bay. The Jurassics on the E. coast of Kuhn Island are Marls and fine-grained Sandstones, with a Fauna very nearly allied to that of the Russian Jurassics. On the S. side they are coarse-grained Sandstones and shell-breccias, with Coal-seams. They rest on crystalline rocks, forming a high central ridge with glaciers. *Aucella concentrica* in five varieties, connected by intermediate forms, is frequent on the E. coast of Kuhn Island. The genus *Aucella* is frequent in all the Russian Jurassics, extending from the Lower Volga northward to the mouth of the Petschora, and is known to occur in Spitzbergen (*Aucella Mosquensis*). Its westernmost limit is the E. coast of Kuhn Island. The other organic remains

* *See above*, pp. 378, 432, &c.

from this locality are: *Cyprina*, sp. (*Cypr. Syssolæ*, Keys.); several *Belemnites*, among them well-preserved specimens of *Bel. Panderianus*, d'Orb.; *Bel. absolutus*, Fisch.; and indeterminable fragments of a third species. The *Ammonites* are represented only by two specimens; one indeterminable, highly involute, with a nearly circular transversal section; the other a new species, *Perisphinctes Payeri*, somewhat similar to, but evidently different from, *Ammon. involutus*, Quenst., and *Ammon. striolaris*, Rein.

The Jurassics (Middle Dogger ?) of the S. coast are brownish micaceous Sandstones, and a seam of bituminous fissile Coal, associated with indistinct and indeterminable fragments of plants. Possibly these beds may prove equivalent to the Carbonaceous Jurassics of Brora, Mull, and Sky (Scotland). These fine-grained Sandstones include an abundance of a middle-sized *Ostrea*, an incomplete cast of *Goniomya v-scripta* (Sow.), moulds and casts of *Myacites*, sp. indet.; *Modiola*, sp., reminding us of *Mod. Strajeskiana* (d'Orb.); *Avicula*, sp., possibly *Avic. Muensteri* (Goldf.), and an undetermined Belemnite. A coarse-grained variety of these Sandstones, abundant in shells, includes an Ostrea-like Bivalve with fibrous shell (*Trichites*, Lycett ?), two species of *Patella*, a *Nerita* (*Ner. hemisphærica*, Roem. ?), moulds and casts of *Trochus*, and spines of *Echinida*.

The Sandstone on the southern coast of False Bay is very calcareous and light-coloured, with cavities including crystals of calcareous spar. The only hand-specimen collected includes a *Rhynchonella* perfectly agreeing with *Rh. fissicostata*, Suess, a characteristic Rhætian form,—a young individual of a smooth and equivalve *Terebratula*,—a small, nearly circular, smooth *Pecten*,—some indistinct casts of Bivalves,—and abundant sections of Cidarite spines.

Palæozoic strata seem to be widely extended on the N. coast of Francis-Joseph Fjord, in the form of red, brown, blueish, and greenish, somewhat calcareous clay-slates, without any trace of organic remains, and of grey or black, white-veined compact Limestones. Possibly these may be analogous to the Carboniferous Limestones.

Fine-grained Gneisses, frequently separable into laminæ, prevail among the crystalline rocks. Those from the Francis-Joseph Fjord contain Garnet (Almandine) in distinct rhombic dodecahedra up to the size of the fist. Oligoclase-gneisses occur at Payer's Peak (7,000 feet altitude), in the westernmost part of the same fjord, together with Gneissoïd Micaschists. A fine Amphibolic Gneiss, with Amphibole crystals, two inches in length, appears on the north point of Shannon Island, farther northward at Haystack, and at Cape Schuhmacher, S.E. point of Kuhn Island. Granitic rocks are rather subordinate, as gneissic Granite near Cape Koner, and the large-grained Granitite between Bessels Bay and Cape Seebach (N. of Hochstetter Foreland), a compound of white and reddish Felspar, Quartz, and black Mica in large lamellæ. Crystalline Dolomite is imbedded in the Gneiss of False

Bay. It offers two varieties: the one nearly compact and yellowish-white, the other coarsely crystalline and blueish-white, with abundant interspersed lamellar Graphite.

2.—LIST of MESOZOIC FOSSILS from EAST GREENLAND. Extracted from Dr. FRANZ TOULA'S Memoir in "Die zweite deutsche Nordpolarfahrt," vol. ii., 1874.

P. 498. *Jurassic Fossils from the East Coast of Kuhn Island.*

Perisphinctes Payeri, n. sp.
Ammonites, sp. ind.
Belemnites Panderianus, d'Orb.
B. absolutus, Fisch.
B. Volgensis, d'Orb. ?
B., sp. ind.
Aucella concentrica, Keys. (non Fisch.).
" " var. rugosa, K.
" " var. crassicollis, K.
" " var. sublævis, K.
" " var. rugossima, nov.
Cyprina, sp.

Pp. 505–6. *Jurassic Fossils from the Dogger of Kuhn Island.*

Ostrea.
Goniomya v-scripta (Sow.).
Myacites, sp. ind.
Modiola Strajeskiana (d'Orb.) ?
Avicula Muensteri (Goldf.) ?
Belemnites fusiformis (Quenst.) ?
Trichites, sp.
Patella Aubentonensis, d'Arch. ?
Patella, sp.
Nerita hemisphærica, Rom. ?
Trochus, sp.
Echinite spines.

3.—ANALYSES of ROCKS, by HERR J. STINGL; pp. 508–510.

Dolomite from False Bay.
Crystalline Limestone from False Bay.
Labradorite from Dolerite in Sabine Island.
Coal from the South coast of Kuhn Island.

4.—On the FOSSIL PLANTS from EAST GREENLAND. By Prof. Dr. O. HEER. "Die zweite deutsche Nordpolarfahrt," vol. ii., pp. 512, &c. 1874. [Extracts.]

Herr Payer and Dr. Copeland collected Fossil Plants at three different places:—

1. East side of Kuhn Island. Brown-coal, with obscure plant-remains, associated with rocks containing *Belemnites*, *Ostræa*, and *Rhynchonella fissicostata*.

2. Hochstetter's Promontory. Coal and sandstone without plant-remains.
3. Sabine Island. In shales and sandstones on the east slopes of Hasenberg and Germaniaberg, both about 74⅓° N. lat.
   Taxodium distichum miocænum, Hr., the most abundant fossil plant in Sabine Island.
   Populus arctica, Hr.
   Diospyros brachysepala, Al. Br.
   Celastrus, sp.

These are of Miocene age, and belong to a plant-bearing formation like that in W. Greenland and Spitzbergen. Of the four species above mentioned, three are found also in W. Greenland. Prof. Heer remarks (p. 515), "I have stated in my 'Miocene Flora " 'of Spitzbergen' that Spitzbergen has 25 species of Miocene " plants in common with W. Greenland; and I have mentioned " these species in particular, *Taxodium distichum*, *Sequoia* " *brevifolia*, *Populus arctica*, *P. Richardsoni*, *P. Zaddachi*, " *Corylus Macquarrii*, *Quercus platania*, *Q. Grœnlandica*, *Pla-* " *tanus aceroides*, *Andromeda protogæa*, *Viburnum Whymperi*, " *Cornus hyperborea*, *Hedera Macclurii*, *Rhamnus Eridani*, " *Paliurus Colombi*, and *Nordenskiœldia borealis*. As these " species had been found in W. Greenland at 70° N. lat., and " in Spitzbergen at 78°, I remarked that very probably they once " spread over a great intermediate land and the whole of North " Greenland, and that they would be found in the Tertiary beds " of East Greenland (Heer's 'Flora Spitsberg.,' p. 12). This anti- " cipation has been fulfilled by the finding of *Populus arctica* and " *Taxodium distichum* on Sabine Island, where probably further " search would discover the other above-mentioned species. The " *Tax. distichum* is of the greatest interest, as it was certainly " the most abundant tree in Miocene Spitzbergen and West " Greenland, and exists still as far as the Southern States of " North America." *See above*, Prof. Heer's later researches, pp. 378, *et seq.*

CVIII.—On Rock-specimens from South Greenland, collected by Prof. G. C. Laube. By Dr. Karl Vrba.

[Sitzungsberichte d. k. Akad. Wissenschaften, Wien, Math.-Nat. Cl., vol. 69, pp. 91–123. 1874. With three plates.]

Dr. K. Vrba describes the following, both as to constitution and structure:—

1. *Gneiss* from Illuidlek Island (Lat. 61° N., Long. 42½° W.), associated with a coarse-grained hornblendic gneissose granite. It is fine and sometimes loosely grained mixture of black-brown mica, yellowish-grey felspar, greyish quartz, and some garnet.

2. The *Granite* of the south end of Greenland, from along the east coast, on King-Christian-IV. Land, Sedlevik Island, the district of Friedrichsthal and Lichtenau, and elsewhere. It is hornblendic at Munarsoit. It contains apatite at the South Cape of Christian's Island.
3. *Eudialyte-syenite* of the Kittisut Islands, W. of Friedrichsthal. Elæolite in the Island of Kikkertarsursoak.
4. *Orthoclase-porphyry*, piercing the hornblendic granite and over-lying red sandstone of Igalliko Fjord.
5. The *Diorite* of the many greenstone-dykes in the granite of both East and West Greenland is mostly a coarse-grained mixture of hornblende and plagioclase.
6. *Diabase*, of plagioclase and augite, occurs on King-Christian-IV. Land, and is not common.
7. The *Gabbro* of the Lichtenau Fjord consists of plagioclase, diallage, and pinchbeck-brown mica.
8. "*Weichstein*" ("soft stone") of the Greenlanders:—1. *Serpentine*, at Unortok Fjord, bordering a dyke of dioritic porphyry. 2. *Clinochlore*, (*loc. incog.*).

For some of these localities, compare GIESECKÉ's List, *above*, p. 349.—EDITOR.

CIX.—SPITZBERGEN.

Besides the references to the Natural History and Geology of Spitzbergen in this "Manual" (*see* Index, &c.), nothing can be given here except references to, and brief notes from, some of the more important memoirs and books concerning the country, such as:—

I. (1.) PAUL GAIMARD. Voyage en Islande et au Groenland, sur la Corvette "La Recherche," en 1835–36, &c. In parts, 8vo. Paris, 1838–40 (?); and Atlas, fol.

Histoire de la Voyage, &c. (I., 1838). Instructions for Sweden, Norway, and other parts of Scandinavia; for Spitzbergen (p. 412), Bear Island (p. 414), Jan Mayen (p. 424). At p. 415, Elie de Beaumont notices the possible relation of the Carboniferous plants of Cherry (or Bear) Island, with those of Melville Island and Ingloolik Bay (Parry). Full geological descriptions and illustrations of Iceland are given.

In the Atlas, the plate of *Cardium Grœnlandicum* by Beck (for both Iceland and Greenland), is the only Nat. Hist. illustration for Greenland, except some Fish.

The "Recherche" had in 1836 visited Iceland, and rounded S. Greenland, touching at Fredrickshaab, of which place and its people the Atlas contains some views. The "Lilloise," which the "Recherche" went to seek, had visited the East Greenland coast in Lat. 69° N., and Long. 27° W. in 1833.

(2.) PAUL GAIMARD. Voyages de la Commission Scientifique du Nord, en Scandinavie, en Laponie, au Spitzberg et aux Feröe en 1838–40, &c. In parts, 8vo., Paris, 1842–48 (?), and Atlas, fol.

Durocher, geology, metallurgy, and chemistry; and *Eugène Robert*, geology, mineralogy, and metallurgy, for Scandinavia, Feröe, Spitzbergen, Nova Zembla, Poland, &c. *A. Bravais, Ch. Martins*, &c., Physical Geography, Botany, &c. *Bivalet*, &c., Birds. *Thornam, Petersen*, &c., Fishes. *Kröyer, Crustacea. Boeck*, Rhizopods, &c. &c.

Many good views of cliffs, snow, ice, places, and people, &c.

II. Reisen nach dem Nordpolarmer in den Jahren, 1870–71, &c., Mit Beiträge zur Fauna, Flora, und Geologie von Spitzbergen und Novaja Semlja, von M. TH. VON HEUGLIN. 8vo. 3 vols. 1873–74. 8vo. Brunswick.

Zoology: Vertebrata, pp. 3–229. Invertebrata, pp. 229–269.

Botany: Phanerogams of Spitzbergen, p. 269, &c. Cryptogams of Spitzbergen, p. 282, &c.

Phanerogams of Novaya Zemlya and Waigatsch, p. 286, &c. Cryptogams of Novaya Zemlya and Waigatsch, p. 307, &c.

Endophytic Fungi of Spitzbergen and Novaya Zemlya, p. 317, &c.

Geology: Geology of Spitzbergen, p. 325, &c. Elevation of the Coast, p. 328, &c. Glaciers, p. 339, &c. Drift, &c., p. 342, &c.

Geological notice of Novaya Zemlya, p. 348, &c.

III. Among the Zoological works also are the memoirs by Prof. ALFRED NEWTON, Proceed. Zoolog. Soc. London, Nov. 8, 1864, pp. 494-502; by the Rev. A. E. EATON, in the "Zoologist," ser. 2, vol. viii., pp. 3762-72; ix., pp. 3805-22, 1873-4; by Dr. O. A. L. MÖRCH, on the Molluscs of Spitzbergen, in the "Annales de la Société Malacologique de Belgique," iv., 1869, pp. 8–32, in which references also to former notices of the Molluscs will be found. Also OTTO TORELL'S "Bidrag till Spitsbergens Molluskfauna jemte, en allmän Öfversigt af Arktiska Regionens naturförhållanden och forntida Utbredning." 8vo. Stockholm. 1859.

IV. In Sir JOHN RICHARDSON'S "Polar Regions" (8vo. London), 1861, are some notices of the Nat. Hist., Geology, &c. of Spitzbergen, at pp. 66, 201, 203, 204, 210, 259.

V. J. LAMONT. Notes about Spitzbergen; with appendix by Woodward, Prestwich, Salter, Horner, and T. R. Jones, Quart. Journ. Geol. Soc., vol. xvi. pp. 428–444, 1860. Some of the information in the above is repeated also in J. LAMONT'S "Seasons with Sea-horses." 8vo. Hurst and Blacket, London, 1861.

VI. The Geology of Spitzbergen has been largely treated of by Prof. A. E. NORDENSKIÖLD, in his "Geografisk och geognostik Beskrifning öfver Nordöstra Delarne af Spitzbergen," &c. Kongl. Vets.-Akad. Handl. vol. iv., No. 7. Translated into English, "Sketch of the Geology of Spitzbergen, by A. E. Nordenskjöld."

With two plates, including coloured map and sections. 8vo. Stockholm. 1867.

Prof. Nordenskiöld enumerates the researches of Parry (1827), Keilhau (chiefly Bear Isl., 1827), Lovèn (1837), Robert (1838), Torell and Nordenskiöld (1858), Lamont (1858–9), Blomstrand (1861), Nordenskiöld and Malmgren (1864); and the results of the palæontological work by Von Buch, De Koninck, Salter, Lindström, and O. Heer.

This work affords the following condensed view of the geological series represented by the rocks and strata of Spitzbergen (pp. 50, 51.)

I. **Crystalline Rocks.**

1. Granite and gneiss.
2. Vertical and contorted strata of mica and hornblende-schist, with beds of quartzite, crystalline limestone, and dolomite.

II. **Hecla-Hook Formation.**

1. Strata, at least 1,500 feet in thickness, with fossils, consisting of red and green clay-slate, grey white-veined limestone, and quartzite.
2. Red ferruginous slates and conglomerates, without fossils; of less extent and unascertained thickness.

III. **Mountain-limestone.**

1. Ryss-Island limestone, or rather dolomite, non-fossiliferous; traversed by beds of quartzite and flint. Thickness about 500 feet.
2. Cape-Fanshawe strata, containing Corals; 1,000 feet at the most in thickness.
3. Layers of hyperite.
4. Upper part of the Mountain-limestone, consisting of calcareous sandstone, limestone, gypsum, and flint, abounding in fossils, and 2,000 feet in thickness. Between the strata are extensive seams of hyperite.
5. A very extensive and regular bed of hyperite, stretching from Mount Edlund to the Thousand Islands.

IV. **Triassic Formation.**

Black bituminous shale, stratified hyperite, limestone, coprolite beds, and sandstone, with remains of Saurians,* Nautilus, Ammonites, &c.; about 1,500 feet in thickness.

V. **Jurassic Formation.**

Shales, limestone, and sandstone, abounding in pyrites, and traversed by a small seam of hyperite. At Mount Agardh the thickness amounts to 1,200 feet.

---

* Described and named as *Ichthyosaurus polaris*, *Icht. Nordenskiœldii*, and *Acrodus Spitsbergensis*, by J. W. Hulke, Esq., F.R.S., &c., in the "Bihang "till k. Svenska Vet. Akad. Handlingar, vol. i. No. 9; 8vo. Stockholm, "1873."

VI. Miocene Formation.

A freshwater formation in Bell Sound, 1,500 feet thick, and consisting of conglomerates, shale, limestone, and sandstone, almost devoid of animal fossils, but containing coal-seams and fine impressions of Plants.

VII. Recent Deposits.

Deposits belonging to the glacial period, present glacier detritus, &c.

If *Pliocene* or *Post-Tertiary* beds exist at Spitzbergen, they are most likely to be met with either in the interior of the peninsula between Ice Sound and Bell Sound, the only tract of any extent uncovered by ice, or on the eastern shore of Barents Land and Stans Foreland. A complete exploration of the first-named place, especially, would be of great interest. Doubtful traces of Post-pliocene beds were found on the shore south of the entrance to Bell Sound.

VII.—Ch. Martins, Le Spitzberg: Tableau d'un archipel à l'époque glaciaire. "Bullet. Soc. Géolog. de la France, sér. 2, vol. xxii. pp. 336–348. 1865.

VIII.—J. C. Wells, "The Gateway to the Polynia; A Voyage to Spitzbergen." 8vo., London. 1873.

In the Appendix to this book there are some notes on the Natural History of Spitzbergen. Of Insects it is said—

There are no *Coleoptera* in Spitzbergen; while 21 species are recorded as having been found in W. Greenland.

Of *Hymenoptera* there are 13 species, and only three have been noticed in W. Greenland. In the latter country *Lepidoptera* are relatively abundant, 26 having been described; while but one specimen has, as yet, been noticed in the former.

Of the *Diptera*, however, there are 49 in Spitzbergen, and only 26 in W. Greenland; and no *Hemiptera*, against four W.-Greenland species.

At pp. 351–355, the following résumé of Prof. O. Heer's researches on the Miocene Plants, &c., of Spitzbergen is given.

The fossil plants belonging to the Miocene period of Spitzbergen have been ably treated by the distinguished naturalist, Oswald Heer, in the "Kongliga Svenska Vetenskaps-Academiens "Handlingar," for 1869. The results fully prove that a warmer climate existed in Spitzbergen when these plants flourished. In his list of the Miocene Flora he identifies—

3	species of	Fungi.	14	species of	Grasses.
1	,,	Algæ.	10	,,	Cyperaceæ.
1	,,	Mosses.	1	,,	Rushes.
2	,,	Ferns.	2	,,	Aroideæ.
1	,,	Equiseta	1	,,	Typhacea.
5	,,	Cypress.	2	,,	Alismaceæ.
17	,,	Poplar.	2	,,	Irideæ.
3	,,	Taxinieæ.	4	,,	Salicineæ.
1	,,	Ephedrineæ.	3	,,	Betulaceæ.

5 species of Cupuliferæ.		2 species of Ranunculaceæ.		
1	,, Plataneæ.	2	,,	Nymphæaceæ.
1	,, Polygoneæ.	2	,,	Tiliaceæ.
1	,, Chenopodiaceæ.	2	,,	Rhamneæ.
1	,, Elæagneæ.	1	,,	Juglandeæ.
2	,, Lynantheræ.	2	,,	Pomaceæ.
1	,, Ericaceæ.	1	,,	Rosaceæ.
1	,, Oleaceæ.	1	,,	Amygdaleæ.
2	,, Caprifoliaceæ.	1	,,	Leguminosæ.
8	,, Araliaceæ.	21	,,	Dubiæ sedis.

It will be seen from this list that the plants of a temperate region once existed there; and individuals of the same species are found in the districts named below in the following order—

25 in Greenland.
8 in Iceland.
5 on the Mackenzie River.
7 in Alaska.

30 are recognised as belonging to the Arctic Flora.			
10	,,	,,	,, Baltic Flora.
5	,,	,,	Schosonetz.
2	,,	,,	the Bonn Coal.
8	,,	,,	,, Wetterau.
8	,,	,,	Bilin.
11	,,	,,	Switzerland.
5	,,	,,	France.
8	,,	,,	Italy.
2	,,	,,	Kumi (Greece).

The Insect-fauna is represented by 23 species.

Of the 9 families of the *Coleoptera*, the list contains—

Carabus; 2.
Dytiscidæ; 1.
Sylphidæ; 1.
Hydrophilidæ; 1.
Elateridæ; 2.
Serropalpæ; 1.
Donacidæ; 2.
Chrysomelidæ; 2.
Curculionidæ; 4.

Of the *Orthoptera*, *Blatta hyperborea* is the sole representative; and of the *Hymenoptera* there are two species.

Of the Marine Fauna in the Miocene period there are—

Terebratula grandis.
Dentalium incrassatum.
Dentalium, sp.
Pecten, sp.
Corbula Heukelinsi.
Corbula, sp.
Ostrea, sp.
Perna, sp.
Turbo, sp.?
Buccinum, sp.
Natica phasianella.

Of the *Polyzoa*, a new species of *Lunulites*.

CX.—Franz-Joseph Land.

(1.)—Notes on the Land discovered by the Austro-Hungarian Expedition under Lieut. Weyprecht and Lieut. Payer, in 1872–4.—By Julius Payer.

[By Permission, from the R. Geograph. Soc. Proceedings, November 1874, xix. pp. 17, &c.]

At p. 24 Lieut. Payer says "The newly discovered country (Franz-Joseph Land) equals Spitzbergen in extent, and consists of several large masses of land—Wilczek Land in the east, Zichy Land in the west, which are intersected by numerous fjords and skirted by a large number of islands. A wide sound—Austria Sound—separates these masses of land. It extends north from Cape Hansa to about lat. 82° N., where Rawlinson Sound forks off towards the north-east. The latter we were able to trace with the eye as far as Cape Buda-Pest.

"The tide rises about 2 feet in Austria Sound, and exercises but a small effect, merely causing the bay-ice to break near the coasts.

"Dolerite is the prevailing rock. Its broad, horizontal sheets, and the steep table-mountains, which recall the Ambas of Abyssinia, impart to the country its peculiar physiognomy. Its geological features coincide with those of portions of North-eastern Greenland. A Tertiary coal-bearing sandstone occurs in both; but only small beds of Brown-coal were discovered. On the other hand amygdaloid rocks, which are so frequent in North-eastern Greenland, were not met with in Francis-Joseph Land; and, whilst the rocks in the south were frequently aphanitic in their texture, and resembled true basalt, those in the north were coarse-grained and contained nepheline.

"It is an established fact that portions of North-eastern Greenland, Novaya Zemlya, and Siberia, are being slowly upheaved; and it was therefore very interesting to meet with Raised Beaches along the shores of Austria Sound, which attested that a similar upheaval was taking place there.

"The mountains, as a rule, attain a height of 2,000 or 3,000 feet, and only towards the south-west do they appear to attain an altitude of 5,000 feet. The extensive depressions between the mountain-ranges are covered with glaciers, of those gigantic proportions only met with in the Arctic Regions. Only in a few instances were we able to determine the daily motion of the glaciers by direct measurements. On the coast they usually form mural precipices, 100 to 200 feet in height. The Dove Glacier on Wilczek Land is undoubtedly one of the most considerable of the Arctic Regions. The glaciers visited by us were characterised by their greenish-blue colour, the paucity of crevasses, and extraordinarily coarse-grained ice, a small development of moraines, slow motion, and the considerable thickness of the annual layers. The *névé*, or glacial region above the snow-line, was much more elevated above the sea than in Greenland or Spitzbergen.

"Another peculiarity which characterises all the low islands in the Austria Sound is their being covered by a glacial cap.

"The vegetation is far poorer than that of Greenland, Spitzbergen, or Novaya Zemlya; and, excepting in the Antarctic Regions, no country exists on the face of the earth which is poorer in that respect. The general physiognomy of the Flora (but not that of the species) resembles that met with in the Alps at an altitude of 9,000 or 10,000 feet. The season during which we visited the country was certainly that in which vegetable life first puts forth its appearance, and most of the slopes were still covered with snow; but even the most favoured spots near the sea-level, which were no longer covered with snow, were unable to induce us to arrive at a different conclusion. On level spots even we scarcely met with anything but poor and solitary bunches of Grass, a few species of Saxifrage, and *Silene acaulis*. Dense carpets of Mosses and Lichens were more abundant, but most abundant of all was a Lichen—the winterly *Umbilicaria arctica*.

"Drift-wood, mostly of an old date, was met with on many occasions, but only in very small quantities. We once saw lying only a trifle higher than the water-line the trunk of a Larch, about a foot thick, and some 10 feet in length. The Drift-wood, like our vessel, had probably been carried to these latitudes by the winds, in all likelihood from Siberia, and not by currents.

"The country, as might have been supposed, has no Human inhabitants, and in its southern portion scarcely any animals excepting Ice-Bears are met with.

"Many portions of the newly discovered country are exceedingly beautiful, though it bears throughout the impress of Arctic rigidity."

Lieut. Payer remarks (p. 24.) that in January 1874, on Count Wilczek Island, south of Franz-Joseph Land, "the visits of Bears were as frequent then as they had been at other seasons of the year: they came close up to the ship, and were killed by regular volleys fired from deck. The Bears here are certainly much less ferocious than those we met with in Eastern Greenland, where they not unfrequently attacked us, and, on one occasion, they even carried one of the crew out of the ship: here they generally took to flight as soon as we made our appearance.

"As regards the disputed question whether Bears pass the winter in a dormant state or not, we observed that amongst the great number shot by us during two winters, there was not a single female; and during our second sledge-expedition in the spring of 1874, we even discovered a tunnel-shaped winter hole in a snow cone lying at the foot of a cliff, which was inhabited by a female Bear and her cubs.

"On encountering Bears, we found it generally most advantageous to fire after they had approached within a distance of 50 or 80 paces.

". . . . . . We frequently encountered Ice-Bears while in the Rawlinson Sound [much further to the north]. They came towards us whenever they caught sight of us, and fell an easy prey to our rifles." (p. 29.)

In Rudolph Land an increase of temperature was noticed (as in Smith Sound by the Americans), and he says (p. 30), "We had previously noticed the flight of Birds from the north, here we found the rocks covered with thousands of Auks and Divers. They rose before us in immense swarms, and filled the air with the noise of their vehement whirring, for breeding-time had arrived. Traces of Bears, Hares, and Foxes were met with everywhere, and Seals reposed sluggishly upon the ice. . . . .

"We rounded Auk Cape, which resembled a gigantic aviary, and reached the two lonely rocky towers of the Cape of Columns. Here we first found open water extending along the coast."*

(2.)—On the Newly Discovered Franz-Joseph Land. By Lieut. Julius Payer.

[From the Proceedings of the Imperial Academy of Sciences, Vienna, 17 December 1874.]

(Communicated by Count A. G. von Marschall, F.C.G.S., &c.)

This land has been surveyed by means of eleven determinations of latitude, the compass, and by geodetic operations as far as possible under the existing circumstances.

---

* "It is, perhaps, not premature to hazard the conclusion that in this new land (Franz-Joseph Land), we have the eastern continuation of the abrupt and mountainous coast of East Greenland, which, as geographers of Petermann's school are of opinion, trends away to the north-east, slightly beyond Parry's furthest point of Spitzbergen. The scarcity of animal life at first sight appears important. Around the shores of Smith Sound, Hayes tells us, the whole region teems with animal life, and one good hunter would feed twenty mouths; the sea abounds in Walrus, Seal, Narwhal and White Whale; the land in Reindeer, Foxes, Eider-duck, Wild Geese, Snipe, and Gulls of various description, and the ice is the roaming-ground of Bears. Again, to the north in Thank-God Bay, Hall's party found the plain free from snow, a creeping herbage covering the ground, on which numerous herds of Musk-oxen found pasture, while Rabbits and Lemmings abounded. The wild flowers were brilliant, and large flocks of Birds came northward. Francis-Joseph Land would, on the other hand, appear to be less favoured as regards climate. This may be attributed to the presence of colder currents than are met with in Smith Sound. But it is not too much to assume that animal life may have existed there though not seen. Admiral Osborn, that staunch advocate of Arctic research, points out that in the Parry Group there is scarcely a single island where properly organised hunting parties could not have largely added to the resources of the English who wintered there. No limit has as yet been discovered to the existence of animal life within the Arctic Circle, and it is most improbable that the failure to discover it immediately in this new region has more than a passing significance."—*Academy*, Sept. 12, 1874.

West Greenland is a high and monotonous glacier-plateau. East Greenland offers beautiful Alpine scenery, with a rich development of animal and vegetable life. Spitzbergen and Nowaja-Zemlya have the aspect of high mountain-groups, rising about 9,000 feet above the sea. Both are far less impressed with the type of Arctic severity than Franz-Joseph Land is; the latter, by its enormous glaciers and many plateaux, reminds one of West Greenland, and in the low level of the limit of its congealed snow (Firn), and in its volcano-like peaks, it bears some resemblance to Victoria Land of the Antarctic regions.

The average height of its mountains is between 2,000 and 3,000 feet; 5,000 feet in the S.W. There are great volcanic "massifs," with recent deposits in their hollows. The prevailing rock is crystalline and fine-grained Dolerite, identical with that of Greenland. Amygdaloids have not been met with, though common in Greenland. In the south the Dolerite becomes aphanitic, passing into real Basalt. To the north they are coarse-grained, and contain Nepheline. Erratic blocks are scarce.

Some of the newly discovered islands of this group must be of considerable extent, as they bear enormous glaciers, whose abrupt slopes, sometimes 200 feet high, generally border the coast. The greater part of Franz-Joseph Land seems to be buried under snow. The vesicular form of the glaciers on all the smaller islands is remarkable and peculiar.

The *Tides* upheaving the ice in the bays, and breaking it against the shore, do not reach above two feet along the coasts of Austria Sound.

The *Vegetation* of Franz-Joseph Land is extremely scanty, far more so than that of Greenland. Its general character is that of an Alpine Flora of between 9,000 and 10,000 feet above the sea. The hollows, free from snow, and even in the most favourable situations, offer no richer aspect. Level places are scantily beset with Grasses, some few species of *Saxifraga* and *Silene acaulis*. *Cerastium* and Poppy are of rare occurrence. Mosses and Lichens are more frequent. Among them *Umbilicaria arctica* prevails, which is found very high up in Greenland.

Drift-wood of old date, is common, though in small quantity. It was probably driven thither by winds.

No traces of human habitations were met with. In the southern part White Bears and passing Birds were the only representatives of Animal Life, excepting the Walrus, seen only twice.

CXI.—On the GLACIATION of the POLES of the EARTH. By HENRY WOODWARD, ESQ., F.R.S., Pres. Geol. Assoc., &c.

[I. With Permission; from the Proceedings of the Geologists' Association, vol. iv., pp. 17–24. 1875.]

[In connection with the Glacial Accumulations at the Polar Regions, and their varying conditions, the following extracts from Mr. H. WOODWARD'S *Presidential Address* to the GEOLOGISTS' ASSOCIATION, November 6th, 1874, supply a résumé of much that is known, and bear upon the migration of Faunæ and Floræ, such as that treated of (*above*) in Dr. HOOKER'S "Memoir on the Distribution of Arctic Plants," at pages 200, 207, &c.—EDITOR.]

. . . . . If the obliquity of the Earth were increased, a greater and yet greater area would be brought under Arctic conditions. When the obliquity reached 35½°, the Arctic Circle would extend to lat. 54½°, and the Tropics to 35½°, reducing the Temperate Zones from their present width, of 43° each, to 19°, one-half of the decrease being added to the Tropics, and the other half to the Arctic Zone. The half near the Equator would have its

FIG. 8.

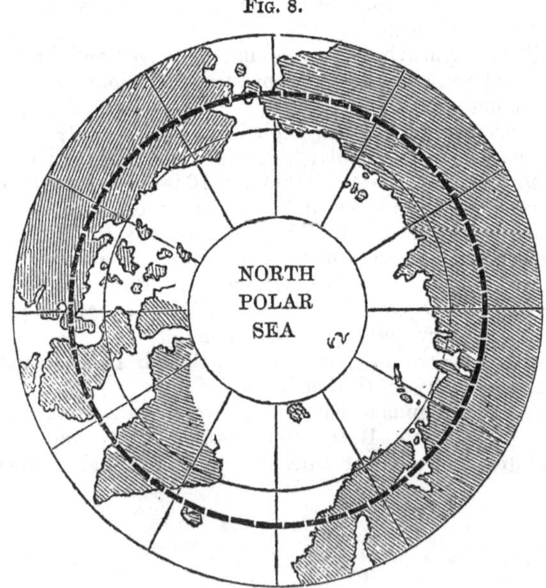

Fig. 8.—Diagram of North-Polar Sea, to contrast with the South Pole (Fig. 9). The South Pole has a land-surface surrounded on all sides by water; whereas the North Pole appears to be a marine area almost entirely surrounded by land.

[These woodcuts (Figs. 8 and 9) have been kindly lent by the President of the Geologists' Association.]

temperature greatly raised, and become an additional area of evaporation; whilst that nearest the Poles would be an additional area of condensation.

As the ice extended and consolidated, it would gradually shut off the warm currents of the ocean, now extending nearly to the Pole; and the heat so diverted would tend to assist and increase evaporation, and thus precipitation would also be proportionably increased.

It is a fundamental part of Mr. CROLL's theory, that the glaciation of the two hemispheres should take place separately and alternately.

It is an equally fundamental principle of Mr. BELT's theory that they should be glaciated simultaneously.

In support of his views, Mr. BELT reminds us—

(*a.*) That glacial conditions due to the obliquity of the Ecliptic exist at the present time around *both Poles.*

That more ice is heaped on the Antarctic regions than upon the Arctic is explicable by the fact that a larger area for evaporation is afforded in the southern hemisphere by its more extended ocean-basins; whereas, in the north, not only are the Arctic regions almost encircled by lands, and so offer a far less area for

FIG. 9.

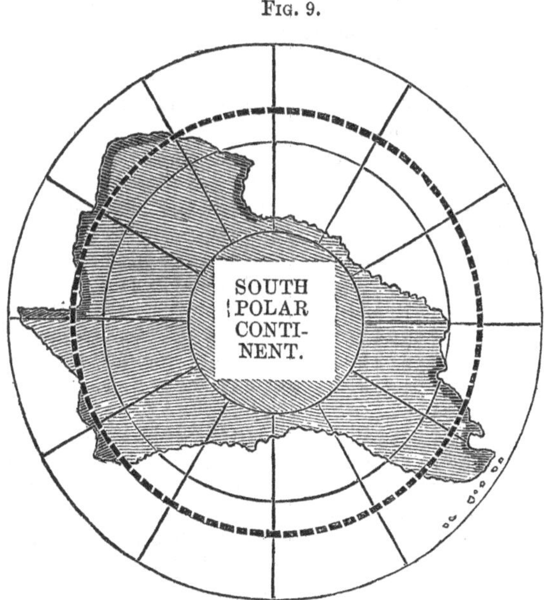

Fig. 9.—Diagram of the South-Polar Continent, to illustrate the vast land-area forming the gathering ground for the great Antarctic Icebergs, referred to by Mr. James Croll.

The darker parts along the margins of the South-Polar Land indicate those portions which have been more or less accurately surveyed, such as South-Victoria Land (with Mount Erebus), Enderby Land, Trinity Land, &c.

evaporation, but much of the snow which would doubtless reach the Polar regions is intercepted by mountain-chains and added to local glacier-systems, such as the Alps, the Himalayas, and other high northern ranges. (Compare Diagrams, Figs. 8 and 9.)

That the difference is due to lessened precipitation, and not to differences of temperature, will be seen by following the isotherm of 30° around each hemisphere.

In the southern hemisphere it deviates but little from the line of 60° S. lat. In the north it is much more irregular, but the mean is again about 60° N. lat., proving that if the precipitation were the same, there would be as much snow and ice north of lat. 60° N. as there is south of lat. 60° S.

(*b.*) Accumulations of snow have only been observed on one planet viz., Mars, which, with an obliquity of 31½°, is *glaciated at both Poles* at the same time.

Judging, then, from analogy, we might expect the Glacial Period to have been *contemporaneous* in both hemispheres.

(*c.*) Many plants and animals are found in both the northern and southern Temperate Zones, having close affinities, and even pointing to a common origin, yet separated by the whole width of the Tropics, which they cannot now pass.

Mr. DARWIN has explained their presence by supposing that during the Glacial Period they were driven to the high lands of the Tropics by the advancing ice, and that on its retreat they followed it north and south.

A glacial period in one hemisphere only would not, Mr. BELT thinks, afford this means of migration. The plants and animals driven south by the northern ice would always have a hot zone to the south of them which they could not pass.

(*d.*) Mr. ALFRED TYLOR has suggested that the piling up of ice in the nothern hemispheres during the Glacial Period would lower the level of the general ocean 600 feet.

Mr. CROLL, on the contrary, in his recent papers on the subject (*Geol. Mag.*, July and August, 1874), shows that, if each hemisphere were glaciated alternately, the level of the ocean would be raised and *not* lowered in the one in which the ice accumulated (as already explained) by the melting of the ice at the opposite pole, and the shifting of the centre of the earth's gravity towards that covered by an ice-cap.

Mr. BELT concludes that one Polar ice-cap could not materially lower the general ocean, if the opposite pole were freed from ice; but, on the contrary, if the Glacial Period in the two hemispheres were contemporaneous, then the water abstracted from the sea and frozen into ice at the Poles, together with that impounded in the great lakes of North America, North Europe, and North Asia, by the blockade of the northern drainage of the continents with ice, must have lowered the general level of the ocean to a great extent.

From data gathered in Central America, and in Siberia, Mr. BELT estimates the lowering of the sea-level by the accumulation of ice at the Poles could not have been less than 2,000 feet, and may have been much more.

Applying this result to the solution of the problem of insular floras and faunas, he points out that their derivation has usually been explained by supposing such islands to have been at one time joined to the continents nearest to them, and to each other in Post-Tertiary times. In every case, therefore, the last movement has been one of depression, the bridge by which they passed over has been submerged and destroyed . . . .

It is a significant fact, first noticed by WALLACE that, whilst all islands having *shallow channels*, however broad, separating them from each other, and from not distant continents, give evidence of a former connection in Post-Tertiary times ; on the other hand, islands surrounded by *deep water* are marked by peculiar faunas. Thus Madagascar, though near the coast of Africa, is separated by a deep sea, and its fauna and flora are singularly distinct. The Galapagos islands have also a peculiar fauna, indicating the antiquity of their insulation.

A uniform lowering of the sea-level would afford a satisfactory solution of these difficulties which cannot so readily be explained by local subsidences and elevations. . . . .

A rise of the sea, owing to the increase of its volume—liberated after the Glacial Period from the ice of the Polar regions—would produce the same effect as the lowering of the land, which might not occur *generally* with the same precision of level as that of the waters of the ocean.

The continent of South America, and indeed the mountain-chains generally, all over the world, testify to the fact that they have been glaciated to lower levels upon their slopes than that to which ice and snow now descend, or where they occur perennially. A general lowering of the sea-level would produce precisely the same effect on climate as the raising of the continental surfaces ; the atmosphere, following down, would bring the freezing point so much lower than it otherwise would be.

An increase in the present obliquity of the Ecliptic would not only permit a greatly increased accumulation of ice on Circum-Polar lands, but it would be the cause of lowering the mean temperature of the Tropics ; so that the snow-line would even descend still further from increased precipitation, due to the greater evaporation. . . . .

If, as Mr. BELT concludes, the melting as well as the accumulation of the ice of the Glacial Period must have occupied thousands of years in accomplishment, this is quite in accordance with the gradual growth of coral-reefs, and the silting up of deltas filled with freshwater deposits. But probably some sudden rises did from time to time occur, causing local floods and the inundation of vast low-lying tracts of country.

Since the close of the great Ice-age, there is clear and indisputable *evidence of the Rise of Land* * both towards the North and the South Pole, which is probably even now in progress. . . .

---

* *See above*, Mr. HOWORTH on the Elevation of the Circumpolar Regions. —EDITOR.

A part of the coast of Greenland is said to be in a state of subsidence; but Greenland *is now undergoing intense glaciation*, and is buried in snow and ice piled mountains high. All the geological evidence is in favour of these lands having been above the sea in Pre-glacial times. The lands around the Poles must then have sank down after they were covered with ice, and they have been slowly rising since it melted away.

Mr. BELT believes the *cause of the depression* was the piling up of so vast a weight of ice around the Poles, and that the *cause of the elevation* was the removal of that vast weight by the gradual melting of the ice. . . . .

That the movement of elevation still continues in some places only shows that the earth is a (more or less) rigid body, and gives way but slowly to great strains.

In considering the theories which have been put forward in order to account for the cold of the Glacial Epoch, one important point to be borne in mind is the *relative time* assumed to be occupied by the respective operations of eccentricity of the Earth's orbit, on the one hand, and greater obliquity of the Ecliptic, on the other.

Thus, Mr. CROLL's theory of the eccentricity of the earth's orbit requires a period of 200,000 years to elapse. On the other hand, Mr. BELT's theory of the greater obliquity of the Ecliptic would not require much more than 20,000 years, or one-tenth of the time demanded by Mr. CROLL. . . . .

If our Glacial period merely necessitated, as Mr. CROLL supposes, the heaping up of snow and ice around the North Pole, the only result would be a slight shifting of the centre of gravity of the Earth northwards; but, if it was contemporaneous in the two hemispheres (as Mr. BELT assumes), the figure of the Earth would be changed, its Polar diameter would be lengthened, its mean Equatorial diameter shortened, and a series of strains would be set up, tending to restore its figure to a state of equilibrium. And if, during the Glacial Epoch, that state had been arrived at by the sinking-down of the circumpolar land, and the rising of land in the Tropics, then, when the ice melted away, the Polar diameter would be shortened, the mean Equatorial diameter lengthened, and forces would be set in operation tending to lower the land at the Tropics, and raise that around the Poles.

That the deepening of tropical seas, as evidenced by the growth of Coral islands, is due to an actual sinking of the bed of the ocean is perfectly in accord with this theory. But whether by the deepening of the sea by addition to its volume, or by depression of its bed, the cause (according to Mr. BELT) has been the gradual melting of the ice piled up during the Glacial Epoch, which, by its liberation, has disturbed the equilibrium of the figure of the Earth. . . . .

PART II.—PHYSICS.

SYNOPSIS OF CONTENTS.

INTRODUCTION.

The following account of the physical work done in the Arctic regions by past Expeditions must be regarded as a collection of *rough hewn facts*, which have been hurriedly brought together at the request of the Arctic Committee of the Royal Society in about *six weeks* during the months of March and April. In consequence of the short time given for collating and arranging, it will, no doubt, be found that many of these facts have not fallen into their proper places, so as to present a clear and complete account of the state of our knowledge on each separate branch of the subject, and that some important results are altogether wanting; nor has it been possible, for the same reason, to discuss the observations or to form them into one harmonious whole, so as to show their relations to one another.

As most of the temperatures are given in Fahrenheit degrees, the observations made by German observers, which are expressed in degrees Reaumur, have been converted into degrees Fahrenheit; had time permitted, it would probably have been better to have converted all the observations of temperature into degrees Centigrade, especially as the observations on board the "Challenger" are so expressed. In some cases, especially in the accounts of the earlier voyages, where the physical observations of all kinds are not numerous, and are scattered throughout the narrative, it has been impossible to classify them under their several headings; but the general arrangement of this, the second part of the "Manual," is according to subjects, and in each subject as far as possible a comparison has been made of the observations and conclusions arrived at by the several observers, and other important matters added bearing on the subject.

My thanks are due to R. H. Scott, Esq., F.R.S., for his kind interest in the work and for the loan of valuable papers; and I have also to thank numerous Authors, Publishers, and Societies for permission, of which I have freely availed myself, to make use of valuable papers, relating to Arctic scientific work; these are specially referred to in the "Manual."

If the curves of temperature of the air and of the surface of the sea be laid down graphically on the same sheet, there will generally be seen to be a remarkable correspondence between them, and their irregularities may readily be compared. It would be convenient to have sheets of regularly ruled papers (similar to Letts' divisional papers, with *ten* divisions to the inch) arranged with two or more separate divisions, one for the barometric curves, *one* for the temperature of the air in *black* ink, and for the temperature of the sea-surface in *red* ink; another might be added for the dry and wet bulbs, or Regnault's hygrometer and the hair hygrometer, to be distinguished in the same manner. It would be also of great service if the changes in the barometric and thermometric curves were laid down graphically on a sheet in such a way as to show their relation to the winds. The clearness of the tables and graphical illustrations contribute not a little to the value of the record of the Expedition to East Greenland under Captain Koldewey, as published by the Committee of the German Arctic Expedition.

Since all descriptions of phenomena which may be called *appearances*, such as the forms and character of clouds, and especially the display of the aurora, depend not only on the phenomena, but also on the sense of relief felt by the observer in having something to interest him beyond mere routine, it would be well that the description of such appearances should be accompanied by sketches. For clouds reference should be made to the excellent sketches of clouds in the Report of the Proceedings of the Conference on Maritime Meteorology, 1874; sketches of the Aurora similar to these would be of very great service. Good sketches of Aurora may be made in Chinese white and indigo on grey paper. (Such sketches have been made by M. Capron, who suggests that drawings should be made through a grating of wires or threads upon ruled paper.)

The excellent papers of M. Lemström and M. Wijkander, taken in connexion with the discoveries of M. Angström and other observers on the Spectrum of the Aurora, seem to show clearly that the Aurora is an Electrical effect, closely resembling the discharge in a vacuum tube; whilst the important results of M. Wijkander for determining the electrical state of the air point to some very important questions with regard to the conduction of electricity by air at different temperatures. Some light would probably be thrown on these questions by measurements of Earth currents in the Arctic regions.

It does not appear, from M. Wijkander's results on the Electricity of the Air, that there is any distinct relation between the conducting power of air and its temperature: for we find that the air insulates well at $-15°$ and $-16°$; whilst at another time at $-17°$, or between $-17°$ and $-13°$, the charge instantly disappears: we also find that although at lower temperatures air usually insulates well, yet at times the insulation is not good at $-27°$ or again at $-31°$.

When the temperature of the air is or has been increasing, the electrometer may be somewhat colder than the surrounding air; in such a case moisture will be condensed on the insulating supports and the charge will be lost; whereas when the temperature of the air is diminishing, and the electrometer warmer than the air, no moisture will be condensed upon the supports and the insulation will be good.

This may explain M. Wijkander's results, and is in accordance with the results of experiments by Sir W. Thomson, that Air, whether dry or moist, always insulates perfectly at all ordinary temperatures.

If, as M. Wijkander supposes, Air at a low temperature becomes a good conductor of electricity, then it will be well to employ his method of using the Quadrant Electrometer to detect it; the quadrants may be kept at a constant difference of potential by employing Mr. Latimer Clark's constant cells. It might also be possible to test the direct effect of the Aurora on the needle of a galvanometer, by placing an insulated copper plate or ball above the mast-head, or in an elevated position above the Observatory, with an insulated wire connecting it to the galvanometer, the other galvanometer-wire being connected with a metal plate buried at some depth in the ground or in contact with water. For this purpose and for detecting Earth currents, a delicate thick-wire galvanometer would be required.

As far as lines of equal cold can be laid down from the observations made in the most extreme latitudes on the temperature, the dryness and the direction of the winds, it would seem that a pole of greatest cold may possibly be met with in the region to the north-west of Melville Island and Wellington Channel; and supposing that the tides which meet in Wellington Channel are two branches of the Baffin's Bay tide, it seems probable that in the neighbourhood of the pole of greatest cold there is also a pole or region where the three tides meet which enter the Arctic regions by Behring Strait, Baffin's Bay, and the North Atlantic or Spitzbergen Sea. What relation these facts may have to the supposition that cold upper currents descend to the Earth's surface at the pole of greatest cold remains to be considered. Of true currents independent of the tides there seem to be few. The Gulf Stream flows to the north on the west and east of Spitzbergen, and from the north or north-east come two currents divided by North Greenland, the westerly current probably flowing through Robeson Channel and Lady Franklin Strait westward, beyond which we know nothing of it; but we find in Barrow Strait and Bellot Strait and Lancaster Sound a return current flowing from the west into Baffin's Bay.

As these currents seem to have nothing to give rise to them but the Gulf Stream and the Siberian rivers, may not the current on the north of Greenland flowing westward be regarded as the continuation of this stream, which, flowing till it strikes against Kellett Land, is reflected back to the south of Melville Island?

There is one very remarkable fact with regard to the direction and temperature of the winds. On the east and north of Spitzbergen it has been found by Lieutenant Weyprecht and others that the prevailing winds are from the south-east and east, driving the ice round the north of Spitzbergen. As we go westward to Polaris Bay and Kennedy Channel, and almost everywhere immediately west of Greenland except at Van Rensselaer, the prevailing winds are from the north-east and east, *except in summer*, and they are invariably warm winds loaded with vapour. On the east of Greenland the prevailing winds, *except in summer*, are from the north; they also are the warmest winds, and are loaded with vapour. This seems to show that the prevailing winds to the east of Spitzbergen, blowing over the region of the Gulf Stream, are warmed by its heat, and take up vapour; retaining their warmth and vapour, they become the north-east warm winds of the regions west of Greenland and the north winds of the East Greenland coast.

Also their direction both on the east and on the west of Greenland is precisely the direction of the currents which may be regarded as the continuation of the Gulf Stream.

From the evidence of Parry and others it appears that sounds travel very long distances in the Arctic regions during winter, when the atmosphere is no doubt very homogeneous ; it would be interesting to test whether sounds can be heard equally and with equal clearness over the same distance in opposite directions, or whether at times sounds seem to travel farther in one direction than in the opposite.

Making allowance for the wind, the theory of Sound requires that any sound shall travel with equal speed and equal intensity between two points, whether it be going in one direction or in the opposite. It may be expected that, in experiments over the ice-floe, spaces of open water intervening will tend to deaden sounds by producing changes of density of the atmosphere, even when to the eye the atmosphere may be perfectly clear. That sounds are sometimes deadened in a clear atmosphere is clearly shown by the experiments of Dr. Tyndall in the English Channel, which also show that even in a thick fog which light cannot penetrate sounds may be distinctly heard for long distances ; under such circumstances the atmosphere may probably be regarded as consisting of a homogeneous fog. The mists which sometimes prevail over open water will probably afford opportunities for testing these interesting questions.

W. G. ADAMS.

Physical Laboratory, King's College,
May 1875.

The under-mentioned Physical Subjects are treated incidentally in Part I.

1. J. TAYLOR. Temperature of the Kickertine Islands, in Cumberland Gulf (Inlet or Sound), Davis Strait, p. 243.
2. P. C. SUTHERLAND. (1.) Glacial Conditions and Temperature of Davis' Strait, Baffin's Bay, and Barrow Strait, p. 356. (2.) Glaciers of their Coasts, and Remarks on Ice, p. 358. (3.) Icebergs and Coast-ice, p. 360. (4.) Polar Currents, and Specific Gravity of Sea-water, p. 363.
3. D. WALKER. Temperature of the Soil at Port Kennedy, Boothia, p. 528.
4. W. FLIGHT. Auroræ, p. 465.

SUMMARY OF VARIOUS EXPEDITIONS, SOME NOTICE OF WHICH IS CONTAINED IN THE "MANUAL."

I. Parry's First Voyage in search of a North-west Passage, 1819–20:

Observations on Tides at Melville Island.
Determination of the three magnetic elements.
Observations on Sound.
Existence of twilight throughout the year.
Pendulum observations.
Description of Aurora Borealis.

II. Parry's Second Voyage in search of a North-west Passage.

III. Parry's Third Voyage for the discovery of a North-west Passage.

IV. Parry's Voyage to Spitzbergen and approach to the Pole.

V. Sir John Ross's Second Voyage in 1829–33:

Meteorological observations.
Observations on refraction.
Auroras.
Magnetic observations.

VI. Sir John Franklin's Second Expedition to the Polar Regions:

Observations on velocity of sound.
Auroras.

VII. Expedition of Capt. Maguire, in H.M.S. "Plover," to Point Barrow:

Magnetic observations.
Auroras.

VIII. Sir Edward Belcher's Expedition in search of Sir John Franklin:

Meteorological observations.
Tides.
Rate of formation of floe-ice.
Composition of floe-ice.
Remarkable refraction, "Paraselenæ."
Observations of Aurora.

IX. McClintock's voyage of the "Fox":

Meteorological observations.
Tidal observations.
Ice observations.
Observations on refraction.
Magnetic observations.
Auroras.

X. Second Grinnell Expedition of Dr. Kane in 1853–55 to Smith Sound and Kennedy Channel:

Meteorological observations.
Tidal observations.
Sea temperatures.
Magnetic observations.

XI. Expedition of Dr. Hayes, 1860–61.
- Meteorological observations.
- Tidal observations.
- Magnetic observations.
- Pendulum observations.
- Measurement of glaciers.
- Auroras.

XII. Observations in Polaris Bay, made by the Expedition under Capt. Hall, from a letter by Dr. Bessels, to the French Geographical Society.

A small observatory was built in Polaris Bay (lat. 81° 31′ N., long. 61° 44′ W.) at a height of 34 feet above the level of the sea. The instruments were—

(1.) A transit instrument.
(2.) A pendulum.
(3.) A theodolite.
(4.) A sextant by Gambey, graduated to 10″.

Meteorological and magnetic observations were also made.

The magnetic instruments were—

(1.) A unifilar magnetometer.
(2.) A dipping needle.
Several prismatic compasses.

A great part of the work was lost when the vessel was wrecked in Baffin's Bay, but some of the meteorological records were saved. They include hourly observations on—

- Temperature of the air.
- The barometer, anemometer, and hygrometer.
- Terrestrial and solar radiation.
- The Aurora and on Ozone.

Twenty series of pendulum experiments were made to determine the figure of the earth. Dr. Bessels says that the magnetic observations were more complete than any of those made up to the present time in the Polar regions.

The declination was 96° 00′ W., and the absolute declination [? inclination] 84° 23′.

NOTE.—As Dr. Bessels gives this value as a correction to the statement made in America that the *inclination* is 45°, the word *declination* in his letter is probably a misprint. Hourly observations for declination were made for five months, and on three days in each month they were made every six minutes.

Dr. Bessels also says that the observations on magnetic intensity, &c. were lost during the storm in the ice. Although he does not say so, yet there is great fear from the way in which he speaks of them, that the pendulum observations are also lost. His account of the scientific work of the Expedition is not yet published, but he says that the first volume on the physical observations will be ready very shortly.

XIII. Physical work of the second German Expedition: Meteorology and Hydrography.

1. Observations and full discussion of the temperature of the Air at Sabine Island and on the East Greenland coast.

2. Temperature and thickness of ice, and the wind and weather.
3. The barometer.
4. The temperature of the sea and currents.
5. The tides.
6. Specific gravities of salt water.

Geodetic measurements.
Astronomical observations along the coast.
Magnetic observations for the three magnetic elements.
Remarks on the Aurora Borealis.
Measurement of the velocity of sound.
Measurement of the motion of glacier in Franz-Joseph-Fiord.

XIV. The second Swedish Expedition under Professor Nordenskiöld.

In the first Swedish Expedition to Spitzbergen in 1868, in addition to the meteorological, magnetic, and other observations, attempts were made by M. Lemström to determine the spectrum of the Aurora, and also the electrical state of the air. His apparatus was not sufficiently delicate to give satisfactory results; but to M. Lemström we owe the discovery of a spectroscope which has revealed several of the lines of the spectrum of the Aurora, not only to himself at home, but also to M. Wijkander in the Arctic regions, on the second Swedish Expedition. An account of these results will be found under Aurora (p.737).

Among the general results of this Expedition, which are not yet published, are contained:

I. Hourly Meteorological observations.
II. Astronomical positions of stations, &c.
III. Pendulum and refraction observations in great cold.
IV. Hourly Magnetic observations, besides observations every five minutes on two term days a month in connexion with observations at Upsala.
V. Tides and Currents.

The record of the meteorological observations will be ready very shortly.

XV. Austro-Hungarian Expedition under Lieut. Weyprecht:

Meteorological observations, winds, and ice-drifts.
Formation and melting of ice.
Astronomical observations.
Magnetic observations, including disturbances of all three magnetic elements.
The Aurora.

The "Manual" also contains an account of a discussion of the temperatures of places in the Arctic regions by M. Dove, to which more recent observations have been added. It also treats somewhat fully of the Aurora, and its relation to the earth's magnetism, to the electrical state of the air, and to observed phenomena of sun spots.

THE PHYSICS OF THE ARCTIC REGIONS.

I.—METEOROLOGICAL OBSERVATIONS.

It will probably be unnecessary to discuss at length the records of all those ordinary observations which have been made by previous expeditions to Polar regions, and which are usually included in the ship's log.

These include—

(1.) Temperature of the air.
(2.) Direction and force of the wind; and the weather.
(3.) The barometer.
(4.) The temperature of the sea at the surface.
(5.) Latitude and longitude.

These observations have generally been made every one or two hours, instead of every watch.

In Dr. Hayes' expedition they were made every two hours, but hourly observations were made for part of the winter, and the mean of hourly observations of temperature differed from the bi-hourly mean by less than 0°·04.

Also in McClintock's expedition the hourly and two-hourly series of observations of the barometer give precisely the same mean.

Observations have also generally been made for—

(6.) The temperature of the sea at different depths.
(7.) The specific gravity of sea water.
(8.) Tides and currents.

There are some few points on the relative temperature of different places, and on the connexion between the temperature and the direction of the wind to which it may be well to draw attention.

1. Meteorological Observations by Sir Leopold McClintock.

The meteorological records of McClintock's expedition in the "Fox" are discussed in the fourth number of Meteorological Papers of the Board of Trade, 1860.

They are also fully discussed in the "Smithsonian Contributions to Knowledge" (Vol. 13). The following facts relating to this expedition are drawn from these sources or from McClintock's Voyage of the "Fox."

Temperature.

"Among the principal features of this Arctic register will be found instances of temperature 48° below zero; a mercurial column of 31 inches (very nearly) at the sea level; notices of warm winds not only from the south-eastward, but also from the north-westward (in the high latitude of 74°), and of several cyclones."

On December 28th, when drifting on the ice down Baffin's Bay, McClintock says :—

"We have been in expectation of a gale all day. This evening there is still a doubtful truce among the elements. Barometer down to 28·83 ; thermometer up to +5°, although the wind has been strong and steady from the N. for twenty-four hours, low scud flying from the E., snow constantly falling. An hour ago the wind suddenly changed to S.S.E. ; the snowing has ceased ; thermometer falls and barometer rises."

Northerly winds prevailed throughout December and January, and McClintock attributes the drift down Baffin's Bay entirely to the winds. He says, "During December we drifted 67 miles: "We move before the wind in proportion to its strength ; we "remain stationary in calm weather ; neither surface or submarine "current has been detected."

A table of the winds and ice-drifts is given on p. 110.

Capt. Allen Young, in his summary of the voyage, says :—

"Dec. 26th, 74° N., 66° W. During Divine Service the wind increased, and towards the afternoon we had a gale from the *north-westward,* attended by an unusual *rise of temperature.* The gale continued on the 28th with a *warm wind* from the north north-west."

"The Danish settlers at Upernavik are at times startled by a similar sudden rise of temperature. During the depth of winter, when all nature has been long frozen, rain sometimes falls in torrents. This rain comes with the *warm south-east wind.*"

McClintock notices the formation of icebergs going on in Melville Bay in August 1857, when the mean temperature of the air is +35° F. when drifting in the pack down Baffin's Bay.

The mean temperature for February 1858 was −16° F., for June +36°, and for July +40°.

At *Port Kennedy.*—The mean temperature for January 1859 was −33° ; for February, −32° ; for March, −3°.

The minimum temperature of the earth 2 feet below the surface was found to be half a degree above 0° F. during the winter of 1858–59.

The greatest cold recorded was −48° F., in January and again in February at Port Kennedy ; and the maximum temperature of those months, −14° and −12° F.

June *was the warmest,* and January the coldest month.

The greatest cold at Port Kennedy was generally recorded during five days after full moon. The five-daily means of McClintock's observations, both in Baffin's Bay and Port Kennedy, bear out the statements of Artic travellers that it is coldest about the time of the full moon. This is also borne out by the observations of Sir John Ross and of Dr. Kane.

A comparison of the temperature with the direction of the wind shows that in *Baffin's Bay* the S.E. wind is the warmest, next to it the N.E., and the S.W. the coldest.

At Port Kennedy the East wind is warmest, next to it the South, and the N.W. the coldest.

McClintock also notices a dense and continued mist over Bellot Strait, caused by considerably warmer water than the air above it; and strong local winds.

The prevailing winds in Baffin's Bay and at Port Kennedy are about N.W. by N., whereas the prevailing winds at Van Rensselaer are S. by W., or S.S.W.

The average velocity of the resulting wind in miles per hour was:—In Baffin's Bay, 6 miles per hour; at Port Kennedy, 11·4; and at Van Rensselaer, 4·5 miles per hour.

At Port Kennedy the average velocity of most of the winds was about 10 or 12 miles an hour; but of the S. and S.E. winds the average velocity was from 19 to 20 miles an hour.

The Barometer.

The more remarkable storms at Port Kennedy were accompanied by sudden falls of the barometer.

Readings were generally taken with the aneroid barometer, but observations were made with both the aneroid and an excellent mercurial barometer during the stay at Port Kennedy.

The discussion of the observations with the barometers shows that the hourly and bi-hourly series of observations give the same mean, and that the mean deduced from observations every four hours does not differ much from them.

	Inches.
Mean of 24 observations a day -	30·049
„ 12 „ „ -	30·049
„ 6 „ „ -	30·047

The diurnal variation of the barometer in *Baffin's Bay*, in latitude 72°·5 N., shows an increase from 4 a.m. to 6 p.m., and a more rapid diminution from 6 p.m. to 4 a.m.; the pressure having its mean value at midnight and at 10 a.m.

At *Port Kennedy*, from 10 p.m. to 4 a.m., there is a rapid fall of the barometer, then a rapid rise to the mean value at 9 a.m., still a rise until noon, then the barometer remains pretty steady, but on the whole rising a little until 10 p.m.

The range of fluctuations is:—

	Inches.
In Baffin's Bay - - -	0·028
Port Kennedy - - -	0·048
Van Rensselaer - - -	0·010
Port Foulke - - - -	0·014

---

2. Record of Dr. Kane's Meteorological Observations. "Smithsonian Contributions," vol. 11.

The *Meteorological* Observatory was a wooden structure, 140 yards from the ship, on the open ice floe, latticed and pierced with holes on all sides, and firmly frozen to the ice. To keep out drift, a series of screens were placed at right angles to each other, so as to surround the inner chamber in which the thermometers were suspended. They were read and illuminated through a lens and pane of glass without going inside the screens.

The Temperature.

The temperature was lower in winter at the Astronomical Observatory and in the outer channel than at the Meteorological Observatory.

The following notes, as to the amount of daylight, are added to the tables :—

"Nov. 10. Thermometers could not be read without a lantern.

"Nov. 22. Barely able to read at midday.

"Jan. 19. Able to read large type by illumination from southern sky.

"Jan. 22, 1854. Thermometer read at 12 o'clock without a lantern.

"April 19. The sun was refracted above the horizon at midnight.

"May 20. Sun's power felt. Black tarpaulin sinks on the snow about 2 inches in a day.

"Nov. 19, 1854. Could not read Parry's type at noon.

"Nov. 29. Mercury congealed at $-42°\cdot7$, and melts at $-38°$.

"Rise of temperature during fall of snow; maximum fall of snow during the period of new moon, as seen in tables; minima of cold near the time of new moon.

"Superior maximum of Dec. 27th to Jan. 1st, 1853–4, during which time it snowed 44 hours."

This should be taken in connexion with the direction of wind, which on Dec. 28th blew a gale from S.E. (magnetic), during which the temperature rose nearly 23° F. in five hours, from $-6°\cdot4$ F. at midnight to $+16°\cdot5$ at 5 a.m.

During December 1853 there are clear indications of a rise of temperature accompanying S.E. wind, and so often does this occur that the following remark is appended to the table of force of wind, p. 65 :—"The connexion of the rise of temperature with "the wind is embarrassing." (Sm. Con., vol. ii., p. 10.)

Under table of temperatures, Dec. 1st, is note, "Temperature "falling $21°\cdot8$ in eight hours." On referring to the table of wind it is found that this is during a wind from the N.E., followed by a calm. With the S. and S.E. wind following, there is a rise of temperature. So again on Dec. 4th and 5th.

Again, on Feb. 7th, 1854, there is a rise of temperature of 35° F. in 12 hours, during a calm of four hours, followed by wind from E. and S. during eight hours, during which thermometer rose to $20°\cdot5$ F.

To this there is the remark :—"These warm changes are very "trying to the health, and *curious in their relation to the winds.*"

On the next day one hour's north wind sent the temperature down 10°, but three hours after the wind from the S.E. sent it up 10°.

"March 9. Between 1 a.m. and 2 a.m. the temperature rose $12°\cdot5$. At the close of the watch there was a fresh breeze from the eastward.

"March 22. Temperature rises $11°\cdot1$ between 6 and 7 a.m. Wind S.E."

July 23. 10 and 11 a.m. Highest temperature of the season, +51°·0 after a calm for two days.

"The warmest month is July, and the coldest is March, but December is almost as cold, and December is colder than March during the second winter."

As other observers have noted, so Dr. Kane often remarks :—

"There is a seeming connexion between the increasing cold and the increasing moonlight."

"The full moon season, with cloudless nights, is always in correspondence with the lowest mean temperatures."

"p. 59.—Nov. 28th, 1853. After 8 a.m. the wind, which had been previously from the S., set in from due S.E. The thermometer instantly rose 2°·1, and by 6 p.m. gave the extraordinary temperature of −5°·4. This, when compared with the record of the 24th, of −41°·8, shows a change of 36°·4. This effect is due to the S.E. wind, and is sometimes much more excessive."

Snow fell for 12 or 13 hours after this, during which the temperature increased to +1° F., although the wind was from the S. and S.W.

At 10 a.m. on the 29th the snow ceased, and the temperature began to fall rapidly, reaching −24° at 11 p.m. In two hours, at 1 a.m. on the 30th, snow began to fall, and the temperature had risen to −9°, and continued to rise to −3°·5. The snow continued to fall until 8 p.m., after which the temperature fell. Dr. Kane remarks ("Narrative," vol. i., p. 154) :—

"The temperature on the floes was always somewhat higher than in the island, the difference being due, as I suppose, to the heat conducted by the sea water, which was at a temperature of +29°, the suspended instruments being affected by radiation." Also (p. 267), "Upon the ice floes, commencing with a surface temperature of −30°, I found at 2 feet deep a temperature of −8°, and at 4 feet +2°, and at 8 feet +26°."

This was in mid-winter, on the largest floe in the open way off Cape Stafford.

With Parry and other more recent Arctic travellers, Dr. Kane notes the effect of the vessel in increasing the temperature in the neighbourhood, causing a difference of sometimes 2°, and also the heating effect due to the approach of the observer or a lantern.

For very low temperature spirit thermometers are not trustworthy. Instruments agreeing within 1·8 down to −40° were found to differ from 15° to 20° at a temperature of about −60°. The freezing point of mercury varied from −38°·5 to −41°·5; after freezing its contraction is very uniform, and the column was seen to descend to −44°.

Sir E. Belcher observed the mercury to descend to −46°.

Spirit thermometers were found in some cases to agree well together, but to differ considerably from others, and from the most probable temperatures. In such cases they were probably companion thermometers, made with the same preparation of spirits.

From April to September 1854, observations were made with a black bulb thermometer to determine the radiation of heat by the

Sun, and corresponding notes were made on the Solar light, *i.e.* on the amount of cloud obscuring the Sun.

Discussion of Winds.

The mean true direction of the wind is from the eastward. In the month of June the wind veers round to the west of south. In winter it is E.N.E. (true), and in summer S.E. by S.

The S.W. and N.W. winds (magnetic), *i.e.*, S.S.W. and E.S.E. (true), blow most violently, the mean velocity being 10 miles an hour, the average velocity of the other winds being about five miles an hour.

The calms greatly predominate, there being more hours of calm than hours of wind, a circumstance quite characteristic of the locality.

All gales come from the S.W. in the summer; and the S.E. (magnetic) in winter, *i.e.*, over South Greenland from the true E.S.E. in summer, and from the true N.N.E. in winter.

NOTE.—Gales appear to come with the warm winds. Snow comes either from the N.W. (magnetic), *i.e.*, from the upper part of Baffin's Bay, or from the S.E. or S.W. *i.e.*, from between N.N.E. and E.S.E. (true), or from the direction of the Spitzbergen Sea.

Barometric Observations.

In the diurnal changes of pressure there is an increase from 2 p.m. (rather rapid) until the maximum is reached at 10 p.m.; then the pressure diminishes until 3 a.m., remains pretty steady until 9 a.m., and then gradually diminishes to a minimum at 1 p.m.; the average amount of this change being about ·01 of an inch.

There are much greater irregularities both in pressure and temperature in winter than in summer. There seems to be very little, if any, change of the barometric pressure caused by the fall of snow.

During S. and S.W. (magnetic) winds the barometer rises above its mean value. For other winds it is depressed.

In the annual changes there is a maximum in April, then a fall (rather rapid) until August, then a gradual rise until November, after which there is little change until February, followed by a sharp rise to the maximum in April.

---

3. DR. HAYES' METEOROLOGICAL OBSERVATIONS in the ARCTIC SEAS. "Smithsonian Contributions," vol. 15, p. 167.

The Temperature.

Port Foulke.—The annual fluctuations of the temperature of the air are exceedingly regular, being lowest at the end of January and increasing very uniformly until July, and then diminishing as uniformly from July until January; the range being from -26° F. to $+42^\circ$ F.

The diurnal fluctuation of temperature through the year is a minimum at 2.30 a.m. and a maximum at about 2.30 p.m., passing through the mean value about 8 a.m. and 8 p.m.; the whole daily range being about 4° F.

The observations at Port Foulke do not bear out the statements of other observers, that it is coldest about the time of full moon.

Daily Range of Barometer.

The barometer is at its lowest between midnight and 4 a.m., and then rises slowly until 8 a.m., but descends again until noon; it then rises to its highest point at 4 p.m., and then descends to its lowest value at midnight.

The yearly maximum occurs in April at Port Foulke and Port Kennedy, and in May at Van Rensselaer. The minimum occurs in October at Port Foulke and at Port Kennedy, and in September at Van Rensselaer. At Port Foulke another maximum was also observed in November, and a smaller maximum at Port Kennedy in the same month.

"The resulting direction of the wind at Port Foulke is N.E. "(true), which agrees with the general movement of the atmosphere "in Arctic regions, but the resulting direction at Van Rensselaer "is S.S.W. nearly." (Sm. Con., vol. 15, p. 237.) In June, July, and August the winds on the West Greenland coast are variable, but throughout the other nine months of the year they are mostly from the north-east. The winds which prevail the least are the south-west in January and the north-west in July.

In proceeding north on the West Coast of Greenland, the difference of temperature between the extreme seasons gradually increases thus:—

The Relative Range of Temperature in West Greenland.

Place.	Minimum.	Mean.	Maximum.	Mean.	Yearly Range of Temperature.	Latitude.
					°	° ′
Jacobshaven - -	—	—	—	—	41·6	69 12
Omenak -	Feb. 10	April 30	July 12	Nov. 2	45·8	70 41
Upernavik - -	Feb. 11	April 26	July 12	Nov. 2	47·7	72 47
Wolstenholme Sound	—	—	—	—	66·7	76 33
Port Foulke - -	—	—	—	—	58·0	78 18
Van Rensselaer -	March 1	April 29	July 8	Oct. 12	62·0	78 37
Port Kennedy -	Jan. 19	April 23	July 20	Oct. 22	72·4	72 01
Sabine Island -	Feb. 23	May 4	July 13	Oct. 1	66·4	74 32

Port Kennedy and Sabine Island are added for comparison, and it will be seen that the range at Wolstenholme Sound and at Sabine Island is nearly the same.

The mean temperatures for Port Foulke, Van Rensselaer, and Port Kennedy are:—

—	Spring.	Summer.	Autumn.	Winter.	Year.
	° ′	° ′	° ′	° ′	° ′
Port Foulke - -	− 3 19	+ 36 82	+ 11 01	− 21 22	+ 5 86
Van Rensselaer Harbour	− 10 59	+ 33 38	− 4 03	− 28 59	− 2 46
Port Kennedy -	− 2 04	+ 37 40	+ 7 09	− 35 04	+ 1 85

Simultaneous observations at Port Foulke and Van Rensselaer Harbour in 1861 showed that during four days in March the mean difference of temperature was 26° F.; Van Rensselaer being 26° colder than Port Foulke. At or near Port Foulke the sea does not freeze over entirely during the winter; the Esquimaux reside here during the winter, and animals abound. At Van Rensselaer, which is exposed to the north, the climate is much more severe, and there is scarcely any animal life.

At Port Foulke the *south-east, south,* and *south-west* winds are warmest, and the *north-east* and *north* winds are coldest.

In his summer journey northwards in 1861 Dr. Hayes made observations of temperature at several stations. At Jensen's Camp, latitude 80° 48′ N., it was 4°·8 colder than at Port Foulke.

Across Smith Sound, and up the west coast of Kennedy Channel it was on the average 10°·7 colder than at Port Foulke.

Dr. Hayes left Mr. Knott at Jensen's Camp on May 16th, who made observations during his absence; he himself went north to Cape Lieber, in latitude 81° 37′ N., longitude 69¼ W., which he reached on May 18th. [No longitude seems to be given in the tables for Jensen's Camp, or anywhere north of Cape Hawks Camp, in latitude 79° 44′ N., longitude 73° 6′ W.]

The fluctuations of atmospheric pressure for the month and year, and the extremes as well as the direction and force of wind, and the state of weather, are fully shown in the tables and curves accompanying the record given in the 15th volume of the "Smithsonian Contributions," which also contains an excellent map of Dr. Hayes' discoveries.

The remarkable fact, which so much puzzled Dr. Kane at Van Rensselaer Harbour, that the true N.N.E. wind is a warm wind, and is laden with moisture, seems to show that the northern part of Greenland does not extend northwards very far beyond his position, but that the winds obtain their moisture from an extensive sea to the north or north-east of North Greenland, and that they derive their high temperature from the waters of the Gulf Stream, since that seems to be the only warm current flowing from low latitudes to the Polar regions.

---

4. Meteorological Observations by Sir John Ross, in Baffin's Bay in 1818, and at Boothia Felix (1829–33).

In the appendix to Sir John Ross's account of his first voyage in 1818, the temperature of the air and the surface temperature of the sea are laid down graphically on the same diagram sheet, showing a close correspondence between them. On the same sheet are contained curves showing the variations of

The barometer.
The sympiesometer.
Kater's hygrometer.
The latitude and longitude.
The winds.
The specific gravity of sea-water.
The magnetic variation.

The appendix to his "Second Voyage" contains a complete register of temperatures, direction and force of the wind, and the state of the weather at various hours, and the daily and monthly means, also the register of the barometer, and an abstract of the meteorological observations.

The prevailing winds through the first winter, 1829–30, were N.W and next N.E. South-easterly winds were light, and the total average of the wind was much greatest from the northward. This was also true of the summer months. With regard to winds and currents Sir John Ross remarks :—" This, together with the " numerous and large rivers which discharge themselves into the " Gulf of Boothia, must account for the strong current which Sir " E. Parry found running to the eastward in Hecla and Fury " Strait. During the second winter the wind prevailed from the " north-west ; but north-east winds were not so prevalent as " during the former winter, south-west winds being the next ; " this may account for the winter being so severe, as there can " be no doubt that the wind came from a colder quarter. These " winds brought vast quantities of ice into the gulf."

In the account of his first voyage to discover a North-west Passage Parry remarks :—

" The wind and the thermometer rose together on more than " one occasion. On Dec. 31 the wind freshened from the north- " east, and the thermometer rose from $-28°$, the wind changed to " the S.S.E., and the thermometer still continued to rise and " reached 5° F. Other instances also appear in the register where " a change of wind from north to north-east or east was accom- " panied by a great increase of temperature. The north-east, east, " and south-east are the warmest winds." The prevailing winds appear to have been from the north.

In Sir John Franklin's second expedition to the Polar regions in 1825–27, when he went overland to the Mackenzie River, in addition to the usual meteorological record, observations were made on the amount of solar radiation by means of a blackened thermometer at Port Franklin, latitude 62° 12′ N., longitude 125° 12′ W. ; and at Carlton House in the following year, latitude 52° 51′ N., longitude 106° 13′ W.

---

5. Meteorological Observations by Sir Edward Belcher. (The Last of the Arctic Voyages, &c.)

From Sir Edward Belcher's Record of Temperatures it seems very clear that the winters of 1852–53 and 1853–54 were very severe.

On January 6, 1853, the temperature was $-51{\cdot}5°$ and continued to fall; on the 12th, at 9 p.m., the temperature was $-62{\cdot}5°$, and the indices of four thermometers the next morning at 8 a.m. read $-62{\cdot}0°$, $-61{\cdot}6°$, $-66{\cdot}0°$, $-63{\cdot}2°$; on the 14th the temperature had not risen above $-46°$ for four days, and the mean for the last three days had been $-55{\cdot}61°$.

From the 5th to the 15th the mean temperature was $-48{\cdot}88°$ F. The temperature did not rise above $-46°$ for $6\frac{1}{2}$ days, or above

− 52° for 3½ days. For 14 hours it continued between − 58° and − 62·5°. Yet with these low temperatures Sir Edward says "We leave our warm cabins at 50° and rush after science or bruin" with a change of temperature of about 99° F. "without damage."

On referring to Vol. II., p. 100, it is seen that the low temperatures of − 55° and − 57° were much more keenly felt during the second winter, thus illustrating what has been remarked at Moscow, that English residents scarcely notice the cold during the first winter, but during the second they feel it much more acutely.

Sir James Ross is said to have registered − 60° F., and on the 15th of February 1820 the thermometer stood at − 54° for 15½ hours (Belcher, I. p. 202).

The mercurial thermometers were observed to register correctly below the temperature − 39·5° F., and towards the end special attention was given to this, and the comparison with the standard spirit thermometer showed that down to − 46° the mercurial thermometers rise and fall with the spirit thermometer very regularly.

Experiments on the freezing of mercury showed that exposed for 20 minutes to a temperature of − 47·7° the mercury began to crystallize, the circumference became very convex at the edges of contact with the conical vessel in which it was contained, and the centre raised to a point when it had actually congealed. When placed solid into water at −47·5° the mercury became encased in ice, but flowed on the bottom at a temperature of −44·5° F. Previous to freezing, or at the instant of returning to fluidity, the mercury assumed a very active motion, resembling lively polypi, parts moving in circles with great velocity. An attempt was made to determine the form of crystallization (p. 208). Allsopp's ale was found to freeze at 22·5°; the frozen part was found to be very insipid, the remainder concentrated forming an excellent liqueur.

NOTE.—McClintock froze Allsopp's ale when the temperature of the air was −35°. When its temperature had risen to 17° it was almost all thawed, at 22° it was completely so.

P. 209. In describing experiments on the expansion of alcohol at low temperatures a very useful caution is given, the neglect of which may be very serious, to take care of the fingers in handling good conductors or good absorbers of heat at low temperatures.

During the early part of March 1853, the temperature still continued low, the mean on some days being as low as − 46°, −52·8°, − 52°, and the mean of 10 days being − 43·23° F. After the 17th of March a most decided change took place in the weather.

The hole for the tide-gauge was cleared, the ice being 7 feet thick, and the whole thickness of ice and snow was 21 feet within 20 yards of the ship. The ice towards the shore grounds in 16 feet of water, the thickness of off-shore floe formed in one season probably does not exceed six feet in thickness. As soon as the hole under the stern for the tide-gauge had been completed, the water rushed up similar to an artesian spring; this is singular since the fire-hole abreast of the ship had been kept open all the winter.

Although the winter of 1852–53 was a very cold one it appears from the records for November 1853 and January and February 1854, that the second winter 1853–54 was much colder than the first.

Thus at the end of January for 84 continuous hours the mean temperature is − 54·9°, the maximum being − 50°, and for 48 continuous hours the mean is −57·12° and the maximum −55°.

The mean of 10 days at beginning of February, − 47·145°, is the coldest on record for that time of year.

This was the winter during which Dr. Kane wintered at Van Rensselaer Harbour, when he experienced such intense cold.

From the Meteorological Tables at the end of Vol. II. it will be seen that the warm winds in Wellington Channel are either from the N.E. or from the S.E.

At Beechey Island strong southerly gales are experienced on the southern side, but no such breeze appears to extend up Wellington Channel or even into Union Bay.

In the table which gives a comparison of the temperatures observed in various Arctic voyages from 1819 to 1855, it is shown that there is a remarkable coincidence in the times of low temperatures and in the mean amount of cold throughout the whole range.

In the northern part of Wellington Channel the warm winds flow from the east (I. p. 127), and the gales come from this quarter with increasing temperature.

"Gales inevitably accompany any undue rise of temperature."

Gales on the N.E. of Melville Island were strong from the N.E.; also animals frequent the *eastern* side of the north-east point of Melville Island, showing that it is milder on the north-east side (II. p. 48).

---

6. Observations of Temperature of Air and Sea Water by Capt. Markham (as given in Map).

—	Lat.	Longitude.	Temp. of Air.	Sea at Surface.	Deep Sea.	Depth.
	°		° ′	° ′	° ′	
18 May -	62 N.	63° W. - -	39 0 F.	34 0 F.	30 0 F.	210 fathoms.
7 June -	75 N.	62° W. (Melville Bay)	32 0	29 0	29 5	200 (edge of land ice).
12 " -	74 N.	76° W. (C. Byam-martin.)	28 5	29 5	29 5	190 fathoms
13 " -	74 N.	76° 30′ W. (nearer shore.)	30 0	37 0	32 0	190 "
15 July -	73 N.	74° W. (off Pond's Inlet.)	41 9	35 5	32 0	200 "
17 " -	73 N.	73° W. (farther off coast.)	40 0	38 2	30 5	200 "
21 " -	74 N.	82°30′W. (Elwin Bay)	37 0	34 0	29 5	190 (bottom).
1 Aug. -	73 N.	92° W. (Fury Beach)	36 4	33 0	—	—

In addition to these, which include the deep-sea temperatures observed, Capt. Markham made observations on temperature of air and of surface water frequently throughout July and August in Baffin's Bay and Lancaster Sound, mostly near the coast of Baffin Land.

---

7. Second German Expedition under Captain Koldewey.

Temperature.

The winter quarters of the Expedition were at Sabine Island, on the East Greenland coast, in latitude 74° 32′ N., longitude 18° 49′ W., but observations were made at various points of the coast towards the north and the south.

On comparing observations on the East and West Coast of Greenland it is found that, except in the month of December, the mean diminution of temperature for one degree of increase of latitude is nearly the same on the East and West Greenland coast.

The changes of temperature throughout the year at Sabine Island are very regular, and can be said to have only one maximum and one minimum. The mean temperature is lowest in January, but changes very little until the middle of March, being about −13° F.; then it increases somewhat rapidly for three months, until the end of June, to about 39° F. In August it descends again, not quite so rapidly as it ascended, until November, and with a slight check in December it again reaches its minimum toward the end of January.

The curves representing the mean temperatures for different hours of the day are also very regular.

In winter the coldest part of the day is from 2 p.m. to 2 a.m., and the temperature almost remains constant throughout these hours. There is very little change throughout the day.

Throughout *each* season, the temperature at 8 a.m. and at 8 p.m. is exceedingly near to the mean temperature for the day.

The greatest changes of temperature during the day take place in the spring, and both in spring and summer the temperature is lowest at 2 a.m., and increases very steadily until 2 p.m., and again diminishes very steadily until 2 a.m.

In the *autumn* there is no change of temperature from 8 p.m. to 4 a.m., then a gradual rise until 1 o'clock in the day, and an almost equally gradual fall until 8 p.m.

The coldest month is January, but February and March are very nearly as cold, the lowest cold registered occurring in February. The warmest month is July, and the highest temperature 55·6° F., on the 1st of July at 2 p.m.

Temperature and Wind at Sabine Island.

It appears that (p. 589) in *autumn* the north wind is the warmest, and next to it the north-east and the north-west winds, and that the south wind is the coldest.

During east wind it is perfectly clear, and also during the north-west wind.

Two-thirds of the whole of the snow came from the north with the warm winds.

Winter.—The south-east wind is warmest with a very high barometer, and the north-west the coldest. During calms it was cold and cloudy.

Spring.—The south and south-east winds are warmest, and the west the coldest, but the difference of temperature is very small for even the extremes.

Summer.—There is very little variation in the temperature throughout the summer, the mean value being 39·67° F.

The force of the wind was very much greater from the *north* than from any other quarter; with the exception of the winds between the north and north-east all the winds were very light.

Taking the means for the different winds through the year, it is seen that a change in the direction of the wind affects the barometer very little. It also appears that in autumn the *north* wind is warmer than any other, and that, taking the mean of the four seasons, the south-east, the south, the north-east, and the north are the warmest winds.

The Barometer.

The mean daily changes of the barometer show that—

In *winter* the barometer is highest at 10 a.m., and diminishes very regularly through the day and until midnight, after which hour it rises as steadily until 10 a.m.; the whole variation being about ·03 inches.

In *spring* there is a gradual rise of the barometer from 4 a.m. to 12 noon, then a gradual fall from noon until 10 p.m.; between 10 p.m. and 4 a.m. the pressure remains constant, and the whole variation is about ·02 inches.

In *summer* there are two daily maxima, at 12 noon and 12 midnight, and two minima, at 4 p.m. and at 2 a.m.; but the total change is less than ·01 inches.

In *autumn* the barometer is highest about 11 a.m., and falls until 4 p.m., then rises and is steady from 6 to 8 p.m., then falls again to its lowest level at 2 a.m.; but the whole variation is ·01 inches.

In considering the monthly changes, it appears that the barometer rises rather rapidly (about ·3 inches) from January to March, being then at its highest; then falls rapidly to April and keeps steady until June, then falls rapidly to its lowest in July and as rapidly rises to August, then slowly falls again about ·1 inches from August until January, its second minimum

The barometer was very little affected by storms. On Dec. 5, whilst a very severe storm from the north was coming on, the barometer remained quite steady; but while its force was at the greatest the barometer rose gradually, and continued to rise as the storm abated. When it became calm and the wind changed to south for an hour, the barometer fell a little, but afterwards continued to rise.

The temperature continued steadily to increase from the beginning of this gale, and when the wind changed increased more rapidly, and came down again with the south wind, to increase again as the wind returned to the north.

---

8. TEMPERATURES of the POLAR REGIONS.

The CLIMATOLOGISCHE BEITRÄGE (Dove 1857) contains a discussion of the temperatures of the Polar Regions with especial reference to the existence and position of Poles of cold.

The north of Asia is warmer in summer than the regions to the north of America, and the changes of temperature through the year are very much greater, being quite as low or lower in winter than any place where temperatures are observed in the Polar regions, and very much higher in summer. The Asiatic summer is entirely absent from the Greenland seas, and in summer these seas have the character of a sea climate.

Thus comparing Ustjansk, Jakutsk, and Van Rensselaer for the four seasons we get:—

——	Ustjansk.	Jakutsk.	Van Rensselaer Harbour.
Winter - - -	−35·95 F.	−36·69 F.	−29·56 F.
Spring - - -	−0·62	14·83	−11·47
Summer - -	46·78	58·39	32·99
Autumn - -	1·02	12·27	−4·83

showing the difference in character, as well as the greater differences at an inland station than on the coast of Siberia.

Comparing the means of the same month of the year at various places in the American Polar Sea, we find the places of greatest cold for the several months of the year as follows:—

January, Northumberland Sound -	−40·00
February, Disaster Bay - - -	−40·24
March, Van Rensselaer - - -	−38·09
April, ,, - - -	−14·00
May, Disaster Bay - - -	9·34
June, Winter Island - - -	23·17
July, ,, - - -	35·36
August, Port Bowen - - -	29·72
September, Van Rensselaer - -	9·81
October, ,, - -	−10·54
November, ,, - -	−23·03
December, Northumberland Sound -	−35·51

As these values are obtained in different years, no allowance can be made for mild or severe seasons, and therefore the table formed on so few observations may be somewhat misleading.

We shall probably get a closer approximation to the region of greatest cold, if we compare observations made at several places during the same season.

If we take the winter of 1852-53 we get the greatest cold at—

Mercy Bay in October - - -	−5·6
Melville Island in November - -	−20·1
Northumberland Sound in December -	−35·51
Mercy Bay in January - - -	−43·87
Melville Island in February - -	−40·8
,, in March - - -	−31·7
Northumberland Sound in April - -	−8·60

Taking the following year in the same way we may compare Beechey Island, Disaster Bay, and Van Rensselaer.

Disaster Bay in September 1853 -	17·00
Van Rensselaer in October - -	0·55
„ in November - -	−23·01
Disaster Bay in December - -	−28·08
„ in January 1854 - -	−37·38
„ in February - -	−40·24
Van Rensselaer in March - -	−38·09
„ in April - -	−8·6
Disaster Bay in May - - -	9·34
„ in June - - -	27·91
„ in July - - -	38·12
Van Rensselaer in August - -	31·35

These results show that the pole of greatest cold (if there is one) in the American Polar Sea is nearly equally distant from Mercy Bay, the south-west side of Melville Island, Northumberland Sound, and Van Rensselaer, except in so far as the extreme cold at any of these places may be due to local causes.

If we consider this question in connexion with the three ocean tides from Behring Straits, Baffin's Bay, and around the north coast of Greenland, we shall see that the line joining places where these tides meet coincides very closely indeed with the line of greatest and nearly equal cold as already observed. This suggests the consideration whether there is not a direct connexion between the temperature of the air in the Polar regions and the meeting of the ocean tides. *See* "Tides and Currents."

A table of monthly mean temperatures of Polar stations is added, chiefly taken from Dove's work, with the addition of the results of those expeditions which have been sent out since 1854, as far as they could be obtained.

A small table of the monthly mean temperatures at stations on the West Greenland coast is added for comparison, this includes the mean of the temperatures for several years. It also appears from tables given by Dr. Rink in his account of the meteorology of West Greenland that out of 14 years observations at Upernavik the mean temperature for January as well as for March was lower in 1853 than in any other year, being as low as 33° F. in each month. This was the winter when Sir E. Belcher was in Northumberland Sound.

Mean Temperature of Polar Stations (Fahrenheit).

Place.	Jan.	Feb.	March.	April.	May.	June.	July.	August.	Sept.	Oct.	Nov.	Dec.	Year.
Fort Hope, 60° 32′ N., 86° 56′ W.	..	..	..	..	..	..	..	..	28·57	12·56	0·68	−19·27	1846
	−29·32	−26·68	−28·10	−3·95	17·88	31·38	41·46	46·9	..	..	..	..	1847
Winter Island, 66° 11′ N., 83° 11′ W.	..	..	..	..	..	..	35·36	36·86	31·61	13·25	7·88	−14·24	1821
	−23·17	−23·99	−10·72	6·47	23·29	23·17	..	..	..	..	..	..	1822
Igloolik, 69° 21′ N., 81° 53′ W.	..	..	..	..	..	..	..	33·83	25·09	13·72	−18·65	−28·25	1822
	−16·13	−19·58	−19·01	−0·85	25·14	32·16	39·09	..	..	..	..	..	1823
Boothia Felix, 69° 59′ N., 92° 1′ W.	..	..	..	..	..	..	..	..	..	7·94	−3·58	−23·08	1829
	−33·13	−29·9	−20·93	1·37	15·27	36·76	44·57	40·87	27·42	10·95	−11·45	−20·24	1830
	−25·43	−32·46	−34·74	−6·44	16·02	31·56	37·94	36·51	23·4	8·32	−1·23	−23·96	1831
	−27·52	−33·69	−31·37	..	..	..	..	..	..	..	..	..	1832
Mean	−28·69	−32·02	−28·68	−2·59	15·65	34·16	41·26	38·69	25·41	9·07	−5·41	−22·43	..
Port Bowen, 73° 14′ N., 88° 56′ W.	..	..	..	..	..	..	35·81	29·72	25·88	10·85	−5·00	−19·05	1824
	−28·91	−27·32	−28·38	−6·50	17·57	36·12	37·29	31·36	..	..	..	..	1825
Batty Bay, 73° 12′ N., 91° 10′ W.	..	..	..	..	..	..	..	..	..	..	−5·43	−21·54	1851
	−20·95	−19·21	−18·11	2·12	..	..	..	..	..	..	..	..	1852
Port Leopold, 73° 50′ N., 90° 20′ W.	..	..	..	..	..	..	..	..	..	9·7	−14·5	−22·8	1848
	−35·7	−35·2	−22·8	−10·0	..	..	..	..	..	..	..	..	1849
Beechey Island, 74° 5′ N., 91° 51′ W.	..	..	..	..	..	..	..	..	18·46	−1·40	−6·64	−35·51	1852
	−31·63	−17·95	−12·97	1·83	19·0	36·8	39·4	34.5	18·5	7·39	15·63	−24·12	1853
	−31·41	−32·97	..	..	..	..	..	..	..	..	..	..	1854
Assistance Bay, 74° 14′ N., 94° 16′ W.	..	..	..	..	..	..	..	..	21·3	1·5	−6·7	−21·4	1850
	−29·0	−29·8	−22·4	−3·2	12·1	34·3	37·8	35·6	..	..	..	..	1851
Griffith Island, 74° 40′ N., 95° 0′ W.	..	..	..	..	..	..	..	..	..	−0·6	−7·5	−22·9	1850
	−31·0	−32·5	−25·7	−7·31	8·96	32·27	..	..	..	..	..	..	1851
Northumberland Sound, 76° 52′ N., 97° 0′ W.	..	..	..	..	..	..	..	..	18·46	−1·40	−6·64	−35·51	1852
	−40·00	−29·58	−17·71	−8·60	14·73	29·84	35·69	33·80	..	..	..	..	1853
Disaster Bay, 75° 31′ N., 92° 10′ W.	..	..	..	..	..	..	..	..	17·00	9·51	−18·33	−28·08	1853
	−37·38	−40·24	−30·86	4·84	9·34	27·91	38·12	36·20	..	..	..	..	1854
Dealy Island, 74° 56, N., 108° 40′ W.	..	..	..	..	..	..	..	..	..	−1·19	−10·83	−26·48	1852
	−35·84	−31·18	−21·9	..	..	..	..	..	..	..	..	..	1853
Melville Island, 74° 47′ N., 110° 48′ W.	..	..	..	..	..	..	..	..	22·52	−2·83	−21·14	−21·62	1819
	−31·28	−32·45	−18·19	−8·21	16·82	36·21	42·45	32·59	..	..	..	..	1820
75° 0′ N., 109° [illegible]′ W.	..	..	..	..	..	..	..	..	..	4·1	−20·1	−32·45	1852
	−36·58	−40·8	−31·7	−5·4	..	..	..	..	..	..	..	..	1853
Mercy Bay, 74° 6′ N. 117° 54′ W.	..	..	..	..	..	..	..	..	..	3·3	−15·2	−20·0	1851
	−27·3	−25·8	−28·4	−1·4	10·2	31·5	36·7	33·2	20·1	−5·6	−16·5	−26·1	1852
	−43·87	−38·5	−25·09	..	..	..	..	..	..	..	..	..	1853
Prince of Wales Strait, 72° 47′ N., 117° 44′ W.	..	..	..	..	..	..	..	(36·5)	20·2	0·2	−10·2	−23·4	1850
	−32·5	−37·7	−28·8	−4·8	18·9	36·1	37·5	37·6	(24·6)	..	..	..	1851

Mean Temperature of Polar Stations—*continued.*

Place.	Jan.	Feb.	March.	April.	May.	June.	July.	August.	Sept.	Oct.	Nov.	Dec.	Year.
Point Providence, 64° 14′ N., 165° W.	..	..	..	..	..	..	..	..	..	25·5	17·5	3·75	1848
	−20·5	−16·0	−6·25	21·5	29·5	38·0	(44·42)	(42·75)	..	..	..	..	1849
Choris Peninsula, 66° 58′ N., 173° 3′ W.	..	..	..	..	..	..	..	45·00	42·75	25·00	1·25	5·25	1849
	−12·0	−15·5	−6·0	14·5	30·0	..	..	..	..	..	..	..	1850
Point Clarence, 60° 45′ N., 165° W.	..	..	..	..	..	..	47·74	44·91	38·34	22·08	−3·25	3·41	1850
	−10·28	9·43	2·57	17·66	33·71	40·06	51·91	(46·47)	43·03	23·19	4·57	−2·78	1851
	−12·05	−7·93	6·62	5·34	31·96	40·78	..	..	..	..	..	..	1852
Wolstenholme Sound, 76° 30′ N., 68° 58′ W.	..	..	..	..	..	..	..	33·67	26·76	11·32	−18·60	−27·05	1849
	−25·07	−34·02	−17·47	−3·74	25·82	39·73	40·52	..	..	..	..	..	1850
Van Rensselaer Harbour, 78° 37′ N., 70° 40′ W.	..	..	..	..	..	..	(37·83)	(33·41)	17·16	0·55	−23·01	−25·99	1853
	−30·24	−33·60	−38·09	−8·60	12·89	29·23	38·40	31·35	9·81	−10·54	−23·03	−37·74	1854
	−28·61	−21·21	−33·97	−14·00	..	..	..	..	..	..	..	..	1855
	−29·42	−27·40	−36·03	−11·30	12·89	29·23	38·40	31·35	13·48	−5·00	−23·02	−31·86	Mean.
Port Foulke, 78° 18′ N., 73° 00′ W.	..	..	..	..	..	..	..	..	22·60	7·60	2·84	−12·81	1860
	−25·97	−24·88	−22·32	−11·01	23·77	33·85	40·54	..	..	..	..	..	1861
Port Kennedy, 72° 01′ N., 94° 14′ W.	..	..	..	..	..	..	..	..	..	7·54	−11·29	−32·97	1858
	−33·54	−36·03	−17·78	−2·45	15·42	35·52	40·13	..	..	..	..	..	1859
Polaris Bay	..	..	..	..	..	..	..	..	..	..	..	..	..
	..	..	..	..	..	..	..	..	..	..	..	..	..
Sabine Island, 74° 32′ N., 18° 49′ W.	..	..	..	..	..	..	..	33·21	24·22	7·12	−0·98	1·15	1869
	−11·47	−10·86	−9·98	2·28	22·24	36·07	38·84	..	..	..	..	..	1870

Mean Temperature on West Greenland Coast.

Place.	Jan.	Feb.	March.	April.	May.	June.	July.	August.	Sept.	Oct.	Nov.	Dec.
Godthaab	14·50	16·14	20·21	24·96	34·29	41·07	46·44	44·44	38·82	31·33	20·28	15·10
Jacobshaven	2·44	−0·28	8·13	18·82	32·49	41·47	45·36	42·39	34·59	25·16	12·56	7·48
Upernavik	−7·15	−11·65	−4·45	6·60	24·98	36·88	39·96	37·92	32·02	22·03	10·94	−1·05

GREATEST observed COLD in the POLAR REGIONS during the several months, in Fahrenheit degrees.

Place.	Jan.	Feb.	March.	April.	May.	June.	July.	August.	Sept.	Oct.	Nov.	Dec.	Date.
Port Hope	..	..	..	..	..	..	..	..	16·0	−15·0	−24·9	−40·0	1846
	−47·0	−42·0	−44·9	−24·9	− 4·0	12·0	29·1	34·0	..	..	..	..	1847
Winter Island	..	..	..	..	..	..	29·1	28·0	20·1	13·1	−20·0	−29·0	1821
	−36·8	−36·4	−35·0	−11·9	− 4·9	20·1	..	..	..	..	..	..	1822
Igloolik	..	..	..	..	..	..	..	27·1	11·1	− 8·9	−31·9	−42·9	1822
	−44·9	−42·9	−40·9	−24·9	− 8·0	7·9	30·0	..	..	..	..	..	1823
Boothia Felix	..	..	..	..	..	..	..	..	..	−16·4	−37·1	−37·1	1829
	−44·9	−46·9	−42·0	−20·9	− 1·1	25·9	32·0	32·9	5·0	−11·9	−40·9	−46·9	1830
	−59·3	−49·0	−51·0	−24·9	−16·0	14·0	32·0	24·1	6·1	−22·9	−42·0	−42·0	1831
	−46·9	−44·0	−48·5	..	..	..	..	..	..	..	..	..	1832
Port Bowen	..	..	..	..	..	..	..	..	16·0	−12·0	−26·0	−35·0	1824
	−42·5	−45·0	−47·5	−37·0	− 7·5	23·0	12·0	25·0	..	..	..	..	1825
Port Leopold	..	..	..	..	..	..	..	..	..	−14·0	−37·5	−56·5	1848
	−50·5	−60·0	−51·0	−29·4	− 9·0	..	..	..	..	..	..	..	1849
Assistance Bay	..	..	..	..	..	..	..	..	6·1	−13·9	−24·0	−37·1	1850
	−40·9	−44·9	−40·9	−31·0	−20·0	16·0	29·1	31·1	..	..	..	..	1851
Griffith Island	..	..	..	..	..	..	..	..	..	−19·0	−31·0	−39·5	1850
	−45·0	−46·0	−44·5	−29·0	−20·0	10·0	31·1	..	..	..	..	..	1851
Northumberland Sound	..	..	..	..	..	..	29·1	21·0	1·0	−22·0	−30·2	−46·7	1852
	−62·0	−47·0	−55·5	−27·0	− 9·0	20·0	26·6	22·6	..	..	..	..	1853
Disaster Bay	..	..	..	..	..	..	..	..	0·1	−15·0	−37·0	−46·5	1853
	−59·25	−55·75	−49·62	−37·50	−19·0	15·0	32·0	28·0	..	..	..	..	1854
Melville Island	..	..	..	..	..	..	..	..	− 1·1	−28·0	−47·0	−43·0	1819
	−47·0	−50·0	−40·0	−32·0	− 4·0	28·0	32·0	22·1	..	..	..	..	1820
	..	..	..	..	..	..	..	..	..	−17·0	−39·0	−46·0	1852
	−55·0	−56·0	−46·0	−36·0	..	..	..	..	..	..	..	..	1853
Prince of Wales St.	..	..	..	..	..	..	..	27·1	− 1·1	−22·9	−31·9	−40·0	1850
	−51·0	−51·0	−51·0	−31·9	− 4·9	27·1	32·0	21·0	1·0	..	..	..	1851
Mercy Bay	..	..	..	..	..	..	..	..	..	−22·0	−40·0	−44·0	1851
	−51·0	−47·0	−51·9	−38·0	−26·9	11·1	30·0	19·0	− 4·0	−33·0	−42·9	−47·9	1852
	−64·9	−57·0	−58·0	..	..	..	..	..	..	..	..	..	1853
Port Foulke	..	..	..	..	..	..	..	..	7·3	−12·0	−17·0	−27·0	1860
	−43·0	−38·0	−37·0	−27·5	− 1·0	16·0	31·0	..	..	..	..	..	1861
Van Rensselaer	..	..	..	..	..	..	..	..	− 0·7	−22·4	−41·8	−44·0	1853
	−58·3	−66·4	−54·2	−41·7	− 7·5	18·4	28·0	20·2	− 8·8	−37·8	−47·9	−59·9	1854
	−66·5	..	..	..	..	..	..	..	..	..	..	..	1855
Sabine Island	..	..	..	..	..	..	..	21·0	10·9	− 8·9	−13·7	−17·9	1869
	−28·5	−40·4	−29·2	−24·7	− 1·3	25·3	29·8	..	..	..	..	..	1870

II.—TEMPERATURE OF THE SEA, &c.

1. Tables of Temperatures of the Sea at Various Depths below the Surface, taken between 1749 and 1868. Collated by Professor Prestwich, F.R.S.

The first experiments were made by bringing up water by means of a bucket with valves; this was employed by Scoresby, who showed that in the seas around Spitzbergen, while at the surface the temperature varied from 29° to 42°, at depths of from 2,000 to 4,000 feet the temperature was generally about 34° or 36°. The apparatus was improved by Lenz, and remarkable results were obtained by him in Kotzebue's expedition in 1823. Sir John Ross sometimes used Six's thermometers, and at others took the temperatures of the silt from the bottom. Six's thermometers were also used by Parry and on most subsequent expeditions.

The observations made in the Mediterranean with these instruments, which were very probably protected by an outer coating, have a remarkably close agreement with those recently made by Dr. Carpenter.

The experiments of Scoresby, Martins, and Ross show that in Polar regions the temperature at depths is higher than the average surface temperature, and this applies to Arctic and Antarctic regions. From the experiments of Ross, Sabine, and Parry, this rule is not found to hold good in Baffin's Bay, where the temperature falls from 30° and 32° at the surface to 29° and 28°·5 at the greatest depths attained.

The facts show that at depths in the tropical seas the temperature is about 34° or 35° F., and Lenz has shown that this could only be maintained by a constant slow under current from the poles to the equator, and has also shown, by making observations, that a belt of cooler water exists at the equator, and that the temperature at equal depths is lower at the equator than a few degrees to the north or south of it. He concluded that this arose from the circumstance that the deep-seated Polar waters there met and rose to the surface.

As bearing out this, he showed that the waters in the same zone were of lower specific gravity, which was also noticed by Humboldt:—

"Lenz shows that in the Arctic Ocean the bathymetrical isotherm of 35° is deepest on the west of Spitzbergen, while nearer Greenland, and again nearer Norway, the deep waters are colder. The several isothermal surfaces of 40°, 50°, 60°, 70°, and 80° are then traced southward, attaining their maximum depth between 50° and 40° latitude, and rising thence towards the equator.

"Taking another zone on the western side of the Atlantic, from Baffin's Bay to the Equator, he shows that the highest isotherms are not prolonged so far north as on the first line, and that the water at the bottom of the Bay is colder than in the Spitzbergen seas, approaching much nearer that of its maximum density and of its point of congelation; whence he concludes that this is the main source of supply of the deep-seated cold waters in the Atlantic, which, after attaining their greatest depths between latitudes 40° to 50° N., are found 3,000 to 4,000 feet nearer the surface on approaching the Equator."

2. Specific Gravity of Sea-Water.—J. Y. Buchanan.

In No. 160 of the Proceedings of the Royal Society (just published) is a paper on the determination at sea of the specific gravity of sea-water which well deserves attention, being the result of Mr. Buchanan's experience on board H.M.S. Challenger. He draws attention to the great care required in the means used for collecting the water and making the necessary measurements, and for reducing the results, and describes the special instrument (on the principle of Nicholson's hydrometer) which had been made for him, and which he had used and tested by comparison with other methods.

His mode of reduction by the graphical method in which he lays down isothermal lines and forms a chart by means of which the specific gravity of sea-water at any temperature may be at once read off from its observed specific gravity at any other temperature, is also well worthy of adoption.

3. Forchhammer's Researches on Sea-water and Currents. (Report of British Association, 1846, p. 90.)

The greatest quantity of saline matter is found in the tropical regions far from land where there is 36·6 parts of salt per 1,000 of salt water. The proportion diminishes on the western side of the Gulf Stream and near any coast. Towards the north-east it decreases slightly, but is pretty constant over the North Atlantic, about 35·7 per 1,000.

More than 100 miles S. of Greenland the proportion of salt is 35·0 per 1,000 parts.

In Davis Straits, 40 miles from land, 32·5 per 1,000, and this is nearly the proportion in the Polar current.

In latitude 43½° N. and longitude 46⅓° W., it is 33·8 per 1,000. This points to the fact that the vapour rising in tropical regions is condensed in Polar regions and flows back in the form of Polar currents, so that more water flows away from than towards the poles.

[This seems clearly indicated by Parry's rapid drift to the southward off Spitzbergen, where some portion of the influence of the Gulf Stream might be supposed to extend to prevent such a southerly drift.]

The water of different seas may contain more or less salt, but the *relative* proportion of its constituent saline parts changes very little.

Comparing the sea in different parts, the proportion between *chlorine* and *sulphuric acid*—

In the Atlantic is - - -	10,000 to 1,188
In the sea between the Faröe Isles, Iceland, and Greenland - -	10,000 to 1,193
In the German Ocean - -	10,000 to 1,191
In Davis Strait - - -	10,000 to 1,220
In the Kattegat - - -	10,000 to 1,240

The proportion of sulphuric acid increases near shores.

4. Surface Temperature of Sea-Water.

The mean monthly values from hourly observations in Van Rensselaer Harbour, at a depth of 4 feet were:—

1853, September, 29°·1, October 28°·8.
From November to March, 28°·7.
1854, April, 28°·8, May, 28°·9.
„ June, 30°·3, July, 32°·3.
„ August, 31°·8, September, 31°·3, October, 30°·9.

After November 1854 the readings remained constant at 29°·0.

5. Measuring the Specific Gravity of Sea Water. (Die zweite deutsche Nordpolarfahrt, vol. ii. p. 667.)

A very complete series of determinations of the specific gravity of sea-water at the surface and at various depths was made by Dr. Börgen with a very delicate glass hydrometer. A full account of the results is given in the excellent record of M. Koldewey's Expedition to East Greenland. The instruments were tested at Göttingen by Professor Kohlrausch before the Expedition sailed. In working out the results the temperature +15° C., or 59° F., was taken as the standard temperature, and the tables of Dr. Gerlach on the specific gravities of salt solutions of different strengths and temperatures (Freiberg, 1859) were employed. For other temperatures than +15° C., a correction was required for the change of volume of the glass.

As these tables will be useful for working out the results of any observations which may be made on the specific gravity of sea-water, and will at once give the relative strength of the salt-water when its temperature is known; the portion of them which may be required is added to the Manual, and also the table of corrections for the change of volume of glass for different temperatures.

To make an observation, the reading of the hydrometer must be taken and also the temperature of the water; multiplying the observed reading of the hydrometer by the correction for the observed temperature, and substracting the result from the reading of the hydrometer, we get the specific gravity at 15° C.; on referring to the table of strengths of solution the per-centage of salt in the sea is obtained.

STRENGTH of SALT SOLUTION.

Temperature, (Centigrade Scale.)	Distilled Water.	Salt Solutions.					Volume of Glass Hydrometer.
		1 per cent.	2 per cent.	3 per cent.	4 per cent.	5 per cent.	
0	1·0007	1·0083	1·0159	1·0235	1·0312	1·03877	0·999612
1	1·0007	1·0083	1·0159	1·0235	1·0311	1·0386	0·999638
2	1·0008	1·0084	1·0159	1·0234	1·0310	1·0385	0·999664
3	1·0009	1·0084	1·0159	1·0234	1·0309	1·0384	0·999690
4	1·00092	1·0084	1·0159	1·0233	1·0308	1·0383	0·999716
5	1·0009	1·0083	1·0158	1·0232	1·0307	1·03815	0·999741
6	1·0008	1·0083	1·0157	1·0231	1·0305	1·0380	0·999767
7	1·0007	1·0082	1·0156	1·0230	1·0304	1·0378	0·990793
8	1·0007	1·0081	1·0155	1·0229	1·0303	1·0377	0·999819
9	1·0006	1·0080	1·0154	1·0228	1·0301	1·0375	0·999845
10	1·00058	1·0079	1·0153	1·0226	1·0300	1·03737	0·999871
11	1·0005	1·0078	1·0151	1·0225	1·0298	1·0371	0·999897
12	1·0004	1·0077	1·0150	1·0223	1·0296	1·0369	0·999922
13	1·0002	1·0075	1·0148	1·0221	1·0294	1·0367	0·999948
14	1·0001	1·0074	1·0147	1·0219	1·0292	1·0365	0·999974
15	1·0000	1·0072	1·0145	1·0217	1·0290	1·03623	1·000000

The following table contains the corrections for the change of volume of the Hydrometer:

Temperature.	Reduction.	Temperature.	Reduction.	Temperature.	Reduction.
0°	−0·0016	5°	−0·0013	10°	−0·0009
1	16	6	13	11	7
2	16	7	12	12	5
3	15	8	11	13	3
4	14	9	09	14	2

The observations are divided into four sets—

(1.) Those in the North Sea 60° N. lat.
(2.) The North Atlantic.
(3.) The Polar Sea to the border of the ice.
(4.) The Ice-bound sea.

The mean specific gravities and corresponding strengths of solution were—

(1) 1·02545 3·511 (2) { 1·02594 3·578 } (3)1 { 1·02493 3·439 } (4) { 1·02411 3·326 }

Observations were also made at various points along the coast of East Greenland in July and August 1870, which showed that after the melting of the snow the water collected contained only about 2 per cent. of salt.

Observations were also made on the increase of specific gravity with the depth, which show some remarkable changes in different parts of the Atlantic and the Polar Sea.

A useful table is also given in Die zweite deutsche Nordpolarfahrt, vol ii., p. 697, comparing the summer temperatures at almost all Polar stations in Greenland, Arctic America, and Russia where observations have been made.

6. Specific Gravity of Sea Water (McClintock).

(*Appendix to Record of Voyage of "Fox,"* p. iii.)

In this Appendix are contained the specific gravities of sea water; remarks on the state of the ice and on aurora and atmospheric phenomena.

The sp. gr. of surface water fell from 1·027 on the 9th of Aug. 1857 to 1·0208 on the 10th, when the yacht was surrounded by icebergs.

			Temp.
On Sept. 5th sp. gr. of sea	at surface	- 1·0265	28·8
" "	in 25 fathoms	- 1·0290	29·0
" "	50 "	- 1·0292	29·0
" "	88 "	- 1·0302	29·0

McClintock refers to freezing salt water and to evaporation through ice (pp. 77, 78).

On a calm day the temperature of the external air being −33°, but within a snow hut the thermometer stood 17° higher, the difference being due to the transmission of heat through the ice from the sea beneath. Evaporation goes on through ice from the water underneath it.

On the coast of Lapland, much salt is made from the *sea* or floe ice, *not* from fresh water, glacier or *pack-ice.*

In the drift down Baffin's Bay, McClintock remarks :—

Sept.—Melville Bay. Ice drift affected by local cause.

Dec.—In mid-channel a tendency to drift westward.

Feb. and March.—Drifted parallel to the Greenland shore.

April.—Ice deflected eastward by land to the west. Rapid increase of drift. Could not ascertain the existence of current.

A table of the winds and ice-drifts is given on p. 110.

III.—PHYSICAL PROPERTIES OF ICE.

---

1. First Grinnell Expedition under Commander De Haven (1852). Forms of Ice. By Dr. Kane.

De Haven, in his drift through Lancaster Sound and Baffin's Bay, met with some very singular cases of the bending of large blocks of ice.

In one case, a table of ice, 4 feet thick, 18 feet long, and 15 broad, was curved so as to form a well-arched bridge across a water chasm. It had evidently reared high in the air, and then gradually bent over.

In another case, a straight block of rectangular section and 50 feet long resting on other blocks near its two ends, was bent down by its own weight so as to form the arc of a circle.

[The change of form produced in ice by pressure or by its own weight is due to the fact that the temperature of the melting point of ice is different under different pressures. Under pressure ice is melted below 32° F.; so that if pieces of ice rather below 32° F. be placed in a mould and pressed, the portions of the ice which press against one another will melt, and so the pressure will be relieved. In consequence of the removal of the pressure, the water will freeze again, since its temperature is below 32° F., and the pieces will be frozen together and take up a new form; this may be again moulded in the same way into new forms in consequence of the alternate melting of the ice by pressure, and its regelation when the pressure is removed.]

(Reference should be made to Tyndall's "Glaciers of the Alps" on this and on other questions connected with glacier-ice.)

In his narrative of the voyage, Dr. Kane says:—

"In Wellington Channel our ice had not acquired its full firmness and tenacity; its structure was granular and almost spongy, its mass infiltrated with salt water, and its plasticity such that it crumbled and moulded itself to our form under pressures which would otherwise have destroyed us.

"By the time we had reached the middle of Barrow's Strait, and the winter's midnight of December had darkened around us, our thermometers indicating a mean of 15° and 20° below zero, the ice attained a thickness of three feet, with an *almost flinty hardness*, and *a splintery fracture* at right angles to its horizontal plane. Such ice was at its surface completely fresh, and, when tested with nitrate of silver, gave not the slightest discolouration."

(Chapters XLII. and XLIII. contain an account of remarkable changes in the floe-ice in Baffin's Bay).

Dr. Kane discovered that the floes "which had formed in " mid-winter at temperatures below −30 were still fresh and pure, " while the floes of slower growth, or of the early and late portions " of the season, were distinctly saline. Indeed, ice which only " two months before I had eaten with pleasure was now so salt " that the very snow which covered it was no longer drinkable."

"Another element in the disintegration of the floes, of which this was but a preliminary process, struck me forcibly a little later in the season. The invasion of the capillary structure of the ice by salt water from below would act both chemically and mechanically in destroying its structure; but I am led to believe that, in addition to the actions of simple infiltration, forces allied to endosmosis are called into play."

"The infiltration of saline water through the ice assists the process of disintegration. The water formed by surface or sun thaw is, by the peculiar endosmitic action which I believe I have mentioned elsewhere, at once rendered salt, as was evident from Baumé's hydrometers and the test of the nitrate of silver. The surface crust bore me readily this evening at a temperature of 21° and 19°, giving no evidences of thaw. Beneath, for two inches it was crisp and fresh. As I tried it lower, cutting carefully with my bear-knife, it became spongy and brackish; at eight inches markedly so; and at and below twelve, salt-water paste. On the other hand, all my observations, and I have made a great many, prove to me that cold, if intense enough, will, by its unaided action, independent of percolation, solar heat, depending position, or even depth of ice, produce from salt water a fresh, pure, and drinkable element."

---

2. Ice Observations. By Sir Edward Belcher.

The Formation of Floe-Ice. (Sir Edward Belcher.)

Some interesting experiments were made on the rate and mode of formation of the floe or salt-water ice by inserting a wooden tube, with two opposite sides partially open, into the ice. The results from November to March, recorded on page 123, Vol. I. show an average increase of ice of half an inch a day solely from below. Observations of thickness were made every 10 days, and compared with the recorded temperatures.

After the ice-gauge was raised in March 1854, observations were made on the in-shore ice to discover when the ice-crystals ceased to attach themselves to the under surface. From November 5th to March 25th the thickness had increased from 18 to 68 inches.

Again on April 9th (p. 161) the thickness of the ice was found to be 66 inches. The ice had not only ceased to form, but the *lower* portion of two inches in depth was entirely composed of loosely cohering separate crystals yielding easily to the pressure of the finger. The ice in contact with these crystals was also in a rotten state.

By experiments made by rapidly lowering bottles fitted with a plug of loose cotton to a considerable depth so as to bring up water from that depth, the water was found to be full of crystalline stars of ice.

These results entirely agreed with experiments made on sea-water in glass cylinders at 50° F., and submitted to a temperature of – 24° F. The crystallization commenced from below. As the freezing point was approached, peculiar stars were produced, and rose to the surface where they became attached to and formed the general mass. These stars were perfect detached crystals, similar to those met with in the atmosphere.

Experiments are also described in Vol. I. (p. 150) intended to measure the amount of evaporation from ice, and cubes of ice from different layers were exposed to the upper-deck temperature during winter, and weighed at intervals to determine the changes. The water which thawed from these cubes was bottled for future examination, and the atmospheric air in well-dried bottles, and covered with leather and bladder, was also obtained.

In the Appendix at the end of Vol. II. of the Last of the Arctic Voyages will be found records of experiments to ascertain the amount of evaporation from cubes of salt-water ice from different depths in the floe; also an Analysis of the Water, and a most interesting Account of Observations on the Forms of and Changes in Crystals of Ice and Snow, with magnificent pictures of the remarkable and, in some cases, complicated forms which were observed.

---

3. The COMPOSITION of SEA-WATER and of SALT-WATER or FLOE-ICE. (Vol. II. p. 293.)

In the Appendix are also tables giving the composition of sea-water and of the melted sea-water or floe-ice from different depths of the floe in Wellington Channel, taken during the period at which it became solid for the season.

The waters contain large quantities of carbonates of lime and magnesia, no doubt due to the presence of those substances in rocks in the neighbourhood. Water taken at spring-tide from a depth of 10 fathoms, contained more chloride of sodium and less sulphate of lime than the water 4 feet below the ice, and contains sulphate of magnesia in place of chloride or bromide of magnesia. The water of the ice is obviously derived from land sources, but the amount of common salt shows that some sea salts are intermixed with the natural salts derived from the land.

ANALYSIS of the ARCTIC FLOE : in 1,000 grains of Water.

——	No. 1.	No. 3.	No. 4.	No. 6.
Specific gravity -	1·004	1·006	1·007	1·007
Sésquioxide of iron and phosphate -	·010	·050	·010	·008
Insoluble lime - - -	trace	·009	·032	·025
Soluble lime - - -	·078	·142	·113	·103
Insoluble magnesia - - -	·024	·041	·043	·052
Soluble magnesia -	·124	·363	·471	·360
Sodium - - -	1·739	1·933	2·708	2·286
Potassium - - -	·594	·096	·128	·096
Chlorine - - -	2·723	3·476	4·424	3·950
Sulphuric acid - -	·372	·472	·575	·497
Carbonic acid - - -	·052	·128	·140	·154

From Northumberland Sound, lat 76° 52′ N., long. 97° W.
No. 1.—From the under 6 inches of the floe.
No. 3.—From the upper 6 inches of the floe.
All collected on October 24, 1852.

The COMPOSITION according to the preceding Table.

——	No. 1.	No. 3.	No. 4.	No. 6.
Sesquioxide of iron and phosphate -	·010	·050	·010	·008
Carbonate of lime - - -	trace	·016	·058	·046
Sulphate of lime - -	·189	·344	·274	·250
Carbonate of magnesia - -	·050	·110	·090	·110
Chloride of magnesium - -	·289	·859	1·122	·855
Sulphate of potash - -	·133	·220	·284	·220
Sulphate of soda - - -	·355	·300	·504	·442
Chloride of sodium - - -	4·132	4·668	6·129	5·454
Total residue - -	5·158	6·567	8·471	7·385
Residue by evaporation -	5·400	6·600	8·340	7·280

ANALYSIS of the ARCTIC SEA WATER.

—	No. 2.	No. 7.
Sesquioxide of iron and phosphate	·010	·022
Insoluble lime	·027	·091
Soluble lime	1·260	·474
Insoluble magnesia	·110	·081
Soluble magnesia	·571	·491
Sodium	10·853	—
Potassium	·384	·432
Chlorine	17·950	17·585
Sulphuric acid	1·986	2·040
Carbonic acid	·284	·260

No. 2.—Four feet below ice.
No. 7.—Ten fathoms, at spring tide.

COMPOSITION of the SEA WATER of WELLINGTON CHANNEL, calculated from the preceding Table.

—	No. 2.	No. 7.
Sesquioxide of iron and phosphate	·010	·022
Carbonate of lime	·048	·091
Sulphate of lime	2·711	1·149
Chloride of calcium	·283	—
Carbonate of magnesia	·232	·171
Sulphate of magnesia	—	1·380
Chloride of magnesium } Bromide of magnesium }	1·353	·097
Sulphate of potash	·854	·965
Chloride of sodium	27·621	28·916
	33·112	32·791
Specific gravity	1·026	1·027

A comparison of the results of the analysis of the floe-ice water with that from the Straits of Dover, analysed by Professor Guthrie, shows that the amount of solid residue after evaporation of the Dover water is nearly the same in amount as from the upper 6 inches of floe-ice in Wellington Channel.

In May 1854 experiments were made to ascertain the protecting power of snow from cold. Thermometers at 32° F. were inserted horizontally into holes close to the floe in 4 feet of snow and 3 feet above it at a depth of 1 foot from the top of the snow bank. After 10 days that at the bottom of the bank in contact with the ice indicated, + 14°; the other indicated + 2°; while the external temperature had been −19°, and the mean + 2·5°, so that one foot of snow was equivalent to an increase of temperature of 21° and four feet of snow to 33° F. After 10 days more, at one foot deep, the temperature was + 8°, and at four feet deep it was 16°; the minimum had been −11°, and the mean + 11·19°.

4. Ice Observations. By Dr. Walker, during the Voyage of the "Fox." (From the Journal of the Royal Dublin Society, Jan. 1860.)

My attention was directed to the analysis of ice, and the changes produced in sea-water when freezing, by the various and contradictory opinions held by different authorities. Sir Charles Lyell, in his "Principles of Geology," p. 96, states that sea-water ice is *fresh*, having lost its salt by the decomposing process of freezing. Dr. Sutherland, in his "Journal of Penny's Voyage," says that sea-water ice is salt, containing about a quarter part of the salt of the original water, the proportion depending upon the temperature; it being very probable that the lower the temperature is the *more* salt the ice will contain. At variance with both of these, Dr. Kane asserts in his "Arctic Expedition," pp. 377, 385, 392, that, although at some temperatures and at some small thicknesses the ice *may* be salt, yet, "if the cold be sufficiently intense, independent of percolation, solar heat, depending position, or even "depth of ice, there *will* be formed from salt water a *fresh, pure, and drinkable element.*" Baron Wrangell ("Le Voyage au Nord de la Siberie") mentions that *the salt left by evaporation* on the surface of the ice is mixed with the snow that falls upon it, and is eaten as salt with food, though bitter and aperient.

Five different sets of observations were carried out during the winters of 1857–8 as we drifted down Baffin's Bay, Davis' Strait.

1st. To observe the changes in specific gravity of sea-water ice formed under different degrees of cold.

2nd. Forming fresh water from salt (or sea) water, by freezing the latter, and thawing and re-freezing the ice thus formed, continued one, two, or three times, as necessary, at different temperatures.

3rd. Converse of last, forming brine from sea-water by freezing at different temperatures, and re-freezing the residue.

4th. Specific gravity, &c., of saline efflorescence at different temperatures expressed from sea-water ice in its early process of formation.

5th. Specific gravity of ordinary ice of winter formation at close of winter, taking specimens at different depths.

On the 7th of September 1857, we were frozen in whilst endeavouring to pass through Melville Bay, and until the month of April 1858 our ship was immoveably fixed in the ice. During all the intervening time I have watched and tested the growth and condition of the ice formed, taking advantage of all opportunities which offered, and filling up every blank, as far as possible, by artificial means. An abstract of the changes produced are here given, with tables of the observations.

When the temperature of the water and air falls below 28°·5, sea-water exposed will in a short time be covered with a thin and almost pellucid pellicle of ice, of a very plastic nature, allowing of a great amount of bending, curving, and such like accommodations to external circumstances. In proportion to the temperature, this covering becomes thicker, and presents a vertically striated appearance, identical with that of sal ammoniac, gradually disappearing as the mass thickens and gets more compact; still, the lowest portion, or that most recently formed, always presents this aspect. When this pellicle or covering becomes of the thickness of a quarter of an inch, or more, small white crystals appear on its surface, at first sparse and widely separated, but gradually forming into tufts, and ultimately covering the whole surface. This may be called "efflorescence." This also differs as to its appearance, quantity, and quality, according to the temperature of the air at the time; its true nature and the purpose it fulfils will be seen presently. For the first few hours or days the increase of the thickness of the ice is rapid, on account of the comparative exposure to the coldness of the atmosphere, but afterwards the rate of growth is much slower and more uniform.

In this way is formed the great body of ice known as "the floe," or "great pack" extending approximately from 78° to 65° N. latitude, about 900 miles in length and 200 miles broad, enclosing an area of about 500,000 square miles. Dr. Kane states that the average thickness of this ice in 1850–51 was 8 feet; during our drift in 1857–58 it was 5 feet. This body of ice is not all level; its otherwise uniform level is disturbed by hummocks, these being produced by the piling up of the tables of the floe by pressure; such accumulations not only appear *above* the surface of the floe, but exist also *underneath* it to a far greater degree, so that a larger quantity of ice is formed annually than the mere superficial extent of the floe would lead one to suppose, and at the same time, on account of their abundant presence, materially increasing its average thickness. The ice thus formed varies in specific gravity according to the temperature at which it was frozen, the density decreasing in proportion to the degree of cold; such decrease being more apparent from the freezing point of sea-water down to zero, and from that becoming less evident, yet still present. Yet in no case (and my observations extend from below the freezing point to —42°) could I obtain fresh water, the purest being of specific gravity 1·005, and affording abundant evidence of the presence of salts, especially chloride of sodium, rendering it unfitted for culinary purposes, much less for photographic use. Table No. 1 gives the observations on the changes of specific gravity of the ice formed under different degrees of cold; this is surface ice, but as this ice increases in thickness and becomes exposed to greater degrees of cold, it gradually becomes more compact and close, and part of the remaining salt is expressed, so that, by the termination of the season, this same ice is somewhat fresher than when it was formed. This is shown by the observation on the 15th of January and the 22nd of March 1858. All the floe which annually drifts out of Davis' Straits is not formed at the same temperature, as wide cracks and lanes appear abundantly at times throughout the whole winter, so that parts of the surface are formed at + 10° or so, and others — 20° or — 40°. I say *parts of the surface*, because that portion first formed acts as a covering to the remainder from the cold, so that, proportionately, there is but little formed at a low temperature; but, no matter at what temperature formed, a part of the remaining salt is expressed as the ice contracts and becomes closer.

Having thus found that sea-water ice loses some of its saline constituents, I endeavoured to find out what became of the remainder. I have above spoken of the white crystals formed on the surface, called efflorescence; on testing this, I found that it contained a considerable portion of the ejected salts. This efflorescence formed at *all* temperatures, from a little below the freezing point of water to the lowest temperature observed; this also in character, density, and amount, varied with the temperature at which formed, its density *increasing* according to the degree of cold. In the act of freezing, some of the saline particles were squeezed out of the ice and forced to the surface, there becoming crystallized if the cold were sufficiently intense, but, if not, only part forming crystals, the remainder forming a pasty semi-fluid understratum. This efflorescence appeared, sooner or later, according to the temperature, but generally commenced when the ice was about $\frac{3}{8}$ths of an inch thick; from this period until the ice was from 4 to 9 inches thick, more or less was forced out; when it attained the latter thickness, this exudation ceased, as the upper portion of the mass had now become too compact and too consolidated to allow of its protrusion. At temperatures above zero, the amount of efflorescence was small, not always well crystallized, and scattered over the surface, being slow of appearance, and taking a long time to crystallize; whereas at low temperatures there was an abundant crop, well crystallized, and covering thickly the whole extent of surface. To test the residuary mass of water after the ice had been formed, so as to determine the amount, if any, of salts expressed into it, set No. 3 was instituted. By exposing sea-water to different temperatures in a large tub, and testing the residue after the ice was taken off, I found it contained some considerable portion of the expressed salts; this portion being dependent on the temperature and length of exposure. On many of these occasions I found on the surface of the ice a pasty fluid, being the liquid efflorescence not yet crystallized. Baron Wrangell, in his Travels, noticed this extra saltness of the crystals found on the surface, but attributed it to evaporation. Now, this cannot be the case, as evaporation is less in proportion to the cold, whereas the amount and density of the efflorescence increased in proportion to the cold. By exposing the residuum of the first freezing of salt-water to a second, a residue was obtained, containing much more salt, and by a continuance of freezing this resldue or its residue, a concentrated brine was obtained, no limit being apparent, so far as my experiments were made. This set of observations led to another, No. 2. Finding that some of the salts were precipitated, I tried to form fresh water by a similar exposure of sea-water, and, thawing and exposing the ice thus formed three or more times, I succeeded in obtaining water of a density of 1·0025 and 1·0020, these results being dependent on the time of exposure andtem perature. Opportunities not always occurring during the winter of examining the ice formed naturally, I endeavoured to fill up the blanks by artificially freezing sea water. The results given in the table may not seem as uniform as would be expected, consequent on the amount of efflorescence not being always crystallized, and adhering to the ice at the time of thawing and testing. Still, from the results obtained, and the composition of the efflorescence and the residue, the conclusion may with propriety be educed, *that, as the temperature decreases from* +28°·5 *to the lowest amount of natural cold, the ice formed will, in equal times, also decrease in density and increase in thickness.* Most likely there will be a point at which this decrease in density will be in a degree limited, due to the rapid increase of thickness under *extreme* cold (say −60° or −70°) in a short space of time.

I have before stated that we were frozen in on the 7th of September 1857; the new ice then forming around us was of specific gravity 1·0235 (30°); on the 15th of January 1858 this same surface ice was specific gravity 1·0102 (30°); on the 22nd of March 1858 specific gravity 1·0078 (30°); so that this surface ice, which was exposed for 6½ months to a temperature varying from +28·2 to −47°, had, by contraction, squeezed out so much extra of its remaining saline constituents.

Set No. 4 gives the results obtained from the testing of specific gravity of efflorescence formed at different temperatures. If the temperaturo at which the efflorescence was expressed were not sufficiently low to freeze it, it remained for a time semi-fluid and pasty, only producing crystals on its surface; but as the cold increased, all the quantity squeezed out became crystalline. At first, the spots of emergence of those crystals were few and widely separated, but as the ice increased in thickness, the efflorescence also increased in quantity, and, after collecting in tufts, ultimately spread over the whole; except the temperature were low, these crystals generally presented a moist appearance, but if below −10° they were always dry and hard. These crystals, from below +28°·5 to −25°, presented almost the same form, the changes being very slight and dependent on the amount of aggregation, the original being in the form of a broad feather, with plumes and secondary plumes branching off these; length varying from ¼ to 2 inches, breadth from ¼ to 1 inch. But at and below the temperature of −25°, the crystals presented a very different form, having the appearance of long acicular fibres, varying from ¼ to 2 inches in length, of no definite structure, but always fibrous, from $\frac{1}{16}$ to $\frac{1}{24}$ of an inch in diameter, sprouting up from a basic tuft very like pure crystals of "*caffeine.*" Whether this change in form in the crystal is dependent on the cold, or greater amount of saline ingredients present, I cannot say.

When the sun-rays in spring become sufficiently intense, the exposed surfaces of snow, ice, or efflorescence, begin to melt. During the winter, as cracks and lanes of water appear in the floe, these are covered with ice, efflorescence, and some little snow. The melting point of this mixture of efflorescence and snow is much lower than that of sea-water ice, the degree depending upon the amount of salts contained. Consequently, in these lanes does the disintegrating action first commence; the sun's rays acting on this mass, thaw it, and form a kind of sludge on top; afterwards exosmosis and endosmosis commence, the ice acting as the membrane between the salt water below and the sludgy pools above; in time this intervening ice becomes permeated and infiltrated with salt, less buoyant, and thus soon becomes a prey to the eroding action of the water and the thawing power of the sun. Thus the new lanes and bay-ice are first broken up, allowing more play to the water on the remaining thicker mass, and at the same time giving greater advantage to the action of the wind, which drifts it to a warmer situation, where it is soon reduced in bulk, and gradually wastes away. But not only do these lanes experience the effect of the efflorescence, but also the main body of the floe, but not to so great an extent, as there is more snow mixed with it, and the intervening membrane is much thicker. The portions of the hummocks which were below the general under surface of the floe having become water-eaten, honeycombed, and detached, now make their appearance (under the name of tongue-pieces), and act the part of wedges, keeping separate the different fields and masses of ice. I have said these tongue pieces are water-eaten and honeycombed. The salts retained in the ice are not equally dispersed through it, but are (if I might so speak) contained in cells, having aggregated at the time of freezing; as these cells are exposed to the thawing power of the water beneath, they become melted much sooner than the mass of the ice, less heat being required, and thus present the honeycombed appearance seen by every one as they sail through nips in the floe. The needle-ice spoken of by Parry and other Arctic travellers over ice is due to the same cause, although that distinguished navigator says it depends on the drops of rain. I have often seen it in early summer, before one drop of rain had fallen, but when the sun had commenced to act on the surface of the ice.

A word or two on the appearances seen when sea-water was frozen in a capacious vessel. After, say, 12 hours' exposure, when the ice was removed, it was found vertically striated, and oftentimes divisible into two or more layers, the under-surface presenting a curious aspect, small lines of about half an inch long, and in groups of four or six, reticulating, seemed to cover the entire under-surface.

This occurred at all temperatures, and independent of length of exposure and consequent thickness of ice. From the bottom of the vessel thin plates of ice were seen protruding into the residuary water, from ½ to 2½ inches long, and proportionately wide. These were found to be much fresher than the ice formed on top, or the residue below; they completely studded the bottom of the vessel. The appearance of these plates was not at all modified by the varied density of the residue in which they were formed, or even the density of the primary solution; but of course their density varied with that of the solution in which they were found. On breaking this artificial sea-water ice, as well as that naturally formed whilst under two inches thick, I observed that it easily separated into perpendicular laminæ, thin, but not long; when viewed with a magnifying glass, presenting the same appearance as that of the plates just mentioned; the markings noticed on the under-surface being but the edges of these laminæ, which, being placed side by side, presented a sal ammoniac appearance: in fact, a mere lateral assemblage or building up of thin laminæ of ice, the increase being from below on the lower end and under edge of these plates.

Perhaps the statement of Dr. Kane, that sea-water ice under certain circumstances is completely free from salt, may be explained by the following facts and experiments. After our winter preparations had been commenced, and the pool of fresh water (from melting snow) had been frozen over, the men sent out to bring in snow for culinary purposes brought in some ice instead; this they obtained from some hummocks near the ship, these hummocks being part of the formation of the previous winter's pack in which we were caught. This ice turned out to be sufficiently fresh for all the purposes of domestic use. On several occasions the party sent out for this ice, digging too deep into the hummock, and not content with the surface pieces, found that the ice was no longer fresh, but quite salt—this ice being a continuation of the same hummock, and also of the previous winter's growth.

On the 12th and 13th of August 1857, whilst lying off Browne Islands, and within about four miles of the glacier, surrounded by bergs, I noticed an appearance like oil on the surface of the water; on closer inspection and testing, this proved to be fresh water floating on the surface of the salt to the depth of two or three inches. The sun, beaming down upon the bergs, had melted the ice and snow; this, running off, floated on the surface and remained separate, so long as there was no wind to mix and agitate the fluids of different densities. To a combination of such circumstances, with an after freezing of this surface water, do these fresh hummocks owe their origin; the water, being frozen in this state, and afterwards the ice elevated into hummocks afforded us a "drinkable element" during the winter; and when the men had exhausted the supply of top-pieces, they, supposing that all was alike, continued their labours, but were disappointed in obtaining salt-water ice instead of fresh. To make sure of this, I took a tub of salt-water, and poured upon its surface, on a glass plate, fresh water, to the thickness of two inches, and allowed the whole to freeze. On testing the upper two inches of ice formed, I found it quite fresh, whilst the under portion was salt.

No. 1.

Manner of Exposure.	Specific Gravity when exposed.	Hours of Exposure.	Temperature at which exposed.	Depth of Water in Vessel at first.	Thickness of Ice formed.	Specific Gravity of Ice formed.	Specific Gravity of Residue.
				in.	in.		
A tub of salt water	1·0260 (28°.5)	—	+20·0	—	—	1·0220 (36°)	
„ „ -	1·0260 (28 ·5)	12	+18·0	6⅞	⅝	1·0200 (30)	1·0322 (28°)
„ „ -	1·0265 (28 ·5)	12	+10·0	6	⅞	1·0180 (40)	1·0350 (28)
A hole in the floe -	1·0278 (28 ·5)	—	+ 3·0	—	1⅝	1·0160 (32)	
A tub of sea water -	1·0278 (28 ·5)	12	Zero.	5½	2	1·0135 (30)	1·0375 (28)
„ „ -	1·0270 (28 ·5)	12	− 5·5	5½	2⅜	1·0130 (35)	1·0560 (27)
„ „ -	1·0265 (28 ·5)	12	− 7·5	9	2	1·0135 (40)	1·0580 (28)
A hole in the floe -	1·0265 (28 ·5)	12	− 7·0	—	2½	1·0130 (35)	
A tub of sea water -	1·0265 (28 ·5)	12	− 8·0	6⅞	3¾	1·0160 (40)	1·080 (28)
A hole in the floe -	1·0265 (28 ·5)	—	− 8·0	—	3½	1·0105 (30)	
A tub of sea water -	1·0270 (28 ·5)	12	− 9·0	5½	3	1·0140 (30)	1·0575 (30)
A hole in the floe -	1·0270 (28 ·5)	24	−17·0	—	3¾	1·0105 (40)	
A tub of sea water -	1·0270 (28 ·5)	12	−17·5	5½	2¾	1·0098 (30)	1·0625 (28)
„ „ -	1·0270 (28 ·5)	12	−18·0	3⅞	3⅛	1·0165 (40)	1·0830 (28)
A hole in the floe -	1·0270 (28 ·5)	4	−23·0	—	½	1·0140 (30)	
A tub of sea water -	1·0270 (28 ·5)	12	−23·0	5½	3	1·0155 (40)	1·0770 (30)
„ „ -	1·0270 (28 ·5)	12	−29·0	5½	2⅝	1·0130 (30)	1·0595 (30)
A hole in the floe -	1·0270 (28 ·5)	—	−30·0	—	2½	1·0115 (30)	
A tub of sea water -	1·0270 (28 ·5)	12	−32·0	6¼	2½	1·0150 (35)	1·0508 (29)
„ „ -	1·0270 (28 ·5)	10	−35·0	6	3¼	1·0095 (35)	1·0545 (28)
A hole in the floe -	1·0270 (28 ·5)	24	−35·0	—	4	1·0105 (30)	
A tub of sea water -	1·0270 (28 ·5)	6	−36·0	6	2¾	1·0085 (32)	1·040 (30)
A hole in the floe -	1·0270 (28 ·5)	15	−40·0	—	3⅝	1·0140 (30)	
A tub of sea water -	1·0270 (28 ·5)	12	−40·0	5½	3	1·0095 (40)	1·0660 (30)

No. 2.

Manner of Exposure.	Specific Gravity when exposed.	Hours of Exposure.	Temperature at which exposed.	Depth of Water in the Vessel.	Thickness of Ice formed.	Specific Gravity of Ice formed.	Specific Gravity of Residue.
				in.	in.		
Exposed sea water	1·0278 (28°·5)	12	Zero.	5½	2	1·0135 (30°)	1·0375 (28°)
„ residue of former experiment.	1·0375 (28)	5	+ 3°	3½	—	1·0230 (29)	1·050 (30)
Exposed sea water	1·0278 (28°·5)	12	− 5·5	5½	2⅜	1·0030 (30)	1·0560 (27)
„ residue of former experiment.	1·0560 (27)	5	−11	3	1	1·040 (40)	1·0850 (28)
Exposed sea water	1·0278 (28°·5)	12	− 9	5½	3	1·0140 (30)	1·0575 (28)
„ residue of former experiment.	1·0575 (28)	5	− 2	3	1½	1·0475 (40)	1·102 (30)
Exposed sea water	1·0270 (28°·5)	12	−17·5	5½	2¾	1·0098 (30)	1·0625 (28)
„ residue of former experiment.	1·0625 (38)	5	−10	3	1	1·0375 (40)	1·0925 (30)
Exposed sea water	1·0270 (28°·5)	12	−18	3⅞	3⅛	1·0165 (40)	1·0830 (28)
„ residue of former experiment.	1·0830 (28)	2	−18	1¾	1	1·051 (40)	1·1225 (26)
Exposed sea water	1·0270 (28°·5)	12	−23	5½	3	1·0155 (40)	1·0770 (30)
„ residue of former experiment.	1·0770 (30)	5	−27	2	1⅛	1·0535 (30)	1·120 (40)

No. 2—*continued.*

Manner of Exposure.	Specific Gravity when exposed.	Hours of Exposure.	Temperature at which exposed.	Depth of Water in the Vessel.	Thickness of Ice formed.	Specific Gravity of Ice formed.	Specific Gravity of Residue.
				in.	in.		
Exposed sea water	1·0270 (28°·5)	12	−29	5½	2⅜	1·0130 (30)	1·0595 (30)
„ residue of former experiment.	1·0595 (30)	5	−25	3	1⅜	1·0480 (30)	1·1070 (30)
Exposed sea water	1·0265 (28°·5)	10	−35	6	3¼	1·0095 (40)	1·0545 (28)
„ residue of former experiment.	1·0545 (28)	3	−35	—	¾	1·0165 (50)	1·0750 (28)
Exposed residue of former experiment.	1·0750 (28)	3	−36	—	¾	1·0460 (30)	1·0910 (28)
Exposed residue of former experiment.	1·0910 (28)	2	−36	—	—	1·070 (30)	1·1225 (26)
Exposed sea water	1·0265 (28°·5)	0	−36	6	2¾	1·0085 (30)	1·040 (28)
„ residue of former experiment.	1·040 (28)	3	−36	—	1	1·0225 (30)	1·0490 (28)
Exposed residue of former experiment.	1·0490 (28)	3	−36	—	—	1·0340 (30)	1·061 (28)
Exposed residue of former experiment.	1·061 (28)	5	−35	—	—	1·050 (30)	1·1175 (26)

No. 3.

Manner of Exposure.	Specific Gravity when exposed.	Hours of Exposure.	Temperature at which exposed.	Depth of Water in the Vessel.	Thickness of Ice formed.	Specific Gravity of Ice formed.	Specific Gravity of Residue.
				in.	in.		
A hole cut through the floe.	1·0265 (28°)	—	− 8°	—	3½	1·0105 (30°)	
Ice off last - -	1·0105 (30)	12	− 9	7⅛	2⅝	1·0055 (60)	1·0315 (28°)
„ „ - -	1·0055 (60)	6	− 1	—	—	1·0045 (35)	1·0105 (28)
„ „ - -	1·0045 (35)	3½	+ 5	—	—	1·0025 (30)	1·008 (29)
A hole cut through the floe.	1·0270 (28·5)	24	−17	—	3¾	1·0110 (30)	
Ice off last - -	1·0110 (30)	6	−20	5¼	1¼	1·0055 (40)	1·0195 (28)
„ „ - -	1·0055 (40)	2	−20	2½	¾	1·0045 (40)	1·0120 (30)
„ „ - -	1·0045 (40)	1	−20	¾	¼	1·0025 (30)	1·0080 (29)
A hole cut through the floe.	1·0270 (28·5)	—	−28	—	3½	1·0130 (30)	
Ice off last - -	1·0130 (30)	12	−20	3¾	2¾	1·0095 (30)	1·0450 (30)
„ „ - -	1·0095 (30)	3	−14	2	½	1·0020 (32)	1·0125 (30)
A hole cut through the floe.	1·0270 (28·5)	—	−20	—	3	1·0105 (70)	
Ice off last - -	1·0105 (70)	18	−31	5⅛	2½	1·0070 (70)	1·050 (28)
„ „ - -	1·0070 (70)	7	−31·5	—	—	1·0020 (32)	1·0140 (30)
Tub of sea water -	1·0265 (28·5)	12	−32	6¼	2½	1·0150 (35)	1·0508 (29)
Ice off last - -	1·0150 (35)	7	−32	—	—	1·0097 (30)	1·0210 (29)
„ „ - -	1·0097 (30)	2½	−33	—	—	1·0030 (32)	

Monthly Measurements of Ice and Snow.

Date.	Ice Thickness.	Increase during Month.	Snow Thickness.	Increase during Month.
	ft. in.	ft. in.	in.	in.
1857. October 16 -	1 3¾	—	2½	—
—— November 16 -	2 0½	8¾	5¼	2¾
—— December 16 -	3 0	11½	6½	1¼
1858. January 16 -	3 7½	7½	8½	2
—— February 16 -	3 9	1½	9½	1
—— March 16 -	4 3½	6½	9½	0
—— April 16 - -	3 1	decrease 1·2½	10½	1

Small Measurements of Increase of Ice.

Hours.	Thickness.	Temperature.
	in.	
3	⅞	−29°
6	1½	−26
18	3⅛	−25
24	3⅞	−22·5

No. 4.

—	Number of Hours in forming.	Temperature.	Specific Gravity.
Efflorescence -	18	− 2°	1·0375 (70°)
„ „ -	36	− 2	1·0410 (44)
„ „ -	—	− 3·5	1·0495 (34)
„ „ -	24	− 7	1·0625 (45)
„ „ -	—	−23	1·0650 (28)
„ „ -	—	−28	1·0665 (40)
„ „ -	—	−30	1·0710 (30)
„ „ -	48	−35	1·0945 (30)
„ „ -	24	−38	1·0775 (28)
„ „ -	—	−40	1·0880 (30)

Date.	Ice formed (3 inches thick) at Temperature of +25.	Specific Gravity.
1857. September 7 and 8 -	- - - - - -	1·0235 (26°)
1858. January 15 - -	Same ice, upper 3 inches - - -	1·0102 (29)
—— March 22 - -	„ 0 to 6 inches from surface -	1·0078 (30)
„ „ - - -	„ 6 to 12 „ „ -	1·0065 (35)
„ „ - - -	„ 12 to 18 „ „ -	1·0052 (32)
„ „ - - -	„ 18 to 24 „ „ -	1·0050 (30)
„ „ - - -	„ 24 to 30 „ „ -	1·0050 (30)
„ „ - - -	„ 30 to 36 „ „ -	1·0058 (32)
„ „ - - -	„ 36 to 42 „ „ -	1·0055 (30)
„ „ - - -	„ 42 to 48 „ „ -	1·0050 (32)
„ „ - - -	„ 48 to 54 „ „ -	1·0050 (35)
	Total depth, 4 feet 6 inches.	

Port Kennedy. Latitude, 72° 1′ N.; longitude, 94° 15′ W.

1858, October 7. Surface ice formed at temperature +21°. Specific gravity 1·0230 (30°)
1859, June 3. " " " (upper 6 inches) " " 1·0055 (30°)

Salts in 100 Grains.	No. 0.	No. 1.	No. 2.	No. 3.	No. 4.	No. 5.
Sulphate and carbonate of lime	0·5658	0·1170	0·1496	0·1382	0·1126	0·1439
Sulphate of magnesia	0·3666	0·2316	0·2616	0·2594	0·2762	0·2296
Chloride and bromide of magnesium	0·9766	0·2470	0·2412	0·2118	0·2184	0·3695
Chlorides of sodium and potassium	12·6192	2·4158	2·7064	2·8212	2·7796	2·7825
Solid contents	14·4282	3·0114	3·3588	3·4306	3·3868	3·5255

No. 5.

Salts; 100 grammes.	6.	7.	—
Sulphate and carbonate of lime.	0·0378	0·0358	6.—Portion of ice taken from middle of perpendicular block of ice—forming part of floe—after being exposed in situ to a varying temperature during eight months; this floe had drifted 1,200 miles. Specific gravity, 1·005 (30°). 7.—In surface ice which was formed at temperature of 21°, specific gravity, 1·0230. After being exposed in situ to a temperature between +20° and −48° during nine months. Specific gravity, 1·0055 (30°).
Sulphate of magnesia	0·0610	0·0548	
Chloride and bromide of magnesium.	0·0884	0·0698	
Chlorides of potassium and sodium.	0·3506	0·2474	
Solid contents	0·5378	0·4078	

5. Physical Properties of Ice. By Dr. J. Rae. (From the Proceedings of the Physical Society.)

If a saturated solution of salt is frozen, and the ice so formed is fresh, it is evident that the salt that has been "rejected" must be deposited or precipitated in a crystalline or some other solid form, because the water, if any, that remains unfrozen, being already saturated, can hold in solution no more salt than it already contains.

Could not salt be obtained readily and cheaply by this means from sea-water in cold climates?

During several long journeys on the Arctic coast, in the early spring before any thaw had taken place, the only water to be obtained was by melting snow or ice. By experience I found that a kettleful of water could be obtained by thawing ice with a much less expenditure of fuel, and in a shorter time, than was required to obtain a similar quantity of water by thawing snow. Now, as we had to carry our fuel with us, this saving of fuel and of time was an important consideration, and we always endeavoured to get ice for this purpose. We had another inducement to test the sea-ice frequently as to its freshness or the reverse.

I presume that almost every one knows that to eat snow when it is very cold, tends to increase thirst, whereas a piece of ice in the mouth is refreshing and beneficial, however cold it may be; we were consequently always glad to get a bit of fresh ice whilst at the laborious work of hauling our heavy sledges; yet with these strong inducements we were never able to find sea-ice, *in situ**, either eatable when solid or drinkable when thawed, it being invariably much too salt. The only exception (if it may be called one) to this rule, was when we found rough ice, which, from its wasted appearance and irregular form, had evidenty been the formation of a previous winter. This old ice, if projecting a foot or two above the water-level, was almost invariably fresh, and, when thawed gave excellent drinking-water. It may be said that these pieces of fresh ice were fragments of glaciers or icebergs; but this could not be so, as they were found where neither glaciers nor icebergs are ever seen.

How is this to be accounted for? Unfortunately I have only a theory to offer in explanation.

When the sea freezes by the abstraction of heat from its surface I do not think that the saline matter, although retained in and incorporated with the ice, assumes the solid state, unless the cold is very intense, but that it remains fluid in the form of a very strong brine enclosed in very minute cells. So long as the ice continues to float at the same level, or nearly the same level, as the sea, this brine remains; but when the ice is raised a little above the water-level, the brine by its greater specific gravity, and probably by some solvent quality acting on the ice, gradually drains off from the ice so raised; and the small cells, by connecting one with another downwards, become channels of drainage.

There may be several other requisites for this change of salt ice into fresh, such as temperature raised to the freezing-point, so as to enable the brine to *work out* the cell-walls into channels or tubes—that is, if my theory has any foundation in fact, which may be easily tested by any expedition passing one or more winters on the Arctic, or by anyone living where ice of considerable thickness is formed on the sea, such as some parts of Norway.

All that is required, as soon as the winter has advanced far enough for the purpose, is to cut out a block of sea-ice (taking care not to be near the outflow of any fresh-water stream) about 3 feet square, remove it from the sea to some convenient position, test its saltness at the time, and at intervals repeat the testing both on its upper and lower surfaces, and observe the drainage if any.

The result of the above experiment, even if continued for a long while, *may* not be satisfactory, because the fresh ice, that I have described must have been formed at least 12 months, perhaps 18 months, before.

The Transposition of Boulders from below to above the Ice.

When boulders, small stones, sand, gravel, &c. are found lying on sea-ice, it is very generally supposed that they must have rolled down a steep place or fallen from a cliff, or been deposited by a flow of water from a river or other source. There is, however, another way, in which boulders, &c. get upon floe-ice, which I have not seen mentioned in any book on this subject.

---

* What I mean by ice *in situ* is ice lying flat and unbroken on the sea, as formed during the winter it is formed in.

During the spring of 1847, at Repulse Bay on the Arctic shores of America, I was surprised to observe, after the thaw commenced, that large boulders (some of them 3 or 4 feet in diameter) began to appear on the surface of the ice; and after a while, about the month of July, they were wholly exposed, whilst the ice below them was strong, firm, and something like 4 feet thick.

There were no cliffs or steep banks near from which these boulders could have come; and the only way in which I could account for their appearance, was that which by subsequent observation I found to be correct.

On the shores of Repulse Bay the rise and fall of the tide are 6 or 8 feet, sometimes more. When the ice is forming in early winter, it rests, when the tide is out, on any boulders, &c. that may be at or near low-water mark. At first, whilst the ice is weak, the boulders break through it; but when the ice becomes (say 2 or 3 feet) thick, it freezes firmly to the boulder, and when the tide rises, is strong enough to lift the boulder with it. Thus, once fastened to the ice, the stone continues to rise and fall with the rise and fall of each tide, until, as the winter advances, it becomes completely enclosed in the ice, which by measurement I found to attain a thickness of more than 8 feet.

Small stones, gravel, sand, and shells may be fixed in the ice in the same way.

In the spring, by the double effect of thaw and evaporation, the upper surface of the ice, to the extent of 3 feet or more, is removed, and thus the boulders, which in autumn were lying at the bottom of the sea, are now on the ice, while it is still strong and thick enough to travel with its load, before favourable winds and currents to a great distance.

The finding small stones and gravel on ice out at sea does not always prove that such ice has been near the shore at some time or other.

I have noticed that wherever the Walrus in any numbers have been for some time lying either on ice or rocks, a not inconsiderable quantity of gravel has been deposited, apparently a portion of the excreta of that animal, having probably been taken up from the bottom of the sea and swallowed along with their food.

---

6. Some EXPERIMENTS with SEA-WATER, by DR. F. GUTHRIE. (From the Proceedings of the Physical Society.)

Sec. 32. *Freezing Sea-water.*—The sea-water with which the following experiments were performed was procured from Dover. After filtration, it was found to have at 760 millims. the boiling point 100°·6 C., while the temperature of its vapour was 100°·2. This sea-water began to freeze at $-2°$ C. On evaporation on a water-bath and keeping at 100° C. for two hours, the per-centage of solid residue was 6·5786. A large beaker of this sea-water was cooled to 0° C. A tin vessel was supported inside the beaker so that its bottom just touched the surface of the water; and a freezing-mixture was placed in the tin vessel. When about $\frac{1}{100}$ of the whole had solidified, the solid was removed and divided into two parts: one was allowed to melt, and its per-centage of solid matter was determined as above; the other was broken up and frequently pressed between linen and flannel in a screw press, being allowed to melt as little as possible. The per-centage of solid matter in this also was determined. The following numbers show the result of this examination:—

	Per cent. at 100° of solid residue.
Sea-water - - - -	6·5786
Frozen sea-water - - -	5·4209
Frozen and pressed sea-water - -	0·4925

It appears, then, that under these conditions the freezing of sea-water is little more than the freezing of ice, and that the almost undiminished saltness of the unpressed ice is due, as suggested by Dr. Rae, to the entanglement amidst the ice-crystals of a brine richer in solid constituents than the original water itself. Such brine, which is here squeezed out in the press, drains in nature down from the upper surface of the ice-floe by gravitation, and also is replaced by osmic action by new sea-water which again yields up fresh ice; so that while new floes are porous and salt, old ones are more compact and much fresher, as the traveller observed.

---

7. Observations on Sea-water Ice. By J. Y. Buchanan.

These observations were made on board H.M.S. Challenger in the broken pack ice in the Antarctic regions. An analysis of the melted ice showed the presence of lime, magnesia, and sulphuric acid, and on the average 0·1723 gramme of chlorine per litre. Another piece, when pounded and melted in a beaker and examined, gave 0·0520 gramme of chlorine per litre, showing that the lump of ice was not homogeneous, as might be expected from the different ways in which it may be formed in the Polar regions.

The melting point of the pack ice was determined. The fresh ice began to melt at $-1°$ C.; after twenty minutes the thermometer had risen to $-0·9$ C., then to $-0·4$ C., where it remained constant for an hour, and then to $-0·3$ C.

"These determinations show that the salt in sea-water ice is not "contained in it only in the form of mechanically enclosed brine, "but exists in the solid form either as a single crystalline sub- "stance, or as a mixture of ice and salt crystals." The melting point of ice crystals, formed by freezing salt water in a bucket, was found to be $-1·3°$ C. (Royal Society Proceedings, vol. xxii., p. 431.)

---

8. The Temperature of Ice and Snow. (Sir John Ross.)

The thickness of ice was measured regularly every month, and increased until the end of May, when it was 10 feet in the sea, and

11 feet in a lake of fresh water. In February and March, with the air temperature at −50° F., the temperature of the ice gradually diminished between the surface and the water which was immediately below the ice at a temperature of − 27° F.

The change of temperature in 12 feet depth of snow being the same as in 7 feet of ice.

Effect of Cold on Mercury.

"Every year the mercury which had been used froze at a higher temperature until it reached −31° F., being eight degrees higher than the usual point, while mercury which had not been exposed retained its purity."

---

9. The Second German Expedition under M. Koldewey.

Temperature and Thickness of the Ice.

Observations were made on the rate of formation and on the temperature of the ice at various depths in the harbour at Sabine Island, with the following results:—

On Sept. 28 the thickness of the ice was 7 English inches.

On Nov. 11 the thickness was 31 inches.

Temperature of the air −12°·3 F.

Temperature of ice at the surface			− 6°·7 F.
„	„	8 inches deep	− 0°·8 F.
„	„	12 „	5°·9 F.
„	„	18 „	13°·3 F.
„	„	24 „	22°·6 F.

The level of the water was $3\frac{1}{4}$ inches under the surface of the ice, and the temp. of water +28° F.

Nov. 24.—Thickness of ice 36·5 inches.

Temperature of the air −2°·9 F.

Temperature of ice at the surface			0°·1 F.
„	„	$7\frac{1}{2}$ inches	7°·0 F.
„	„	13 „	12°·4 F.
„	„	18·5 „	16°·5 F.
„	„	23 „	19°·9 F.
„	„	27·5 „	21°·4 F.
„	„	30·0 „	23°·9 F.

The level of the water was $3\frac{1}{2}$ inches under the surface of the ice, and the temperature of the water was 28°·2 F.

Jan. 20.—Thickness of the ice 53 inches.

Temperature of the air −5°·3 F.

Temperature of ice at the surface			− 0°·2 F.
„	„	10 inches	1°·2 F.
„	„	14 „	3°·4 F.
„	„	20·5 „	5°·5 F.
„	„	26 „	7°·9 F.
„	„	31 „	11°·3 F.
„	„	37·5 „	13°·3 F.
„	„	44 „	19°·2 F.

The level of the water was 5·5 inches under the surface of the ice, and temperature of water 28°·2.

Feb. 18.—Thickness of the ice 57 inches.

Temperature of the air − 14°·8.

Temperature of the ice at the surface			− 16°·8 F.
"	"	14·4 inches	− 7°·1 F.
"	"	19 "	− 3°·8 F.
"	"	35·9 "	− 7°·5 F.
"	"	45·6 "	10°·0 F.
"	"	50·6 "	15°·8 F.

The level of the water was 6·1 inches under the surface of the ice.

May 21.—Thickness of the ice 79 inches.

Temperature of the air 29°·1 F.

Temperature of the ice at the surface			26°·8 F.
"	"	17 inches	23°·5 F.
"	"	26·5 "	26°·2 F.
"	"	31 "	26°·2 F.
"	"	38 "	26°·8 F.

This was the greatest thickness of the ice. From April to May the increase in thickness was at the rate of 7 inches a month, whilst from January to February it was only 4 inches a month. In winter the snow lying on the ice protects it, whilst in spring the evaporation and the storms which sweep the surface of the ice quite clear of snow keep down the mean temperature, and tend to increase the rate of formation of ice.

IV.—TIDES AND CURRENTS.

1. Parry's First Voyage for the Discovery of North-west Passage (1819–20).

Latitude of anchorage in Melville Island -	74° 47′ 19″·36 N.
Longitude - - - -	110° 48′ 29″·2 W.
Dip of magnetic needle - - -	88° 43′
Variation - - - - -	127° 47′ 50″ E.

The mean time of high water on full and change days (luni-tidal interval) was 1h. 29m. Usually the ice cracked from the high spring tides about *two* days after new and full moon, but in December the ice did not crack from the beach until the 22nd, *i.e.*, 5 days 8 hours after the new moon. This retardation of the tides may, perhaps, have arisen from the circumstance of the moon and sun having both had their greatest south declination at the usual time of the highest spring tide. There were also fresh gales from the eastward on December 17th and 18th. During the three months May, June, and July the night spring tides were higher than the day spring tides, showing that the two tides in the day are not equal. Tides were observed during May, June, to the 20th, and in July 1820.

From this it appears that at Parry's winter quarters the tides came from the east through Barrow Strait.

Height of the Tides.

——	Maximum.	Minimum.	Mean.
May -	4 ft. 2 in.	0 ft. 10 in.	2 ft. 6½ in.
June -	3 „ 7 „	1 „ 4 „	2 „ 7 „
July - -	3 „ 9 „	1 „ 5 „	2 „ 8¼ „

At the quarters of the moon the low tides were much greater during the day than during the night. Also the evening high tides reach their minimum 1½ days before the morning high tides.

2. Parry's Second Voyage.—Observations on the Tides at Winter Island and Igloolik.

The observations on tides were carried on at Winter Island from October 1821 to May 1822, and those at Igloolik from November 18, 1822, to April 1823. The heights of the tides were measured by means of a tide pole which was let through the ice and moored to the bottom by a heavy weight.

At Winter Island, the mean time of high water on full and change days is 12h. 11m.

The highest spring tide occurred on October 13, and was 15 ft. 8 in. high.

The lowest neap tide on March 18 was 3 ft. 1 in.

At Igloolik the mean time of high water on full and change days was 7h. 28m.

The highest spring tide on January 27 was 9 ft. 8 in., and the lowest neap tide on February 5 was 0 ft. 5 in.

3. Tides at Port Bowen. (Parry's Third Voyage, &c.)

"The great depth of water in which we lay at Port Bowen prevented our observing the rise and fall of the tides during the winter, by the usual method of a pole moored to the bottom. In the spring, however, when the fire-hole alongside the ship could not be kept constantly open, we adopted another plan, which it may be useful to describe. A stove of about 3 cwt. was let down the fire-hole to the bottom, having a whale-line attached to it. The line was rove through a block fixed to an outrigger from the ship's side, and to its other end was fastened a weight of 50 lbs. By this means the line was kept quite tight, and a marked pole being attached to it, served to indicate with great accuracy the perpendicular rise and fall of the water. The observations being given at length in the tide-table, I shall only here mention the fact, that during nine weeks in the months of April, May, and June, the morning tides were found, almost invariably, to rise several inches higher than those of the evening."

From the tables of tides it appears that when high water occurred at about 8 to 9 a.m., and 9 to 10 p.m., then the two heights were nearly equal; and if we call these hours the limits of day and night, then in almost all cases it is found that throughout April, May, and June (the period during which tides were observed), the night high tides are higher than the day tides. The two high tides were equal about four or five days after the half-moons, and the time of these equal high tides was a little earlier in the day as the summer advanced.

After new moon both high tides continued to increase for three or four days, and attained their maximum nearly together; but after full moon, the day tide began to diminish on the day after the full moon, whilst the night high tide still continued to increase for two or three days.

After *new* moon the height (*i.e.* the rise from the last low water) of the day tide is from 3 to 4 ft., and of the night tide is about 5 ft.; after *full* moon the height of the day tide is from 3 to 4 ft. and of the night tide about 6 ft.

If, on the other hand, we measure the heights of the tides by the *fall from the last high water*, then about three days after new and full moon the depths are equal and occur about 7 p.m. and 8 a.m., and they continue to be very nearly equal, except in the interval between three or four days after the half moons and the next new or full moons.

The greatest fall is about 4 feet at new moon and about 5 feet at full moon.

The mean time of high water at full and new moon is 11h. 10m.

The greatest rise of the tide is 6 ft. 4 in. and the least 1 ft. 1½ in.

---

4. Tidal Observations by Sir Edward Belcher.

The pack-ice to the northward of Jones' Sound in the neighbourhood of North Cornwall was acted on by a strong tide, and Sherard Osborn accounts for the disturbed state of the floe-ice as early as the 18th of May by supposing that a strong tidal wave flows in an east and west direction.

Again, on the north coast of Melville Island, on the east side of Hecla and Griper Bay, Commander Richards found a piece of pine wood which had been drifted by the sea.

Vol. II. page 154. "The distinct increased dimensions of all "the floe-ice noticed to the westward, as well as northward, of "the Queen's Channel, is to my mind satisfactory proof that it "belongs to another sea, and has no connexion with Baffin's Bay, "Lancaster Sound, or the Wellington Channel; and although it "may have considerable *motion* in summer, yet I believe that in "the Victoria Archipelago, as in the case between Banks' Land "and Melville Island, it remains unbroken for years, even "ages."

In Vol. I. p. 116, in speaking of the region near the western part of Jones' Sound at Pell Point (lat. 77° 29′ N., long. 95° W., declination 141° 18′ W.), Admiral Belcher says: — "In this " region, where the tides or currents are scarcely obstructed by " islands, and run with some velocity, ripping up the floe like " paper, much open water must of necessity prevail, and possibly " still more so to the northward."

Again, (II. p. 132):—"North of our present position (Northum- " berland Sound) the flood tide sets in from the Polar Sea, and " brings its *warmer oceanic water;* southerly the flood has to " pass up Lancaster Sound, there to be deflected up this channel, " and makes high water somewhere between this and Beechey " Island."

Vol. I. p. 244. Strong breezes from the westward and from the north-west are found to have a surprising influence on the tides, and show that there must be vacant space somewhere; "nothing " but open water to the northward or westward can effect such " motions." It was found by Commander Richards that "the " main-tide channel at the north end of Wellington Channel, " between Pioneer Island and Village Point, was open for half " a mile; and here the gales ranged from north to north-north- " west."

Early in May a large sheet of water was found at Village Point about one mile in length, and extending nearly across the Strait.

Heights of Tides at various Points in Arctic Regions.

In Behring Straits, at Kotzebue Sound, the flood tide from the Pacific rose 2 feet, and at Point Barrow at flood tide the rise was 7 inches.

In the Prince of Wales Straits M'Clure found that the flood tide came from the south with only 3 feet rise and fall on spring tides.

At the Bay of Mercy, Banks Island, as at places generally in Barrow Strait, the flood came from the east with a rise of about 2 feet.

In Jones' Strait the flood tide also flows from the east.

The observations of Sir Edward Belcher at Northumberland Sound show either that the tides from Lancaster Sound and from Jones' Strait meet in Wellington Channel; or if the distance by Jones' Strait is too great to admit of this explanation, then, taking them in connexion with the observations of Dr. Bessels in Kennedy Channel, they seem to show that the tides from the open Atlantic Ocean to the north-east of Greenland, entering through Robeson Channel, and possibly through Lady Franklin Straits and other openings more to the north, meet the tides from Baffin's Bay about Cape Fraser and also in Wellington Channel.

Difficulties in Measuring the Height of Tides.

Notwithstanding the very great care taken by Admiral Belcher to register the tides as accurately as possible (as shown in Vol. I.

p. 141), after a manner similar to, but somewhat more elaborate than that since adopted by Captain Koldewey in East Greenland, and described, p. 664, he says (p. 163):—"Nov. 28. The "increased thickness of the ice, and consequent gradual rise of "the ship, prevents the tide-gauge from acting correctly, unless "indeed the entire frozen surface does not admit the due flow of "tide. The difficulty, and incessant labour also, of breaking "away the constantly-forming ice, is too much for the men; I "have therefore put it out of gear until spring."

Also, page 155:—"The gale of November 9th came on with "the thermometer at 20°; on the 10th the temperature ranged "from 0° to 9°, rising according to the strength of the wind. "The tide-gauge rose one foot above its scale, and I have reason "to believe that some movement of ice unperceived by us shook "the observatory. A heavy snow-bank formed on our port side; "water flowed above the ice."

The Amount of Rise of the Tide.—(II. p. 181.)

"The rise and fall of the tide is apparent, not only on the tide-pole, but also on the ice, proving that until the floe becomes entirely free from the shore it does not rise and fall to the *extent* to which the water indicates it *should*. Thus, in addition to the rise and fall, as exhibited by the true index (the tide gauge secured to the bottom), we notice a rise and fall between shelf cakes of ice deposited at high and low water, a distance of 18 inches, fully proving a resistance in rising due to floatation if free. This is specially evident at the in-shore cracks, where the communication is impeded at high water by thin sludgy ice and water."

On May 14, 1854, strong tidal action was observed, and experiments were made by inserting a tide-pole through the fire-hole. The depth was 21 feet 8 inches amidships, and the tide fell one foot within the hour.

(Vol. II. p. 203.)

June 25.—"We have noticed that the weather here is more "influenced at the actual moments of the moon's quartering than "at the spring tides, which is opposed to my experience in other "parts of the world. To-day, however, the moon changed at noon, "but the wind, which has prevailed strong, still continues in heavy "gusts; about 8 p.m. it abated, and at midnight ceased. But the "tide does not appear to coincide to-day with its natural move-"ments, not rising at noon by 6 feet to its natural height, at mid-"night it flowed 6 feet above."

Some interesting and important facts are recorded by Sir Edward Belcher (I. p. 146) on the production of raised terraces or beaches by the ice being driven up the shore by the tides, and on the formation of fissures in the ice along the line of coast.

TIDES of NORTHUMBERLAND SOUND and REFUGE COVE, discussed by PROFESSOR HAUGHTON. (Proceedings of Royal Society, vol. xxiii., p. 2.)

—	Northumberland Sound.	Refuge Cove.
Semi-diurnal tide.		
Mean luni-tidal interval for high water	+ 0h. 7·05m.	– 0h. 26·7m.
" " " for low water	6h. 35·35m.	6h. 1·1m.
Difference between true luni-tidal and soli-tidal intervals.	·38m.	—
Diurnal tide.		
True soli-tidal interval - -	7h. 49m.	—
True luni-tidal " - - -	—	20h. 48m.
True solar coefficient - -	4·7 inches.	—

---

5. The TIDES of PORT KENNEDY. (Proceedings of Royal Society, vol. xxiii., p. 299.)

The tidal observations made by Sir Leopold McClintock in July 1859 have been discussed by Professor Haughton, with the following results :—

The heights of the tides were observed every hour for 23 days. They are remarkable for—

(1.) The magnitude of the diurnal tide.
(2.) The solar diurnal tide is greater than the lunar diurnal tide.

Solar Diurnal Tide.

The true soli-tidal interval is 5h. 12m. 7½s.

The coefficient 23·4 inches.

The true luni-tidal interval of the lunar diurnal tide is 0h. 33m. 50s.

Coefficient - { 18·4 inches (in time).
23·37 " (in height.

The age - { 1d. 4h. 14½m. (time).
4d. 6h. 20½m. (height).

Lunar Semi-diurnal Tide.

True luni-tidal interval 23h. 18m. 1s.

Uncorrected ratio of solar and lunar coefficients - } $\frac{S''}{M''}$ = { 0·412 inches (height).
0·549 " (time).

---

6. Record of Dr. Kane's Tidal Observations, in Winter Quarters at Van Rensselaer Harbour. "Smithsonian Contributions," vol. xiii., p. 1, &c.

An attempt was first made to determine the height of the tides by means of a tide-staff, and afterwards by a pulley-gauge attached to the vessel and rising with it. The graduations were on the arc of the pulley.

The difficulties arising from the slow movement of the vessel, the softness of the bottom, the alteration in length of rope, and especially the slipping of the rope over the pulley, which often froze to the axis, render the series of tidal observations very defective, and entirely break their continuity.

The observations were continued, with several breaks, from October 1853 to October 1854, and part of these series, from their regularity, may probably be regarded as trustworthy. A comparison of the tides at Port Foulke and Van Rensselaer is given in connexion with the account of Dr. Hayes' Expedition.

P. 81. The observations show that when the sea is partially open the wind has an effect on the tides; thus:—

Aug. 17, 1853. With a heavy gale from the southward, "our "flood rose 2 feet above any previous register, overflowing the "ground ice, and our last ebb or outgoing tide was hardly per-"ceptible."

When the snow melts in spring, the ice cracks open and water rises through the ice, making large basins at every tide.

The tide at Van Rensslaer is derived from the Atlantic through Baffin's Bay, as shown by the following table:—

—	Lat.	Long.	High Water at F. and C.	Highest Spring Tide.
	° ′	° ′	h. m.	feet.
Julianshaab - -	60 35	46 5	5 6	7
Fredrickshaab - -	62 0	50 5	6 3	12½
Holsteinborg Harbour -	66 56	53 42	6 30	10
Whalefish Islands - -	68 59	53 13	8 15	7½
Godhavn (Disco) - -	69 12	53 28	9 0	7½
Upernavik - -	72 47	56 03	11 0	8
Wolstenholme Sound -	76 33	68 56	11 8	7 to 7½
Port Foulke -	78 18	73 0	11 24	9·9
Van Rensselaer Harbour -	78 37	70 53	11 52	10·8
Polaris Bay - -	81 38	61 44	12 13	8

The value 12h. 3m. is given by Dr. Bessels as the establishment of Polaris Bay. At other places in W. Greenland an addition of 10m. to the mean establishment gives the luni-tidal interval, or the time of high water at full and change. This addition gives 12h. 13m. for the luni-tidal interval.

It is estimated that the whole distance, 770 knots, from Holsteinborg to Van Rensselaer is travelled by the tide in 6½ hours, which would correspond to a mean depth of 220 fathoms.

Comparing Upernavik and Van Rensselaer, the velocity corresponds to a depth of 800 fathoms. These are probably extreme values.

---

7. Observations by Dr. Hayes at Port Foulke, in 1860–61. "Smithson. Contrib.," vol. xv. Consisting of two series.

In the first, for 17 days in November and December, half-hourly observations of height of tide were made, and in the second series, in June and July 1861, observations were made at intervals of 10 minutes about the time of high and low water only.

The readings are regular, and are not affected by agitation of the surface because of the surrounding ice. The graphical method of recording the tides was adopted, and was found to be the best. By means of it the actual times of high and low water can be more exactly determined.

The tide-gauge consisted of a tripod mounted over a hole in the ice, and supporting a pulley, over which passed a rope; the rope was fastened to a heavy weight at the bottom, and was kept stretched by a counterpoise.

The rope was graduated, and the exact readings of the lengths of the portions of it measured.

The following are some of the difficulties in the way of making accurate observations:—

(1.) The weight at the bottom may drag along the bottom from currents or ice-motion.

(2.) The rope may stretch or may contract from becoming soaked with water.

(3.) The ice-field may have a slow motion, and incline the rope.

From these causes the zero level may be lost.

Note.—With an index on the rope, and a scale fixed vertically on the tripod, any dragging or inclination of the rope may be detected.

There is great advantage in referring all tidal observations to half-tide level—

(1.) For testing for any change of zero.

(2.) For measuring any secular change of level, *i.e.*, relative level of sea and land.

(3.) For separating and measuring the changes due to the lunar and solar tides.

To construct a diagram showing the half-tide level, the mean of two successive high tides is placed opposite the reading of the intermediate low tide, and the mean of these gives one point showing the half-tide level. In the same way the mean of two successive low tides is compared with the intermediate high tide, and their mean gives another point showing the half-tide level. In this way the half-tide level curve is traced out.

Superposed on the semi-diurnal tides at Port Foulke is the diurnal tide. The diurnal inequality in height of high water amounts to about two feet, and in the November observations is very regular.

In discussing the observations in June 1861, the diurnal inequality in times of high water and of low water are somewhat irregular, but their periods agree with the computed inequality in time of low water.

In November and December 1860 the day high tides were higher than the night high tides, whereas in June 1861 the night high tides were higher than the day high tides. The range of the half-monthly inequality in time amounts to 1h. 26m. At Van Rensselaer it amounted to 1h. 52m., a very large value.

General Character of Port Foulke Tides.

The general character of the half-monthly and diurnal inequalities is very much the same as at Van Rensselaer Harbour; the establishment is half-an-hour less at Port Foulke. The average range of the tide is about the same at the two places, and the diurnal inequality in the height of high water is greater than in the height of low water.

NOTE.—Comparing the heights of the highest spring tides in Baffin's Bay, generally about $7\frac{1}{2}$ feet, with the height in Van Rensselaer Harbour, 11 feet, it would appear possible that the tide at Van Rensselaer may result from both the southern and northern tides. The fact that the duration of the fall of the tide is less at Van Rensselaer than at Port Foulke, would also seem to show that Van Rensselaer is nearer to the open ocean than Port Foulke. Also at Van Rensselaer Harbour the extreme fluctuation of low water is very much greater than at Port Foulke.

---

8. THEORY of TIDES. HALF-MONTHLY INEQUALITY. Phil. Trans., 1834 and 1836.

The Theory of Bernouilli.—The pole of the fluid spheroid follows the pole of the spheroid of equilibrium at a distance (at the hour angle λ), and the spheroid of equilibrium corresponds to the configuration of the sun and moon, not at the moment of the tide, but at a previous moment at which the right ascension of the moon was less by a quantity α.

$$\text{Thus } \tan 2(\theta_1-\lambda_1)=-\frac{\sin 2(\varphi-\alpha)}{\frac{h_1}{h}+\cos 2(\varphi-\alpha)}.$$

θ_1 is the hour angle of the place of high water from the moon's place; φ the hour angle of the moon from the sun; h, h_1, the heights of solar and lunar tides; λ_1 the hour angle by which the tide follows the pole of equilibrium; α the retardation or diff. of R.A. due to the age of the tide.

The value of $(\theta_1 - \lambda_1)$ is a maximum or minimum when

$$\cos 2(\varphi - \alpha) = -\frac{h}{h_1}$$

hence $\tan 2(\theta_1 - \lambda_1) = -\dfrac{\frac{h}{h_1}}{\sqrt{1-\left(\frac{h}{h_1}\right)^2}}$ or $\sin 2(\theta_1 - \lambda_1) = \dfrac{h}{h_1}$.

The height y of the tide above the mean surface is given by the equation $y = \sqrt{h^2 + h_1{}^2 + 2\,h\,h_1 \cos 2(\varphi - \alpha)}$

To determine the variations in the mean level accurately requires very great care, and the observations made in Dr. Kane's or Dr. Hayes' Expeditions do not at all determine them, as the zero was several times shifted during the series of observations.

The height of the tide is affected by the barometric pressure, and also by the wind.

The effect of pressure may be tested by grouping the mean levels for days below the average pressure, and those for days above the average pressure, in two separate columns; the differences from the average value are then taken, and the mean of the whole series gives the change of level due to a given change in the height of the barometer.

Similarly the effects due to opposite local winds may be determined by an arrangement of the observations according to the direction of the wind.

---

9. Observations on Tides and Currents in "Polaris Bay" by Dr. Bessels (Bullétin Soc. Géographique, March 1875).

As soon as the ice was strong enough an Observatory was erected on it for the observations on the sea and on ice.

Observations on the tides were generally made hourly and extended over seven lunations; during three or four weeks they were made every ten minutes, and the following results were obtained.

The Establishment of Polaris Bay is 12h. 3m.

The highest spring tide -	- 8·0 English feet.
The lowest spring tide -	- 2·5 ft.
Mean height of low tides	- 3·8 ft.
Mean height of high tides	- 5·47 ft.
Mean of neap tides -	- 1·83 ft.

Observations were also made on the specific gravity of the sea, the depth, and the temperature at different depths.

After entering Smith's Sound a current to the south was met with varying from 1·5 to 5 miles, and this current brought drift wood into Polaris and Newmann Bays. The wood was pine-wood, and the different layers showed that it had grown in northern latitudes.

The floe ice was always in motion in the direction of the currents and the wind. The ice in Robeson Strait was divided by lanes of open water, which were too broad to admit of travelling on sledges, but too narrow to admit of navigation.

---

10. Currents on the East Coast of Greenland (Koldewey).

The observations on currents show that between the latitudes of 70° N. and 75° N., there are two currents towards the south.

Outside the ice and in the drift ice quite up to the ice fields which form its inner barrier there is a current of 8 to 10 knots a day, which shifts eastwards or westwards according to the wind and the ice-drift. Close to the coast there is a current of about 4·6 knots in 24 hours, which almost ceases in summer when the southerly winds are stronger and the northerly winds weaken.

There is no true current attending the ebb and flow of the tide, although the height of the spring tide was about 5 feet: even in the neighbourhood of the Kaiser-Franz-Joseph Fiord there was no current.

---

11. Observations of Tides by Capt. Koldewey at Sabine Island, on the East Coast of Greenland, during the Winter of 1869–70. "Die zweite deutsche Nordpolarfahrt," vol. ii., part 4, 1874.

Apparatus.

At the end of one of the davits was fastened (in a vertical position) a wooden scale, divided into feet and tenths. Vertically below it, a stone sunk in the ground, and fastened by an iron bar, was attached to a light rope, which passed over a pulley on the davit, and had a counterpoise at the other end.

Attached to the rope was an index, which marked the position of some point of the scale.

As the ship and scale sank and rose with the ice during the ebb and flow of the tide, the position of the fixed index on the scale showed how much the scale had fallen or risen, and in this way the height and times of high and low water were recorded.

Capt. Koldewey's method of registering is better than that employed by Dr. Kane, viz., graduating the arc of the pulley, for he found that often the pulley was frozen on its axis, in which case the rope slid over it.

Dr. Kane tried to correct this defect by taking soundings, but these soundings are somewhat uncertain, because the vessel moves with the ice, and very numerous soundings at various times would be required.

The difficulties as to shifting of position are well shown by a note in Dr. Kane's log, of Feb. 3, 1854. He says:—

"The enormous elevation of the land ice by the tides has raised a barrier of broken tables 72 feet wide and 20 feet high between the brig and the islands. This action has caused a secession of the main floe; *our vessel has changed her position 20 feet within the last two spring tides*, and the hawser connected with Butler Island parted with the strain."

NOTE.—Under these circumstances it seems hardly necessary to make allowance for the rise of the ship in consequence of the consumption of food by the crew during the winter.

—	Port Foulke.	Van Rensselaer.	Sabine Island.
Average rise and fall of tide	7·7 ft.	7·9 ft.	3·11 ft.
Extreme fluctuation of high water.	7·3 "	8·4 "	—
Ditto, of low water - -	5·2 "	9·0 "	—
Ditto, in water level - -	13·8 "	16·6 "	—
Mean establishment of high water (λ_1).	11h. 13·8m.	11h. 43·3m.	10h. 53·7m.
Ditto, of low water - -	17h. 19·5m.	17h. 48·0m.	17h. 6·1m.
Retardation of R.A. (α) for high water.	9° 36′	5° 15′	16° 45′
Ditto, for low water - -	10° 30′	12° 30′	13° 0′
Value of $\frac{h}{h}$ for high water -	·3624	·367	·4305
Ditto, low water - -	·3665	·471	·4019
Range of half-monthly diurnal inequality in time.	1h. 26m.	1h. 52m.	1h. 42m.
Ditto, in height - -	2·4 ft.	3·0 ft.	1·38 ft.
Cotidal hour and minute -	15h. 43m.	16h. 04m.	11h. 47m.
High water (moon's tide) interval.	11h. 24m.	11h. 52m.	—
Duration of fall of tide (=difference of establishments).	6h. 05·7m.	6h. 04·7m.	—
Duration of rise of tide -	6h. 18·7m.	6h. 19·7m.	—

To obtain the cotidal time, the mean establishment must be added to the longitude west of Greenwich, and a correction of 1m. in half-an-hour made for the moon's motion.

In this way the cotidal time of places in W. Greenland are as follows:—

—	Mean Establishment.	Correction for Moon.	Long. W.	Cotidal Time.	Latitude.
	h. m.	m.	h. m.	h. m.	° ′
Julianshaab - -	4 56	− 9	3 04	7 51	60 35
Fredrikshaab - -	5 53	−12	3 20	9 1	62 0
Holsteinborg - -	6 20	−13	3 35	9 42	66 56
Whalefish Islands -	8 5	−16	3 33	11 22	68 59
Godhavn - -	8 50	−18	3 34	12 6	69 12
Upernavik - -	10 50	−22	3 44	14 12	72 47
Wolstenholme Sound -	10 58	−22	4 36	15 12	76 33
Port Foulke - -	11 14	−23	4 52	15 43	78 18
Van Rensselaer Harbour -	11 43	−23	4 44	16 4	78 37
Polaris Bay - -	12 3	−24	4 7	15 46	81 38

COTIDAL TIMES of Places in E. GREENLAND (Koldewey, p. 665).

—	Mean Establishment.	Correction for Moon.	Long. W.	Cotidal Time.	Latitude.
	h. m.	m.	h. m.	h. m.	° ′
Nubarbik - -	6 30	−13	2 48	9 5	63 24
Eleonoren Bay - -	10 45	−21	1 40	12 4	73 27
Cape Broer Ruys -	10 51	−22	1 20	11 49	73 28
Jackson Island - -	11 3	−22	1 20	12 1	73 54
Sabine Island - -	11 14	−23	1 15	12 6	74 32
Klein Pendulum - -	11 21	−23	1 14	12 12	74 37
Cape Philip Broke -	11 28	−23	1 11	12 16	74 56
Cape Bergen - -	12 7	−24	1 12	12 55	75 26

The difference in the establishments of high and low water is the duration of the fall of the tide.

In places receiving the direct ocean tide, the duration of rise is greater than the duration of the fall of the tide; but shallow water tends to diminish the duration of the rise of the tide, and so makes the rise shorter than the fall.

The law of the velocity of tides as depending on the depth may be well tested in Baffin's Bay by observations in different parts from Cape Farewell northwards, and for this purpose soundings should be taken at all points at which the time of high or low water is observed.

It is important to have observations on both shores of the bay during two lunations.

The heights of high water are different in two successive tides, because of the action of the diurnal and semi-diurnal inequalities combined, and the difference depends on the declination of the moon.

Near full moon the night tides are higher than the day tides in the winter, but the night tides may be less than the day tides for other positions of the moon.

---

12. AVERAGE DEPTH of DAVIS STRAIT, BAFFIN'S BAY, and SMITH SOUND.

—	Distance.	Diff. in Cotidal hours.	Velocity.	Depth.
	miles.	h.		ft.
Julianshaab to Whalefish Island.	680	3·5	194m. per h.	418
Whalefish Island to Port Foulke.	770	4·35	177 „	349
Port Foulke to Van Rensselaer Harbour.	55	·35	157 „	277

Hence the average depth of Davis Strait and Baffin's Bay (according to theory) is about 383 fathoms, the length of the free tide wave nearly 2,300 miles, with a height between trough and crest of about 7½ feet.

[By comparison of these results with other similar results obtained by the Expedition and with soundings, some important conclusions may be arrived at as to the theory of the tides.

A table is given comparing the different tidal observations at points on the East Coast of Greenland by M. Koldewey, with those on the West Coast by Drs. Kane and Hayes, showing the progress of the tides from the south along those two coasts. The average height of high water at spring tides on the east side is only about 4 feet, and at neap tides is about 2 feet. It will be seen that the tides in "Polaris" Bay, which are 5 feet at spring tides and 2 feet at neap tides, are nearly of the same amount as those at Sabine Island. At some places M. Koldewey had great difficulty in determining within half-an-hour the time of high water, even by his apparatus, which was more perfect than that used by Dr. Hayes.

From the account of the observations of tides made by Dr. Kane and Dr. Hayes, it is seen that there are great defects in the apparatus, and even with the improvement adopted by Capt. Koldewey, viz., with a straight scale fixed to the vessel, and rising and falling with it, and an index fixed to the rope, the observations were liable to serious and unknown errors, either (1), from the dragging of the weight; (2), from the motion of the ice-field; (3), and especially from the freezing up of the hole through which the rope passes. This is a serious objection to observations made by cutting holes in the ice, for the rope will be frozen to the ice and be lifted or dragged with it, and so the zero of the scale may be altered either by the motion of the ice or even in breaking the ice before making the next observations.

From Admiral Belcher's observations it is clear that the objections to a tidal pole are even greater than the objections to the above method.

There is one method by which these difficulties may be diminished or entirely got rid of, but the ships will require to be specially adapted for the purpose.

An opening may be constructed in the bottom of the ships, through which direct communication may be had with the water below the ice when the ships are frozen up in winter quarters. Through these openings not only the observations on tides could be made, but also on the temperature and on the specific gravity of sea water at various depths, as well as observations on currents.

Some of the evident advantages of such a plan are,—

(1.) Observations made within the ship will be far more accurate, because the observer is not exposed to cold and the weather.

(2.) The rope does not get frozen to the ice, and therefore will not be subject to a sudden pull, as it would in breaking the ice round it.

(3.) Valuable observations could be made on currents, and proper apparatus could be fitted in the ships for the purpose.

Usually no observations have been made on the currents in the Arctic regions, except where the sea is clear of ice, because of the want of apparatus.

(4.) Much more perfect apparatus for registering and even recording the tides may be fitted within the ships; there seems to be no reason why the height of the tides should not be continually registered on a revolving cylinder as in various ports, according to the plan adopted by Sir Wm. Thomson. Fixing the tidal apparatus within the ship, the rope, which is usually attached to a float, must be attached to a stone at the bottom, and must have a pencil or index attached to it, as in M. Koldewey's apparatus. Thus the pencil will remain fixed, and the cylinder on which the tidal curves are traced will rise and fall with the tide. Also the currents in the sea might be registered and recorded by means of an apparatus similar in construction to the apparatus devised by Dr. Robinson for registering the rate of the wind, but with the cups at the bottom instead of at the top of the shaft. The apparatus consists of four hemispherical cups placed symmetrically in a horizontal plane, and attached to a vertical revolving shaft. By means of proper gearing, the rate of revolution of the shaft is recorded by a pencil on a revolving cylinder. There seems to be no reason why the rate of currents of water should not be registered by this apparatus quite as accurately as the rate of the wind by means of Dr. Robinson's apparatus.

Even if the delicate apparatus for recording the currents continously cannot be conveniently applied on board the ships, it will still be of very great importance to employ some such apparatus as the revolving cups above described, and to apply to them an index which shall enable the observer to record the rate of revolution, and so to determine the rate of the current.]

---

13. VOYAGE of the "POLARIS" under Captain HALL.

A comparison of the facts relating to this voyage are gathered from Captain A. H. Markham's "Whaling Cruise to Baffin's Bay," and Clement Markham's "Threshold of the Unknown Regions," with the records of the discoveries of Dr. Kane and Dr. Hayes in Kennedy Channel.

p. 189.—In August 1871 the "Polaris" went from Cape Shackleton (in latitude 74°, near Upernavik) to her extreme northern point in 82° 16′ N. (a distance of about 600 miles) in five days (part of the way against a current of 1 knot an hour), and was stopped by an insignificant stream of ice which might have been easily passed through, while there was a magnificent water-sky to the northward. The floes in Smith Sound were small, and no icebergs of any size were seen north of 80° N., *i.e.*, north of Cape Fraser or the Humboldt Glacier. Like Dr. Hayes, Captain Hall went northwards on the west side of Kane Basin and Kennedy Channel, whereas Dr. Kane was proceeding on the eastern side when he was caught in the ice in Van Rensselaer Harbour, where he experienced very severe cold.

This suggests that probably the extreme cold of Van Rensselaer, as experienced by Dr. Kane and also by Dr. Hayes in 1861, may be due to the extensive glaciers on the high lands to the east as well as to the exposure to the N.W.

The Humboldt Glacier, in latitude 79° 12′ N., with a sea face 45 miles long and from 330 to 500 feet high, bounded by precipices of Old Red Sandstone and Silurian limestone from 100 to 1,400 feet high, sends off icebergs in lines which fill up the eastern side of Kane Basin, and are drifted by the prevailing winds (and the current, if there be any, in Kennedy Channel) into Smith's Sound and Baffin's Bay.

A comparison of the six maps of the Smith's Sound route in "The Threshold of the Unknown Regions" scarcely bear out Mr. Clement Markham's statements with regard to the accuracy of Morton's statements on page 165, or the statements of Captain Markham in his "Whaling Cruise," page 192, as to the coast line laid down by Hayes in 1853–54 and again in 1860–61. These remarks are of course founded on conversations with Dr. Bessels on the voyage.

Morton stated that he went 76 miles farther (than Dr. Kane ?). Dr. Kane did not cross the front of the glacier; thus, according to Morton's statement, he would only have gone about 30 miles, or half a degree of latitude, beyond Cape Jackson, which is rightly placed by Morton in Kane's map in latitude 80° N. and longitude nearly 66° W.

From his extreme point about a distance of 90 miles, or 1° 30′ of latitude, directly north, is marked in Dr. Kane's map as open sea up to Mount Parry. This entirely agrees with the discoveries of the "Polaris." Throughout the whole of this region, according to Dr. Bessels, there is an open sea or basin extending over a space of 3° of longitude by 1° of latitude, with four broad channels leading out in four directions, where the ice was insignificant, and where there was a magnificent water-sky to the northward; where also the ice floes seldom exceeded 5 feet in thickness, and were all of one year's growth. He also states that the heavier ice probably drifts up Lady Franklin Strait.

This seems to agree exceedingly well with Morton's statement that he found open water extending in an iceless channel to the western shores, and the fact that the northern tide reaches southwards to Cape Fraser favours Morton's statement that the heavy surf was beating against the rocks. Morton was very near to, and rather to the north of, the point where the northern and southern flood tides meet, and in an open sea there would be that disturbance of the water which was noticed by Morton, but which was not noticed at Cape Fraser by the "Polaris" or by Dr. Hayes along the western shore, because of the ice.

There is one point in which Morton may have been wrong, viz., he may have overestimated the distance he had travelled in a direct line northwards; but with this exception it seems that Morton's statements are fairly trustworthy. According to Dr. Bessels' observations, Morton must have gone about 76 miles beyond Dr. Kane to reach Cape Constitution.

As the prevailing winds in Kennedy Channel are from the north-east, and this would cause the drift of the ice, it does not seem certain, as Dr. Bessels' results have not yet been published, that there is any true current *anywhere to the south of Lady Franklin Bay*, and in that case a channel can hardly have been cut in the ice by the strong current during the days of midsummer. The meeting of the tides and consequent disturbance would tend to keep the sea clear of ice provided there were no heavy ice. The current through Robeson Channel, in which heavier ice was seen drifting southwards, may proceed up Lady Franklin Strait, carrying with it the heavy ice, which according to Dr. Bessels did not come farther to the south.

De Haven, McClintock, and other observers all attribute the drift down Baffin's Bay entirely to the winds.

Except perhaps in Barrow Strait, there seems to be no evidence from De Haven's Drift, in Wellington Channel, Lancaster Sound, and Baffin's Bay, that any true current exists. At first, in Wellington Channel, the drift was in opposite directions, according as the wind changed, while in Barrow Strait, at times with a north wind, the drift was westward, the ice being driven by the tide from the eastward through Lancaster Sound and Baffin's Bay; the amount and direction of the drift seemed to depend almost entirely on the direction of the wind.

On the N.E. of Spitzbergen it has been found by Lieut. Weyprecht that the prevailing winds are from E. to W., causing the great ice drift which drifted Parry to the S.W., and which brings down the ice towards East Greenland. The rate of this drift varies from 8 to 13 miles a day. These winds, which may partly be caused by the warm Gulf Stream, towards which they blow, to the N. and W. of Spitzbergen, will be warmed as they pass over that region, since the temperature of the sea water in the whole of that region is found to be high, and flowing on still westward will pass over the sea to the west, and over the north of Greenland, and so to the shores of Kennedy Channel and Smith Sound.

This would also seem to be borne out by the fact reported from the observations of Captain Hall in the "Polaris," that the tide round the north of Greenland met the tide up Smith Sound at Cape Fraser, on the west side of Kennedy Channel, in latitude 80° N.

With regard to the existence of an open sea north of Smith Sound, Kane says:—"Whether it exists simply as a feature of "the immediate region, or as a part of a great and unexplored "area communicating with a polar basin, may be questions for "men skilled in scientific deductions; whether it does or does "not communicate with a polar basin we are without facts to "determine." He adds:—"The influence of rapid tides and "currents in destroying ice by abrasion can hardly be realised by "those who have not witnessed their action. It is not uncom- "mon to see such tidal sluices remain open in the midst of "winter."

The observations of Dr. Bessels with regard to the climate of the winter quarters of the "Polaris" in 81° 38′ N., on the east side of Hall Basin or Kennedy Channel, that *it is much milder than it is several degrees to the south*, and that *a current of a knot an hour flows down Robeson Channel from the north, and carries the ice out into Baffin's Bay*, also agree with the view that the ice to the north of Spitzbergen, Nova Zembla, and Siberia is drifted to the west and north-west by the winds, and divides to the north of Greenland into two streams, one current flowing down along East Greenland, and the other flowing through Robeson Channel to the south-west.

These observations must also be taken in connexion with Admiral Belcher's observations at the north end of Wellington Channel, that the tides from the north meet the tides from the south in that channel.

Sir Edward Belcher says, in "The Last of the Arctic Voyages," page 219, vol. ii. :—"I had well proved, by experiments conducted " at Beechey Island, Cape Bowden, and the late winter quarters " (at the north end of the Wellington Channel), that the flood- " course did not run from Lancaster Sound *through* Wellington " and the Queen's Channel; but that the northern flood from the " Arctic Ocean *met* that from Lancaster Sound, as nearly as I " could determine, at Cape Bowden, and much in the same manner " as the Channel and North Sea tides meet about Dover."

"Hence it was clear to my mind that without a great effort of nature to clear away the ice *northerly* as well as *southerly* of that *parallel*, antagonistic forces must continue to compress any loose floes together, and perfect a solid barrier in that the narrowest portion of the Wellington Channel."

He also says, on p. 222 :—"We know of no southern drift in " Wellington Channel; we have never experienced northern " gales ; we do possess facts to prove the reverse; portions of " the 'Breadalbane' reached our winter quarters, and that *with- " out a gale*;" on the 4th of September 1853. "The 'Breadal- " bane' was wrecked 55 miles to the south on the 21st of " August."

That there is a "permanent easterly current on the *south* of " Melville Island, and in Barrow Strait," towards Lancaster Sound, is clear from the evidence of Captain Kellett from observations between Sept. 10 and Nov. 12, 1873.

---

14. Remarks on Tides and Currents.

If, as seems not unlikely, the tide from the north of Greenland passing through Lady Franklin Channel to the south-west meets the Baffin's Bay tide near the north end of Wellington Channel, then there may be no one point where the three tides actually meet, but Melville Island will probably be washed by all three tides, and it seems not improbable to expect that the pole of greatest cold will be not far from Melville Island.

But it seems more probable, and this should be made clear from the character of the tides near Wellington Channel, that the tides meeting in that channel are the two branches of the Baffin's Bay tide flowing through Lancaster Sound and through Jones' Sound. The shortest way to the north-west end of Wellington Channel being probably through Jones' Sound.

The flowing of the three tides into the region between Melville Island, Grinnell Land, and Kellett Land, the Behring Strait tide from the S.W., the Baffin's Bay tide from the S.E. through Jones' Sound, and the North Greenland tide through Lady Franklin Strait, and other straits to the north of it, will fully account for the heavy nature of the ice in that region as far as it has been observed, and for the fact that it is never carried out of this Arctic sea through the somewhat narrow openings into it. In this case there will be some point in the open ocean to the north of Melville Island where these three tides meet, forming, as it were, a pole of tides. McClure was prevented from passing from Prince of Wales Strait to Barrow Sound by the ice collected where two opposing tides meet, and was afterwards obliged to abandon his ship in Mercy Bay probably for the same reason. Professor Haughton has shown that where heavy ice exists, as, for instance, on the west of Boothia Felix, the meeting of the tides tends to consolidate the ice and to prevent its drifting, and that it was owing to this meeting of the tides that Sir John Franklin's ships could not escape.

In the neighbourhood of this pole of tides we should look for the pole of greatest cold, since we may expect that the tides and currents in the air and its temperature will be regulated to some extent by the distribution of sea and land, so that where we get the pole of ocean tides there we may also expect the pole of aerial tides.

The correspondence between the temperature of the air and the surface temperature of the sea is also in favour of this hypothesis.

Such a position may be expected to be one of comparative rest, so that the colder air above will more readily descend to the surface of the earth and produce the cold winds.

Judging from the direction of the cold winds as observed, the pole of cold may be expected to be somewhere to the N.W. of Wellington Channel.

The fact that "the oscillations of the barometer are much "greater in the neighbourhood of water," although the surface temperature over water is more nearly constant, shows that the distribution of land and water must have an important effect on, and to some extent regulate, the aerial tides.

May not the flow of the tide through Lady Franklin Bay to the westward, nearly in the direction of the prevailing N.E. winds, be connected with Admiral Belcher's observations in Wellington Channel, which would be about 400 miles to the south-west? Dr. Bessels saw large ice-fields in latitude 82° 16′ which were drifting south, but which were never seen to the *south* of Lady Franklin Bay, and it is his impression that they go up that sound or strait. Does not this also seem to show that the current of one knot an hour in Robeson Channel flows through Lady Franklin Strait rather than down Kennedy Channel?

It is also stated by Sir Edward Belcher that the east winds are the warm winds which are laden with snow.

Dr. Bessels states that the coast of Greenland extends only as far as 82° 30′ N., and then trends away to the eastward.

Pieces of pine-wood, up to 4 feet in length, were picked up in 82° N., drifting with the stream from the north, and the Esquimaux, Jem, told them that plenty of wood came from the north, and was washed up along the shore of Grinnell Land, on the west side of Kennedy Channel.

" To the south of Cape Fraser the tide rose to a greater height during the night, as is the case along the coast of Greenland; whereas to the north of Cape Fraser there was no perceptible difference between the day and night tides." This will depend on the age of the moon.

Captain Markham, p. 191.—At times of full moon, *i.e.*, at the highest spring tides, which must take place at Cape Fraser about 12 o'clock, this will be true ; but at other times, when the tides are small, it will probably not be found to be true even south of Cape Fraser. Where the whole rise of the neap tide is only 2 feet, there may be some difficulty in finding the difference in the height of two tides.

On referring to the map (p. 159) showing Kane's and Hayes' discoveries, it will be seen that the latitudes of all places observed by Hayes up to his most northern point on the shore of Lady Franklin Strait, in latitude 81° 35′ N., agrees with the latitudes as determined by Dr. Bessels; also that he agrees almost entirely with Dr. Bessels in the longitude of Cape Fraser and Cape Barrow (lat. 80° N., long. 70° W.), and in the position and dimensions of Hayes Sound.

In going north from Cape Barrow, according to Dr. Bessels, the coast line bears rather more to the east than according to Hayes; so that while Hayes places his extreme northern point in the same longitude as Cape Barrow, Dr. Bessels places it rather more than 2°, *i.e., about* 20 *miles*, to the east of Cape Barrow.

[From the discussion of Dr. Hayes' Observations, I cannot find that any longitudes were determined to the north of Cape Barrow.]

Hayes found the channel frozen over in May 1861, but the ice much decayed. The coast on the west side was lined with a heavy ridge of pressed-up ice, in some cases 60 feet high, and far up on the beach.

Captain Markham's " Whaling Cruise," p. 190.—The north coast of Greenland is steep and precipitous, and is free from land ice, whilst the shores of Grinnell Land are low and shelving, and have fast ice attached to them. This agrees with the observations of Dr. Hayes, p. 194. The lowest temperature registered was —48° F., with very little wind blowing at the time. The prevailing winds were from the N.E. (It would be interesting to know whether the highest temperatures were registered during these winds. The season seems to have been a mild one.) " The fall of snow " was remarkably small."

p. 196.—In latitude 81° 36′ N., "on the 21st of December there "were six hours of twilight, which rose at noon to an altitude "of 10°."

Dr. Bessels visited Port Foulke, and ascended to a height of 4,181 feet on the glacier visited by Hayes. His observations here are not given.

p. 190.—The north coast of Greenland is in latitude 82° 30′ N. The land on the west side of Robeson Channel is a little farther to the north, and is bounded by a channel trending towards the west, while to the north, at a distance of 60 miles (estimated), is seen another northern land. This would give a latitude of 83° 30′ for this northern land.

While Kane was in Smith Sound his Danish interpreter, Petersen, conversed with the Esquimaux who had been to a large island called Umingmuk (Musk-ox) Isle, where there was open water, with Walrus there, and some of their people formerly lived on the island.

"Whaling Cruise," p. 194.—In Hall Land, the winter quarters of the 'Polaris,' Musk-oxen were met with, and Foxes and Lemmings; also three kinds of Seals, but no Walruses.

p. 192.—Traces of Esquimaux were found in latitude 82° N., a couple of ribs of Walrus (used as sledge-runners), and a small piece of wood (the back of a sledge); also remains of a summer encampment. This would seem to show that Umingmuk Isle and Hall Land are identical, the Petermann Fjord dividing it off from the mainland at least for a considerable distance from the mouth of the fjord, as shown by Dr. Bessels.

Captain Sabine met with Esquimaux in East Greenland in 1823, which were believed to have come from the north; and M. Koldewey met with Musk-oxen, which probably came from the north, since none are found to the south of Wolstenholme Sound.

p. 193.—"The 'Polaris' drifted to the southward from her most "northern point, in consequence of the prevailing north-east "winds." There was a current of 1 knot an hour down the channel.

Can this current down Robeson Channel be the effect of the tides coming from a larger area of water beyond into the smaller channel, or do the winds cause the ice to move in that direction and so cause a surface current in the channel, or is there a true current down this channel from the N.E. which flows into Lady Franklin Strait? These are interesting questions to determine.

15.—The Austro-Hungarian North-Polar Expedition of 1872–74. (A Lecture by Lieutenant Weyprecht.)

Meteorology and Hydrography.

(Petermann's "Geograph. Mittheilungen," 1875; and "Nature," March 11th and 18th, 1875.)

"In the preliminary expedition to the east of Spitzbergen in 1871, the observers had found the sea completely free from ice as far as 78° N. lat., north of Nowaja Semlja, and their intention at the second expedition was to investigate this sea in an easterly direction, taking the Siberian coast as basis, and depending on the influence of the great Siberian rivers, whose great quantities of comparatively warm water probably free the coast from ice almost every summer.

"Unfortunately the year 1872 was one of the most unfavourable ever seen. Already in 74·5° N. lat. the explorers found ice; they could only reach Cape Nassau with great difficulty, and were finally blocked up by packed ice in a locality where, in the previous and following years, there was no ice for one hundred German miles round. They never got within the reach of the Siberian rivers, and the uncertainty with regard to their influence upon the ice along the Siberian coast is still the same as ever. But one point is clearly proved, namely, that the conditions of ice between Spitzbergen and Nowaja Semlja are highly variable from year to year; this circumstance, more than any other, speaks against the advisability of future expeditions to be made on the basis of Franz-Joseph's Land. In 1874 the explorers found the ice again in the same position as in 1871; there is perhaps a certain periodicity in this.

"Lieut. Weyprecht formerly thought that marine currents were the principal cause of the general motion of the ice in Arctic regions; he is now of a different opinion, as he maintains that during the drift of their vessel, which was frozen in, in packed ice, and drifted in this state for over fourteen months, the influence of currents was imperceptible compared to that of winds upon the drifting ice. The existence of Gulf-stream water in the great area between Norway, Spitzbergen, and Nowaja Semlja is undeniable; the current cannot, however, be traced directly by its course, but rather by the unproportionally high sea-temperatures in those high latitudes. As a natural consequence of this, the Gulf stream does not regulate the limits of ice, but the ice, set in motion by winds, regulates the limits of the warmer Gulf-stream water, depriving the same of the last degrees of heat which it contains. A comparison of the *Hansa* drift with the winds would show whether on the east coast of Greenland the drift of ice is only produced by the latter; Sir L. McClintock proves with figures that this decidedly is the case in Baffin's Bay. The speed of the drift of course depends upon the force of the winds, local conditions, vicinity of coasts, and the more or less open water. The great influence of the wind upon the ice-fields is explained by their ruggedness; each projecting block represents a sail.

"In the vicinity of coasts it is somewhat different; immense currents are often perceived there, originating through the tides, or perhaps through the motion of the ice itself and the winds.

"There is a decided general tendency in the ice to move southward during the summer; the reason of this may be the flowing off of melted water in all directions, which causes a breaking-up of the whole Arctic mass of ice. But all other influences upon the motion of the ice are nearly imperceptible when compared to that of winds, and can only be traced in their most general effects.

It is quite certain, however, that in the south of Franz-Joseph's Land there is a constant flow of ice from east to west, *i.e.* from the Siberian sea. If the field of ice which held Lieutenant Weyprecht's ship a prisoner had not attached itself to Wilczek Island, it would have drifted toward the northern end of Spitzbergen; he arrives at this conclusion from observing the winds of last winter.

"To the influence of winds Lieutenent Weyprecht also ascribes the existence of open water near all west coasts in those regions: he found the main direction of winter-storms in Franz-Joseph's Land to be E.N.E.; the ice under west coasts is therefore constantly broken up. Lieutenant Payer, on the northernmost point he reached, was stopped from extending his sledge journeys further by open water near a west coast, upon which he was travelling.

"Also, with regard to quality, the ice in those seas is very variable. While in the summer of 1873 the explorers could not see the end of the field in which their ship was frozen in, they never met fields of such an extent during their retreat; also, with regard to thickness, there was great variety. In 1873 their field formed an irregular frozen mass, with high ice walls in all directions and immense protuberances; in 1874 they found much greater evenness, and although thawing had begun so late that they almost perished with thirst during a month and a half, the ice was so thin in some places at the end of July that they often broke through while drawing their sledges. During the drift the whole mass was doubtless packed very closely; the field, in spite of the constant drifting motion, did not turn round, the bow of the ship pointing always in the same direction; only in September, when the field was greatly reduced, it began to turn; in October and November large holes were seen in it in the vicinity of the coast towards the south.

"Whether Franz-Joseph's Land can again be reached by ship, Lieutenant Weyprecht thinks mainly dependent on favourable conditions of weather and ice; in any case a very warm summer will be necessary, and then it could be done only late in the year. As to the most favourable point to start from in such an expedition, he indicates 45° East long., as here he found the barrier of ice in 1871 to be 50 German miles more to the north than in 60° E. long.

"In the preliminary expedition of 1871, Lieut. Weyprecht found sure signs of the vicinity of land in 43° E. long. and 78° 75′ N. lat., and accordingly he proposed this unknown land as the basis for future expeditions sent to reach the pole. The mysterious Gillis-Land is situated upon 30° E. long. The south coast of Franz-Joseph's Land was seen by Payer at least as far as 50° E. long. Lieut. Weyprecht now thinks he may be permitted to conclude that these three points are connected. Thus Franz-Joseph's Land would become greatly extended in a western direction. Numerous icebergs floating along the coast seem to confirm this idea, and it is hardly necessary to point out how much the interest in Arctic investigation would be increased by this idea proving a correct one.

"During a year and a half the explorers had constant opportunities closely to observe the behaviour and formation of pack-ice. The phenomenon is instructive, as it is the same in the whole of the Arctic regions. With the exception of land-ice, which clings to the coasts and never reaches far out into the sea, all ice—icebergs as well as fields—is in constant motion, winter and summer; and this, as has been shown, is through the influence of winds. The motion, however, is a different one almost with every field, and thus a certain pressure results wherever two fields touch; this naturally leads to the breaking up of the fields, and the contraction of the ice during sudden low temperatures plays its part in a similar way. If one considers the great extent of the fields, sometimes of many miles, and their enormous masses, one can easily imagine the colossal forces which are active in these phenomena, and the greatness of their effects. When two fields meet, a combat body to body ensues, often lasting only a few minutes, but sometimes even for days and weeks. The edges are then turned up on both sides, upwards and downwards, an irregular wall of ice consisting of wildly-mixed blocks begins to build itself, the pressure increases more and more, masses of ice eight feet long and broad are lifted 30 to 40 feet high, and then fall to make room for others. At last one of the fields begins to shift itself for some distance underneath the other one; often they separate for a while only to renew the struggle. But the end of it always is that the intense cold unites all into one solid mass; a single field results from the two, and the next storm or quick change of temperature cracks the new fields in some other direction, the pieces renewing the old struggle. This is the origin of the ice-fields, which are quite irregular above and below, sometimes only consisting of blocks that have frozen together, and filling up the whole Arctic region as so-called pack-ice.

"During winter, snow-storms fill up all smaller irregularities completely. As soon as the sun begins its action, the crushing of the ice decreases, the winterly ice walls diminish considerably, immense masses of ice and snow are melted, and the resulting sweet water forms large lakes on all the lower even parts of the field. During the summer, about four feet of ice are thus melted down from above; of course the whole field and everything upon it—the explorer's ship, for instance—is raised so much higher. In the following winter it grows below in the same ratio, and thus the whole of the ice is in an uninterrupted process of renovation, from below upwards; we may conclude that all the old pack-ice is replaced by new in the course of two years.

"The spaces of open water which naturally occur during the great crushes are soon again covered by fresh ice in winter; the intense cold keeps repairing the broken field of ice. Lieut. Weyprecht observed that within 24 hours, and with a temperature of −30° to 40° R. (37·5–50° C.,) the new crust becomes about a foot thick. The salt of the sea-water has not time to be displaced entirely, the formation of ice going on too quickly, and a considerable quantity freezes into the upper strata of the ice; this quantity decreases downwards as the ice takes more time to form. Beginning at a certain thickness, the ice is almost free from salt. The upper strata, however, on account of the salt they contain, attract moisture in a great degree, and form a tough, leathery mass which bends under foot without breaking. This, however, is only the case with new ice, as after a short time the salt crystallizes out of the ice, and the surface covers itself with a snowy layer of salt, sometimes reaching two inches of thickness. Even in the most intense cold this layer retains so much moisture that it makes the impression of a thaw; only little by little, evaporation and drizzling snow do their work, and the ice itself becomes brittle.

"In this way almost all the salt, which was frozen in, crystallizes out, and is washed off and back into the sea by the melted water in the next summer. The melted water at the end of the summer is, therefore, almost free from salt, and has a specific gravity of 1·005. It is evident that a smooth plane of ice, as is found on sweet water, is a very rare occurrence in Arctic regions.

"The astronomical observations while the ship was still drifting were confined to determinations of latitude and longitude, the latter by chronometers and correction of clocks, by lunar distances, as often as opportunity served. In this only a sextant and a prism circle with artificial horizon were used. When the ship was lying still, a little 'universal' instrument was erected, and the determinations of time, latitude, and azimuth were made with this. The longitude was calculated from the mean of as many lunar distances as could be observed during the winter; they were 210 in number. The azimuth of a base line 2,171 metres long, measured by Lieut. Weyprecht with a Stampfer levelling instrument, was determined with the universal instrument of the magnetic theodolite. All this work was done by Ensign Orel, Lieut. Weyprecht only taking a share in measuring lunar distances. The determinations of locality were made without regard to temperatures; if the mercury of the artificial horizon was frozen, blackened oil of turpentine was used instead. The winter quarters were in latitude 79° 51′, and longitude 58° 56′ E.

"Of the results of the meteorological observations, only some general ideas can be given, as here figures alone decide. They were begun on the day the explorers left Tromsö, and were only discontinued when they left the ship; thus they were made during twenty-two months. Readings were taken every two hours, and also at 9 a.m. and 3 p.m., therefore fourteen times daily. The observers were Lieut. Brosch, Ensign Orel, Capt. Lusina, Capt. Carlsen, Engineer Krisch (from autumn 1872 till spring 1873), and Dr. Kepes (during the last two months only). The direction as well as force of winds were noted down without instruments. Lieut. Weyprecht thinks this method by far the best in Arctic regions, as errors are more or less eliminated, while when using instruments the constant freezing, drifting snow, &c., produce errors that cannot be determined nor controlled; besides, any one who has been to sea for a short time will soon acquire sufficient exactness in these observations.

"Until the autumn of 1873 winds were highly variable. In the vicinity of Nowaja Semlja many S.E. and S.W. winds were observed; in the spring these veered more to N.E. A prevalent direction of winds was only recognized when in the second winter the Expedition was near Franz-Joseph's Land. There all snowstorms came from E.N.E., and constituted more than 50 per cent. of all winds. They generally produced clouded skies, and the clouds only dispersed when the wind turned to the north. The explorers never met with those violent storms from the north, from which the *Germania* party had so much to suffer on the east coast of Greenland, and which seem to be the prevalent winds in the Arctic zone. Altogether, they never observed those extreme forces of wind which occur regularly in our seas several times in every winter (for instance, the "Bora" in the Adriatic). Every Arctic seaman knows that the ice itself has a calming effect upon the winds; very often white clouds are seen passing with great rapidity, not particularly high overhead, while there is an almost perfect calm below.

"One peculiarity must here be mentioned. Lieut. Weyprecht made the remarkable discovery that the ice never drifted straight in the direction of the wind, but that it always deviated *to the right*, when looking from the centre of the compass; with N.E. wind it drifts due W. instead of S.W.; with S.W. wind it drifts due E. instead of N.E.; in the same manner it drifts to the north with S.E. wind, and to the south with N.W. wind. There was no exception to this rule, which cannot be explained by currents nor by the influence of the coasts, as with these causes there would be opposite results with opposite winds. Another interesting phenomenon in both years was the struggle between the cold northern winds and the warmer southern ones in January, just before the beginning of the lasting and severe cold; the warm S. and S.W. winds always brought great masses of snow and produced a rise in the temperature amounting to 30–35° R. within a few hours.

"Little can at present be said on the result of the barometer readings, without a minute comparison of the long tables of figures, although very extreme readings occurred at times. The explorers had three mercury and four aneroid barometers; by way of control, Ensign Orel took the readings from five of these instruments every day at noon, while the intermediate observations were made with an aneroid.

"The thermometers were suspended about four feet from the surface of the snow, in the open air, and perfectly free on all sides, about twenty-five yards from the vessel. Excepting the maximum thermometers, they were all spirit thermometers, made by Cappeller of Vienna. They were often compared with a very exact normal thermometer of the same make. Readings from a minimum thermometer were noted daily at noon; during the summer a black bulb thermometer was exposed to the rays of the sun; during the winter frequent observations were made with exposed and covered minimum thermometers to ascertain the nightly radiation at low temperatures. In both winters February was the coldest month, while January both times showed a rise in the temperature when compared either with December or February. In winter the temperature was highly variable, and sudden rises or falls were frequent; in the three summer months, however, the temperature was very constant, and changes very rare. July was the warmest month. The lowest reading was $-37\frac{1}{2}°$ R. (nearly $-47°$ C.) The influence of extremely low temperatures upon the human body has often been exaggerated; there are tales of difficulty in breathing, pains in the breast, &c., that are caused by them. Lieut. Weyprecht and his party did not notice anything of the kind; and although many of them had been born in southern climes, they all bore the cold very easily indeed; there were sailors amongst them who never had fur coats on their bodies. Even in the greatest cold they all smoked their cigars in the open air. The cold only gets unbearable when there is wind with it, and this always raises the temperature. Altogether, the impression cold makes upon the body differs widely according to personal disposition and the quantity of moisture contained in the air; the same degree of frost produces a very uncomfortable effect at one time, while at another one does not feel it.

"To determine the quantity of moisture in the atmosphere, an ordinary psychrometer, a dry and a wet thermometer, were used. But the observations with these instruments are not reliable at low temperatures, and had to be given up altogether during winter, as the smallest errors give great differences in the absolute quantity of moisture in the air. In order to determine approximately the evaporation of ice during winter, Lieut. Weyprecht exposed cubes of ice that had been carefully weighed to the open air, and determined the loss of their weight every fourteen days."

"During winter the air seemed always to contain particles of ice; this was seen not only by parhelia and paraselenæ when the sky was clear, but also in astronomical observations. The images of celestial objects were hardly ever as clear and well defined as they are at home, although the actual moisture in the atmosphere was far less. It happened very often that with a perfectly clear sky needles of ice were deposited in great quantities upon all objects. It was quite impossible to determine the quantity of atmospheric deposits, as during the snow-storms no distinction could be made between the snow actually falling and that raised from the ground by the storm; it was remarkable, however, that during the first winter the quantity of snow was small compared with that of the second winter, when the snow almost completely buried the ship (this happened near Franz-Joseph's Land). The same proportion was repeated in the quantity of rain during the first and second summer; in the first only a little rain fell late in the year, while in July 1874 it rained in torrents for days.

"Clouds are naturally of a very different character from those seen at home; nimbus and cumulus are never seen. The form of cloud is either that uniform melancholy grey of an elevated fog, or cirrus: the latter consists of round but undefined masses of fog at but a small elevation, therefore somewhat different from the cirrus of the temperate zone. Instead of clouds, fogs are prevalent, now higher, now lower, and 24 hours of clear weather rarely occur during the summer; generally the sun is seen for a few hours, then to disappear again behind the thick fogs. Melancholy and depressing as these fogs are, they are nevertheless necessary for the general conditions of the ice; they prevent the escape of the heat of the sun's rays, and melt more ice than the direct rays.

"Parhelia and paraselenæ were often observed; they always were certain indications of coming snow-storms. A new phenomenon was only observed once, when, besides the double system of parhelia, two other mock suns appeared on the same altitude with the real sun.

"On the whole path which the vessel described soundings were made constantly, and the depth of the sea was found to increase towards the east; on the easternmost point, 73° E. long., there were 400 metres of water, and the depth steadily decreased towards the west. In front of Franz-Joseph's Land there is a bank which seems to reach as far as Nowaja Semlja; beyond it the depth increases again. The whole area east of Spitzbergen rarely exceeds 300 metres in depth. Lieut. Hopfgarten specially constructed an instrument to fetch up dredgings, which was frequently done. The deep-sea temperatures were measured with Casella's minimum and maximum thermometer, and these measurements were continued throughout the winter. They showed a slight increase in the temperature at the bottom. The per-centage of salt in the sea-water at different depths was also determined. Until the ship was blocked up the surface temperatures of the sea were also measured. Lieut. Weyprecht thinks that, as a rule, too much importance is attached to these, as the state of the weather is not taken into account, and it is just that which has the greatest influence upon the surface temperature; it is quite wrong to imagine the existence of currents from observations of this kind."

V.—GEODESY AND PENDULUM EXPERIMENTS TO DETERMINE THE FIGURE OF THE EARTH.

---

1. Parry's Pendulum Observations in Melville Island.

These experiments were made by setting up a clock containing the pendulum and observing the number of vibrations, the pressure and temperature of the air, the arcs described, and the height above the sea.

Two clocks were used, and were the same which Captain Cook took round the world.

The pendulums were prepared by Captain Kater, being each cast in one piece of solid brass, vibrating on a knife edge of hard steel on hard agate, in the form of a hollow cylinder.

The number of vibrations made by the clocks in London was ascertained before the departure and after the return of the Expedition.

				London.	Mean.
Clock I.	Jan.	1819	-	86392·5673	86392·4513
,,	Dec.	1820	-	86392·3353	
Clock II.	March	1819	-	86496·997	86496·9855
,,	Dec.	1820	-	86496·9741	

The daily acceleration of a pendulum, from the mean of the results with both pendulums, is 74·734 vibrations between the latitudes 51° 31′ 08·4″ and 74° 47′ 14·36″ N., giving the ellipticity of the earth $\frac{1}{312\cdot 6}$.

---

2. Observations in Spitzbergen by Captain Sabine.

Similar observations were made by Sabine at Spitzbergen in latitude 79° 50′ N., in the year 1823.

Experiments were also made by Sabine at Hammerfest in latitude 79° 40′ N.

Comparing together these results with those which have been obtained by Sabine at Greenwich and at Paris, we get—

—	Length of Seconds Pendulum.	Measure of the Force of Gravity.
At Spitzbergen - - -	39·21469	32·2528
,, Hammerfest - - -	39·19475	32·2363
,, Greenwich - - -	39·13983	32·1912
,, Paris - - -	39·12851	32·1819

---

3. Parry's Experiments at Port Bowen. (Third Voyage.)

These experiments on the time of vibration of one of Kater's Pendulums, were made at Port Bowen, in lat. 73° 13′ 39·4″ N., long. 88° 54′ 48·55″ W., and compared with experiments with the same pendulum at Greenwich, before the Expedition left England, and also after its return. The number of vibrations made at Greenwich in 24 hours mean time reduced to the level of the sea, and in vacuo at the temperature of 50° Fahr. in April 1824, was 86159·315, and in November 1825, was 86159·554; the difference 0·24 is probably due to the wear of the knife edges, which gives the average 86159·434 at Greenwich, to compare with 86230·242, the mean of two series of observations made at Port Bowen. Taking these values, the ellipticity of the earth is found to be $\frac{1}{309\cdot 16}$, and the comparison of the force of gravity at the two places is at once determined. The length of the Pendulum being invariable, the value of the acceleration due to the force of gravity, is directly proportional to the square of the number of vibrations; or, which amounts to the same thing, the force of gravity is directly proportional to the length of the second's Pendulum. If g be the acceleration at Greenwich, and g_1 the acceleration at Port Bowen, then $\frac{g_1}{g}=\left(\frac{86230\cdot242}{86159\cdot434}\right)^2=1\cdot0016445$.

---

4. EXPERIMENTS made by DR. HAYES at PORT FOULKE in September and October 1860, to determine the Figure of the Earth.

A series of experiments on the time of vibration of a Pendulum were compared with a series made with the same Pendulum at Cambridge, Massachusetts, before the Expedition sailed. The number of vibrations in 24 hours at Cambridge was 86421·14, and the number of vibrations at Port Foulke was 86550·72.

These values would give a compression or flattening of the pole $= \frac{1}{372}$.

This value is smaller than that derived from Parry's experiments and by other methods. Baily's Pendulum experiments gave the ellipticity $\frac{1}{285}$, but other smaller values have been obtained. The value given by geodetic measurements is $\frac{1}{293}$.

The ratio of the value of the force of gravity at Port Foulke, and at Cambridge, from these Pendulum experiments, is 1·003 very nearly.

---

5. GEODESY. (Die zweite deutsche Nordpolarfahrt, vol. ii., p. 761.)

A base line was measured on the coast near Germania Harbour, and connected by a system of triangulation with three points in Sabine Island, at some distance from the harbour and from one another.

The sides of this triangle were then used as bases of other systems of triangulation extending to other islands, and to various points on the mainland, extending over nearly one degree of latitude.

The comparison of these measurements, with the astronomical observations for the elevation of the Pole Star at the different places, gives, by the usual formulæ for the length of a degree of latitude, 72751·429 metres. The length of a degree of latitude as deduced from theoretical considerations is 72937·627 metres, the difference being about one four-hundredth part of the whole distance.

The heights of a great many points along the coast were also determined by trigonometrical measurement, as well as by the barometer.

---

6. MEASURING the ADVANCE of the GLACIER in FRANZ-JOSEPH-FIORD. (R. COPELAND.)

Two signal stations were taken on the glacier, and their positions determined with reference to two points on the side of the glacier, 80 metres apart, in a line parallel to the motion of the glacier.

The motion of the marks was measured on August 11, after 19 hours, from 4 p.m. to 11 a.m., and again on August 12, after another interval of 18 hours, from 11 a.m. to 5 a.m. the next morning.

From these observations the daily advance of the glacier was found to be $0{\cdot}164 \pm 0{\cdot}019$ metres.

---

7. MEASUREMENT of MOTION of GLACIER by DR. HAYES. (Smithson Contrib., vol. xv.)

In the autumn of 1860 Dr. Hayes made a survey of glacier in the neighbourhood of Port Foulke, the end of it being about two miles from the sea. A base line was measured along its axis, and bearings taken to fixed objects on the mountain on each side. After eight months the measurements indicated a motion of 94 feet.

On the interesting subject of the motion of the Swiss glaciers, reference may be made to papers of Prof. Forbes, in the Edinburgh New Philosophical Journal (Oct. 1842), who discovered that the central portion of a glacier moves more rapidly than the sides, and that a glacier is continually advancing ; also to the works of De Saussure, about the same period, in the "Comptes Rendus," and especially to the experiments of Dr. Tyndall, on the motion of the Mer de Glace. (Tyndall's Glaciers of the Alps, p. 275.)

VI.—ATMOSPHERIC REFRACTION AT LOW TEMPERATURES.

---

1. APPENDIX to SIR JOHN ROSS'S NORTH-WEST PASSAGE.

"Captain Scoresby gives some extraordinary instances of both land and ships seen at immense distances, and on our first voyage it is recorded that Cape Clarence was seen from the deck at the distance of 120 miles, the ship being at that time 2° of latitude south of the cape. The most remarkable circumstance which occurred was the uneven current of refraction raising an intermediate body (an iceberg or island) above the more distant land, which at the time of no refraction was considerably higher."

---

2. OBSERVATIONS at PORT BOWEN. (Parry's Third Voyage, &c.)

(Page 56.) The amount of atmospheric refraction at low temperatures was the subject which, next to magnetism, appeared the most interesting to investigate. The extreme difficulty attending the use of the repeating circle during intense cold, rendering it next to impossible to obtain with that instrument observations of a star *having quick motion* with the minute accuracy requisite for this purpose, we were led to adopt the simple method of observing the setting of a star behind a horizontal board fixed edgewise on a neighbouring hill, the altitude of the board being obtained at leisure by repeated observations with the circle. These observations, the details of which are given in the *Philosophical Transactions*, make the refraction at low temperatures, and from the altitude of 4½° to 7½°, as computed from the table in the Nautical Almanack, considerably *in defect*.

The instruments used were a theodolite and a small repeating circle with two telescopes which were fixed at 122 feet above the sea-level on casks filled with sand and firmly frozen to the ground.

It will be seen from the observations made by Sir John Ross that the accuracy of the Nautical Almanack tables cannot be tested by the method adopted.

---

3. Capt. Parry's Third Voyage by Spitzbergen, p. 99.

Fog Bows.—At half-past 5 p.m. on 23rd July 1827, in lat. 82° 43′, a beautiful phenomenon, consisting of six fog bows, was seen in a direction opposite to the sun. First one white bow was seen, then it became tinged with prismatic colours and five other complete coloured arches were formed within it, the interior ones gradually narrowing.

The larger bow and the one next within it had the red on the outer or upper part of the circle, the others on the inner side. Altitude of the outer arc, 20° 45′, its extent at the horizon, 72½°, the altitude of the sun being 20° 40′.

The smaller bows were seen for 20 minutes.

In these travels the southerly drift of ice was more than four miles a day for the last five days of the journey, so that after immense labour in making a good 23 miles journey there was found to be only a gain of one mile northwards.

Lat. 82°·45′, long. 19°·25′ E., dip, 82°·21′·6, and declination 18°·10′ W. No bottom at a depth of 500 fathoms; the specific gravity of water at that depth was 1·034 at 37° F.

At 6 fathoms the specific gravity of water was 1·0225.

On July 27th, at 3 a.m., saw another fog bow, the sun's altitude being 12½°, that of the centre of the outer arch 28°, and its extent at the horizon, 77½°.

Three smaller fog bows were seen within the large one.

Aug. 2nd.—In lat. 82° 6′, and long. 17° 45′ E., we met with snow, tinged with some red colouring matter to the depth of several inches, giving a rose tint and in some places a salmon coloured tint to the snow when pressed.

Aug. 5th.—Appearances of the raising of ice in the horizon by refraction until it resembles a perpendicular wall, were generally taken to be an indication of open water, but were not thought by Parry an infallible one.

---

4. Scoresby's Observations of Fog Bows.

Scoresby in his Greenland Voyage (pp. 96, 144, 164) also describes remarkable effects of refraction by layers of air of different densities; and (p. 275) remarkable coronæ or fog-bows, with a number of prismatic circles, seen from the mast-head 105 feet above the sea-level in a direction opposite to the sun.

Round the centre was—

1. A bow from 3° to 4° in diameter.
2. A bow 9½° in diameter.
3. A bow 13° in diameter.

These at times gave the colours of the spectrum.

4. A circle about 5° broad, and mean radius 38° 50′.

The upper edge of this circle was about 6° above the horizon. Another outer circle was seen on another occasion.

The sun's altitude was 35° 42′. Many shining spikes or needles, supposed to be crystals of snow, were seen floating in the breeze, but they were so small that they could not be caught and examined under a microscope.

The sun shone above the fog with great splendour, and the shadows of the masts, sails, &c. were clearly thrown on the water.

p. 351. Passed through brown coloured patches of sea-water, the colour being due to animalcules covering the sea in streaks, as shown by examining the water under a microscope. They must be exposed to an average temperature 15° below the freezing point in a sea perpetually covered with ice.

Water of a similar appearance is also noticed by Parry on his entrance into Davis Strait. See Journal of Voyage for Discovery of N.W. Passage.

Sometimes these streaks are of a yellowish green colour.

---

5. Observations on Refraction. Paraselenæ. (Belcher.)

On November 30, 1852, Admiral Belcher saw a beautifully defined paraselena (see illustration in Vol. I. p. 169) about N.N.E. true. The interior circle had a radius of 22° 30′, the exterior 44°, the moon's altitude at the time 30°.

There were two concentric halos, incomplete near the horizon, accompanied by a strong vertical and a horizontal beam passing through the moon. It was also accompanied by arcs of other concentric circles, having their common centre at a point within the zenith. The greater of these intersected the moon and outer halos, forming at their contact luminous spots.

At the points of intersection of these halos, bright paraselenæ, forming five on the lower and two on the upper arcs, presented themselves; the moon and the intersections by the vertical ray exhibiting the most luminous. The moon was four days past the full.

On the same evening, between 9 p.m. and midnight, the magnetometer was deflected westwards 10° more than usual. The next night the magnetometer was again unusually disturbed. There was also a fainter display of paraselenæ.

On December 1st a beautiful illustration of the effect of intense cold, fixing, as it were, the clouds and currents of air, and giving them rounded outlines. Probably another remarkable effect due, like the paraselenæ, to the floating of crystals or needles of ice in the air.

"The distant land on the southern shore, near Cape Lady Franklin, was peculiarly refracted, and appeared to be considerably nearer than our positive knowledge would warrant, displaying very beautifully its particular features." A delicate salmon tint pervaded the horizon.

---

6. Halos, &c.—McClintock, p. 74.

Dec. 4, Dawn at 11h. 0m. a.m. A well-marked halo and several paraselenæ between 7h. and 10h. p.m., consisting of five false moons, three arcs of halos, and a horizontal belt of light, round the heavens and passing through the moon.

---

7. Twilight throughout the Winter.

With regard to the existence of twilight through the winter, McClintock says:—"On the mid-winter day (December 21st) at "noon we could just read type similar to the leading article of the "'Times.' Few people could read more than two or three lines "without their eyes aching."

Parry in his first voyage to Melville Island says:—"Up to the "shortest day of the year the return of each successive day was "very decidedly marked by a considerable twilight for some time "about noon, that on the shortest day being sufficient to enable "us to walk out very comfortably for nearly two hours. There "was usually a beautiful arch of bright red light in the southern "horizon for an hour or two."

---

8. Parry's Second Voyage to discover a North-west Passage.

Analysis of Air, p. 240.

Bottles of air collected at Igloolik and packed in oakum were examined by Prof. Faraday, whose results showed that the air contained 20·5885 per cent. of oxygen, whilst the air of the Laboratory at the Royal Institution contained 21·9625 per cent.

Expansion of Air, p. 245.

A glass cylinder, 1·7 inch in diameter and 2·6 inches in length, was used; into this was fitted a long tube or stem 10 inches in length and nearly half an inch in diameter, graduated into 140 equal parts. The cylinder was also fitted with a ground glass stopper. A drop of mercury was placed in the stem, and the vessel carried from a cold atmosphere to a warm one, and then the temperature taken by removing the stopper. Again, the vessel was carried from the warm room to the cold, and the results showed that 2059·854 volumes of air at 55°·5 F. are reduced to 1682·24 volumes at 34°·5 F.

---

9. EXPERIMENTS and OBSERVATIONS on SOUND. (Parry's First Voyage, &c.)

The distance at which sounds were heard during intense cold constantly afforded matter for surprise.

" We have often heard people distinctly conversing in a com-
" mon tone of voice at the distance of a mile." A man was heard singing to himself as he walked along the beach at even a greater distance than this.

[These facts, recorded by Parry, are no doubt due to the singularly homogeneous nature of the air over the whole region.

The experiments of Prof. Tyndall in the English Channel show that whether the weather be foggy or clear, the air will convey sound well, provided it is homogeneous; but that if there are layers of air of different densities, or if there are dry and moist currents of air alternating, then the sound will be reflected or diffused at the common surface of two layers, and will be lost. The acoustic transparency of the air and its optical transparency are entirely independent of one another.]

---

10. The VELOCITY of SOUND. (Franklin's Second Journey.)

Observations on the velocity of sound at different temperatures were made by Lieutenant E. N. Kendall during October, November, and December 1825.

The experiments, divided into five series, according to the temperatures at the time of observation, give the following results:—

Temperature.	Velocity of Sound.
° °	
28·0 to 27·0 F. -	- 1112 feet per second.
10·0 to 3·0 -	- 1089 ,,
– 2·0 to – 4·0 -	- 1079 ,,
–36·0 to –35·5 -	- 1036·1 ,,
–41·0 - -	- 1030·3 ,,

---

11. Parry's Experiments on Sound at Winter Island. (Second Voyage, &c.).

(p. 237.) Base lines were determined between two stations, and the time between the flash and the report of a six-pounder was taken by pocket chronometers, &c., by each observer. The gun had an elevation of 10°, and was directed towards the observers. The results are arranged according to temperatures.

Therm.	Mean Velocity of Sound.
°	Feet.
−41·3	985·9
−33·3	1011·2
−27·2	1009·2
−21·0	1031·0
− 2·0	1039·8
+33·3	1069·9

From these observations the velocity of sound is seen to increase with an increase of temperature, at the rate of 1·126 feet for 1° Fahr.

On Feb. 9th, with barometer 28·96, and thermometer 25° Fahr., the officer's word of command "fire" was several times distinctly heard about three-eighths of a second *after* the report of the gun.

---

12. Experiments on the Velocity of Sound at Port Bowen. (Parry's Third Voyage to discover a Northw-est Passage.)

(p. 58.) Lieutenant Foster having occasion to send a man from the Observatory to the opposite shore of the harbour, a measured distance of 6,696 feet, or about one statute mile and two-tenths, in order to fix a meridian mark, had placed a second person half way between, to repeat his directions; but he found on trial that this precaution was unnecessary, as he could without difficulty keep up a conversation with the man at the distant station. The thermometer was at this time 18°, the barometer 30·14 inches, and the weather nearly calm, and quite clear and serene.

(Appendix, p. 86.) The following experiments were made with a six-pounder brass gun placed on the beach at the head of Port Bowen, and fired by signal from the Hecla, the interval elapsed between the flash and report being carefully noted by the beats of a pocket chronometer held to the ear of each observer. The initials in the columns of interval are those of Capt. Parry and Lieut. Foster, and the result in the last column of the table is deduced from the mean of both. The distance of the muzzle of the gun from the place of observation, as measured trigonometrically by Lieut. Foster, was 12,892·82 feet, and by Capt. Parry (by a different series of triangles), 12,892·96. The mean distance employed in the calculation is 12,892·89 feet. The bearing of the gun was S. 71° 48′ E. The observations appear to indicate a decided decrease of velocity, with an increased density of the atmosphere, the rate of travelling decreasing from 1,098 feet per-second, at a pressure of 30·118 in., and temperature +33·5°, to 1,014 feet per second, at a pressure of 30·398 in., and temperature −38·5°, all other circumstances being alike. The last experiment in the table shows a still greater velocity at a high atmospheric temperature, which, however, might have been influenced by a stronger breeze blowing from the direction of the gun at that time.

Date 1824.	Barometer. Inches.	Thermometer.	Wind.		Weather.	No. of Guns.	Interval in Seconds between Flash and Report.			Rate of Travelling in ft. per second.
			Direction.	Force.			P.	F.	Mean.	
							s.	s.	s.	
Nov. 24 -	29·841	−7°	E.S.E.	Light	Overcast -	5	12·3525	12·430	12·3912	1040·49
Dec. 9 -	22·561	−9	N.N.E.	Squally	Very clear	6	12·331	12·5266	12·4288	1037·34
Jan. 10, 1825	30·268	−37	E.S.E.	Light	Clear -	4	12·5889	12·4700	12·5290	1029·04
Feb. 7 -	29·647	−24·5	N.E.	Light	Very clear	6	12·639	12·6167	12·6278	1020·99
Feb. 17 -	29·598	−18	Calm.		Overcast -	6	12·372	12·440	12·406	1039·25
Feb. 21 -	29·735	−37·5	Calm.		Overcast -	6	12·8167	12·7067	12·7617	1010·28
March 2 -	30·398	−38·5	E.	Light	Little overcast.	6	12·640	12·780	12·710	1014·39
March 22 -	30·258	−21·5	W.	Light	Very clear and fine.	6	12·40	12·7167	12·5583	1026·64
June 3 -	30·118	+33·5	E.	Light	Very clear	6	11·7333	11·744	11·7387	1098·32
June 4 -	30·102	+35	S.E.	Strong and squally	Clear -	6	11·5889	11·4733	11·5311	1118·10

---

13. Determination of the Velocity of Sound at Low Temperatures. (Second German Expedition.)

Observations were made at a point of Walross Island and at the cairn containing the records of the Expedition at Sabine Island, the distance being 2091·21 metres.

One observer was stationed at each point with a gun and a chronometer, and each observer noted the instant at which the gun was fired, and the time at which the sound was heard.

The difference in the times corrected for the personal equation of the observer by taking repeated observations will be the time Sound takes to travel the distance.

On January 28, 1870, the velocity of sound at a temperature of $-21\cdot2°$ C. or $-6\cdot2°$ F. was 321·58 metres, and on February 20, at a temperature of $-33\cdot9°$ C. or $-29°$ F., the velocity of sound was 317·52 metres per second.

Assuming the ordinary law for the reduction of the observations to 0° C. to be true, the mean of the two results at 0° C. or 32° F. is 337·0 metres or 1,105 feet per second.

VII.—TERRESTRIAL MAGNETISM.

---

1. Observations in Davis Strait and Baffin's Bay, by Sir James Ross and Captain Sabine. Phil. Trans. 1819, and Voyage of Sir John Ross in 1818.

The observations on the dip and magnetic intensity were made by Sir James Ross and Captain Sabine at Waygate Islands, and at several points on ice in Baffin's Bay, by observing the time of 100 vibrations of the dipping-needle both *in* and *perpendicular to* the magnetic meridian. There was no opportunity for using instruments which required fixing during the voyage.

The dipping needle used was one by Nairne and Blunt, and the balance of the needle was truly adjusted by screws on the cross-wires attached to its axis, so that no alteration took place in the indication of the dip on reversing the poles.

The intensity of the magnetic force was determined with the dipping needle, the needle being drawn to a horizontal position by a magnet and then allowed to oscillate freely. Observations of the time of 100 vibrations, and the extent of arc traversed at every tenth vibration, were made both in and perpendicular to the magnetic meridian.

The observations in the magnetic meridian, as well as those for the dip, were repeated with the face of the instrument towards the east and towards the west.

These observations were compared with observations made with the same instruments at Regent's Park in April 1818 by Captain Kater, and in March 1819 at the same spot and also at Brassa Island, Shetland, by Sabine.

The magnetic declination was determined by means of Captain Kater's azimuth compasses, described in "Instructions for the use "of instruments furnished to the Northern Expeditions." The compass was placed on a copper-fastened stool, and carefully levelled by means of a spirit level. Each altitude and azimuth is a mean of several observations, the compass being removed and levelled afresh between every one, thus making each faithfully distinct.

The latitudes and longitudes are deduced by the ship's log from the nearest observed. "The influence of the ship's iron on "her compasses, increasing as the directive power of magnetism "diminished, produced irregularities that rendered observations "on board ship of little or no value towards a knowledge of the "true variation."

Abstract of the times in which 100 vibrations of dipping needle were performed:

Place.	Lat.	Long. W.	In Meridian.	First Arc.	Perpendicular to the Meridian.	First Arc.
	° ′	° ′	M. S.	°	M. S.	°
Regent's Park - -	51 31	0 08	—	—	8 18·3	90
Shetland - - -	60 09	1 12	7 49¾	74	7 59·5	90
Davis Strait -	68 22	53 50	7 20	83	7 33	90
Hare Island -	70 26	54 52	7 21	83	7 26	90
Baffin's Bay -	75 05	60 23	7 27½	84	7 26	90
" - -	75 51½	63 06	7 23½	84	—	—
" - -	76 45	76 00	7 15	85	7 26	90
" - -	76 08	78 21	7 16	85	7 18	90
Davis Strait - -	70 35	66 55	7 16	83	7 18·5	90
Regent's Park -	51 31	0 08	8 02	70	8 18	90

2. PARRY'S FIRST VOYAGE in search of a NORTH-WEST PASSAGE.

Magnetic Observations at Observatory, Melville Island.—(Made by Captain Sabine.)

Variation - 127° 47′ 50″ E.
Dip - - 88° 43′ (on July 17, 18, 19, 1820).

Dipping Needle.—To adjust the needle to the magnetic meridian, a box containing an horizontal needle and card was fitted to the instrument. The variation is the mean of a great number of observations extending from October to July.

The intensity of the magnetic force was determined by the dipping needle in July 1820, by determining the time of 100 vibrations *in* and *perpendicular to* the magnetic meridian, observing the amount of swing after every ten vibrations, and the results were compared with determinations made in Regent's Park in March 1819 and December 1820:—

The time of 100 vibrations in Regent's Park, 8m. 2s., in the meridian.

The time of 100 vibrations in Regent's Park, 8m. 18·5s., perpendicular to the meridian.

The time of 100 vibrations in Melville Island, 7m. 26·25s., in the meridian.

The time of 100 vibrations in Melville Island, 7m. 26·4s., perpendicular to meridian.

The time of 100 vibrations in Regent's Park, 8m. 00s., in magnetic meridian.

Observations for Horizontal Intensity were also made with bars of steel suspended horizontally in a stirrup by a silk line, and allowed to vibrate on each side of the magnetic meridian, the readings being taken at each 10 vibrations.

These were compared with observations made at Sheerness. Comparing the results obtained, we get—

—	Dip.	Observed Force.	Calculated.	—
London - -	70° 33·3′	1	1	By dipping needle.
Melville Island -	88° 43·5′	1·163	1·153	

—	Dip.	Observed Horizontal Force.	Calculated Horizontal Force.	—
Sheerness - -	69° 55′	13·33	13·275	From several observations with three needles.
Melville Island -	88° 43·5′	1	1	

Showing a remarkable agreement between theory and experiment.

A description is given of the special means employed to keep the agate planes horizontal, and also of a better method of releasing the magnet instead of employing another magnet for the purpose.

From observing the variation 128° 58′ W., in longitude 91° 48′ W., and then 165° 50′ 9″ E. in longitude 103° 44′ 37″ (where the dip was 88° 25·58′), Parry estimated that he was north of the magnetic pole in longitude 100° W.

---

3. Appendix to Parry's Second Voyage in the "Fury" and "Hecla."

(p. 97.) Observations were made to determine the Declination on ice and on shore. On a floe 132 yards, and even at 200 yards from the ship, the ship's influence on the needle was found to be considerable, which is shown by the effect of a change in the position of the ship: the floes of ice to which it was attached had a slow motion in azimuth. Observations were afterwards made at a quarter of a mile from the ship.

(p. 276.) Magnetic observations at Winter Island and at Igloolik were generally made at 9 a.m. and 3 p.m. throughout the winter. The mean of all the observations gave:

Place.	Lat.	Long.	Declination.	Dip.	Force.	Time of 100 Vibrations.
	° ′	° ′	° ′	° ′ ″		′ ″
At Winter Island - -	66 11	82 54 W.	57 24⅓	87 51 9	125	4 57·3
At Igloolik -	69 21	81 37 W.	83 1½	88 9 49	135	4 46·7

The force in London, as determined by the same dipping needle, being 100.

The time of 100 vibrations in London was 5′ 30″ before the Expedition started, and 5′ 36·5″ after its return: giving a mean value of 5′ 33·3″.

With regard to the diurnal changes in all the magnetic elements contrary results were obtained at the two winter stations.

At Winter Island the declination W., the dip and the force were each greater in the morning than in the afternoon; but at Igloolik each of the three magnetic elements was greater in the afternoon than in the morning.

At both stations the magnetic force increased as summer advanced.

---

4. Parry's Journal of a Third Voyage for the Discovery of a North-West Passage. (1824–25.)

Magnetic Observations at Port Bowen.

The dipping needles employed were three by Jones, $7\frac{3}{4}$ inches in length; one of which, belonging to the same instrument, was employed for intensity, and to another each one of three brass spheres could be attached, by means of which its moment of inertia could be determined.

There were also two dipping needles by Dollond, one of which was employed exclusively for intensity.

"The observations for intensity, by means of the time in which the needles performed 100 vibrations in the meridian, are deduced from the mean of 400 vibrations obtained with the face of the instrument on each side of the vertical, and the needles reversed on their axes in the two positions."

These times are compared with the times during which the same needle made 100 vibrations on "Woolwich Common," at 11.30 a.m. on Dec. 3, 1825.

Three observations for intensity were made at Port Bowen on Nov. 8, 1824, Jan. 10, and June 27, 1825, and the ratio of the force at Port Bowen to that at Woolwich was found to be 1·296, 1·298, and 1·286 in the three experiments.

The observations for the dip gave:

Date.	No. of Observations.	Mean Dip.			
		°	′	″	
Nov. 1–12	10	87	5	29	Mean for all the observations, 88° 01′ 23″ (Nov. 1–12 to June 27, 28)
Jan. 4–8	6	88	05	31	
April 26, 27	3	88	13	02	
June 2	3	88	08	12	
June 27, 28	4	87	44	41	
Aug. 30, 1825	1	88	08	25	At Neill's Harbour.
Dec. 4, 5, 1825	3	70	0	4	At Woolwich.

Observations were made for the variation with Gilbert's Azimuth compass, and also with Kater's compass.

A great many observations were made in November 1824, and a few in December and January.

The mean variation was found to be 123° 21′ 55″ at Port Bowen.

At Neill's Harbour the mean variation was 118° 48′ W. on June 4, 1825.

Since 1819 the variation had increased at Port Bowen from 114° to 123°.

The maximum variation westerly was observed to occur between the hours of 10 a.m. and 1 p.m.; and the minimum between 8 p.m. and 2 a.m.; the quantity being seldom less than 1½° or 2°, and sometimes amounting to 5, 6, or even 7°.

In connexion with these observations a regular series of hourly experiments was made on the magnetic intensity with a *suspended* needle, which admitted of the intervals of vibration being observed with minute accuracy. There was found to be a diurnal change of intensity, generally showing a regular increase of intensity from the morning to the afternoon, and as regular a decrease from the afternoon to the morning. The changes seemed to depend on the sun and the relative position of the sun and moon. The diurnal maximum Variation occurred at 11h. 49m., and the minimum Variation at 10h. 1m. p.m.

[The importance of having needles delicately *suspended* instead of *supported* on a point for observing the magnetic disturbances is noticed by Parry, and also by McClintock in his observations at Port Kennedy. Corresponding disturbances in the dip were not observed, probably because the dipping needles were too heavy and not delicately suspended.

For measuring these disturbances the dipping needle may be made very light, and be *suspended* by two threads attached to two points in a horizontal axis, passing through the centre of gravity of the needle. If the magnetic axis of the needle is at right angles to the axis of suspension, the disturbances will be indicated, even though the centre of gravity should not be accurately on the axis of suspension. The importance of having very light needles is shown by M. Weyprecht in his account of his Arctic scientific work (*see* p. 709).]

Observations were made hourly for Horizontal Intensity.

The maximum intensity of the horizontal needle uniformly took place at 7 p.m., but the minimum was not so well marked.

The needle for estimating the diurnal changes in the horizontal needle was frequently observed to vibrate in very small arcs in its passage to and fro westward from 2 to 6 p.m., and also in its passage eastward. During the same interval the intensity of the horizontal needle was observed to be very changeable, and the action of the suspended needles very irregular.

Feb. 22, 2. a.m.—Aurora appeared in an arch from north to west by compass, with bright streamers towards the zenith; the needle was not affected in any way whatever.

A comparison of the changes of intensity of the horizontal force with the changes of intensity of the total force seems to show that the change of intensity in the horizontal force is due principally to a change in the dip, rather than to a real change of intensity in the Earth's magnetic force.

Lieutenant Foster believes he accounts for the diurnal variations of direction and intensity by supposing that the magnetic axis of the earth revolves round and is inclined to the earth's axis at an angle of 2′ or 2½′.

---

5. Magnetic Observations by Sir John Ross.

The magnetic observatory at Felix Harbour was built of snow, 200 yards from any metal.

Observations for Variation were made every two hours in April and May.

In the Diurnal Variation instrument (by Dollond) the needle was nicely suspended by a single fibre of New Zealand flax. Sir John Ross says, "It was materially disturbed by the approach of " any metallic substance, by any sudden increase or decrease of " light, and by alteration in the temperature." The increase of temperature caused by looking through the microscopes had a considerable effect, and the needle had a "constant tremulous motion."

Sir John Ross also says, "When too dark to observe without " a light, I was obliged to take a paper lantern, and even then, " when the lighted candle shining through the paper was held " for the purpose of reading off the arc, it produced a horizontal " motion in the needle corresponding to the direction and " strength of the light."

The position was near the magnetic pole, in latitude 69° 59′ and longitude 92° W., where the variation was 89° 45′ W. and the dip 89° 55′. "During the winter it was proved that the needle was " disturbed by and followed the light of a candle; that it was " materially disturbed by a brilliant Aurora Borealis, particularly " when that was of a deep red."

Note.—This passage shows the delicacy of suspension of the needle, and may possibly have some bearing on the beautiful experiments of Mr. Crookes on delicately suspended bodies subject to the action of a candle, although the effects may probably be due to the fact that some kinds of paper are strongly magnetic.

The dip was determined by three dipping needles, and was the same with each, viz., 89° 55′.

At Spence Bay the dip was 89° 56′, and the variation 68° 35′ W.

On the shores of the sea to the west of Boothia Felix, Sir James Ross made observations of declination and dip, and calculated the position of the magnetic pole.

On reaching the calculated position, attempts were made to determine the magnetic meridian by suspending a dipping needle first by means of a single fibre of silk, and afterwards by a single fibre of flax, also by finding in what direction a given number of vibrations was made in the shortest time, but no result could be obtained. Six sets of observations were taken in what seemed to be the most probable direction, and at 45° and 90° to it, and the mean of all the observations gave the dip 89° 59′.

The latitude of the position was 70° 5′ 17″ N.
The longitude „ „ 96° 45′ 48″ W.

---

6. MAGNETIC OBSERVATIONS in the ARCTIC SEAS, by DR. E. K. KANE, during the SECOND GRINNELL EXPEDITION in 1853–55, at VAN RENSSELAER HARBOUR. ("Smithsonian Contributions," vol. x.)

Observations for changes of magnetic declination were made with a unifilar magnetometer. The magnet was suspended by a silk string 9½ inches in length, and on several trials the effect of torsion was found to be exceedingly small. The Expedition not being well provided with instruments, the same instrument was employed for observing absolute declinations. Mr. Sonntag obtained a few values in winter quarters by detaching the box containing the magnet from the circle which bears the telescope. The telescope was then moved in azimuth until a well-defined object within the range of its vertical motion could be observed. The same instrument was also employed to determine the Earth's horizontal magnetic force.

The instrument was perched on a pedestal of frozen gravel, the contents of two barrels, and this mounting was as stable as the rock underneath.

At Van Rensselaer Harbour the Observatory was placed upon the northernmost of the rocky group of islets that formed the harbour. It is 76 English feet from the highest and northernmost salient point of this island, in a direction S. 14° E., or in one with said point and the S.E. projection of the southernmost islet of the group. A natural face of gneiss rock formed the western wall of the Observatory. A crevice in this rock had been filled with melted lead, in the centre of which is a copper bolt. Eight feet from this bolt, and in the direction indicated by the crevice, stood the magnetometer. This was called the Fern Rock Observatory. The highest point of the island was about 30 feet above the mean tide level of the harbour.

On the 9th of June 1854 Mr. Sonntag examined the instrument in reference to local disturbance, and found no sensible deviation arising from such a source. "The local deviation seems "to have corrected itself; the iron in our comfortless little cell "seems to have been so distributed that our results were not "affected by it." Hourly observations for diurnal variation were made during January, February, and March 1854, generally at intervals of three or four days, as well as on term days. The term days were January 18, 19, February 24, 25, March 22, 23, April 19, 20, May 26, 27, June 21, 22, 1854. Readings (the mean of two extremes during a vibration when the magnet is in motion) were taken every sixth minute on term days, beginning between 4 and 5 o'clock in the afternoon, so as to correspond with the times of observations at Göttingen. Each set of observations extends over 24 hours, and the times of observations are given in the tables, beginning at 4h. 37m. 34s. mean Fern Rock time, *i.e.* at 10 p.m. Göttingen time, the difference of longitude being assumed to be equal to 5h. 22m. 26s.

The winter quarters at Van Rensselaer Harbour were in latitude 78° 37′ N. and longitude 70° 40′ W. The Observatory was of stone, 10 feet square, with a wooden floor and roof, and with a copper fire-grate.

The results of the observations of changes of magnetic declination as given in the tables show that the total easterly deflection is greatest at 2 a.m., when it is equal to 29′, and the greatest westerly deflection, which amounts to 37′·8, occurs at 12 noon, the needle being in its mean position about 6·45 p.m. and 7 a.m.

The observations for changes of magnetic declination at Fern Rock have been compared with those at Greenwich and Washington, and the results of these comparisons are given in the tables.

---

7. Magnetic Observations in the Arctic Seas, by Dr. Hayes, at Port Foulke, in 1860–61. ("Smithsonian Contributions, vol. xv., p. 42.)

The Observatory was erected on the first of a series of terraces which lay north-east from the anchorage, and its foundation was 38 feet above the mean tidal level. The rock on which it stood was a dark reddish-brown syenite, which rose on either side of the harbour into hills from 600 to 800 feet high. It faced the S.W., its axis being nearly in the magnetic meridian.

Winter quarters in Port Foulke were in latitude 78° 17′ 39″ and longitude 73° 0′ 0″ west of Greenwich, 20 miles south of the latitude of Rensselaer Harbour, and distant from it by the coast-line about 55 miles.

The unifilar magnetometer was mounted in the centre of a room 8 feet square and 7 feet high, on a stand made of two kegs, whose heads being removed, and the ends carefully fitted together, were filled with beans and water; the lower keg was placed on the solid rock through a hole cut in the floor.

For diurnal variation of magnetic declination, hourly observations were recorded on 15 days between November 26, 1860, and March 4, 1861, at intervals of from six to eight days.

The diurnal variations at Port Foulke and Van Rensselaer Harbour through the winter are very similar in character, both positions being northward of the magnetic pole.

A maximum *west* deflection occurs about 1 p.m., and a maximum *east* deflection between 2 and 3 a.m., the needle being in its mean position at 6.30 p.m. and 7 a.m.

The range of motion at Fort Foulke in 1860–61 was 42′, and at Van Rensselaer Harbour in 1854 was 69′; the horizontal force at Fort Foulke is less than the horizontal force at Van Rensselaer Harbour.

A maximum *west* deflection occurs at 1 p.m. at Godhavn (as

observed by Lieutenant Foster at Whalefish Islands, lat. 68° 59′, long. 53° 13′, in 1824; also at Port Kennedy, observed by McClintock 1858–59, lat. 72° 1′, long. 94° 19′, very near the magnetic pole, magnetic declination 135° 47′).

The time of maximum *west* deflection appears to be 1 p.m. for all places in the north magnetic hemisphere.

The time of the maximum *east* deflections appears to be subject to disturbances which cause it to vary from 2 a.m. to 9 a.m.

A westerly magnetic motion of the north end of the needle at Port Foulke or Van Rensselaer Harbour means a change farther from the north on the western side, *i.e.*, a change from a point geographically nearly S.W. to a point more towards the south.

Observations for absolute declination were made at 14 stations in various parts of the Polar regions west of Greenland, and the results are recorded in the following table.

8. Declinations observed by Dr. Hayes.

Place.	Latitude.	Longitude.	Declination.	Date.
	° ′	° ′	° ′	
Pröven - - -	72 23	55 33	83 24	Aug. 1860.
Starr Island - -	78 18	73 6	109 45	Oct. ,,
Cairn Point - -	78 31	72 59	110 9	April 1871.
Foggy Camp - -	79 55	71 28	106 53	May ,,
Camp Hawks - -	79 44	73 6	115 38	,, ,,
Câche-on Floe -	79 30	72 53	113 52	,, ,,
Scouse Camp - -	79 29	72 53	112 6	,, ,,
Potato Camp - -	79 4	72 30	105 34	,, ,,
Camp Separation - -	78 53	72 8	105 4	,, ,,
Last Camp - -	78 38	72 8	108 36	,, ,,
Port Foulke - -	78 18	73 0	111 40	July ,,
Northumberland Island -	77 11	72 20	106 0	Aug. ,,
Netlik - -	77 8	71 22	106 49	,, ,,
Upernavik - -	72 47	56 3	72 12	,, ,,

Observations for the Horizontal Magnetic Force were made at five stations by Dr. Hayes and at three stations by Dr. Kane, and the results are as follows:—

Values of Horizontal Force of the Earth. (Sm. Con., vol. xv., p. 66. Drs. Kane and Hayes.)

—	Lat.	Long.	Hor. Force.	Date.
	° ′	° ′		
Cambridge, Massa. - -	42 23	71 07	3·607	July 1860
Godhavn - - -	69 12	53 28	1·762	Sept. 1861
Pröven - - -	72 23	55 33	1·576	Aug. 1860
Upernavik - - -	72 47	56 03	1·358	,, 1861
Netlik - - -	77 08	71 22	1·110	,, ,,
Port Foulke - -	78 18	73 00	1·084	July ,,
Cape York - - -	76 03	68 00	1·573	1855
Hakluyt Island -	77 23	73 10	1·344	1855
Van Rensselaer Harbour -	78 37	70 53	1·139	1854

These observations of Horizontal Force made by Drs. Kane and Hayes are compared with what seems to be the *absolute* value of the Horizontal Force, 3·607, for the same period at Cambridge, Massachusetts; in which case the Total Force at Port Foulke is 12·56, and at Van Rensselaer 12·38 British absolute units, according to the system adopted by Sir Edward Sabine in his Magnetic Survey of the Earth.

The inclination was obtained at five stations in the neighbourhood of Smith Sound by means of a dip circle, and the determinations were made with each of two needles. In the following table they are compared with three determinations from Dr. Kane's Expedition in 1853–54, and with observations made at Port Kennedy and at Polaris Bay.

OBSERVED INCLINATION.

Date.	Locality.	Lat.	Long.	Observed Dip.	Declination.	Hor. Force.
		° ′	° ′	° ′	° ′	
Aug. and Sept. 1861	Godhavn, Disco Island.	69 12	53 28	81 51	—	1·762
July 1861 -	Port Foulke, Smith Str.	78 18	73 0	85 3	111 40W.	1·084
„ -	Littleton Island	78 22	73 30	84 43	—	—
„ -	Gale Pt., Cadogan Inlet.	78 11	76 28	85 21	—	—
Aug. 1861 -	Hakluyt Island, off Whale Sound.	77 23	73 10	85 0	—	1·344 (?)
„ -	Netlik, Whale Sound	77 8	71 22	84 58	106 49W.	1·110
Aug. 1853 -	Cape Grinnell - -	78 34	71 34	85 8	—	—
Sept. „ -	Marshall Bay - -	78 51	68 54	84 49	—	—
June 1854 -	Van Rensselaer Harbour.	78 37	70 53	84 46	108 12	1·139
Nov. 1858 -	Port Kennedy - -	72° 0′ 49″	94° 19′ W.	88° 27′ 4″	135 47	—
—	Polaris Bay - -	81° 31′	61° 44′	84° 23′	96° 00′	—

---

9. HOURLY OBSERVATIONS of MAGNETIC DECLINATION, by CAPT. R. MAGUIRE and the OFFICERS of H.M.S. "PLOVER," in 1852–54, at POINT BARROW. Phil. Trans., 1857, p. 497.

The instruments used were two dip circles, one of 9½ inches, the other of 6 inches, each with two needles; and a portable declinometer having a perforated magnet of 3 inches in length, carrying a collimator scale in the interior, the divisions of which were read by a detached telescope.

The Observatory was composed of an outer house of ice 12 feet square and 7 feet high, within which was another of sealskin 7 feet by 6. Two posts 23 inches apart were then sunk, and being firmly frozen into the earth served as supports for the declinometer and telescope, whilst another post was placed in the N.W. corner for the chronometer, and a pedestal was placed outside the Observatory for the dip circle, which was afterwards removed to another house 15 feet N.W. of the Observatory.

The dip was observed twice a week. Two dipping needles were broken through awkward handling owing to the extreme cold.

Latitude of place of observation, 71° 21′ N., long. 156° 15′ W. A heavy gale on December 17th stopped observations for five days, raised the water 3 feet round the wall of the observatory, and on December 21st the water had come within 6 inches of the top of the pedestal.

There were two series of hourly observations of declination from November 1852 to June 1853, and from October 1853 to June 1854. The influence of the disturbances at Point Barrow was to occasion a small mean deflection of the needle towards the east, thus slightly increasing the easterly declination.

The mean declination at Point Barrow is 41° E.

There are two epochs of maximum disturbance, the principal at from 7 to 9 a.m., when the proportion reaches twice the average amount, and a lesser maximum at from 11 p.m. to 1 a.m. The principal minimum is from 3 to 6 p.m., and a smaller minimum at 3 or 4 a.m.

The amount of disturbance at the different hours as compared with Toronto shows that it is regulated by a law which has reference to the hours of solar time; but the easterly disturbances at Point Barrow correspond to the westerly disturbances at Toronto.

The principal maximum disturbance at 7 to 9 a.m. is occasioned chiefly by easterly disturbances, and the lesser maximum by westerly disturbances; the minimum for both easterly and westerly disturbances being from 3 to 6 p.m. The easterly are in all respects as to time and range more strongly marked than the westerly disturbances, but each probably consists of a single progression in the 24 hours, and the double maxima and minima would seem to be capable of being resolved into two single progressions (easterly and westerly), having different hours of maximum and minimum.

For the further elucidation of these relations of disturbances and deflections at different stations, observations should include variations of the magnetic *force* as well as of the magnetic *direction*.

The mean dip in each year was found to be 81° 36′, being greatest in February, March, and April in 1852–53, and in February, March, and June in 1853–54, and least in November and December 1852–53, also in December and January 1853–54; the range throughout the year being about 10′.

Both at Toronto and Point Barrow the north end of the magnet is at its easterly extreme at 8 a.m.; in returning towards the west its motion is more rapid than at any other part of the 24 hours; that it passes its mean position at 11 a.m., and reaches its westerly extreme at 1 p.m. The subsequent motion to the east is slower at Point Barrow than at Toronto, and at both stations is checked by a small retrogression towards the west, which would seem to be due to the disturbances. (Phil. Trans., 1857, p. 508.)

---

10. McCLINTOCK'S VOYAGE of the "FOX." Hourly Observations of Magnetic Declination at Port Kennedy in the Winter of 1858–59, and Results compared with Observations made at Point Barrow in 1852–54. (Phil. Trans., 1863.)

Port Kennedy is in lat. 72° 0′ 49″, and long. 94° 19′ W. The Observatory was built on a large hummock of old ice about 220 yards magnetic south from the ship and about 400 yards from land.

Ice was cut into blocks and built to enclose a space 7 feet square. The roof was of loose planks covered and cemented by a mixture of snow and water. A slab of marble placed on slabs of ice in the centre of the room was adjusted and frozen to the ice. The declinometer was mounted and levelled on a tripod table-top on brass grooves, and the table was frozen to the slab and the brass levelling screws to the table.

The declinometer was supported on a point and steadied, and its weight relieved by a silk thread, but it did not work satisfactorily until the magnet was supported *only* by the silk thread which had been employed to relieve its weight.

Within the observatory were only a wooden candlestick, a copper lamp, and a board upon which the observation-paper was fastened by copper tacks.

Accumulations of snow-drift on one side or the other of the observatory occasionally slightly altered the level of the ice and of the instrument.

Auroras were of frequent occurrence.

The extreme easterly disturbances are found to occur approximately at the same *absolute* time at Port Kennedy and at Port Barrow; and the principal maximum of westerly disturbance at the same *local* time at the two stations. The secondary maxima of westerly disturbance do not agree, and are stronger at Port Kennedy. The positions differ about four hours in time, and are nearly in the same latitude, both being about 2° north of the magnetic pole.

The mean value of the dip at Port Kennedy in October and November was 88° 27′·4 N., and hence the horizontal force was small. Notwithstanding this, it appears that the energy of the disturbing force is greater at Point Barrow.

It is also a fact that the aurora is very much more frequent at Point Barrow than at Port Kennedy.

At Port Kennedy the aurora is seen about one day in four. At Point Barrow it was seen about six days in seven, which must be considered in connexion with the fact that 1853–54 were years of minimum disturbance, and 1858–59 were years of maximum disturbance, so that many more might have been expected at Port Kennedy.

Several of the auroras at Port Kennedy were seen over a space of water open throughout the winter towards the S.W.

When the disturbances are eliminated it is found that the solar diurnal variation produces a maximum easterly deflection magnetically at 8 a.m., and a maximum westerly deflection at 2 p.m., at Port Rennedy as at other places of observation.

Before elimination these effects are marked by the greater disturbance deflections at these times in the opposite directions.

At Port Kennedy what is magnetically north is geographically S.W., and at Point Barrow magnetic north is about N.E.

---

11. MAGNETIC OBSERVATIONS made in the Second GERMAN EXPEDITION on the EAST COAST of GREENLAND, in 1869–70, by CAPTAIN KOLDEWEY, 1874.

The instruments used by this expedition were—

(1.) A portable magnetometer.

(2.) An inclination-needle.

(3.) An induction magnetometer by Weber, consisting of a coil which could be rotated through 180° about an axis which could be placed either vertical or horizontal. The galvanometer to be used with this coil was an astatic needle, but this was found to be too delicate, and one of the needles was taken off, and only one needle employed.

(4.) A small needle was employed for observations of horizontal intensity.

By means of the theodolite of the magnetometer the bearing of a small cairn on the highest point of Walross Island was determined, and the azimuth was found to be N. 123° 11′ 10″.

The Declination.

Observations for determining the absolute declination were made generally on two consecutive days about twice a month, during December 1869, January, February, March, and June 1870. In all these observations the declination was nearly 45° W., and the mean value is 45° 6′ 7″ W. Generally it was found on trial that there was no correction required in the declination for the torsion of the thread.

Observations on Changes of Declination on Term Days.

It had been previously arranged with Professors Klinkerfus and Kohlrausch at Göttingen, that on the 21st of the month, for 24 hours, terminal observations of changes of declination should be made.

For the first 10 minutes of every half-hour (Copenhagen time) observations were to be made every 2 minutes, and during the remainder of each hour at every 5 minutes.

From these observations the hourly and daily means are deduced and recorded in the tables.

The observations were made on the term days in December, January, February, and March; also on January 5, February 11, and March 4.

On January 5, at 9 o'clock (Copenhagen time), the aurora was seen, and there was, about 9.30 p.m., very great disturbance of the magnetometer.

On Feb. 11, another of the term days, there was great disturbance of the magnetometer, but there seemed great doubt whether the aurora had any connexion with it. This was eight days after the sun had returned.

[The results are recorded in divisions of the scale employed, and the reading 318 corresponds to the mean declination: on days when there were no remarkable disturbances, the change in the deflection was from 10 to 15 divisions of the scale.

On Jan. 5, at 9 p.m., the reading was 298, and increased to 326 at 9.30 p.m., after which the change was 20 divisions in 5 minutes, and 16 more in the next 2 minutes to 362, then back to 315 in the next 2 minutes, and to 298 in the next 2 minutes.

Again, on Feb. 11, from 7.30 to 8, the changes in declination are from 298·57 to 213·40. For some hours the disturbances in both directions are very great. From 10h. to 13h. the change is from 210 to 360.

Each division of the scale is equivalent to 2′·2, so that this change of deflection amounts to 5° 30′.

On March 21, again, there are great disturbances at 18h. and 19h., but not so sudden nor so great as on the 11th of February; they are chiefly in the opposite direction.]

"The observers were troubled with cloudy weather, which often may have prevented them from seeing the aurora. Still it must be considered that the declination needle was not generally greatly disturbed by the aurora."

Dr. Börgen remarks, with regard to the return of the sun being accompanied by magnetic disturbance, that Dr. Kane's observations also show the greatest disturbance in the magnetic declination on Feb. 16, soon after the return of the sun.

The maximum westerly declination occurs at 6 to 7 p.m., and the minimum declination at 4 to 5 a.m., the needle being in its mean position at 10 or 11 in the morning and in the evening (Copenhagen time). The difference of time between Copenhagen and Sabine Island is 2h. 6m.

In November 1869, as the sun sinks towards the horizon, the movement of the needle was observed to increase, whilst Dec. 21 shows very little disturbance of the needle; the disturbance again increases as the sun returns towards the horizon. The changes which take place in the Polar regions are much greater than corresponding changes which are also observed in lower latitudes.

"It would seem that with the approach of the observer to the magnetic pole the disturbances are increased."

[As Dr. Börgen observes also:—"This is scarcely borne out, "at any rate not to the same extent, by Dr. Kane's observa- "tions." Nor does it seem to be borne out by the observations of English expeditions at Point Barrow, and at points nearer the magnetic pole; nor again by the results of the Austro-Hungarian Expedition.]

The Dip.

The magnetic dip was determined on Aug. 6, 1869, by the dipping needle, and on Aug. 7 by the induction magnetometer with the astatic galvanometer.

This galvanometer was found to be too delicate, and only one needle was afterwards used for the observations which were made on Jan. 7, 11, 15, 16, and June 25, 26, 27, 28.

The mean value of the dip from these determinations is 79° 42′ 4″.

Hourly observations for changes of dip were also made with the dipping needle at 4 minutes past the hour on all term days.

On Feb. 11, from 6h. to 12h., there was found to be first a great increase, and then a great diminution in the dip, accompanied by great oscillations. The change of dip in 2 hours amounted to about 2°.

On March 21, especially at 18h. and 19h., there is great diminution in the dip.

[On comparing the tables it seems that magnetic disturbances which cause an increase in the reading of the scale of the magnetometer also cause a diminution in the dip, and those which diminish the reading of the scale increase the dip.

This is true of most of the simultaneous disturbances observed on the term days, February 11 and March 21.

Since an increase in the reading on the scale of the magnetometer means a diminution in the westerly declination, we see, by comparing the observations for declination and dip, that a disturbance which causes an *increase in the westerly declination* also causes an *increase in the dip.*]

---

12. Method of Observation with the Induction Magnetometer.

A circular coil of copper wire, to the ends of which the galvanometer wires are attached, is capable of rotating through 180° about an axis which may be placed in a horizontal or in a vertical position in the magnetic meridian. When the coil is rotated at a given rate, it cuts the lines of terrestrial magnetic force, and a current is induced in the wire, the strength of the current being proportional to the number of lines of force cut during the half revolution. The number of lines of force cut by the revolving coil will depend on the angle between their direction (*i.e.*, the direction of the dip) and the axis of the coil, being proportional to the sine of half the angle.

If φ and φ_1 be the deflections of the galvanometer when the axis is horizontal and vertical respectively, and i the dip, then

$$\tan i = \frac{\sin \frac{\varphi}{2}}{\sin \frac{\varphi_1}{2}}.$$

If x be the observed reading on a reflecting galvanometer, when the scale is at a distance r from the centre, then

$$\tan 2\,\varphi = \frac{x}{r}.$$

If the angle φ is small, then

$$2\,\varphi = \frac{x}{r} - \frac{1}{3}\cdot\frac{x^3}{r^3}$$

$$\text{and } \frac{\varphi}{2} = \frac{x}{4r} - \frac{1}{12}\cdot\frac{x^3}{r^3}$$

$$\text{So that } \sin\frac{\varphi}{2} = \frac{1}{4r}\left(x - \frac{11}{32}\cdot\frac{x^3}{r^2}\right)$$

$$\text{hence } \tan i = \frac{x - \frac{11}{32}\cdot\frac{x^3}{r^2}}{x_1 - \frac{11}{32}\cdot\frac{x_1{}^3}{r^2}}$$

The correction $-\frac{11}{32}\cdot\frac{x^3}{r^2}$ must be applied to the observed reading of the scale.

The needle being at rest, the coil was turned through 180°, causing a deflection of galvanometer, at the instant when the galvanometer needle was swinging back through its point of rest, the coil was turned back again, thus increasing the swing, and at each instant when the needle is passing its point of rest, the coil is turned through 180°, each time increasing the swing of the needle until the increase in the swing becomes very small.

Taking the mean of all these successive corrected deflections of the needle on both sides for the horizontal and vertical positions of the axis, their ratio will give the tangent of the dip.

Horizontal Intensity.

Observations of horizontal intensity were made in the usual way, by measuring the moment of inertia of a magnet, its time of vibration, and the amount of deflection which it produces in a small needle placed at given distances from it.

Observations to determine the values of the magnetic elements were also made at various points of the coast with the several instruments, and the results are contained in the following table :—

Place.	Time.	Lat.	Long.	Declination.	Inclination.	Hor. Force.
		° ′	° ′	° ′	° ′	
	20 Apr. 1870	76 23	19 36	45 28		
	7 „ „	76 18	19 41	45 30		
	14 Aug. 1869	75 30·5	17 40	45 24		
Cape Bergen - -	28 July 1870	75 26	18 1	—	79 50·1	0·9608
— Bremen - - -	30 May „	74 58·4	19 57	45 48·2	79 57	0·9830
Phil. Broke - -	—	74 55·8	17 40	45 25·5	78 50·6	1·1325
Klein Pendulum - -	29 Aug. 1869	74 37·6	18 29	43 3·6	80 10·4	1·0614
Sabine Island. - -	—	74 32·3	18 49	45 6·7	79 42·4	1·0489
Cape Broer Ruys -	4 Aug. 1870	73 27·7	20 43	43 52·5	79 54	0·9832
Franz Jos. Fiord -	12 „ „	73 11·6	25 58	50 30·2	79 59·7	0·9587

The Inclination observed at Sabine Island, in East Greenland, in 1823, by Sabine was 80° 11′, and the Total Intensity observed by Sabine 11·54 in absolute British units.

The values obtained by Koldewey in 1869 are, Inclination 79° 42′, and the Total Intensity 5·8690 Göttingen units, that is, 11·585 British units.

In most of the expeditions to the Arctic regions the unifilar magnetometer has been employed to determine the declination, and the changes of declination, and also the earth's horizontal magnetic force. Let F stand for the magnet which is employed to determine the earth's horizontal force. For this purpose it is necessary to determine the time of oscillation of the magnet F, employed with its usual mounting, and to determine it again when loaded with a ring or weight of known moment of inertia.

If t and t_1 be the times of vibration of the magnet when free and when loaded respectively, and I, I_1, be the moments of inertia of the magnet and the weight added to it, then the observation of the times t and t_1 will give the moment of inertia of the magnet by the formula

$$I = I_1 \frac{t^2}{t_1^2 - t^2}.$$

If M be the magnetic moment of the magnet F, and X the horizontal force of the earth, then the observation of the time of vibration t will give the products MX, by the formula

$$MX = \frac{\pi^2 I}{t^2}.$$

The ratio $\frac{M}{X}$ is given by the deflections produced by the magnet F when employed to produce deflections in another magnetic needle.

The formula is $\frac{M}{X} = \frac{\frac{1}{2} r^3 \sin u}{1 + \frac{p}{r^2} + \frac{q}{r^4}}$ where u is the deflection of the suspended magnet, and r the distance between their centres.

To determine the values p and q it is necessary to take observations for deflection at *three* different distances. (For details as to the method see "Lloyd's Magnetism," p. 91, and "Admiralty "Manual of Scientific Inquiry," p. 96.)

A less approximate value of $\frac{M}{X}$ is obtained by neglecting the fraction $\frac{q}{r^4}$ in the above expression, in which case

$$\frac{M}{X} = \frac{\frac{1}{2} r^3 \sin u}{1 + \frac{p}{r^2}}$$

$$\text{or } \frac{M}{X} = \tfrac{1}{2} r^3 \sin u \left(1 - \frac{p}{r^2}\right), \text{ where } \frac{p}{r^2} \text{ is small.}$$

For this formula it is only necessary to take deflections at *two* different distances; but the results obtained will not be accurate unless $\frac{q}{r^4}$ is very small.

If r and r_1 be the two distances between the centres of the magnets, and u, u_1, the deflections, then combining the two formulæ we get

$$-\frac{p}{r^2}=\frac{r_1^5 \sin u_1 - r_1^2 r^3 \sin u}{r_1^5 \sin u_1 - r^5 \sin u}$$

$$\frac{\frac{\sin u_1}{\sin u}-\left(\frac{r}{r_1}\right)^3}{\frac{\sin u_1}{\sin u}-\left(\frac{r}{r_1}\right)^5}=\frac{\sigma-\rho^3}{\sigma-\rho^5}$$

$$\text{so that } \frac{M}{X}=\tfrac{1}{2} r^3 \sin u \frac{\rho^3-\rho^5}{\sigma-\rho^5}$$

$$\text{where } \rho=\frac{r}{r_1} \text{ and } \sigma=\frac{\sin u_1}{\sin u}.$$

These were the formulæ employed in discussing the observations of Dr. Kane and Dr. Hayes in the Arctic regions; and the values of M and X determined by Dr. Kane in 1854–55 at three stations, and by Dr. Hayes in 1860–61 at six stations, are given on p. 699.

[The great differences in the value of $\frac{p}{r^2}$, obtained by Dr. Hayes at Port Foulke in the course of one week's observations in July, differences which range from $+0{\cdot}0044$ to $-0{\cdot}0851$, would seem to show that if this method is employed to determine the Earth's magnetic force, it will be necessary to take the more exact formula, and to determine the deflections for *three* different distances between the centres of the magnets.

In the neighbourhood of the magnetic pole the results obtained by this method are liable to large errors because of the smallness of the Earth's horizontal force, and hence one of the methods with the dipping needle should be adopted. The early Arctic voyagers, Sir John Ross, Sir James Ross, Sabine, and Parry, all made use of a dipping needle to determine the Earth's force as compared with its value in London, determined by the same needle before starting and after returning.]

[For methods of using the dipping needle see Admiralty Manual of Scientific Inquiry, p. 103 to 106.]

The absolute value of the Earth's magnetic force may be determined by observing the inclination of the dipping needle, 1st, under the action of the Earth's magnetism alone; and 2nd, under the combined action of magnetism and gravity, by attaching a weight to the needle as in Fox's dip circle. (See Lloyd's Magnetism, p. 95).

If F stand for the dipping needle employed to determine the Earth's force; if θ be the inclination of the needle F without the weight, η the inclination of the needle F with the weight, and η_1 the inclination of another needle which is deflected by the needle F, and r the distance between their centres, R the Earth's force, W the weight on the axis of radius a, then

$$R^2 \sin(\theta-\eta)\sin(\theta-\eta_1)=UW a$$

$$\text{where } U=\frac{2}{r^3}\left(1+\frac{p}{r^2}+\frac{q}{r^4}\right).$$

U depends on the distance between the centres of the two needles, and its value will be determined by using the needle F as a deflector at *three* different distances.

Dr. Lloyd thus sums up the advantages of this method (p. 100):—

"1. It is applicable with equal accuracy at all parts of the globe.

"2. It dispenses with the employment of a separate instrument for the determination of the magnetic intensity, and with the separate adjustments required in placing it.

"3. The constants to be determined, *i.e.*, the magnitude of the added weight, and the radius of the pulley by which it acts, can be ascertained with more ease and certainty than those with which we have to deal in the method of vibrations, and are less liable to subsequent change.

"4. The observations themselves are less varied in character than the usual ones, and may be completed in a shorter time."

---

13. Magnetic Observations, made during the Austro-Hungarian Expedition, by Lieut. Weyprecht.

(From Petermann's "Geograph. Mittheilungen," Jan. 1875, and "Nature," March 18, 1875.)

"In close connexion with the Aurora are the Magnetic disturbances; while these are in our regions the exception, in northern latitudes they are the normal condition; there the needle scarcely ever remains still at rest. This is the case with Declination as well as Intensity and Inclination needles. As long as the ship drifted, which was till the October of the second year, it was, of course, impossible to put up the fixed Variation instruments. We did, however, often carry out absolute observations with Lamont's magnetic theodolite, but it was evident already in Nova Zembla, that, in consequence of continuous disturbances, all these observations, without the contemporary reading of the Variation instruments were of very little worth.

"In November 1873, as soon as it was certain that we were anchored for the winter, I had some snow huts built close together, in one of which the Variation instruments, in the other the magnetic theodolite and the Inclinometer for the absolute observations, as well as the Astronomical instruments were put up. The three Variation instruments for declination, horizontal intensity, and inclination were supplied by Dr. Lamont, the Director of the Observatory in Munich, on the pattern of those instruments which are in use there.

"After the first series of observations, it was evident that the earlier methods of observation, *i.e.*, simple readings at certain hours, at least in these regions, were worthless, as they depended merely on the accidental amount of the momentary disturbances. They give neither a true mean, neither do they give a picture of the movement of the needles. In former expeditions the recorded observations are much too far apart to render it possible to draw correct conclusions on the magnetic conditions.

"Under these circumstances I took quite a different course. Every third day, at an interval of four hours, I caused minute readings for an hour to be taken at all three instruments, and in such a way that different hours were chosen in each day. Besides this, we observed in order to obtain an insight into the whole daily proceedings, twice in the month, at interval of 5 minutes through 24 hours. To obtain the observations as nearly as possible at the same exact time, the three instruments whose telescopes were all fixed on the same axis, were read off one after another as quickly as possible (on an average within 8 to 10 seconds). These observations were continued from the middle of January till the end of April, in all 32 observation-days, and I think that put together they will give a true picture of the continuous changes as regards the direction and intensity of the Magnetic forces in the Arctic regions."

"In order to confirm the connexion between the auroræ and the action of the needles, a second observer, independently of the others, observed the changes and motion of the auroræ. Absolute determinations of the three constants were made as often as circumstances permitted, to control the Variation instruments.

"Apart from the Swedish Expedition, whose observations are not yet published, Lieut. Weyprecht points out that his are the first regular and simultaneous observations that were ever made in the Arctic districts. Moreover, he thinks that all former observations were made with the ordinary heavy needles, and that he was the first to use the light Lamont needles. For observations, however, under such conditions as the normal ones near the pole prove to be, heavy needles are perfectly useless; even the comparatively light intensity needle of Lamont's theodolite oscillated so violently, on account of its unproportionally great moment of inertia, and even with moderate disturbances, that the readings became quite illusory. Almost on each magnetic day some disturbances were so great that the image of the scale could no longer be brought into the field of the telescopes on account of deflection; in order to ascertain also these greatest effects, Lieut. Weyprecht constructed an apparatus by which he could measure them at least approximately. He owns that as a matter of course his observations could not possibly be as perfect as those made at home, but thinks that it will be easy to modify Lamont's instruments on the basis of his experience, so that on a future expedition, where there is a greater staff of observers, results could be obtained of any desired exactness. Altogether Lieut. Weyprecht's party of observers, consisting besides himself only of Lieut. Brosch and Ensign Orel, have taken about 30,000 readings from their different magnetic instruments, and the principal results are the following:—

"The magnetic disturbances in the district visited are of extraordinary frequency and magnitude. They are closely connected with the aurora borealis: the quicker and more fitful the motion of the rays of the aurora, and the more intense the prismatic colours, the greater are the magnetic disturbances. Quiet and regular arcs, without changing rays or streamers, exercise almost no influence upon the needles. With all disturbances the declination needle moved towards the east, and the horizontal intensity decreased, while the inclination increased. Movements in an opposite direction, which were very rare, can only be looked upon as movements of reaction.

"The instrument upon which Lieut. Weyprecht placed the greatest expectations, namely, the Earth-current galvanometer, gave no results at all, through the peculiar circumstances in which the explorers were placed. He had expected to be able to connect the auroræ with the galvanic Earth currents. But as the ship was lying two-and-a-half German miles from land, he could not put the collecting plates into the ground, but was obliged to bury them in the ice. Now, as ice is no conductor, the plates were insulated, and the galvanometer needle was but little affected. Prof. Lamont had supplied these excellent instruments also; the conducting wires were 400 feet long. Later on, Lieut. Weyprecht tried to obtain some results by connecting a collector for atmospheric electricity with the multiplier of the galvanometer, but failed, doubtless for the same reason."

---

14. Magnetic Survey of the North Polar Regions.

In the Philosophical Transactions for the year 1872, Part II., is contained a "Contribution" on this subject from Sir Edward Sabine. It contains the Magnetic Survey of the region from 40° N. to the most northern limits for the three magnetic elements, for the epoch from 1842 to 1845. The observations are arranged in eight zones of successive 5° of latitude.

The observations for Magnetic Force to which, prior to Gauss' improvement, have always been expressed according to some arbitrary scale having reference to the force at a *base* station, have in this paper been referred to *absolute* British units. The arbitrary value which used commonly to be adopted for London was 1·372, and for this the value in absolute measure at Kew, viz., 10·28, has been substituted. From 1830 to 1869 the value of Kew has been gradually increasing from 10·27 to 10·31 units.

In the American Polar regions there is more than ordinary difficulty in respect to secular change especially of declination; but the facts observed seem to point to a probable reversal in the direction of the secular change at some interval between 1818 and 1860, such as has been proved to take place about 1842 at York Fort by the observations of Franklin, Lefroy, and Blakiston. These changes may be connected with the easterly progression in north-eastern Asia, showing the approach of the present Asiatic point of maximum force to the American continent.

Future researches alone can clear up these difficulties.

Since the early part of the 17th century the dip on the coasts of Norway and in the Spitzbergen Sea has greatly diminished; in some cases the recorded observation gave 80° and even 86° in localities where the inclination is now 10° less.

These valuable tables drawn up by Sir Edward Sabine, and the lines of equal declination, inclination, and magnetic intensity given in the three maps, supply all the information with regard to the absolute values of the magnetic elements in the Polar regions which had been obtained previous to the German and Swedish Expeditions.

Sir Edward Sabine has shown that the theory of Halley, which was put forth in 1683, is the theory which is supported by recent observations, viz., that "the globe of the Earth may be regarded as "one great magnet, having four magnetic poles, two of them near "each pole of the equator, and that in those parts of the world "which lie near any of those magnetic poles the needle is chiefly "governed thereby, the nearest pole being always predominant "over the more remote."

The work of Hansteen, "Magnetismus der Erde," published in 1819, of which an abstract was published in British Association Report for 1835 by Sir Edward Sabine, and this work of Sabine just published complete our knowledge of the magnetic conditions of the northern regions up to the present time.

The results which have been obtained are collected at the end of the paper in groups for convenient comparison with the phenomena which may be observed at future periods.

VIII.—THE AURORA BOREALIS.

---

1. Observations of Aurora at Melville Island. (Parry's First Voyage, &c.)

A brilliant display of Aurora seen on Jan. 15 is described by Captain Sabine. It began with an arch nearly north and south, a little east of the zenith. "Towards the southern horizon was "the ordinary Aurora, giving a pale light, apparently issuing "from behind an obscure cloud, at from six to twelve degrees of "altitude."

"The luminous arch broke into irregular masses, streaming "in different directions, varying in shape and intensity, and "always to the east of the zenith, and was most vivid to the "E.S.E."

"The various masses seemed to arrange themselves in two "arches, one passing near the zenith, and a second midway "between the zenith and the horizon both north and south, but "*curving towards* each other." "At one time a part of the arch "near the zenith was bent into convolutions like a snake in "motion, and undulating rapidly." The light was estimated as equal to that of the moon when a week old. Besides the pale light, which resembled the combustion of phosphorus, a very slight tinge of *red* was noticed when the Aurora was most vivid, but no other colours were visible.

On the next day it blew a fresh gale from N.N.W.

The Aurora had the appearance of being very near us, but no sound could be heard. The Aurora was repeatedly seen on the following day, assuming the shape of a long low arch, from 3° to 12° high in the centre, extending from south to north-west The temperature at this time was −36° F.

On Feb. 18 an Aurora is described as "of a pale yellow, at " other times white, excepting to the southward, in which direc- " tion a dull red tinge was now and then perceptible."

The fresh gale which blew at the time from the N.N.E. appeared to have no effect on the Aurora, which streamed directly to windward, and this with great velocity.

---

2. Observations of Aurora. By Sir John Franklin.

Observations were made in Sir John Franklin's Expeditions for magnetic declination and the daily changes of declination, and for the dip.

Sir John Franklin sums up the results of his experiences of the Aurora in the following general conclusions :—

1. Brilliant and active coruscations of the Aurora Borealis cause a deflection of the needle almost invariably if they appear through a hazy atmosphere, and if the prismatic colours are exhibited in the beams or arches. When, on the contrary, the atmosphere is clear, and the Aurora presents a steady dense light of a yellow colour, and without motion, the needle is often unaffected by its appearance.

2. The Aurora is generally most active when it seems to have emerged from a cloud near the earth.

3. When the Aurora is very active, a haziness is very generally perceptible about the coruscations, though the other part of the sky may be free from haze or cloud.

4. The nearest end of the needle is drawn towards the point from whence the motion of the Aurora proceeds, and that its deflections are greatest when the motion is most rapid; the effect being the same whether the motion flows along a low arch or one that crosses the zenith.

5. That a low state of temperature seems favourable for the production of brilliant and active coruscations; it being seldom that we witnessed any that were much agitated, or that the prismatic tints were very apparent when the temperature was above zero.

6. The Aurora was registered at Bear Lake 343 times without any sound having been heard to attend its motions.

7. The gold-leaf electrometer was never affected by any appearance of the Aurora.

8. On four occasions the coruscations of the Aurora were seen very distinctly before the daylight had disappeared, and we often perceived the clouds in the day-time disposed in streams and arches such as the Aurora assumes.

A brilliant Aurora was seen by Dr. Richardson on April 23, while Lieut. Kendall, who was watching at the time, by agreement, at a distance of 20 miles off, did not see any coruscation.

[In his observations of the two kinds of Aurora, Sir John Franklin agrees with M. Angström, who finds by the spectroscope that the Auroræ are of two kinds.

The results do not seem to agree with those arrived at by Parry at Port Bowen, that the Aurora does not influence the magnetic needle. At the same time, it must be remembered that the Auroræ at Port Bowen never exhibited the vivid prismatic colours or rapid streams of light, so that Parry only saw that kind of Aurora which, according to Franklin, does not affect the magnetic needle.]

---

3. Parry's Third Voyage.—Winter at Port Bowen.

The Aurora was observed and recorded 47 times. It usually consisted of an arch sometimes continuous, but more frequently broken into irregular masses of light extending from W. to S.E. (true). Its termination to the S.E. was not visible, as land intervened. The altitude of a permanent arch seldom exceeded 15°, and from this arch streamers were generally observed shooting towards the zenith. "The lower edge of the arch was generally well "defined and unbroken, and the sky beneath it appeared by con- "trast so exactly like a dark cloud (to me often of a brownish "colour), that nothing at the time of viewing it could well convince "one to the contrary, if the stars shining there with undiminished "lustre did not discover the deception."

Note.—This description closely resembles the account of the Aurora as seen by M. Koldewey in the Second German Polar Expedition in East Greenland. (*See* p. 719.)

A few of the more important Auroras are described by Parry (pp. 60, 61), one or two seen just over the land to the S.E. or S. appeared as a single compact mass of brillant yellow light constantly varying in intensity, as of numerous streamer-like clouds overlaying one another. He says (p. 62), "While Lieutenants "Sherer and Ross and myself were admiring the extreme "beauty of this phenomenon from the Observatory, we all simul- "taneously uttered an exclamation of surprise at seeing a bright "ray of the Aurora shoot suddenly downward from the general "mass of light, *and between us and the land*, which was there "distant only 3,000 yards. Had I witnessed this phenomenon "by myself, I should have been disposed to receive with caution "the evidence even of my own senses as to this last fact; but "the appearance conveying precisely the same idea to three "individuals at once, all intently engaged in looking towards "the spot, I have no doubt that the ray of light actually passed "within that distance of us."

On several occasions during Auroras the gold leaf electroscope was applied to a chain attached by glass rods to the sky-sail mast head, with the pointed end of the last link considerably above the mast head and 115 feet above the level of the sea, but not the slightest perceptible effect was observed.

The variation needles, which were extremely light, suspended in the most delicate manner, and subject to weak directive energy, were never in a single instance sensibly affected by the Aurora, although the needles were visited every hour for some months.

From the 8th to the 14th of December, but especially on the 12th of December about 11 p.m., several meteors were seen passing from near β Tauri towards the Pleiades. Sudden changes of wind occurred about the times of these phenomena on the 12th and 14th of December, and Parry says, "There appeared to be an evident "coincidence between the occurrence of the meteors and the "changes of the weather at the time."

"Feb. 22, 2 a.m.—Aurora appeared in an arch from north to west by compass, with bright streamers towards the zenith; the needle was not affected in any way whatever."

---

4. Ross's Observations of Aurora.

(*Second Arctic Voyage.*)

P. 223. Nov. 24, 1829.—Brilliant Aurora to the S.W., extending its *red* radiance as far as the zenith. On the following evening there was a still more brilliant one increasing in splendour until midnight. "It constituted a bright arch, the extremities of which "rested on two opposite hills, while its colour was that of the "full moon, and itself seemed not less luminous," with the dark and somewhat blue sky behind it.

On Dec. 3rd, magnificent arch of an Aurora. The colour was a light yellow, and it emitted rays; day calm and sky clear, with a cloudy horizon.

Dec. 17th.—Another beautiful Aurora obscured by clouds.

Dec. 20th.—Brilliant Aurora with bright flashes

The observations were made in Felix Harbour, lat. 69° 59′ N., long. 92° W., where the declination was 89° 45′ W., and dip, 89° 55′.

At the magnetic pole, *i.e.*, where all the declination needles remain in any position in which they are placed, the dipping needle showed an inclination of 89° 59′, in latitude 70° 5′ 17″, longitude 96° 46′ 45″ W.

In his account of his Antarctic voyage he says: "Whilst our "ships lay rolling amidst the foam and spray to windward of the "berg, a beautiful phenomenon presented itself worthy of notice, "as tending to afford some information on the causes of the "exhibition of Auroral light. The unfrequency of the appearance "of this meteor during the present season rather surprised us; "and therefore, to observe its bright light forming a range of "vertical beams along the top of the icy cliff, marking and par-"taking of all the irregularities of its figure, was the more "remarkable."

Captain Ross suggests that this singular appearance was produced by electrical action taking place between the vaporous mist thrown upwards by the dashing of the waves against the berg, and the colder atmosphere with which this latter was surrounded.

From the observations of Captain Back, who observed a large number of Auroras in 1833, and also from the observations of MM. Lottin and Bravais at Bossekop, in Lapland (where the appearances were very frequent), as well as from those of many other observers, it appears that the degree of disturbance of the needle varies with the intensity of the Aurora, and that when the Aurora is faint and diffused or low down in the horizon the magnetic needle is not disturbed at all, but on the appearance of beams and brilliant and coloured streamers the needle is disturbed for some minutes, and oscillates through several degrees.

It has also been noticed by several observers that a grand display of Aurora is often preceded or accompanied by an extraordinary motion of the needle to the westward. This has been noticed at places which are far distant from one another, for instance, by Sir Edward Belcher in Wellington Channel, by Captain Maguire at Point Barrow, and by MM. Lottin and Bravais at Bossekop.

---

5. Observations on Aurora. (McClintock.)

In preface, p. xi., to the discussion of McClintock's Observations in Smithsonian Contrib., vol. 13, it is stated that McClintock says the beams of the Aurora were most frequently seen in the direction of open water, or else in that of places where vapour is rising. In some cases patches of light could be plainly seen a few feet above a small mass of vapour over an opening in the ice.

It is also stated by Dr. Walker that the Aurora was seen by him on more than one occasion at Port Kennedy between himself and the land about three miles off towards the W. or S.W.

This observation is in accordance with a deduction from an examination of a large number of notices of the Aurora in the voyages of Peter Force, published in Smith. Contrib., vol. 8, viz.: —"That on the Atlantic Ocean and other open water the Aurora "is most frequent and most brilliant."

Dec. 17, 18. Bright Aurora S. to N.E., lasting through the night, followed by Aurora during the next night.

Jan. 9th, 8 p.m. Bright Aurora from West to East (magnetic).

Jan. 28, Dawn at 8.25 a.m. Sun's upper limb appeared at $11^h\ 25^m$, disappeared at $1^h\ 0^m$, dusk at $3^h\ 45^m$, showing the length of twilight.

Aurora in the form of arches, sometimes double and treble, patches and streamers seen frequently in February and March, especially on Feb. 2nd, March 2nd and 6th.

Feb. 2, 10 p.m. Auroral arch in S.E., visible for one hour, faint from S.E. to E.N.E., the extremities of the arch touching [meeting ?] the horizon, the S.E. extremity was brightest, with an occasional stream towards the zenith.

The Aurora seems often to have appeared in the S.E. and E. as seen from Baffin's Bay, but sometimes also towards the S.W.

Prismatic halos and parhelia also several times seen.

March 7. Double prismatic halo (red external) about the sun, diameters 45° and 90°; occasional inner halo of same altitude as the sun (16°), and a portion of an inverted arch above the outer halo.

On six occasions of Aurora in Baffin's Bay the electroscope was strongly affected, and on three occasions of Aurora at Port Kennedy, when the Aurora extended from the horizon to the zenith, the electroscope was strongly affected; on all these occasions the electricity was positive.

McClintock's Remarks on the Aurora, p. 79.

On Dec. 18 the Aurora was visible in the morning until eclipsed by the day-dawn at 10 o'clock; afterwards very thin clouds occupied its place, through which, as through the Aurora, stars appeared scarcely dimmed in lustre. He adds: "I do not "imagine that Aurora is ever visible in a perfectly clear atmo-"sphere. I often observe it just silvering or rendering luminous "the upper edge of low fog or cloud banks, and with a few "vertical rays feebly vibrating."

"Dr. Walker called me to witness his success with the electro-"meter. The electric charge was so very weak that the gold "leaves diverged at regular intervals of four or five seconds. "Some hours afterwards it was strong enough to *keep* them "diverged." The temperature ranged from −12° to −25° F.

[These observations entirely agree with the observations of M. Lemström on the First Swedish Expedition in 1868, and M. Wijkander on the Second Swedish Expedition in 1872–73.]

Dr. Walker's Account of the Aurora.—(Appendix to McClintock's Voyage of the "Fox.")

During our drift down Baffin's Bay (1857–58) the Aurora was noticed on 43 nights; of these, 18 were observed in a direction where water or water sky had been seen during the day. The general direction of the remainder was between N.E. and S.E. At times pulsations were noticed in the patches and bands of light; these were often contrary to the surface wind. Once only was there noticed a connexion between cirrus clouds and the Aurora. Of the 42 Auroras observed at Port Kennedy during 1858–59, 24 were in the direction of a space of open water or of the vapour rising from it.

On five occasions the Aurora was observed to cause an agitation of the magnetic needle; once there was a deflection of 15°. Four of these were when the Aurora was from south to north, passing through the zenith.

A fine wire was attached to the foreyard-arm by insulated supports, and led to a snow house, with a connexion through the floe to the water beneath. Here I was enabled to observe the presence of free electricity in the atmosphere and the influence of the Aurora on the instrument. There appeared to occur two periods of minimum electrical disturbance, about 9 p.m. and noon.

OBSERVATION of AURORA. By W. R. GROVE.

Grove in his Correlation of Physical Forces (p. 448) says:— "I remember about seven or eight years ago seeing an Aurora at "Chester when the flashes appeared close to the observer, so that

" gleams of light continuous with the streamers could be seen " *between the houses of the town and myself* like the portions of " a rainbow intervening between terrestrial objects and the ob- " server. I tried to ascertain if there was any reflection or other " cause of optical illusion, but could not see it as other than a real " effect; I seemed to be in the Aurora."

---

6. The Aurora, Sir Edward Belcher..

December 2nd.—An aurora was seen at 9.20 p.m. A light narrow streak through the zenith, S. by E. true. Four cumulus-shaped masses over the mast heads. Electrometer was not affected.

From 4 p.m. to midnight there was great westerly disturbance of magnetometer amounting to 137·80°, equal to 27·60° of deflection. At 8 p.m. it was 117·30°; at 9. p.m. 116·50°; at 10 p.m. 120·60°.

The barometer, during the interval between 8 p.m. and midnight, suddenly changed from 29·86 to 29·65, rising again to 29·90, when the magnetometer showed 107·90° at 4 a.m.

On the 5th, 6th, 9th, and 10th of December there were further exhibitions of aurora, preceded or followed by slight disturbances of the magnetometer.

On the 12th, at 3 a.m., the aurora was reported as very brilliant. The magnetometer exhibited the most unmistakeable signs of disturbance, moving instantaneously from 114° to 128°, and up to 150°, returning at 4 a.m. to 117·90°.

About the 14th and 15th of February 1853 the weather underwent unexpected changes, the temperature rising as high as 2·5° above zero. This was a day or two before the return of the sun, and the magnetometer exhibited sudden and incomprehensible disturbances.

---

7. Dr. Hayes' Observations of Aurora.

Auroras.

In the winter of 1860–61 (*i.e.*, when the 10 or 11 year inequality was at its maximum), only three auroras were seen and recorded, and they were feeble and short.

Jan. 6, 11 a.m. Red aurora seen in the north from the horizon to the zenith; lasted about 15 minutes. Aurora seen again in the evening.

Jan. 11. Heavy mist on the ice; 3 p.m., aurora to the west; extended to the zenith; lasted 10 minutes.

Feb. 16. An aurora at 9 p.m. in the west; lasted 10 minutes; 25° to 30° high.

Last two auroras seen in direction of north end of magnetic needle, over an area of open water present throughout the winter.

---

8. Observations of Aurora, by Captain Maguire. Point Barrow, 1852–54.

The aurora was found to be connected with the movements of the declination needle; the brighter the aurora, the quicker the magnetic changes became, and from repeated observations it appeared that "the appearance of the aurora in the *south* was " connected with the motion of the magnet to the *east* of the " magnetic north, and if in the *north* to the *west* of the same."

In addition to these disturbances, considerable irregularities took place in the daytime, generally in the forenoon, and always in cloudy and misty weather. It was no uncommon occurrence for the magnet to go out of the field of the telescope, not returning again for several minutes, and it was generally to the eastward on these occasions.

Of 40 instances in which the deflection was beyond the scale, 30 were deflections to the east, and 22 of these were between 7 and 9 a.m., *i.e.*, the time of the maximum easterly deflection; and six of the westerly disturbances were between 11 p.m. and 1 am., the time of the greatest westerly deflection.

The aurora was seen six days out of seven, and at the time of the hourly observations on 1,079 occasions, *i.e.*, nearly one-third of the whole number of observations.

The aurora was very seldom seen between 9 a.m. and 5 p.m., but increased from 5 p.m. rapidly and pretty regularly until 1 a.m., then diminished in the same way until 9 a.m.

Not a single display of aurora occurred between 10 a.m. and 4 p.m. Thus there is a close resemblance between the display of aurora and the *westerly* disturbances at Point Barrow.

---

9. Observations of Aurora made in the Second German Expedition.

The aurora was of very frequent appearance, almost every day, yet we have not succeeded well on account of cloudy weather.

The aurora was always first seen in the south-east, *i.e.*, in the direction of the magnetic meridian.

After their first appearance remarkable changes took place in the bands of light, which always stretched from west to east, and often reached from the horizon to the zenith. The higher the bands of light or streamers are above the horizon, the wider they are.

The appearance extends by degrees over the northern half of the sky, which at first was clear of streamers, and the beams now reach their greatest brilliancy and are seen to converge toward the magnetic pole, *i.e.*, in direction of dipping needle. The phenomenon then fades away, first becoming paler in the west and south.

The arc in the south-east remains to the last, and sometimes another display follows from it during the night.

No greater darkness than usual was ever observed before a display of the aurora, such as has been observed in lower latitudes for hours before a display.

The dark appearance between the display was absolutely an effect of contrast.

A spectroscope showed only one clear line at the limit of yellow and green—the same line which has been so often seen.

Dr. Börgen says, "We have never seen a trace of the weak "lines in blue and red, which were observed so distinctly with "the same spectroscope on October 25, 1870, after the return "of the Expedition."

In accordance with this we have never seen the aurora of any other colour than that of greenish yellow.

Frequently during the aurora, observations were made to determine whether there were any striking disturbances of the declination accompanying the display. Only once, during the term day of the 5th January, at same time with aurora, there was an important disturbance, and another term day showed an extraordinary disturbance of the needle, but there was great doubt whether the aurora was the cause of it.

Exact and extended observations on the simultaneous disturbances of all three elements of the earth's magnetism within the Polar regions as well as in mean latitudes would greatly add to our knowledge of the aurora.

---

10. Observations of Aurora, made during the Austro-Hungarian Expedition, by Lieut. Weyprecht.

(From Petermann's Geograph. Mittheilungen, Jan. 1875.)

"The most beautiful and interesting phenomenon in these parts, the only change in the solitude of the long winter night, is the Aurora Borealis; no pen or brush can describe the glory and beauty of this phenomenon in its greatest intensity. In February 1874 we had an Aurora that stretched like a grand broad stream of fire over the zenith from west to east, and with the velocity of lightning waves of light of considerable intensity in prismatic colours darted from one side of the horizon to the other. At the same time it lightened and flashed from the southern horizon up to the magnetic pole; it was the most magnificent firework display that nature has ever here presented to our sight.

"I shall later give proofs, with data, that the intensity of the Aurora phenomena differs in the different parts of the Arctic regions independently of the geographical latitude, and that the region in which we were is a region of greatest intensity. In a clear sky one could almost uninterruptedly see at least traces of it. In the second winter I caused the officers who were making meteorological observations to keep an 'Aurora journal;' this, however, produced very little positive information, and was therefore discontinued.

"The phenomenon defies all description and any systematic classification, it shows constantly new forms and changes every instant. In spite of persevering and most eager endeavour, I have never succeeded in describing the formation of the Aurora; the phenomenon is there, but how or whence it comes it is impossible to say.

"Speaking quite generally, three forms are distinguishable: quiet regular bows, which rise from the southern horizon, and moving slowly over the zenith pale at the northern horizon; then bands of light, which, twisted in themselves, change place and form continuously, and either consist of decided rays or else of light matter; lastly, the phenomenon of the corona, *i.e.*, the direction of the rays from or towards the magnetic pole. Generally the colour is a powerful white with a greenish tinge; when the motion and intensity are greater, particularly when the flashing motion of the rays takes place, the prismatic colours often appear in considerable intensity.

"I spent much time and trouble on the spectral observations of the Aurora, but the spectroscope we took with us was much too small. I was never able to observe more than the well-known green line; in comparison with the spectral observations of the Swedish Expedition with more perfect instruments, ours are worthless.

"Most of us made the observation that the Aurora was connected with the state of the weather; storms mostly follow strong, and particularly the flaming Aurora. However the comparison of the meteorological data alone, can decide whether this view is a right one. I myself have come to the conviction, without being able to give any positive weighty reasons, but from the observation of hundreds of Auroræ, that the Aurora is an atmospheric phenomenon, combined with the meteorological conditions."

---

11. Comparison of Auroras, Magnetic Disturbances, and Sun Spots.—(Prof. Loomis.) (From the American Journal of Science, vol. v., April 1873.)

A comparison between the mean daily range of the magnetic declination and the number of Auroras observed in each year, and also with the extent of the black spots on the surface of the sun, establishes a connexion between these phenomena, and indicates that auroral displays, at least in the middle latitudes of Europe and America, are subject to a law of periodicity; that their grandest displays are repeated at intervals of about 60 years; and that there are also other fluctuations less distinctly marked which succeed each other at an average interval of about 10 or 11 years, the times of maxima corresponding quite remarkably with the maxima of solar spots.

Professor Loomis gives reasons for thinking that in high northern latitudes the inequality of auroral displays in different years depends more on their unequal brilliancy than on their frequency, and on this account, and because the observations are only made in northern latitudes for single years occasionally, they have been omitted in making the comparisons.

The portion of the earth's surface selected for comparison includes the whole area from which there are long continued series of observations, except from a few places in the north, and the results of this comparison are shown in a plate giving the curve for each of the three phenomena.

There is a very close correspondence of the curves, and the coincidence in the times of maximum and minimum is remarkable. The auroral maximum generally occurs a little later than the magnetic maximum, and it appears that the connexion between the auroral and magnetic curves is more intimate than between the auroral and sun-spot curves.

Date of Maximum.			Date of Minimum.		
Sun Spots.	Magnetic Declination.	Auroras.	Sun Spots.	Magnetic Declination.	Auroras.
1778	1777	1778	1784	1784	1784
1788·5	1787	1787·5	1798	1799·5	1798
1804	1803	1804·5	1810	Wanting.	1811
1816·5	1817·5	1818	1823	1823·5	1823
1829·5	1829	1830	1833·5	Wanting.	1834·5
1837	1838	1840	1843·5	1844	1843·5
1848·5	1848·5	1850·5	1856	1856	1856
1860	1859·5	1859·5	1867	1867	1867
1870	1870·5	1870·5			

From the close connexion between the times of maxima and minima Professor Loomis concludes—" That the black spot is a " result of a disturbance of the sun's surface, which is accom- " panied by an emanation of some influence from the sun, which " is almost instantly felt upon the earth in an unusual disturbance " of the earth's magnetism and a flow of electricity, developing the " auroral light in the upper regions of the earth's atmosphere."

12. The Geographical Distribution of Aurora.

M. Petermann's Mittheilungen, vol. 20, 1874, IX., contains a paper on this subject by Prof. Fritz, with map, from which it appears that the northern limit of Auroras chosen by Prof. Loomis nearly coincides, except in England, with the line of frequency represented by 10 in Prof. Fritz's paper. This line in Prof. Fritz's paper nearly passes through Toronto, Manchester, and St. Petersburgh. Prof. Loomis places it as far north as Edinburgh. On a line across Behring's Straits and coming down below 60° N. in America and the Atlantic, and passing just north of the Hebrides to Dröntheim, and including the most northern points of Siberia, the frequency is represented by 100.

Within this another zone is drawn, indicating the zone of greatest frequency and intensity.

This zone passes just south of Point Barrow, in latitude 72° N., and by the Great Bear Lake to Hudson's Bay, where it reaches a latitude of 60°, then on to the coast of Labrador and to the south of Cape Farewell, then bending sharper to the northward it passes between Iceland and the Faroe Isles near to the North Cape, and on to the Northern Ice-sea to Nova Sembla and Cape Tschelyuskin, and on just to the north of the Siberian coast to the south of Kellett Land and Point Barrow. The grouping of the magnetic meridians is very well shown in the map which accompanies the paper.

---

13. On a Definite Arrangement and Order of the Appearance and Progress of the Aurora Borealis, and on its Height above the Surface of the Earth, by Rev. James Farquharson.—(Phil. Trans. 1829.)

The following results had been arrived at from the observation of a great number of Auroras in Aberdeenshire (lat. 57° 15′ N.) in 1823, and were communicated to the Philosophical Magazine :—

" That the Aurora Borealis has in all cases a determinate arrange- " ment and figure, and follows an invariable order in its appear- " ance and progress; that the streamers (pencils of rays) of the " Aurora generally appear first in the north, forming an arch from " east to west, having its vertex at the line of the magnetic meri- " dian; that when this arch is yet only of low elevation it is of " considerable breadth from north to south, having the streamers " of which it is composed placed crosswise in relation to its " own line, and all directed towards a point a little south of the " zenith; that the arch moves forward towards the south, con- " tracting its lateral dimensions as it approaches the zenith, " and increasing in intensity of light by the shortening of the " streamers near the magnetic meridian, and the gradual shifting " of the angles which the streamers near the east and west ex- " tremities of the arch make with its own line, till at length these " streamers become parallel to that line, and then the arch is seen " as a narrow belt 3° or 4° only in breadth, stretching across the " zenith at right angles to the magnetic meridian; that it still " makes progress southwards; and after it has reached several " degrees south of the zenith again enlarges in breadth, by ex- " hibiting an order of appearances the reverse of that which had " attended its progress towards the zenith from the north; and " that the only conditions that can explain and reconcile these " appearances are, that the pencils of rays (streamers) of the Aurora " Borealis are vertical, or nearly so, and form a deep fringe, which " stretches a great way from east to west at right angles to the " magnetic meridian, but which is of no great thickness from " north to south, and that the fringe moves southward, preserving " its direction at right angles to the magnetic meridian."

A remarkable Aurora was seen on Nov. 22, 1825.

When first seen it had already formed two distinct and separate arches in the north and north-eastern parts of the heavens; the continuity of each was only interrupted by a few detached masses of low clouds coming, with a gentle breeze, slowly from the north, and brightly illuminated by the moon.

The most southerly arch approached within about 25° of the zenith. It was abruptly terminated at its west extremity about 35° above the horizon; as will be afterwards more particularly described in discussing the question of the height above the earth. This west abrupt extremity was a little to the north of the west. Its east extremity was near the horizon in the north-east. The streamers at the vertex of this arch were very short and compact, and parallel to the magnetic meridian. From this point towards both extremities the streamers gradually increased in length, and being all directed to a point apparently 10° or 15° south of the zenith, all formed angles with the general line of the arch, which were more acute in proportion to the distance from the vertex.

The arch might be about 10° broad, and speedily moved southward, maintaining a parallelism with its first position. Its lateral dimensions became gradually contracted. The streamers near the zenith shortened into dense bundles, like sheaves of light, parallel to the magnetic meridian, and consequently at right angles to the general line of the arch; and those towards the extremities gradually diminished the angles which they made with that line and approached to a parallelism with it. At length after reaching the zenith the arch became diminished in breadth to about 3° or 4° and coincided in its whole extent with the prime vertical to the magnetic meridian, and the light at its vertex exhibited a nebulous or mottled appearance, and that of the extremities of long streamers or pencils of rays, now parallel to the arch itself. I had no opportunity, upon the present occasion, to witness the enlargement of the breadth again, and the unfolding of the parallel streamers at the vertex, which I had observed in former arches when they got considerably beyond the zenith, for this arch gradually faded and became extinct, about 10° or 12° southwards of the zenith.

I now proceed to the question of the height of the Aurora Borealis above the surface of the Earth. In the paper in the Edinburgh Philosophical Journal, 1823, I had inferred, from the bright phosphorescent light of a cloud apparently under an Aurora, that they were in contact, or nearly so, with each other. Another similar appearance, of a still more decided character, in the autumn of 1825, but the precise date of which I have not noted, confirmed in my mind the justness of the inference. In a dark evening, without moon, an extended mass of clouds stretching along the N. and N.E. quarter, not much raised above the visible horizon, and having a clear sky above it, in which there was playing a fine Aurora of vertical streamers with their lower extremities apparently touching it, was observed giving out at its upper side a fitful but bright white light, more vivid and conspicuous amidst the darkness than if it had been illuminated by the rising moon. Similar clouds in other parts of the horizon exhibited no such light. It was impossible for a spectator to refer the Aurora to a distance more remote than that of the mass of clouds, or to believe that the former and the light of the latter were not part of the same phenomenon. Mr. Otley (Phil. Trans. l. c.) appears to have witnessed a similar phenomenon. "About 7 p.m. a dense cloud appeared in the horizon to the

" N.N.W., bounded by a bright line, the rest of the heavens being " starry. Presently beams of an Aurora began to shoot towards " the Great Bear."

On the evening of Nov. 22, 1825, besides the small detached clouds of the eastern and zenith part of the heavens coming slowly from the north, another of quite different character extended along the whole western part of the sky, to about 25° or 30° above the horizon. It was one dense sheet of stratum, comparatively, with the other clouds, very dark below, waved or furrowed from north to south, and cut off at its east side in an apparently straight edge, trending nearly north and south. It was coming on very slowly towards the east, and had before next morning prevailed over the other clouds, covering the heavens, and accompanied with a fresh westerly breeze, after a frosty night which the 22nd of November was. This large sheet of cloud was much more elevated than the small detached ones, as was fully proved by some of the latter being projected in perspective on its dark under surface, and there appearing as white masses fully enlightened by the moon.

Now the two arches of Aurora of that evening were abruptly terminated at the points where they appeared over the eastern edge of the large cloud; and the abrupt terminations increased their azimuth distances from the north as the arches came southwards, still appearing in their new positions over the east edge of the cloud. The lower extremities of the streamers, which were as long at these terminations as at any other parts of the arches, appeared even in contact with the cloud, and I sometimes conceived that they stretched before its eastern edge, but that part being considerably illuminated by the moon prevented me from being quite positive. Independently, however, of this uncertainty, the appearances are surely decisive of the fact, that the Aurora did not extend into the region occupied by the western cloud; and being seen over it at an angle not much higher than its own, occupied therefore a region of nearly equal elevation above the surface of the Earth.

I should have estimated the height of the phosphorescent clouds above described as so much as 2,000 feet above the surface, or twice the height of some of the neighbouring hills; but while the lower ends of the vertical streamers were at this height their upper might be 2,000 or 3,000 feet more. I have seen the Aurora, however, when the clouds certainly occupied a much more elevated region.

[The additional observations made by Mr. Farquharson have led him to think that the point to which the streamers are directed is a little farther south of the zenith than he had supposed, and that the luminous belt is sometimes a little broader than he had estimated it at its maximum (5°). Also the extremity of the zenith arch sometimes descends to the horizon. Another observer describes a display as consisting of "a bow or arch " of silvery light passing a few degrees south of the zenith, while " waves of light seemed to run along the arch."]

Conclusions.—1. The Aurora Borealis always presents itself in definite and very curious relations to the lines of magnetism, indicated by the needle.

2. The streamers in the direction of their length coincide with the plane of the dip of the needle or nearly so, and that each individual streamer is, in fact, parallel to the dipping needle.

3. They form a thin fringe, stretching often a great way from E. to W., at right angles to the magnetic meridian.

4. The fringe moves away from N. magnetic pole, by the extinction of streamers at its northern face, and the formation of new ones contiguous to its southern face.

5. The invariable regularity of its appearance, as seen by so many observers, when it comes fully within command of the eye, near the zenith, shows the apparent irregularities, when it is seen either more northerly or southerly, to be only optical illusions.

6. The region which it occupies is above and contiguous to that of the clouds, or that in which they are about to form.

---

14. On the Origin of Atmospheric Electricity, by J. Becquerel, Senr., reviewed by M. de la Rive. (Archives des Sciences, No. 41, 1871.)

According to M. Becquerel, Solar spots, which are sometimes 16,000 leagues in extent, appear to be cavities by which hydrogen and various substances escape from the Sun's photosphere. But hydrogen, which appears here to be only the result of decomposition, takes with it positive electricity, which spreads into planetary space even to the earth's atmosphere and to the Earth itself, always diminishing in intensity because of the bad conducting power of the successive denser layers of air and of the crust of the Earth. That would then only be negative, as being less positive than the air. The diffusion of electricity through planetary space would be limited by the diffusion of matter, since it cannot spread in a vacuum.

That gaseous matter extends farther through space than the distance which is generally assigned to the Earth's atmosphere will be proved by the fact that Auroras, which are due to electric discharges, are produced at heights of 100 and 200 kilomètres, where some gaseous matter must exist.

M. de la Rive agrees with M. Becquerel as to the electrical origin of the Aurora, but considers that the Earth is charged with negative electricity, and is the source of the *positive* atmospheric electricity, the atmosphere becoming charged by the aqueous vapour rising in tropical seas. The action of the Sun, he considers, is an indirect action which varies with the state of the Sun's surface, as shown by the coincidence in the periods of Aurora and Sun spots.

In the accounts of travellers in Norway we often read of their being enveloped in the Aurora, and perceiving a strong smell of sulphur, which must be attributed to the presence of ozone. M. Paul Rollier, the aëronaut, who descended on a mountain in Norway 1,300 mètres high, saw brilliant rays of Aurora across a thin mist which glowed with a remarkable light. To his astonishment, an incomprehensible muttering caught his ear; when this ceased he perceived a very strong smell of sulphur almost suffocating him.

---

15. Observations of Aurora in Italy, during April 1871, by M. Denza. (Archives des Sciences, No. 41, 1871.)

Auroras were remarkably frequent in Italy, as elsewhere, in April 1871. On the 9th a remarkable Aurora was seen, occupying an extensive region between Perseus and Cassiopea, and lasting until midnight. During the day, at mid-day, there had been a beautiful solar halo about 35° in diameter, changing from white to red and other colours, followed by a thunderstorm, with a falling barometer and the declinometer deflected more than usual towards the east. On the morning of the 10th the declinometer continued to be deflected, and at mid-day *ninety-seven* spots were counted on the Sun's disc, whereas on the 9th there had only been *sixty-three.*

On the 18th of April there was a similar Aurora, but consisting of two phases, the second more brilliant than the first and lasting through the day up to 10 at night. From this time until the 23rd of April the Aurora appeared every day, giving a reddish tinge in the north and north-west. It made a brilliant contrast with the zodiacal light, which at the same time shone out very brilliantly.

On the evening of the 23rd the Aurora shone out brilliantly, at first at 8.15 with a rosy tint about 20° on the horizon, then soon after 9 o'clock it shone out very brilliantly for five or six minutes and then waned. Before and after the display the northern sky was covered with a reddish-white light.

On the evenings when the Aurora appeared, the magnetometers were agitated throughout Italy, and the disturbances ended by a very violent agitation during the whole of the 24th of the month. During this time numerous sun spots were observed at Rome, Palermo, and at Moncalieri; but the greatest number was observed on the days of the Aurora.

There were also displays of Aurora seen at Moncalieri in June, the most brilliant on June 18, which was accompanied by a very violent magnetic disturbance.

[The winter of 1870 was remarkable, not only for the number of Auroras, but also for their great brilliancy and their very great extent. The third volume of "Nature" contains accounts of several displays seen in various parts of the world.

The displays of Oct. 24 and 25 were remarkably brilliant in England and in America, and the Aurora Australis was seen on the same days at Madras; yellowish-white and crimson beams shot up from a bank of light in the southern horizon. These displays were seen in the daytime in England and America as patches or coronæ of white light with streamers stretching upwards from them.

The spectra were observed by Mr. Alvan Clark in America, and in England by Mr. H. R. Procter, who observed a red line in the spectrum. These displays were accompanied by great magnetic disturbances, and by remarkable Earth-currents both in England and also on the Madras-Bombay lines in India.

Another brilliant display was seen at 6.30 p.m. and again at 7 p.m. on Dec. 19, 1870, in the Mediterranean, on the east coast of Sicily. Towards the north, north-east, and east brilliant pink streamers shot up out of a bank of faint hazy light on the horizon. The planet Jupiter was clearly seen through some of the most brilliant streamers. The sky became covered with a pinkish-mauve colour. Toward the west a pale, steady white light, the zodiacal light, was clearly seen during the evening. This display was followed in Sicily by a falling barometer and very stormy winds, with thunder and lightning, and by very destructive storms in Italy, causing the overflow of the Tiber and the flooding of the city of Rome to a depth which had been scarcely ever known.]

---

16. The Spectrum of the Aurora Borealis.

The following account of Professor Angström's paper on this subject is taken from "Nature," Vol. 10, No. 246 (for July 16, 1874):—

"It may be assumed that the spectrum of the aurora is composed of *two* different spectra, which, even although appearing sometimes simultaneously, have in all probability different origins.

"The one spectrum consists of the homogeneous yellow light which is so characteristic of the aurora, and which is found even in its weakest manifestations. The other spectrum consists of extremely feeble bands of light, which only in the stronger auroræ attain such an intensity as enables one to fix their position, though only approximatively.

"As to the yellow lines in the aurora, or the one-coloured spectrum, we are as little able now as when it was first observed to point out a corresponding line in any known spectrum. True Piazzi Smyth (*Comptes Rendus*, lxxiv., 597) has asserted that it corresponds to one of the bands in the spectrum of hydrocarbons; but a more exact observation shows that the line falls into a group of shaded bands which belong to the spectrum, but almost midway between the second and third Herr Vogel has observed that this line corresponds to a band in the spectrum of rarefied air (Pogg. Ann., cxlvi., 582). This is quite right, but, in Angström's opinion, is founded on a pure misconception. The spectrum of rarefied air has in the green-yellow part seven bands of nearly equal strength; and that the auroral line corresponds with the margin of one of these bands, which is not even the strongest, cannot be anything else than merely accidental.

"Observations on the spectrum have not hitherto agreed with each other; partly, perhaps, because of the weak light of the object, but partly also, it may be, on account of the variability of the aurora. The red does not always appear, and when it does is often so weak that it cannot be observed in the spectroscope. If now it be assumed that the aurora has its final cause in electrical discharges in the upper strata of the atmosphere, and that these discharges, whether disruptional or continuous, take place sometimes on the outer boundary of the atmosphere, and sometimes near to the surface of the earth, this variability will easily show in the appearance of the spectrum what the observations appear to confirm.

"If we consider the conditions under which the electric light appears on the boundary of the atmosphere, moisture in that region must be set down as nil, and consequently the oxygen and hydrogen there must alone act as conductors of electricity. Angström has tried to reproduce these conditions on a small scale. Into a flask, the bottom of which is covered with a layer of phosphate, the platinum wires are introduced, and the air is pumped out to the extent of several millimetres. If the inductive current of a Ruhmkorff coil be sent through the flask, the whole flask will be filled, as it were, with that violet light which otherwise only proceeds from the negative pole, and from both electrodes a spectrum is obtained consisting chiefly of shaded violet bands.

"If this spectrum be compared with that of the aurora, Angström thinks that the agreement between the former and some of the best established bands of the latter is satisfactory.

Lines.		Wave-lengths.		
Of the aurora spectrum	According to Barker	431	470·5	
	" " Vogel	—	469·4	523·3
	" " Angström	—	472	521
	" " Lemström	426·2	469·4	523·5
	Mean	428·6	470·3	522·6
Of the spectrum of the violet light		427·2	470·7	522·7

"In the neighbourhood of the line 469·4 Herr Vogel has moreover observed two weak light bands, 466·3 and 462·9. The spectrum of the violet light has also two corresponding shaded bands, 465·4 and 460·1.

"Should the aurora be flamy and shoot out like rays, there is good reason for assuming a disruptive discharge of electricity, and then there ought to appear the strongest line in the line-spectrum of the air, the green, whose wave-length is 500·3. Precisely this has been actually observed by Vogel, and has moreover been seen by Angström and others.

"Finally, should the aurora be observed as it appears at a less height in the atmosphere, then are recognised both the hydrogen lines and also the strongest of the bands of the dark-banded air-spectrum, as *e.g.*, 497·3. There are found also again nearly all the lines and light bands of the weak aurora spectrum, whose position has with any certainty been observed.

"There still remains the line in the red field, the wave-length of which, according to Vogel, may be valued at 630. Angström has chanced to see it only a single time, while on various occasions, when the aurora has shown red lights, he has found it impossible to distinguish any lines whatever in this part of the spectrum. The cause of this may be that while the red bands in the spectrum of the negative pole are broad and very feeble in light, the corresponding light in the aurora may be imperceptible in the spectroscope on account of the dispersion of the prism, although it is strong enough to give to the aurora a reddish appearance. Angström does not venture to decide whether the red line observed by Vogel coincides with the strongest of these bands, but so much is at least certain, that it may coincide with more than one of the bands in the red field of Plucker's air-spectrum.

"In general it may be thus assumed that the feeble bands in the aurora spectrum belong to the spectrum of the negative pole, and that the appearance of this spectrum may be changed more or less by additions from the banded air-spectrum or the line-spectrum of the air.

"But by this is not yet explained the one-coloured spectrum or the origin of the yellow line. The only explanation of the origin of this line which, in Angström's opinion, is in any way probable, is that it owes its origin to *fluoresence* or phosphorescence. Since fluorescence is produced by the ultra-violet rays, *an electric discharge may easily be imagined, which, though in itself of feeble light, may be rich in ultra-violet light, and therefore in a condition to cause a sufficiently strong fluorescence.* It is also known that oxygen is phosphorescent, as also several of its compounds.

"There is therefore no need, in order to account for the spectrum of the aurora, to have recourse to the 'very great variability of 'gas spectra according to the varying circumstances of pressure 'and temperature,' a variability which, according to Angström's 20 years' observations, does not exist. Just as little can Angström admit that the way in which a gas may be brought to glow or burn, can alter the nature of the spectrum; since it is an established fact in physics that the state of light and of heat which puts a body into a glowing condition is unconnected in character with that which produces glowing.

"Angström does not entirely deny the possibility that a simple body by glowing in a gaseous condition will offer several spectra. Just as one simple body can form a chemical combination with another, and this body by glowing in a gaseous conditions, so long as it is not decomposed, gives its own spectrum, so must it also be able to form combinations with itself—thus to form isomeric combinations—it being always supposed that it exists in the gaseous form and can maintain itself in a glowing condition without decomposition. In this way it is indeed possible to conceive an absorption for oxygen which belongs to ozone; but since ozone, as is well known, cannot maintain itself in a glowing condition, it is in vain to look for more than one spectrum of oxygen.

There is, however, at least a possibility of obtaining several spectra from sulphur, while again with respect to carbon, which cannot even be exhibited in a gaseous condition, a like assumption, in the author's opinion, wants the support of experience."

---

17. PROF. A. S. HERSCHEL on the SPECTRUM of the AURORA.

The spectrum of the Aurora is no doubt in the main the same as that of the pale blue light round the negative pole in an air or nitrogen vacuum-tube, with the induction-spark passing through it. There are so many well marked lines in this spectrum that, looking at Ångström's representation of them, it is probably owing to the insignificant appearance of that part of the vacuum-spark that its proper spectrum has not been more frequently studied with reference to the Aurora, as Ångström seems to have done by an experiment specially adapted for the purpose.

There are several forms or modes (apparently four or five) of electrical discharge through rarefied gases.* When very much rarefied, air transmits the electricity so as to discharge the Ruhmkorff poles without a spark. In that state there is still a glow of heated air round both poles, which increases in size and length along the tube as the air pressure is increased, faster round the positive than round the negative pole. This has been accounted for by showing that the air offers far greater resistance to the passage of electricity when it surrounds a cathode or negative, than when it surrounds an anode or positively electrified pole. The difference becomes more obvious as the pressure and density of the gas are increased. The negative glow shrinks into a very small space, while the positive brush extends through nearly the whole length of the tube, abolishing at last the dark space that at very low tensions separates the two lights from each other. At pressures not exceeding one or two millimetres the positive glow is stratified; but if the pressure is increased it becomes continuous; and if the air-pressure amounts to that of $\frac{1}{4}$ or $\frac{1}{2}$ an inch of mercury, or upwards, it again gathers into somewhat larger light clouds; and at about 1 or 2 inches of barometric pressure a spark passes between the poles. This spark is red; it scarcely diminishes the strength of the concomitant glow discharge; and it is far less luminous than the white spark which begins to appear at 5 or 6 inches of pressure, and may often be seen at first broken up along its length into parts which are alternately white and red. The spectra, like the general appearances, of these two forms of the spark are quite distinct.

---

* An examination of these with revolving mirrors, by A. Wüllmer, at Aix la Chapelle, appeared in the "Jubelband" of Poggendorff's *Annalen* this year at the same time as Ångström's paper in that volume, which also contains some other tracts (by A. de la Rive and others) tracing the effects of magnets and of metallic vapours in augmenting the discharge through air.

I do not know if these several phases of the *positive* part of the discharge have all been examined spectroscopically. They pass into each other according to the shape and size of the tube or flask, as well as the air-pressure; and it is difficult to say how much of each is concerned in those observations which have been made of air-spark spectra in comparison with the aurora. No one, so far as I know, has compared with it the *negative*-glow spectrum so fully as Ångström has now done; and it seems very probable that its peculiar fitness for the comparison has been overlooked—the feat of filling a bottle with the negative glow discharge being certainly a novelty; if it is really true that he succeeded in obliterating the positive brush entirely in its favour.

The next remarkable novelty in the paper is the way in which he proposes to account for the "citron" line of auroras; for there is evidently nothing of the kind in the negative glow, however well that answers to all the secondary facts of faint blue, red, and greenish lines. If oxygen and its compounds are (as has, I believe, been lately shown) strongly *fluorescent*, Tait and Dewar have also proved, as shown by some of their experiments this year, that they also possess powers of *phosphorescence*—Geissler tubes shining for some time after the spark has passed through them, from the production of ozone during the discharge. When one of the globes of a phosphorescent "garland" tube was heated over a Bunsen flame, that globe which was heated did not shine after the spark had passed, apparently because, as we know, a very little heat is sufficient to destroy ozone. Whatever the way may be in which the ozone or otherwise electrified gas remains self-luminous after the discharge, it seems very reasonable to suppose some action of the same kind (perhaps, as Ångström says, simply fluorescence) as common in all auroras, and that this produces the well-known auroral line.

Pocket spectroscopes can, of course, do nothing further to fix the position of the citron line; nor can they alone fix very exactly the places of any of the fainter ones. But as every Aurora shows this strong monochromatic light, it might be used to bring out a row of punctures transverse to the slit, as a divided scale in the field of view whereby to map the fainter lines, or at any rate to recognise those which appear most frequently. For this purpose they should be made large, and the slit should be a wide one. For ordinary miniature spectroscopes, two holes on the red and five on the blue side of the slit, $\frac{1}{40}$ or $\frac{1}{50}$ of an inch apart, would suffice for recognitions and even for very useful measurements. The jaws of the slit can be cut with a fine saw across the middle about $\frac{1}{10}$ or $\frac{1}{8}$ of an inch deep each way; and a piece of copper foil, provided with the row of holes and a sufficiently wide slit across it, can be fastened to one of them inside, opposite to the crosscut and adjoining the edges of the spectroscope jaws.

Some other means may be found of piercing the jaws of a pocket spectroscope at regular intervals; but, as a simple plan, I have found this very efficient, in finer divisions, for laboratory use. The holes are pierced at $\frac{1}{150}$ of an inch apart; and thirty of them include the whole visible spectrum. Sodium-light, which is common in laboratory flames, exhibits the punctures with admirable distinctness; and each fifth hole being punched double, the scale is very easily read off. There are ten holes on the red, and twenty on the blue side of the slit. If the mechanical difficulty of perforated jaws could be overcome, nothing perhaps could be better suited for examining auroras than a pocket spectroscope so prepared with a few close but clear and tolerably open holes on each side of the slit.

The secondary auroral lines can only be seen (in small spectroscopes) with a pretty broad slit; and the strength of the yellow line might then prove embarrassing. I would abolish it, if so, by a blue* glass nearly covering one half, and a red glass the other half of the slit—the blue and red parts of the spectrum respectively, not in its immediate neighbourhood, being freely transmitted. The slit might also be made longer than usual for auroral study.

I have been here supposing that special spectroscopes would be provided for Arctic observers. But it is quite certain that much may be done with common pocket spectroscopes without any such provision. They should have adjustable slits and good dispersion, as the secondary lines are faint; and though abundant enough in the blue to make the spectrum there pretty luminous, they can only be individualised by varying the slit-aperture. On the only occasion when I have seen this spectrum (in February 1872), they seemed to run into each other, and presented a light so nearly continuous in the blue part that, although the slit of the Browning's pocket spectroscope which I was using was extremely fine, and was focused on the yellow line, no interruption or appearance of lines could be made out. It was probably also through not opening the slit that I missed seeing a red line which another observer, using a similar instrument and looking with me at the Aurora, saw very plainly. Although its red colour was intensely brilliant, I failed to see the slightest trace of light on the red side of the yellow line. Had I opened the slit, or perhaps opened and closed it alternately (as the yellow line, though fine, was still very bright), the result would probably have been different.

I can confirm the appearance of the negative or "cathode" spectrum which Ångström gives, from the results of some examinations of it which I have lately made. On projecting the recorded lines in wave lengths, there is a very exact agreement with the chief lines and shadings as figured in the plate. Some fainter lines, however, are visible, which Ångström has, perhaps, omitted purposely to avoid encumbering the drawing. During the years 1871 and 1872 there were several résumés of the Aurora spectrum, accompanied with new measurements, in Poggendorff's "Annalen" and the "American Journal of Science" (by Vogel, Barker, and others).

---

* Some care would be necessary in selecting the blue glass, as these generally transmit a yellow ray closely corresponding with the auroral line.

I have endeavoured to condense the information on the Aurora lines from Ångström's descriptions of them in the annexed Table, as a guide for further observation.

Professor Piazzi Smyth, who has given much attention to the auroral spectrum, has published, in the introduction to vol. xiii. of the "Edinburgh Astronomical Observations," a set of simple comparison spectra, with notes of *desiderata,* which would be of great service to observers well furnished with instrumental means and applying them to measurements of the Aurora; and something similar would very much assist observers using direct-vision spectroscopes to map their spectra, where auroras are frequent and of great brightness, in high latitudes.

Positions and General Characters of Principal Lines in the Auroral Spectrum, according to Ångström.

Red . . 1. *Brightness* 0 to 4 or 8. *Wave-length* 6300.

Relative position. Atmospheric absorption-line α near C in the solar spectrum.

General description of source and frequency. Seen chiefly, if not only, in red auroras; a clearly defined line, sometimes intense; no other red line visible.

Identification with lines of electrical-air spectra. Coincides with a red band in the negative-glow discharge.

Yellow or citron-green. *a. Brightness* 25. *Wave-length* 5570.

Position. Second separable line in the first or citron band of blue gas-flame spectrum.

Description, &c. The most characteristic auroral line; constant and conspicuous in all auroras not divisible; sharp and bright.

Identification, &c. Not identified; possibly a phosphorescent or fluorescent light emitted when air is subjected to the action of electrical discharge.

Greenish blue or blue.	3.	*Brightness*	2 or 0 ? to 6....	*Wave-length* 5225.
	3*a*.	"	0–8 (increases with red line)	" 5170–5190.
	3*b*.	"	2 or 0 ? to 8....	" 5000.
	3*c*.	"	0 to 4 ?	" 4820–4870.

Position. 3 and 3*a* closely adjoin the solar line *b* and the second or green band of the blue gas-flame spectrum. 3 *b* is at $\frac{1}{2}$ (*b*, F); and the line or lines 3 *c* are near F.

Description, &c., The first three are distinct lines; the first most frequently observed; the second and third less commonly; lines in the fourth place (3*c*) noted by Alvan Clark, jun., Barker, and Ångström.

Identification, &c. 3, 3 *b* coincide with lines in the negative glow, 3 *b* that of nitrogen in the nebulæ; 3 *a* with a constant strong line in the spark-discharge. The latter and 3 *c*, it may be, are only seen in auroral streamers of low elevation.

Full blue. { 4. *Brightness* 3–6 (fainter with red line) ..... } *Wave-length* 4665–4740.
4*a*. „ .... „ 4630–4665.

Position. Middle and latter half of the third or blue band in the blue gas-flame spectrum.

Description, &c. A double band, consisting of two lines; the first rather more frequently noted than the second in auroral spectra.

Identification, &c. The principal line and its companion agree well in position with the principal band in the negative-glow spectrum.

Indigo, blue and violet. } 5. *Brightness* 0–6. *Wave-length* 4285.
6. „ ... „ 4110.

Position, &c. 5 coincides nearly with G and with the fourth or indigo band of the gas-flame spectrum; 6 is between G and H near the hydrogen line *h* in the solar spectrum.

Description, &c. 5 is a frequent line, but somewhat difficult to see; and, from its position, it is possibly the limit of vision for pocket spectroscopes; 6 was measured once by Lemström at Helsingfors.

Identification, &c. 5 corresponds exactly with a strong band in the violet in the negative-glow spectrum.

Remarks and Suggestions.

The general character of the subspectrum appears to be a series of bright lines, bands, and shadings, more or less dimly visible on a faint field of light in the blue region of the spectrum, the greatest concentration occurring apparently most frequently at about the positions stated above. They arise, according to Ångström, from discharges of electricity from the denser to the more rarefied strata of the upper air, producing there on a great scale what is seen in artificial discharges of electricity in rarefied air as a blue cap round the negative pole. The appearance in the aurora of only one red line in the place of the many red bands of the negative glow, scarcely less bright than the principal one, is remarkable; and fresh observations are very desirable to confirm it, or to detect other red lines if they exist. Very red auroras should be examined with a wide slit, covered (if of advantage) with light-red glass to shut off all other light as much as possible.

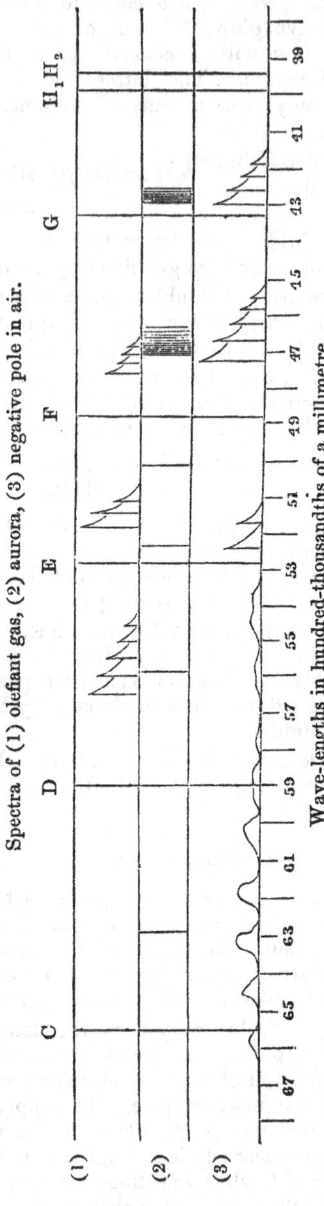

Ångström's representation of the Spectrum of the glow discharge round the negative pole of Air-vacuum Tubes, and its comparison with the Spectrum of the Aurora.

Spectra of (1) olefiant gas, (2) aurora, (3) negative pole in air.

Mixed with the lines of the *negative* glow, Ångström supposes that lines of the positive disruptive spark or brush discharge may appear in flashing auroras, especially near the base or arch as distinguished from the tops of the streamers, giving to the subspectrum a different appearance according to the strength and agitation of the streamers: this may perhaps be traceable in the appearance and disappearance of the lines 3 *a*, 3 *c*, and perhaps of other faint lines, whose positions should be noted. Such lines should also be searched for in quiescent parts, such as stationary auroral bands and the tops of very bright streamers.

A correct recognition of some one or more of the lines described above, other than the citron line, is however, of chief importance in observing with small spectroscopes, as the leading lines themselves must supply the only standard intervals of comparison for eye-estimations of such faint spectra.

18. Comparison of other Spectra with the Aurora Spectrum, by J. R. Capron.—(Phil. Mag., April 1875.)

For comparison with a table of Aurora spectra given in M. Wijkander's paper (*see* p. 742), a table of the results obtained by other observers, as given by J. R. Capron, is added, together with the spectra of hydrogen and oxygen (Vogel), and carbon (Dr. Watts and J. R. Capron).

Aurora Lines and Bands.

Observers.	Red.	Yellow.	Green.				Blue.		Indigo.	Violet.	
Vogel, April 9, 1871.	629·7	556·9	539·0	—	523·3	518·9	500·4	—	469·4	462·9	—
Barker, Nov. 9, 1871.	623·0	562·0	—	—	—	517·0	502·0	482·0	—	—	—
Barker, Oct. 14, 1872	630·0	555·0	—	533·0	—	520·0	505·0 to 499·0	493·0 to 485·0	474·0 to 467·0	—	431·0
A. Clarke, Oct. 24, 1870.	—	569·0	—	532·0	—	—	—	485·0	—	—	435·0
Backhouse, 1873.	606·0	566·0	—	—	—	516·5	501·5	—	—	462·5	430·5
Backhouse, Feb. 4, 1874.	—	557·0	—	—	—	518·0	498·0	483·0	—	464·0	432·0
Hydrogen lines.	—	555·5	542·2	—	—	518·9	500·8	—	—	463·2	—
Oxygen lines	—	560·3	—	—	—	519·5	—	483·4	—	—	—
Carbon (Dr. Watts).	—	562·2	—	—	—	518·9	—	482·9	—	—	—
Carbon (Capron).	—	560·8	—	—	—	519·3	—	482·5	—	—	—

Mr. Capron generally agrees with the results arrived at by M. Ångström, but sees reason to question the statement that "moisture in the region of the Aurora must be regarded as *nil.*" Also he sees no reason for giving to the violet-pole glow any special or distinguished place in comparison with the Aurora. He considers that the fainter lines are partly due to the air-spectrum, in which H. lines play a prominent part, and the spark-spectrum appears nearer the mark than the tube-spectrum; and that the remaining bands or lines (besides the phosphorescent lines) may be due to phosphorus and iron.

For measuring auroral and other lines, a cheap and very effective micrometer is constructed by making the whole slit-plate of the spectroscope traverse the field with a fine micrometer screw, a pointer or pointers being fixed in the eye-piece.

Large aperture, both of prisms and lenses, is almost indispensable in observations of faint Auroral spectra.

---

19. Observations on the Aurora, by M. Lemström.

Description of M. Lemström's Spectroscope.

On a massive foot is fixed a vertical brass tube; another brass tube sliding within it carries a strong plate of metal, which can turn in a vertical plane about an axis at the top of the tube and in a horizontal plane, by the sliding of the two tubes one in the other. To this plate another plate is fixed by a vertical steel axis about which it can turn, and the two are pressed together by a screw and a circular spring. The upper plate carries a horizontal arm of brass. Two prisms having a refracting angle of 60° are fixed on this plate. Their bases are placed against a vertical plate of metal with metal corners so that their bases form an angle of about 166°·5. The prisms are enclosed in a square box, which is fixed by screws. On one side of this box is a tube carrying a slit, and on the other a tube with a telescope. An arm of brass projects through the side of the box on the side of the telescope, and a micrometer screw is attached at the end of it. The rays of light entering by the slit fall on the lens and into the tube, and are then refracted and dispersed by the first prism, and afterwards internally reflected at the base of the second prism. They then enter the telescope. The bases of the prisms are inclined at an angle of 166°·5 to one another. By turning a micrometer screw the prisms are turned slowly so as to bring any ray into the middle of the field. The head of the screw is graduated in hundredth parts, and the scale in intervals corresponding to one turn. A reading of the screw determines the place of a ray of the spectrum, and a comparison with other scales or with wave-lengths may be made by fixing the relative positions and taking the readings for a great number of known lines of the solar spectrum.

The readings of the scale and the positions of the lines may be determined by alternately illuminating and obscuring the scale during the observation, so that the lines of the spectrum and their positions in the scale are blended together.

Spectroscopic Observations.

In the first Swedish Expedition, 1868, some very remarkable observations were made on the appearance of luminous beams around the tops of mountains, and by the spectroscope M. Lemström has showed that these displays were of the same nature as the Aurora.

He mentions also that on the 1st of September, on the Isle of Amsterdam, in the Bay of Smeerenberg, there was a light fall of snow, and the snowflakes were observed falling obliquely; all at once there appeared a luminous phenomenon, which, starting from the earth's surface, shot up vertically, cutting the direction of the falling snowflakes, and the phenomenon lasted for some seconds.

The characteristic yellow line of the Aurora is seen in the light on the tops of the mountains, and its intensity is constantly changing, and its variations are such as to show that the light is discontinuous. This line of the spectrum was also observed, but was of feeble intensity when the slit was directed towards a lake covered with snow, or towards a roof covered with snow, and even in the snow all round the observer.

M. Lemström concluded that an electric discharge, which could only be seen by means of the spectroscope, was taking place on the surface of the ground all round him, and that from a distance it would appear as a faint display of Aurora.

At the end of his very interesting paper on the spectrum of the Aurora and on his Lapland Expedition, M. Lemström sums up the results obtained by himself and other observers thus:—

(1.) The phenomena of pale and flaming light which is sometimes seen on the tops of Spitzbergen mountains appear also in Lapland, and are of the nature of the Aurora.

(2.) Phenomena of the same kind, although with some differences, have been observed in other countries, showing that electric discharges of the nature of the Aurora take place in other than Arctic regions.

(3.) The spectroscope is the surest means of deciding the kind of the phenomena in doubtful cases.

(4.) In Polar regions the electric discharge of thunder lies lower in the atmosphere than elsewhere.

(5.) Earth currents which accompany the Aurora are not induction currents produced by the Aurora, at least not in northern regions.

(6.) In all probability the current of polar light would act on a galvanometer, provided the apparatus which collects the electricity is large enough, or placed high enough in the atmosphere.

(7.) As a rule, positive electricity comes down from the upper regions.

(8.) The corona of the Aurora Borealis is not entirely a phenomenon of perspective, but the rays have a true curvature. (This is explained on the supposition that the rays are currents flowing in the same direction, and therefore attract each other.)

(9.) In the Aurora spectrum there are nine rays, which in all probability agree with lines which belong to the gases of the air.

(10.) The Aurora spectrum can be referred to three distinct types, which depend on the character of the discharge.

---

20. Observations on the Electrical State of the Air during the First Swedish Expedition in 1868, by M. Lemström.—(Archives Sciences Physiques, &c., tome 41, 1871.)

Toward the end of September, at Southgat Strait, between Danskow and Spitzbergen, lat. 79° 39′ 7″ N., long. 11° 7′ E., between mountains 300 mètres (1,000 feet) high on the north and south, the auroral light was seen on the outline of clouds on the mountain, and about 10° or 15° above the mountain, in undulating lines, presenting a diffused yellowish light at their base, from which vertical orange bands shot up, forming a series of very prominent sharp points at the top. The crest of the mountain became enveloped in mist with the wind from the E.N.E. For some minutes after the cloud had passed, the crest continued to be surrounded with a pale glimmer floating along the mountain, and the spectroscope showed the existence in it of the yellow light of the Aurora.

The next day, September 26, a similar effect was seen in the south-west, but the phenomena seemed to be much farther off than on the day before.

The next day (Sept. 27) a beam of yellowish-white light was seen in the morning, and at 11.30 p.m. a faint glimmer was seen gliding along the arête of the mountain, and, from the movement of the mist, was evidently on the arête itself. The light appeared for some seconds under the form of rays of a clear yellow of great brilliancy, following the outline of the mountain.

At other times all the tops and highest ridges of the mountains were enveloped with a pale glimmer, generally when they were covered with a thin veil of mist, the light gradually dying out and disappearing in the upper layers of the mist.

On returning to Tromso, an Aurora was seen on October 21, which commenced in the north and became very brilliant.

The spectroscope showed—

(1.) A yellow line.
(2.) A very clear line in the blue.
(3.) Two lines of a hair's breadth, with very pronounced horizontal striæ on the side of the yellow.

The light of the yellow line was very variable in intensity.

The relative places of the auroral lines were found to be—

(1.) The yellow line at 74·9.
(2.) The blue line at 65·90.

One of the shaded lines at 125·0 and the other about 105·0.

On the entrance of the "Sophie" into the Norwegian Archipelago on the evening of October the 18th, fragments of polar light were seen scattered here and there in all the sky to the north and to the east, which finished by forming a continuous ring around the horizon.

Rays from this ring lengthened gradually and met together suddenly about the zenith, forming during some instants a crown of Aurora of perfect regularity, and presenting the brightest colours.

(p. 157.) I have observed on several occasions discharges accompanied the electric light arising from scattered clouds or from beds of clouds.

In these high latitudes it is not by the clouds only that electricity is discharged; it is also directly by the damp air, as also takes place during winter in the temperate zone. We possess a great number of direct observations verifying the existence of slow discharges of this nature, and we have a remarkable proof of it in an observation by M. Ångström, who on one occasion has verified the presence of the yellow line of the polar light nearly all over the sky. If it is well established that the phenomena of the polar light are due to the electricity of the air, it follows that its appearance depends less on terrestrial magnetism than has been hitherto admitted. This may exercise a direct action upon the electric discharge already formed, but cannot contribute to its formation, which must depend only on the conditions in which the different layers of air are found. Although terrestrial magnetism has an influence upon the position of the luminous arc of the Aurora, it is difficult to admit with Hansteen and Bravais that the position of this arc should be determined only by the magnetic pole. Thus, the summit of the polar arc is rarely in the exact direction of the declination needle. From 226 observations made upon the position of the azimuth of the luminous polar arc we find 36 per cent., which give for this position 30° more to the west ; 32, which give 10° to 20°, 7, 0° to 10°, and 4, 0° to 26° to the east; from which it appears that the position of the arc varies from 25° to 30° and more. These variations are too great to be explained by accidental perturbations in the terrestrial magnetism, as much as about 6° to 7° more than the greatest deviations in the magnetic declination.

We therefore think that terrestrial magnetism plays only a relatively secondary part in the phenomena of the polar light; that this part consists essentially in a direct action upon the rays of this light, and in a movement of rotation exercised upon the rays, circumstances demonstrated positively by the experiences of M. de la Rive.

(p. 160.) The experiments of M. de la Rive, which have shown the influence of magnetism on the electric light in circumstances nearly identical with those which the polar light presents, do not at all furnish the proof that the rays of this light are really united under this influence. The polar light considered as an electric discharge gives the following results :—

(1.) An electric current arising from the discharge itself, which takes place slowly.

(2.) Rays of light consisting of an infinite number of sparks, each spark giving rise to two induction currents going in opposite directions.

(3.) A galvanic current going in an opposite direction to that of the discharge, and having its origin in the electro-motive force discovered by M. Edland in the electric spark.

To be developed, these currents require a closed circuit; it is true that in the phenomena of the Aurora, strictly speaking, it does not exist, but it is not necessary, seeing that in this case the earth and the rarefied air of the upper regions are immense reservoirs of electricity, which produce the same effect as if the circuit were closed.

According to the theory of M. de la Rive, the positive discharge of electricity from the air to the earth produces a current which I shall call the principal current. This is counterbalanced in part by the current due to the electro-motive force of the spark.

We see, by observations made with telegraph wires during the presence of the Aurora, that sometimes one and sometimes the other prevails, the first being in general predominant, the current given by telegraph wires being more frequently from north to south than from south to north.

(p. 163.) The cause of the clouds of the higher regions being discharged in the form of Aurora, and not in thunder and lightning, is the permanent moisture of the air. Hygrometric observations made during the Expedition of the "Sophie" show that the air is constantly saturated with aqueous vapours, which are condensed most frequently in clouds, more rarely in rain. It is clear that this layer of moisture, a good conductor of electricity, causes a slow discharge.

M. de la Rive adds (p. 166):—M. Lemström established by a great number of facts that the Aurora is due to atmospheric electricity, the presence of which in the Polar regions he has proved, often in the region of the clouds, and sometimes nearer the earth. He shows that this light is the consequence of electric discharges, which in these regions, constantly charged with moisture, operate slowly and continuously, instead of suddenly and producing lightning as in the equatorial regions and mean latitudes. He shows with reason that terrestrial magnetism, to which we had attributed an exaggerated importance in the production of Aurora, plays in this phenomena only a very secondary part. This consists simply in giving to the luminous electric streamers a certain direction which they are capable of taking as they are propagated in a gaseous conductor.

The electric discharges which take place in the Polar regions between the positive electricity of the atmosphere and the negative electricity of the earth, are the essential and unique cause of the formation of the polar light--light the existence of which is independent of terrestrial magnetism, which contributes only to give to the polar light a certain direction, and in some cases to give it motion. This is what I have always maintained, contrary to those who believe they see in terrestrial magnetism, or rather in the induction currents which it is capable of developing, the origin of the polar light.

---

21. Observations on the Spectrum of the Aurora made by M. Auguste Wijkander and Lieutenant Parent, of the Swedish Arctic Expedition in 1872–73, under Professor Nordenskiöld.—(Archives des Sciences Physiques et Naturelles, vol. li., p. 25.)

The Swedish Arctic Expedition under Professor Nordenskiöld have made some valuable observations on the spectrum of the Aurora Borealis in Mussel Bay, on the north of Spitzbergen, during the winter of 1872–73.

The instrument employed was a direct-vision spectroscope, constructed by Baron Wrede, and similar to that used by M. Lemström in his observations on the spectrum of the Aurora. The instrument consisted of two prisms, one of which serves to refract and disperse the luminous rays, whilst the other brings them back to the first direction by total reflection. The prisms are so far capable of motion, that the different parts of the spectrum can be brought on the system of crossed spider lines placed at the focus of the eye-piece. By a side opening and a mirror opposite to it the spider lines could be uniformly illuminated when necessary;

with this arrangement the position of the faint aurora-lines could be exactly determined.

For the wave-lengths corresponding to the lines in the spectrum, Ångström's values were made use of, and the wave-lengths have been deduced from the readings of the micrometer screw by the following formula, which is applicable to the rays from D to H:—

$$\lambda = 0{\cdot}00046489 + 0{\cdot}000009540(\alpha - 10) + 0{\cdot}0000004564(\alpha - 10)^2 + 0{\cdot}00000002184(\alpha - 10)^3,$$

where α is the reading of the spectroscope, and λ the length of the wave.

Observations on the solar spectrum were made during the autumn of 1872 and the spring of 1873, and they have shown that during this period no change of the zero took place.

During the winter the instrument was tested by means of the faint part of the flame of a candle. In a few experiments there was an accidental displacement, probably because a small quantity of dust or ice had penetrated into the apparatus, and in these cases the values determined have been enclosed in brackets, and entirely omitted from the final result. A variation of one-tenth of a revolution of the screw corresponds to from 0·0000019 to 0·0000007 of a wave-length in the part of the spectrum where the Aurora lines occur.

The first table contains the numbers found by M. Wijkander for the different series of lines grouped according to the different days. The measures of the most brilliant ray of the Aurora are not included in the table, since it is bright enough to be easily determined by instruments of greater dispersive power. The time when the Aurora was most brilliant was spent in observing the other rays, and the brightest lines have been measured in the intervals. The mean value found for the wave-length of the brightest ray was 0·0005572.

The following are the lengths of wave, with their means and probable errors, taking the millionth of a millimetre as the unit:—

The following Table contains the measurements made by M. Wijkander.

1872.	1.	2.	3.	4.	5.	6.	7.	8.
October 24 - -	536·46	..	..	..	..	470·77	..	428·54
" " - -	534·80	..	..	..	..	468·78	..	..
" " - -	..	..	..	..	..	470·77	..	..
" 30 - -	..	..	..	..	487·13	468·88	..	..
" " - -	..	..	.	..	..	469·08	..	..
November 3 -	535·46	526·89	523·76	..	..	469·37	..	..
" " -	..	..	524·07	..	..	..	..	..
" " -	..	..	526·28	..	..	..	..	..
December 5 -	535·46	527·99	524·85	498·85	..	467·80	436·65	427·86
" " -	537·46	528·63	521·93	503·20	..	..	..	426·60
" " -	..	530·23	522·69	496·29	..	..	..	..
1873.								
January 18 - -	..	531·85	..	497·31	..	468·29	..	429·02
" " - -	..	527·68	..	502·26	..	..	..	..
Mean - -	535·9	528·9	523·9	499·6	487·1	469·2	436·6	428·0
Probable Error -	±0·31	±0·51	±0·43	±0·91	..	±0·25	..	±0·35

The following table contains the measurements made by Lieutenant Parent.

1872.		1.	2.	3.	4.	5.	6.	7.	8.
November 3	-	..	527·84	..	..	..	471·59	..	424·64
" "	-	..	528·47	..	..	..	467·80	..	..
" "	-	..	..	..	..	..	471·79	..	..
" "	-	..	..	..	..	..	470·17	..	..
" 4	-	..	527·84	..	..	..	472·30	..	..
" 8	-	..	..	522·39	..	487·37	471·79	..	430·89
" "	-	..	..	..	..	..	472·30	..	430·75
" "	-	..	..	..	..	..	472·72	..	..
" 29	-	..	..	..	..	..	468·98	..	..
" "	-	..	..	..	..	..	469·27	..	..
December 5	-	..	..	519·05	..	..	..	..	..
Mean -	-	..	528·05	520·72	..	487·37	470·87	..	428·76
Probable Error	-	..	±0·14	±1·11	..	..	±0·54	..	±1·68

[Seeing that the results obtained by Lieutenant Parent are so near to those obtained by Mr. Wijkander, it seems probable that they are quite trustworthy. If we take the mean of all the results we get:

Mean, 535·9, 528·6, 523·1, 499·6, 487·2, 470·1, 436·6, 428·4.]

In these last experiments the crossed spider lines were not placed very exactly in the focus of the eye-piece, and so these have been omitted in making the final determination of wave-lengths. But the agreement between the two sets of observations shows the accuracy of the results.

The appearance of the Aurora was often observed on other days with this spectroscope, when the light was too faint to admit of measurement. With every appearance of the Aurora, the rays have always been visible, except 4, 5, and 7. These have been seen several times, but have often been wanting, not entirely from the faintness of the light, but probably from some change in the nature of the electric discharge, still in that case no indications have been found as to the conditions of this irregularity.

Between the rays 2 and 3 there are probably smaller lines, which form with them a broad band of light of variable intensity. In the same way the ray 6 spreads on the violet side of it into a tolerably broad but somewhat indistinct band. After careful search no red ray has been found, but at the same time no red Aurora was ever seen in winter quarters.

The following table contains the different measures of the lines of the Aurora which have been made with the greatest accuracy :—

Ångström -	..	556·7	..	..	..	521·	501·	487·	472·	..	..	..
Vogel -	629·7	557·1	539·0	..	523·3	518·9	500·4	..	469·4	..	..	..
Lemström -	..	556·9	..	..	523·5	..	496·9	..	469·4	..	426·2	411·2
Wijkander -	..	557·2	535·9	528·9	523·9	..	499·6	487·3	469·2	436·6	428·0	..

The results of the observations entirely agree with M. Ångström's explanation of the origin of the rays. There are only the relatively bright bands 528·9 and 535·9, which cannot be accounted for. The brightest group of iron lines is very near, but does not correspond with them.

---

22. Observations on the Electricity of the Air made in the Swedish Arctic Expedition in 1872–73, by M. Auguste Wijkander.—(Archives des Sciences Physiques et Naturelles, vol. li., p. 31.)

Many attempts have been made to discover the nature of the electricity of the air in high latitudes, because of its great importance, especially with reference to the Aurora Borealis. But the result of all these researches has been that not a trace of electricity has been found until the observations of MM. Bravais and Lottin, in the winter of 1838–39, at Bossekop in Northern Norway. On their way to the north, in experiments made on board during the summer, positive electricity was collected by shooting arrows into the air; but in Northern Norway, and in the month of October, at Bossekop, the experiments were repeated without any result.

On February 25 five, and on March 1 eight, experiments with kites were more successful, giving weak charges of positive electricity.

The observers concluded that the air is charged with positive electricity in those regions as well as in lower latitudes, but they had also found that the air is a much better conductor of electricity in those regions than elsewhere.

In the first Swedish Expedition in 1868, M. Lemström made observations with an electrometer by M. Lamont as well as with a straw electroscope (Snow Harris ?), but these instruments were not sufficiently delicate to detect the slight traces of electricity which are found in these regions, and no traces were obtained.

In the second Swedish Expedition, 1872–73, a modification of Thomson's electrometer was made by M. Holmgren, who also gave excellent directions for making the observations.

The Leyden jar of Thomson's electrometer was replaced by an alcohol battery, the poles of which are attached to the quadrants, the opposite ones being connected together. The needle is insulated from the quadrants and is in connexion with the collecting ball. The whole was sealed almost hermetically by a brass cover, and the air dried by chloride of calcium. In employing an alcohol battery in place of a Leyden jar, the readings made at different times were more nearly on a uniform scale; but, for the better comparison of the different readings with the strength of charge, an alcohol battery of 25 pairs was always connected up to the electrometer, and its effect measured before each observation.

For the winter observations a special room or hut was built by means of sacks filled with northern moss (?) at a distance of a hundred metres from our abode, and in one corner of it the electrometer was placed on a pile of stones, and in the opposite corner the tube and scale. The collecting apparatus could be raised through an opening in the roof. At certain periods a pole 25 feet long was employed, which was supported by stays, and the collector could be raised the whole height of it. When the conduction of the air made it impossible to employ this method, the pole was taken down, and the apparatus was only raised by hand, the ball reaching about three feet above the roof. There were no salient points which could have any disturbing influence. The place of observation was a rather low island in Mussel Bay not far from the 80th degree of latitude. There were no mountains around which were at an elevation of more than three degrees above the horizon.

The collecting apparatus consisted sometimes simply of a hollow knob or ball, of about three inches in diameter, mounted on a rod of ebonite five feet long, at others of a lamp also, which could be screwed on to the top of the ebonite rod by means of a cross piece of ebonite one foot long. The lamp consisted of a cup of metal, with its edges pierced, in the middle of which was a reservoir with a rim of metal to hold the spirit. From the middle of the cup of metal a metal rod a foot and a half long projected vertically downwards, which could be screwed to the ball of the collector.

The lamp was lighted and raised, and, when the spirit was burnt, the lamp was lowered and brought to the electrometer. When the ball alone was employed, it was placed in communication with the Earth by a metallic thread. However, in measuring the electricity of the Earth the lamp with no spirit was often used, because of the inconvenience of taking it off the ball. When there was only a weak charge of electricity, constant contact was made between the collector and the electrometer in order to take the measurements rapidly. In this case a cotton-covered copper wire covered with paraffin connected the electrometer to the collector. The thread was stretched along inside the roof attached to some supports by a thread, with as good insulation as possible, and so as not to disturb the electrometer when the collector was rapidly raised or lowered. Several experiments were made to test this arrangement.

The observations were made by M. Wijkander and Lieutenant Polander.

In the autumn of 1872 experiments were made to discover the electrical state of the air and of the Earth's surface, and there were always feeble traces of positive electricity in the air and negative electricity in the ground. The free air appeared to be a very bad insulator under the circumstances of the experiment. The deviation produced by the battery ceased at once when it was removed. This was not from any defect in the electrometer, for the insulation was always satisfactory in a warm room. The experiments below show that the insulation was also good in free air when it was very cold.

In the month of January 1873 other results appeared. The experiments are given in the order in which they were made, and afterwards follow remarks on the meteorological conditions, as well as the temperature, which gives the best indication of the changes of weather.

The connexion between the electricity of the air and the weather will appear from the meteorological observations:—

Date.	Hour.	Charge from			Temp.	Remarks.
		Air.	Ground.	Battery.		
	h.				°	
Jan. 21	9	+0·5	−0·5	11	−5	The deviation instantly ceased.
„ 22	11	+5	−0·5	22	+3	Heavy storm from S.S.W. to S.
„ 22	21	—	—	—	0	Both ground and air are electrified positively. Deviation small.
„ 23	10	—	—	—	0	
Feb. 6	11	+7	—	11	−1	
„ 7	10–11½	+0·5	−0·5	—	−15	Good insulation. The lamp gave the electricity of Air +5.
„ 10	18	+12	—	—	−26	After strong N.W. wind.
„ 19	17	—	—	—	−32	Good insulation. Great disturbance, so as to prevent measurement.
„ 20	16	+8 +11 +15	−18 −20 −25	7 to 8	−38	Good insulation.
„ 21	10	+12 +14	−70 −80	7 to 8	−31	
„ 25	18	+20 +18	−40 −45	8 to 9	−33	The long pole was employed for the first time for raising the Collector.
„ 26	14	+10 +13	−20 −25	5	−31	Once the lamp gave −12 for Air, after having been raised for some time. Clear, except to the west.
„ 27	10	+25 +30	−35	6 to 8	−33	Almost clear.
„ 27	13	+15 +17	−25 −20	7 to 8	−34	Strong mirage to the west. Clear sky, with stratus to west and south.
„ 27	21	−9 −16	−41 −38	7	−32	Repeated experiments. A slight mist began to rise in the south, which an hour later covered the whole sky.
„ 28	9	−8 −10	−16 −20	8 to 9	−27	Insulation not so good. Light veil of clouds. A few snow flakes.
„ 28	14	−10 −13	−38 −35	8	−28	Clear except to the East. Scattered nimbus clouds.
„ 28	21	−30 −31	−51 −53	7 to 8	−33	Entirely clear.
Mar. 1	10	−5 −7	−10 −13	5 to 6	−33	Clear. Very little wind from E.S.E.
„ 1	16	−1 −2·5	−16 −17	6 to 7	−31	Wind N.E. Sky cloudy to N.E., N., and W., with nimbus and stratus.
„ 1	22	−1 −3	−3 −5	9 to 10	−19	Cloudy.
„ 2	10	—	—	6 to 7	−7	Charge from battery *instantly* falls to zero.
„ 10	14	+22 +24	−25 −30	7 to 9	−16	Good insulation.
„ 20	14	0 −2	−35 −40	7 to 9	−33	
„ 20	21	−3	−45	6 to 7	−32	
„ 21	11	−5	−40	9	−32	Needles of ice floating in the air.
„ 21	15	−8	−38	8	−31	
„ 21	21	−10	−45	—	−32	Wind tolerably strong from the N.W.
„ 22	10	−15	−18	8 to 9	−27	Needles of ice in the air.
„ 23	10	−3	−16	7	−32	Pretty clear. Needles of ice. Almost calm. Some cirro-stratus.
„ 23	14	+10 +19	−30	—	−33	Almost clear. Fog over the sea.
„ 24	10	−9 −6	−29	7	−32	Do. Cirro-stratus to S.W.

Date.	Hour.	Charge from Air.	Charge from Ground.	Charge from Battery.	Temp.	Remarks.
	h.				°	
Mar. 24	16	− 9	−28	6 to 7	−28	Clear and calm.
„ 25	10	{+20 +18 +25}	−48	—	−33	Charge from air suddenly became −32, probably the ball was touched by one of the stays.
„ 25	15	{+30 +35}	−36	10·5	−31	Charge from battery disappeared quickly. Clear. Light E. wind.
„ 25	21	+25	−30	—	−34	Light S.S.E. wind. Clouds gather and temperature is raised during the night from −34° to −15°.
„ 26	10	−15	−12	8 to 9	− 7	Cloudy and calm.
„ 26	15	—	—	—	− 5	Charge from battery immediately disappeared.
„ 27	10	+ 2	−0·5	3 to 4	−17	Wind N., with snow. Charge disappeared instantly.
„ 27	22	+ 1	—	—	−24	With lamp rapidly raised. Rain mixed with snow. Wind W.
„ 28	10	{− 5 − 6}	{−23 −24}	6 to 7	−18	Cloud and calm.
„ 28	15	− 2	− 3	5 to 6	−10	The air once gave +1·5. A little snow. Wind S.E.
„ 30-31	—	—	—	—	−17 to −13	Several experiments. Charge instantly disappeared.

A few more successful experiments were made until May 24, but the charges instantly disappeared, although the insulation of the instrument appeared to be as good as ever.

Observations repeated as spring came on, when the temperature approached zero showed the same results as in the autumn.

There was great difficulty in keeping the electrometer in a good state. The walls of the room allowed the wind and snow to penetrate, and occasionally the instruments were buried in snow, and the room torn with the violent storms.

The observations agree in showing that the air conducts electricity very easily at relatively high temperatures, and to this is due the absence of thunder and the existence of the Aurora.

This is said to be due to the *moisture* in the air in those regions, but the same temperature and the same degree of moisture do not produce this effect in lower latitudes. At the lower temperatures −20°, −30°, and still lower, the air insulates better. Generally the air is electrified positively and the ground negatively. On several occasions the air could only be regarded as itself holding a charge, and not as charged by induction from the Earth.

At certain periods of spring, when the air insulated pretty well, the ground and the air were both charged with negative electricity. A change in the electricity of the air was not a constant result of greater cold, but when the temperature had been low for some time the air seemed to have a tendency to be electrified negatively.

It seems that there is a very natural relation between these facts and the Aurora, as far as conclusions may be drawn from so few observations.

In the months of January and February Auroræ were constantly seen, and they were particularly numerous from 19th to 26th of February, then they ceased, to reappear on March 2nd; at the same time changes were observed in the electricity of the air. The observers were driven to the supposition that the negative electricity, being deprived of the possibility of discharge by means of the Aurora, could not fail to accumulate in the lower layers of the atmosphere, which are comparatively good insulators.

From March 2 to March 11, the Aurora reappeared: during this time the air was a good conductor or, when it insulated comparatively well, was positively electrified. From March 11 the Aurora entirely ceased, and then follows a period of comparatively low temperature, with the electricity of the air generally negative, a period which lasted until the season when the light prevented all further observations of the Aurora.

APPENDIX.

The chief modern Works on Greenland, the Franklin Archipelago, Spitzbergen, &c.

1818. Sir J. Barrow. Chronological History of Voyages into the Arctic Regions, &c.

1819. A. Fisher. Journal of a Voyage to the Arctic Regions in 1818, &c.

1819. Capt. Sir J. Ross. Voyage for inquiring into the Probability of a N.W. Passage. Baffin's Bay and Davis Strait. Appendices by Macculloch, Leach, &c.; and on Meteorology, Soundings, &c. Two editions, 4to and 8vo.

1820. W. Scoresby. Account of the Arctic Regions, &c. Greenland and Spitzbergen.

1821. A. Fisher. A Journal of a Voyage of "Hecla" and "Griper" to the Arctic Regions in 1819–20. Parry Islands.

1821. Parry's Journal of a Voyage for the Discovery of a North-west Passage, &c. Parry Islands.

1823. W. Scoresby. Journal of a Voyage to the Northern Whale Fishery, including Researches on the East Coast of Greenland.

1823. Sir J. Franklin. Narrative of a Journey to the Shores of the Polar Sea in 1819–22, &c. America.

1824. Lyon's Private Journal during Capt. Parry's Second Voyage. Parry Islands.

1824. J. D. Cochrane. Narrative of a Pedestrian Journey to the Frozen Sea, &c.

1824. Sir W. E. Parry. Journal of a Second Voyage for the Discovery of a North-west passage, 1821–23, &c. Parry Islands.

1825. Major-Gen. E. Sabine. Account of Experiments to determine the Figure of the Earth. (Also a brief account of Capt. Clavering's voyage to the Arctic Regions.)

1825. Lyon. Brief Narrative, Repulse Bay, &c. Hudson Strait.

1826. Sir W. E. Parry. Journal of a Third Voyage for the Discovery of a N.W. Passage, 1824–25, &c. Parry Islands. Zoology by Lieut. L. C. Ross; Botany by Prof. Hooker; Geology by Prof. Jameson, &c.

1828. Sir W. E. Parry. Narrative of an Attempt to reach the North Pole in 1827. Spitzbergen. Zoology by Lieut. L. C. Ross; Botany by Prof. Hooker; Geology by Prof. Jameson, &c.

1828. Franklin and Richardson. Second Expedition. America.

1832. W. A. Graah. Undersögelses-Reise til Ostkysten af Grönland, &c. East Greenland. Botany by Hornemann; Miscell., &c.

1834. Anon. Arctic Expeditions from England from 1497–1833.

1834. Sir J. C. Ross. The Position of the North-Magnetic Pole.

1835. R. Huish. The last Voyage of Capt. Sir J. Ross to the Arctic Regions, &c.

1835. S. Braithwaite. Supplement to Sir James Ross's Second Voyage in search of a N.W. Passage. Boothia Felix, &c. Parry Islands.

1835. Capt. Sir J. Ross. Narrative of a Residence in the Arctic Regions during 1829–33, &c. Appendices.

1835. R. Huish. The last Voyage of Capt. Sir J. Ross, for the Discovery of a N.W. Passage.

1835. Capt. Sir J. Ross. Narrative of a Second Voyage in search of a North-west Passage, &c. Appendices.

1835. M. de la Roquette. Sur les Découvertes faites en Grönland, &c.

1836. R. King. Narrative of a Journey to the Arctic Ocean in 1833–35, &c.

1836. Sir G. Back. Narrative of the Arctic Land Expedition to the Great Fish River, &c. America.

1837. W. A. Graah. Narrative of an Expedition to the E. Coast of Greenland, &c. Translation by MacDougal. Appendix.

1838. Sir G. Back. Narrative of an Expedition to the Arctic Shores, &c. America.

1838–40. P. Gaimard. Voyage en Groënland pendant 1835 et 1836, &c. South Greenland and Iceland.

1839. F. von Wrangell. Reise auf dem Eismeere in 1820–24, &c. Asia.

1839. K. E. von Baer. Sur la fréquence des Orages dans les Régions Arctiques.

1839. Capt. Fabvre. Rétour en France de *la Recherche;* Rapport sur la seconde Campagne au Spitzberg.

1840 and 1844. F. von Wrangell. Narrative of an Expedition to the Polar Sea in 1820–23, &c. Siberia.

1843. F. W. Beechey. Voyage towards the North Pole in 1818, &c. Spitzbergen.

1845. C. C. Rafn. Americas Arctiske Landes gamle Geographie, &c.

1846. Sir J. Barrow. Voyages within the Arctic Regions from 1818, &c.

1847. C. C. Rafn. Aperçu de l'ancienne Géographie des Régions Arctiques de l'Amérique.

1848–56. Arctic Expeditions. A Collection of Papers relative to the recent Arctic Expeditions, &c.

1850. Arctic Expeditions. Eskimaux and English Vocabulary.

1850. J. Washington. Eskimaux and English Vocabulary.

1850. R. A. Goodsir. Arctic Voyage to Baffin's Bay. Baffin's Bay and Lancaster Sound.

1850. J. Rae. Narrative of an Expedition to the Shores of the Arctic Sea in 1846–47. N.E. America and southern parts of Parry Island.

1850. H. Kellett and others. The Arctic Expedition of 1849, &c.

1850. W. Scoresby. The Franklin Expedition; or, Considerations for the Discovery of our Countrymen in the Arctic Region.

1850. J. J. Shillinglaw. Narrative of Arctic Discovery from the earliest Period.

1851. Arctic Searching Expeditions of 1850–51, &c.

1851. J. Mangles. Illustrated Geography and Hydrography. Wellington Channel Section.

1851. Sir J. Richardson. Arctic Searching Expedition, &c.

1851. W. P. Snow. Voyage of the "Prince Albert" in search of Sir J. Franklin, &c. Baffin's Bay and Parry Islands.

1852. H. Rink. Om den geographiske Beskaffenhed af de danske Handels-distrikter i Nordgrönland, &c. Greenland.

1852–57. H. Rink. Grönland geographisk og statistisk beskrevet.

1852. Osborn. Stray Leaves from an Arctic Journal. Baffin's Bay and Parry Islands.

1852. P. Force. Remarks on the English Maps of Arctic Discoveries in 1850 and 1851.

1852. Additional Papers relative to the Arctic Expedition (in search of Franklin) under Capt. Austin.

1852. Further Correspondence connected with the Arctic Expedition, &c., in search of Franklin.

1852. Report of the Committee appointed to inquire into the Report on the recent Arctic Expeditions in search of Sir J. Franklin.

1852. A. Petermann. The Search for Franklin. A suggestion, &c.

1852. P. C. Sutherland. Journal of a Voyage in 1850–51. Davis Strait, Baffin's Bay, and Franklin Archipelago.

1853. E. K. Kane. Access to an open Polar Sea, &c.

1853. Greenland Eskimo Vocabulary, &c.

1853. C. R. Markham. Franklin's Footsteps; a Sketch of Greenland, &c.

1853. Kennedy. Second Voyage of the "Prince Albert." South part of Parry Islands.

1853. W. H. Hooper. Ten Months among the Tuski, with an Arctic Boat Expedition, &c.

1853. E. A. Inglefield. A Summer Search for Sir J. Franklin, &c., in the "Isabel" in 1852. Davis Strait and Baffin's Bay. Appendix (Sutherland, Geology, Botany, &c.).

1853. E. K. Kane. The United States Grinnell Expedition, in search of Sir J. Franklin, &c. S.E. Parry Islands. Appendices: currents, meteorology, winds, &c.

1853. B. Seemann. Narrative of Three Cruizes to the Arctic Regions, &c.

1854. J. R. Bellot. Journal d'un Voyage aux Mers Polaires, &c.

1854. Papers relative to the recent Arctic Expeditions in search of Sir J. Franklin, &c.

1854. R. M'Cormick. Narrative of a Boat Expedition up the Wellington Channel in 1852, &c.

1854. C. Irminger. Arctiske Strömning.

1855. Sir E. Belcher. Last of the Arctic Voyages; a Narrative of the Expedition of H.M.S. "Assistance" in search of Sir John Franklin, 1852–54; with notes on the Natural History, by Richardson, Owen, Bell, Salter, and Reeve. Appendix; Snow, &c. Parry Islands.

1855. V. A. Malte-Brun. Coup-d'œil d'ensemble sur les differentes Expéditions Arctiques, &c.

1855. F. Mayne. Voyage in the Arctic Regions.

1856. C. C. Ostergaard and others. Observationes Meteorologicæ per annos 1832–54 in Grœnlandiâ factæ.

1856. E. K. Kane. Arctic Explorations: The Second Grinnell Expedition, &c. Smith Sound, &c. Appendix: Nat. Hist., Physics, &c.

1856. W. Kennedy. A short Narrative of the Second Voyage of the "Prince Albert" in search of Sir J. Franklin.

1856. J. Rae. Voyages and Travels of, in the Arctic Regions. Copy of a letter, &c.

1856. R. le M. M'Clure. Discovery of the North-west Passage, 1850–54.

1856. R. White. On the Open Water at the Great Polar Basin.

1856. S. Osborn. Discovery of the North-west Passage by the "Investigator," &c. Southern part of Parry Islands.

1857. Lord Dufferin. Letters from High Latitudes, being an Account of Iceland, Spitzbergen, &c.

1857. G. F. McDougall. The Voyage of H.M. "Resolute" to the Arctic Regions, 1852–54. Parry Islands.

1857. M'Clure. Discovery of the North-west Passage.

1857. C. Petersen. Erindringer fra Polarlandene, 1850–55, &c.

1857. Armstrong. Personal Narrative of the Discovery of the North-west Passage. Parry Islands (Botany, Hooker).

1857. S. M. Smucker. Arctic Explorations and Discoveries during the Nineteenth Century.

1858. J. Brown. The North-west Passage, &c. South part of Parry Islands.

1859. F. L. M'Clintock. Reminiscence of Arctic Ice-travel, &c. In the Journ. Royal Dublin Soc. Parry Islands.

1859. F. L. M'Clintock. The Voyage of the "Fox" in the Arctic Seas, &c. South-east part of Parry Islands.

1859 and 1861. Sir J. Richardson. Polar Regions.

1860. E. K. Kane. Astronomical Observations made on the North-west Coast of Greenland, &c. ("Smithsonian Contributions.")

1860. E. K. Kane. Tidal Observations in the Arctic Seas, &c. ("Smithsonian Contributions.")

1860. P. Chaix. Explorations Arctiques, &c.

1860. I. I. Hayes. Arctic Boat Journey in 1854, &c. With introductory notice by Norton Shaw, and lists of Arctic Expeditions and Works. Smith Sound.

1860. S. Osborn. The Career, last Voyage, and Fate of Sir J. Franklin.

1860. W. P. Snow. On the lost Polar Expedition, &c.

1860. J. Brown. A Sequel to the North-west Passage.

1861. J. Lamont. Seasons with the Sea Horses. Spitzbergen.

1863. C. Irminger. Notice sur les Pêches du Groenland.

1863. D. G. Lindhagen. Geografiska Ortobestämmningar på Spetsbergen af Prof. A. E. Nordenskiöld, &c.

1863. A. E. Nordenskiöld. Geografisk och geognostisk Beskrifning öfver Nordöstra Delarne af Spetsbergen, &c.

1864. C. C. Rafn. Renseignments sur les premiers habitants de la Côte Occidentale du Groënland, &c.

1864. C. F. Hall. Life with the Esquimaux, &c. Frobisher Bay and Davis Strait.

1867. Dr. I. I. Hayes. The open Polar Sea, &c. Smith Sound.

1867. O. Heer. Ueber die Polarländer.

1867. H. Helms. Grönland und die Grönlander, &c.

1869. Sir F. L. M'Clintock. Fate of Sir J. Franklin. The Voyage of the "Fox" in the Arctic Seas, &c. Parry Islands.

1871. Hayes. Land of Desolation. South Greenland.

1873. Clements R. Markham. Threshold of the unknown Regions. Notices of Arctic Discovery. East Coast of Greenland, &c.

1873. J. C. Wells. Gateway to the Polynia. Spitzbergen. Recent Plants, Insects, and Birds, Fossils, &c.

1874. Th. von Heuglin. Reisen nach dem Nordpolarmeer, &c. Spitzbergen and Novya Zemlya.

1874. H. A. Markham. Whaling Cruise to Baffin's Bay, and Rescue of the Crew of the "Polaris." Baffin's Bay and S.E. Parry Islands. Appendix: Instructions, &c., and Botany and Geology.

1874. Die zweite deutsche Nordpolarfahrt: The German Arctic Expedition of 1869–70, under Koldewey; and Translation by H. Bates. East Coast of Greenland. Appendix: Nat. Hist., Geology, Physics, &c.

1874. Arctic Experiences, containing Capt. G. E. Tyson's Drift on Icefloe, and a general Arctic Chronology. Edited by E. V. Blake.

The following Publications contain Information on Voyages and Matters connected with Arctic Discovery.

ALLGEMEINE HISTORIE. Ice and the Frozen Sea, vols. xviii., xix., xx., xxi.; Spitzbergen, N.W. Passage, Cabot, Frobisher, Davis, vol. xvii.; Cranz, xix.

VOYAGES AND TRAVELS. A. Fisher, vol. i.; Freminville, vol. ii.; Sarylschew, Parry, vol. v.; T. James, vol. x.

CHURCHILL. Monck, vol. i.; La Peyrère, &c., vol. ii.; T. Gatoube, vol. vi.

HAKLUYT. Gudbrandus Thorlacus, N. de Lima, vol. i.; "Sunshine" and "Northstarre," S. Cabot, J. Davis, Sir H. Gilbert, Zeno, vol. iii.; Rundall, vol. v.; G. de Veer, vol. xii.; White, vol. xviii.

HARRIS. Greenland, T. James, vol. ii.

PURCHAS. N.W. Passage, vol. ii.; W. Baffin, D. Blefkens, T. Edge, R. Fotherbye, G. Weymouth, J. Hall, H. Hudson, W. Heley, Knight, Iver Boty, "North Pole," S. Cabot, vol. iii.

PINKERTON. Backstrom, Maupertius' Journey, Pluff, vol. i.; Frobisher, vol. ii.

LA HARPE. Glaces, Climat, Minéraux, &c., vol. xvii.

"ACCOUNT OF SEVERAL LATE VOYAGES," &c. F. Marten, Spitzbergen, 1811; Greenland, 1711.

INDEX OF MANUAL, &c.

(This Index is *complete* only for Part I., §§ I. and II.)

A.

B.

C.

D.

E.

F.

G.

H.

I.

J.

K.

L.

M.

N.

O.

P.

Q.

R.

S.

T.

U.

V.

W.

X.

Y.

Z.

ERRATA ET CORRIGENDA.

Page 133, after line 6 from the top, *insert* Modiolaria faba, O. Fabr.
„ 163, line 3, for *Cendronotus* read *Centronotus*.
„ 167, „ 2, *for* Cythera *read* Cythere.
„ 184, „ 20, for *Chiridota* read *Chirodota*.
„ 205, „ 20, *for* Plantanthera *read* Platanthera.
„ 217, „ 5, *for* its Temperate regions *read* Temperate Greenland.
„ 220, „ 28, *for* Serophularia *read* Scrophularia.
„ 220, „ 34, *for* Eleocharis *read* Elæocharis.
„ 220, „ 43, *for* Pologonium *read* Polygonum.
„ 221, „ 39, for *Polamogeton* read *Potamogeton*.
„ 221, bottom line, *for* Scheuzeria *read* Scheuchzeria.
„ 230, line 32, *for* Artemesia *read* Artemisia.
„ 236, „ 32, *for* Schenchzeri *read* Scheuchzeri.
„ 238, „ 4, *for* Cistopteris *read* Cystopteris.
„ 241, „ 39, *for* Racomitrum *read* Racomitrium.
„ 255, „ 14, *for* Aulacomium *read* Aulacomnium.
„ 255, „ 26, *for* Plocodium *read* Placodium.
„ 257, „ 7 from bottom, in note, *for* Scoreby *read* Scoresby.
„ 272, „ 10, for *Andreæa* read *Andræa*.
„ 273, in the heading, for MOSSES read LICHENS.
„ 279, line 1, for *Holosaccion* read *Halosaccion*.
„ 282, „ 13 from bottom, for CHROOESCCACEÆ read CHROOCOCCACEÆ.
„ 283, in heading, add ALGÆ after FRESHWATER.
„ 308, line 3, *for* Rinodina *read* Rhinodina.
„ 310, „ 22, *for* Arthopyrenia *read* Arthropyrenia.
„ 375, in heading, for ARCIC read ARCTIC.
„ 380, line 19, *for* Lastræa *read* Lastrea.
„ 498, „ 8, *for* vol. x. *read* vol. i., 1857.
„ 508, in the first footnote, *read*, The *Hadena* and *Tipula* next following came from Pond's Bay; but it is not clear whether the Arachnida came from Pond's Bay or from Port Kennedy.
„ 513, line 10, *for* Seppula *read* Serpula.
„ 519, „ 30, for *Arthodesmus* read *Arthrodesmus*.
„ 576, „ 5 from bottom, for *Pedum* read *Ledum*.
„ 582, „ 25, *for* Spachnum *read* Splachnum.
„ 583, „ 8 from the bottom, *for* subuletorum *read* sabuletorum.
„ 587, „ 1, at top, *for* Eudopyreuium *read* Endopyrenium.
„ 592, „ 11, *for* Nordpolarmer *read* Nordpolarmeer.

LONDON:
Printed by GEORGE E. EYRE and WILLIAM SPOTTISWOODE,
Printers to the Queen's most Excellent Majesty.
For Her Majesty's Stationery Office.
[.—750.—5/75.]

Printed in the United States
By Bookmasters